Mathematical Analysis, Optimization and Data Sciences

Series on Computers and Operations Research

ISSN: 1793-7973

Series Editor: Panos M. Pardalos *(University of Florida, USA)*

Published

Vol. 14 *Mathematical Analysis, Optimization and Data Sciences*
edited by P. M. Pardalos and T. M. Rassias

Vol. 13 *Global Optimization, Computation, Approximation and Applications*
edited by P. M. Pardalos and T. M. Rassias

Vol. 12 *Mathematical Analysis, Optimization, Approximation and Applications*
edited by P. M. Pardalos and T. M. Rassias

Vol. 11 *Mathematical Analysis, Differential Equations and Applications*
edited by T. M. Rassias and P. M. Pardalos

Vol. 10 *Analysis, Cryptography and Information Science*
edited by N. J. Daras, P. M. Pardalos and M. Th. Rassias

Vol. 9 *Analysis, Geometry, Nonlinear Optimization and Applications*
edited by P. M. Pardalos and T. M. Rassias

Vol. 8 *Network Design and Optimization for Smart Cities*
edited by K. Gakis and P. M. Pardalos

Vol. 7 *Computer Aided Methods in Optimal Design and Operations*
edited by I. D. L. Bogle and J. Zilinskas

Vol. 6 *Recent Advances in Data Mining of Enterprise Data: Algorithms and Applications*
edited by T. W. Liao and E. Triantaphyllou

Vol. 5 *Application of Quantitative Techniques for the Prediction of Bank Acquisition Targets*
by F. Pasiouras, S. K. Tanna and C. Zopounidis

More information on this series can also be found at https://www.worldscientific.com/series/scor

Series on Computers and Operations Research **Vol. 14**

Mathematical Analysis, Optimization and Data Sciences

Editors

Panos M Pardalos
University of Florida, USA

Themistocles M Rassias
National Technical University of Athens, Greece

World Scientific

NEW JERSEY • LONDON • SINGAPORE • BEIJING • SHANGHAI • HONG KONG • TAIPEI • CHENNAI • TOKYO

Published by

World Scientific Publishing Co. Pte. Ltd.
5 Toh Tuck Link, Singapore 596224
USA office: 27 Warren Street, Suite 401-402, Hackensack, NJ 07601
UK office: 57 Shelton Street, Covent Garden, London WC2H 9HE

Library of Congress Cataloging-in-Publication Data
Names: Pardalos, P. M. (Panos M.), 1954– editor | Rassias, Themistocles M., 1951– editor
Title: Mathematical analysis, optimization and data sciences / editors
Panos M Pardalos, University of Florida, USA, Themistocles M Rassias,
National Technical University of Athens, Greece.
Description: New Jersey : World Scientific, [2027] | Series: Series on computers and operations research, 1793-7973 ; vol. 14 | Includes bibliographical references and index.
Identifiers: LCCN 2025047184 | ISBN 9789819820481 hardcover |
ISBN 9789819820498 ebook | ISBN 9789819820504 ebook other
Subjects: LCSH: Mathematical analysis | Mathematical optimization
Classification: LCC QA300 .M2274 2027
LC record available at https://lccn.loc.gov/2025047184

British Library Cataloguing-in-Publication Data
A catalogue record for this book is available from the British Library.

For any available supplementary material, please visit
https://www.worldscientific.com/worldscibooks/10.1142/14503#t=suppl

Desk Editors: Soundararajan Raghuraman/Steven Patt

Typeset by Stallion Press
Email: enquiries@stallionpress.com

Printed in Singapore

https://doi.org/10.1142/9789819820498_fmatter

Preface

The volume *Mathematical Analysis, Optimization and Data Sciences* publishes research and survey papers in various domains of applied mathematics focusing on the areas of mathematical analysis, optimization and their interplay with various highly applicable problems.

More specifically, the works within this volume investigate classes of operators on Banach lattices, cube-free values of integers, circuit and weight representation of the embedded Markov chain, spherical harmonics for optimal 3D data modeling, a strengthening of a statistical framework by applied mathematical analysis, fluid-structure interaction in arterial hemodynamics, inequalities for normalized entropic determinant of positive operators in Hilbert spaces, ternary derivation-homomorphism functional inequalities, a discrete operator linking adjoint Bernoulli polynomials, extended general quasi variational inclusions and dynamical systems, Hardy-type integral inequalities and equivalent forms, reverse Hardy-type integral inequalities in the whole plane, fractional Hermite–Hadamard-type inequalities, Hardy–Hilbert-type inequalities, exponential stability conditions for differential equations, oscillation and non-oscillation criteria of second-order functional differential equations, angular localization of the eigenvalues of perturbed matrices, non-conventional stacked three relator spaces, PPF-dependent fixed points for multivalued geometric contractions, multiple coercivity, applications of Frobenius–Euler-type polynomials, as well as the use of moments formulas for geometric and negative binomial distributions with special numbers polynomials in data sciences applications.

The chapters have been contributed by experts in their domains from the international community and present the state-of-the-art knowledge on the problems treated. We would like to express our thanks to them for participating in this publication with their valuable chapters. We would also like to express our thanks to the production team of World Scientific for their valuable assistance throughout the preparation process of this book.

https://doi.org/10.1142/9789819820498_fmatter

About the Editors

Panos Pardalos was born in Greece and graduated from Athens University (Department of Mathematics). He received his PhD (Computes and Information Sciences) from the University of Minnesota. He is a Distinguished Emeritus Professor in the Department of Industrial and Systems Engineering at the University of Florida, and an affiliated faculty of Biomedical Engineering and Computer Science & Information & Engineering departments. Panos Pardalos is a world-renowned leader in Global Optimization, Mathematical Modeling, Energy Systems, Financial applications, and Data Sciences. He is a Fellow of AAAS, AAIA, AIMBE, EUROPT, and INFORMS and was awarded the 2013 Constantin Caratheodory Prize of the International Society of Global Optimization. In addition, Panos Pardalos has been awarded the 2013 EURO Gold Medal prize bestowed by the Association for European Operational Research Societies. This medal is the preeminent European award given to Operations Research (OR) professionals for "scientific contributions that stand the test of time." Panos Pardalos has been awarded a prestigious Humboldt Research Award (2018–2019). The Humboldt Research Award is granted in recognition of a researcher's entire achievements to date — fundamental discoveries, new theories, insights that have had significant impact on their discipline. Panos Pardalos is also a Member of several Academies of Sciences, and he holds several honorary PhD degrees and affiliations. He is the Founding Editor of Optimization Letters, Energy Systems, and Co-Founder of the International Journal of Global Optimization, Computational Management Science, and Springer Nature Operations Research Forum. He has published over 600 journal papers, and edited/authored over 200 books. He is one of the most cited authors and has graduated 71 PhD students so far. Details can be found in www.ise.ufl.edu/pardalos.

Themistocles M. Rassias was born in Greece and received his PhD in 1976 from the University of California, Berkeley under the supervision of Fields Medalist Stephen Smale. At Berkeley Shiing-Shen Chern was his academic advisor. Rassias received his PhD as the first degree after highschool, skipping undergraduate studies. Th. M. Rassias is an Emeritus Professor at the National Technical University of Athens, Greece. He has published more than 300 papers, 10 research books and 45 edited volumes in research Mathematics as well as four textbooks in Mathematics (in Greek) for university students. He serves as a member of the Editorial Board of several international mathematical journals. His work extends over several fields of mathematical analysis. It includes nonlinear functional analysis, functional equations, approximation theory, analysis on manifolds, calculus of variations, inequalities, metric geometry and their applications. He has contributed a number of results in the stability of minimal submanifolds, in the solution of Ulam's Problem for approximate homomorphisms in Banach spaces, in the theory of isometric mappings in metric spaces and in complex analysis (Poincaré's inequality and harmonic mappings). He has been awarded the title of Doctor Honoris Causa by several universities. He is a renowned figure in Mathematical Analysis with several terms attributed to him, including the Hyers-Ulam-Rassias stability of functional equations, Alexandrov-Rassias problems, etc. More information on his academic activity can be found in his homepage http://www.math.ntua.gr/~trassias/.

https://doi.org/10.1142/9789819820498_fmatter

Contents

https://doi.org/10.1142/9789819820498_0001

Chapter 1

Some Classes of Operators on Banach Lattices

Halimeh Ardakani[*,¶], **Yeol Je Cho**[†,‖], **Madjid Eshaghi Gordji**[‡,**] **and Themistocles M. Rassias**[§,††]

[*] *Department of Mathematics, Payame Noor University, Tehran, Iran*

[†] *Department of Mathematics Education, Gyeongsang National University, Jinju 52828, Korea*

[‡] *Department of Mathematics, Semnan University, Semnan, Iran*

[§] *Department of Mathematics, National Technical University of Athens, Zografou Campus, 15780 Athens, Greece*

[¶] *ardakani@pnu.ac.ir, halimeh_ardakani@yahoo.com*

[‖] *yjchomath@gmail.com*

[**] *meshaghi@semnan.ac.ir, madjid.eshaghi@gmail.com*

[††] *trassias@math.ntua.gr*

Using weakly p-summable positively limited sequences, the classes of positively limited p-convergent operators, positively limited weakly p-convergent operators and weak* positively p-convergent operators are introduced. Also, by weakly p-summable almost limited sequences, the classes of almost limited p-convergent operators, almost limited weakly p-convergent operators and weak* almost p-convergent operators are studied. Moreover, the positive DP* property of order p and weak DP* property of order p are considered and the behavior of the classes of operators on Banach lattices with the positively limited p-Schur property and almost limited p-Schur property is investigated. In addition, the domination properties of defined operators on Banach lattices are considered.

Keywords: Positively limited set, weakly p-compact set, p-convergent operator, p-Schur property, DP* property.

1. Introduction and Preliminaries

We use E, F for Banach lattices and X, Y for Banach spaces.

If A is a subset of a Banach space X and, for each weak*-null sequence (x_n^*) in X^*, $\lim_{n\to\infty} \sup_{a\in A} |\langle a, x_n^*\rangle| = 0$, then we say that A is *limited.* Each relatively compact set is limited [9, 14, 18]. A subset A of a Banach lattice E is said to be *almost limited* if every disjoint weak*-null sequence (x_n^*) in E^* converges uniformly to zero on A. Each limited set is almost limited. B_{ℓ_∞} is an almost limited set which is not limited [11]. A bounded set $A \subset E$ is *positively limited* if each positive weak*-null sequence (x_n^*) in E^* converges uniformly to zero on A. Each almost limited set is a positively limited set. Also, each order interval in a Banach lattice is positively limited [2].

A sequence $(x_n) \subset X$ is called *weakly p-summable*, where $1 \leq p < \infty$ if, for each $x^* \in X^*$, $(x^*(x_n)) \in \ell_p$. Also, $(x_n) \subset X$ is called *weakly p-convergent* to $x \in X$ if $(x_n - x) \in \ell_p^w(X)$, where $\ell_p^w(X)$ is the space of weakly p-summable sequences of X. For $p = \infty$, weakly p-convergent sequences are exactly weakly convergent sequences. A bounded set $A \subset X$ is called *relatively weakly p-compact* if each sequence in A has a weakly p-convergent subsequence. $A \subset X$ is *weakly p-compact* if the limit point is in A [19].

A Banach space X has the *limited p-Schur property* ($1 \leq p \leq \infty$) if all limited weakly p-compact subsets of X are relatively compact or, equivalently, every limited sequence $(x_n) \in \ell_p^w(X)$ is norm null [13].

Later, the concept of strong limited p-Schur property in Banach lattices was introduced. A Banach lattice E has the *strong limited p-Schur property* if each almost limited weakly p-compact subset of E is relatively compact. Each Banach lattice with the strong limited p-Schur property has the limited p-Schur property too [5].

Recently, the classes of Banach lattices with the positively limited p-Schur property were studied. A Banach lattice E has the *positively limited p-Schur property* if each positively limited weakly p-compact subset of E is relatively compact. Every Banach lattice with the positively limited p-Schur property has the strong limited p-Schur property too [6]. Here, at first the classes of positively limited p-convergent operators, positively limited weakly p-convergent operators and weak* positively p-convergent operators are introduced and some results of them are obtained. As an application, some characterizations of the positively limited p-Schur property and the positive DP* property of order p of E are considered in terms of these operators. In particular, using positively limited p-convergent operators, some operator spaces with the positively limited p-Schur property are obtained.

Next, the classes of almost limited p-convergent operators, almost limited weakly p-convergent operators and weak* almost p-convergent operators are introduced and some results of them are obtained. As an application, some characterizations of the strong limited p-Schur property and the weak DP* property of order p of E are considered in terms of these operators.

We also investigate the domination problem of positively limited p-convergent operators, weak* positively p-convergent operators, almost limited p-convergent operators and weak* almost p-convergent operators.

We recall some definitions and notations as follows:

For a Banach lattice E, $E^+ = \{x \in E : x \geq 0\}$ refers to the positive cone of E. A subset A of E is called *solid* if $|x| \leq |y|$ for some $y \in A$ implies that $x \in A$. The solid hull of A is the set $Sol(A) = \{y \in E : |y| \leq |x| \text{ for some } x \in A\}$. A norm-bounded subset A of E is solid if $|x| \leq |y|$ for some $y \in A$ implies that $x \in A$. If, for every weakly null sequence (x_n) in E, $|x_n| \xrightarrow{w} 0$, then the lattice operations are called weakly sequentially continuous. Also, if, for every weak*-null sequence (x_n^*) in E^*, $|x_n^*| \xrightarrow{w^*} 0$, then the lattice operations are called *weak* sequentially continuous* [1, 26].

All the concepts mentioned above and needed in this chapter are collected as follows:

(1) A Banach lattice E has the following:

  (a) *Schur property* if each relatively weakly compact set in E is relatively compact;
  (b) *Gelfand–Phillips property* (*shortly, GP property*) if each limited set in E is relatively compact;

(c) *DP* property* if each relatively weakly compact set in E is limited [14];
(d) *weak DP* property* if each relatively weakly compact set in E is almost limited [11];
(e) *positive DP* property* if each relatively weakly compact set in E is positively limited [2];
(f) *p-Schur property* if every sequence $(y_n) \in \ell_p^w(E)$ is norm null [31];
(g) *limited p-Schur property*, if every limited sequence $(y_n) \in \ell_p^w(E)$ is norm null [13];
(h) *positively limited p-Schur property* if every positively limited sequence $(y_n) \in \ell_p^w(E)$ is norm null [6];
(i) *strong limited p-Schur property* if every almost limited sequence $(y_n) \in \ell_p^w(E)$ is norm null [5];
(j) *property (d)* if $|x_n^*| \xrightarrow{w^*} 0$ for every weak*-null disjoint sequence (x_n^*) in E^* [17].

(2) An operator $T : E \to X$ is called

(a) *completely continuous* if it carries weakly null sequences to norm null ones [1];
(b) *p-convergent* if it carries weakly p-summable sequences to norm null ones [10];
(c) *disjoint p-convergent* if it carries disjoint weakly p-summable sequences to norm null ones [31];
(d) *limited p-convergent* (shortly, lpc) if it carries limited weakly p-summable sequences to norm null ones [13, 31];
(e) *almost limited p-convergent* (shortly, alpc) if it carries almost limited weakly p-summable sequences to norm null ones [5].

(3) An operator $T : X \to E$ is called *positively limited* if $T(B_X)$ is a positively limited set or equivalently, $\|T^* x_n^*\| \to 0$ for each positive weak*-null sequence (x_n^*) in E^* [2].

We assume that $1 \le p < \infty$, unless otherwise stated.

2. Positively Limited p-convergent Operators

In this section, the class of positively limited p-convergent operators is studied and some characterizations are obtained. Moreover, some necessary and sufficient conditions under which some operator spaces have the positively limited p-Schur property are investigated. We show that a Banach lattice E has the positively limited p-Schur property if and only if, for each Banach space Y, each operator $T : E \to Y$ is positively limited p-convergent. The class of positively limited p-convergent operators was described in [7].

Definition 1. A bounded linear operator $T : E \to X$ is *positively limited p-convergent (shortly, plpc)* if T carries weakly p-summable positively limited sequences of E to norm null sequences of X.

The set of all plpc operators from E into X is denoted by $\mathcal{L}_{plpc}(E, X)$. Clearly, $\mathcal{L}_{plpc}(E, X)$ is a linear subspace of $\mathcal{L}(E, X)$, where $\mathcal{L}(E, X)$ is the class of all bounded linear operators from E to X. Since each almost limited set is positively limited, a plpc operator is lpc and alpc. For the converse, we have the following which are the immediate consequences of [2, Theorems 2.5 and 2.7]:

(1) If $T : E \to X$ is an lpc operator and E^* has the weak*-sequentially continuous lattice operations, then T is a plpc operator.
(2) If $T : E \to X$ is an alpc operator and E has the property (d), then T is a plpc operator.

Example 1.

(1) The identity operator on each Banach lattice with the positively limited p-Schur property, and without the p-Schur property, such as c_0 is plpc, but it is not p-convergent.
(2) The identity operator on each Banach lattice with the limited p-Schur property and without the positively limited p-Schur property, such as $L^1[0, 1]$ is lpc, but it is not plpc for all $p \geq 2$.
(3) The identity operator on each Banach lattice with the strong limited p-Schur property and without the positively limited p-Schur property, such as c is alpc, but it is not plpc.

It is trivial that $T \circ S$ is plpc if $F \xrightarrow{S} X \xrightarrow{T} Y$, where S is plpc and T is a bounded linear operator. It should be noted that order-bounded operators between Banach lattices preserve positively limited sets, see [2, Theorem 2.11]. As a result, if $E \xrightarrow{T} F \xrightarrow{S} X$, where S is plpc and T is order bounded, then $S \circ T$ is also plpc. However, the following example shows that $S \circ T$ need not be plpc if T is not order bounded.

Example 2. Let $T : L^1[0,1] \to c_0$ be defined as

$$Tf = \left(\int_0^1 f(t) r_n(t) dt \right)_{n=1}^{\infty} \qquad \text{for all } f \in L^1[0,1],$$

where $r_n(t)$ is the nth Rademacher function on $[0,1]$. The operator T is not order bounded. Indeed, $(r_n(t))_{n=1}^{\infty}$ is weakly p-summable for all $2 \leq p \leq \infty$ and order bounded ($-\mathbf{1} \leq r_n \leq \mathbf{1}$, $n \geq 1$) and hence positively limited in $L^1[0,1]$, but $(Tr_n(t))_{n=1}^{\infty}$ is not order bounded in c_0. On the other hand, $\|Tr_n\| = 1$ for each $n \geq 1$. Therefore, T is not plpc. Note that the identity operator Id_{c_0} is plpc [2, Theorem 3.6]. However, $Id_{c_0} \circ T = T$ is not plpc.

Theorem 1. *An operator T on E is plpc if and only if, for each positively limited weakly p-compact set $A \subseteq E$, the set $T(A)$ is relatively compact.*

Proof. To show that an operator T is plpc, we assume that (x_n) is a weakly p-summable positively limited sequence in E. For each subsequence of (x_n), which is denoted again by (x_n), the sequence (Tx_n) is relatively compact and so it has a convergent and also weakly p-summable subsequence (Tx_{n_k}). Then $\|Tx_{n_k}\| \to 0$ as $k \to \infty$ which implies that T is plpc.

For the converse, let $A \subset E$ be a positively limited weakly p-compact set and $T : E \to Y$ be a plpc operator. Then every sequence (x_n) in A has a weakly p-convergent subsequence, denoted again by (x_n). On the other hand, the difference set $A - A$ is positively limited. Hence the sequence $(x_n - x_m)$ is positively limited weakly p-summable and, by hypothesis, (Tx_n) is a Cauchy sequence and so it is norm convergent in Y. Thus, $T(A)$ is relatively compact. This completes the proof. □

Clearly, if $T : E \to X$ takes positively limited sets to relatively compact sets, then T is plpc. However, a plpc operator does not necessarily carry a positively limited set to a relatively compact one.

Example 3. We know that every bounded linear operator from ℓ_∞ to c_0 is Dunford–Pettis. Therefore, every operator ℓ_∞ to c_0 is plpc.

However, there exists an operator $T : \ell_\infty \to c_0$ is not compact. This implies that TB_{ℓ_∞} is not relatively compact while B_{ℓ_∞} is positively limited in ℓ_∞.

Example 4. For each $1 \leq p < \infty$, the canonical injection $J_{\infty,p} : L^\infty[0,1] \to L^p[0,1]$ is a Dunford–Pettis operator and so it is plpc which fails to be compact [26, p. 222]. Therefore, $J_{\infty,p}(B_{L^\infty[0,1]})$ is not relatively compact although $B_{L^\infty[0,1]}$ is positively limited.

Example 5. An operator $R : C[0,1] \to c_0$ defined by

$$Rf = \left(\int_0^1 f(t) r_n(t) dt \right)_{n=1}^{\infty} \qquad \text{for all } f \in L^1[0,1],$$

where r_n is the nth Rademacher function on $[0,1]$. The operator R is weakly compact. By the Dunford–Pettis property of $C[0,1]$, R is completely continuous and so it is plpc. However, R is not compact. Hence, $R(B_{C[0,1]})$ is not a relatively compact set in c_0, while the closed unit ball $B_{C[0,1]}$ is a positively limited set.

Indeed, every plpc operator takes positively limited sets to relatively weakly compact sets.

Theorem 2. *Every plpc operator $T : E \to Y$ carries positively limited sets to relatively weakly compact sets. In particular, every plpc operator is order weakly compact.*

Proof. Let (x_n) be an arbitrary order bounded disjoint sequence in E. Then, (x_n) is weakly p-summable [5] and positively limited. Thus, $\|Tx_n\| \to 0$ since T is plpc. Hence, T is order weakly compact. From (cf. Ref. [1, Theorem 5.58]), T admits a factorization through a Banach lattice G with order continuous norm

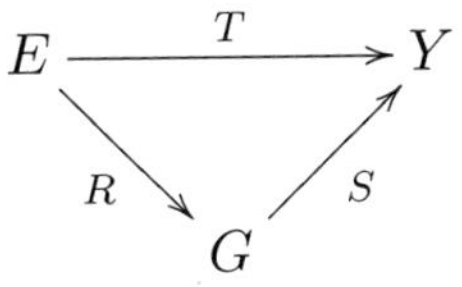

such that $R : E \to G$ is a lattice homomorphism. From Ref. [2, Theorem 2.11], for each positively limited set $B \subset E$, $R(B)$ is also positively limited in G and so it is relatively weakly compact (see Ref. [2, Theorem 3.4]). Therefore, $T(B) = SR(B)$ is relatively weakly compact. This completes the proof. $\square$

The coincidences of these types of operators are formulated as follows:

Remark 1. Let E be a Banach lattice. Then, we have the following:

(1) If E^* has the weak* sequentially continuous lattice operations, then each positively limited set in E is limited [2]. Hence, in this case, three classes of operators lpc, alpc and plpc from E into any Banach space are the same.
(2) If E has the p-DP* property, then each weakly p-compact set in E is limited which implies that three classes of operators lpc, alpc and plpc coincide with the p-convergent operator.

It follows from Theorem 2 that each plpc operator on a Banach lattice is order weakly compact and so it is lcc. Since it admits a factorization through a Banach lattice with order continuous norm with the GP property. So we can draw the following diagram:

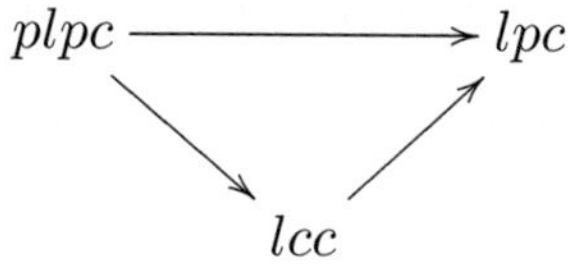

Lemma 1. *The following assertions hold:*

(1) *If (x_n) is a positively limited sequence in E, then the operator $T : \ell_1 \to E$, $T(b) = \sum_{n=1}^{\infty} b_n x_n$, $b = (b_n) \in \ell_1$, is positively limited.*
(2) *If (x_n) is a positively limited weakly p-summable sequence in E, then the operator $T : \ell_{p^*} \to E$, $T(b) = \sum_{n=1}^{\infty} b_n x_n$, $b = (b_n) \in \ell_{p^*}$, is positively limited* $(1 < p < \infty)$.

Proof. (1) Let (f_m) be a positive weak* null sequence in E^*. Since (x_n) is a positively limited sequence in E, we have

$$\sup_{a=(a_j)\in B_{\ell_1}} |f_m(Ta)| = \sup_{a=(a_j)\in B_{\ell_1}} |T^* f_m(a)| = \|T^*(f_m)\|_\infty$$
$$= \sup_n |f_m(x_n)| \to 0.$$

Hence, $T(B_{\ell_1})$ is a positively limited set and so T is a positively limited operator.

(2) For the second part, use the similar techniques of [22, Lemma 3.23] and the fact that the closed unit ball $B_{\ell_{p^*}}$ is weakly p-compact. This completes the proof. □

We then have the following composition result:

Theorem 3. *Let $1 < p < \infty$ and $T : E \to Y$ be an operator. The following are equivalent:*

(1) *T is a plpc operator;*
(2) *for every Banach space Z and any positively limited, weakly p-compact operator $S : Z \to E$, the operator $TS : Z \to Y$ is compact;*
(3) *for every positively limited operator $S : \ell_{p^*} \to E$, the operator $TS : \ell_{p^*} \to Y$ is compact.*

Proof. (1) ⇒ (2) If $S : Z \to E$ is a positively limited, weakly p-compact operator, then $S(B_Z)$ is a positively limited weakly p-compact subset of E. Using the plpc-ness of T and following Theorem 1, we can prove that $TS(B_Z)$ is relatively compact and so the operator $T \circ S$ is compact.

(2) ⇒ (3) It follows easily from the fact that $Id(\ell_{p^*}) \in W_p$ (the class of weakly p-compact operators).

(3) ⇒ (1) Let $(x_n) \in \ell_p^w(E)$ be a positively limited sequence. By Lemma 1, the operator $S : \ell_{p^*} \to E$ defined by $S(b) = \sum_{n=1}^{\infty} b_n x_n$, $b = (b_n) \in \ell_{p^*}$, such that $S(e_n) = x_n$ for all $n \geq 1$ is positively limited (note that $\ell_p^w(E) = L(\ell_{p^*}, E)$). Hence, the operator $T \circ S$ is compact and so $\|Tx_n\| = \|TS(e_n)\| \to 0$. This completes the proof. □

As a consequence of Theorem 3, we obtain the following characterization:

Corollary 1. *Let $1 < p < \infty$. The following are equivalent:*

(1) *E has the positively limited p-Schur property,*
(2) *for every Banach space Z, any positively limited weakly p-compact operator $S : Z \to E$ is compact,*
(3) *each positively limited operator $S : \ell_{p^*} \to E$ is compact.*

The following theorem provides a characterization of the positively limited p-Schur property with respect to plpc operators:

Theorem 4. *For a Banach lattice E, the following are equivalent:*

(1) *E has the positively limited p-Schur property;*
(2) *for each Banach space Y, $\mathcal{L}_{plpc}(E, Y) = \mathcal{L}(E, Y)$;*
(3) *$\mathcal{L}_{plpc}(E, \ell_\infty) = \mathcal{L}(E, \ell_\infty)$.*

Proof. (1) $\Rightarrow$ (2) Let $T : E \to Y$ be an operator and A be a positively limited weakly p-compact subset of E. By the positively limited p-Schur property of E, A and so $T(A)$ are relatively compact sets in Y. Then, by Theorem 1, T is plpc.

(2) $\Rightarrow$ (3) It is obvious.

(3) $\Rightarrow$ (1) Assume, by way of contradiction, that E does not have the positively limited p-Schur property. Then, there is a weakly p-summable positively limited sequence (x_n) in E such that $\|x_n\| = 1$ for all n. Choose a normalized sequence (x_n^*) in E^* such that $|\langle x_n, x_n^* \rangle| = 1$ for all $n \geq 1$. Then, the operator $T : E \to \ell_\infty$ defined by

$$Tx = (\langle x, x_n^* \rangle) \quad \text{for all } x \in E$$

is not plpc (since the sequence (x_n) is weakly p-summable and positively limited and $\|Tx_n\| \geq 1$ for all $n \geq 1$). This leads to a contradiction. This completes the proof. □

Note that, in Theorem 4, one cannot replace the positively limited p-Schur property of the domain E of related operators by their images. The Banach lattice c_0 has the positively limited p-Schur property, but the operator $T : L^1[0, 1] \to c_0$ defined by

$$Tf = \left(\int_0^1 f(t) r_n(t) dt \right)_{n=1}^{\infty} \quad \text{for all } f \in L^1[0, 1]$$

is not plpc for all $p \geq 2$ (since the Rademacher sequence $f_n(t) = r_n(t)$ is weakly 2-summable and positively limited, but $\|Tf_n\| = 1$).

Lemma 2 ([2]). *Let $T : E \to F$ be a positive operator from a Banach lattice E to a Banach lattice F and $A \subset E$ be a positively limited set. Then, $T(A)$ is a positively limited set in F.*

We can show that, in order for the positively limited p-Schur property of the domain E of related operators to be replaced by their images in Theorem 4, it is sufficient for the operators to be positive.

Proposition 1. *If $T : E \to F$ is a positive operator and F has the positively limited p-Schur property, then T is plpc.*

Proof. Note that, from [2], for each positively limited weakly p-compact set $A \subset E$, $T(A)$ is positively limited (and also weakly p-compact) in F. It follows from the positively limited p-Schur property of F that $T(A)$ is relatively compact. This proves that T is plpc. This completes the proof. □

Before doing so, we need to recall almost interval preserving operators. It is remarked in [26] that a positive linear operator $T : E \to F$ between two Banach lattices is *almost interval preserving* if $T[0, x]$ is dense in $[0, Tx]$ for every $x \in E_+$ [26].

Theorem 5. *Let E and F be two Banach lattices and suppose that F has the positively limited p-Schur property. Then, each positive or an almost interval preserving operator $T : E \to F$ is plpc.*

Proof. Let A be a positively limited weakly p-compact set in E and $T : E \to F$ be a positive operator. By [2], $T(A)$ is positively limited and so it is relatively compact (by the positively limited p-Schur property of F). By Theorem 1, T is a plpc operator.

Now, let A be a positively limited weakly p-compact set in E and $T : E \to F$ be an almost interval preserving operator. By Ref. [3, Theorem 3.8], $T(A)$ is positively limited and so it is relatively compact (by the positively limited p-Schur property of F). By Theorem 1, T is a plpc operator. This completes the proof. □

A Banach lattice E has the *dual positive Schur property* if every positive weak*-null sequence in E^* is norm null [29]. One easily verifies that E has the dual positive Schur property if and only if the closed unit ball B_E is a positively limited set.

Proposition 2. *Let E be a weakly p-compact Banach lattice with the dual positive Schur property and F be a Banach lattice with the positively limited p-Schur property. Then, every positive operator $T : E \to F$ is compact.*

Proof. Since E is weakly p-compact with the dual positive Schur property, the closed unit ball B_E is weakly p-compact and positively limited. By Lemma 2 and Theorem 5, $T(B_E)$ is relatively compact and so T is a compact operator. This completes the proof. □

If $R : E \to F$ and $S : F \to X$ are two operators such that R is plpc, then $S \circ R$ is likewise plpc. In fact, if $(x_n) \in \ell_p^w(E)$ is a positively limited sequence and R is an plpc operator, then $\|Rx_n\| \to 0$ and so $\|S(Rx_n)\| \to 0$. Consequently, $S \circ R : E \to X$ is plpc. However, if S is plpc, then $S \circ R$ is not necessarily plpc.

Example 6. An operator $R : L^1[0,1] \to c_0$ defined by

$$Rf = \left(\int_0^1 f(t) r_n(t) dt \right)_{n=1}^{\infty} \quad \text{for all } f \in L^1[0,1],$$

where r_n is the nth Rademacher function on $[0,1]$, is not plpc. Consider an operator $S = Id_{c_0}$. Then, S is plpc, but $S \circ R = R$ is not plpc.

Theorem 6. *Let E and F be two Banach lattices. If $R : E \to F$ and $S : F \to X$ are two operators such that S is plpc and R is positive, then the composition operator $S \circ R$ is plpc.*

Proof. Let $(x_n) \in \ell_p^w(E)$ be a positively limited sequence. Since R is a positive operator, (Rx_n) is a weakly p-summable and positively limited sequence in F. Since $S : F \to X$ is plpc, then $\|S(Rx_n)\| \to 0$ and so $S \circ R$ is plpc. This completes the proof. □

However, one easily verifies that, if S is plpc and R is positive, then $S \circ R$ is plpc. It is enough to note that, in this case, for each positively limited sequence $(x_n) \in \ell_p^w(E)$, (Rx_n) is a weakly p-summable and positively limited sequence. Hence, plpc-ness of S ensures that $\|S(Rx_n)\| \to 0$. This proves that $S \circ R$ is plpc.

In [26, Corollary 2.4.3], the author states that a σ-Dedekind complete Banach lattice F has order continuous norm if and only if ℓ_∞ is not lattice embeddable in F. We use this statement in the following theorem:

Theorem 7. *Let E and F be two Banach lattices with F σ-Dedekind complete. If every bounded linear operator T from E into F is plpc, then either E or F has order continuous norm.*

Proof. If neither E nor F has order continuous norm, then there exists an order bounded disjoint sequence $(x_n)_n^\infty \subset E$ such that $\|x_n\| = 1$ for all $n \geq 1$. Clearly, (x_n) is positively limited and weakly p-summable. We choose a normalized sequence (x_n^*) in E^* such that $|\langle x_n, x_n^* \rangle| = 1$ for all $n \geq 1$ and define the operator $T : E \to \ell_\infty$ by

$$Tx = (\langle x, x_n^* \rangle) \quad \text{for all } x \in E.$$

Since F is a σ-Dedekind complete Banach lattice without order continuous norm, ℓ_∞ lattice embeds in F. Let $j : \ell_\infty \to F$ be the lattice embedding. Then, it is easily verified that $j \circ T : E \to F$ is not plpc since $\|Tx_n\| \geq 1$ for all $n \geq 1$. This completes the proof. □

Remark 2. Example 6 implies that the converse of Theorem 7 does not necessarily hold. However, if E^* has weak*-sequentially continuous lattice operations and either E or F has order continuous norm, then each operator T from E into F is plpc. Indeed, if either E or F has order continuous norm, then either E or F has the limited p-Schur property and hence every operator $T : E \to F$ is lpc. On the other hand, since E^* has the weak*-sequentially continuous lattice operations, T is plpc.

Also, note that the σ-Dedekind completeness of a Banach lattice F cannot be removed. Each operator $T : \ell_\infty \to c$ is plpc, while neither ℓ_∞ nor c has order continuous norm.

Theorem 7 has the following two consequences:

Corollary 2. *If E^* has the weak* sequentially continuous lattice operations, then the following are equivalent:*

(1) *Each operator $T : E \to \ell_\infty$ is plpc.*
(2) *E has order continuous norm.*

The order continuity of the norm is also characterized as follows:

Corollary 3. *Let F be a σ-Dedekind complete Banach lattice. Then, the following are equivalent:*

(1) *Each operator $T : c \to F$ is plpc.*
(2) *F has order continuous norm.*

Recently, the authors introduced the p-positive DP* property in [6]. A Banach lattice E has the *p-positive DP* property* if every relatively weakly p-compact subset of E is positively limited, alternatively, $x_n^*(x_n) \to 0$ for every weakly p-summable sequence (x_n) and every weak*-null sequence $(x_n^*) \subset (E^*)^+$. For our convenience, let $\mathcal{L}_{pc}(X,Y)$ denote the set of all p-convergent operators from X into Y. $\mathcal{L}_{pc}(E,X) \subset \mathcal{L}_{plpc}(E,X)$ holds.

Theorem 8. *For a Banach lattice E, the following assertions are equivalent:*

(1) *E has the p-positive DP* property;*
(2) *for every Banach space X, $\mathcal{L}_{plpc}(E,X) = \mathcal{L}_{pc}(E,X)$;*
(3) *$\mathcal{L}_{plpc}(E,c_0) = \mathcal{L}_{pc}(E,c_0)$.*

Proof. (1) $\Rightarrow$ (2) Let $T \in \mathcal{L}_{plpc}(E,X)$ and $(x_n)_{n=1}^{\infty}$ be a weakly p-summable sequence in E. Since E has the p-positive DP* property, (x_n) is positively limited in E and, by hypothesis, $\|T(x_n)\| \to 0$. This implies that $T \in \mathcal{L}_{pc}(E,X)$

(2) $\Rightarrow$ (3) Obvious.

(3) $\Rightarrow$ (1) Let $(x_n)_{n=1}^{\infty}$ be a weakly p-summable sequence in E and $(x_n^*)_{n=1}^{\infty}$ be a sequence in $(E^*)^+$ satisfying $x_n^* \xrightarrow{w^*} 0$. We define the positive operator $T : E \to c_0$ by

$$Tx = (x_n^*(x)) \text{ for all } x \in E.$$

Since c_0 has the positively limited p-Schur property and the operator T is positive, by Proposition 1, T is plpc. By hypothesis, $T \in \mathcal{L}_{pc}(E,c_0)$ and hence $\|Tx_n\| \to 0$. As a result, $x_n^*(x_n) \to 0$ since the inequality $|x_n^*(x_n)| \leq \|Tx_n\|$ holds for all $n \geq 1$. This implies that E has the p-positive DP* property. This completes the proof. □

The p-positive DP* property is then also characterized by plpc operators as follows. For this, we need the following lemma of [21]:

Lemma 3. *Suppose that E is a Banach lattice. Then, for all $p \geq 2$, each disjoint sequence (z_n) in the solid hull of a bounded set $W \subset E$ is weakly p-summable.*

Theorem 9. *Suppose that E is a Banach lattice. Then, for all $p \geq 2$, the following assertions are equivalent:*

(1) *E has the p-positive DP* property,*
(2) *every plpc operator $T : E \to Y$ is disjoint p-convergent for every Banach space Y,*
(3) *every plpc operator $T : E \to c_0$ is disjoint p-convergent.*

Proof. (1) ⇒ (2) In this case, every plpc operator $T : E \to Y$ is p-convergent, for every Banach space Y.

(2) ⇒ (3) It is clear.

(3) ⇒ (1) Let $(x_n) \in \ell_p^w(E)$ be a disjoint sequence and (x_n^*) be a positive weak*-null sequence in E^*. It is enough to show that $x_n^*(x_n) \to 0$ (see [6, Theorem 3.6]). Consider the positive operator $T : E \to c_0$ defined by $Tx = (\langle x, x_n^* \rangle)$ for all $x \in E$. According to Proposition 1, T is plpc and so it is disjoint p-convergent. Thus $\|Tx_n\| \to 0$ and hence $x_n^*(x_n) \to 0$, as desired. This completes the proof. □

A Banach lattice E has the p-positive Schur property if and only if every disjoint $(x_n) \in \ell_p^w(E)_+$ is norm null [31]. It is obvious that each of the two properties p-positive Schur and weak DP* implies the p-weak DP* property. It is proved in [11] that a Banach lattice E has the positive Schur property if and only if E has the weak DP* property and order continuous norm.

Now, we turn our attention to the domination problem of plpc operators. Recall that a Banach lattice E is *weak p-consistent* $(1 \leq p < \infty)$ if it follows from $(x_n) \in \ell_p^w(E)$ that $(|x_n|) \in \ell_p^w(E)$. For instance, each AM-space with a unit is weak p-consistent [31].

Proposition 3. *Assume that E is a weak p-consistent Banach lattice, F is a Banach lattice and $S, T : E \to F$ are two positive operators satisfying $0 \leq S \leq T$. If T is a plpc operator, then S is a plpc operator.*

Proof. Let $(x_n) \in \ell_p^w(E)$ be a positively limited sequence. Since E is weak p-consistent, $(|x_n|) \in \ell_p^w(E)$ also holds in E. Also, $(|x_n|)$ is positively limited [2] and so $\||T|x_n|\| \to 0$. Our conclusion follows from the inequalities $|S(x_n)| \leq S|x_n| \leq T|x_n|$. This completes the proof. □

If $0 \leq S \leq T : E \to F$ are two positive operators between Banach lattices and T is plpc, is then S necessarily plpc? We pose a question and have to leave it open.

3. Positively Limited Weakly p-convergent Operators

In this section, two classes of operators, limited weakly p-convergent operators and positively limited weakly p-convergent operators, are discussed.

First, recall form [23] that a bounded linear operator $T : X \to Y$ is

(1) *weakly completely continuous* (*shortly, wcc*) if T carries weakly Cauchy sequences of X to weakly convergent sequences of Y;
(2) *limited weakly completely continuous* (*shortly, lwcc*) if T carries weakly Cauchy limited sequences of X to weakly convergent sequences of Y.

A sequence (x_n) in a Banach space X is called a *weakly p-Cauchy sequence* if, for all $m \geq n$, $(x_m - x_n)$ is weakly p-summable.

Definition 2. A bounded linear operator $T : E \to X$ is

(1) *limited weakly p-convergent* (*shortly, lwpc*) if T carries weakly p-Cauchy limited sequences of E to weakly convergent sequences of X;
(2) *positively limited weakly p-convergent* (*shortly, plwpc*) if T carries weakly p-Cauchy positively limited sequences of E to weakly convergent sequences of X.

The set of all lwpc (resp., plwpc) operators from E into X is denoted by $\mathcal{L}_{lwpc}(E, X)$ (resp., $\mathcal{L}_{plwpc}(E, X)$). Clearly, $\mathcal{L}_{lwpc}(E, X)$ and $\mathcal{L}_{plwpc}(E, X)$ are linear subspaces of $\mathcal{L}(E, X)$, where $\mathcal{L}(E, X)$ is the class of all bounded linear operators from E to X.

Since each limited set is positively limited, a plwpc operator is lwpc:

(1) If $T : E \to X$ is an lwcc operator, then T is an lwpc operator.
(2) If $T : E \to X$ is a plwcc operator, then T is an lwpc operator.

A Banach lattice E has the positively BD property if each weakly Cauchy positively limited sequence of E is weakly convergent [2]. Also, a Banach space X has the BD property if each weakly Cauchy limited sequence of X is weakly convergent [23].

Definition 3. We say that a Banach lattice has:

(1) *BD property of order* p (*shortly,* p*-BD property*) if each weakly p-Cauchy limited sequence of E is weakly convergent;
(2) *positively BD property of order* p (*shortly,* p*-positively BD property*) if each weakly p-Cauchy positively limited sequence of E is weakly convergent.

Each Banach lattice with the positively p-BD property has the p-BD property, but the converse is false.

Example 7. Each limited set in a Banach sequence space c is relatively compact and so c has the p-BD property, but c does not have the positively p-BD property.

It is trivial that $T \circ S$ is plwcc if $F \xrightarrow{S} X \xrightarrow{T} Y$, where S is plwcc and T is a bounded linear operator. It should be noted that order bounded operators between Banach lattices preserve positively limited sets, see [2, Theorem 2.11]. As a result, if $E \xrightarrow{T} F \xrightarrow{S} X$, where S is plwcc and T is order bounded, then $S \circ T$ is also plwcc.

A norm bounded set $A \subset X$ is called *weakly* p*-precompact* if each sequence in A has a weakly p-Cauchy subsequence.

Theorem 10. *An operator* T *on* E *is plwpc if and only if, for each positively limited weakly* p*-precompact set* $A \subseteq E$*, the set* $T(A)$ *is relatively weakly compact.*

Proof. To show that an operator T is plwpc, we assume that (x_n) is a weakly p-Cauchy positively limited sequence in E. For each subsequence of (x_n), which is denoted again by (x_n), the sequence (Tx_n) is relatively weakly compact and so it has a weakly convergent subsequence (Tx_{n_k}). On the other hand, (Tx_n) is weakly Cauchy and so (Tx_n) is weakly convergent, which implies that T is plwpc.

For the converse, let $A \subset E$ be a weakly p-precompact positively limited set and $T : E \to Y$ be a plwcc operator. Then every sequence (x_n) in A has a weakly p-Cauchy and positively limited subsequence, denoted again by (x_n) positively limited. Hence, by hypothesis, (Tx_n) is weakly convergent in Y. Thus, $T(A)$ is relatively weakly compact. This completes the proof. □

The following theorem provides a characterization of the positively p-BD property with respect to plwpc operators:

Theorem 11. *For a Banach lattice E, the following are equivalent:*

(1) *E has the positively p-BD property;*
(2) *for each Banach space Y, $\mathcal{L}_{plwpc}(E, Y) = \mathcal{L}(E, Y)$;*
(3) *$\mathcal{L}_{plwpc}(E, \ell_\infty) = \mathcal{L}(E, \ell_\infty)$.*

Proof. (1) ⇒ (2) ⇒ (3) Obvious.

(3) ⇒ (1) Assume to the contrary that E does not have the positively p-BD property. Then, there exists a weakly p-Cauchy and positively limited sequence (x_n) in E, which is not weakly convergent. Let F be the closed subspace of E spanned by this sequence. Since F is separable, F is isomorphic to a closed subspace of ℓ_∞. Let T be such an isomorphic embedding from F to ℓ_∞. Since ℓ_∞ is injective, T has an extension to all of E, denoted again by T. Clearly, T is not plwpc, which leads to a contradiction. This completes the proof. □

Proposition 4. *If $T : E \to F$ is a positive operator and F has the positively p-BD property, then T is plwpc.*

Proof. Note that, from [2], for each positively limited weakly p-precompact set $A \subset E$, $T(A)$ is positively limited (and also weakly p-precompact) in F. It follows from the positively p-BD property of F that $T(A)$ is relatively weakly compact. This proves that T is plwpc. This completes the proof. □

If $R : E \to F$ and $S : F \to X$ are two operators such that R is plwpc, then SR is likewise plwpc. In fact, if (x_n) is a positively limited weakly p-Cauchy sequence in E and R is a plwpc operator, then (Rx_n) is weakly convergent and so $(S(Rx_n))$ is weakly convergent. Consequently, $S \circ R : E \to X$ is plwpc. However, if S is plwpc, then $S \circ R$ is not necessarily plwpc.

However, one easily verifies that, if S is plwpc and R is positive, then $S \circ R$ is plwpc. It is enough to note that, in this case, for each positively limited weakly p-Cauchy sequence (x_n), (Rx_n) is a weakly p-Cauchy sequence and positively limited sequence. Hence, plwpc-ness of S ensures that $(S(Rx_n))$ is weakly convergent, which proves that $S \circ R$ is plwpc.

Theorem 12. *Suppose that E and F are Banach lattices such that F is σ-Dedekind complete. Then the following assertions are equivalent:*

(1) *Every positively limited operator $T : E \to F$ is plwpc.*
(2) *Either E has positively p-BD property or F has order continuous norm.*

Proof. (1) $\Rightarrow$ (2) If E has positively p-BD property, then each weakly p-Cauchy and positively limited sequence (x_n) in E is weakly convergent. Hence each operator $T : E \to F$ is plwpc. Now, assume that F has order continuous norm and (x_n) is weakly p-Cauchy sequence in E. Since $T : E \to F$ is positively limited, (Tx_n) is weakly p-Cauchy and positively limited in F and so weakly convergent, by order continuity of the norm on F. Hence an operator $T : E \to F$ is plwpc.

(2) $\Rightarrow$ (1) We assume to the contrary that E does not have the positively p-BD property and F does not have order continuous norm. Then, there exists a weakly p-Cauchy and positively limited sequence (x_n) in E such that is not weakly convergent. Let W be the closed subspace of E spanned by this sequence. Then W is separable, and hence it is isomorphic to a closed subspace of ℓ_∞. Let Q be such an isomorphic embedding from W to ℓ_∞. Since ℓ_∞ is injective, we can extend Q to all of E. F does not have order continuous norm and it does not have the positively BD property. Hence, F is a σ-Dedekind complete Banach lattice without order continuous norm and there exists a lattice embedding $S : \ell_\infty \to F$ (cf. [26, Corollary 2.4.3]). Since B_{ℓ_∞} is positively limited in ℓ_∞ and S is a positive operator, S is a positively limited operator. Then $T := S \circ Q : E \to F$ is a positively limited operator. However, T not plwpc. This leads to a contradiction. This completes the proof. $\square$

Theorem 12 has the two following consequences:

Corollary 4. *The following are equivalent:*

(1) *Each positively limited operator $T : E \to \ell_\infty$ is plwpc.*
(2) *E has positively p-BD property.*

The order continuity of the norm is also characterized as follows:

Corollary 5. *Let F be a σ-Dedekind complete Banach lattice. Then, the following are equivalent:*

(1) *Each positively limited operator $T : c \to F$ is plwpc.*
(2) *F has order continuous norm.*

Theorem 13. *If E has the p-positive DP* property, then every plwpc operator $T : E \to Y$ is weak p-convergent for every Banach space Y.*

Proof. Let (x_n) be a weakly p-Cauchy sequence. Then, by hypothesis, (x_n) is positively limited. Since $T : E \to Y$ is a plwpc operator, (Tx_n) is weakly convergent and so T is weak p-convergent. This completes the proof. □

Now, we turn our attention to the domination problem of plpwc operators. If $0 \leq S \leq T : E \to F$ are two positive operators between Banach lattices and T is plwpc, then is S necessarily plwpc? We pose a question and have to leave it open.

4. Weak* Positively p-convergent Operators

This section focuses on the so-called weak* positively p-convergent operators and discuss some properties of them related to plpc operators and the positively limited p-Schur property.

Definition 4. An operator $T : X \to E$ is called *weak* positively p-convergent* if, for each sequence $(x_n) \in \ell_p^w(X)$ and positive weak*-null sequence (x_n^*) in E^*, $x_n^*(Tx_n) \to 0$.

Theorem 14. *Let $T : X \to E$ be an operator. The following are equivalent:*

(1) *T is a weak* positively p-convergent operator;*
(2) *T maps each weakly p-compact set in X to a positively limited set in E,*
(3) *for each Banach space Z and each weakly p-compact operator $S : Z \to X$, the operator $T \circ S$ is positively limited;*
(4) *for each operator $S : \ell_{p^*} \to X$, $T \circ S$ is positively limited.*

Proof. (1) $\Rightarrow$ (2) Let W be a weakly p-compact set in X. We show that $T(W)$ is a positively limited set in E, that is, for each positive weak*-null sequence (x_n^*) in E^* and each sequence (x_n) in W, $x_n^*(Tx_n) \to 0$. Since W is weakly p-compact, (x_n) has a weakly p-convergent subsequence to $x \in X$ which is denoted by (x_n) again. Then, $(x_n - x)$ and so $(Tx_n - Tx)$ are weakly p-summable and, by the definition of weak* positively p-convergent operators, $x_n^*(Tx_n) = x_n^*(Tx_n - Tx) + x_n^*(Tx) \to 0$.

(2) $\Rightarrow$ (3) Obvious.

(3) $\Rightarrow$ (4) It is clear (since $B_{\ell_{p^*}}$ is a weakly p-compact set).

(4) $\Rightarrow$ (1) We show that, for each positive weak*-null sequence (x_n^*) in E^* and every sequence $(x_n) \in \ell_p^w(X)$, $x_n^*(Tx_n) \to 0$. An operator $S : \ell_{p^*} \to E$ defined by $S(b) = \sum_{n=1}^{\infty} b_n x_n$, $b = (b_n) \in \ell_{p^*}$ is weakly p-compact, and so TS is positively limited. Then, $TS(B_{\ell_{p^*}})$ is a positively limited set, which implies that $x_n^*(Tx_n) \to 0$. This completes the proof. $\square$

The following result characterizes the class of Banach lattices with the p-positive DP* property by weak* positively p-convergent operators:

Corollary 6. *Let E be a Banach lattice. Then, the following are equivalent:*

(1) *E has the p-positive DP* property,*
(2) *the identity operator on E is weak* positively p-convergent,*

(3) *every weakly p-compact operator T from an arbitrary Banach space Z to E is a positively limited operator,*
(4) *every bounded linear operator $S : \ell_{p^*} \to E$ is positively limited.*

Theorem 15. *Let E be a Banach lattice and $T : X \to E$ be an operator. Then, the following are equivalent:*

(1) *T is a weak* positively p-convergent operator,*
(2) *for each positive operator S from E to a Banach lattice F with the positively limited p-Schur property, the operator ST is p-convergent,*
(3) *for each positive operator $S : E \to c_0$, the operator ST is p-convergent.*

Proof. (1) $\Rightarrow$ (2) Let W be a weakly p-compact set in X. By hypothesis (1), $T(W)$ is a positively limited and weakly p-compact set in E. Since $S : E \to F$ is positive, $ST(W)$ is a positively limited and weakly p-compact set in F. Hence, by the positively limited p-Schur property of F, $ST(W)$ is relatively compact. Thus, $S \circ T$ is p-convergent:

(2) $\Rightarrow$ (3) It is clear.

(3) $\Rightarrow$ (1) We show that, for each positive weak* null sequence (x_n^*) in E^* and every weakly p-summable sequence (x_n) in X, $x_n^*(Tx_n) \to 0$. Consider the operator $S : E \to c_0$ defined by $Sx = (x_n^*(x))$ for all $x \in E$. Then, S is a positive operator and, by hypothesis, $S \circ T$ is p-convergent. Thus, $\|S(T(x_n))\| \to 0$, which implies that $x_n^*(Tx_n) \to 0$. This completes the proof. □

We have immediate consequences of Theorem 15 as follows:

(1) If $T : E \to F$ is weak* positively p-convergent and F has the positively limited p-Schur property, then T is p-convergent.
(2) If $T : E \to F$ is weak* positively p-convergent, then, for each Banach space Y and each plpc operator $S : F \to Y$, the operator $ST : E \to Y$ is p-convergent.

Theorem 16. *Let $T : E \to F$ be an operator. The following are equivalent:*

(1) *T is a weak* positively p-convergent operator,*
(2) *for each positive operator $S : F \to c_0$, the operator $S \circ T$ is p-convergent.*

Proof. (1) $\Rightarrow$ (2) It is clear since the operator $S : F \to c_0$ is plpc (Proposition 1).

(2) $\Rightarrow$ (1) We show that, for each positive weak* null sequence (x_n^*) in F^* and every weakly p-summable sequence (x_n) in E, $x_n^*(Tx_n) \to 0$. Consider the operator $S : F \to c_0$ defined by $Sx = (x_n^*(x))$ for all $x \in F$. S is a positive operator. By hypothesis, $S \circ T$ is p-convergent and so $\|S(T(x_n))\| \to 0$, which implies that $x_n^*(Tx_n) \to 0$. This completes the proof. □

Theorem 17. *Suppose that F is a σ-Dedekind complete Banach lattice with the weakly sequentially continuous lattice operations and E is a Banach lattice. Then, the following are equivalent:*

(1) *Each weak* positively p-convergent operator $T : E \to F$ is plpc.*
(2) *E has the positively limited p-Schur property or F has order continuous norm.*

Proof. (1) $\Rightarrow$ (2) Suppose that E does not have the positively limited p-Schur property and F does not have order continuous norm. Then, there is a weakly p-summable and positively limited sequence (x_n) in E with $\|x_n\| > \epsilon$ for some $\epsilon > 0$. So, there is a sequence (x_n^*) in E^* such that $\|x_n^*\| = 1$ and $x_n^*(x_n) = \|x_n\|$. Consider the operator $S : E \to \ell_\infty$ defined by

$$Sx = (\langle x, x_n^* \rangle) \quad \text{for all } x \in E.$$

Also, F is σ-Dedekind complete without order continuous norm and so there is a lattice embedding $i : \ell_\infty \to F$. Thus an operator $T := i \circ S : E \to F$ defined by $Tx = (\langle x, x_n^* \rangle)$ for all $x \in E$ is weak* (positively) p-convergent, but it is not plpc.

(2) $\Rightarrow$ (1) If E has the positively limited p-Schur property, then clearly each operator $T : E \to F$ is plpc. If F has order continuous norm and $T : E \to F$ is a weak* positively p-convergent operator, then we show that T is p-convergent. Let (x_n) be a weakly p-summable sequence in E. Then, (Tx_n) is a weakly p-summable positively limited sequence in F because T is weak* positively p-convergent. Since E has the weakly sequentially continuous lattice operations and order continuous norm, it is discrete [12, Corollary 2.3]. By [2, Theorem 2.5], (Tx_n) is a weakly p-summable limited sequence in F. Since F has the GP property [30],

$\|Tx_n\| \to 0$ and so the operator T is p-convergent. This completes the proof. □

As some consequences of Theorem 17, we obtain two following characterizations. Note that ℓ_∞ is a σ-Dedekind complete Banach lattice with the weakly sequentially continuous lattice operations.

Corollary 7. *For a Banach lattice E, the following are equivalent:*

(1) *each weak* positively p-convergent operator $T : E \to \ell_\infty$ is plpc.*
(2) *E has the positively limited p-Schur property.*

Corollary 8. *Let F be a σ-Dedekind complete Banach lattice with the weakly sequentially continuous lattice operations. Then; the following are equivalent:*

(1) *each weak* positively p-convergent operator $T : \ell_\infty \to F$ is plpc;*
(2) *F has order continuous norm.*

Following the same arguments as in the proof of Theorem 17, we conclude that, if F is a σ-Dedekind complete Banach lattice with the weakly sequentially continuous lattice operations, then each weak* positively p-convergent operator $T : E \to F$ is p-convergent if and only if E has the p-Schur property or F has order continuous norm.

An operator $T : X \to Y$ is called *weak* p-convergent* if, for each weakly p-summable sequence (x_n) in X and weak* null sequence (x_n^*) in Y^*, $x_n^*(Tx_n) \to 0$, see [31, Definition 4.1.1]. It is clear that each weak* p-convergent operator carries weakly p-compact sets into limited sets. If $T : X \to E$ is an operator and E has the positively limited p-Schur property, then it is easily verified that T is p-convergent if and only if T is weak* p-convergent if and only if T is weak* positively p-convergent.

Proposition 5. *Suppose that $R : E \to F$ is a positive operator. If $S : X \to E$ is a weak* positively p-convergent operator and $T : Y \to X$ is an operator, then the composition operator $R \circ S \circ T : Y \to F$ is a weak* positively p-convergent operator.*

Proof. Let W be a weakly p-compact set in Y. Then, $T(W)$ is a weakly p-compact set in X and so $S(TW)$ is a positively limited set in E (since S is a weak* positively p-convergent operator). Also, $R : E \to F$ is a positive operator and so $R(STW)$ is a positively limited set in F. Hence, $R \circ S \circ T$ is a weak* positively p-convergent operator. This completes the proof. □

We consider a result of the domination property of the weak* positively p-convergent operators.

Theorem 18. *Let $T : E \to F$ be a weak* positively p-convergent operator. If E is weak p-consistent and $0 \leq S \leq T$, then S is a weak* positively p-convergent operator.*

Proof. Let (x_n) be a weakly p-summable sequence in E and (x_n^*) be a positive weak* null sequence in F^*. By assumption, the sequence $(|x_n|)$ is weakly p-summable in E. Then,

$$x_n^*(S(x_n)) \leq x_n^*(S(|x_n|)) \leq x_n^*(T(|x_n|)) \to 0.$$

This proves that S is a weak* positively p-convergent operator. This completes the proof. □

5. Almost Limited p-convergent Operators

A Banach lattice E has the strong limited p-Schur property if and only if each almost limited sequence $(x_n) \in \ell_p^w(E)$ is norm null. Each discrete Banach lattice with order continuous norm has the strong limited p-Schur property [5].

In this section, we are going to study the class of almost limited p-convergent (alpc) operators and show that a Banach lattice E has the strong limited p-Schur property if and only if, for each Banach space Y, each operator $T : E \to Y$ is alpc. Throughout this section, we assume that $1 \leq p < \infty$, unless otherwise stated.

Definition 5. A bounded operator $T : E \to Y$ is said to be *almost limited p-convergent* (*shortly, alpc*) if T carries weakly p-summable almost limited sequences in E to norm null ones and the class of these operators is denoted by $C_{lp}^a(E, Y)$.

It is clear that every p-convergent operator is alpc and so it is lpc. To see that the converse does not hold in general, see the following example.

Example 8. The identity operator on each Banach lattice with the strong limited p-Schur property and, without the p-Schur property such as c_0, is alpc, but it is not p-convergent. Also, the identity operator on each Banach lattice with the limited p-Schur property and, without the strong limited p-Schur property such as $L^1[0,1]$, is lpc, but it is not alpc.

Theorem 19. *A bounded operator T on E is alpc if and only if, for each almost limited weakly p-compact set $A \subseteq E$, the set $T(A)$ is relatively compact.*

Proof. To show that an operator T is alpc, we assume that (x_n) is a weakly p-summable almost limited sequence in E. For each subsequence of (x_n), which is denoted again by (x_n), the sequence (Tx_n) is relatively compact and so it has a convergent and also weakly p-summable subsequence (Tx_{n_k}). Then, $\|Tx_{n_k}\| \to 0$ as $k \to \infty$ and this shows that T is alpc.

For the converse, let $A \subset E$ be an almost limited weakly p-compact set and T be an alpc operator on E. Then, every sequence (x_n) in A has a weakly p-convergent subsequence; denoted again by (x_n). Hence, the sequence $(x_n - x_m)$ is almost limited weakly p-summable and by hypothesis (Tx_n) is Cauchy and so is norm convergent. Thus, $T(A)$ is relatively compact. This completes the proof. □

A bounded operator $T : X \to Y$ between two Banach spaces is called *limited completely continuous* (*shortly, lcc*) if T carries weakly null limited sequences in X to norm null ones in Y. Each weakly compact operator is lcc [28].

The following example shows that two operators lcc and alpc are completely different in general:

Example 9. The identity operator on each Banach lattice with the GP property and without the strong limited p-Schur property such as $L^1[0,1]$ is lcc, but it is not alpc for all $p \geq 2$. Indeed, the Rademacher sequences (r_n) in $L^1[0,1]$ are weakly 2-summable (see [27, Proposition 3.6]) and almost limited (by the weak DP* property), but $\|r_n\| = 1$ for all $n \geq 1$. Also, the identity operator on a Banach lattice with the strong limited p-Schur and, without the GP property, is alpc, but it is not lcc.

In general, for the connection between three operators lcc, lpc and alpc, we have the following theorem:

Theorem 20. *Let E be a Banach lattice. Then we have the following:*

(1) *If E is Grothendieck (i.e. weak* convergent sequences in E^* are weakly convergent) with the property (d), then each alpc operator T from E into any Banach space is lcc.*
(2) *If E is σ-Dedekind complete Grothendieck, then each lpc operator T from E into any Banach space is lcc.*
(3) *If E^* has the weak* sequentially continuous lattice operations, then each lpc operator T from E into any Banach space is alpc. In fact, in this case, each lcc operator is alpc.*

Proof. (1) To prove that the operator T is lcc, it is enough to show that it is weakly compact. For this, we show that the operator T is order weakly compact (since every order weakly compact operator on a Grothendieck space is weakly compact [26, Theorem 5.3.13]). Let (x_n) be an order-bounded disjoint sequence in E. It is useful to know that (x_n) is weakly null (see [1, p. 192]). Actually, it has been shown that, (x_n) is weakly 1-summable and so weakly p-summable, by the well known fact $\ell_p^w(E) \subset \ell_q^w(E)$ for all $1 \leq p \leq q$.

On the other hand, E has the property (d) and so the sequence (x_n) is almost limited. Since the operator T is alpc, $\|Tx_n\| \to 0$. Thus, T is order weakly compact and the proof is completed.

(2) Just note that every order bounded disjoint sequence in a σ-Dedekind complete Banach lattice is limited. The rest of the proof is similar to the previous case.

(3) Let (x_n) be an almost limited weakly p-summable sequence in E. Since E^* has the weak* sequentially continuous lattice operations, by Ref. [2], (x_n) is limited weakly p-summable. So, by hypothesis, $\|Tx_n\| \to 0$ which implies that T is alpc. This completes the proof. $\square$

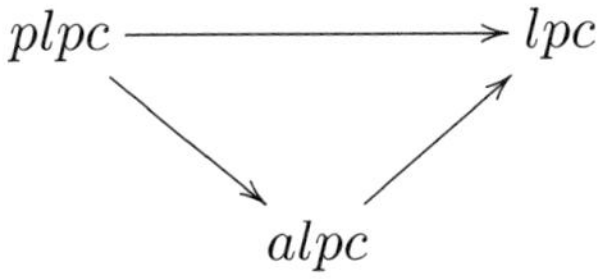

Recall that an operator $T : X \to E$ is called *almost limited* if $T(B_X)$ is an almost limited set in E or, equivalently, $\|T^* f_n\| \to 0$ for each disjoint weak* null sequence (f_n) in E^*.

Lemma 4. *For a Banach lattice E, the following statements hold:*

(1) *If (x_n) is an almost limited sequence in E, then the operator $T : \ell_1 \to E$, $T(b) = \sum_{n=1}^{\infty} b_n x_n$, $b = (b_n) \in \ell_1$ is almost limited.*
(2) *Let $1 < p < \infty$. If (x_n) is an almost limited weakly p-summable sequence in E, then the operator $T : \ell_{p^*} \to E$, $T(b) = \sum_{n=1}^{\infty} b_n x_n$, $b = (b_n) \in \ell_{p^*}$ is almost limited.*

Proof. To prove that $T(B_{\ell_1})$ is an almost limited set, let (f_m) be a disjoint weak* null sequence in E^*. Since (x_n) is an almost limited sequence in E, we have

$$\sup_{a=(a_j)\in B_{\ell_1}} |f_m(Ta)| = \sup_{a=(a_j)\in B_{\ell_1}} |T^* f_m(a)|$$
$$= \|T^*(f_m)\|_\infty = \sup_n |f_m(x_n)| \to 0.$$

Hence, $T(B_{\ell_1})$ is an almost limited set and so T is an almost limited operator.

For the second part use the similar techniques of [22, Lemma 3.23]. Note that the closed unit ball $B_{\ell_{p^*}}$ is weakly p-compact. This completes the proof. □

Theorem 21. *Let $1 < p < \infty$ and $T : E \to Y$ be an operator. The following statements are equivalent:*

(1) *T is an alpc operator.*
(2) *For every Banach space Z and any almost limited, weakly p-compact operator $S : Z \to E$, the operator $TS : Z \to Y$ is compact.*
(3) *For every almost limited operator $S : \ell_{p^*} \to E$, the operator $TS : \ell_{p^*} \to Y$ is compact.*

Proof. (1) ⇒ (2) If $S : Z \to E$ is an almost limited, weakly p-compact operator, then $S(B_Z)$ is an almost limited weakly p-compact subset of E. Since T is alpc, by Theorem 19, $TS(B_Z)$ is relatively compact and so the operator TS is compact.

(2) ⇒ (3) It is obvious (note that, $Id(\ell_{p^*}) \in W_p$).

(3) $\Rightarrow$ (1) Let $(x_n) \in \ell_p^w(E)$ be an almost limited sequence. By Lemma 4, the operator $S : \ell_{p^*} \to E$ defined by $S(b) = \sum_{n=1}^{\infty} b_n x_n$, $b = (b_n) \in \ell_{p^*}$ such that $S(e_n) = x_n$ is almost limited (note that $\ell_p^w(E) = L(\ell_{p^*}, E)$). Hence, the operator TS is compact and so $\|Tx_n\| = \|TS(e_n)\| \to 0$. This completes the proof. $\square$

Corollary 9. *Let $1 < p < \infty$ and E be a Banach lattice. Then the following statements are equivalent:*

(1) *E has the strong limited p-Schur property.*
(2) *For every Banach space Z, any almost limited weakly p-compact operator $S : Z \to E$ is compact.*
(3) *Each almost limited operator $S : \ell_{p^*} \to E$ is compact.*

In the following theorem, we give a characterization of the strong limited p-Schur property, with respect to alpc operators.

Theorem 22. *For a Banach lattice E, the following statements are equivalent:*

(1) *E has the strong limited p-Schur property,*
(2) *for each Banach space Y, $C_{lp}^a(E, Y) = L(E, Y)$,*
(3) *$C_{lp}^a(E, \ell_\infty) = L(E, \ell_\infty)$.*

Proof. (1) $\Rightarrow$ (2) Let $T : E \to Y$ be an operator and A be an almost limited weakly p-compact subset of E. By the strong limited p-Schur property of E, the sets A and $T(A)$ are relatively compact. Hence, by Theorem 19, T is an alpc operator.

(2) $\Rightarrow$ (3) It is obvious.

(3) $\Rightarrow$ (1) If E does not have the strong limited p-Schur property, then there is a weakly p-summable almost limited sequence (x_n) in E such that $\|x_n\| = 1$ for all $n \geq 1$. Choose a normalized sequence (x_n^*) in E^* such that $|\langle x_n, x_n^* \rangle| = 1$ for all $n \geq 1$. Then, the operator $T : E \to \ell_\infty$ defined by

$$Tx = (\langle x, x_n^* \rangle) \quad \text{for all } x \in E$$

is not alpc (since the sequence (x_n) is weakly p-summable and almost limited and $\|Tx_n\| \geq 1$ for all $n \geq 1$). This completes the proof. $\square$

Note that in Theorem 22 one cannot replace the strong limited p-Schur property of the domain E of related operators by their images. In fact, the Banach lattice c_0 has the strong limited p-Schur property, but the operator $T : L^1[0,1] \to c_0$ defined by

$$Tf = \left(\int_0^1 f(t) r_n(t) dt \right)_{n=1}^{\infty} \quad \text{for all } f \in L^1[0,1]$$

is not alpc for all $p \geq 2$ (since the Rademacher sequence $f_n(t) = r_n(t)$ is weakly 2-summable and almost limited, but $\|Tf_n\| = 1$).

Lemma 5 ([2]). *Let $T : E \to F$ be an order bounded operator from a Banach lattice E to a Banach lattice F with the property (d) and $A \subset E$ be an almost limited set. Then $T(A)$ is an almost limited set in F.*

Proposition 6. *Let E and F be two Banach lattices. Then, we have the following:*

(1) *If F has the property (d) and strong limited p-Schur property, then each order bounded operator $T : E \to F$ is alpc.*
(2) *If F has the strong limited p-Schur property, then each almost interval preserving operator $T : E \to F$ is alpc.*

Proof. (1) Let A be an almost limited weakly p-compact set in E and $T : E \to F$ be an order bounded operator. By Lemma 5, $T(A)$ is an almost limited set in F. On the other hand, F has the strong limited p-Schur property and so $T(A)$ is relatively compact. By Theorem 19, T is an alpc operator.

(2) Let A be an almost limited weakly p-compact set in E and $T : E \to F$ be an almost interval preserving operator. By [3, Theorem 3.8], $T(A)$ is an almost limited set and so it is relatively compact (by the strong limited p-Schur property of F). By Theorem 19, T is an alpc operator. This completes the proof. □

If $R : E \to F$ and $S : F \to X$ are two operators such that R is alpc, then $S \circ R$ is likewise alpc. In fact, if $(x_n) \in \ell_p^w(E)$ is an almost limited sequence and R is an alpc operator, then $\|Rx_n\| \to 0$ and so $\|S(Rx_n)\| \to 0$. Consequently, $S \circ R : E \to X$ is alpc. But, if S is alpc, then $S \circ R$ is not necessarily alpc. Indeed, an operator

$R : L^1[0,1] \to c_0$ defined by

$$Rf = \left(\int_0^1 f(t) r_n(t) dt \right)_{n=1}^{\infty} \quad \text{for all } f \in L^1[0,1],$$

where r_n is the n'th Rademacher function on $[0,1]$, is not alpc. If $S = Id_{c_0}$, then S is alpc, but $S \circ R = R$ is not alpc. However, we have the following result:

Theorem 23. *Let E and F be two Banach lattices. Then, the following statements hold:*

(1) *If F has the property (d) and $R : E \to F$ and $S : F \to X$ are two operators such that S is alpc and R is order bounded, then the composition operator $S \circ R$ is alpc.*
(2) *If $R : E \to F$ is an almost interval preserving operator and $S : F \to X$ is an operator such that S or R is alpc, then the composition operator $S \circ R$ is alpc.*

Proof. (1) Let $(x_n) \in \ell_p^w(E)$ be an almost limited sequence. By Lemma 5, (Rx_n) is a weakly p-summable and almost limited sequence in F. Since $S : F \to X$ is alpc, $\|S(Rx_n)\| \to 0$ and so $S \circ R$ is alpc.

(2) Let $(x_n) \in \ell_p^w(E)$ be an almost limited sequence and the operator S be alpc. Since R is an almost interval preserving operator, then the sequence (Rx_n) is a weakly p-summable almost limited sequence in F and so $\|S(Rx_n)\| \to 0$. Hence, the operator $S \circ R$ is alpc. If R is alpc, then clearly the operator $S \circ R$ is alpc. This completes the proof. □

Theorem 24. *Let $T : X \to E$ be a bounded linear operator. Then, for the following assertions,*

(1) *T is an almost limited and weakly p-compact operator,*
(2) *for each Banach space Y and each alpc operator $S : E \to Y$, the operator $S \circ T$ is compact,*
(3) *for each order bounded operator $S : E \to F$, where F has the property (d) and the strong limited p-Schur property, the operator $S \circ T$ is compact,*

it follows that the implications (1) $\Rightarrow$ (2) $\Rightarrow$ (3) are valid.

Proof. (1) $\Rightarrow$ (2) If T is almost limited and weakly p-compact, then $T(B_X)$ is an almost limited weakly p-compact set in E. Since $S : E \to Y$ is alpc, $S(T(B_X))$ is relatively compact, that is, the operator $S \circ T$ is compact.

(2) $\Rightarrow$ (3) It follows from Proposition 6. This completes the proof. $\square$

Theorem 25. *Let E and F be two Banach lattices with the weakly sequentially continuous lattice operations and F be σ-Dedekind complete. Then, the following statements are equivalent:*

(1) *Each positive operator $T : E \to F$ is alpc.*
(2) *E has the strong limited p-Schur property or F has order continuous norm.*

Proof. (1) $\Rightarrow$ (2) Suppose that E does not have the strong limited p-Schur property and F does not have order continuous norm. Then, there is a weakly p-summable almost limited sequence (x_n) in E with $\|x_n\| > \epsilon$ for some $\epsilon > 0$. Since E has the weakly sequentially continuous lattice operations, the sequence $|x_n| \xrightarrow{w} 0$ in E. So, by [16, Corollary 2.6], there is a norm bounded disjoint sequence (x_n^*) in E_+^* such that $|x_n^*(x_n)| > \epsilon$ for all $n \geq 1$. Consider the positive operator $S : E \to \ell_\infty$ defined by

$$Sx = (\langle x, x_n^* \rangle) \text{ for all } x \in E.$$

Also, F is σ-Dedekind complete without order continuous norm and so, by [26, Corollary 2.4.3], there is a lattice embedding $i : \ell_\infty \to F$. Thus, the positive operator $T := i \circ S : E \to F$ defined by $Tx = (\langle x, x_n^* \rangle)$ for all $x \in E$ is not alpc. (Indeed, the sequence (x_n) in E is weakly p-summable almost limited, but $\|Tx_n\| = \|ioS(x_n)\| \nrightarrow 0$.)

(2) $\Rightarrow$ (1) If E has the strong limited p-Schur property, then clearly each operator $T : E \to F$ is alpc. If F has order continuous norm and $T : E \to F$ is a positive operator, then we show that T is alpc. Let (x_n) be a weakly p-summable almost limited sequence in E. Since F has the property (d) and T is positive (and so order bounded), by Proposition 6, (Tx_n) is a weakly p-summable almost limited sequence in F. Since F has the weakly sequentially continuous lattice operations, $|Tx_n| \xrightarrow{w} 0$. On the other hand, F has order

continuous norm and so each disjoint sequence $(x_n^*) \in B_{F_+^*}$ is weak* null and so $x_n^*(Tx_n) \to 0$. Now, by [16, Corollary 2.6], $\|Tx_n\| \to 0$ which implies that T is alpc. This completes the proof. $\square$

Corollary 10. *Let E be a Banach lattice with the weakly sequentially continuous lattice operations. Then, the following statements are equivalent:*

(1) *Each positive operator $T : E \to \ell_\infty$ is alpc.*
(2) *E has the strong limited p-Schur property.*

Corollary 11. *Let F be a σ-Dedekind complete Banach lattice with the weakly sequentially continuous lattice operations. Then the following statements are equivalent:*

(1) *Each positive operator $T : \ell_\infty \to F$ is alpc.*
(2) *F has order continuous norm.*

6. Almost Limited Weakly p-convergent Operators

In this section, we consider another class of operators related to almost limited sets: almost limited weakly p-convergent operators. Recall from [4] that a bounded linear operator $T : E \to X$ is *almost limited weakly completely continuous* (*abbr. alwcc*) if T carries weakly Cauchy almost limited sequences of E to weakly convergent sequences of X.

Definition 6. A bounded linear operator $T : E \to X$ is *almost limited weakly p-convergent* (*shortly, alwpc*) if T carries weakly p-Cauchy almost limited sequences of E to weakly convergent sequences of X.

The set of all alwpc operators from E into X is denoted by $\mathcal{L}_{alwpc}(E, X)$. Clearly, $\mathcal{L}_{alwpc}(E, X)$ is a linear subspace of $\mathcal{L}(E, X)$, where $\mathcal{L}(E, X)$ is the class of all bounded linear operators from E to X.

(1) If $T : E \to X$ is an alwpc operator, then T is an lwpc operator.
(2) If $T : E \to X$ is a plwpc operator, then T is an alwpc operator.

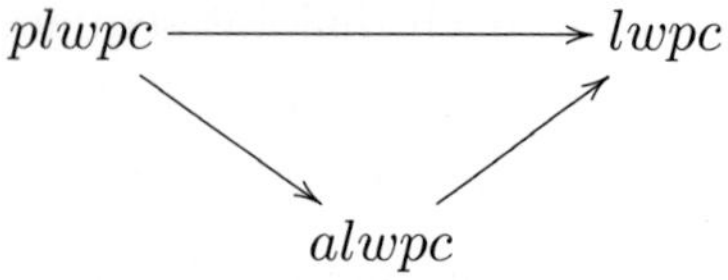

A Banach lattice E has the *strong BD property* if each weakly Cauchy almost limited sequences of E is weakly convergent [4].

Definition 7. A Banach lattice E has the *strong BD property of order p (shortly, p-strong BD property)* if each weakly p-Cauchy almost limited sequences of E is weakly convergent.

Example 10.

(1) The identity operator on each Banach lattice with the p-BD property, and without the p-positive BD property is lwpc, but it is not plwpc.
(2) The identity operator on each Banach lattice with the p-BD property and without the p-strong BD property is alwpc, but it is not plwpc.
(3) The identity operator on each Banach lattice with the p-strong BD property and without the p-positive BD property is lwpc, but it is not alwpc.

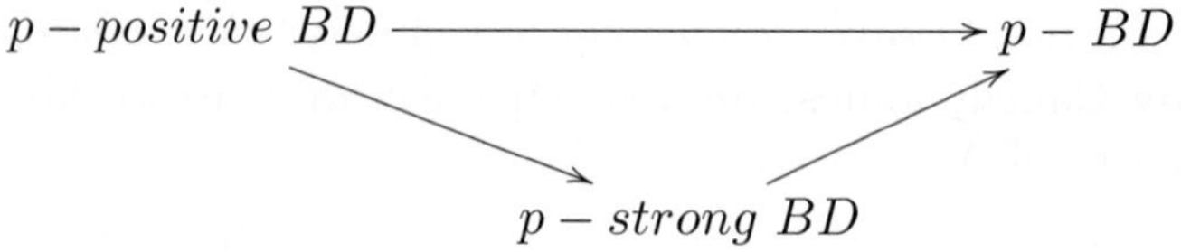

It is trivial that TS is alwpc if $F \xrightarrow{S} X \xrightarrow{T} Y$ where S is alwpc and T is a bounded linear operator. It should be noted that order bounded operators between Banach lattices preserve almost limited sets, see [3]. As a result, if $E \xrightarrow{T} F \xrightarrow{S} X$ where S is alwpc and T is order bounded, then ST is also alwpc.

Lemma 6 ([3]). *Let $T : E \to F$ be a positive operator from a Banach lattice E to a Banach lattice F and $A \subset E$ be an almost limited set. Then, $T(A)$ is a positively limited set in F.*

Theorem 26. *An operator T on E is alwpc if and only if for each almost limited weakly p-precompact set $A \subseteq E$, the set $T(A)$ is relatively weakly compact.*

Proof. To show that an operator T is alwpc, we assume that (x_n) is a weakly p-Cauchy almost limited sequence in E. For each subsequence of (x_n), which is denoted again by (x_n), the sequence (Tx_n) is relatively weakly compact and so it has a weakly convergent subsequence (Tx_{n_k}). On the other hand, (Tx_n) is weakly Cauchy and so on. Then, (Tx_n) is weakly convergent which implies that T is alwpc. For the converse, let $A \subset E$ be a weakly p-precompact almost limited set and $T : E \to Y$ be an alwpc operator. Then, every sequence (x_n) in A has a weakly p-Cauchy and almost limited subsequence, denoted again by (x_n) almost limited. Hence, by hypothesis, (Tx_n) is weakly convergent in Y. Thus, $T(A)$ is relatively weakly compact. This completes the proof. □

The following theorem provides a characterization of the p-strong BD property with respect to alwpc operators.

Theorem 27. *For a Banach lattice E, the following are equivalent:*

(1) *E has the p-strong BD property;*
(2) *for each Banach space Y, $\mathcal{L}_{alwpc}(E, Y) = \mathcal{L}(E, Y)$;*
(3) *$\mathcal{L}_{alwpc}(E, \ell_\infty) = \mathcal{L}(E, \ell_\infty)$.*

Proof. (1) $\Rightarrow$ (2) $\Rightarrow$ (3) Obvious.

(3) $\Rightarrow$ (1) Assume to the contrary that E does not have the p-strong BD property. Then, there exists a weakly p-Cauchy and almost limited sequence (x_n) in E which is not weakly convergent. Let F be the closed subspace of E spanned by this sequence. Since F is separable, F is isomorphic to a closed subspace of ℓ_∞. Let T be such an isomorphic embedding from F to ℓ_∞. Since ℓ_∞ is injective, T has an extension to all of E, denoted again by T. Clearly, T is not alwpc, and this leads to a contradiction. This completes the proof. □

Proposition 7. *If $T : E \to F$ is a positive operator, and F has the p-strong BD property, then T is alwpc.*

Proof. Note that from [3], for each almost limited weakly p-precompact set $A \subset E$, $T(A)$ is almost limited (and also weakly p-precompact) in F. It follows from the p-strong BD property of F that $T(A)$ is relatively weakly compact. This proves that T is alwpc. This completes the proof. □

Theorem 28. *Let E and F be two Banach lattices and F has the p-strong BD property. Then, each positive operator $T : E \to F$ is alwpc.*

Proof. Let A be an almost limited weakly p-precompact set in E and $T : E \to F$ be a positive operator. By [3], $T(A)$ is an almost limited and so it is relatively weakly compact (by the p-strong BD property of F). By Theorem 26, T is an alwpc operator. This completes the proof. □

Theorem 29. *Let E and F be two Banach lattices and F has the p-strong BD property. Then, each almost interval preserving operator $T : E \to F$ is alwpc.*

Proof. Let A be an almost limited weakly p-precompact set in E and $T : E \to F$ be an almost interval preserving operator. By [3, Theorem 3.8], $T(A)$ is almost limited and so it is relatively weakly compact (by the p-strong BD property of F). By Theorem 26, T is an alwpc operator. This completes the proof. □

A Banach lattice E has the *dual Schur property* if every disjoint weak*-null sequence in E^* is norm null [25]. One easily verifies that E has the dual Schur property if and only if the closed unit ball B_E is an almost limited set.

Proposition 8. *Let E be a weakly p-precompact Banach lattice with the dual Schur property and F be a Banach lattice with the p-strong BD property. Then, every positive operator $T : E \to F$ is compact.*

Proof. Since E is weakly p-precompact with the dual Schur property, then the closed unit ball B_E is weakly p-precompact and almost limited. By Theorem 29, $T(B_E)$ is relatively compact and so T is a compact operator. This completes the proof. □

If $R : E \to F$ and $S : F \to X$ are two operators such that R is alwpc, then $S \circ R$ is likewise alwpc. In fact, if (x_n) is an almost limited weakly p-Cauchy sequence in E, and R is an alwpc operator, then (Rx_n) is weakly convergent, and so $(S(Rx_n))$ is weakly convergent.

Consequently, $S \circ R : E \to X$ is alwpc. However, if S is alwpc, then $S \circ R$ is not necessarily alwpc.

However, one easily verifies that if S is alwpc and R is positive, then $S \circ R$ is alwpc. It is enough to note that in this case, for each almost limited weakly p-Cauchy sequence (x_n), (Rx_n) is a weakly p-Cauchy and almost limited sequence. Hence, alwpc-ness of S ensures that $(S(Rx_n))$ is weakly convergent. This proves that $S \circ R$ is alwpc.

Theorem 30. *Let E and F be Banach lattices such that F is σ-Dedekind complete. Then the following assertions are equivalent:*

(1) *Every positively limited operator $T : E \to F$ is alwpc.*
(2) *Either E or F has p-strong BD property.*

Proof. (1) $\Rightarrow$ (2) This follows easily from [3].

(2) $\Rightarrow$ (1) We assume to the contrary that neither E nor F has p-strong BD property. Then, there exists a weakly p-Cauchy and almost limited sequence (x_n) in E such that is not weakly convergent. Let W be the closed subspace of E spanned by this sequence. Then, W is separable, and hence, it is isomorphic to a closed subspace of ℓ_∞. Let Q be such an isomorphic embedding from W to ℓ_∞. Since ℓ_∞ is injective, we can extend Q to all of E. Since F does not have the p-strong BD property, it does not have the strong BD property. Hence, F is a σ-Dedekind complete Banach lattice without order continuous norm; there exists a lattice embedding $S : \ell_\infty \to F$ (cf. [26, Corollary 2.4.3]). Since B_{ℓ_∞} is almost limited in ℓ_∞ and S is a positive operator, S is an almost limited operator. Then, $T := S \circ Q : E \to F$ is an almost limited operator. However, T is not alwpc. This leads to a contradiction. This completes the proof. □

Theorem 30 has the two following consequences:

Corollary 12. *For a Banach lattice E, the following are equivalent:*

(1) *Each positively limited operator $T : E \to \ell_\infty$ is alwpc.*
(2) *E has order continuous norm.*

The order continuity of the norm is also characterized.

Corollary 13. *Let F be a σ-Dedekind complete Banach lattice. Then the following are equivalent:*

(1) *Each positively limited operator $T : c \to F$ is alwpc.*
(2) *F has order continuous norm.*

Recently, the authors introduced the p-weak DP* property in [5]. A Banach lattice E has the *p-weak DP* property* if every relatively weakly p-compact subset of E is almost limited; alternatively, $x_n^*(x_n) \to 0$ for every weakly p-summable sequence (x_n), and every disjoint weak*-null sequence $(x_n^*) \subset (E^*)$. It can be proved that E has the p-weak DP* property if and only if every relatively weakly p-precompact subset of E is almost limited.

For our convenience, let $\mathcal{L}_{wpc}(X, Y)$ denote the set of all weakly p-convergent operators from X into Y. $\mathcal{L}_{wpc}(E, X) \subset \mathcal{L}_{alwpc}(E, X)$ holds.

Theorem 31. *If E has the p-weak DP* property, then every alwpc operator $T : E \to Y$ is weak p-convergent for every Banach space Y.*

Proof. Let (x_n) be a weakly p-Cauchy sequence. Then by hypothesis, (x_n) is almost limited. Since $T : E \to Y$ is an alwpc operator, (Tx_n) is weakly convergent and so T is weak p-convergent. This completes the proof. □

We now turn our attention to the domination problem of alwpc operators. If $0 \leq S \leq T : E \to F$ are two positive operators between Banach lattices and T is alwpc, is S then necessarily alwpc? We pose a question and have to leave it open.

7. Weak* Almost p-convergent Operators

In this section, we introduce and study the class of weak* almost p-convergent operators and discuss some properties of them related to alpc operators and the strong limited p-Schur property.

Definition 8. An operator $T : X \to E$ is called *weak* almost p-convergent* if, for each weakly p-summable sequence (x_n) in X and each disjoint weak* null sequence (x_n^*) in E^*, $x_n^*(Tx_n) \to 0$.

Theorem 32. *Let $T : X \to E$ be an operator. The following statements are equivalent:*

(1) *T is a weak* almost p-convergent operator;*
(2) *T carries weakly p-compact subsets of X to almost limited subsets of E;*
(3) *for any Banach space Z and each weakly p-compact operator S from Z to X, the operator $T \circ S$ is almost limited;*
(4) *for any bounded linear operator $S : \ell_{p^*} \to E$, $T \circ S$ is almost limited.*

Proof. (1) $\Rightarrow$ (2) Let W be a weakly p-compact set in X, we show that $T(W)$ is an almost limited set in E; that is, for each disjoint weak* null sequence (x_n^*) in E^* and each sequence (x_n) in W, $x_n^*(Tx_n) \to 0$. Since W is weakly p-compact, (x_n) has a weakly p-convergent subsequence converging to $x \in X$ which is denoted by (x_n) again. Then, $(x_n - x)$ and so $(Tx_n - Tx)$ are weakly p-summable, and by the definition of weak* almost p-convergent operators, $x_n^*(Tx_n) = x_n^*(Tx_n - Tx) + x_n^*(Tx) \to 0$.

(2) $\Rightarrow$ (3) Obvious.

(3) $\Rightarrow$ (4) It is clear (since $B_{\ell_{p^*}}$ is a weakly p-compact set).

(4) $\Rightarrow$ (1) We show that, for each disjoint weak* null sequence (x_n^*) in E^* and every weakly p-summable sequence (x_n) in X, $x_n^*(Tx_n) \to 0$. An operator $S : \ell_{p^*} \to E$ defined by $S(b) = \sum_{n=1}^{\infty} b_n x_n$, $b = (b_n) \in \ell_{p^*}$ is weakly p-compact and so $T \circ S$ is almost limited. Then, $TS(B_{\ell_{p^*}})$ is an almost limited set which implies that $x_n^*(Tx_n) \to 0$. This completes the proof. □

The following result characterizes the class of Banach lattices with the p-weak DP* property by weak* almost p-convergent operators:

Corollary 14. *Let E be a Banach lattice. Then, the following statements are equivalent:*

(1) *E has the p-weak DP* property;*
(2) *the identity operator on E is weak* almost p-convergent;*
(3) *every weakly p-compact operator T from an arbitrary Banach space Z to E is an almost limited operator;*
(4) *any bounded linear operator $S : \ell_{p^*} \to E$ is almost limited.*

Theorem 33. *Let E be a Banach lattice with the property (d) and $T : X \to E$ be an operator. Then, the following statements are equivalent:*

(1) *T is a weak* almost p-convergent operator;*
(2) *for each positive operator S from E to a Banach lattice F with the property (d) and the strong limited p-Schur property, the operator SoT is p-convergent;*
(3) *for each positive operator $S : E \to c_0$, the operator SoT is p-convergent.*

Proof. (1) ⇒ (2) Let W be a weakly p-compact set in X. By hypothesis (1), $T(W)$ is an almost limited and weakly p-compact set in E. Since $S : E \to F$ is order bounded and F is a Banach lattice with the property (d) and strong limited p-Schur property, by Proposition 6, S is alpc and so $ST(W)$ is relatively compact. Hence, the operator $S \circ T$ is p-convergent.

(2) ⇒ (3) It is clear.

(3) ⇒ (1) We show that for each disjoint weak* null sequence (x_n^*) in E^* and every weakly p-summable sequence (x_n) in X, $x_n^*(Tx_n) \to 0$. Consider the operator $S : E \to c_0$ defined by $Sx = (x_n^*(x))$ for all $x \in E$. Since E has the property (d), $|x_n^*| \xrightarrow{w^*} 0$ and so we can assume that (x_n^*) is positive. Then, S is a positive operator and by hypothesis, $S \circ T$ is p-convergent. Thus, $\|S(T(x_n))\| \to 0$, which implies that $x_n^*(Tx_n) \to 0$. This completes the proof. □

Corollary 15. *Let $T : X \to E$ be a weak* almost p-convergent operator. The following statements hold:*

(1) *If E has the strong limited p-Schur property, then T is p-convergent.*
(2) *For each Banach space Y and each alpc operator $S : E \to Y$, the operator $SoT : X \to Y$ is p-convergent.*

Theorem 34. *Let $T : X \to E$ be an operator. The following statements are equivalent:*

(1) *T is a weak* almost p-convergent operator,*
(2) *for each disjoint operator $S : E \to c_0$, the operator $S \circ T$ is p-convergent.*

Proof. (1) ⇒ (2) Let W be a weakly p-compact set in X, then $T(W)$ is an almost limited set in E. Since S is a disjoint operator by [20, Proposition 2.3], $S(T(W))$ is relatively compact. Hence, the operator $S \circ T$ is p-convergent.

(2) ⇒ (1) We show that for each disjoint weak* null sequence (x_n^*) in E^*, and every weakly p-summable sequence (x_n) in X, $x_n^*(Tx_n) \to 0$. The operator $S : E \to c_0$ defined by $Sx = (x_n^*(x))$ for all $x \in E$ is a disjoint operator. By hypothesis, $S \circ T$ is p-convergent and so $\|S(T(x_n))\| \to 0$, which implies that $x_n^*(Tx_n) \to 0$. This completes the proof. □

Theorem 35. *Let E and F be two Banach lattices such that F is σ-Dedekind complete with weakly sequentially continuous lattice operations. Then, the following statements are equivalent:*

(1) *Each weak* almost p-convergent operator $T : E \to F$ is alpc.*
(2) *E has the strong limited p-Schur property or F has order continuous norm.*

Proof. (2) ⇒ (1) If E has the strong limited p-Schur property, then clearly each operator $T : E \to F$ is alpc. If F has order continuous norm and $T : E \to F$ is weak* almost p-convergent operator, then we show that T is p-convergent. Let (x_n) be a weakly p-summable in E. Then, (x_n) and so (Tx_n) are weakly null sequences. Since F has the weakly sequentially continuous lattice operations, $|Tx_n| \xrightarrow{w} 0$. On the other hand, F has order continuous norm and so each disjoint sequence $(x_n^*) \in B_{F_+^*}$ is weak* null. Also, the operator T is weak* almost p-convergent and so $x_n^*(Tx_n) \to 0$. Now, by [16, Corollary 2.6], $\|Tx_n\| \to 0$ which implies that T is p-convergent.

(1) ⇒ (2) Suppose that E does not have the strong limited p-Schur property and F does not have order continuous norm. Then, there is a weakly p-summable and almost limited sequence (x_n) in E with $\|x_n\| > \epsilon$, for some $\epsilon > 0$. So, there is a sequence (x_n^*) in E^* such that $\|x_n^*\| = 1$ and $x_n^*(x_n) = \|x_n\|$. Consider the operator $S : E \to \ell_\infty$ defined by

$$Sx = (\langle x, x_n^* \rangle) \text{ for all } x \in E.$$

Also, F is σ-Dedekind complete without order continuous norm and so there is a lattice embedding $i : \ell_\infty \to F$. Thus, an operator

$T := ioS : E \to F$ defined by $Tx = (\langle x, x_n^* \rangle)$ for all $x \in E$ is weak* (almost) p-convergent, but it is not alpc. This completes the proof. □

Corollary 16. *For a Banach lattice E, the following statements are equivalent:*

(1) *E has the strong limited p-Schur property.*
(2) *Each weak* almost p-convergent operator $T : E \to \ell_\infty$ is alpc.*

Corollary 17. *Let F be a σ-Dedekind complete with the weakly sequentially continuous lattice operations. Then the following statements are equivalent:*

(1) *F has order continuous norm.*
(2) *Each weak* almost p-convergent operator $T : \ell_\infty \to F$ is alpc.*

By the same method, we conclude with the following proposition:

Proposition 9. *Let E and F be two Banach lattices such that F is σ-Dedekind complete with the weakly sequentially continuous lattice operations. Then, the following statements are equivalent:*

(1) *Each weak* almost p-convergent operator $T : E \to F$ is p-convergent.*
(2) *E has the p-Schur property or F has order continuous norm.*

Proposition 10. *Let $T : X \to E$ be an operator and E has the strong limited p-Schur property. The following statements are equivalent:*

(1) *T is a p-convergent operator.*
(2) *T is a weak* p-convergent operator.*
(3) *T is a weak* almost p-convergent operator.*

Proposition 11. *Suppose that F has the property (d) and $R : E \to F$ is an order bounded operator. If $S : X \to E$ is a weak* almost p-convergent operator and $T : Y \to X$ is an operator. Then the composition operators $R \circ S \circ T : Y \to F$ is a weak* almost p-convergent operator.*

Proof. Let W be a weakly p-compact set in Y, then $T(W)$ is a weakly p-compact set in X and so $ST(W)$ is an almost limited set in E (since S is a weak* almost p-convergent operator). Also, $R : E \to F$ is an order bounded operator and F has the property (d), and so by Proposition 6, $RST(W)$ is an almost limited set in F. Hence, $R \circ S \circ T$ is a weak* almost p-convergent operator. This completes the proof. □

Note that the previous theorem is also valid if R is an almost interval preserving operator or R is a positive, one to one, onto operator. In fact, in this case, almost limited sets are preserved by the operator R.

Theorem 36. *Let $T : E \to F$ be a weak* almost p-convergent operator. If E is weak p-consistent, F has the property (d) and $0 \leq S \leq T$, then S is a weak* almost p-convergent operator.*

Proof. Let (x_n) be a weakly p-summable sequence in E and (x_n^*) be a disjoint weak* null sequence in F^*. By assumption, the sequence $(|x_n|)$ is weakly p-summable in E. Also, F has the property (d) and so $|x_n^*| \xrightarrow{w^*} 0$ in F. Then,

$$x_n^*(S(x_n)) \leq |x_n^*|(S|x_n|) \leq |x_n^*|(T|x_n|) \to 0.$$

This proves that S is a weak* almost p-convergent operator. This completes the proof. □

References

[1] C. D. Aliprantis and O. Burkishaw. *Positive Operators*. Academic Press, New York, 1978.

[2] H. Ardakani and J. X. Chen. Positively limited sets in Banach lattices. *J. Math. Anal. Appl.*, 526(1), 2023, 127220. https://doi.org/10.1016/j.jmaa.2023.127220.

[3] H. Ardakani, S. M. Moshtaghioun, S. M. S. Modarres Mosadegh and M. Salimi. The strong Gelfand-Phillips property in Banach lattices. *Banach J. Math. Anal.*, 10, 2016, 15–26.

[4] H. Ardakani, M. Salimi and S. M. Moshtaghioun. The strong BD property in Banach lattices. *Filomat*, 37, 2023, 3063–3073.

[5] H. Ardakani and K. Taghavinejad. The strong limited p-Schur property in Banach lattices. *OaM*, 16, 2022, 811–825.

[6] H. Ardakani and K. Amjadi. The positively limited p-Schur property in Banach lattices. *Ann. Funct. Anal.*, 16, 2023. https://doi.org/10.1007/s43034-023-00292-y.

[7] H. Ardakani and K. Amjadi. Positively limited p-convergent and weak* positively p-convergent operators. *OaM*, 18, 2024, 361–374.

[8] H. Ardakani and F. Vali. On almost limited p-convergent operators on Banach lattices. *Positivity*, 28, 2024. https://doi.org/10.1007/s11117-024-01040-9.

[9] J. Bourgain and J. Diestel. Limited operators and strict cosingularity. *Math. Nachr.,* 119, 1984, 55–58.

[10] J. Castillo and F. Sanchez. Dunford-Pettis-like properties of continuous vector function spaces. *Rev. Math.*, 6, 1993, 43–59.

[11] J. X. Chen, Z. L. Chen and G. X. Ji. Almost limited sets in Banach lattices. *J. Math. Anal. Appl.*, 412, 2014, 547–563.

[12] Z. L. Chen and A. W. Wickstead. Relative weak compactness of solid hulls in Banach lattices. *Indag. Math. N.S.*, 9, 1995, 187–196.

[13] M. B. Dehghani, S. M. Moshtaghioun and M. Dehghani. On the limited p-Schur property of some operator spaces. *IJAA*, 16, 2018, 50–61.

[14] J. Diestel. *Sequences and Series in Banach Spaces.* Graduate Texts in Mathematics, Vol. 92. Springer–Verlag, Berlin, 1984.

[15] J. Diestel. *Absolutely Summing Operators.* Cambridge University Press, 1995.

[16] P. G. Dodds and D. H. Fremlin. Compact operators in Banach lattices. *Isr. J. Math.*, 34, 1979, 287–320.

[17] A. Elbour. Some characterizations of almost limited operators. *Positivity*, 21, 2017, 865–874.

[18] G. Emmanuele. On Banach spaces with the Gelfand–Phillips property, III. *J. Math. Pures Appl.*, 72, 1993, 327–333.

[19] J. H. Fourie and E. D. Zeekoei. DP*-properties of order p on Banach spaces. *Quaest. Math.*, 37, 2014, 349–358.

[20] P. Galindo and V. C. C. Miranda. Grothendieck-type subsets of Banach lattices. *J. Math. Anal. Appl.*, 506, 2022, 1–14.

[21] P. Galindo and V. C. C. Miranda. Some properties of p-limited sets. *Proc. Am. Math. Soc.*, 152(2), 2024, 749–763.

[22] I. Ghenciu. Some classes of Banach spaces and complemented subspaces of operators. *Adv. Oper. Theory*, 4, 2019, 369–387.

[23] I. Ghenciu. The BD property in spaces of compact operator. *Ann. Funct. Anal.*, 7, 2016, 635–645.

[24] M. Hajji and M. Mahfoudhi. LW-compact operators and domination problem. *Positivity*, 25, 2021, 1959–1972.

[25] N. Machrafi, A. Elbour and M. Moussa. Some characterizations of almost limited sets and applications, 2013. arXiv:1312.2770v1.

[26] P. Meyer-Nieberg. *Banach Lattices.* Universitext. Springer-Verlag, Berlin, 1991.

[27] C. Palazuelos, E. A. Sanchez Perez and P. Tradacete. Maurey-Rosenthal factorization for p-summing operators and Dodds-Fremlin domination. *J. Oper. Theory*, 68, 2012, 205–222.

[28] M. Salimi and S. M. Moshtaghioun. The Gelfand-Phillips property in closed subspaces of some operator spaces. *Banach J. Math. Anal.*, 2, 2011, 84–92.

[29] W. Wnuk. On the dual positive Schur property in Banach lattices. *Positivity*, 2, 2012, 759–773.

[30] W. Wnuk. *Banach Lattices with Order Continuous Norms.* Polish Scientific Publishers PWN, Warsaw, 1999.

[31] E. Zeekoei. Classes of Dunford-Pettis-type operators with applications to Banach spaces and Banach lattices. Ph.D. Thesis, 2017.

https://doi.org/10.1142/9789819820498_0002

Chapter 2

Pairs of Cube-Free Values of the Type $[n^c \tan^\theta(\log n)]$, $[n^c \tan^\theta(\log n)] + 1$

S. I. Dimitrov

Faculty of Applied Mathematics and Informatics,
Technical University of Sofia, Bulgaria
Institute of Biophysics and Biomedical Engineering,
Bulgarian Academy of Sciences, Bulgaria
sdimitrov@tu-sofia.bg
xyzstoyan@gmail.com

Let $[\cdot]$ be the floor function. In this chapter, we prove that when $1 < c < \frac{3}{2}$ and $\theta > 1$ is a fixed, then there exist infinitely many cube-free pairs of the form $[n^c \tan^\theta(\log n)]$, $[n^c \tan^\theta(\log n)] + 1$.

1. Notations

Let x be a sufficiently large positive number. By ε we denote an arbitrary small positive constant, not the same in all appearances. We denote by $\mu(n)$ the Möbius function and by $\tau(n)$ the number of positive divisors of n. As usual, $[t]$ and $\{t\}$ denote the integer part, respectively, the fractional part of t. Moreover, $e(y) = e^{2\pi iy}$. Instead of $m \equiv n \pmod k$, we write for simplicity $m \equiv n\,(k)$. Throughout this chapter, we suppose that $1 < c < \frac{3}{2}$ and $\gamma = \frac{1}{c}$. Assume that $\theta > 1$ is a fixed. Denote

$$z = x^{\frac{2c-1}{4}}; \tag{1.1}$$

$$\psi(t) = \{t\} - 1/2; \tag{1.2}$$

$$\Delta_1 = e^{\pi\left[\frac{\log x}{\pi}\right] + \arctan 1}; \tag{1.3}$$

$$\Delta_2 = e^{\pi\left[\frac{\log x}{\pi}\right] + \arctan 2}; \tag{1.4}$$

$$S_c(x) = \sum_{\substack{\Delta_1 \leq n < \Delta_2 \\ [n^c \tan^\theta(\log n)],[n^c \tan^\theta(\log n)]+1 \text{ are cube-free}}} 1. \tag{1.5}$$

2. Introduction and Statement of the Result

The problem for the consecutive cube-free numbers arose in 1932 when Carlitz [1] showed that

$$\sum_{\substack{n \leq x \\ n,\, n+1 \text{ are cube-free}}} 1 = \prod_p \left(1 - \frac{2}{p^3}\right) x + \mathcal{O}\big(x^{\lambda+\varepsilon}\big), \tag{2.1}$$

where $\lambda = \frac{1}{2}$. Afterward, the reminder term of (2.1) was improved by Mirsky [2], Meng [3] and Brandes [4]. The best result up to now belongs to Reuss [5] with $\lambda = \frac{169}{432}$.

Another interesting problem we know in number theory is consecutive k-free numbers of the form $[n^c], [n^c]+1$. In 2018, the author [6] using the method of Cao and Zhai [7] proved that when $1 < c < \frac{22}{13}$, then

$$\sum_{\substack{n \leq x \\ [n^c],\, [n^c]+1 \text{ are square-free}}} 1 = \prod_p \left(1 - \frac{2}{p^2}\right) x + \mathcal{O}(x^{1-\delta^2/2}),$$

where $0 < \delta < \frac{22-13c}{5(14c-1)}$. In the same year, the author [8] using the method of Zhang and Li [9] showed that when $1 < c < \frac{31}{17}$, then

$$\sum_{\substack{n \leq x \\ [n^c],\, [n^c]+1 \text{ are cube-free}}} 1 = \prod_p \left(1 - \frac{2}{p^3}\right) x + \mathcal{O}(x^{1-\delta^2/2}),$$

where $0 < \delta < \min\{\frac{31-17c}{9c-9}, 10^{-10}\}$. Recently, the author [10] proved that there exist infinitely many prime numbers of the form $[n^c \tan^\theta(\log n)]$. More precisely, we showed that when $1 < c < \frac{12}{11}$ and $\theta > 1$ is a fixed, then the lower bound

$$\sum_{\substack{\Delta_1 \leq n < \Delta_2 \\ [n^c \tan^\theta(\log n)] = p}} 1 \gg \frac{x}{\log x}$$

holds. Here, Δ_1 and Δ_2 are denoted by (1.3) and (1.4).

Motivated by these results, we prove that there exist infinitely many cube-free pairs of the form $[n^c \tan^\theta(\log n)]$, $[n^c \tan^\theta(\log n)] + 1$. More precisely we establish the following theorem.

Theorem 1. *Let* $1 < c < \frac{3}{2}$ *and* $\theta > 1$ *is a fixed. Then, for the sum* (1.5), *the lower bound*

$$S_c(x) \gg x \tag{2.2}$$

holds.

3. Preliminary Lemmas

Lemma 1. *Let* $H \geq 1$. *Then, for the function defined by* (1.2), *we have*

$$\psi(t) = \sum_{1 \leq |h| \leq H} a(h) e(ht) + \mathcal{O}\left(\sum_{|h| \leq H} b(h) e(ht) \right),$$

where

$$a(h) \ll \frac{1}{|h|}, \quad b(h) \ll \frac{1}{H}. \tag{3.1}$$

Here, $\psi(t)$ *is denoted by* (1.2).

Proof. See Ref. [11]. □

Lemma 2. *Let* $q \geq 0$ *be an integer. Suppose that* $f(t)$ *has* $q + 2$ *continuous derivatives on* I *and that* $I \subseteq (N, 2N]$. *Assume also that* *there is some constant F such that*

$$|f^{(r)}(t)| \asymp FN^{-r} \tag{3.2}$$

for $r = 1, \ldots, q+2$. Let $Q = 2^q$. Then,

$$\left|\sum_{n\in I} e(f(n))\right| \ll F^{\frac{1}{4Q-2}} N^{1-\frac{q+2}{4Q-2}} + F^{-1}N.$$

The implied constant depends only upon the implied constants in (3.2).

Proof. See Ref. [12, Theorem 2.9]. □

4. Beginning of the Proof

Using (1.1), (1.5) and the characteristic function of cube-free numbers

$$\sum_{d^3|n} \mu(d) = \begin{cases} 0, & \exists\, m \text{ s.t. } m^3 \,|\, n, \\ 1, & \text{otherwise,} \end{cases}$$

we write

$$\begin{aligned} S_c(x) &= \sum_{\Delta_1 \le n < \Delta_2} \sum_{d^3|[n^c \tan^\theta(\log n)]} \mu(d) \sum_{t^3|[n^c \tan^\theta(\log n)]+1} \mu(t) \\ &= \sum_{\substack{d,t \\ (d,t)=1}} \mu(d)\mu(t) \sum_{\substack{\Delta_1 \le n < \Delta_2 \\ [n^c \tan^\theta(\log n)] \equiv 0\, (d^3) \\ [n^c \tan^\theta(\log n)]+1 \equiv 0\, (t^3)}} 1 \\ &= S_c^{(1)}(x) + S_c^{(2)}(x), \end{aligned} \tag{4.1}$$

where

$$S_c^{(1)}(x) = \sum_{\substack{dt \le z \\ (d,t)=1}} \mu(d)\mu(t) \sum_{\substack{\Delta_1 \le n < \Delta_2 \\ [n^c \tan^\theta(\log n)] \equiv 0\, (d^3) \\ [n^c \tan^\theta(\log n)]+1 \equiv 0\, (t^3)}} 1, \tag{4.2}$$

$$S_c^{(2)}(x) = \sum_{\substack{dt > z \\ (d,t)=1}} \mu(d)\mu(t) \sum_{\substack{\Delta_1 \le n < \Delta_2 \\ [n^c \tan^\theta(\log n)] \equiv 0\, (d^3) \\ [n^c \tan^\theta(\log n)]+1 \equiv 0\, (t^3)}} 1. \tag{4.3}$$

5. Estimation of $\mathbf{S_c^{(1)}(x)}$

From (4.2), we obtain

$$\begin{aligned}
S_c^{(1)}(x) &= \sum_{\substack{dt \le z \\ (d,t)=1}} \mu(d)\mu(t) \sum_{\substack{(\Delta_1^c \tan^\theta(\log \Delta_1)-1)d^{-3} < k < \Delta_2^c \tan^\theta(\log \Delta_2)d^{-3} \\ kd^3+1 \equiv 0\,(t^3)}} \\
&\quad \times \sum_{\substack{\Delta_1 \le n < \Delta_2 \\ [n^c \tan^\theta(\log n)] = kd^3}} 1 \\
&= \sum_{\substack{dt \le z \\ (d,t)=1}} \mu(d)\mu(t) \sum_{\substack{(\Delta_1^c \tan^\theta(\log \Delta_1)-1)d^{-3} < k < \Delta_2^c \tan^\theta(\log \Delta_2)d^{-3} \\ kd^3+1 \equiv 0\,(t^3)}} \\
&\quad \times \sum_{\substack{\Delta_1 \le n < \Delta_2 \\ kd^3 \le n^c \tan^\theta(\log n) < kd^3+1}} 1 \\
&= \sum_{\substack{dt \le z \\ (d,t)=1}} \mu(d)\mu(t) \sum_{\substack{(\Delta_1^c \tan^\theta(\log \Delta_1)-1)d^{-3} < k < \Delta_2^c \tan^\theta(\log \Delta_2)d^{-3} \\ kd^3+1 \equiv 0\,(t^3)}} \\
&\quad \times \sum_{\Delta_1' \le n < \Delta_2'} 1 + \mathcal{O}(x^\varepsilon) \\
&= \sum_{\substack{dt \le z \\ (d,t)=1}} \mu(d)\mu(t) \sum_{\substack{(\Delta_1^c \tan^\theta(\log \Delta_1)-1)d^{-3} < k < \Delta_2^c \tan^\theta(\log \Delta_2)d^{-3} \\ kd^3+1 \equiv 0\,(t^3)}} \\
&\quad \times ([-\Delta_k'] - [-\Delta_k'']) + \mathcal{O}(x^\varepsilon) \\
&= S_c^{(3)}(x) + S_c^{(4)}(x) + \mathcal{O}(x^\varepsilon),
\end{aligned} \tag{5.1}$$

where

$$[\Delta_k', \Delta_k'') \subset [\Delta_1, \Delta_2) \tag{5.2}$$

and the interval $[\Delta_k', \Delta_k'')$ is a solution of the system inequalities

$$\left| \begin{array}{l} n^c \tan^\theta(\log n) \ge kd^3 \\ n^c \tan^\theta(\log n) < kd^3 + 1 \end{array} \right. \tag{5.3}$$

and where

$$S_c^{(3)}(x) = \sum_{\substack{dt\leq z\\(d,t)=1}} \mu(d)\mu(t) \sum_{\substack{(\Delta_1^c \tan^\theta(\log\Delta_1)-1)d^{-3}<k<\Delta_2^c\tan^\theta(\log\Delta_2)d^{-3}\\ kd^3+1\equiv 0\,(t^3)}} \times (\Delta_k'' - \Delta_k'), \tag{5.4}$$

$$S_c^{(4)}(x) = \sum_{\substack{dt\leq z\\(d,t)=1}} \mu(d)\mu(t) \sum_{\substack{(\Delta_1^c \tan^\theta(\log\Delta_1)-1)d^{-3}<k<\Delta_2^c\tan^\theta(\log\Delta_2)d^{-3}\\ kd^3+1\equiv 0\,(t^3)}} \times (\psi(-\Delta_k'') - \psi(-\Delta_k')). \tag{5.5}$$

5.1. *Lower bound for* $\mathbf{S_c^{(3)}(x)}$

Consider the function $t(y)$ defined by

$$t = y^c \tan^\theta(\log y) \quad \text{for } y \in \left[\Delta_1', \Delta_2'\right]. \tag{5.6}$$

Using (5.3), (5.4), (5.6), Abel's summation formula and the mean-value theorem, we get

$$\begin{aligned}
S_c^{(3)}(x) &= \sum_{\substack{dt\leq z\\(d,t)=1}} \mu(d)\mu(t) \sum_{\substack{(\Delta_1^c \tan^\theta(\log\Delta_1)-1)d^{-3}<k<\Delta_2^c\tan^\theta(\log\Delta_2)d^{-3}\\ kd^3+1\equiv 0\,(t^3)}} \\
&\quad \times (y(kd^3+1) - y(kd^3)) \\
&= \sum_{\substack{dt\leq z\\(d,t)=1}} \mu(d)\mu(t) \Bigg[(y(\Delta_2^c \tan^\theta(\log\Delta_2)+1) - y(\Delta_2^c\tan^\theta(\log\Delta_2))) \\
&\quad \times \sum_{\substack{(\Delta_1^c \tan^\theta(\log\Delta_1)-1)d^{-3}<k<\Delta_2^c\tan^\theta(\log\Delta_2)d^{-3}\\ kd^3+1\equiv 0\,(t^3)}} 1
\end{aligned}$$

$$- \int_{(\Delta_1^c \tan^\theta(\log \Delta_1)-1)d^{-3}}^{\Delta_2^c \tan^\theta(\log \Delta_2)d^{-3}} \left(\sum_{\substack{(\Delta_1^c \tan^\theta(\log \Delta_1)-1)d^{-3}<k\le u \\ kd^3+1\equiv 0\,(t^3)}} 1 \right)$$

$$\left. \times \frac{\partial(y(ud^3+1) - y(ud^3))}{\partial u}\, du \right]$$

$$= \sum_{\substack{dt\le z \\ (d,t)=1}} \mu(d)\mu(t) \left[y'(\xi_1) \right.$$

$$\times \left(\frac{\Delta_2^c \tan^\theta(\log \Delta_2) - \Delta_1^c \tan^\theta(\log \Delta_1) + 1}{d^3 t^3} + \mathcal{O}(1) \right)$$

$$- \int_{(\Delta_1^c \tan^\theta(\log \Delta_1)-1)d^{-3}}^{\Delta_2^c \tan^\theta(\log \Delta_2)d^{-3}} \left(\frac{u - (\Delta_1^c \tan^\theta(\log \Delta_1) - 1)d^{-3}}{t^3} + \mathcal{O}(1) \right)$$

$$\left. \times \frac{\partial\big(y(ud^3+1) - y(ud^3)\big)}{\partial u}\, du \right]$$

$$= \sum_{\substack{dt\le z \\ (d,t)=1}} \frac{\mu(d)\mu(t)}{t^3} \int_{(\Delta_1^c \tan^\theta(\log \Delta_1)-1)d^{-3}}^{\Delta_2^c \tan^\theta(\log \Delta_2)d^{-3}} (y(ud^3+1) - y(ud^3))\, du$$

$$+ \mathcal{O}(zx^\varepsilon(|y'(\xi_1)| + |y'(\xi_2)|))$$

$$= \sum_{\substack{dt\le z \\ (d,t)=1}} \frac{\mu(d)\mu(t)}{d^3 t^3} \int_{\Delta_1^c \tan^\theta(\log \Delta_1)-1}^{\Delta_2^c \tan^\theta(\log \Delta_2} (y(z+1) - y(z))\, dz$$

$$+ \mathcal{O}(zx^\varepsilon(|y'(\xi_1)| + |y'(\xi_2)|))$$

$$= \sum_{\substack{dt\le z \\ (d,t)=1}} \frac{\mu(d)\mu(t)}{d^3 t^3} \int_{\Delta_1^c \tan^\theta(\log \Delta_1)-1}^{\Delta_2^c \tan^\theta(\log \Delta_2} y'(\zeta_z)\, dz + \mathcal{O}(zx^\varepsilon(|y'(\xi_1)|$$

$$+ |y'(\xi_2)|)), \tag{5.7}$$

where

$$\left|\begin{array}{l} y(\xi_1) \asymp \xi_1^\gamma \\ y(\xi_2) \asymp \xi_2^\gamma \\ y(\zeta_z) \asymp \zeta_z^\gamma \\ z < \zeta_z < z+1 \\ \quad \Delta_2^c \tan^\theta(\log \Delta_2) < \xi_1 < \Delta_2^c \tan^\theta(\log \Delta_2) + 1 \\ \quad \Delta_1^c \tan^\theta(\log \Delta_1) - 1 < \xi_2 < \Delta_1^c \tan^\theta(\log \Delta_1) \end{array}\right. . \tag{5.8}$$

For the implicit function y defined by (5.6), the first derivative with respect to t is

$$y' = \frac{y^{1-c}}{\big(c\tan(\log y) + \theta \sec^2(\log y)\big)\tan^{\theta-1}(\log y)}. \tag{5.9}$$

Now, (1.1), (1.3), (1.4), (5.2), (5.6)–(5.8) and the well-known asymptotic formula

$$\sum_{\substack{dt\le z \\ (d,t)=1}} \frac{\mu(d)\mu(t)}{d^3t^3} = \prod_p \left(1 - \frac{2}{p^3}\right) + \mathcal{O}(z^{\varepsilon-2})$$

give us

$$S_c^{(3)}(x) \gg x. \tag{5.10}$$

5.2. *Upper bound for* $\mathbf{S_c^{(4)}(x)}$

By (5.5) we have

$$S_c^{(4)}(x) \ll \sum_{dt\le z} |S_c^{(5)}(x)|, \tag{5.11}$$

where

$$S_c^{(5)}(x) = \sum_{\substack{(\Delta_1^c \tan^\theta(\log \Delta_1)-1)d^{-3} < k < \Delta_2^c \tan^\theta(\log \Delta_2) d^{-3} \\ kd^3+1\equiv 0\,(t^3)}} (\psi(-\Delta_k'') - \psi(-\Delta_k')). \tag{5.12}$$

Now, (1.2), (5.12) and Lemma 1 yield

$$S_c^{(5)}(x) = S_c^{(6)}(x) + S_c^{(7)}(x) + S_c^{(8)}(x), \tag{5.13}$$

where

$$S_c^{(6)}(x) = \sum_{\substack{(\Delta_1^c\tan^\theta(\log\Delta_1)-1)d^{-3}<k<\Delta_2^c\tan^\theta(\log\Delta_2)d^{-3}\\ kd^3+1\equiv 0\,(t^3)}} \times \sum_{1\le|h|\le H} a(h)(e(-h\Delta_k'')-e(-h\Delta_k')), \tag{5.14}$$

$$S_c^{(7)}(x) \ll \sum_{\substack{(\Delta_1^c\tan^\theta(\log\Delta_1)-1)d^{-3}<k<\Delta_2^c\tan^\theta(\log\Delta_2)d^{-3}\\ kd^3+1\equiv 0\,(t^3)}} \sum_{|h|\le H} b(h)e(-h\Delta_k'), \tag{5.15}$$

$$S_c^{(8)}(x) \ll \sum_{\substack{(\Delta_1^c\tan^\theta(\log\Delta_1)-1)d^{-3}<k<\Delta_2^c\tan^\theta(\log\Delta_2)d^{-3}\\ kd^3+1\equiv 0\,(t^3)}} \sum_{|h|\le H} b(h)e\big(-h\Delta_k''\big). \tag{5.16}$$

Put

$$H = \frac{x^{\frac{2c-1}{4}}}{dt}. \tag{5.17}$$

We first estimate $S_c^{(7)}(x)$. From (1.3), (1.4), (3.1) and (5.15), we deduce

$$\begin{aligned} &S_c^{(7)}(x)\\ &\ll \frac{1}{H}\sum_{|h|\le H}\left|\sum_{\substack{(\Delta_1^c\tan^\theta(\log\Delta_1)-1)d^{-3}<k<\Delta_2^c\tan^\theta(\log\Delta_2)d^{-3}\\ kd^3+1\equiv 0\,(t^3)}} e(-h\Delta_k')\right|\\ &\ll \frac{x^c}{Hd^3t^3}+\frac{1}{H}\sum_{1\le h\le H}\left|\sum_{\substack{(\Delta_1^c\tan^\theta(\log\Delta_1)-1)d^{-3}<k<\Delta_2^c\tan^\theta(\log\Delta_2)d^{-3}\\ kd^3+1\equiv 0\,(t^3)}} e(-h\Delta_k')\right|\\ &\ll \frac{x^c}{Hd^3t^3}+\frac{1}{H}\\ &\quad\times\sum_{1\le h\le H}\left|\sum_{\Delta_1^c\tan^\theta(\log\Delta_1)t^{-3}<l\le(\Delta_2^c\tan^\theta(\log\Delta_2)+1)t^{-3}} e(-h\Delta'_{(lt^3-1)d^{-3}})\right|. \end{aligned} \tag{5.18}$$

It is easy to see that (1.3) and (1.4) imply

$$\frac{2\big(\Delta_1^c \tan^\theta(\log \Delta_1)\big)}{t^3} < \frac{\Delta_2^c \tan^\theta(\log \Delta_2) + 1}{t^3}. \tag{5.19}$$

Taking into account (5.18) and (5.19), we split the range of l into dyadic subintervals to obtain

$$S_c^{(7)}(x) \ll \frac{x^c}{Hd^3t^3} + \frac{\log x}{H} \sum_{1\leq h\leq H} \left| \sum_{L<l\leq 2L} e(-h\Delta'_{(lt^3-1)d^{-3}}) \right|, \tag{5.20}$$

where

$$\left| \begin{array}{c} L \asymp \frac{x^c}{t^3} \\ \frac{\Delta_1^c \tan^\theta(\log \Delta_1)}{t^3} \leq L \leq \frac{\Delta_2^c \tan^\theta(\log \Delta_2)+1}{2t^3} \end{array} \right. . \tag{5.21}$$

According to (5.2), (5.3), (5.20) and (5.21), the variable $\Delta'_{(lt^3-1)d^{-3}}$ is an implicit function of l defined by

$$y^c \tan^\theta(\log y) = lt^3 - 1 \quad \text{for } l \in (L, 2L] \tag{5.22}$$

and

$$y \subset [\Delta_1, \Delta_2). \tag{5.23}$$

Bearing in mind (1.3), (1.4), (5.6), (5.9), (5.21), (5.22) and (5.23), we find

$$y' \asymp L^{\gamma-1}t^{3\gamma}. \tag{5.24}$$

Proceeding in the same way, we derive

$$y'' \asymp L^{\gamma-2}t^{3\gamma}. \tag{5.25}$$

Now, (5.20), (5.21), (5.24), (5.25) and Lemma 2 with $q = 0$ lead to

$$\begin{aligned} S_c^{(7)}(x) &\ll \frac{x^c}{Hd^3t^3} + \frac{\log x}{H} \sum_{1\leq h\leq H} \left(h^{\frac{1}{2}} L^{\frac{\gamma}{2}} t^{\frac{3\gamma}{2}} + h^{-1}L^{1-\gamma}t^{-3\gamma}\right) \\ &\ll x^c H^{-1} d^{-3} t^{-3} + L^{\frac{\gamma}{2}} H^{\frac{1}{2}} t^{\frac{3\gamma}{2}} \log x + L^{1-\gamma} H^{-1} t^{-3\gamma} \log^2 x \end{aligned}$$

$$\ll x^c H^{-1} d^{-3} t^{-3} + x^{\frac{1}{2}} H^{\frac{1}{2}} + x^{c-1} H^{-1} t^{-3} \log^2 x$$
$$\ll x^c H^{-1} d^{-3} t^{-3} + x^{\frac{1}{2}} H^{\frac{1}{2}}. \tag{5.26}$$

Arguing in the same way for the sum defined by (5.16), we get

$$S_c^{(8)}(x) \ll x^c H^{-1} d^{-3} t^{-3} + x^{\frac{1}{2}} H^{\frac{1}{2}}. \tag{5.27}$$

It remains to estimate the sum $S_c^{(6)}(K)$. Working similar to estimation of $S_c^{(7)}(x)$ from (3.1) and (5.14), we have

$$S_c^{(6)}(x) \ll \sum_{1\le|h|\le H} \frac{1}{h} \left| \sum_{L<l\le 2L} \Phi_h(l) e\left(-h\Delta'_{(lt^3-1)d^{-3}}\right) \right|, \tag{5.28}$$

where L satisfies (5.21) and

$$\Phi_h(u) = e(h(\Delta'_{(ut^3-1)d^{-3}} - \Delta''_{(ut^3-1)d^{-3}})) - 1. \tag{5.29}$$

From (1.3), (1.4), (5.6), (5.9), (5.21), (5.22), (5.23), (5.29) and the mean-value theorem it follows

$$|\Phi_h(u)| \le 2\left|\sin\left(\pi h(\Delta'_u - \Delta''_u)\right)\right| \ll |h|\left|\Delta'_u - \Delta''_u\right| \ll |h| x^{1-c}, \tag{5.30}$$

$$\Phi'_h(u) = 2\pi i h \left(\frac{\partial \Delta'_u}{\partial u} - \frac{\partial \Delta''_u}{\partial u} \right) e\big(h(\Delta'_u - \Delta''_u)\big) \ll |h| x^{1-2c} \tag{5.31}$$

for $u \in [L, 2L]$. Now, (5.21), (5.22), (5.28), (5.30), (5.31) and Abel's summation formula give us

$$S_c^{(6)}(x) \ll \sum_{1\le|h|\le H} \frac{1}{h} \left| \Phi_h(2L) \sum_{L<l\le 2L} e\left(-h\Delta'_{(lt^3-1)d^{-3}}\right) \right|$$
$$+ \sum_{1\le|h|\le H} \frac{1}{h} \int_L^{2L} \left| \Phi'_h(u) t^3 \sum_{L<l\le u} e\left(-h\Delta'_{(lt^3-1)d^{-3}}\right) \right| du$$
$$\ll x^{1-c} \left| S_c^{(9)}(L_1) \right|, \tag{5.32}$$

where

$$S_c^{(9)}(L_1) = \sum_{1\le h\le H} \left| \sum_{L<l\le L_1} e\Big(h\Delta'_{(lt^3-1)d^{-3}}\Big) \right| \tag{5.33}$$

for some number $L_1 \in (L, 2L]$. By (5.21), (5.24), (5.25), (5.33) and Lemma 2 with $q = 0$, we obtain

$$\begin{aligned} S_c^{(9)}(L_1) &\ll \sum_{1\le h\le H} (h^{\frac{1}{2}} L^{\frac{\gamma}{2}} t^{\frac{3\gamma}{2}} + h^{-1} L^{1-\gamma} t^{-3\gamma}) \\ &\ll L^{\frac{\gamma}{2}} H^{\frac{3}{2}} t^{\frac{3\gamma}{2}} + L^{1-\gamma} t^{-3\gamma} \log H \\ &\ll x^{\frac{1}{2}} H^{\frac{3}{2}} + x^{c-1} t^{-3} \log H. \end{aligned} \tag{5.34}$$

Now, (5.32) and (5.34) imply

$$S_c^{(6)}(x) \ll x^{\frac{3-2c}{2}} H^{\frac{3}{2}} + t^{-3} \log H \ll x^{\frac{3-2c}{2}} H^{\frac{3}{2}}. \tag{5.35}$$

Taking into account (1.1), (5.11), (5.13), (5.17), (5.26), (5.27) and (5.35), we deduce

$$\begin{aligned} S_c^{(4)}(x) &\ll \sum_{dt\le z} \Big(x^c H^{-1} d^{-3} t^{-3} + x^{\frac{1}{2}} H^{\frac{1}{2}} + x^{\frac{3-2c}{2}} H^{\frac{3}{2}} \Big) \\ &\ll x^{\frac{2c+1}{4}} \sum_{dt\le z} (dt)^{-2} + x^{\frac{2c+3}{8}} \sum_{dt\le z} (dt)^{-\frac{1}{2}} + x^{\frac{9-2c}{8}} \sum_{dt\le z} (dt)^{-\frac{3}{2}} \\ &\ll x^{\frac{2c+1}{4}} + x^{\frac{2c+3}{8}} \sum_{n\le z} \tau(n) n^{-\frac{1}{2}} + x^{\frac{9-2c}{8}} \\ &\ll x^{\frac{2c+1}{4}} + x^{\frac{2c+3}{8}+\varepsilon} z^{\frac{1}{2}} + x^{\frac{9-2c}{8}} \\ &\ll x^{\frac{2c+1}{4}+\varepsilon} + x^{\frac{9-2c}{8}}. \end{aligned} \tag{5.36}$$

5.3. *Lower bound for* $\mathbf{S_c^{(1)}(x)}$

Bearing in mind (5.1), (5.10) and (5.36), we derive

$$S_c^{(1)}(x) \gg x. \tag{5.37}$$

6. Upper Bound for $\mathbf{S_c^{(2)}(x)}$

Splitting the range of d and t into dyadic subintervals from (4.3) we obtain

$$\begin{aligned} S_c^{(2)}(x) &\ll \sum_{\substack{dt>z \\ (d,t)=1}} \sum_{\substack{(\Delta_1^c \tan^\theta(\log \Delta_1)-1)d^{-3}<k<\Delta_2^c \tan^\theta(\log \Delta_2)d^{-3} \\ kd^3+1\equiv 0\,(t^3)}} 1 \\ &\ll (\log x)^2 \sum_{D\le d<2D} \sum_{T\le t<2T} \sum_{\substack{(\Delta_1^c \tan^\theta(\log \Delta_1)-1)d^{-3}<k<\Delta_2^c \tan^\theta(\log \Delta_2)d^{-3} \\ kd^3+1\equiv 0\,(t^3)}} 1, \end{aligned} \tag{6.1}$$

where

$$\frac{1}{2} \le D, T \le \sqrt[3]{\Delta_2^c \tan^\theta(\log \Delta_2) + 1}, \quad DT > \frac{z}{4}. \tag{6.2}$$

On the one hand, (1.3), (1.4) and (6.1) give us

$$\begin{aligned} S_c^{(2)}(x) &\ll x^\varepsilon \sum_{T\le t<2T} \sum_{l\le(\Delta_2^c \tan^\theta(\log \Delta_2)+1)T^{-3}} \sum_{D\le d<2D} \\ &\quad\times \sum_{\substack{(\Delta_1^c \tan^\theta(\log \Delta_1)-1)d^{-3}<k<\Delta_2^c \tan^\theta(\log \Delta_2)d^{-3} \\ kd^3=lt^3-1}} 1 \\ &\ll x^\varepsilon \sum_{T\le t<2T} \sum_{l\le(\Delta_2^c \tan^\theta(\log \Delta_2)+1)T^{-3}} \tau(lt^3-1) \\ &\ll x^\varepsilon \sum_{T\le t<2T} \sum_{l\le(\Delta_2^c \tan^\theta(\log \Delta_2)+1)T^{-3}} 1 \\ &\ll x^{c+\varepsilon}T^{-2}. \end{aligned} \tag{6.3}$$

On the other hand, (1.3), (1.4) and (6.1) imply

$$\begin{aligned} S_c^{(2)}(x) &\ll x^\varepsilon \sum_{D\le d<2D} \sum_{(\Delta_1^c \tan^\theta(\log \Delta_1)-1)d^{-3}<k<\Delta_2^c \tan^\theta(\log \Delta_2)d^{-3}} \sum_{T\le t<2T} \\ &\quad\times \sum_{\substack{l\le(\Delta_2^c \tan^\theta(\log \Delta_2)+1)T^{-3} \\ kd^3+1=lt^3}} 1 \end{aligned}$$

$$\ll x^{\varepsilon} \sum_{D \leq d < 2D} \sum_{k \leq \Delta_2^c \tan^{\theta}(\log \Delta_2) D^{-3}} \tau(kd^3 + 1)$$

$$\ll x^{\varepsilon} \sum_{D \leq d < 2D} \sum_{k \leq \Delta_2^c \tan^{\theta}(\log \Delta_2) D^{-3}} 1$$

$$\ll x^{c+\varepsilon} D^{-2}. \tag{6.4}$$

Now, (1.1) and (6.2)–(6.4) yield

$$S_c^{(2)}(x) \ll x^{c+\varepsilon} z^{-1} = x^{\frac{2c+1}{4}+\varepsilon}. \tag{6.5}$$

7. The End of the Proof

Summarizing (4.1), (5.37) and (6.5), we establish the lower bound (2.2).

This completes the proof of Theorem 1.

References

[1] L. Carlitz. On a problem in additive arithmetic II. *Quart. J. Math.*, 3, 1932, 273–290.

[2] L. Mirsky. On the frequency of pairs of square-free numbers with a given difference. *Bull. Am. Math. Soc.*, 55, 1949, 936–939.

[3] Z. Meng. Some new results on k-free numbers. *J. Number Theory*, 121, 2006, 45–66.

[4] J. Brandes. Twins of s-free numbers, 2013. arXiv:1307.2066.

[5] T. Reuss. *The Determinant Method and Applications.* PhD thesis, University of Oxford, 2015.

[6] S. I. Dimitrov. Consecutive square-free numbers of the form $[n^c]$, $[n^c]+1$. *JP J. Algebra Number Theory Appl.*, 40, 2018, 945–956.

[7] X. Cao and W. Zhai. The distribution of square-free numbers of the form $[n^c]$, II. *Acta Math. Sinica (Chin. Ser.)*, 51, 2008, 1187–1194.

[8] S. I. Dimitrov. Consecutive cube-free numbers of the form $[n^c]$, $[n^c]+1$. *Proceedings of the 44th International Conference on Applications of Mathematics in Engineering and Economics, AIP Conference Proceedings*, 2048, 2018, 050004.

[9] M. Zhang and J. Li. Distribution of cube-free numbers with form $[n^c]$. *Front. Math. China*, 12, 2017, 1515–1525.

[10] S. I. Dimitrov. Prime numbers of the form $[n^c \tan^\theta(\log n)]$. *Indian J. Pure Appl. Math.*, 55, 2024, 1198–1209.

[11] J. D. Vaaler. Some extremal problems in fourier analysis. *Bull. Am. Math. Soc.*, 12, 1985, 183–216.

[12] S. W. Graham and G. Kolesnik. *Van der Corput's Method of Exponential Sums.* Cambridge University Press, New York, 1991.

https://doi.org/10.1142/9789819820498_0003

Chapter 3

Inequalities for Normalized Entropic Determinant of Positive Operators in Hilbert Spaces

Silvestru Sever Dragomir

Applied Mathematics Research Group, ISILC, Victoria University, Melbourne City, MC 8001, Australia

Department of Mathematical and Physical Sciences, La Trobe University, Plenty Road, Bundoora, Melbourne VIC 3086, Australia

sever.dragomir@vu.edu.au

http://rgmia.org/dragomir

For positive invertible operators A on a Hilbert space H and a fixed unit vector $x \in H$, define the normalized determinant by $\Delta_x(A) := \exp[\langle \ln Ax, x\rangle]$ and the normalized entropic determinant by

$$\eta_x(A) := \exp[-\langle A \ln Ax, x\rangle].$$

In this chapter, we show among others that, if $A_j > 0$ and $p_j \geq 0$, $j \in \{1, \ldots, n\}$ with $\sum_{j=1}^n p_j = 1$, then

$$\prod_{j=1}^n [\eta_x(A_j)]^{p_j} \leq \left(\prod_{j=1}^n [\Delta_x(A_j)]^{p_j}\right)^{-\langle \sum_{j=1}^n p_j A_j x, x\rangle}$$

for all $x \in H$ with $\|x\| = 1$. In particular, we have

$$\eta_x(A) \le [\Delta_x(A)]^{-\langle Ax,x\rangle}$$

for all $x \in H$ with $\|x\| = 1$.

Keywords: Positive operators, normalized determinants, inequalities.

1. Introduction

Let $B(H)$ be the space of all bounded linear operators on a Hilbert space H, and I stands for the identity operator on H. An operator A in $B(H)$ is said to be positive (in symbol: $A \ge 0$) if $\langle Ax, x\rangle \ge 0$ for all $x \in H$. In particular, $A > 0$ means that A is positive and invertible. For a pair A, B of selfadjoint operators, the order relation $A \ge B$ means as usual that $A - B$ is positive.

In 1998, Fujii *et al.* [6, 7], introduced the *normalized determinant* $\Delta_x(A)$ for positive invertible operators A on a Hilbert space H and a fixed unit vector $x \in H$, namely, $\|x\| = 1$, defined by $\Delta_x(A) := \exp\langle \ln Ax, x\rangle$ and discussed it as a continuous geometric mean and observed some inequalities around the determinant from this point of view.

Some of the fundamental properties of normalized determinant are as follows, Ref. [6].

For each unit vector $x \in H$, see also Ref. [9], we have the following:

(i) **continuity:** the map $A \to \Delta_x(A)$ is norm continuous;
(ii) **bounds:** $\langle A^{-1}x, x\rangle^{-1} \le \Delta_x(A) \le \langle Ax, x\rangle$;
(iii) **continuous mean:** $\langle A^p x, x\rangle^{1/p} \downarrow \Delta_x(A)$ for $p \downarrow 0$ and $\langle A^p x, x\rangle^{1/p} \uparrow \Delta_x(A)$ for $p \uparrow 0$;
(iv) **power equality:** $\Delta_x(A^t) = \Delta_x(A)^t$ for all $t > 0$;
(v) **homogeneity:** $\Delta_x(tA) = t\Delta_x(A)$ and $\Delta_x(tI) = t$ for all $t > 0$;
(vi) **monotonicity:** $0 < A \le B$ implies $\Delta_x(A) \le \Delta_x(B)$;
(vii) **multiplicativity:** $\Delta_x(AB) = \Delta_x(A)\Delta_x(B)$ for commuting A and B;
(viii) **Ky Fan type inequality:** $\Delta_x((1-\alpha)A + \alpha B) \ge \Delta_x(A)^{1-\alpha}$ $\Delta_x(B)^{\alpha}$ for $0 < \alpha < 1$.

We define the logarithmic mean of two positive numbers a, b by

$$L(a,b) := \begin{cases} \frac{b-a}{\ln b - \ln a} & \text{if } b \neq a, \\ a & \text{if } b = a. \end{cases} \tag{1.1}$$

In Ref. [6], the authors obtained the following additive reverse inequality for the operator A which satisfy the condition $0 < mI \leq A \leq MI$, where m, M are positive numbers,

$$\begin{aligned} 0 &\leq \langle Ax, x\rangle - \Delta_x(A) \\ &\leq L(m,M)\left[\ln L(m,M) + \frac{M\ln m - m\ln M}{M-m} - 1\right] \end{aligned} \tag{1.2}$$

for all $x \in H$, $\|x\| = 1$.

We recall that *Specht's ratio* is defined by [13]

$$S(h) := \begin{cases} \frac{h^{\frac{1}{h-1}}}{e\ln(h^{\frac{1}{h-1}})} & \text{if } h \in (0,1)\cup(1,\infty), \\ 1 & \text{if } h = 1. \end{cases} \tag{1.3}$$

It is well known that $\lim_{h\to 1} S(h) = 1$, $S(h) = S\left(\frac{1}{h}\right) > 1$ for $h > 0$, $h \neq 1$. The function is decreasing on $(0,1)$ and increasing on $(1,\infty)$.

In Ref. [7], the authors obtained the following multiplicative reverse inequality as well

$$1 \leq \frac{\langle Ax, x\rangle}{\Delta_x(A)} \leq S\left(\frac{M}{m}\right) \tag{1.4}$$

for $0 < mI \leq A \leq MI$ and $x \in H$, $\|x\| = 1$.

For the entropy function $\eta(t) = -t\ln t$, $t > 0$, the operator entropy has the following expression:

$$\eta(A) = -A\ln A$$

for positive A.

For $x \in H$, $\|x\| = 1$, we define the *normalized entropic determinant* $\eta_x(A)$ by

$$\eta_x(A) := \exp(-\langle A \ln Ax, x\rangle) = \exp\langle \eta(A)x, x\rangle. \tag{1.5}$$

Let $x \in H$, $\|x\| = 1$. Observe that the map $A \to \eta_x(A)$ is *norm continuous* and since

$$\begin{aligned}
&\exp(-\langle tA \ln(tA)x, x\rangle) \\
&\quad = \exp(-\langle tA(\ln t + \ln A)x, x\rangle) = \exp(-\langle (tA \ln t + tA \ln A)x, x\rangle) \\
&\quad = \exp(-\langle Ax, x\rangle t \ln t) \exp(-t\langle A \ln Ax, x\rangle) \\
&\quad = \exp \ln(t^{-\langle Ax,x\rangle t})[\exp(-\langle A \ln Ax, x\rangle)]^{-t},
\end{aligned}$$

hence

$$\eta_x(tA) = t^{-t\langle Ax,x\rangle}[\eta_x(A)]^{-t} \tag{1.6}$$

for $t > 0$ and $A > 0$.

Observe also that

$$\eta_x(I) = 1 \text{ and } \eta_x(tI) = t^{-t} \tag{1.7}$$

for $t > 0$.

In Ref. [2], we showed that, if A, $B > 0$, then for all $x \in H$ with $\|x\| = 1$ and $t \in [0, 1]$,

$$\eta_x((1-t)A + tB) \geq (\eta_x(A))^{1-t}(\eta_x(B))^t.$$

We also have the bounds

$$\left(\frac{\langle A^2x, x\rangle}{\langle Ax, x\rangle}\right)^{-\langle Ax,x\rangle} \leq \eta_x(A) \leq \langle Ax, x\rangle^{-\langle Ax,x\rangle},$$

where $A > 0$ and $x \in H$ with$\|x\| = 1$.

If $y \in H$, $y \neq 0$, then we can extend the definition of the normalized entropic determinant as follows:

$$\breve{\eta}_y(A) := \exp(-\langle A \ln Ay, y\rangle) = \exp\langle \eta(A)y, y\rangle.$$

Also, we can consider

$$\breve{\Delta}_y(A) := \exp\langle \ln Ay, y\rangle$$

for $y \in H$, $y \neq 0$.

We observe that

$$\breve{\eta}_y(A) = \left[\eta_{\frac{y}{\|y\|}}(A)\right]^{\|y\|^2} \quad \text{and} \quad \breve{\Delta}_y(A) = \left[\Delta_{\frac{y}{\|y\|}}(A)\right]^{\|y\|^2}$$

for $y \in H$, $y \neq 0$.

Motivated by the above results, in this chapter, we show among others that, if $A_j > 0$ and $p_j \geq 0$, $j \in \{1, \dots, n\}$ with $\sum_{j=1}^{n} p_j = 1$, then

$$\prod_{j=1}^{n} [\eta_x(A_j)]^{p_j} \leq \left(\prod_{j=1}^{n} [\Delta_x(A_j)]^{p_j}\right)^{-\langle \sum_{j=1}^{n} p_j A_j x, x\rangle}$$

for all $x \in H$ with $\|x\| = 1$. In particular, we have

$$\eta_x(A) \leq [\Delta_x(A)]^{-\langle Ax, x\rangle}$$

for all $x \in H$ with $\|x\| = 1$.

2. Main Results

We say that the functions f, $g : [a, b] \longrightarrow \mathbb{R}$ are *synchronous (asynchronous)* on the interval $[a, b]$ if they satisfy the following condition:

$$(f(t) - f(s))(g(t) - g(s)) \geq (\leq) 0 \quad \text{for each } t, \quad s \in [a, b].$$

It is obvious that, if f, g are monotonic and have the same monotonicity on the interval $[a, b]$, then they are synchronous on $[a, b]$, while if they have opposite monotonicity, they are asynchronous.

Theorem 1. *Assume that A, $B > 0$ and x, $y \in H$ with $\|x\| = \|y\| = 1$, then*

$$\eta_x(A)\eta_y(B) \leq [\Delta_x(A)]^{-\langle By, y\rangle} [\Delta_y(B)]^{-\langle Ax, x\rangle}. \tag{2.1}$$

In particular, we have

$$\eta_x(A)\eta_x(B) \leq [\Delta_x(A)]^{-\langle Bx, x\rangle} [\Delta_x(B)]^{-\langle Ax, x\rangle}, \tag{2.2}$$

$$\eta_x(A)\eta_y(A) \leq [\Delta_x(A)]^{-\langle Ay, y\rangle} [\Delta_y(A)]^{-\langle Ax, x\rangle} \tag{2.3}$$

and

$$\eta_x(A) \leq [\Delta_x(A)]^{-\langle Ax, x\rangle}. \tag{2.4}$$

Proof. In Ref. [1], we obtained, between others, the following two operators and two vectors inequality: If A and B are selfadjoint operators and $\mathrm{Sp}(A)$, $\mathrm{Sp}(B) \subseteq [m, M]$, then for any continuous synchronous (asynchronous) functions f, $g : [m, M] \longrightarrow \mathbb{R}$ we have the more general result

$$\begin{aligned}&\langle f(A)g(A)x, x\rangle + \langle f(B)g(B)y, y\rangle \\ &\geq (\leq) \langle f(A)x, x\rangle\langle g(B)y, y\rangle + \langle f(B)y, y\rangle\langle g(A)x, x\rangle\end{aligned} \tag{2.5}$$

for each x, $y \in H$ with $\|x\| = \|y\| = 1$.

If we write (2.5) for the asynchronous functions $f(t) = -t$, $g(t) = \ln t$, $t > 0$, then we obtain

$$-\langle A \ln Ax, x\rangle - \langle B \ln By, y\rangle \leq -\langle Ax, x\rangle\langle \ln By, y\rangle - \langle By, y\rangle\langle \ln Ax, x\rangle \tag{2.6}$$

for each x, $y \in H$ with $\|x\| = \|y\| = 1$.

If we take the exponential in (2.6), then we get

$$\begin{aligned}&\exp[-\langle A \ln Ax, x\rangle] \exp[-\langle B \ln By, y\rangle] \\ &\leq \exp[-\langle Ax, x\rangle \langle \ln By, y\rangle] \exp[-\langle By, y\rangle\langle \ln Ax, x\rangle] \\ &= [\exp\langle \ln By, y\rangle]^{-\langle Ax,x\rangle} [\exp\langle \ln Ax, x\rangle]^{-\langle By,y\rangle}\end{aligned}$$

for each x, $y \in H$ with $\|x\| = \|y\| = 1$, which is equivalent to (2.1). □

We also have the following:

Theorem 2. *Assume that* $A_j > 0$ *and* $x_j \in H$, $j \in \{1, \ldots, n\}$ *with* $\sum_{j=1}^{n} \|x_j\|^2 = 1$, *then*

$$\prod_{j=1}^{n} \left[\eta_{\frac{x_j}{\|x_j\|}}(A)\right]^{\|x_j\|^2} \leq \left(\prod_{j=1}^{n} \left[\Delta_{\frac{x_j}{\|x_j\|}}(A_j)\right]^{\|x_j\|^2}\right)^{-\sum_{j=1}^{n}\langle A_j x_j, x_j\rangle}. \tag{2.7}$$

Proof. In Ref. [1], we also obtained the following result: Let A_j be selfadjoint operators with $\mathrm{Sp}(A_j) \subseteq [m, M]$ for $j \in \{1, \ldots, n\}$ and for some scalars $m < M$. If f, $g : [m, M] \longrightarrow \mathbb{R}$ are continuous and *synchronous (asynchronous) on* $[m, M]$, *then*

$$\sum_{j=1}^{n} \langle f(A_j)g(A_j)x_j, x_j\rangle \geq (\leq) \sum_{j=1}^{n}\langle f(A_j)x_j, x_j\rangle \sum_{j=1}^{n}\langle g(A_j)x_j, x_j\rangle, \tag{2.8}$$

for each $x_j \in H$, $j \in \{1, \ldots, n\}$ with $\sum_{j=1}^{n} \|x_j\|^2 = 1$.

If we write the inequality (2.8) for the asynchronous functions $f(t) = -t$, $g(t) = \ln t$, $t > 0$, then we obtain

$$\sum_{j=1}^{n} -\langle A_j \ln A_j x_j, x_j\rangle \leq -\sum_{j=1}^{n}\langle A_j x_j, x_j\rangle \sum_{j=1}^{n}\langle \ln A_j x_j, x_j\rangle, \tag{2.9}$$

for each $x_j \in H$, $j \in \{1, \ldots, n\}$ with $\sum_{j=1}^{n} \|x_j\|^2 = 1$.

By taking the exponential, we derive

$$\exp\left(\sum_{j=1}^{n} - \langle A_j \ln A_j x_j, x_j\rangle\right) \leq \exp\left(-\sum_{j=1}^{n}\langle A_j x_j, x_j\rangle \sum_{j=1}^{n}\langle \ln A_j x_j, x_j\rangle\right).$$

Observe that

$$\exp\left(\sum_{j=1}^{n} - \langle A_j \ln A_j x_j, x_j\rangle\right) = \prod_{j=1}^{n} \exp\left[-\langle A_j \ln A_j x_j, x_j\rangle\right] = \prod_{j=1}^{n} \breve{\eta}_{\frac{x_j}{\|x_j\|}}(A_j) = \prod_{j=1}^{n}\left[\eta_{\frac{x_j}{\|x_j\|}}(A)\right]^{\|x_j\|^2}$$

and

$$\exp\left(-\sum_{j=1}^{n}\langle A_j x_j, x_j\rangle \sum_{j=1}^{n}\langle \ln A_j x_j, x_j\rangle\right) = \left[\exp\left(\sum_{j=1}^{n} \langle \ln A_j x_j, x_j\rangle\right)\right]^{-\sum_{j=1}^{n}\langle A_j x_j, x_j\rangle}$$

$$= \left[\prod_{j=1}^{n} \exp\left(\langle \ln A_j x_j, x_j\rangle\right)\right]^{-\sum_{j=1}^{n}\langle A_j x_j, x_j\rangle}$$

$$= \left[\prod_{j=1}^{n} \breve{\Delta}_{\frac{x_j}{\|x_j\|}}(A_j)\right]^{-\sum_{j=1}^{n}\langle A_j x_j, x_j\rangle}$$

$$= \left(\prod_{j=1}^{n} \left[\Delta_{\frac{x_j}{\|x_j\|}}(A_j)\right]^{\|x_j\|^2}\right)^{-\sum_{j=1}^{n}\langle A_j x_j, x_j\rangle}$$

for each $x_j \in H$, $j \in \{1, \dots, n\}$ with $\sum_{j=1}^{n} \|x_j\|^2 = 1$.

By utilizing (2.9), we derive (2.7). □

Corollary 1. *Let $A_j > 0$ and $p_j \geq 0$, $j \in \{1, \dots, n\}$ with $\sum_{j=1}^{n} p_j = 1$. Then,*

$$\prod_{j=1}^{n} [\eta_x(A_j)]^{p_j} \leq \left(\prod_{j=1}^{n} [\Delta_x(A_j)]^{p_j}\right)^{-\langle \sum_{j=1}^{n} p_j A_j x, x\rangle} \tag{2.10}$$

for all $x \in H$ with $\|x\| = 1$.

Proof. If we choose in Theorem 2 $x_j = \sqrt{p_j}x$, $j \in \{1, \dots, n\}$, where $p_j \geq 0$, $j \in \{1, \dots, n\}$, with $\sum_{j=1}^{n} p_j = 1$ and $x \in H$, with $\|x\| = 1$ then a simple calculation shows that inequality (2.7) becomes (2.10). The details are omitted. □

Remark 1. The case of two operators is as follows:

$$[\eta_x(A)]^{1-t} [\eta_x(B)]^t \leq \left([\Delta_x(A)]^{1-t} [\Delta_x(B)]^t\right)^{-\langle (1-t)A + tBx, x\rangle}, \tag{2.11}$$

for A, $B > 0$, $t \in [0, 1]$ and $x \in H$ with $\|x\| = 1$.

For $t = 1/2$, we get

$$\eta_x(A)\eta_x(B) \leq [\Delta_x(A)\Delta_x(B)]^{-\langle \frac{A+B}{2}x, x\rangle},$$

for A, $B > 0$ and $x \in H$ with $\|x\| = 1$.

If we take $B = A$, we recapture (2.18).

Theorem 3. *Assume that* A, $B > 0$ *and* x, $y \in H$ *with* $\|x\| = \|y\| = 1$. *If* $q \in \mathbb{R} \setminus \{0\}$ *and* $p > 0$ *with* $p + q \neq 0$, *then*

$$[\eta_x(A^{p+q})]^{\frac{1}{p+q}\langle B^q y,y\rangle}[\eta_y(B^{p+q})]^{\frac{1}{p+q}\langle A^p x,x\rangle} \leq [\eta_y(B^q)]^{\frac{1}{q}\langle A^{p+q}x,x\rangle}[\eta_x(A^p)]^{\frac{1}{p}\langle B^{p+q}y,y\rangle}. \tag{2.12}$$

If $p < 0$ *and* $p + q \neq 0$, *then*

$$[\eta_x(A^{p+q})]^{\frac{1}{p+q}\langle B^q y,y\rangle}[\eta_y(B^{p+q})]^{\frac{1}{p+q}\langle A^p x,x\rangle} \geq [\eta_y(B^q)]^{\frac{1}{q}\langle A^{p+q}x,x\rangle}[\eta_x(A^p)]^{\frac{1}{p}\langle B^{p+q}y,y\rangle}. \tag{2.13}$$

Proof. We consider only the case of synchronous functions. In this case, we have then

$$f(t)g(t) + f(s)g(s) \geq f(t)g(s) + f(s)g(t)$$

for each t, $s \in [m, M]$.

Assume that h is positive if we multiply this inequality by $h(t)\,h(s) \geq 0$ we get

$$h(t)f(t)g(t)h(s) + h(t)h(s)f(s)g(s) \geq h(t)f(t)h(s)g(s) + h(s)f(s)h(t)g(t)$$

for each t, $s \in [m, M]$.

If we fix $s \in [m, M]$ and use the continuous functional calculus for operators with spectra in $[m, M]$ over t, we get

$$h(s)h(A)f(A)g(A) + h(s)f(s)g(s)h(A) \geq h(s)g(s)h(A)f(A) + h(s)f(s)h(A)g(A),$$

and by taking the inner product over $x \in H$ with $\|x\| = 1$, we get

$$h(s)\langle h(A)f(A)g(A)x,x\rangle + h(s)f(s)g(s)\langle h(A)x,x\rangle \geq h(s)g(s)\langle h(A)f(A)x,x\rangle + h(s)f(s)\langle h(A)g(A)x,x\rangle.$$

If we apply again the functional calculus over B, then we get

$$\langle h(A)f(A)g(A)x,x\rangle h(B) + \langle h(A)x,x\rangle h(B)f(B)g(B) \geq \langle h(A)f(A)x,x\rangle h(B)g(B) + \langle h(A)g(A)x,x\rangle h(B)f(B).$$

If we take the inner product for $y \in H$ with $\|y\| = 1$, then we get

$$\begin{aligned}\langle h(A)f(A)g(A)x, x\rangle \langle h(B)y, y\rangle + \langle h(A)x, x\rangle \langle h(B) f(B)g(B)y, y\rangle \\ \geq \langle h(A)f(A)x, x\rangle\langle h(B)g(B)y, y\rangle \\ + \langle h(A)g(A)x, x\rangle\langle h(B)f(B)y, y\rangle,\end{aligned} \tag{2.14}$$

which holds for x, $y \in H$ with $\|x\| = \|y\| = 1$.

If f and g are asynchronous, then inequality (2.14) reverses.

For $p > 0$, $f(t) = t^p$ is increasing, $g(t) = -\ln t$ is decreasing and $h(t) = t^q$ is positive on $(0, \infty)$. By (2.14), we derive

$$\begin{aligned}&-\langle A^{p+q} \ln Ax, x\rangle\langle B^q y, y\rangle - \langle A^p x, x\rangle\langle B^{p+q} \ln By, y\rangle \\ &\leq -\langle A^{p+q}x, x\rangle \langle B^q \ln By, y\rangle - \langle A^p \ln Ax, x\rangle\langle B^{p+q}y, y\rangle,\end{aligned}$$

namely,

$$\begin{aligned}&-\frac{1}{p+q}\langle A^{p+q} \ln A^{p+q}x, x\rangle\langle B^q y, y\rangle - \frac{1}{p+q}\langle A^p x, x\rangle\langle B^{p+q} \ln B^{p+q}y, y\rangle \\ &\leq -\frac{1}{q}\langle A^{p+q}x, x\rangle\langle B^q \ln B^q y, y\rangle - \frac{1}{p} \langle A^p \ln A^p x, x\rangle \langle B^{p+q}y, y\rangle,\end{aligned}$$

for x, $y \in H$ with $\|x\| = \|y\| = 1$.

By taking the exponential, we get

$$\begin{aligned}&(\exp[-\langle A^{p+q} \ln A^{p+q}x, x\rangle])^{\frac{1}{p+q}\langle B^q y,y\rangle} \\ &\quad\times \left(\exp\left[-\langle B^{p+q} \ln B^{p+q}y, y\rangle\right]\right)^{\frac{1}{p+q}\langle A^p x,x\rangle} \\ &\leq (\exp\left[-\langle B^q \ln B^q y, y\rangle\right])^{\frac{1}{q}\langle A^{p+q}x,x\rangle} \\ &\quad\times (\exp[-\langle A^p \ln A^p x, x\rangle])^{\frac{1}{p}\langle B^{p+q}y,y\rangle},\end{aligned}$$

which is equivalent to (2.12). □

Corollary 2. *With the assumptions of Theorem 3, and $p > 0$, we have the particular inequalities*

$$\begin{aligned}&[\eta_x(A^{p+q})]^{\frac{1}{p+q}\langle B^q x,x\rangle}[\eta_x(B^{p+q})]^{\frac{1}{p+q}\langle A^p x,x\rangle} \\ &\leq [\eta_x(B^q)]^{\frac{1}{q}\langle A^{p+q}x,x\rangle}[\eta_x(A^p)]^{\frac{1}{p}\langle B^{p+q}x,x\rangle},\end{aligned} \tag{2.15}$$

$$[\eta_x(A^{p+q})]^{\frac{1}{p+q}\langle A^q y,y\rangle}[\eta_y(A^{p+q})]^{\frac{1}{p+q}\langle A^p x,x\rangle}$$
$$\leq [\eta_y(A^q)]^{\frac{1}{q}\langle A^{p+q}x,x\rangle}[\eta_x(A^p)]^{\frac{1}{p}\langle A^{p+q}y,y\rangle} \tag{2.16}$$

and

$$[\eta_x(A^{p+q})]^{\frac{1}{p+q}[\langle A^p x,x\rangle+\langle A^q x,x\rangle]}$$
$$\leq [\eta_x(A^q)]^{\frac{1}{q}\langle A^{p+q}x,x\rangle}[\eta_x(A^p)]^{\frac{1}{p}\langle A^{p+q}x,x\rangle}. \tag{2.17}$$

If $p < 0$, *then we have the particular inequalities*

$$[\eta_x(A^{p+q})]^{\frac{1}{p+q}\langle B^q x,x\rangle}[\eta_x(B^{p+q})]^{\frac{1}{p+q}\langle A^p x,x\rangle}$$
$$\geq [\eta_x(B^q)]^{\frac{1}{q}\langle A^{p+q}x,x\rangle}[\eta_x(A^p)]^{\frac{1}{p}\langle B^{p+q}x,x\rangle}, \tag{2.18}$$

$$[\eta_x(A^{p+q})]^{\frac{1}{p+q}\langle A^q y,y\rangle}[\eta_y(A^{p+q})]^{\frac{1}{p+q}\langle A^p x,x\rangle}$$
$$\geq [\eta_y(A^q)]^{\frac{1}{q}\langle A^{p+q}x,x\rangle}[\eta_x(A^p)]^{\frac{1}{p}\langle A^{p+q}y,y\rangle} \tag{2.19}$$

and

$$[\eta_x(A^{p+q})]^{\frac{1}{p+q}[\langle A^p x,x\rangle+\langle A^q x,x\rangle]} \geq [\eta_x(A^q)]^{\frac{1}{q}\langle A^{p+q}x,x\rangle}[\eta_x(A^p)]^{\frac{1}{p}\langle A^{p+q}x,x\rangle}. \tag{2.20}$$

If we take $q = p > 0$ in (2.17), then we get

$$[\eta_x(A^{2p})]^{\frac{1}{2p}\langle A^p x,x\rangle} \leq [\eta_x(A^p)]^{\frac{1}{p}\langle A^{2p}x,x\rangle} \tag{2.21}$$

for $A > 0$ and $x \in H$ with $\|x\| = 1$.

Remark 2. If we take $p = 1-t$, $q = t \in (0,1)$ in (2.12), then we get

$$[\eta_x(A)]^{\langle B^t y,y\rangle}[\eta_y(B)]^{\langle A^{1-t}x,x\rangle} \leq [\eta_y(B^t)]^{\frac{1}{t}\langle Ax,x\rangle}[\eta_x(A^{1-t})]^{\frac{1}{1-t}\langle By,y\rangle}, \tag{2.22}$$

for all $t \in (0,1)$, A, $B > 0$ and x, $y \in H$ with $\|x\| = \|y\| = 1$.

In particular, for $B = A$ and $y = x$, we derive

$$[\eta_x(A)]^{\langle (A^t+A^{1-t})x,x\rangle} \leq [\eta_x(A^t)]^{\frac{1}{t}\langle Ax,x\rangle}[\eta_x(A^{1-t})]^{\frac{1}{1-t}\langle Ax,x\rangle}, \tag{2.23}$$

for $x \in H$ with $\|x\| = 1$.

3. Related Results

We also have the following:

Theorem 4. *Assume that* $A_j > 0$ *and* $x_j \in H$, $j \in \{1, \ldots, n\}$ *with* $\sum_{j=1}^{n} \|x_j\|^2 = 1$, *then for all* $p \geq 1$,

$$\frac{\left(\prod_{j=1}^{n}\left[\eta_{\frac{x_j}{\|x_j\|}}(A_j^p)\right]^{\|x_j\|^2}\right)^{1/p}}{\left[\prod_{j=1}^{n}\left(\Delta_{\frac{x_j}{\|x_j\|}}(A_j)\right)^{\|x_j\|^2}\right]^{\sum_{j=1}^{n}\langle A_j^p x_j, x_j\rangle}}$$

$$\geq \left(\frac{\sum_{j=1}^{n}\langle A_j x_j, x_j\rangle}{\prod_{j=1}^{n}\left(\Delta_{\frac{x_j}{\|x_j\|}}(A_j)\right)^{\|x_j\|^2}}\right)^{\sum_{j=1}^{n}\langle A_j^p x_j, x_j\rangle - \left(\sum_{j=1}^{n}\langle A_j x_j, x_j\rangle\right)^p} \geq 1. \tag{3.1}$$

Proof. Let A_j be selfadjoint operators with $\mathrm{Sp}(A_j) \subseteq [m, M]$ for $j \in \{1, \ldots, n\}$ and for some scalars $m < M$. In Ref. [1], we also obtained the following result: If f, g are a*synchronous on* $[m, M]$, then

$$\sum_{j=1}^{n}\langle f(A_j)x_j, x_j\rangle \sum_{j=1}^{n}\langle g(A_j)x_j, x_j\rangle - \sum_{j=1}^{n}\langle f(A_j)g(A_j)x_j, x_j\rangle$$
$$\geq \left[\sum_{j=1}^{n}\langle f(A_j)x_j, x_j\rangle - f\left(\sum_{j=1}^{n}\langle A_j x_j, x_j\rangle\right)\right]$$
$$\times \left[\sum_{j=1}^{n}\langle g(A_j)x_j, x_j\rangle - g\left(\sum_{j=1}^{n}\langle A_j x_j, x_j\rangle\right)\right] \tag{3.2}$$

for each $x_j \in H$, $j \in \{1, \ldots, n\}$ with $\sum_{j=1}^{n} \|x_j\|^2 = 1$. Moreover, if either both of them are convex or both of them are concave on $[m, M]$, then the right-hand side of (3.2) is non-negative as well.

Assume that A_j are positive definite and $p \geq 1$. Then, by writing the inequality (3.2) for the functions $f(t) = t^p$ and $g(t) = -\ln t$, $t > 0$,

$$\begin{aligned}
&\sum_{j=1}^{n} \langle A_j^p \log A x_j, x_j \rangle - \sum_{j=1}^{n} \langle A_j^p x_j, x_j \rangle \sum_{j=1}^{n} \langle \log A_j x_j, x_j \rangle \\
&\quad \geq \left[\sum_{j=1}^{n} \langle A_j^p x_j, x_j \rangle - \left(\sum_{j=1}^{n} \langle A_j x_j, x_j \rangle \right)^p \right] \\
&\quad\quad \times \left[\log \left(\sum_{j=1}^{n} \langle A_j x_j, x_j \rangle \right) - \sum_{j=1}^{n} \log \langle A_j x_j, x_j \rangle \right] \\
&\quad \geq 0
\end{aligned} \tag{3.3}$$

for each $x_j \in H$, $j \in \{1, \ldots, n\}$ with $\sum_{j=1}^{n} \|x_j\|^2 = 1$.

By taking the exponential in (3.3), we get

$$\begin{aligned}
&\exp \left[\sum_{j=1}^{n} \langle A_j^p \log A x_j, x_j \rangle - \sum_{j=1}^{n} \langle A_j^p x_j, x_j \rangle \sum_{j=1}^{n} \langle \log A_j x_j, x_j \rangle \right] \\
&\geq \left(\exp \left[\log \left(\sum_{j=1}^{n} \langle A_j x_j, x_j \rangle \right) \right.\right. \\
&\quad\quad \left.\left. - \sum_{j=1}^{n} \log \langle A_j x_j, x_j \rangle \right] \right)^{\sum_{j=1}^{n} \langle A_j^p x_j, x_j \rangle - \left(\sum_{j=1}^{n} \langle A_j x_j, x_j \rangle \right)^p} \\
&\geq 1.
\end{aligned}$$

Observe that

$$\begin{aligned}
&\exp \left[\sum_{j=1}^{n} \langle A_j^p \log A x_j, x_j \rangle - \sum_{j=1}^{n} \langle A_j^p x_j, x_j \rangle \sum_{j=1}^{n} \langle \log A_j x_j, x_j \rangle \right] \\
&\quad = \exp \left[\frac{1}{p} \sum_{j=1}^{n} \left\langle A_j^p \log A_j^p x_j, x_j \right\rangle - \sum_{j=1}^{n} \langle A_j^p x_j, x_j \rangle \sum_{j=1}^{n} \langle \log A_j x_j, x_j \rangle \right]
\end{aligned}$$

$$
\begin{aligned}
&= \frac{\exp\left(\frac{1}{p}\sum_{j=1}^{n}\left\langle A_j^p \log A_j^p x_j, x_j\right\rangle\right)}{\exp\left[\sum_{j=1}^{n}\left\langle A_j^p x_j, x_j\right\rangle \sum_{j=1}^{n}\langle \log A_j x_j, x_j\rangle\right]} \\
&= \frac{\left(\prod_{j=1}^{n}\exp\left\langle A_j^p \log A_j^p x_j, x_j\right\rangle\right)^{1/p}}{\left[\exp\left(\prod_{j=1}^{n}\langle \log A_j x_j, x_j\rangle\right)\right]^{\sum_{j=1}^{n}\langle A_j^p x_j, x_j\rangle}} \\
&= \frac{\left(\prod_{j=1}^{n}\breve{\eta}_{\frac{x_j}{\|x_j\|}}(A_j^p)\right)^{-1/p}}{\left[\prod_{j=1}^{n}\breve{\Delta}_{\frac{x_j}{\|x_j\|}}(A_j)\right]^{\sum_{j=1}^{n}\langle A_j^p x_j, x_j\rangle}} \\
&= \frac{\left(\prod_{j=1}^{n}\left[\eta_{\frac{x_j}{\|x_j\|}}(A_j^p)\right]^{\|x_j\|^2}\right)^{-1/p}}{\left[\prod_{j=1}^{n}\left(\Delta_{\frac{x_j}{\|x_j\|}}(A_j)\right)^{\|x_j\|^2}\right]^{\sum_{j=1}^{n}\langle A_j^p x_j, x_j\rangle}}
\end{aligned}
$$

and

$$
\begin{aligned}
&\exp\left[\log\left(\sum_{j=1}^{n}\langle A_j x_j, x_j\rangle\right) - \sum_{j=1}^{n}\log\langle A_j x_j, x_j\rangle\right] \\
&= \frac{\exp\log\left(\sum_{j=1}^{n}\langle A_j x_j, x_j\rangle\right)}{\exp\left(\sum_{j=1}^{n}\log\langle A_j x_j, x_j\rangle\right)} = \frac{\sum_{j=1}^{n}\langle A_j x_j, x_j\rangle}{\prod_{j=1}^{n}\breve{\Delta}_{\frac{x_j}{\|x_j\|}}(A_j)} \\
&= \frac{\sum_{j=1}^{n}\langle A_j x_j, x_j\rangle}{\prod_{j=1}^{n}\left(\Delta_{\frac{x_j}{\|x_j\|}}(A_j)\right)^{\|x_j\|^2}}
\end{aligned}
$$

for each $x_j \in H$, $j \in \{1, \ldots, n\}$ with $\sum_{j=1}^{n}\|x_j\|^2 = 1$ and the inequality (3.1) is thus proved. □

Corollary 3. *Let $A_j > 0$ and $p_j \geq 0$, $j \in \{1, \ldots, n\}$ with $\sum_{j=1}^{n} p_j = 1$. Then,*

$$\frac{\left(\prod_{j=1}^{n}\left[\eta_x\left(A_j^p\right)\right]^{p_j}\right)^{-1/p}}{\left[\prod_{j=1}^{n}[\Delta_x(A_j)]^{p_j}\right]^{\langle\sum_{j=1}^{n} p_j A_j^p x, x\rangle}} \geq \left(\frac{\langle\sum_{j=1}^{n} p_j A_j x, x\rangle}{\prod_{j=1}^{n}[\Delta_x(A_j)]^{p_j}}\right)^{\langle\sum_{j=1}^{n} p_j A_j^p x, x\rangle - \langle\sum_{j=1}^{n} p_j A_j x, x\rangle^p} \geq 1 \quad (3.4)$$

for all $x \in H$ with $\|x\| = 1$.

Remark 3. For $p = 1$, we get

$$\frac{\left(\prod_{j=1}^{n}\left[\eta_x\left(A_j\right)\right]^{p_j}\right)^{-1}}{\left[\prod_{j=1}^{n}\left[\Delta_x(A_j)\right]^{p_j}\right]^{\langle\sum_{j=1}^{n} p_j A_j x, x\rangle}} \geq 1, \quad (3.5)$$

while for $n = 1$, we obtain

$$\frac{[\eta_x(A^p)]^{-1/p}}{[\Delta_x(A)]^{\langle A^p x, x\rangle}} \geq \left(\frac{\langle Ax, x\rangle}{\Delta_x(A)}\right)^{\langle A^p x, x\rangle - \langle Ax, x\rangle^p} \geq 1 \quad (3.6)$$

provided that $A > 0$, $x \in H$ with $\|x\| = 1$ and $p \geq 1$.

For $p = 1$ in (3.6), we recapture (2.4), while for $p = 2$ in (3.6), we derive

$$\frac{[\eta_x(A^2)]^{-1/2}}{[\Delta_x(A)]^{\langle A^2 x, x\rangle}} \geq \left(\frac{\langle Ax, x\rangle}{\Delta_x(A)}\right)^{\langle A^2 x, x\rangle - \langle Ax, x\rangle^2} \geq 1 \quad (3.7)$$

for all $x \in H$ with $\|x\| = 1$.

References

[1] S. S. Dragomir. Čebyšev's type inequalities for functions of selfadjoint operators in Hilbert spaces. *Linear Multilinear Algebra*, 58(7–8), 2010, 805–814. Preprint *RGMIA Res. Rep. Coll.*, 11(e), 2008, Article 9. https://rgmia.org/papers/v11e/CebysevOperators.pdf.

[2] S. S. Dragomir. Some basic results for the normalized entropic determinant of positive operators in Hilbert spaces. Preprint *RGMIA Res. Rep. Coll.*, 25, 2022.

[3] S. S. Dragomir. Some reverses of the Jensen inequality for functions of selfadjoint operators in Hilbert spaces. *J. Inequal. Appl.*, 2010, 15 p. Article ID 496821. DOI: 10.1155/2010/496821.

[4] S. S. Dragomir. Some inequalities for convex functions of selfadjoint operators in Hilbert spaces. *Filomat*, 23(3), 2009, 81–92. Preprint *RGMIA Res. Rep. Coll.*, 11, 2008. Supplement, Article 11. https://rgmia.org/papers/v11e/ConvFuncOp.pdf.

[5] T. Furuta, J. Mićić Hot, J. Pečarić and Y. Seo. *Mond-Pečarić Method in Operator Inequalities. Inequalities for Bounded Selfadjoint Operators on a Hilbert Space.* Element, Zagreb, 2005.

[6] J. I. Fujii and Y. Seo. Determinant for positive operators. *Sci. Math.*, 1, 1998, 153–156.

[7] J. I. Fujii, S. Izumino and Y. Seo. Determinant for positive operators and Specht's theorem. *Sci. Math.*, 1, 1998, 307–310.

[8] T. Furuta. Precise lower bound of $f(A)-f(B)$ for $A > B > 0$ and non-constant operator monotone function f on $[0,\infty)$. *J. Math. Inequal.*, 9(1), 2015, 47–52.

[9] S. Hiramatsu and Y. Seo. Determinant for positive operators and Oppenheim's inequality. *J. Math. Inequal.*, 15(4), 2021, 1637–1645.

[10] J. Mićić, Y. Seo, S.-E. Takahasi and M. Tominaga. Inequalities of Furuta and Mond-Pečarić. *Math. Ineq. Appl.*, 2, 1999, 83–111.

[11] D. S. Mitrinović, J. E. Pečarić and A. M. Fink. *Classical and New Inequalities in Analysis.* Kluwer Academic Publishers, Dordrecht, 1993.

[12] B. Mond and J. Pečarić. Convex inequalities in Hilbert space. *Houston J. Math.*, 19, 1993, 405–420.

[13] W. Specht. Zer Theorie der elementaren Mittel. *Math. Z.*, 74, 1960, 91–98.

[14] H. Zuo and G. Duan. Some inequalities of operator monotone functions. *J. Math. Inequal.*, 8(4), 2014, 777–781.

https://doi.org/10.1142/9789819820498_0004

Chapter 4

Inequalities for the Normalized Determinant of Positive Operators in Hilbert Spaces via Jensen and Slater's Results

Silvestru Sever Dragomir

Applied Mathematics Research Group, ISILC, Victoria University, Melbourne City, MC 8001, Australia
Department of Mathematical and Physical Sciences, La Trobe University, Plenty Road, Bundoora, Melbourne VIC 3086, Australia
http://rgmia.org/dragomir

For positive invertible operators A on a Hilbert space H and a fixed unit vector $x \in H$, define the normalized determinant by $\Delta_x(A) := \exp\langle \ln Ax, x\rangle$. In this paper we prove among others that,

$$1 \leq \frac{\langle Ax, x\rangle}{\Delta_x(A)} \leq \exp\left[\langle Ax, x\rangle \left\langle A^{-1}x, x\right\rangle - 1\right]$$

for $A > 0$ and $x \in H$ with $\|x\| = 1$.

Keywords: Positive operators, normalized determinants, inequalities.

1. Introduction

Let $B(H)$ be the space of all bounded linear operators on a Hilbert space H, and I stands for the identity operator on H. An operator A in $B(H)$ is said to be positive (in symbol: $A \geq 0$) if $\langle Ax, x\rangle \geq 0$ for all $x \in H$. In particular, $A > 0$ means that A is positive and invertible. For a pair A, B of selfadjoint operators the order relation $A \geq B$ means as usual that $A - B$ is positive.

In 1998, Fujii *et al.* [5, 6], introduced the *normalized determinant* $\Delta_x(A)$ for positive invertible operators A on a Hilbert space H and a fixed unit vector $x \in H$, namely $\|x\| = 1$, defined by $\Delta_x(A) := \exp\langle \ln Ax, x\rangle$ and discussed it as a continuous geometric mean and observed some inequalities around the determinant from this point of view.

Some of the fundamental properties of normalized determinant are as follows, Ref. [5].

For each unit vector $x \in H$, see also Ref. [8], we have the following:

(i) **continuity:** the map $A \to \Delta_x(A)$ is norm continuous;
(ii) **bounds:** $\langle A^{-1}x, x\rangle^{-1} \leq \Delta_x(A) \leq \langle Ax, x\rangle$;
(iii) **continuous mean:** $\langle A^p x, x\rangle^{1/p} \downarrow \Delta_x(A)$ for $p \downarrow 0$ and $\langle A^p x, x\rangle^{1/p} \uparrow \Delta_x(A)$ for $p \uparrow 0$;
(iv) **power equality:** $\Delta_x(A^t) = \Delta_x(A)^t$ for all $t > 0$;
(v) **homogeneity:** $\Delta_x(tA) = t\Delta_x(A)$ and $\Delta_x(tI) = t$ for all $t > 0$;
(vi) **monotonicity:** $0 < A \leq B$ implies $\Delta_x(A) \leq \Delta_x(B)$;
(vii) **multiplicativity:** $\Delta_x(AB) = \Delta_x(A)\Delta_x(B)$ for commuting A and B;
(viii) **Ky Fan type inequality:** $\Delta_x((1-\alpha)A + \alpha B) \geq \Delta_x(A)^{1-\alpha}\Delta_x(B)^{\alpha}$ for $0 < \alpha < 1$.

We define the logarithmic mean of two positive numbers a, b by

$$L(a,b) := \begin{cases} \dfrac{b-a}{\ln b - \ln a} & \text{if } b \neq a, \\ a & \text{if } b = a. \end{cases}$$

In Ref. [5], the authors obtained the following additive reverse inequality for the operator A which satisfy the condition $0 < mI \leq$

$A \le MI$, where m, M are positive numbers,

$$0 \le \langle Ax, x\rangle - \Delta_x(A) \le L(m, M) \times \left[\ln L(m, M) + \frac{M \ln m - m \ln M}{M - m} - 1\right] \tag{1.1}$$

for all $x \in H$, $\|x\| = 1$.

We recall that *Specht's ratio* is defined by [9]

$$S(h) := \begin{cases} \dfrac{h^{\frac{1}{h-1}}}{e \ln\left(h^{\frac{1}{h-1}}\right)} & \text{if } h \in (0,1) \cup (1, \infty), \\ 1 & \text{if } h = 1. \end{cases} \tag{1.2}$$

It is well known that $\lim_{h \to 1} S(h) = 1$, $S(h) = S\left(\frac{1}{h}\right) > 1$ for $h > 0$, $h \ne 1$. The function is decreasing on $(0,1)$ and increasing on $(1,\infty)$.

In Ref. [6], the authors obtained the following multiplicative reverse inequality as well

$$1 \le \frac{\langle Ax, x\rangle}{\Delta_x(A)} \le S\left(\frac{M}{m}\right) \tag{1.3}$$

for $0 < mI \le A \le MI$ and $x \in H$, $\|x\| = 1$.

Motivated by the above results, in this chapter, we prove among others that

$$1 \le \frac{\langle Ax, x\rangle}{\Delta_x(A)}, \frac{\Delta_x(A)}{\langle A^{-1}x, x\rangle^{-1}} \le \exp\left[\langle Ax, x\rangle \langle A^{-1}x, x\rangle - 1\right]$$

for $A > 0$ and $x \in H$ with $\|x\| = 1$.

2. Main Results

Our first main result is as follows:

Theorem 1. *For all* $A > 0$ *and* $x \in H$, $\|x\| = 1$,

$$1 \le \frac{\langle Ax, x\rangle}{\Delta_x(A)} \le \exp\left[\langle Ax, x\rangle \langle A^{-1}x, x\rangle - 1\right]. \tag{2.1}$$

Proof. In Ref. [1], we proved the following reverse of Jensen's inequality:

$$0 \leq \langle f(A)x, x\rangle - f(\langle Ax, x\rangle) \leq \left\langle f'(A)Ax, x\right\rangle - \langle Ax, x\rangle \left\langle f'(A)x, x\right\rangle,$$

where $f : I \to \mathbb{R}$ is a convex and differentiable function on $\mathring{I}$ (the interior of I) whose derivative f' is continuous on $\mathring{I}$, A is a selfadjoint operator with spectrum $\mathrm{Sp}(A) \subset I$ and $x \in H$, $\|x\| = 1$.

If we take in this inequality $f(t) = -\ln t$, $t > 0$, then we get [1]

$$0 \leq \ln\langle Ax, x\rangle - \langle \ln Ax, x\rangle \leq \langle Ax, x\rangle \left\langle A^{-1}x, x\right\rangle - 1$$

for $A > 0$ and $x \in H$, $\|x\| = 1$.

Therefore,

$$\ln\langle Ax, x\rangle \leq \langle \ln Ax, x\rangle + \langle Ax, x\rangle \left\langle A^{-1}x, x\right\rangle - 1$$

and by taking the exponential, we get

$$\begin{aligned}\exp(\ln\langle Ax, x\rangle) &\leq \exp\left[\langle \ln Ax, x\rangle + \langle Ax, x\rangle \left\langle A^{-1}x, x\right\rangle - 1\right] \\ &= \exp\langle \ln Ax, x\rangle \exp\left[\langle Ax, x\rangle \left\langle A^{-1}x, x\right\rangle - 1\right],\end{aligned}$$

namely,

$$\langle Ax, x\rangle \leq \Delta_x(A) \exp\left[\langle Ax, x\rangle \left\langle A^{-1}x, x\right\rangle - 1\right],$$

which is equivalent to (2.1). □

Corollary 1. *Assume that* $0 < mI \leq A \leq MI$, *where* m, M *are positive numbers, then*

$$\begin{aligned}1 &\leq \frac{\langle Ax, x\rangle}{\Delta_x(A)} \leq \exp\left[\langle Ax, x\rangle \left\langle A^{-1}x, x\right\rangle - 1\right] \\ &\leq \exp\left[\frac{(M-m)^2}{4mM}\right]\end{aligned} \tag{2.2}$$

for all $x \in H$, $\|x\| = 1$.

Proof. We use Kantorovich inequality, see, for instance, Ref. [7, p. 30]

$$\langle Ax, x\rangle \left\langle A^{-1}x, x\right\rangle \leq \frac{(m+M)^2}{4mM}$$

for all $x \in H$, $\|x\| = 1$.

Then,

$$\langle Ax, x\rangle \left\langle A^{-1}x, x\right\rangle - 1 \leq \frac{(m+M)^2}{4mM} - 1 = \frac{(M-m)^2}{4mM}$$

and the third inequality in (2.2) is obtained. □

Corollary 2. *With the assumptions of Corollary* 1,

$$\begin{aligned} 1 &\leq \frac{\langle Ax, x\rangle}{\Delta_x(A)} \leq \exp\left[\langle Ax, x\rangle \left\langle A^{-1}x, x\right\rangle - 1\right] \\ &\leq \exp\left[\frac{\left(\sqrt{M}-\sqrt{m}\right)^2}{mM}\langle Ax, x\rangle\right] \leq \exp\left[\frac{\left(\sqrt{M}-\sqrt{m}\right)^2}{m}\right] \end{aligned} \tag{2.3}$$

for all $x \in H$, $\|x\| = 1$.

Proof. We use the additive inequality [7, p. 30]

$$\left\langle A^{-1}x, x\right\rangle - \langle Ax, x\rangle^{-1} \leq \frac{\left(\sqrt{M}-\sqrt{m}\right)^2}{mM}$$

for all $x \in H$, $\|x\| = 1$.

If we multiply by $\langle Ax, x\rangle > 0$, we get

$$\langle Ax, x\rangle \left\langle A^{-1}x, x\right\rangle - 1 \leq \frac{\left(\sqrt{M}-\sqrt{m}\right)^2}{mM}\langle Ax, x\rangle \leq \frac{\left(\sqrt{M}-\sqrt{m}\right)^2}{m},$$

which proves the last part of (2.3). □

Theorem 2. *For all* $A > 0$ *and* $x \in H$, $\|x\| = 1$,

$$1 \leq \frac{\Delta_x(A)}{\left\langle A^{-1}x, x\right\rangle^{-1}} \leq \exp\left[\langle Ax, x\rangle \left\langle A^{-1}x, x\right\rangle - 1\right]. \tag{2.4}$$

Proof. Assume that A is a selfadjoint operator with spectrum $\operatorname{Sp}(A) \subset I$ and $x \in H$, $\|x\| = 1$. In Ref. [2], we proved the following reverse of Slater's inequality:

$$\begin{aligned} 0 &\leq f\left(\frac{\langle Af'(A)x, x\rangle}{\langle f'(A)x, x\rangle}\right) - \langle f(A)x, x\rangle \\ &\leq f'\left(\frac{\langle Af'(A)x, x\rangle}{\langle f'(A)x, x\rangle}\right) \frac{\langle f'(A)Ax, x\rangle - \langle Ax, x\rangle \langle f'(A)x, x\rangle}{\langle f'(A)x, x\rangle}, \end{aligned}$$

where $f : I \to \mathbb{R}$ is a convex and differentiable function on $\mathring{I}$ (the interior of I) whose derivative f' is continuous on $\mathring{I}$ and

$$\frac{\langle Af'(A)x, x\rangle}{\langle f'(A)x, x\rangle} \in \mathring{I} \text{ for any } x \in H, \|x\| = 1.$$

Now, if we write this inequality for the convex function $f(t) = -\ln t$, $t > 0$, then we have [2]

$$0 \leq \langle \ln Ax, x\rangle - \ln\left(\langle A^{-1}x, x\rangle^{-1}\right) \leq \langle Ax, x\rangle \langle A^{-1}x, x\rangle - 1$$

for $A > 0$ and $x \in H$, $\|x\| = 1$.

Therefore,

$$\langle \ln Ax, x\rangle \leq \ln\left(\langle A^{-1}x, x\rangle^{-1}\right) + \langle Ax, x\rangle \langle A^{-1}x, x\rangle - 1$$

and by taking the exponential, we get

$$\begin{aligned} \Delta_x(A) &\leq \exp\left[\ln\left(\langle A^{-1}x, x\rangle^{-1}\right) + \langle Ax, x\rangle \langle A^{-1}x, x\rangle - 1\right] \\ &= \exp\left[\ln\left(\langle A^{-1}x, x\rangle^{-1}\right)\right] \exp\left[\langle Ax, x\rangle \langle A^{-1}x, x\rangle - 1\right] \\ &= \langle A^{-1}x, x\rangle^{-1} \exp\left[\langle Ax, x\rangle \langle A^{-1}x, x\rangle - 1\right], \end{aligned}$$

which proves the second part of (2.4). □

Finally, we have the following:

Corollary 3. *With the assumptions of Corollary* 1,

$$1 \leq \frac{\Delta_x(A)}{\langle A^{-1}x, x\rangle^{-1}} \leq \exp\left[\langle Ax, x\rangle \langle A^{-1}x, x\rangle - 1\right] \leq \exp\left[\frac{(M-m)^2}{4mM}\right] \quad (2.5)$$

and

$$1 \leq \frac{\Delta_x(A)}{\langle A^{-1}x, x\rangle^{-1}} \leq \exp\left[\langle Ax, x\rangle \langle A^{-1}x, x\rangle - 1\right] \leq \exp\left[\frac{\left(\sqrt{M}-\sqrt{m}\right)^2}{mM}\langle Ax, x\rangle\right] \leq \exp\left[\frac{\left(\sqrt{M}-\sqrt{m}\right)^2}{m}\right] \quad (2.6)$$

for all $x \in H$, $\|x\| = 1$.

Theorem 3. *Assume that* $0 < mI \leq A \leq MI$, *where* m, M *are positive numbers, then*

$$1 \leq \frac{\Delta_x(A)}{m^{\frac{M-\langle Ax,x\rangle}{M-m}} M^{\frac{\langle Ax,x\rangle - m}{M-m}}} \leq \exp\left[\frac{\langle (MI-A)(A-mI)x, x\rangle}{mM}\right] \leq \exp\left[\frac{((M-\langle Ax, x\rangle)(\langle Ax, x\rangle - m))}{mM}\right] \leq \exp\left[\frac{(M-m)^2}{4mM}\right] \quad (2.7)$$

for all $x \in H$, $\|x\| = 1$.

Proof. Let $f : [m, M] \to \mathbb{R}$ be a convex function with finite lateral derivatives $f'_+(m)$ and $f'_-(M)$ and A a selfadjoint operator satisfying the condition $mI \leq A \leq MI$. We have the following generalized trapezoid inequality [3]

$$0 \le \frac{f(m)(MI - A) + f(M)(A - mI)}{M - m} - f(A)$$
$$\le \frac{f'_-(M) - f'_+(m)}{M - m}(MI - A)(A - mI)$$

in the operator order.

If we write this inequality for the convex function $f(t) = -\ln t$, $t > 0$, then we get

$$0 \le \ln A - \frac{\ln m(MI - A) + \ln M(A - mI)}{M - m} \le \frac{(MI - A)(A - mI)}{mM}$$

for all $x \in H$, $\|x\| = 1$.

Therefore,

$$\langle \ln Ax, x\rangle \le \frac{\ln m(M - \langle Ax, x\rangle) + \ln M(\langle Ax, x\rangle - m)}{M - m} + \frac{\langle (MI - A)(A - mI)x, x\rangle}{mM}$$

and by taking the exponential, we get

$$\exp\langle \ln Ax, x\rangle \le \exp\left[\frac{\ln m(M - \langle Ax, x\rangle) + \ln M(\langle Ax, x\rangle - m)}{M - m}\right] \times \exp\left[\frac{\langle (MI - A)(A - mI)x, x\rangle}{mM}\right]$$
$$= m^{\frac{M - \langle Ax, x\rangle}{M - m}} M^{\frac{\langle Ax, x\rangle - m}{M - m}} \times \exp\left[\frac{\langle (MI - A)(A - mI)x, x\rangle}{mM}\right]$$

for all $x \in H$, $\|x\| = 1$, which proves the second inequality in (2.7).

The function $g(t) = (M - t)(t - m)$ is concave on $[m, M]$, and by Jensen's inequality

$$\langle g(A)x, x\rangle \le g(\langle Ax, x\rangle), \ x \in H, \|x\| = 1,$$

we have

$$\langle (MI - A)(A - mI)x, x\rangle \le ((M - \langle Ax, x\rangle)(\langle Ax, x\rangle - m))$$

for all $x \in H$, $\|x\| = 1$, which proves the third inequality in (2.7).

The last part is obvious. □

3. Related Results

We have the following:

Theorem 4. *For all* $A,\ B > 0$ *and* $x \in H,\ \|x\| = 1$,

$$\begin{aligned}
&\frac{2}{\langle A^{-1}x, x\rangle + \langle B^{-1}x, x\rangle} \\
&\quad \leq \left\langle \left[\int_0^1 ((1-t)A + tB)^{-1}\, dt\right] x, x\right\rangle^{-1} \\
&\quad \leq \int_0^1 \Delta_x((1-t)A + tB)dt \leq \left\langle \frac{Ax + Bx}{2}, x\right\rangle. \qquad (3.1)
\end{aligned}$$

Proof. By utilizing the inequalities from (ii), we get

$$\begin{aligned}
&\left\langle ((1-t)A + tB)^{-1}\, x, x\right\rangle^{-1} \\
&\quad \leq \Delta_x((1-t)A + tB) \leq \langle ((1-t)A + tB)\, x, x\rangle
\end{aligned}$$

for all $A,\ B > 0,\ x \in H,\ \|x\| = 1$ and $t \in [0, 1]$.

By taking the integral over $t \in [0, 1]$, we get

$$\begin{aligned}
\int_0^1 \left\langle ((1-t)A + tB)^{-1}\, x, x\right\rangle^{-1} dt &\leq \int_0^1 \Delta_x((1-t)A + tB)dt \\
&\leq \int_0^1 \langle ((1-t)A + tB)\, x, x\rangle\, dt \\
&= \left\langle \frac{Ax + Bx}{2}, x\right\rangle
\end{aligned}$$

for all $A,\ B > 0,\ x \in H,\ \|x\| = 1$. This proves the last part of (3.1).

Using Jensen's integral inequality for the convex function $g(u) = u^{-1},\ u > 0$, namely,

$$\int_0^1 [f(t)]^{-1} \geq \left(\int_0^1 f(t)\, dt\right)^{-1},$$

we obtain

$$\begin{aligned}
&\int_0^1 \left\langle ((1-t)A+tB)^{-1}x,x\right\rangle^{-1} dt \\
&\quad \geq \left(\int_0^1 \left\langle ((1-t)A+tB)^{-1}x,x\right\rangle dt\right)^{-1} \\
&\quad = \left\langle \left[\int_0^1 ((1-t)A+tB)^{-1} dt\right] x,x\right\rangle^{-1}
\end{aligned}$$

for all A, $B > 0$, $x \in H$, $\|x\| = 1$.

The function $g(u) = u^{-1}$, $u > 0$ is operator convex and by Hermite–Hadamard inequality for operator convex functions, we have, see Ref. [4],

$$\left(\frac{A+B}{2}\right)^{-1} \leq \int_0^1 ((1-t)A+tB)^{-1} dt \leq \frac{1}{2}\left(A^{-1}+B^{-1}\right)$$

for A, $B > 0$.

Therefore,

$$\left\langle \left[\int_0^1 ((1-t)A+tB)^{-1} dt\right] x,x\right\rangle \leq \frac{1}{2}\left\langle \left(A^{-1}+B^{-1}\right)x,x\right\rangle$$

$x \in H$, $\|x\| = 1$, which gives

$$\begin{aligned}
&\left\langle \left[\int_0^1 ((1-t)A+tB)^{-1} dt\right] x,x\right\rangle^{-1} \\
&\quad \geq \left(\frac{1}{2}\left\langle \left(A^{-1}+B^{-1}\right)x,x\right\rangle\right)^{-1} = \left(\frac{1}{2}\left[\left\langle A^{-1}x,x\right\rangle + \left\langle B^{-1}x,x\right\rangle\right]\right)^{-1} \\
&\quad = \frac{2}{\langle A^{-1}x,x\rangle + \langle B^{-1}x,x\rangle}
\end{aligned}$$

for $x \in H$, $\|x\| = 1$. This proves the first part of (3.1). □

Theorem 5. *For all* A, $B > 0$ *and* $x \in H$, $\|x\| = 1$, *then*

$$\int_0^1 \Delta_x((1-t)A + tB)dt \geq L(\Delta_x(A), \Delta_x(B)). \tag{3.2}$$

Also,

$$\ln \Delta_x\left(\frac{A+B}{2}\right) \geq \int_0^1 \ln \Delta_x((1-t)A + tB)dt \geq \frac{1}{2}[\ln \Delta_x(A) + \ln \Delta_x(B)], \tag{3.3}$$

or, equivalently,

$$\Delta_x\left(\frac{A+B}{2}\right) \geq \exp\left[\int_0^1 \ln \Delta_x((1-t)A + tB)dt\right] \geq \sqrt{\Delta_x(A)\Delta_x(B)}. \tag{3.4}$$

Proof. Let $x \in H$, $\|x\| = 1$. From (viii), we have by taking the integral and assuming that $\Delta_x(A) \neq \Delta_x(B)$,

$$\int_0^1 \Delta_x((1-t)A + tB)dt \geq \int_0^1 \Delta_x(A)^{1-t}\Delta_x(B)^t dt = \frac{\Delta_x(A) - \Delta_x(B)}{\ln \Delta_x(A) - \ln \Delta_x(B)} = L(\Delta_x(A), \Delta_x(B)).$$

If $\Delta_x(A) \neq \Delta_x(B)$, then inequality (3.2) also holds.

If we take the log in (viii), then we get

$$\ln \Delta_x((1-t)A + tB) \geq (1-t)\ln \Delta_x(A) + t \ln \Delta_x(B) \tag{3.5}$$

for all A, $B > 0$ and $t \in [0,1]$.

By taking the integral over $t \in [0,1]$ in (3.5), we get

$$\int_0^1 \ln \Delta_x((1-t)A + tB)dt \geq \int_0^1 [(1-t)\ln \Delta_x(A) + t \ln \Delta_x(B)]\, dt = \frac{1}{2}[\ln \Delta_x(A) + \ln \Delta_x(B)],$$

which proves the second inequality in (3.3).

From (3.5), we get

$$\ln \Delta_x \left(\frac{C+D}{2}\right) \geq \frac{1}{2}\left[\ln \Delta_x(C) + \ln \Delta_x(D)\right] \tag{3.6}$$

for all $C, D > 0$ and $x \in H$, $\|x\| = 1$.

Now, if we take $C = (1-t)A + tB$ and $D = tA + (1-t)B$ in (3.6), then we get

$$\ln \Delta_x \left(\frac{A+B}{2}\right) \geq \frac{1}{2}\left[\ln \Delta_x((1-t)A + tB) + \ln \Delta_x(tA + (1-t)B)\right] \tag{3.7}$$

for all $A, B > 0$ and $t \in [0,1]$.

If we take the integral over $t \in [0,1]$ in (3.7), we deduce

$$\begin{aligned}
&\ln \Delta_x \left(\frac{A+B}{2}\right)\\
&\geq \frac{1}{2}\int_0^1 \left[\ln \Delta_x((1-t)A + tB) + \ln \Delta_x(tA + (1-t)B)\right] dt\\
&= \frac{1}{2}\left[\int_0^1 \ln \Delta_x((1-t)A + tB)dt + \int_0^1 \ln \Delta_x(tA + (1-t)B)dt\right]\\
&= \int_0^1 \ln \Delta_x((1-t)A + tB)dt,
\end{aligned}$$

which proves the first part of (3.3). □

We also have the following:

Proposition 1. *Assume that* $0 < mI \leq A, B \leq MI$, *where* m, M *are positive numbers, then*

$$\begin{aligned}
0 &\leq \left\langle \frac{Ax+Bx}{2}, x\right\rangle - \int_0^1 \Delta_x((1-t)A + tB)dt \\
&\leq L(m,M)\left[\ln L(m,M) + \frac{M\ln M - m\ln M}{M-m} - 1\right]
\end{aligned} \tag{3.8}$$

and

$$1 \leq \frac{\left\langle \frac{Ax+Bx}{2}, x\right\rangle}{\int_0^1 \Delta_x((1-t)A + tB)dt} \leq S\left(\frac{M}{m}\right) \tag{3.9}$$

for $x \in H$, $\|x\| = 1$.

Proof. Since $0 < mI \leq A,\ B \leq MI$, hence $0 < mI \leq (1-t)\,A + tB \leq MI$ for all $t \in [0,1]$. By using (1.1), we get

$$\begin{aligned} 0 &\leq \langle ((1-t)\,A + tB)\,x, x\rangle - \Delta_x((1-t)\,A + tB) \\ &\leq L\,(m, M) \left[\ln L\,(m, M) + \frac{M \ln M - m \ln M}{M - m} - 1\right], \end{aligned}$$

and by taking the integral over $t \in [0,1]$, we get (3.8).

The inequality (3.9) follows in a similar way from (3.9). □

Proposition 2. *With the assumptions of Proposition* 1*, we have*

$$1 \leq \frac{\left\langle \frac{Ax+Bx}{2}, x\right\rangle}{\int_0^1 \Delta_x((1-t)\,A + tB)dt} \leq \exp\left[\frac{(M-m)^2}{4mM}\right] \tag{3.10}$$

and

$$1 \leq \frac{\left\langle \frac{Ax+Bx}{2}, x\right\rangle}{\int_0^1 \Delta_x((1-t)\,A + tB)dt} \leq \exp\left[\frac{\left(\sqrt{M} - \sqrt{m}\right)^2}{m}\right] \tag{3.11}$$

for $x \in H$, $\|x\| = 1$.

Proof. If we use the inequality (2.2), then we can state for $x \in H$, $\|x\| = 1$ that

$$\langle ((1-t)\,A + tB)\,x, x\rangle \leq \Delta_x\left(((1-t)\,A + tB)\right) \exp\left[\frac{(M-m)^2}{4mM}\right]$$

for all $t \in [0,1]$.

If we take the integral over $t \in [0,1]$, then we get (3.10).

Inequality (3.11) follows by (2.3). □

Proposition 3. *With the assumptions of Proposition* 1, *we have*

$$1 \le \frac{\int_0^1 \Delta_x((1-t)A+tB)\,dt}{\int_0^1 \left\langle ((1-t)A+tB)^{-1}x,x\right\rangle^{-1} dt} \le \exp\left[\frac{(M-m)^2}{4mM}\right] \tag{3.12}$$

and

$$1 \le \frac{\int_0^1 \Delta_x((1-t)A+tB)\,dt}{\int_0^1 \left\langle ((1-t)A+tB)^{-1}x,x\right\rangle^{-1} dt} \le \exp\left[\frac{\left(\sqrt{M}-\sqrt{m}\right)^2}{m}\right] \tag{3.13}$$

for $x \in H$, $\|x\| = 1$.

Proof. By (2.5), we have

$$\Delta_x((1-t)A+tB) \le \left\langle (((1-t)A+tB))^{-1}x,x\right\rangle^{-1} \exp\left[\frac{(M-m)^2}{4mM}\right]$$

for all for all $t \in [0,1]$.

If we take the integral over $t \in [0,1]$, then we get

$$\int_0^1 \Delta_x((1-t)A+tB)\,dt \le \int_0^1 \left\langle ((1-t)A+tB)^{-1}x,x\right\rangle^{-1} dt \exp\left[\frac{(M-m)^2}{4mM}\right]$$

for $x \in H$, $\|x\| = 1$, which proves (3.12).

Inequality (3.13) follows by (2.6). □

Proposition 4. *With the assumptions of Proposition* 1, *we have for* $x \in H$, $\|x\| = 1$ *that*

$$\begin{aligned} &\exp\left[\frac{M\ln m - m\ln M}{M-m} + \frac{\ln M - \ln m}{M-m}\right] L(A,B,x,m,M) \\ &\le \int_0^1 \Delta_x(((1-t)A+tB))\,dt \\ &\le \exp\left[\frac{M\ln m - m\ln M}{M-m} + \frac{\ln M - \ln m}{M-m} + \frac{(M-m)^2}{4mM}\right] \\ &\quad \times L(A,B,x,m,M), \end{aligned} \tag{3.14}$$

where

$$L(A,B,x,m,M) := \begin{cases} \dfrac{M-m}{(\ln M - \ln m)\langle (B-A)x,x\rangle} \\ \quad \times \left[\exp\left[\frac{\ln M - \ln m}{M-m}\langle (B-A)x,x\rangle\right] - 1\right] \\ \textit{if } \langle (B-A)x,x\rangle \neq 0, \\ 1 \textit{ if } \langle (B-A)x,x\rangle = 0. \end{cases}$$

Proof. From (2.7), we get

$$\exp\left[\frac{\ln m\left(M - \langle((1-t)A+tB)x,x\rangle\right) + \ln M\left(\langle((1-t)A+tB)x,x\rangle - m\right)}{M-m}\right]$$
$$\leq \Delta_x\left(((1-t)A+tB)\right)$$
$$\leq \exp\left[\frac{\ln m\left(M - \langle((1-t)A+tB)x,x\rangle\right) + \ln M\left(\langle((1-t)A+tB)x,x\rangle - m\right)}{M-m}\right]$$
$$\times \exp\left[\frac{(M-m)^2}{4mM}\right],$$

namely,

$$\exp\left[\frac{M\ln m - m\ln M}{M-m}\right]\exp\left(\frac{\ln M - \ln m}{M-m}\langle((1-t)A+tB)x,x\rangle\right)$$
$$\leq \Delta_x\left(((1-t)A+tB)\right) \leq \exp\left[\frac{M\ln m - m\ln M}{M-m}\right]$$
$$\times \exp\left(\frac{\ln M - \ln m}{M-m}\langle((1-t)A+tB)x,x\rangle\right)\exp\left[\frac{(M-m)^2}{4mM}\right]$$

for $x \in H$, $\|x\| = 1$.

This is equivalent to

$$\exp\left[\frac{M\ln m - m\ln M}{M-m} + \frac{\ln M - \ln m}{M-m}\right]$$
$$\times \exp\left(t\frac{\ln M - \ln m}{M-m}\langle (B-A)x,x\rangle\right)$$

$$\leq \Delta_x\left(((1-t)A+tB)\right)$$
$$\leq \exp\left[\frac{M\ln m - m\ln M}{M-m} + \frac{\ln M - \ln m}{M-m}\right]$$
$$\times \exp\left(t\frac{\ln M - \ln m}{M-m}\left\langle (B-A)x, x\right\rangle\right)$$
$$\times \exp\left[\frac{(M-m)^2}{4mM}\right]$$

for $x \in H$, $\|x\| = 1$.

If we take the integral over $t \in [0,1]$, then we get

$$\exp\left[\frac{M\ln m - m\ln M}{M-m} + \frac{\ln M - \ln m}{M-m}\right]$$
$$\times \int_0^1 \exp\left(t\frac{\ln M - \ln m}{M-m}\left\langle (B-A)x, x\right\rangle\right) dt$$
$$\leq \int_0^1 \Delta_x\left(((1-t)A+tB)\right) dt$$
$$\leq \exp\left[\frac{M\ln m - m\ln M}{M-m} + \frac{\ln M - \ln m}{M-m}\right]$$
$$\times \int_0^1 \exp\left(t\frac{\ln M - \ln m}{M-m}\left\langle (B-A)x, x\right\rangle\right) dt$$
$$\times \exp\left[\frac{(M-m)^2}{4mM}\right]$$

for $x \in H$, $\|x\| = 1$.

Now, observe that

$$\int_0^1 \exp\left(t\frac{\ln M - \ln m}{M-m}\left\langle (B-A)x, x\right\rangle\right) dt = L\left(A, B, x, m, M\right)$$

and the proof is completed. □

References

[1] S. S. Dragomir. Some reverses of the Jensen inequality for functions of selfadjoint operators in Hilbert spaces. *J. Inequalities Appl.*, 2010, 15 p. Article ID 496821. DOI: 10.1155/2010/496821.
[2] S. S. Dragomir. Some Slater's type inequalities for convex functions of selfadjoint operators in Hilbert spaces. *Rev. Un. Mat. Argentina*, 52(1), 2011, 109–120.
[3] S. S. Dragomir. Vector and operator trapezoid type inequalities for continuous functions of selfadjoint operators. *Electron. J. Linear Algebra*, 22, March 2011, 161–178.
[4] S. S. Dragomir. Hermite–Hadamard's type inequalities for operator convex functions. *Appl. Math. Comput.*, 218(3), 2011, 766–772.
[5] J. I. Fujii and Y. Seo. Determinant for positive operators. *Sci. Math.*, 1, 1998, 153–156.
[6] J. I. Fujii, S. Izumino and Y. Seo. Determinant for positive operators and Specht's theorem. *Sci. Math.*, 1, 1998, 307–310.
[7] T. Furuta, J. Mičić-Hot, J. Pečarić and Y. Seo. *Mond-Pečarić Method in Operator Inequalities.* Published by Element in Zagreb, Croatia, in 2005.
[8] S. Hiramatsu and Y. Seo. Determinant for positive operators and Oppenheim's inequality. *J. Math. Inequal.*, 15(4), 2021, 1637–1645.
[9] W. Specht. Zer Theorie der elementaren Mittel. *Math. Z.*, 74, 1960, 91–98.

https://doi.org/10.1142/9789819820498_0005

Chapter 5

Circuit and Weight Representation of the Embedded Markov Chain in the M/G/1 Queuing System

Chrysoula Ganatsiou
Department of Digital Systems, University of Thessaly,
Gaiopolis Campus, Larissa-Trikala Ring Road,
41500 Larissa, Greece
chganats@uth.gr
ganatsio@otenet.gr

By using the circuit representation theory of Markov processes, we investigate a representation by directed circuits and weights of the corresponding embedded Markov chain in the M/G/1 queue in fixed environments.

Keywords: M/G/1 queue, embedded Markov chain, circuit chain, directed weighted circuit, fixed environment.

1. Introduction

It is known that in queuing theory [6, 23, 24, 34], a discipline associated with the mathematical theory of probability ([4, 5]), the most common queuing systems are single-server queue models to which customers arrive according to a Poisson process and service times must be exponentially distributed (e.g. M/M/1 queue). In many cases, Poisson arrivals are a fairly realistic model for the arrival processes, but exponential service times are not very common in practice. Since in many systems the coefficient of variation of the service times must be smaller (or greater) than 1, it is essential to extend the theory to the case of generally distributed service times. To this direction, one of the most classical and best-understood queuing system is the M/G/1 queue, a single-server queue model requiring generally distributed service times and unlimited waiting space while the arrivals of the customers are Markovian following a Poisson process. The model written in Kendall's notation is an extension of the M/M/1 queue, where service times are exponentially distributed [1, 2, 33]. In general, the M/G/1 queue has many applications, but the most classic one is to model performance of a fixed head hard disk. Taking into account that the M/G/1 queue is a stochastic process with state space $S = \{0, 1, 2, 3, \ldots\}$, where the elements of S correspond to the number of customers in the queue including any being served, we may say that a state of the M/G/1 queue at time t can be described by two parameters (n, r), where n denotes the number of customers in the system and r the service time already received by the customer in service, that is, a two-dimensional state, where the first dimension is still discrete and the second one is continuous. This complication in the analysis of the M/G/1 queues may be eliminated by studying the system just after departures, where the state description can be simplified to n only since $r = 0$ for the new customer in service. The whole study of the queue-length process leads to a discrete, time *Markov chain embedded* on the *departure epochs* (or departure instants) with infinitely many cases, which is irreducible and aperiodic under conditions [26, 27, 29, 31, 32].

In parallel, in recent years, a systematic research has been developed by Kalpazidou [25], MacQueen [28], Qian Minping and Qian Min [30], Zemanian [35] and others in order to investigate representations of the finite-dimensional distributions of Markov processes (with discrete or continuous parameter) having an invariant measure, as decompositions in terms of the *circuit passage functions*

$$J_c(i, j) = \begin{cases} 1, & \text{if } i, j \text{ are consecutive states of } c, \\ 0, & \text{otherwise} \end{cases}$$

for any directed sequence $c = (i_1, i_2, \ldots, i_v, i_1)$ of states, called a *circuit*, $v = 1, 2, 3, \ldots$, of the corresponding Markov process. This research has stimulated a motivation toward the representation of Markov processes through directed circuits and weights in terms of circuit passage functions in fixed or random environments as well as the study of specific problems associated with Markov processes in a different way. The representations are called *circuit representations,* while the corresponding discrete parameter Markov chains generated by directed weighted circuits are called *circuit chains* [3, 7–22].

The main purpose of this work is to bring together the two subjects, queue models and circuit chains, by discussing their interconnection. In particular, following the context of the theory of Markov processes' circuit representation, the present work arises as an attempt to study the circuit and weight representation of the discrete-time Markov chain embedded in the queue-length process, especially on the departure epochs in the M/G/1 queuing system in the case of unlimited waiting room capacity in fixed environments, giving a new perspective in the study of specific problems associated with queuing systems in another way through directed weighted circuit representations.

The work is organized as follows. In Section 2, we give a brief account of certain concepts of circuit representation theory of Markov processes that we shall need throughout the chapter. In Section 3, some auxiliary results regarding the Markov chain embedded on departure epochs in the M/G/1 queue model are presented in order to make the presentation of the work more comprehensible. Finally, in Section 4, a representation by directed circuits and weights of the above-mentioned discrete-time embedded Markov chain (circuit chain) is investigated in fixed environments, motivating the study of specific problems associated with the above-mentioned Markov chain in the M/G/1 queue through directed weighted circuits. Throughout the chapter, we shall use the following notations: $\mathbb{N} = \{0, 1, 2, \ldots\}$, $\mathbb{N}^* = \{0, 1, 2, 3, \ldots\}$, $\mathbb{Z} = \{\ldots, -1, 0, 1, \ldots\}$.

2. Preliminaries

Let S be a denumerable set and also C be a class of overlapping directed circuits in S such that for any $i \in S$, there exists a loopcircuit

(i, i). The directed sequence $c = (i_1, i_2, \ldots, i_v, i_1)$ modulo the cyclic permutations, where $i_1, i_2, \ldots, i_v \in S, v \in \mathbb{N}^*$, completely defines a *directed circuit* in S. The ordered sequence $\hat{c} = (i_1, i_2, \ldots, i_v)$ associated with the given directed circuit c is called a *directed cycle* in S. A directed circuit may be considered as a periodic function from $\mathbb{Z}$ to S. The smallest positive integer $p \equiv p(c) \geq 1$ satisfying the equation $c(m+p) = c(m)$, for all $m \in \mathbb{Z}$, is called the *period* of the circuit c. A directed circuit c such that $p(c) = 1$ is called a *loop*. Then the circuit c is completely determined by its points $c(m), m \in \mathbb{Z}$, while the directed pairs $(c(m+r-1), c(m+r)), r = 1, 2, \ldots, p-1$, are the *directed edges* of c. Furthermore, the equivalence class $c = \{c' \mid c' \text{ is a circuit in } S \text{ such that } c(m+i) = c'(m) \text{ for some } i \in \mathbb{Z} \text{ and any } m \in \mathbb{Z}\}$ can be characterized by a period $p(c)$ which is the common period of all the circuits belonging to C. This allows us to call a circuit the corresponding classcircuit. Next, we give the following definitions due to MacQueen [28] and Kalpazidou [25]:

Definition 1. Let c be a directed circuit with period $p(c) > 1$. Then we define by

$$J_c^{(n)}(i,j) = \begin{cases} 1, & \text{if there exists an } m \in \mathbb{Z} \text{ such that } i = c(m), \\ & \quad j = c(m+n), \\ 0, & \text{otherwise} \end{cases}$$

the *n-step passage function* associated with the directed circuit c, for any $i, j \in S, m \in \mathbb{N}^*$.

For $n = 1$, we have

$$J_c^{(1)}(i,j) = \begin{cases} 1, & \text{if there exists an } m \in \mathbb{Z} \text{ such that } i = c(m), \\ & \quad j = c(m+1), \\ 0, & \text{otherwise.} \end{cases}$$

Definition 2. Let c be a directed circuit with period $p(c) > 1$. Then we define by

$$J_c(i) = \begin{cases} 1, & \text{if there exists an } m \in \mathbb{Z} \text{ such that } i = c(m), \\ 0, & \text{otherwise} \end{cases}$$

the *passage function* associated with the directed circuit c for any $i \in S$.

The n-step passage functions $J_c^{(n)}(i,j)$ satisfy the following equations:

$$J_c^{(n+1)}(i,j) = \sum_{k \in S} J_c^{(n)}(i,k) \cdot J_c^{(1)}(k,j), \quad n \in \mathbb{N}^*, \tag{2.1}$$

$$J_c^{(n+1)}(i,j) = \sum_{k \in S} J_c^{(1)}(i,k) \cdot J_c^{(n)}(k,j), \quad n \in \mathbb{N}^*, \tag{2.2}$$

$$J_c^{(n)}(i,j) = \sum_{i_1, i_2, \ldots, i_{n-1} \in S} J_c^{(1)}(i,i_1) \cdot J_c^{(1)}(i_1,i_2) \cdots J_c^{(1)}(i_{n-1},j),$$
$$n = 2, 3, \ldots, \tag{2.3}$$

$$J_c^{(n+t)}(i,j) = \sum_{k \in S} J_c^{(n)}(i,k) \cdot J_c^{(t)}(k,j), \quad n,t \in \mathbb{N}^* \tag{2.4}$$

(Chapman–Kolmogorov-type equations)

$$J_c(i) = \sum_{j \in S} J_c^{(n)}(i,j) = \sum_{k \in S} J_c^{(n)}(k,i), \quad n \in \mathbb{N}^*, \tag{2.5}$$

(balance properties) for any $i, j \in S$. For the proofs of the above-mentioned equations, see the work by Kalpazidou [25].

Given a denumerable set S and an infinite denumerable class C of overlapping directed circuits with distinct points (except for the terminals) in S such that all the points of S can be reached from one another following paths of circuit edges, that is, for each two distinct points i and j of S, there exists a finite sequence $c_1, c_2, \ldots, c_k, k \in \mathbb{N}^*$, of circuits of C such that i lies on c_1 and j lies on c_k and any pair of consecutive circuits (c_n, c_{n+1}) have at least one point in common. We may also assume that the class C contains, among its elements, circuits with period greater or equal to 2. With each directed circuit, let us associate a *strictly positive weight* w_c which must be independent of the choice of the representative of c, that is, it must satisfy the consistency condition $w_{c \circ t_k} = w_c, k \in \mathbb{Z}$, where t_k is the translation of length k.

For a given class C of overlapping directed circuits and for a given sequence $(w_c)_{c \in C}$ of weights, we may define by

$$p_{ij} = \frac{\sum_{c \in C} w_c \cdot J_c^{(1)}(i,j)}{\sum_{c \in C} w_c \cdot J_c(i)} \tag{2.6}$$

the elements of a Markovian transition matrix on S if and only if $\sum w_c \cdot J_c(i) < +\infty$ for any $i \in S$. This means that a given Markov transition matrix $P = (p_{ij}), i, j \in S$, can be represented by directed circuits and weights if and only if there exists a class of overlapping directed circuits C and a sequence of positive weights $(w_c)_{c \in C}$ such that the formula (2.6) holds. In this case, the representation of the distribution of Markov process (with discrete or continuous parameter) having an invariant measure as decomposition in terms of the circuit passage functions is called *circuit representation,* while the corresponding discrete parameter Markov chain generated by directed circuits is called *circuit chain* with Markov transition matrix P given by (2.6) and unique stationary distribution p (a solution of $p \cdot P = p$) defined by

$$p(i) = \sum_{c \in C} w_c \cdot J_c(i), \quad i \in S.$$

It is known that the following classes of Markov chains may be represented uniquely by directed circuits and weights: (i) the recurrent Markov chains ([30]) and (ii) the reversible Markov chains.

3. Auxiliary Results

A *queuing system* may be defined by (i) *the input,* (ii) *the queuediscipline* and (iii) *the servicemechanism.* Let us consider that successive customers demand service at the time instants, $t_{n-1}, t_n, t_{n+1}, \ldots$ In this case, $(t_n)_{n \in \mathbb{Z}}$ is a chain with non-negative independent and identically distributed increments. Suppose that F is the distribution function of the increments $(t_n - t_{n-1}), n \in \mathbb{Z}$. Then

(1) the *input* is said to be

(i) *general independent* (with no assumptions about F except for the condition that there is zero probability of two customers arriving simultaneously),

(ii) *deterministic or regular*
$$\left(F(x) = \begin{cases} 0, & 0 \le x < a \\ 1, & x \ge a \end{cases}, \quad a > 0 \right),$$

(iii) *random* or *Poisson* $(F(x) = 1 - e^{-x/a}, x \ge 0)$,

(iv) *Erlang* (E_k) $\left(F'(x) = \frac{(k/a)^k}{\Gamma(k)} x^{k-1} \cdot e^{-\frac{kx}{a}}, x \geq 0\right)$.

In case (ii) the customers arrive at regular intervals of time such that the interval-arrival time being fixed and equal to a while in (iii) the customers arrive at "random" according to a Poisson process.

(2) Regarding the *queuediscipline,* it will be supposed that the customers form up into a single queue in the usual way such that the customer at the head of the queue is usually served as soon as a sever is free to attend to him (FIFO → first comefirst outserved), taking into account that there will be s servers in general ($s = 1, 2, 3, \ldots$),

(3) The *service mechanism* is defined by the assertions that the servicetimes are non-negative random variables independent of one another and of the input (implying that the existence of a long queue is supposed to have no effect on the speed of service) and that for all customers the servicetime has an arbitrary distribution function with finite mean value and with the condition that there is a zero probability of a customer receiving instantaneous service.

Moreover, a standardized shorthand is used for identifying queuing systems by giving them labels such as M/G/1 (that is, Poisson input, no special assumption about the service-time distribution, one server).

In general, the queuing theory is primarily concerned with a continuous parameter stochastic process $\xi = (\xi(t))_{t\in\mathbb{R}}$, where $\xi(t)$ denotes the number of people waiting or being served at time $t \in \mathbb{R}$, which does not constitute a Markov process except in the M/M/1 queue model. In many cases, its study can be simplified by taking the so-called *embedded Markov chain.* To this direction, we denote by A_t a subset of the set of functions $\mathbb{N}^{(-\infty,t]}$, $t \in \mathbb{R}$, E the random set of those values of $t \in \mathbb{R}$ for which A_t contains the element $(\xi(\tau))_{-\infty<\tau\leq t}$ and $n(t) = f_t\{(\xi(\tau))_{-\infty<\tau\leq t}\}$, $t \in E$, where f_t is a specified functional with domain A_t. Supposing that $\{A_t, f_t/t \in \mathbb{R}\}$ have been chosen such that

(i) E has almost surely no finite point of accumulation (in this case, we may write its members in an increasing order, $\tau_{k-1}, \tau_k, \tau_{k+1}, \ldots$ *(departure epochs)*,

(ii) if $n_k \equiv n(\tau_k)$, then the distribution $\{n_{k+1}/n_k, n_{k-1}, \ldots\}$ is equivalent to the distribution $\{n_{k+1}/n_k\}$, for all $k \in \mathbb{Z}$. In this case, the variables, $n_{k-1}, n_k, n_{k+1}, \ldots$, are said to constitute an *embedded Markov chain.*

Let us now consider the *Markov chain embedded* on departure epochs in the M/G/1 queue model, where the time points selected are immediately after the moment of departure. Since the customer being served (if there is one) has received zero seconds of service, we may conclude that between departures, there can be $0, 1, 2, 3, \ldots$ arrivals. Therefore, from state i, the chain can move to state $i-1$, i, $i+1$, $i+2$, $i+3, \ldots$. In particular, transitions from state i to state $(i+1)$ represent the arrival of a new customer taking into account that the times between such arrivals have an exponential distribution, while transitions from state i to state $(i-1)$ represent a customer who has been served, finishing being served and departing such that the length of time required for serving an individual customer has a general distribution function. Here, the length of times between arrivals and of service periods are random variables assumed to be statistically independent.

More strictly, let us define $\xi(\cdot)$ to be continuous to the right at its points of discontinuity. If A_t is the set of N-valued functions h such that $h(t) = h(t-) - 1$ then the set $E = \{t/\xi(t) = \xi(t-) - 1\}$ consists of *the epochs of departure*. By setting $f_t(h) = h(t), h \in A_t$, we have that $n(t), t \in E$, denotes the number of customers left behind by a departing customer (including the customer, if any, whose service is just starting). Kendall [26] managed to show that this *embedded chain* is Markov with state space $S = \{0, 1, 2, 3 \ldots\}$ and with a well-defined transition matrix $P = (p_{ij})_{i,j \in S}$.

The specification of the transition probabilities p_{ij} of the above-mentioned embedded Markov chain following an algorithmic approach ([31]) to the queue-length distribution is given as follows. Supposing that E has no finite point of accumulation, we may write its members as, $\tau_{k-1}, \tau_k, \tau_{k+1}, \ldots$ in an increasing order (*departure epochs*). Then the embedded Markov chain $n_k \equiv n(\tau_k)$, where $n_k, k \in \mathbb{N}$, expresses the queuelengths immediately after the departure epoch τ_k evolves as follows:

$$n_{k+1} = \begin{cases} n_k - 1 + U, & n_k \geq 1, \\ U, & n_k = 0, \end{cases} \tag{3.1}$$

where U is the number of arrivals of new customers during the service time of the customer $(k+1)$. Denote by

- $a_i = P(U = i)$,
- $f_T(x) =$ the probability distribution function of the service time T (that is, the service time of the customer starting the busy period).

Then following the law of the total probability, it holds

$$\begin{aligned} a_i = P(U = i) &= \int_0^{+\infty} P(U = i/T = x) f_T(x) dx \\ &= \int_0^{+\infty} \frac{(\lambda x)^i}{i!} e^{-\lambda x} \cdot f_T(x) dx, \quad i \in \mathbb{N}, \end{aligned}$$

where λ is the arrival intensity, that is, the arrival rate of the Poisson arrival process. (It is worth mentioning that in a Poisson process, the interarrival times are distributed according to an exponential distribution with parameter λ). Regarding the evolution equation (3.1) of the chain $(n_k)_k$, the transition probabilities p_{ij}, $i, j \in S$, can be determined as follows:

$$p_{ij} = P(n_{k+1} = j/n_k = i) = \begin{cases} P(U = j - i + 1) = a_{j-i+1}, & i \geq 1, \\ P(U = j - i) = a_{j-i}, & i = 0. \end{cases} \tag{3.2}$$

Equivalently, the transition matrix $P = (p_{ij})_{i,j \in S}$ of the embedded Markov chain $(n_k)_k$ is given by

$$P = \begin{array}{c} \\ 0 \\ 1 \\ 2 \\ 3 \\ \vdots \end{array} \!\!\! \begin{array}{c} \begin{array}{ccccc} 0 & 1 & 2 & 3 & \cdots \end{array} \\ \begin{pmatrix} \alpha_0 & \alpha_1 & \alpha_2 & \alpha_3 & \cdots \\ \alpha_0 & \alpha_1 & \alpha_2 & \alpha_3 & \cdots \\ 0 & \alpha_0 & \alpha_1 & \alpha_2 & \cdots \\ 0 & 0 & \alpha_0 & \alpha_1 & \cdots \\ \vdots & \vdots & \vdots & \vdots & \cdots \end{pmatrix} \end{array}, \tag{3.3}$$

as shown in Figure 1. Here, $k_i = a_{i+1} + a_{i+2} + a_{i+3} + \cdots = P(U > i)$ is the probability that during the service time T there are at least $(i+1)$ arrivals, $i \in \mathbb{N}$.

It is obvious that $p_{ij} = 0$ for all $j < i - 1$, while p_{ij} for all $j \geq i - 1$ gives the probability that exactly $(j - i + 1)$ customers arrived during

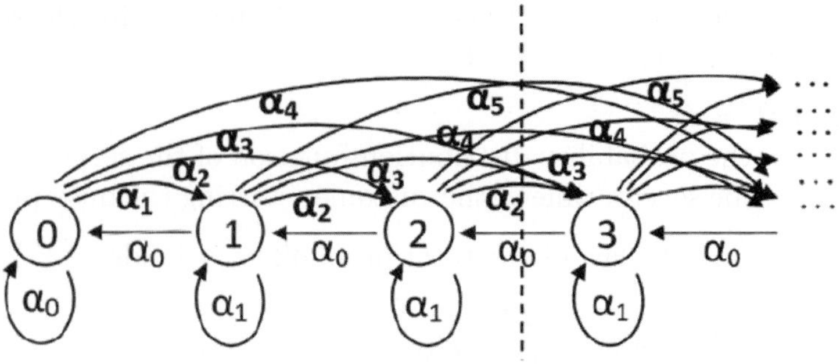

Fig. 1. The embedded Markov chain in the M/G/1 queuing system.

the service time of a customer, $i \in \mathbb{N}^*$. Since in state 0 a customer leaves behind an empty system, it is clear that p_{0j} is the probability that during the service time of the next customer, exactly j customers arrived.

Remark
The discrete-time process $(n_k)_k$ constitutes a discrete-time Markov chain but not a birth–death type-process. In the following, we shall suppose that $a_i > 0, i \in \mathbb{N}$, such that $\sum_{i\in\mathbb{N}} a_i = 1$, which implies that the embedded Markov chain is irreducible and aperiodic.

4. Circuit and Weight Representation of the Embedded Markov Chain in the M/G/1 Queue in Fixed Environments

Let us consider the S-state discrete-time Markov chain $(n_k)_k$ on $\mathbb{N}$, $S = \{0, 1, 2, \ldots\}$ (n_k expresses the queuelength immediately after the departure epoch τ_k, that is, the number of customers in the system immediately after the k_{th} departure including the customer, if any, whose service is just starting), that is, the *embedded Markov chain* in the M/G/1 queue model with transitions $i \to (i-1), i \in \mathbb{N}^*, i \to i, i \to (i+1), i \to (i+2), i \to (i+3), \ldots, i \in \mathbb{N}$, in a fixed environment, whose elements of the corresponding Markov transition matrix (transition probabilities) are defined by (3.3) as follows:

$$P(n_{k+1} = i + r/n_k = i)$$
$$= p_{i,i+r} \equiv \begin{cases} a_{r+1}, & \text{if } i = m = 1, 2, \ldots,\ r = i - m, i - m + 1, \ldots, \\ a_r, & \text{if } i = m = 0,\ r \in \mathbb{N}, \end{cases}$$
$$P(n_{k+1} = i - 1/n_k = i) = p_{i,i-1} \equiv a_0, \quad i \in \mathbb{N}^*,$$
$$P(n_{k+1} = i - r/n_k = i) = p_{i,i-r} = 0, \quad i = 2, 3, 4, \ldots,$$
$$r = 2, 3, 4, \ldots, i,$$

such that

$$\sum_{r=0}^{+\infty} p_{0_r} = 1,$$

$$\sum_{r=2}^{i} p_{i,i-r} + \sum_{r=i-m}^{+\infty} p_{i,i+r} + p_{i,i-1} = 1, \quad i = m = 1, 2, \ldots.$$

Assume that $(p_{i,i+r})_{i,r}$, $i = m = 0, 1, 2, \ldots$, $r = i - m, i - m + 1, i - m + 2, \ldots$, and $(p_{i,i-1})_i, i \in \mathbb{N}^*$ are arbitrary fixed sequences with $p_{i,i+r} > 0$ for every $i = m = 0, 1, 2, \ldots, r = i - m, i - m + 1, i - m + 2, \ldots$ and $p_{i,i-1} > 0$, for every $i \in \mathbb{N}^*$. If we consider the directed circuits $c_i^{(r)} = (i, i + r, i + r - 1, i + r - 2, i + r - 3, \ldots, i)$, and the collection of weights $(w_{c_i^{(r)}})_{i,r}$, $i = m = 0, 1, 2, \ldots$, $r = i - m, i - m + 1, i - m + 2, \ldots$, then we may obtain the corresponding transition probabilities

$$p_{0_r} = \frac{w_{c_0^{(r)}} \cdot J^{(1)}_{c_0^{(r)}}(0, r)}{\sum_{u=0}^{+\infty} \left[w_{c_0^{(u)}} \cdot J_{c_0^{(u)}}(0) \right]}, \quad r \in \mathbb{N}, \tag{4.1}$$

$$p_{i,i+r} = \frac{w_{c_i^{(r)}} \cdot J^{(1)}_{c_i^{(r)}}(i, i + r)}{\sum_{u=0}^{+\infty} \left[w_{c_i^{(u)}} \cdot J_{c_i^{(u)}}(i) + w_{c_{i-1}^{(u+1)}} \cdot J_{c_{i-1}^{(u+1)}}(i) + w_{c_{i-2}^{(u+2)}} \cdot J_{c_{i-2}^{(u+2)}}(i) + \ldots \right]},$$

$$i = m = 1, 2, \ldots, r = i - m, i - m + 1, i - m + 2, \ldots, \tag{4.2}$$

$$p_{i,i-1} = \frac{\sum_{u=0}^{+\infty}\left[w_{c_{i-1}^{(u+1)}} \cdot J^{(1)}_{c_{i-1}^{(u+1)}}(i,i-1) + w_{c_{i-2}^{(u+2)}} \cdot J^{(1)}_{c_{i-2}^{(u+2)}}(i,i-1) + \cdots ,\right]}{\sum_{u=0}^{+\infty}\left[w_{c_i^{(u)}} \cdot J_{c_i^{(u)}}(i) + w_{c_{i-1}^{(u+1)}} \cdot J_{c_{i-1}^{(u+1)}}(i) + w_{c_{i-2}^{(u+2)}} \cdot J_{c_{i-2}^{(u+2)}}(i) + \ldots ,\right]}, \quad i \in \mathbb{N}^*, \tag{4.3}$$

$$p_{i,i-r} = 0, \quad i = 2,3,\ldots, \quad r = 2,3,\ldots,i, \tag{4.4}$$

if and only if

$$\sum_{u=0}^{+\infty}\left[w_{c_0^{(u)}} \cdot J_{c_0^{(u)}}(0)\right] < +\infty,$$

$$\sum_{u=0}^{+\infty}\left[w_{c_{i-1}^{(u+1)}} \cdot J^{(1)}_{c_{i-1}^{(u+1)}}(i,i-1) + w_{c_{i-2}^{(u+2)}} \cdot J^{(1)}_{c_{i-2}^{(u+2)}}(i,i-1) + \ldots ,\right] < +\infty,$$

$$\sum_{u=0}^{+\infty}\left[w_{c_i^{(u)}} \cdot J_{c_i^{(u)}}(i) + w_{c_{i-1}^{(u+1)}} \cdot J_{c_{i-1}^{(u+1)}}(i) + w_{c_{i-2}^{(u+2)}} \cdot J_{c_{i-2}^{(u+2)}}(i) + \ldots ,\right] < +\infty,$$

$$i = m = 0,1,2,\ldots, \quad r = i-m, i-m+1, i-m+2, \ldots,$$

such that

$$\sum_{r=0}^{+\infty} p_{0_r} = 1,$$

$$\sum_{r=2}^{i} p_{i,i-r} + \sum_{r=i-m}^{+\infty} p_{i,i+r} + p_{i,i-1} = 1, \quad i = m = 1,2,\ldots.$$

Here, the class C contains the directed circuits $c_i^{(r)} = (i, i+r, i+r-1, i+r-2, \ldots, i)$, $i = m = 0,1,2,\ldots$, $r = i-m, i-m+1, i-m+2, \ldots$. As a consequence, the transition matrix $P = (p_{ij})_{i,j}$, $i,j \in S$, with elements given by the formulas (4.1)–(4.4), where $J^{(1)}_{c_i^{(r)}}(i,i+r) = 1$ if $i, i+r$ are consecutive points of the circuit $c_i^{(r)}, J^{(1)}_{c_{i-1}^{(u+1)}}(i, i-1), J^{(1)}_{c_{i-2}^{(u+2)}}(i, i-1), \ldots$ equals 1 if $i, i-1$ are consecutive points of the circuits $c_{i-1}^{(u+1)}, c_{i-2}^{(u+2)}, \ldots$, respectively, and $J_{c_i^{(u)}}(i), J_{c_{i-1}^{(u+1)}}(i), \ldots,$ equals 1 if i is a point of the circuit $c_i^{(u)}, c_{i-1}^{(u+1)}, \ldots$, respectively, for every $u \in \mathbb{N}$, $i = m = 0, 1, 2, \ldots$, $r = i - m, i - m + 1, i - m + 2, \ldots$, expresses a representation of the embedded Markov chain $(n_k)_k$ in the M/G/1 queue by directed circuits and weights.

References

[1] J. Abate, G. L. Choudhury, W. Whitt. Calculating the M/G/1 busy-period density and LIFO waiting-time distribution by direct numerical transform inversion. *Oper. Res. Lett.*, 18(3), 1995, pp. 113–119.

[2] J. N. Daigle. The basic M/G/1 queueing system. In *Queueing Theory with Applications to Packet Telecommunication*, Springer, 2005, pp. 159–223.

[3] Yv. Derriennic. Random walks with jumps in random environments (examples of cycle and weight representations). In V. Paulauskas, V.A. Statulevicius, B. Grigelionis, J. Kubilius and H. Pragarauskas (eds.), *Probability Theory and Mathematical Statistics: Proceedings of the 7th Vilnius Conference 1998*, pp. 199–212, De Gruyter, 2020.

[4] W. Feller. *An Introduction to Probability Theory and Its Applications*, Vol. I, 3rd edn. (Wiley, Hoboken, NJ, 1968).

[5] W. Feller. *An Introduction to Probability Theory and Its Applications*, Vol. II, 2nd edn. (Wiley, Hoboken, NJ, 1971).

[6] N. Gautam. *Analysis of Queues: Methods and Applications* (CRC Press, 2012).

[7] Ch. Ganatsiou. On the application of Markov chains cycle representation to a random walk with jumps in a fixed, random environment. In *Proceedings of the 33th International Conference on Stochastic Processes and Their Applications*, Berlin, 2009.

[8] Ch. Ganatsiou. On cycle representations of random walks in fixed, random environments. In *Proceedings of the 58th World Congress of the International Statistical Institute*. Dublin, 2011.

[9] Ch. Ganatsiou. On the study of transience and recurrence of the Markov chain defined by directed weighted circuits associated with a random walk in fixed environment. *J. Prob. Stat.*, Article ID 424601, 2013.

[10] Ch. Ganatsiou. On cycle representations of discrete-time birth-death processes. In *Proceedings of the 29th European Meeting of Statisticians*. Budapest, 2013.

[11] Ch. Ganatsiou. On the study of circuit chains associated with a random walk with jumps in fixed, random environments. In *Proceedings of the 7th European Congress of Mathematics*, Berlin, 2016.

[12] Ch. Ganatsiou. On the expected extinction time for the adjoint birth-death circuit chains in random environments. In *Proceedings of the 32nd European Meeting of Statisticians*, Palermo, Italy, 2019.

[13] Ch. Ganatsiou. On the study of circuit chains associated with a random walk with jumps in fixed, random environments: Criteria of recurrence and transience. In T. Rassias and P. Pardalos (eds.), *Mathematical Analysis and Applications*, Springer Optimization and Its Applications, Vol. 154, Springer, Berlin, 2019, pp. 185–203.

[14] Ch. Ganatsiou. On the study of extinction of circuit chains describing random walks in random environments. In *Proceedings of the 63rd World Congress of the International Statistical Institute (Virtual)*, Hague, 2021.

[15] Ch. Ganatsiou. On irreducible denumerable continuous parameter circuit chains: Analysis of a generalized sample path case. *Quaest. Math.*, 45(1), 2022, pp. 41–53.

[16] Ch. Ganatsiou. On the approximation of extinction time for the discrete-time birth-death circuit chains in random environments. In T. Rassias and N. J. Daras (eds.), *Approximation and Computation in Science and Engineering*, (Springer, 2022), pp. 333–348.

[17] Ch. Ganatsiou. On the expected extinction time of the adjoint circuit chains associated with a random walk with jumps in random environments. In A. Nikeghbali, P. Pardalos, A. Raigorodskii, M. Th. Rassias (eds.), *High Dimensional Optimization and Probability: With a View Towards Data Sciences*, Springer, 2022, pp. 219–239.

[18] Ch. Ganatsiou and Ilias K. Savvas. On the study of circuit chains describing a random walk with jumps in fixed environments: A generalized sample path case. In *Global Optimization, Computation, Approximation and Applications*, World Scientific, 2025, pp. 499–521.

[19] Ch. Ganatsiou. On the stochastic properties of birth-death circuit chains for a generalized sample path case. In *Mathematical Analysis, Optimization, Approximation and Applications*, World Scientific, 2025, pp. 325–345.

[20] Ch. Ganatsiou. On birth-death circuit chains in fixed environments: Analysis of a generalized sample path case. In *Proceedings of the 34th European Meeting of Statisticians (EMS-2023)*, Warsaw, 3–7 July, 2023.

[21] Ch. Ganatsiou, Ilias K. Savvas and A. Xenakis. On the study of cycle chains representing non-reversible Markov chains associated with random walks with jumps in fixed environments. In A. Nikeghbali, P.M. Pardalos, M.T. Rassias (eds.), *Optimization, Discrete Mathematics and Applications to Data Sciences.* Springer Optimization and Its Applications, Vol. 220, Springer, Cham, 2025, pp. 91–103.

[22] Ch. Ganatsiou and Ilias K. Savvas. On the study of cycle chains associated with non-reversible Markov chains describing a random walk with jumps in random environments. In *Trends in Applied Mathematical Analysis*, Springer (to appear).

[23] B. V. Gnedenko and I. N. Kovalenko. *Introduction to Queueing Theory*, Birkhäuser, 1989, pp. 167–169.

[24] S. Asmussen. *Applied Probability and Queues*, 2nd ed. (Springer, 2003).

[25] S. Kalpazidou. *Cycle Representations of Markov Processes* (Springer, New York, NY, 1995).

[26] D. G. Kendall. Some problems in the theory of queues (with discussion). *J. Roy. Stat. Soc. B*, 24, 1951, pp. 151–185.

[27] D. G. Kendall. Stochastic processes occurring in the theory of queues and their analysis by the method of the imbedded Markov chain, *Ann. Math. Stat.*, 24, 1953, pp. 338–354.

[28] J. MacQueen. Circuit processes. *Ann. Probab.*, 9, 1981, pp. 604–610.

[29] B. Meini. Solving M/G/1 type Markov chains: Recent advances and applications. *Commun. Stat. Stoch. Models,* 14(1–2), 1998, pp. 479–496.

[30] Oian Minning and Qian Min. Circulation for recurrent Markov chains, *Z.Wahrsch. Verw. Gebiete,* 59(2), 203–210 (1982).

[31] M. F. Neuts. *Matrix-Geometric Solutions in Stochastic Models: An Algorithmic Approach* (Johns Hopkins University Press, 1981).

[32] N. U. Prabhu and U. N. Bhat. Some first passage problems and their application to queues. *Sankhya*, A, 25, 1963, pp. 281–292.

[33] J. Riordan. *Stochastic Service Systems* (John Wiley and Sons, Inc., 1962).

[34] W. J. Stewart. *Probability, Markov Chains, Queues, and Simulation: The Mathematical Basis of Performance Modelling* (Princeton University Press, 2009).

[35] H. Zemanian. *Infinite Electrical Networks* (Cambridge University Press, Cambridge, 1991).

https://doi.org/10.1142/9789819820498_0006

Chapter 6

Exponential Stability Conditions for Differential Equations with Several Delays and Unbounded Operators in a Banach Space

Michael Gil'

Department of Mathematics, Ben Gurion University of the Negev, Beer-Sheva 84105, Israel
gilmi@bezeqint.net

Let A be a generator of an exponentially stable semigroup in a Banach space, B_j $(j = 1, \ldots, m)$ and $B_0(\tau)$ $(0 \leq \tau \leq \eta < \infty)$ be bounded operators. We consider the equation

$$u'(t) = Au(t) + \sum_{k=1}^{m} B_k u(t - h_k) + \int_0^{\eta} B(\tau)u(t - \tau)d\tau \quad (t \geq 0),$$

where $0 < h_k = \text{const} < \eta$ $(k = 1, \ldots, m)$. We derive explicit delay-dependent and delay-independent exponential stability conditions for that equation. Applications of these conditions to partial integro-differential equations with delays are also discussed.

Keywords: Banach space, differential-delay equations, stability, integro-differential equations with delays.

1. Introduction and Notations

In this chapter, we suggest delay-dependent and delay-independent stability conditions for a class of differential equations in a Banach space with unbound operators and several discrete and distributed delays.

The basic method for the stability analysis of functional differential equations is the Lyapunov–Krasovskii method [3, 10]. By that method, many great results have been obtained. Recently, that method has been extended to functional differential equations in a Hilbert space, cf. Refs. [5, 12, 15] and references given therein. It should be noted that finding the Lyapunov–Krasovskii-type functionals or solving the corresponding operator inequalities are often connected with serious mathematical difficulties, especially regarding equations with many delays. To the contrary, the conditions presented in this chapter are explicitly formulated in terms of the coefficients and delays. The literature on the delay-dependent stability criteria is rather rich, but mainly equations in a finite dimensional space are considered, cf. Refs. [2, 10], etc.

In the papers [6, 8], the delay-dependent stability conditions for non-autonomous equations in a Banach space with *bounded operators* have been derived. In Ref. [9], the delay-dependent stability test for an autonomous equation in a Banach space with an unbounded operator and one discrete delay has been obtained. In this chapter, we generalize the main result from Ref. [9] to equations with many delays.

Introduce the notations: $\mathcal{X}$ is a complex Banach space with a norm $\|\cdot\|_{\mathcal{X}} = \|\cdot\|$ and the unit operator $I_{\mathcal{X}} = I$. By $\mathcal{B}(\mathcal{X})$, we denote the set of all bounded linear operators in $\mathcal{X}$. For a linear operator T, $D(T)$ is the domain, and $\|T\|_{\mathcal{X}} = \|T\|$ is the operator norm of T if it is bounded; $C(J, \mathcal{X})$ is the space of $\mathcal{X}$-valued functions f defined and continuous (in the strong topology) on a finite or infinite real segment J, and equipped with the finite norm $\|f\|_{C(J)} = \|f\|_{C(J,\mathcal{X})} := \sup_{t\in J} \|f(t)\|_{\mathcal{X}}$. Put $R_+ = [0, \infty)$ and $R_\eta = [-\eta, \infty)$ for a finite $\eta > 0$.

Everywhere below A is a generator of a strongly continuous semigroup e^{At} in $\mathcal{X}$, B_j $(j = 1, \ldots, m)$ are bounded operators mapping $D(A)$ into itself, and $B_0(\tau)$ is a piece-wise strongly continuous function defined on $[0, \eta]$ with values in $\mathcal{B}(\mathcal{X})$ mapping $D(A)$ into itself and having a finite number of gaps (if they are).

Define on $C(R_\eta, \mathcal{X})$ the operator E by

$$(Eu)(t) = \sum_{k=1}^{m} B_k u(t-h_k) + \int_0^\eta B_0(\tau)u(t-\tau)d\tau \ (u \in C(R_\eta, \mathcal{X}); t \geq 0),$$

where $0 \leq h_k = \text{const} \leq \eta$ $(k = 1, \ldots, m)$. Our main object is the equation

$$dv(t)/dt = Av(t) + (Ev)(t) \quad (t \geq 0) \tag{1.1}$$

with the initial condition

$$v(t) = \phi(t) \quad (-\eta \leq t \leq 0), \tag{1.2}$$

where ϕ is a given function from $C([-\eta, 0], \mathcal{X})$ with values in $D(A)$.

A solution of problem (1.1), (1.2) is a function $v : R_\eta \to \mathcal{X}$ having a strong derivative $v'(t) \in \mathcal{X}$ for each finite $t > 0$, a right strong derivative at zero, and satisfying (1.1) and (1.2).

Let $Y \subseteq C([-\eta, 0], \mathcal{X})$ be a normed space of functions defined on $[-\eta, 0]$ with a norm $\|.\|_Y$ and values in $D(A)$.

Equation (1.1) is said to be *exponentially stable in the norm of* Y if there are positive constants m_Y, δ_Y independent of the initial function, $\phi \in Y$, such that

$$\|v(t)\|_{\mathcal{X}} \leq \|\phi\|_Y m_Y e^{-\delta_Y t} \quad (t \geq 0)$$

for any solution $v(t)$ of problem (1.1), (1.2).

A few words about the contents. In Section 2, we check the existence of solutions to (1.1) under consideration. In Section 3, delay-independent stability conditions are obtained. The main results of this chapter are Theorem 1 and its corollary are presented in Section 4. In Section 5, we discuss applications of our results to partial integro-differential equations.

2. Existence of Solutions

Lemma 1. *Let* $f \in C([0, t_0], \mathcal{X})$ $(0 < t_0 < \infty)$ *and* $f(t) \in D(A)$ *for all* $0 \leq t \leq t_0$. *In addition, let* $K(t, s)$ *be a function defined on* $0 \leq$

$s \le t \le t_0$ with values in $\mathcal{B}(\mathcal{X})$ mapping $D(A)$ into itself, continuous in t and piece-wise continuous in $s \le t$. Then, the equation

$$u(t) = \int_0^t K(t,s)u(s)ds + f(t) \quad (0 \le t \le t_0) \tag{2.1}$$

has a unique solution $u \in C(0,t_0)$ with values in $D(A)$. Moreover,

$$\|u\|_{C(0,t_0)} \le e^{m_0 t_0}\|f\|_{C(0,t_0)}, \tag{2.2}$$

where

$$m_0 := \sup_{0 \le s \le t \le t_0} \|K(t,s)\|.$$

Proof. Put

$$(Vu)(t) = \int_0^t K(t,s)u(s)ds.$$

Then, for any finite natural n,

$$(V^n u)(t) = \int_0^t K(t,t_1) \int_0^{t_1} K(t_1,t_2)... \int_0^{t_{n-2}} K(t_{n-2},t_{n-1}) \times \int_0^{t_{n-1}} K(t_{n-1},t_n)u(t_n)dt_n dt_{n-1}...dt_1.$$

Hence,

$$\|((V^n u)(t)\|_{C(0,t_0)} \le m_0^n \|u\|_{C(0,t_0)} \int_0^{t_0}\int_0^{t_1} ... \int_0^{t_{n-1}} dt_n dt_{n-1}...dt_1 = \frac{m_0^n t_0^n}{n!}\|u\|_{C(0,t_0)}.$$

Thus,

$$\|V^n\|_{C(0,t_0)} \le \frac{m_0^n t_0^n}{n!}$$

and

$$(I-V)^{-1} = \sum_{n=0}^{\infty} V^n.$$

Since K maps $D(A)$ into itself, $(I - V)^{-1}$ maps $D(A)$ into itself. Moreover,

$$\|(I - V)^{-1}\|_{C(0,t_0)} \le \sum_{n=0}^{\infty} \frac{m_0^n t_0^n}{n!} = e^{m_0 t_0}.$$

The required solution $u(t)$ is given by $u = (I - V)^{-1} f$ and therefore, (2.2) is valid. This proves the lemma. □

Now, let us consider the integral equation

$$v(t) = e^{At}\phi(0) + \int_0^t e^{(t-s)A}(Ev)(s)ds \quad (t \ge 0). \tag{2.3}$$

A solution of this equation is a continuous function $v : R_\eta \to D(A)$ satisfying (2.3) and condition (1.2).

Lemma 2. *Let A generate a continuous semigroup; B_k ($k = 1, \ldots, m$) and $B_0(\tau)$ ($0 \le \tau \le 1$) be bonded linear operators mapping $D(A)$ into itself and $\phi \in C(-\eta, 0)$ have values in $D(A)$. Then, equation (2.3) with condition (1.2) has a unique solution with values in $D(A)$.*

Proof. We have

$$\int_0^t e^{(t-s)A}(Ev)(s)ds = \sum_{j=0}^{m} W_j(t) \quad (t \ge 0), \tag{2.4}$$

where

$$W_0(t) = \int_0^t e^{(t-s)A} \int_0^\eta B_0(\tau)v(s-\tau)d\tau\, ds$$
$$= \int_0^\eta \int_0^t e^{(t-s)A} B_0(\tau)v(s-\tau)d\tau\, ds$$

and

$$W_k(t) = \int_0^t e^{(t-s)A} B_k v(s - h_k)ds \quad (k = 1, \ldots, m).$$

Obviously,

$$W_k(t) = \int_{-h_k}^{t-h_k} e^{(t-s_1-h_k)A} B_j v(s_1) ds_1$$
$$= \int_0^{t-h_k} e^{(t-s_1-h_k)A} B_k v(s_1) ds + \int_{-h_k}^{0} e^{(t-s_1-h_k)A} B_k \phi(s_1) ds_1.$$

So,

$$W_k(t) = \int_0^t Q_k(t,s) v(s) ds + b_k(t),$$

where

$$Q_k(t,s) = e^{(t-s-h_k)A} B_k \quad \text{for } 0 \le s < t - h_k \quad \text{and}$$
$$Q_k(t,s) = 0 \quad \text{for } t - h_k \le s \le t,$$

and

$$b_k(t) = \int_{-h_k}^{0} e^{(t-s-h_k)A} B_k \phi(s) ds \quad (t \ge 0; k = 1, \dots, m).$$

Furthermore,

$$W_0(t) = \int_0^{\eta} \int_{-\tau}^{t-\tau} e^{(t-s_1-\tau)A} B_0(\tau) v(s_1) ds_1 \, d\tau$$
$$= \int_0^{\eta} \left(\int_{-\tau}^{0} + \int_0^{t-\tau} e^{(t-s_1-\tau)A} B_0(\tau) v(s_1) ds_1 \right) d\tau$$
$$= b_0(t) + \int_0^t Q_0(t,s) v(s) ds,$$

where

$$b_0(t) = \int_0^{\eta} \int_{-\tau}^{0} e^{(t-s_1-\tau)A} B_0(\tau) \phi(s_1) ds_1 \, d\tau$$

and

$$Q_0(t,s) = \int_0^{\eta} \hat{Q}(t,s,\tau) d\tau.$$

Here,

$$\hat{Q}(t,s,\tau) := e^{(t-s-\tau)A}B_0(\tau) \quad \text{for } 0 \le s < t-\tau \quad \text{and}$$

$$\hat{Q}(t,s,\tau) = 0 \quad \text{for } t-\tau \le s \le t \quad (t \ge 0).$$

Put

$$\hat{K}(t,s) = \sum_{k=0}^{m} Q_k(t,s) \quad \text{and} \quad b(\phi,t) = \sum_{k=0}^{m} b_k(t).$$

According to (2.4), equation (2.3) with (1.2) taken into account can be written as (2.1) with

$$K(t,s) = \hat{K}(t,s) \quad \text{and} \quad f(t) = b(\phi,t).$$

Now, the existence and uniqueness of solutions to (2.3) (1.2) are due to Lemma 1. □

Due to the variation of constant formula [4, Section III.1], (2.3) is equivalent to equation (1.1). We thus arrive at the following:

Corollary 1. *Under the hypothesis of Lemma* 2 *problem* (1.1), (1.2) *has a unique solution.*

About the recent existence results, see Refs. [1, 7, 13, 14].

3. Delay-Independent Stability Conditions

Assume that e^{At} exponentially stable. Then,

$$\gamma(A) := \int_0^\infty \|e^{At}\| dt < \infty.$$

Put

$$\chi(E) := \sum_{k=1}^{m} \|B_k\| + \int_0^\eta \|B_0(\tau)\| d\tau.$$

Simple calculations show that

$$\|(Ef)\|_{C(0,t)} \le \chi(E)\|f\|_{C(-\eta,t)} \quad (f \in C(-\eta,t), 0 < t \le \infty). \quad (3.1)$$

For the brevity, put

$$|f|_t := \sup_{0\le s\le t} \|f(s)\|_{\mathcal{X}} \quad (f \in C([0,t];\mathcal{X})).$$

Let v be a solution of (1.1), (1.2) and

$$a_0 := \sup_{t\ge 0} \|e^{At}\|_{\mathcal{X}}. \tag{3.2}$$

From (2.3), it follows

$$|v|_t \le a_0\|\phi(0)\| + \int_0^t \|e^{(t-s)A}\|ds|Ev|_{C(-\eta,0))} \quad (t \ge 0).$$

According to (1.2),

$$\|v\|_{C(-\eta,t))} \le |v|_t + \|\phi\|_{C(-\eta,0))}.$$

Thus, (3.1) yields

$$|v|_t \le a_0\|\phi(0)\| + \int_0^\infty \|e^{sA}\|ds \quad |Ev|_t \le a_0\|\phi(0)\|$$
$$+ \int_0^\infty \|e^{sA}\|ds\ \chi(E)\|v\|_{C(-\eta,t)}$$
$$\le a_0\|\phi(0)\| + \gamma(A)\chi(E)(|v|_t + \|\phi\|_{C(-\eta,0)}).$$

So,

$$|v|_t \le c_1\|\phi\|_{C(-\eta,0)} + \gamma(A)\chi(E)|v|_t.$$

Here and in the following, $c_j (j = 1, 2, \dots)$ are constants independent of ϕ. If

$$\gamma(A)\chi(E) < 1, \tag{3.3}$$

then

$$|v|_t \le c_1(1 - \gamma(A)\chi(E))^{-1}\|\phi\|_{C(-\eta,0)}.$$

Now, letting $t \to \infty$ we get the following:

Lemma 3. *Let B_k $(k = 0, \dots, m)$ be the same as in Lemma 2. Let A generate an exponentially stable semigroup and condition (3.3) hold. Then, problem (1.1), (1.2) with $\phi \in C(-\eta, 0)$ and values in $D(A)$ has a unique solution $v(t)$, and*

$$\|v\|_{C(R_+)} \le c_2\|\phi\|_{C(-\eta,0)},$$

where $c_2 = const \ge 1$ does not depend on ϕ.

This lemma implies the following:

Corollary 2. *Under the hypothesis of Lemma* 3 *equation* (1.1) *is exponentially stable in the norm of* $C([-\eta,0],\mathcal{X})$.

Indeed, substitute

$$v_\epsilon(t) = e^{\epsilon t} v(t) \tag{3.4}$$

with $\epsilon > 0$ into (1.1). We obtain the equation

$$v_\epsilon'(t) = (A + \epsilon I)v_\epsilon(t) + (E_\epsilon v_\epsilon)(t) \quad (t > 0), \tag{3.5}$$

where

$$(E_\epsilon f)(t) = \sum_{k=1}^m B_k e^{\epsilon h_k} f(t-h_k) + \int_0^\eta B_0(\tau) e^{\epsilon\tau} f(t-\tau)d\tau \quad (f \in C(R_\eta)).$$

The operator E_ϵ acting from $C(R_\eta)$ into $C(R_+)$ satisfies $E_\epsilon \to E$ in the strong topology as $\epsilon \to 0$. For a small enough $\epsilon > 0$, the hypothesis of Lemma 3 holds, a solution v_ϵ of (3.5) with the initial function $\phi \in C(-\eta, 0)$ satisfies the inequality $\|v_\epsilon\|_{C(R_+)} \le c_\epsilon \|\phi\|_{C(-\eta,0)}$, where c_ϵ does not depend on ϕ. Hence, (3.4) yields the required result. □

4. The Main Result and Delay-Dependent Stability Conditions

In this section, it is assumed that the bounded operators B_k $(k = 1,\ldots,m)$ *and* $B_0(\tau)$ $(0 \le \tau \le \eta)$ *map* $\mathcal{X}$ *into* $D(A)$. So, due to the Banach theorem [11, Section 2], the operators AB_k $(k = 1,\ldots,m)$, $AB_0(\tau)$ and therefore, AE are bounded. Besides, as above $B_0(\tau)$ is piece-wise continuous.

Put

$$M = A + \sum_{k=1}^m B_k + \int_0^\eta B_0(\tau)d\tau,$$

$$\zeta(E) := \sum_{k=1}^m \|B_k\|h_k + \int_0^\eta \|B_0(\tau)\|\tau d\tau$$

and

$$\psi(A, E) = \sum_{k=1}^{m} \|AB_k\| + \int_0^\eta \|AB_0(\tau)\| d\tau.$$

Furthermore, introduce the space $W(-\eta, 0)$ of functions f with values in $D(A)$ defined and strongly continuously differentiable on $[-\eta, 0]$ and equipped with the norm

$$\|f\|_{W(-\eta,0)} := \|f'\|_{C([-\eta,0],\mathcal{X})} + \|f\|_{C([-\eta,0],\mathcal{X})}.$$

Now, we are in a position to formulate and prove the main result of this chapter.

Theorem 1. *Let A generate an exponentially stable semigroup. In addition, let the conditions*

$$\gamma(M) := \int_0^\infty \|e^{Mt}\|_{\mathcal{X}} dt < \infty$$

and

$$\gamma(M)\zeta(E)\ [\gamma(A)\psi(A, E) + \chi(E)] < 1 \qquad (4.1)$$

hold. Then, problem (1.1), (1.2) *with $\phi \in W(-\eta, 0)$ has a unique solution $v(t)$, and $\|v\|_{C(R_+)} \leq \hat{c}\|\phi\|_{W(-\eta,0)}$, where $\hat{c} = const \geq 1$ does not depend on ϕ.*

Proof. The existence and uniqueness of solutions is due to Corollary 1. As it was above noted, (1.1) is equivalent to the equation

$$v(t) = e^{At}\phi(0) + \int_0^t e^{A(t-s)}(Ey)(s) ds.$$

Consequently, in view of (1.1),

$$\begin{aligned} dv(t)/dt &= Av(t) + (Ev)(t) \\ &= A(e^{At}\phi(0) + \int_0^t e^{A(t-s)}(Ev)(s) ds) + (Ev)(t). \qquad (4.2) \end{aligned}$$

Since AE is bounded, the integral

$$\int_0^t e^{A(t-s)}(AEv)(s) ds \quad (0 < t < \infty)$$

converges and

$$A \int_0^t e^{A(t-s)} (Ev)(s) ds = \int_0^t e^{A(t-s)} (AEv)(s) ds.$$

Thus, (4.2) can be written as

$$\frac{dv}{dt} = Ae^{At}\phi(0) + \int_0^t e^{A(t-s)} (AEv)(s) ds + (Ev)(t) \quad (t \geq 0).$$

Hence,

$$|v'|_t \leq a_0 \|A\phi(0)\| + \int_0^t \|e^{A(t-s)} (AEv)(s)\| ds + \sup_{-\eta \leq s \leq t} \|(Ev)(s)\|. \tag{4.3}$$

Recall that

$$|f|_t := \sup_{0 \leq s \leq t} \|f(s)\| \quad (0 < t < \infty, f \in C(0,t)) \text{ and } a_0 := \sup_{t \geq 0} \|e^{At}\|.$$

Take into account that

$$\begin{aligned}
\|(AEv)(s)\| &= \left\| \sum_{k=1}^m AB_k v(s - h_k) + \int_0^\eta AB_0(\tau) v(s-\tau) d\tau \right\| \\
&\leq \sum_{k=1}^m \|AB_k\| \|v(s-h_k)\| + \int_0^\eta \|AB_0(\tau)\| \|v(s-\tau)\| d\tau \\
&\leq \psi(A,E) \|v\|_{C(-\eta,t)} \leq \psi(A,E)|v|_t + \psi(A,E)\|\phi\|_{C(-\eta,0)} \\
&\quad (0 \leq s \leq t).
\end{aligned}$$

Due to (3.1),

$$\|(Eu)(s)\|_{\mathcal{X}} \leq \chi(E)|u|_t + \chi(E)\|\phi\|_{C(-\eta,0)} \quad (0 \leq s \leq t),$$

and therefore, from (4.3), we get

$$\begin{aligned}
|v'|_t &\leq a_0\|A\phi(0)\| + \left(\int_0^\infty \|e^{At}\|_{\mathcal{X}} dt \; \psi(A,E) + \chi(E) \right) \\
&\quad \times (\|\phi\|_{C(-\eta,0)} + |v|_t).
\end{aligned}$$

That is,

$$|v'|_t \le a_0\|A\phi(0)\| + (\gamma(A)\psi(A,E) + \chi(E))(\|\phi\|_{C(-\eta,0)} + |v|_t). \quad (4.4)$$

From (1.1) and (1.2), it follows

$$\phi'(0) = A\phi(0) + (E\phi)(0)$$

and

$$\|A\phi(0)\| \le \|\phi'(0)\| + \|(E\phi)(0)\| \le c_2\|\phi\|_{W(-\eta,0)}.$$

Now, (4.4) yields

$$|v'|_t \le c_4\|\phi\|_{W(-\eta,0)} + (\gamma(A)\psi(A,E) + \chi(E))|v|_t. \quad (4.5)$$

Furthermore, rewrite (1.1) as

$$v'(t) = Mv(t) + (Fv)(t), \quad (4.6)$$

where

$$(Fv)(t) = \sum_{k=1}^{m} B_k(v(t-h_k) - v(t)) + \int_0^{\eta} B_0(\tau)(v(t-\tau) - v(t))d\tau \quad (t > 0).$$

Since

$$\|v(t) - v(t-h)\|_{\mathcal{X}} = \left\|\int_{t-h}^{t} v'(s_1)ds_1\right\|_{\mathcal{X}} \le h\|v'\|_{C(-\eta,t)} \quad (0 \le h \le \eta),$$

we get

$$\|(Fv)(t)\|_{\mathcal{X}} \le \|v'\|_{C(-\eta,t)}\zeta(E). \quad (4.7)$$

Due to the above mentioned Variation of Constants formula, (4.6) is equivalent to the equation

$$v(t) = e^{Mt}\phi(0) + \int_0^t e^{M(t-s)}(Fv)(s)ds.$$

Now, (4.7) yields

$$|v|_t \le m_0\|\phi(0)\|_{\mathcal{X}} + \int_0^t \|e^{M(t-s)}\|_{\mathcal{X}} ds\ \zeta(E)\|v'\|_{C(-\eta,t)}$$
$$\le m_0\|\phi(0)\|_{\mathcal{X}} + \gamma(M)\zeta_M(E)\|v'\|_{C(-\eta,t)},$$

where $m_0 := \sup_{t\ge 0}\|e^{Mt}\|_{\mathcal{X}}$. But

$$\sup_{-\eta\le s\le t}\|v'(s)\|_{\mathcal{X}} \le |v'|_t + \|\phi'\|_{C(-\eta,0)}.$$

Consequently,

$$|v|_t \le c_5\|\phi\|_{W(-\eta,0)} + \gamma(M)\zeta(E)|v'|_t. \tag{4.8}$$

Substituting (4.5) into this inequality, we get

$$|v|_t \le c_5\|\phi\|_{W(-\eta,0)} + \gamma(M)\zeta(E)[c_4\|\phi\|_{W(-\eta,0)}$$
$$+ (\gamma(A)\psi(A,E) + \chi(E))|v|_t]$$

or

$$|v|_t \le c_6\|\phi\|_{W(-\eta,0)} + \gamma(M)\zeta(E)[\gamma(A)\psi(A,E) + \chi(E)]|v|_t].$$

Hence, due to condition (4.1),

$$|v|_t \le \frac{c_6\|\phi\|_{W(-\eta,0)}}{1-\gamma(M)\zeta(E)[\gamma(A)\psi(A,E)+\chi(E)]}.$$

Letting $t \to \infty$, we get the required result. □

Corollary 3. *Under the hypothesis Theorem* 1, *equation* (1.1) *is exponentially stable in the norm of* $W(-\eta, 0)$.

Indeed, substitute (3.4) with $\epsilon > 0$ into (1.1). We obtain equation (3.5). We have $E_\epsilon \to E$ in the strong topology as $\epsilon \to 0$. The hypothesis of Theorem 1 holds, for a small enough $\epsilon > 0$, and a solution v_ϵ of (3.5) with the initial function $\phi \in W(-h, 0)$ satisfies the inequality $\|v_\epsilon\|_{C(R_+)} \le c_\epsilon\|\phi\|_{W(-h,0)}$, where c_ϵ does not depend on ϕ. Hence, (3.4) yields

$$\|v(t)\|_{C(R_+)} \le c_\epsilon e^{-\epsilon t}\|\phi\|_{W(-h,0)} \quad (t \ge 0),$$

as claimed. □

Corollary 3 is sharp: If $\eta = 0$, then the condition $\gamma(M) < \infty$ is necessary, and condition (4.1) automatically holds.

5. Example

We need the following simple result.

Lemma 4. *Let C be a closed linear operator with a dense domain in a Hilbert space $\mathcal{H}$ with a scalar product $\langle ..\rangle_{\mathcal{H}}$ and the norm $\|.\|_{\mathcal{H}} = \sqrt{\langle .,.\rangle_{\mathcal{H}}}$. Let $D(C) = D(C^*)$ and C generate a strongly continuous semigroup e^{Ct}. If, in addition,*

$$\nu(C) := \frac{1}{2} \sup_{z \in D(C), \|z\|_{\mathcal{H}}=1} \langle (C + C^*)z, z\rangle < \infty,$$

then $\|e^{Ct}\|_{\mathcal{H}} \le e^{\nu(C)t}$ $(t \ge 0)$.

Proof. Put $w(t) = e^{tC}w(0)$ with $w(0) \in D(C)$. Then,

$$\begin{aligned}\frac{d}{dt}(w(t), w(t)) &= (w'(t), w(t)) + (w(t), w'(t)) \\ &= (Cw(t), w(t)) + (w(t), Cw(t)) \\ &= ((C + C^*)w(t), w(t)) \le 2\nu(C)(w(t), w(t)).\end{aligned}$$

Hence,

$$\frac{d}{dt}\|w(t)\|^2 \le 2\nu(C)\|w(t)\|^2.$$

Solving this inequality, we get the required result. □

Furthermore, let $\mathbf{C}^2$ be the two-dimensional Euclidean space of vectors $z = (z_1, z_2)^T, w = (w_1, w_2)^T$ with the scalar product $\langle z, w\rangle_2 = z_1\overline{w}_1 + z_2\overline{w}_2$ and norm $\|z\|_2 = \sqrt{\langle z, z\rangle_2}$. In this section, $\mathcal{X} = L^2([0,1], \mathbf{C}^2)$, where $L^2([0,1], \mathbf{C}^2) = L^2$ is the Hilbert space of vector-valued functions $f(x) = (f_1(x), f_2(x))^T, g(x) = (g_1(x), g_2(x))^T$ defined on $[0,1]$ with values in $\mathbf{C}^2$, the scalar product

$$\langle f, g\rangle_{L^2} = \int_0^1 \langle f(x), g(x)\rangle_2 dx \quad (f, g \in L^2)$$

and the norm $\|f\|_{L^2} = \sqrt{\langle f, f\rangle_{L^2}}$. Consider the coupled system

$$\frac{\partial u_1(t,x)}{\partial t} = \frac{\partial^2 u_1(t,x)}{\partial x^2} + b_1 u_2(t,x) + \int_0^1 K_1(x,s) u_1(t-h_1,s)ds,$$
$$\frac{\partial u_2(t,x)}{\partial t} = \frac{\partial^2 u_2(t,x)}{\partial x^2} + b_2 u_1(t,x) + \int_0^1 K_2(x,s) u_2(t-h,s)ds,$$
$$(0 \le x \le 1;\ t \ge 0;\ b_j = \text{ const} > 0;\ 0 < h_j < \infty, j = 1,2), \tag{5.1}$$

where $K_j(x,s)$ $(j = 1,2)$ are defined on $[0,1] \times [0,1]$, twice continuously differentiable in $x \in [0,1]$, and bounded and measurable in s, and $K_j(0,s) = K_j(1,s) = 0$ $(s \in [0,1])$.

Take the boundary conditions

$$u(t,0) = u(t,1) = 0 \quad (t \ge 0). \tag{5.2}$$

Here, $u(t,x) = (u_1(t,x), u_2(t.x))^T$. Put

$$D(A) = \{f \in L_2^2 : f'' \in L_2^2, f(0) = f(1) = 0\} \quad (f = (f_1, f_2)^T).$$

Besides, A and B_j $(j = 1,2)$ are defined by

$$(Af)(x) = (f_1''(x) + b_1 f_2(x), f_2''(x) + b_2 f_1(x))^T \quad (f \in D(A))$$

and

$$(B_1 f)(x) = \left(\int_0^1 K_1(x,s) f_1(s)ds, 0\right)^T,$$
$$(B_2 f)(x) = \left(0, \int_0^1 K_2(x,s) f_2(s)ds\right)^T \quad (f \in L_2^2).$$

Clearly, B_j maps $L^2([0,1], \mathbf{C}^2)$ into $D(A)$. In addition,

$$(AB_1 f)(x) = \left(\int_0^1 \frac{\partial^2 K_1(x,s)}{\partial x^2} f_1(s)ds + b_1 f_2(x), 0\right)^T,$$
$$(AB_2 f)(x) = \left(0, \int_0^1 \frac{\partial^2 K_2(x,s)}{\partial x^2} f_2(s)ds + b_2 f_1(x)\right)^T.$$

We can write $A = A_1 + A_2$ where A_1 is the self-adjoint operator $\frac{d^2}{dx^2}$ on $D(A)$ and A_2 is defined by $A_2 f = bf$, where

$$b = \begin{pmatrix} 0 & b_1 \\ b_2 & 0 \end{pmatrix}.$$

Simple calculations show that the largest eigenvalue of the self-adjoint operator $\frac{d^2}{dx^2}$ on $D(A)$ is $-\pi^2$ and the eigenvalues of b are $\pm i\sqrt{b_1 b_2}$, $\|e^{bt}\|_2 = 1$. Hence, $\|e^{A_2 t}\|_{L^2} = 1$. Since A_1 and A_2 commute,

$$\|e^{At}\|_{L^2} = \|e^{A_1 t} e^{A_2 t}\|_{L^2} \le e^{-\pi^2 t} \quad (t \ge 0). \tag{5.3}$$

Besides,

$$\gamma(A) = \int_0^\infty \|e^{At}\|_{L^2} dt \le 1/\pi^2. \tag{5.4}$$

Furthermore, under consideration, we have $M = A + B_1 + B_2$ and

$$\nu(M) \le -\pi^2 + \nu(B_1) + \nu(B_2).$$

Due to Lemma 4, $\|e^{Mt}\|_{L^2} \le e^{\nu(M)t}$ $(t \ge 0)$. If $\nu(M) < 0$, then

$$\gamma(M) = \int_0^\infty \|e^{Mt}\|_{L^2} dt \le 1/|\nu(M)|.$$

In addition, $\chi(E) = \|B_1\|_{L^2} + \|B_2\|_{L^2}, \zeta(E) = \|B_1\|_{L^2} h_1 + \|B_2\|_{L^2} h_2$, and

$$\psi(A, E) = \|AB_1\|_{L^2} + \|AB_2\|_{L^2}.$$

Making use of Corollary 3, we arrive at the following:

Corollary 4. *Let the conditions* $\nu(M) < 0$ *and*

$$\frac{\zeta(E)}{\nu(M)} \left(\frac{\psi(A, E)}{\pi^2} + \chi(E) \right) < 1$$

hold. Then (5.1), (5.2) *is exponentially stable in the norm of* $W(-\eta, 0)$ *with* $\eta = \max\{h_1, h_2\}$.

References

[1] D. Andrica and T. M. Rassias (eds.). *Differential and Integral Inequalities.* Springer Optimization and Its Applications, Vol. 151. Springer Nature, Switzerland, 2019.

[2] L. Berezansky, J. Diblik, Z. Svoboda and Z. Šmarda. Simple tests for uniform exponential stability of a linear delayed vector differential equation. *IEEE Trans. Autom. Control*, 67(3), 2022, 1537–1542.

[3] C. Corduneanu, Y. Li and M. Mahdavi. *Functional Differential Equations. Advances and Applications.* Pure and Applied Mathematics. John Wiley & Sons, Inc., Hoboken, NJ, 2016.

[4] K.-J. Engel and R. Nagel. *A Short Course on Operator Semigroups.* Universitext. Springer, New York, 2006.

[5] E. Fridman and Y. Orlov. Exponential stability of linear distributed parameter systems with time-varying delays. *Automatica*, 45, 2009, 194–201.

[6] M. I. Gil'. Delay-dependent stability conditions for non-autonomous functional differential equations with several delays in a Banach space. *Nonauton. Dyn. Syst.*, 8(1), 2021, 168–179.

[7] M. I. Gil'. On Dyson-Phillips type approach to differential-difference equations in a Banach space. *Discuss. Math. Differ. Incl. Control Optim.*, 42(2), 2022, 1–8.

[8] M. I. Gil'. Delay-dependent stability conditions for differential- difference equations with small commutators in a Banach space. *Bull. Math. Sci.*, 2350009, 2023, 1–12.

[9] M. I. Gil'. Delay-dependent stability conditions for differential-delay equations with unbounded operators in a Banach space. *Electron. J. Differ. Equ.*, 2024(76), 2024, 1–8.

[10] V. Kolmanovskii and A. Myshkis. *Applied Theory of Functional Differential Equations.* Kluwer, Dordrecht, 1999.

[11] S. G. Krein. *Linear Equations in Banach Spaces.* Boston, Birkhauser, 1982.

[12] S. G. Luo and F. Q. Deng. LMI-based approach of robust control for uncertain distributed parameter control systems with time-delay. *Control Theory Appl.*, 23, 2006, 318–324.

[13] P. M. Pardalos and T. M. Rassias (eds.). *Mathematics without Boundaries.* Surveys in Interdisciplinary Research, Vol. VIII. Springer, NY, 2014.

[14] P. M. Pardalos and T. M. Rassias (eds.). *Mathematical Analysis, Optimization, Approximation an Applications.* World Scientific, New Jersey, 2025.

[15] L. Wang and Y. Wang. LMI-based approach of global exponential robust stability for a class of uncertain distributed parameter control systems with time-varying delays. *J. Vib. Control*, 15, 2009, 1173–1185.

https://doi.org/10.1142/9789819820498_0007

Chapter 7

Angular Localization of the Eigenvalues of Perturbed Matrices

Michael Gil'

Department of Mathematics, Ben Gurion University of the Negev, Beer-Sheva 84105, Israel

gilmi@bezeqint.net

Let A and $\tilde{A}$ be matrices. We consider the following problem: Let the eigenvalues of A lie in some angular sector. In what sector the eigenvalues of $\tilde{A}$ lie if the norm of $A - \tilde{A}$ is sufficiently small? We also discuss applications of our results to the theory of polynomials. Illustrative examples are also presented.

1. Introduction and Preliminaries

Let $\mathbf{C}^n$ be the n-dimensional complex Euclidean space with a scalar product $(.,.)$, the norm $\|.\| = \sqrt{(.,.)}$ and unit operator I. As usual, $\mathbf{C}^{n\times n}$ denotes the set of complex $n \times n$-matrices. For an $A \in \mathbf{C}^{n\times n}$, A^* is the adjoint matrix, $\|A\|$ is the spectral norm: $\|A\| = \sup_{x\in\mathbf{C}^n, \|x\|=1} \|Ax\|$, $\lambda_k(A)$ $(k = 1, \ldots, n)$ are the eigenvalues of A taken with their multiplicities and $\sigma(A)$ is the spectrum.

Let A and $\tilde{A} \in \mathbf{C}^{n\times n}$. We investigate the following problem: Let $\sigma(A)$ lie in some angular sector. In what sector $\sigma(\tilde{A})$ lies if $q = \|A - \tilde{A}\|$ is sufficiently small?

Not too much works are devoted to the angular localizations of spectra. The papers [1, 2, 13, 16] should be mentioned. In particular, in the papers by Anderson *et al.* [1, 2], it is shown that the test to determine whether all eigenvalues of a complex matrix of order n lie in a certain sector can be replaced by an equivalent test to find whether all eigenvalues of a real matrix of order $4n$ lie in the left half plane. The results from Ref. [1] have been applied by Hostetter [12] to obtain an improved test for the zeros of a polynomial in a sector. Krein [13] announces two theorems concerning the angular localization of the spectrum of a multiplicative operator integral. The so-called sector matrix function [11] also should be mentioned. It is deeply connected with angular localization of eigenvalues, but perturbation of that function was not explored.

Note that although the perturbation theory of matrices is very rich, cf. Refs. [3, 10, 15, 17], and the references given therein, to the best of our knowledge, the just pointed problem was not considered in the available literature, except in the papers [8, 9], where some special classes of operators have been considered.

For a Hermitian $Y \in \mathbf{C}^{n\times n}$ we write $Y > 0$ if Y is positive definite, i.e. $\inf_{x\in\mathbf{C}^n,\|x\|=1}(Yx, x) > 0$.

It is assumed that

$$\beta(A) := \min_{k=1,\dots,n} \text{Re}\, \lambda_k(A) > 0. \tag{1.1}$$

By the Lyapunov theorem, cf. Ref. [4, Theorem I.5.1], this condition implies that there exists a positive definite $Y \in \mathbf{C}^{n\times n}$ such that $(YA)^* + YA > 0$. Define the quantity $\tau(A, Y)$ by

$$\cos\tau(A, Y) := \inf_{x\in\mathbf{C}^n,\|x\|=1} \frac{\text{Re}\,(YAx, x)}{|(YAx, x)|}.$$

Let $\lambda = re^{it}$ $(r > 0, 0 \le t < 2\pi)$ be an eigenvalue of A and d the corresponding eigenvector: $Ad = \lambda d$. Then,

$$\frac{\text{Re}\,(YAd, d)}{|(YAd, d)|} = \frac{\text{Re}\;\, re^{it}(Yd, d)}{r(Yd, d)} = \cos t.$$

We thus get the following:

Lemma 1. *For an* $A \in \mathbf{C}^{n\times n}$*, let condition* (1.1) *hold and* Y *be a positive definite matrix such that* $(YA)^* + YA > 0$*. Then, any eigenvalue of* A *lies in the sector*

$$S(A,Y) := \{z \in \mathbf{C} : |\arg z| \leq \tau(A,Y)\}.$$

Example 1. Let $A = A^* > 0$. Then, condition (1.1) holds. For any $Y > 0$ commuting with A (for example, $Y = I$), we have $(YA)^* + YA = 2YA$ and $\mathrm{Re}\,(YAx, x) > 0$. That is, $\mathrm{Re}\,(YAx, x) = |(YAx, x)|$ and $\cos\tau(A,Y) = 1$, and $S(A,Y) = \{z \in \mathbf{C} : \arg z = 0\}$.

So, Lemma 1 is sharp.

Remark 1. Suppose A is non-singular. Recall that the quantity $\mathrm{dev}(A)$ defined (in the finite dimensional case) by

$$\cos \mathrm{dev}(A) := \inf_{x\in\mathbf{C}^n, x\neq 0} \frac{\mathrm{Re}\,(Ax, x)}{\|Ax\|\|x\|}$$

is called the angular deviation of A, cf. Ref. [4, Chapter 1, Exercise 32]. For example, for a positive definite Hermitian matrix A, one has

$$\cos\ \mathrm{dev}(A) = \frac{2\sqrt{\lambda_M(A)\lambda_m(A)}}{\lambda_M(A) + \lambda_m(A)},$$

where $\lambda_M(A)$, $\lambda_m(A)$ are the boundary of the spectrum of A (see Ref. [4, Chapter 1, Exercise 33]). Besides, in Exercise 32, it is pointed that the spectrum of A lies in the sector $|\arg z| \leq \mathrm{dev}(A)$. Since $|(Ax, x)| \leq \|Ax\|\|x\|$, Lemma 1 refines the just pointed assertion.

2. The Main Result

By the above-mentioned Lyapunov theorem, under condition (1.1), there exists a positive definite $X \in \mathbf{C}^{n\times n}$ solving the Lyapunov equation:

$$2\mathrm{Re}\,(AX) = XA + A^*X = 2I. \tag{2.1}$$

Hence,

$$\cos \tau(A, X) = \inf_{x \in \mathbf{C}^n, \|x\|=1} \frac{(x, x)}{|(XAx, x)|} = \frac{1}{\sup_{x \in \mathbf{C}^n, \|x\|=1} |(XAx, x)|}$$
$$\geq \frac{1}{\|AX\|}. \tag{2.2}$$

Put

$$J(A) = 2 \int_0^\infty \|e^{-At}\|^2 dt.$$

In Section 3, we suggest an estimate for $J(A)$. Now, we are in a position to formulate our main result.

Theorem 1. *Let* $A, \tilde{A} \in \mathbf{C}^{n \times n}$, *condition* (1.1) *hold and* X *be a solution of* (2.1). *Then, with the notation* $q = \|A - \tilde{A}\|$ *one has*

$$\cos \tau(\tilde{A}, X) \geq \cos \tau(A, X) \frac{(1 - qJ(A))}{(1 + qJ(A))} \geq \frac{(1 - qJ(A))}{\|AX\|(1 + qJ(A))}, \tag{2.3}$$

provided

$$qJ(A) < 1. \tag{2.4}$$

The proof of this theorem is based on the following lemma:

Lemma 2. *Let* $A, \tilde{A} \in \mathbf{C}^{n \times n}$, *let condition* (1.1) *hold and* X *be a solution of* (2.1). *If, in addition,*

$$q\|X\| < 1, \tag{2.5}$$

then

$$\cos \tau(\tilde{A}, X) \geq \cos \tau(A, X) \frac{(1 - q\|X\|)}{(1 + q\|X\|)} \geq \frac{1 - q\|X\|}{\|AX\|(1 + q\|X\|)}.$$

Proof. Put $E = \tilde{A} - A$. Then, $q = \|E\|$ and due to (2.1), with $\|x\| = 1$ we obtain

$$\text{Re } (X(A + E)x, x) \geq \text{Re } (XAx, x) - |(XEx, x)|$$
$$= (x, x) - |(XEx, x)|$$
$$\geq (x, x) - \|X\|\|E\|\|x\|^2 = 1 - \|X\|q. \tag{2.6}$$

In addition,

$$|(X(A+E)x,x)| \leq |(XAx,x)| + \|X\|\|E\|\|x\|^2$$
$$= |(XAx,x)| \left(1 + \frac{\|X\|q}{|(XAx,x)|}\right) \ (\|x\|=1).$$

But

$$|(XAx,x)| \geq |\text{Re}\ (XAx,x)| = \text{Re}\ (XAx,x) = (x,x) = 1.$$

Hence,

$$|(X(A+E)x,x)| \leq |(XAx,x)| \left(1 + \frac{\|X\|q}{|(XAx,x)|}\right)$$
$$\leq |(XAx,x)|(1+\|X\|q).$$

Now, (2.2) and (2.6) yield

$$\frac{\text{Re}\ (X\tilde{A}x,x)}{|(X\tilde{A}x,x)|} \geq \frac{1}{|(XAx,x)|}\frac{(1-\|X\|q)}{(1+\|X\|q)} \ (\|x\|=1),$$

provided (2.5) holds. Since

$$\cos\tau(\tilde{A},X) = \inf_{x\in\mathbf{C}^n, \|x\|=1} \frac{\text{Re}\ (X\tilde{A}x,x)}{|(X\tilde{A}x,x)|},$$

according to (2.2), we arrive at at the required result. □

Proof of Theorem 1. Note that X is representable as

$$X = 2\int_0^\infty e^{-A^*t}e^{-At}dt \tag{2.7}$$

Reference [4, Section 1.5]. Hence, we easily have $\|X\| \leq J(A)$. Now, the latter lemma proves the theorem. □

3. An Estimate for $J(A)$

Let $\|A\|_F$ be the Frobenius (Hilbert–Schmidt) norm of A: $\|A\|_F = (\text{trace}\ (AA^*))^{1/2}$. Introduce the quantity (the departure from normality)

$$g(A) = \left[\|A\|_F^2 - \sum_{k=1}^n |\lambda_k(A)|^2\right]^{1/2}.$$

The following relations are checked in Ref. [7, Section 3.1]:

$$g^2(A) \leq \|A\|_F^2 - |\text{trace } A^2| \text{ and } g^2(A) \leq \frac{\|A - A^*\|_F^2}{2} = 2\|A_I\|_F^2,$$

where $A_I = (A - A^*)/2i$. If A is a normal matrix $AA^* = A^*A$, then $g(A) = 0$.

If A_1 and A_2 are commuting matrices, then

$$g(A_1 + A_2) \leq g(A_1) + g(A_2).$$

By the inequality between geometric and arithmetic mean values,

$$\left(\frac{1}{n}\sum_{k=1}^{n} |\lambda_k(A)|^2\right)^n \geq \left(\prod_{k=1}^{n} |\lambda_k(A)|\right)^2.$$

So, $g^2(A) \leq \|A\|^2 - n(|determinant(A)|)^{2/n}$.

Lemma 3. *Let condition* (1.1) *hold. Then,* $J(A) \leq \hat{J}(A)$, *where*

$$\hat{J}(A) := \sum_{j,k=0}^{n-1} \frac{g^{j+k}(A)(k+j)!}{2^{j+k}\beta^{j+k+1}(A)(j!\ k!)^{3/2}}.$$

Proof. By virtue of Example 3.2 from Ref. [7],

$$\|e^{-At}\| \leq \exp[-\beta(A)t] \sum_{k=0}^{n-1} \frac{g^k(A)t^k}{(k!)^{3/2}} \quad (t \geq 0).$$

Then,

$$\begin{aligned} J(A) &\leq 2\int_0^\infty \exp[-2\beta(A)t]\left(\sum_{k=0}^{n-1} \frac{g^k(A)t^k}{(k!)^{3/2}}\right)^2 dt \\ &= 2\int_0^\infty \exp[-2\beta(A)t]\left(\sum_{j,k=0}^{n-1} \frac{g^{k+j}(A)t^{k+j}}{(j!k!)^{3/2}}\right) dt \\ &= \sum_{j,k=0}^{n-1} \frac{2(k+j)!g^{j+k}(A)}{(2\beta(A))^{j+k+1}(j!\ k!)^{3/2}} = \hat{J}(A), \end{aligned}$$

as claimed. □

If A is normal, then $g(A) = 0$, and, taking $0^0 = 1$, we have $\hat{J}(A) = \frac{1}{\beta(A)}$.

The latter lemma and Theorem 1 imply the following:

Corollary 1. *Let $A, \tilde{A} \in \mathbf{C}^{n\times n}$ and the conditions* (1.1) *and $q\hat{J}(A) < 1$ hold. Then,*

$$\cos\tau(\tilde{A}, X) \geq \cos\tau(A, X)\frac{(1 - q\hat{J}(A))}{(1 + q\hat{J}(A))} \geq \frac{(1 - q\hat{J}(A))}{\|A\|\hat{J}(A))(1 + q\hat{J}(A))}.$$

4. Matrices "Close" to Triangular Ones

Let A_+ be the upper triangular part of A. That is, $A_+ = (a^+_{jk})^n_{j,k=1}$, where $a^+_{jk} = a_{jk}$ if $j \leq k$ and $a^+_{jk} = 0$ for $j > k$. To illustrate our results, apply Corollary 1 with A instead of $\tilde{A}$ and with A_+ instead of A.

Since A_+ is triangular, we have $\lambda_j(A_+) = a_{jj}$ $(j = 1, \ldots, n)$,

$$g(A_+) = g_+(A) := \left(\sum_{k=2}^{n}\sum_{j=1}^{k-1} |a_{jk}|^2\right)^{1/2}$$

and $\beta(A_+) = \beta_+(A) := \min_k \operatorname{Re} a_{kk}$. Assuming that $\beta_+(A) > 0$, we can write

$$\hat{J}(A_+) = \hat{J}_+(A) := \sum_{j,k=0}^{n-1} \frac{g_+^{j+k}(A)(k+j)!}{2^{j+k}\beta_+^{j+k+1}(A)(j!\ k!)^{3/2}}.$$

In addition, $q = q_+ = \|A - A_+\|$. Now, Corollary 1 implies.

Corollary 2. *Let $\beta_+(A) > 0$ and the condition*

$$q_+\hat{J}_+(A) < 1$$

hold. Let the diagonal entries of A lie in the sector $|\arg z| \leq \phi$ $(\phi < \pi/2)$. Then, its eigenvalues lie in the sector $|\arg z| \leq \psi$ with ψ satisfying

$$\cos\psi \geq \frac{(1 - q_+\hat{J}_+(A))}{(1 + q_+\hat{J}_+(A))}\cos\phi$$

Example 2. Consider the matrix

$$A = \begin{pmatrix} 4+2i & 0.1 \\ 0.2 & 8+4i \end{pmatrix}.$$

Then,

$$A_+ = \begin{pmatrix} 4+2i & 0.1 \\ 0 & 8+4i \end{pmatrix}.$$

We have $\arg a_{11} = \arg a_{22} = \phi$, where $\phi = \arctan(1/2)$, and therefore, $\cos\phi = \frac{2}{\sqrt{5}}$. In addition, $q_+ = 0.2$, $\beta_+(A) = 4$, $g_+(A) = 0.1$ and consequently,

$$\hat{J}_+(A) \approx 0.3125 \le 0.313.$$

Hence,

$$\frac{1 - q_+\hat{J}_+(A)}{1 + q_+\hat{J}_+(A)} \ge \frac{1 - 0.2 \cdot 0.313}{1 + 0.2 \cdot 0.313} \approx 0.8821.$$

Now, Corollary 2 implies that the eigenvalues of the considered matrix A lie in the sector $|\arg z| \le \psi$ with ψ satisfying

$$\cos\psi \ge 0.882 \cos\phi = \frac{2}{\sqrt{5}} 0.882 \approx 0.787.$$

The direct calculations show that $\cos\psi \approx 0.893$.

5. Perturbation of Diagonalizable Matrices

In this section, we are going to refine Lemma 3 in the case when all the eigenvalues of A are different:

$$\delta := \min_{j,k=1,\ldots,n;\ k\neq j} |\lambda_j - \lambda_k| > 0. \tag{5.1}$$

As is well known, under this condition, A is a diagonalizable matrix, i.e. there is an invertible matrix T, such that $\hat{D} = T^{-1}AT$ is a normal one. Besides, $\kappa_T := \|T\|\|T^{-1}\|$ is called a condition number.

Introduce the notations

$$\theta(A) := \sum_{k=0}^{n-2} \frac{g^{k+1}(A)}{\delta^{k+1}}$$

and

$$\gamma(A) := \left(1 + \frac{\theta(A)}{\sqrt{n-1}}\right)^{2(n-1)}.$$

Due to Ref. [6, Theorem 1.1], we can write

$$\kappa_T \le \gamma(A). \tag{5.2}$$

About the recent results on condition numbers, see, for example, Ref. [5] and references which are given therein.

For any function $f(A)$ of A, (5.2) yields the inequality

$$\|f(A)\| \le \gamma(A) \max_j |f(\lambda_j(A))|.$$

In particular,

$$\|e^{-At}\| \le \gamma(A) e^{-\beta(A)t} \quad (t \ge 0). \tag{5.3}$$

Recall that $\beta(A) = \min_j \operatorname{Re} \lambda_j(A)$. From (5.3), we have

$$J(A) \le \frac{\gamma^2(A)}{\beta(A)},$$

provided $\beta(A) > 0$. Making use of Theorem 1, we arrive at the following:

Corollary 3. *Let all the eigenvalues of A be different, and*

$$\beta(A) > q\gamma^2(A).$$

Then,

$$\begin{aligned}\cos\tau(\tilde{A}, X) &\ge \cos\tau(A, X) \frac{(\beta(A) - q\gamma^2(A))}{(\beta(A) + q\gamma^2(A))} \\ &\ge \frac{\beta(A)(\beta(A) - q\gamma^2(A))}{\gamma^2(A)(\beta(A) + q\gamma^2(A))\|A\|}.\end{aligned}$$

6. Perturbations of Polynomials

For a finite integer $n \geq 2$, let us consider the polynomials

$$P(\lambda) = \sum_{k=0}^{n} c_k \lambda^{n-k} \text{ and } Q(\lambda) = \sum_{k=0}^{n} b_k \lambda^{n-k} \quad (c_0 = b_0 = 1;\ \lambda \in \mathbf{C}).$$

As is well known, P and Q are characteristic values of the matrices

$$A = \begin{pmatrix} -c_1 & -c_2 & \cdots & -c_{n-1} & -c_n \\ 1 & 0 & \cdots & 0 & 0 \\ 0 & 1 & \cdots & 0 & 0 \\ . & . & \cdots & . & . \\ 0 & 0 & \cdots & 1 & 0 \end{pmatrix} \tag{6.1}$$

and

$$\tilde{A} = \begin{pmatrix} -b_1 & -b_2 & \cdots & -b_{n-1} & -b_n \\ 1 & 0 & \cdots & 0 & 0 \\ 0 & 1 & \cdots & 0 & 0 \\ . & . & \cdots & . & . \\ 0 & 0 & \cdots & 1 & 0 \end{pmatrix},$$

respectively. So,

$$q = \|A - \tilde{A}\| = \left(\sum_{k=1}^{n} |c_k - b_k|^2 \right)^{1/2}$$

and

$$N_2(A) = \left(n - 1 + \sum_{k=1}^{n} |c_k|^2 \right)^{1/2}.$$

Denote the zeros of P and Q by $z_j(P)$ and $z_j(Q)$ $(j = 1, \ldots, n)$, respectively, and assume that $\beta_P := \min_j \operatorname{Re} z_j(P) = \min_j \operatorname{Re} \lambda_j(A) > 0$. We have $g(A) = g_P$, where

$$g_P = \left[n - 1 + \sum_{k=1}^{n} |c_k|^2 - \sum_{k=1}^{n} |z_k(P)|^2 \right]^{1/2}.$$

Simple calculations show that

$$\sum_{k=1}^{n} z_k^2(P) = \text{trace } A^2 = c_1^2 - 2c_2,$$

and therefore,

$$g_P \le \left[n - 1 + \sum_{k=1}^{n} |c_k|^2 - |c_1^2 - 2c_2| \right]^{1/2}. \tag{6.2}$$

If $\beta_P > 0$, then

$$\hat{J}(A) = J_P := \sum_{j,k=0}^{n-1} \frac{g_P^{j+k}(k+j)!}{2^{j+k}\beta_P^{j+k+1}(j!\ k!)^{3/2}}.$$

Moreover, for any $x = (x_k) \in \mathbf{C}^n$ and $y = (y_k) = Ax$, we have $y_1 = \sum_{k=1}^{n} c_k x_k$, $y_k = x_{k-1}$. So,

$$\|y\|^2 = |y_1|^2 + \sum_{k=2}^{n} |y_k|^2 = \left| \sum_{k=1}^{n} c_k x_k \right|^2 + \sum_{k=2}^{n} |x_{k-1}|^2$$
$$\le \left| \sum_{k=1}^{n} |c_k|^2 \right| \sum_{j=1}^{n} |x_j|^2 + \left| \sum_{k=1}^{n} c_k x_k \right|^2 \le s_P^2 \sum_{j=1}^{n} |x_j|^2,$$

where

$$s_P := \left(\sum_{k=1}^{n} |c_k|^2 + 1 \right)^{1/2}.$$

Thus, $\|A\| \le s_P$. Now, Corollary 1 yields the following result.

Corollary 4. *Let* $\beta_P > 0$ *and* $qJ_P < 1$. *Then,*

$$\cos \arg(z_j(Q)) \ge \frac{1 - qJ_P}{s_P J_P (1 + qJ_P)} \quad (j = 1, \ldots, n).$$

Now, let us assume that all the zeros of P are different:

$$\delta_P := \min_{j,k} |z_j(P) - z_k(P)| > 0.$$

Then, we obtain

$$\theta(A) = \psi_P := \sum_{k=0}^{n-2} \frac{g_P^{k+1}}{\delta^{k+1}}$$

and

$$\gamma(A) = \left(1 + \frac{\psi_P \theta_P}{\sqrt{n-1}}\right)^{2(n-1)}.$$

Now, Corollary 3 yields the following:

Corollary 5. *Let all the zeros of P be different, and*

$$\beta_P > q\gamma_P^2.$$

Then,

$$\cos(\arg z_j(Q)) \geq \cos(\arg z_j(P)) \frac{(\beta_P - q\gamma^2(A))}{(\beta_P + q\gamma^2(A))}$$
$$\geq \frac{(\beta_P - q\gamma^2(A))}{s_P\psi_P(\beta(A) + q\gamma^2(A))}.$$

The literature devoted to the theory of polynomials is very rich, cf. Ref. [14] and the references given therein, but to the best of our knowledge, the angular localization of the zeros of perturbed polynomials are not considered in the available literature.

References

[1] B. D. O. Anderson, N. K. Bose and E. I. Jury. A simple test for zeros of a complex polynomial in a sector. *IEEE Trans. Autom. Control (Tech. Notes Corresp.)*, 19, 1974, 437–438.

[2] B. D. O. Anderson, N. K. Bose and E. I. Jury. On eigenvalues of complex matrices in a sector. *IEEE Trans. Autom. Control (Tech. Notes Corresp.)*, 20, 1975, 433.

[3] R. Bhatia. *Perturbation Bounds for Matrix Eigenvalues.* Classics in Applied Mathematics, Vol. 53. SIAM, Philadelphia, 2007.

[4] Y. L. Daleckii and M. G. Krein. *Stability of Solutions of Differential Equations in Banach Space.* American Mathematical Society, Providence, R. I., 1974.

[5] N. J. Daras, M. T. Rassias and N. B. Zographopoulos (eds.). *Exploring Mathematical Analysis, Approximation Theory, and Optimization.* Springer Nature, Switzerland, 2023.
[6] M. I. Gil'. A bound for condition numbers of matrices. *Electron. J. Linear Algebra*, 27, 2014, 162–171.
[7] M. I. Gil. *Operator Functions and Operator Equations.* World Scientific, New Jersey, 2018.
[8] M. I. Gil'. On angular localization of spectra of perturbed operators. *Extracta Math.*, 35(2), 2020, 197–204.
[9] M. I. Gil'. Perturbation bounds for angular sectors of spectra of unbounded operators. *Ann. Univ. Ferrara*, 67(1), 2021, 73–81.
[10] M. I. Gil'. Bounds for the relative and absolute spectral variations of matrices. *Results Math.*, 79(4), 2024, 10 p. Paper No. 174.
[11] N. G. Higham. *Functions of Matrices.* SIAM, Philadelphia, 2008.
[12] G. H. Hostetter. An improved test for the zeros of a polynomial in a sector. *IEEE Trans. Autom. Control (Tech. Notes Corresp.)*, 20, 1975, 433.
[13] M. G. Krein. The angular localization of the spectrum of a multiplicative integral in Hilbert space. *Funkcional. Anal. i Prilozhen*, 3(1), 1969, 89–90 (Russian).
[14] G. V. Milovanović, D. S. Mitrinović and T. M. Rassias. *Topics in Polynomials: Extremal Problems, Inequalities, Zeros.* World Scientific, Singapore, 1994.
[15] I. N. Parasidis and T. M. Rassias (eds.). *Mathematical Analysis in Interdisciplinary Research.* Springer Optimization and Its Applications, Vol. 179. Springer, Cham, 2021.
[16] G. V. Rozenblyum. Angular asymptotics of the spectrum of operators that are close to normal. *J. Sov. Math.*, 45(3), 1250–1261 (1989).
[17] G. W. Stewart and J. G. Sun. *Matrix Perturbation Theory.* Academic Press, New York, 1990.

https://doi.org/10.1142/9789819820498_0008

Chapter 8

Spherical Harmonics for Optimal 3D Data Modeling

Malika Jallouli[*,**], **Anouar Ben Mabrouk**[†,‡,§,††], **Michael Rassias**[¶,‡‡] and **Carlo Cattani**[‖,§§]

[*]*LATIS Laboratory of Advanced Technology and Intelligent Systems, Université de Sousse, Ecole Nationale d'Ingénieurs de Sousse, Sousse 4023, Tunisia*

[†]*Laboratory of Algebra, Number Theory and Nonlinear Analysis, Department of Mathematics, Faculty of Sciences, University of Monastir, 5019 Monastir, Tunisia*

[‡]*Department of Mathematics, Higher Institute of Applied Mathematics and Computer Science, University of Kairouan, 3100 Kairouan, Tunisia*

[§]*Department of Mathematics, Faculty of Sciences, University of Tabuk, 71491 Tabuk, Saudi Arabia*

[¶]*Department of Mathematics and Engineering Sciences, Hellenic Military Academy, 16673 Vari Attikis, Greece*

[‖]*Engineering School (DEIM), Tuscia University, Viterbo, Italy*

[**]*jallouli.malika3@gmail.com*

[††]*anouar.benmabrouk@fsm.rnu.tn*

[‡‡]*mrassias@sse.gr*

[§§]*cattani@unitus.it*

The processing of big data such as 3D images is nowadays a very interesting subject in many scientific fields, such as medicine, physics, chemistry and computer science. However, the evolution of acquisition equipment means that the advanced quality of the data processed is bulky and difficult to analyze and process. Therefore, the development of appropriate tools ensuring effective treatment is a necessity. This chapter concerns the whole topic of data analysis, and modeling, and aims to review some recent methods applied in 3D data processing based on the frequency decompositions. We are concerned especially to apply the concept of spherical harmonics due to two main advantages, such as the optimization of the modeling process in the sense of complexity reduction, as well as the reduction of the representing coefficients number, and the entropy measure. Concrete applications on 3D data objects are provided, such as Benchmark models and medical images.

1. Introduction

The processing of 3D images constitutes today a very interesting subject, in many scientific fields, such as medicine, physics, chemistry and computer science. However, the progress of acquisition tools, modeling and visualization of 3D objects has caused an increase in the number of 3D models available on the web or in specialized databases.

Efficient processing of geometric objects requires, as in all other areas of computer science, the design of data structures appropriate. For each specific problem in the geometry treatment, one can identify a set of operations by which the computation is dominated, and consequently, the development of appropriate tools guaranteeing treatment effective is a necessity.

This work concerns the whole subject of the treatment of three-dimensional objects based on the frequency decompositions applied to the modeling 3D surfaces. Among these decompositions, we notice that spherical harmonics (SHs) dominate all other transformations above 3D spherical compacts.

SHs are important in many theoretical applications and practices, such as the representation of electrostatic fields and multipolar electromagnetic fields, the calculation of orbital electron configurations and atomic fields, the representation of gravitational fields, the reconstruction of fibers to estimate the path and location of neuronal axons the properties of water diffusion by diffusion-weighted MRI imaging for aerodynamic tractography, the magnetic fields of planetary bodies and stars, and the characterization of background radiation cosmic microwave.

In 3D computer graphics, spherical harmonics play a role in a wide variety of subjects, including indirect lighting (ambient occlusion, global illumination, pre-calculated luminance transfer, etc.) and modeling of 3D shapes.

The different three-dimensional acquisition techniques (medical or multimedia) have evolved considerably over the last decades. They certainly provide high resolution and true-to-life viewing. However, their mathematical manipulation (invariance, geometric transformation, etc.) is complicated given the density of the data, on the one hand, and the lack of structuring, on the other. Hence, the need to create a three-dimensional model that allows the faithful representation of the initial object while guaranteeing compactness and relevance.

The spherical harmonics represent a family of compact, relevant and rotation-invariant descriptors. Indeed, this 3D sphere modeling method makes it possible to extend the properties of the Fourier transform on the surfaces of $\mathbb{R}^3$ of genus zero. This spectral analysis relative to the basis of spherical harmonics has the advantage of the possibility of constructing a family of invariant descriptors by rotations of $\mathbb{R}^3$ thanks to the generalized delay theorem.

On the other hand, the decomposition on the basis of SHs allows a compact and relevant representation of the initial surface because a mesh formed of hundreds of thousands of points and facets can be represented by a model faithful to the initial surface and formed of a few tens of SH coefficients. These coefficients will be a way to do more in-depth calculations, such as deformation or motion estimation.

The challenge is to revisit the method of definition and calculation of the basis and the coefficients of the spherical harmonics in order to optimize the complexity and the processing time in the case of complex surfaces:

- Mathematically, the spherical harmonics are special functions obtained as particular solutions of the Laplace equation generated by the Legendre polynomials. By resorting to the three-level recurrence relation of these polynomials, a new recursive basis of spherical harmonics has been proposed. This leads to a similar recursive computation of the spherical harmonic coefficients. These new recurrence relations allow a considerable contribution in terms of complexity and optimization of processing time.

- The classical notion of Shannon's entropy has been exploited in order to define a new 3D entropy through which the reconstruction order is defined in an automatic and accurate way without even having to resort to reconstructing the spherical harmonic model at each iteration of the process.
- In order to validate our research work, the first experiment was made on synthetic surfaces, theoretical methods have been proposed for modeling a specific 3D surface to demonstrate the efficiency and the scalability of creating accurate SH models for large 3D surfaces. The tests were performed on the well-known medical image of the human brain. Then, a medical application was made on the left ventricle of the heart which aims to estimate the overall deformation of this anatomical structure over time by resorting to modelization by spherical harmonics.

This chapter is organized as follows: Section 2 is devoted to the description of some representations and models of 3D objects. In Section 3, we present the spherical harmonics tool. Section 4 is devoted to the application of the approach based on spherical harmonics on 3D objects of Benchmark models and medical imaging cases. Finally, Section 5 is a general conclusion.

2. 3D Data Modeling Review

For a decade, the flow of 3D objects is increasingly used in all areas. Representations of 3D models can take many forms. The choice of a particular representation strongly depends on the context in which it will be used. Indeed, the development of computer graphics and acquisition techniques has given rise to many fields, such as mechanics, architecture, medical imaging, software, 3D games, 3D videos and computer-aided design, to benefit from this new category of data. In this chapter, we subdivide 3D representation techniques into two broad categories: volume representation and surface representation. Indeed, a volume representation can be either by voxel or a volume mesh, whereas a surface representation can be either by a cloud of points or by a surface mesh.

The representation of 3D surfaces by polygonal meshes remains an efficient means for storage and visualization. However, this method requires a large storage space, especially when it reaches high resolution. On the other hand, it is difficult to carry out specific mathematical processing on this type of data, such as geometric transformations and deformation calculation. Hence, the interest of three-dimensional modeling which is compact and relevant for representing 3D data generally, usually dense and unstructured. In this chapter, we present the surface models most cited in the literature by classifying them into global, local and hybrid models. We also explain the particularities of each type of model as well as its fields of application.

In the literature, different representations of 3D data have been applied such as voxelization, point cloud, and mesh. Briefly, a voxel represents a value on a regular grid in three-dimensional space. A point cloud is a collection of data points in a three-dimensional coordinate system. A mesh is a geometric data structure composed of vertices in space, linked together by edges. Figure 1 illustrates these different representations.

2.1. *Voxelic representation*

Volume representation is frequently used in the visualization and analysis of medical and scientific data. Some volumetric displays use voxels to describe their resolution. This representation results from

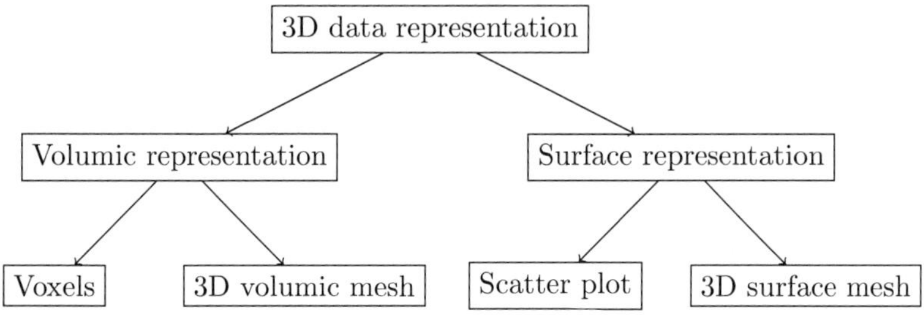

Fig. 1. A flowchart for different illustrative 3D data representations.

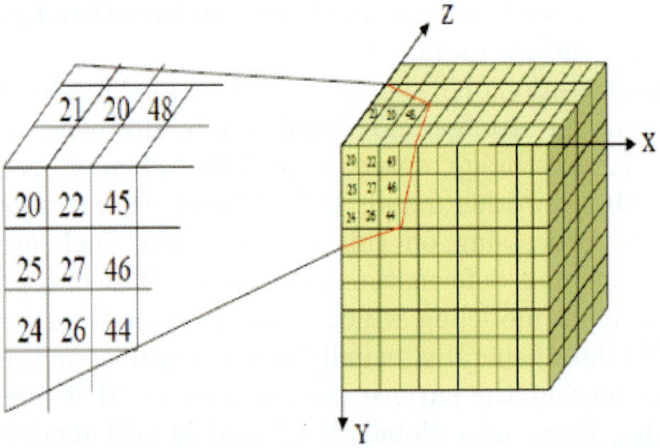

Fig. 2. An example of voxelic 3D representation.

a regular division of the 3D image and the position of the voxels (see Figure 2).

2.2. *Scatter plot*

In the representation of 3D data, scatter plots look like a set of points produced by 3D scanners, taken from the external surfaces of the objects surrounding them. They are used for a variety of purposes, including creating 3D CAD models for manufactured parts, for metrology and quality control, visualization, ..., etc.

The information is stored by the relative position of each point with respect to each other. Figure 3 shows an example of a scatter plot representing a dragon. The model is taken from the Stanford Computer Graphics Laboratory. The colors in Figure 3(b) represent the distance from the point displayed to the camera, going from blue (closest) through green to red (farthest).

A scatter plot containing only points; this representation has the advantage of being very easy to store and display. The main problem comes from the fact that the surface represented is unknown. Another major problem concerns the rendering of the point cloud. See Figure 3.

2.3. *3D mesh*

This type of 3D data representation does not use a regular division of space. However, it offers a satisfactory compromise between

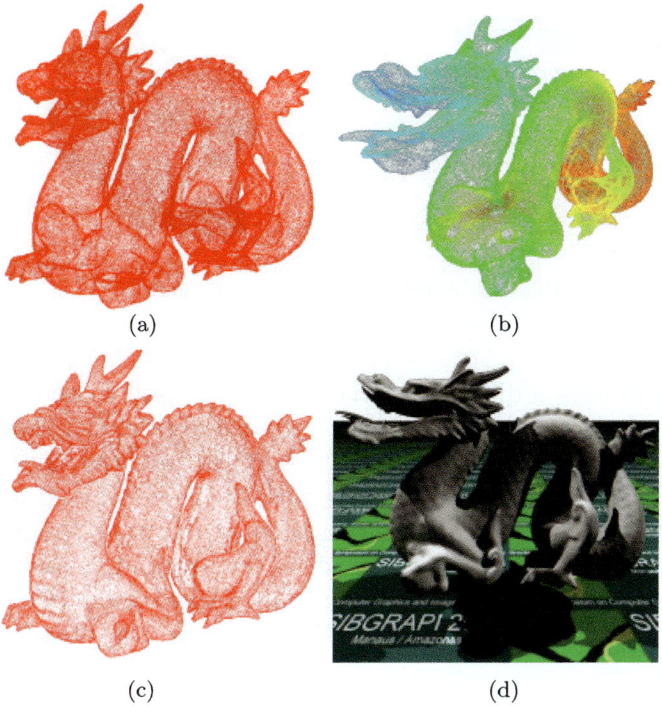

Fig. 3. (a) The scattered plot, (b) the reconstructed dragon relative to the distance of each point to the camera, (c) the scattered object with less scattering points, and (d) A scatter plot for Dragon – Stanford Computer Graphics Laboratory.

precision, storage space and display speed. The most important information in this type of data is the 3D position in space of the set of points describing the 3D model. In this second type, we distinguish two sub-categories called surface mesh and volume mesh. Surface meshes describe only the surface of the 3D model while volume meshes describe the entire volume of the object. Figure 4 illustrates the two types of 3D meshes.

2.3.1. *3D mesh regularity*

After acquiring a 3D model using scanners, a scatter plot is obtained. This scatter plot can be triangulated in several ways using for example the Delaunay triangulation [15, 16, 29] or the Voronoi diagram [4, 7] or any triangulation [8, 13, 26]. Each triangulation method will generate a different topology using the same geometry where each

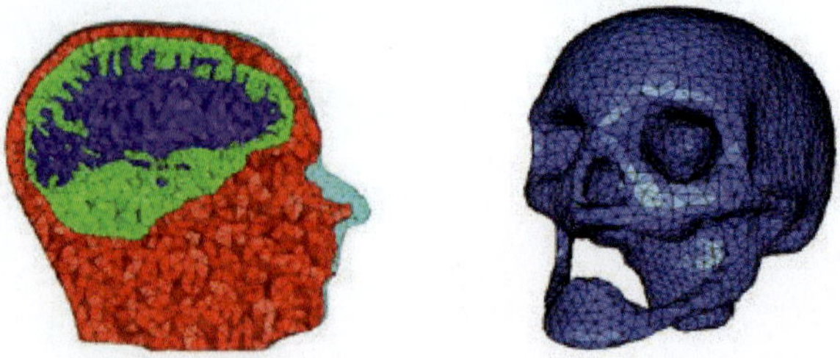

Fig. 4. A volume and a surface mesh [25].

vertex will have a different valence from one mesh to another, i.e. a different number of incident edges at each vertex. Consequently, a mesh is said to be regular if all the vertices have the same valence which is generally equal to 6. this type of mesh is found especially in the field of CAD where the regularity of the mesh facilitates calculations and simulations. A mesh is semi-regular if each triangle has a regular mesh. A mesh is irregular if there are no Consistency in the way vertices are connected. Therefore, they have different valences.

2.3.2. *Some mesh types*

A mesh can be mono or multi-resolution. In the case of mono-resolution meshes, a 3D object is represented in a single level of resolution. All polygons have the same importance. Therefore, to be able to display it on the screen, the whole object must be loaded into memory.

In the case of multi-resolution meshes, the 3D object is represented at different levels of detail. It can be modeled by a coarse mesh or in detail by a new mesh when the application requires it (Figure 5).

This multiresolution mesh representation has several interests:

- The evolution of the resolution allowing a visualization, a treatment or a progressive transmission (Figure 5).
- The adaptation of the resolution according to the concerned media.
- If broken during such processing, unlike single-resolution meshes, the object can still be viewed globally.

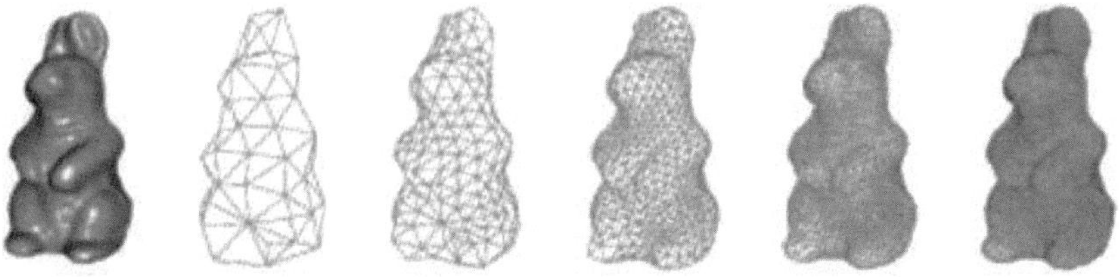

Fig. 5. A multiresolution representation.

The polygonal mesh is the simplest form of representation of 3D objects. However, this technique does not offer the possibility of knowing either the geometry or the shape of the 3D image. For this, it is necessary to determine the mathematical model of the object. Hence, the need for the notion of 3D modeling.

2.4. *3D data modeling*

Global surface models are described by a small number of parameters which allow characterizing the shape and possibly the global movements of any 3D surface. They are useful in cases of imprecise data or small sizes.

Fourier surfaces make it possible to model any parameterized 3D surface (torus, closed surface, tube). They were presented by Staib and Duncan in a general framework devoted to the extraction of contours in 3D images [31]. The general objective is to model types of objects that can be encountered in medical imaging. This modeling is used to be able to create a correspondence between simple geometric surfaces and three-dimensional medical images such as Magnetic Resonance Imaging (MRI).

The 3D Fourier representation can be adapted to objects of different shapes. It allows modeling four types of surfaces, namely, tori, open surfaces, closed surfaces and tubes.

Surprisingly, closed surfaces homeomorphic to a sphere are the most difficult to represent with the model presented by Staib and Duncan. They considered the surface as a tube whose two extremities are closed and point-like. The authors consider separately the parametrization of the three coordinates x, y, z and limit themselves to a surface with rectilinear axis z.

3. Spherical Harmonics

The base of trigonometric functions is universal and allows representing any surface. However, for a star surface, it is more interesting to represent it in spherical coordinates.

A well-adapted orthonormal basis is then that of spherical harmonics. In Ref. [14], the authors have studied the problem of modeling the epicardium from the bifurcation points of the coronary artery detected on a series of 10 pairs of angiography images. The points are first approximated by a parametric surface $r_s(\theta, \varphi)$ and represented in spherical coordinates.

The basis of spherical harmonics has been the subject of several research works. The real forms existing in nature have in the majority of cases a spherical topology. In order to optimize the process of reconstruction by spherical harmonics, many researchers have focused on the choice of a method of parametrization allowing to pass to the space of projection (the unit sphere) by having a uniform mesh faithful to the starting shape. See, for example, Ref. [21].

In order to obtain a homogeneous distribution of the quadrilaterals on the parametrization space, the initial projection is transformed into an optimization problem having these two constraints:

- Preservation of areas: Each region of the object is associated with a region of proportional area.
- The minimization of distortions: Each quadrilateral must transform into a spherical quadrilateral in the parametrization space.

The choice of poles as well as the resolution of an optimization problem seems to be a suitable way to migrate to the projection space, but this remains very expensive in terms of time especially for a dense mesh ($\geq$5000 vertex). On the other hand, the meshes most used in medical imaging are often triangular meshes.

SHs are rotation invariant descriptors. Modeling by SHs makes it possible to reduce the dimensionality of the shape descriptor and consequently to reduce both the storage space and the comparison time (property have been essential for the implementation of interactive pattern retrieval systems). Our research work falls within this framework.

3.1. SH recursivity

One of the main challenges in the area of shape matching is that in many applications a model and its image by a transformation are considered to be identical. Therefore, the challenge of comparing two shapes is to find the best measure of similarity on the space of transformations. The need for efficient retrieval makes it difficult to explicitly query all transformations.

In the mathematical literature, SHs are special functions defined on the surface of a sphere [3, 9, 10, 22, 23].

The spherical coordinate system is a system for representing geometric figures in three-dimensional space using three coordinates (r, θ, φ), where r is the distance from the origin, $\theta \in [0, \pi]$ is the angle with the z axis and $\varphi \in [0, 2\pi]$ is the angle with the x axis in the (x, y) plane. Sometimes, θ and φ are called co-latitudinal and azimuth angles, respectively.

Using the Laplace operator in the spherical coordinate system, the SHs will be defined by the angular parts of the Y solutions of the Laplace equation $\Delta^2 Y = 0$ in the spherical coordinate system.

Let M be a point on the unit sphere $\mathbb{S}^2$ with coordinates $\mathbb{R}^3$:

$$(x, y, z) = (\sin\theta\cos\varphi, \sin\theta\sin\varphi, \cos\theta).$$

The idea follows from the fact that any function supported on the sphere must be a linear combination of basis functions satisfying Laplace's equation. To introduce such a basis, the solutions Y are represented by the following form:

$$Y(r, \theta, \varphi) = f(r)g(\theta)h(\varphi). \tag{3.1}$$

We get

$$f(r) = Cr^l,\ l \in \mathbb{N}, \quad \text{and} \quad h(\varphi) = e^{im\varphi},\ m \in \mathbb{Z}. \tag{3.2}$$

Next, by setting $x = \cos\theta$ and $P(x) = g(\theta)$ above, we get the Legendre equation

$$(1 - x^2)P''(x) - 2xP'(x) + \left(l(l+1) - \frac{m^2}{1 - x^2}\right)P(x) = 0, \tag{3.3}$$

where the solution is given by the famous Legendre polynomials defined on $[-1, 1]$ by

$$P_{l,m}(x) = \frac{(-1)^m}{2^l l!}(1 - x^2)^{\frac{m}{2}} \frac{d^{l+m}}{dx^{l+m}}(x^2 - 1)^l, \tag{3.4}$$

where $l \in \mathbb{N}$ and $m \in \mathbb{Z}$ satisfying $|m| \leq l$. Applying for example (3.4), we get

$$Y_{l,m}(\theta, \varphi) = K_{l,m} P_{l,m}(cos\theta) e^{im\varphi}, \quad l \in \mathbb{N}, \;\; m \in \mathbb{Z}; \; |m| \leq l, \quad (3.5)$$

where $K_{l,m}$ is the normalization constant

$$K_{l,m} = (-1)^m \sqrt{\frac{(2l+1)(l-m)!}{4\pi(l+m)!}}. \quad (3.6)$$

The parameter l is known as the degree of the SH $Y_{l,m}$, and m is its order.

Any element $S \in L^2(\mathbb{S}^2)$ is decomposed into a series known as the SH series decomposition of S as

$$S(\theta, \varphi) = \sum_{l=0}^{\infty} \sum_{m=-l}^{l} S_{l,m} Y_{l,m}(\theta, \varphi). \quad (3.7)$$

Mathematically, the $S_{l,m}$ are the SHs coefficients of S with respect to the basis $(Y_{l,m})_{l,m}$ and are evaluated by

$$S_{l,m} = \int_0^{2\pi} \int_0^{\pi} S(\theta, \varphi) \overline{Y_{l,m}}(\theta, \varphi) \sin\theta d\theta d\varphi. \quad (3.8)$$

The functional SH basis has several useful mathematical properties for modeling and pattern recognition:

- $\mathbb{S}^2$ being a compact group, the Fourier transform is therefore represented by the Fourier coefficients with respect to the associated Legendre basis.
- The SH basis forms a complete system on the surface of the unit sphere.

Denote for the next

$$a_l = \frac{2l+1}{l+1} \quad \text{and} \quad b_l = \frac{l}{l+1}, \quad (3.9)$$

and also

$$\begin{cases} \alpha_{l,m} = a_{l-1}\sqrt{\frac{(2l+1)(l-m)}{(2l-1)(l+m)}}, \\ \beta_{l,m} = m a_{l-1}\sqrt{\frac{(2l+1)}{(2l-1)(l+m)(l+m-1)}}, \\ \gamma_{l,m} = -b_{l-1}\sqrt{\frac{(2l+1)(l-1)(l-m-1)}{(2l-3)(l+m)(l+m-1)}}. \end{cases} \quad (3.10)$$

We obtain the three-level induction rule for the $Y_{l,m}$ of the SHs:

$$Y_{l,m} = \alpha_{l,m}\cos\theta Y_{l-1,m} + e^{i\varphi}\beta_{l,m}\sin\theta Y_{l-1,m-1} + \gamma_{l,m}Y_{l-2,m}. \quad (3.11)$$

We consequently deduce a three-level induction rule for the evaluation of the SH coefficients.

The SHs of degree l generate a subspace of dimension $(2l+1)$ in $L^2(\mathbb{S}^2)$ corresponding to the spherical functions of width of band l. The induction rule allows a gain in terms of complexity and time calculation.

An important interest is then particularly linked its ability as a mathematical/physical basis to approach with precision 3D data with minimal storage costs, memory requirements and the possibility of obtaining computationally efficient algorithms to process. Theoretically, we know that in the Hilbert space $L^2(\mathbb{S}^2)$, we always have

$$\left\| S - \sum_{k=0}^{l}\sum_{m=-k}^{k} S_{k,m}Y_{k,m} \right\| \to 0 \text{ as } L \to +\infty.$$

A major problem is to know how to define automatically and with precision the optimal reconstruction order $L = L_{\min}$. We then proposed a new variant of entropy based on SH which can be a good black box allowing to stop the procedure with the best guaranteed approximation and to precisely determine the optimal order of reconstruction. Recall that the entropy in general is a global measure of invariance of the studied system. Its value should be consequently stationary as the multi-resolution level increases.

In Ref. [30], Claude Shannon introduced the revolutionary notion of information entropy as the average rate at which information is produced by a stochastic source of data. The measure of information entropy associated with each possible data value is the negative logarithm of the probability mass function $-\log(p(x))$, where $p(x)$ is the probability of presence or occupation of the information of a position or state x in the state space.

In the context of 2D data, the concept of entropy is somehow new. It was recently introduced in 2D image analysis such as segmentation in Ref. [18] using maximum entropy thresholding. In Ref. [27], algorithms have been proposed to analyze the irregularity at spatial scales and to extend the multientropy 1D scales in the 2D case. In Ref. [32], the entropy measure has been applied to determine an automatic localization of positions.

The concept of entropy based on wavelets remains still recent and the studies on it remain few. This motivates the development of more works about it. In Ref. [27], a wavelet-based approach has been developed to measure Shannon entropy in the context of spatial point models. Spatial heterogeneity and complexity of spatial point models were analyzed using of multiscale anisotropic wavelet entropy.

In the case of 3D data, especially spherical images, to our knowledge, the concept of entropy has not yet been widely known, especially with SHs. Let $S \in L^2(\mathbb{S}^2)$ expressed as in (3.7)–(3.8). We introduce the energy in the detail m and the level l of S is defined by

$$\mathrm{SHs}E(l,m) = |S_{l,m}|^2. \tag{3.12}$$

Next, as for the wavelet entropy, the energy at level l is the mean value of all the details m, $|m| \leq l$, and, it is given by

$$\mathrm{SHs}E(l) = \frac{1}{N} \sum_{|m|\leq l} \mathrm{SHs}E(l,m). \tag{3.13}$$

The total energy is the sum over all multiresolution levels

$$\mathrm{SHs}E = \sum_{l} \mathrm{SHs}E(l) = \frac{1}{N} \sum_{l} \sum_{|m|\leq l} \mathrm{SHs}E(l,m). \tag{3.14}$$

In practice, it is difficult to compute all levels l, so we usually estimate the energy by means of a finite approximation as given by the following expression:

$$\mathrm{SHs}E_L = \sum_{l=0}^{L} \mathrm{SHs}E(l) = \frac{1}{N} \sum_{l=0}^{L} \sum_{|m|\leq l} \mathrm{SHs}E(l,m). \tag{3.15}$$

It is subsequently called the L approximation of the energy of SHs. Then, as for the Shannon entropy, we introduced the distribution of SHs as the probability density of SHs at a level l by

$$\mathrm{SHs}P(l) = \frac{\mathrm{SHs}E(l)}{\mathrm{SHs}E} = \frac{\sum_{|m|\leq l} \mathrm{SHs}E(l,m)}{\sum_l \sum_{|m|\leq l} \mathrm{SHs}E(l,m)}, \tag{3.16}$$

or equivalently,

$$\mathrm{SHs}P(l) = \frac{\sum_{|m|\leq l} |S_{i,m}|^2}{\sum_l \sum_{|m|\leq l} |S_{j,m}|^2}. \tag{3.17}$$

Finally, the SH entropy is defined as

$$\mathrm{SHsEnt} = -\sum_l \mathrm{SHs}P(l) \log(\mathrm{SHs}P(l)). \tag{3.18}$$

In practice, as previously, we may speak about an L-estimation of the entropy as

$$\mathrm{SHsEnt}_L = -\sum_{l=0}^{L} \mathrm{SHs}P(l) \log(\mathrm{SHs}P(l)). \tag{3.19}$$

4. Applications to 3D Data Modeling

4.1. *3D synthetic surface modeling*

The general shape of a synthetic surface S parametrized by the spherical coordinates (θ, φ) is written as [5]

$$\begin{pmatrix} X(\theta,\varphi) \\ Y(\theta,\varphi) \\ Z(\theta,\varphi) \end{pmatrix} = \mathcal{A}(\theta,\varphi) \begin{pmatrix} \cos(\theta)\cos(\varphi) \\ \cos(\theta)\sin(\varphi) \\ \sin(\theta) \end{pmatrix}, \tag{4.1}$$

where

$$\mathcal{A}(\theta,\varphi) = \sum_{n\in N} \sum_{m=0}^{n} (a_n^m \cos(m\varphi) + b_n^m \sin(m\varphi)) P_{n,m}(\cos(\theta)). \tag{4.2}$$

a_n^m and b_n^m are arbitrary constants, and $P_{n,m}$ is an associated Legendre polynomial. In Figure 6, we have presented some examples of these synthetic surfaces.

(a) (b) (c) (d) (e)

Fig. 6. Some synthetic surfaces: (a) $l = 3; m = 1$, (b) $l = 4; m = 3$, (c) $l = 9$; $m = 3$, (d) $l = 13; m = 12$ and (e) $l = 23; m = 10$.

By applying the SH, we obtain a recurrence formula for the associated SH coefficients:

$$S_{l,m} = a_{l-1}S_{l-1,m} + ma_{l-1}S_{l-1,m-1} - b_{l-1}S_{l-2,m}. \quad (4.3)$$

This recursive formula with three levels is in fact independent of the mesh used and which makes it possible to calculate all the coefficients for all the levels from the first two consecutive coefficients. Figure 7 shows the stages of the reconstruction of the different synthetic surfaces by applying recursivity:

- The first spherical model A has been reconstructed at $L = 3$.
- The second model B was rebuilt for $L = 5$.
- The third C model was made for $L = 10$.
- Model D has been rebuilt at level $L = 14$.
- Model E was obtained at level $L = 24$.

It is worth recalling that the classical method based on the direct calculation of the coefficients SHs via the integral expression must calculate a number $L(L + 2)$ of coefficients for each level L. Table 1 shows that, for the same initial surface comprising the same number of points, the recursive method gives a minimal number of elements of basis of the SHs and therefore a minimum number of coefficients of the associated SHs calculated directly to validate the model.

We note that the evaluation of the recursion of SHs considerably simplifies the calculation algorithms, which has direct repercussions on their complexity. Indeed, the complexity is estimated at $O(L(L+2)N^2)$ for the classical method of modeling SHs.

(a) (b) (c) (d) (e)

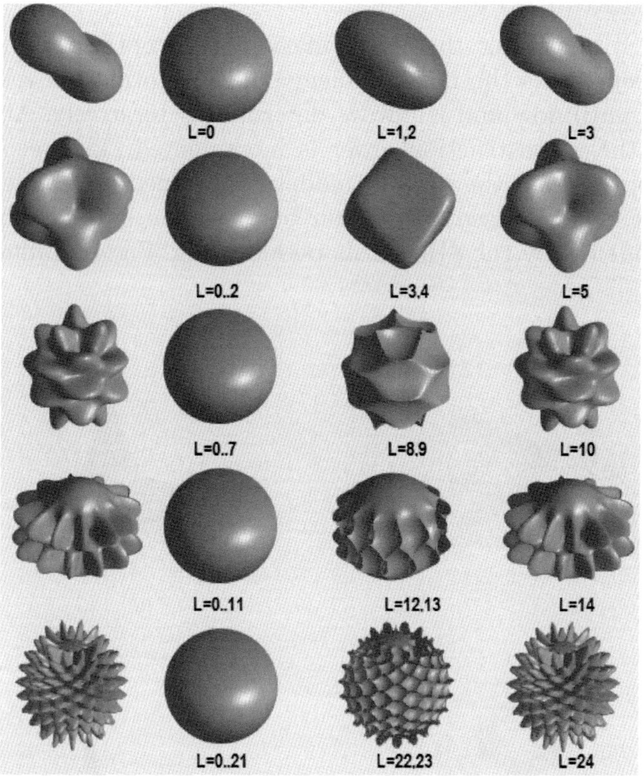

Fig. 7. Reconstruction of 3D models due to Figure 6 with the recursive SHs.

Table 1. A minimal number of coefficients of the SHs directly calculated to reconstruct the 3D surfaces of Figure 7.

Surface	Number of Points	Order L	Direct SH Method	Recursive SH Method
A	2601	4	24	2
B	3721	7	63	2
C	4489	11	143	2
D	5041	13	195	2
E	7569	17	323	2

4.2. *SH entropy for 3D data modeling*

In the SH modeling, the determination of the optimal reconstruction order represents the main challenge. This information remains subjective until now and is always determined iteratively. At each iteration, the reconstruction order is incremented. The reconstruction is stopped when the desired model is obtained. In this section, we present the proposed correlation between the new 3D entropy and the modeling by the SHs for shape modeling. Figure 8 illustrates the

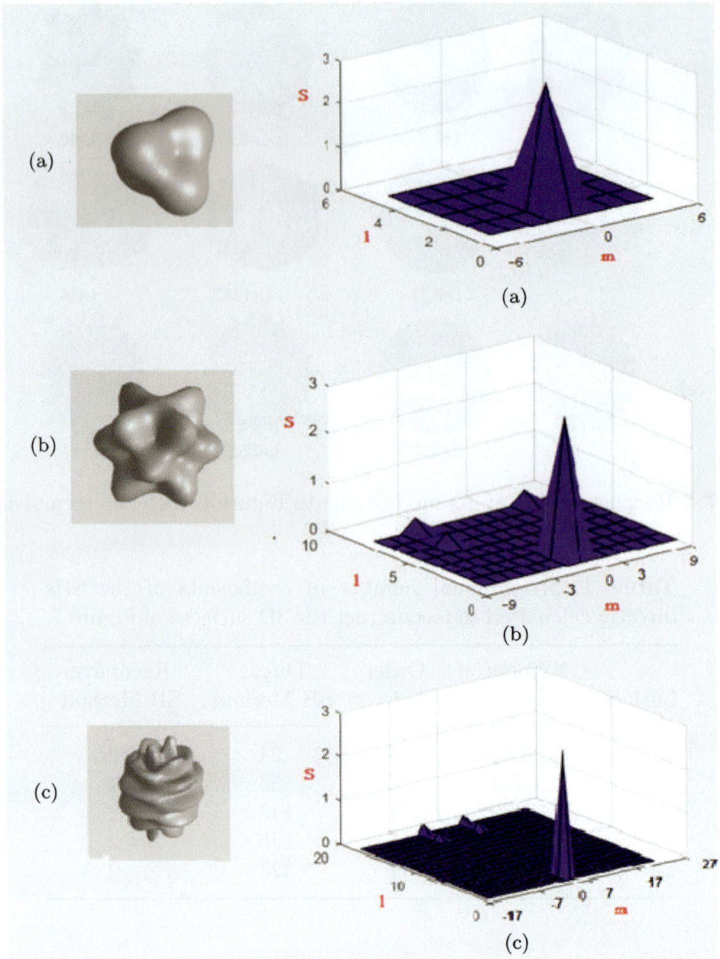

Fig. 8. SH coefficients spectrum. (a) degree 4, (b) degree 7, and (c) degree 17.

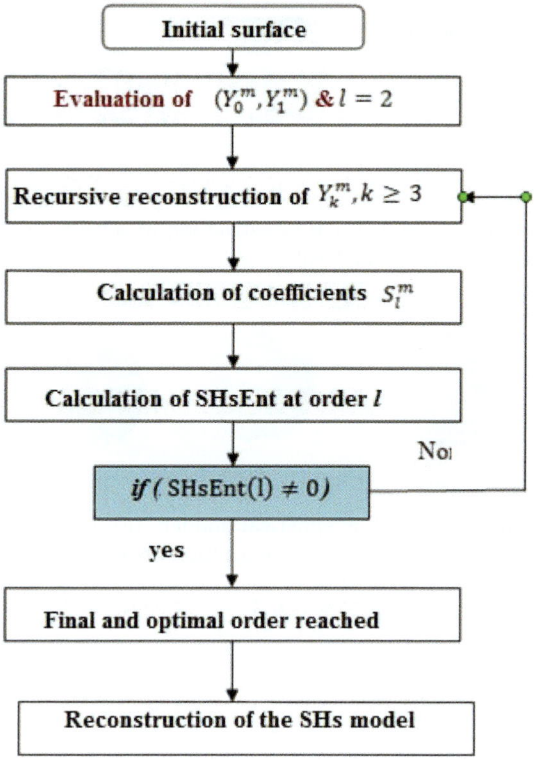

Fig. 9. SH entropy processing.

spectra of the values of the SHs according to the orders L and m for the synthetic surfaces A, B and C.

These spectra show that the information resides around zero, then as L increases, the details of the initial object are represented. The diagram in Figure 9 shows that at each iteration of the process of calculating the SHs coefficients, the value of the entropy is determined. The optimal order is deduced when the value of the entropy stabilizes.

Figures 10 and 11 show an example of an experiment carried out on some synthetic images. The entropy of the SHs can then be a good black box allowing to stop the procedure with the best guaranteed approximation.

Figures 12 and 13 illustrate the curves of the entropy calculated on two samples of synthetic surfaces of genus zero. It should be

(a) (b) (c) (d) (e)

Fig. 10. Reconstruction of a spherical model: (a) Initial synthetic object, (b) Reconstruction at level 4, (c) Reconstruction at level 6, (d) Reconstruction at level 7, (e) Reconstruction at level 8.

(a) (b) (c) (d) (e)

Fig. 11. Reconstruction of a spherical model: (a) Initial synthetic object, (b) Reconstruction at level 4, (c) Reconstruction at level 15, (d) Reconstruction at level 17, (e) Reconstruction at level 18.

(b)

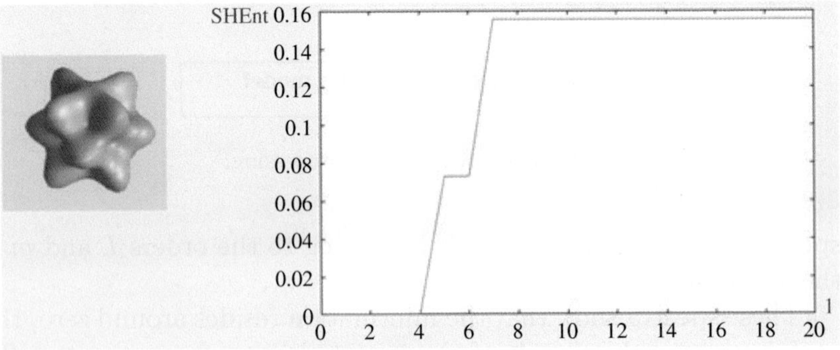

Fig. 12. SHsEnt due to Figure 10.

noted that the SHsEnt values become stationary from a level L. This order will be considered as the optimal level L for the reconstruction process.

It appears from Figures 12 and 13 that the entropy of the SHs is an efficient way to predict the best optimal reconstruction order.

(c)

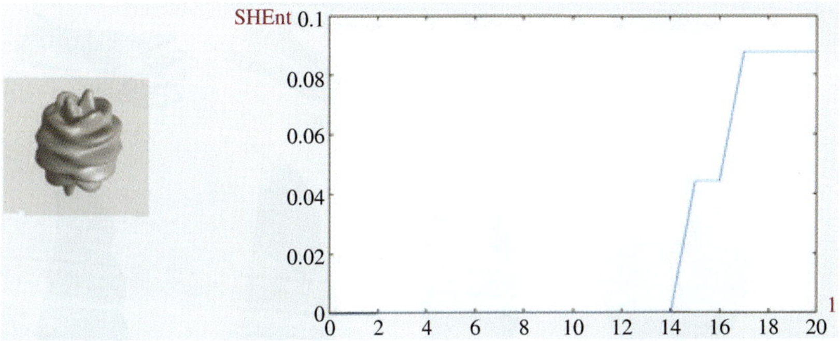

Fig. 13. SHsEnt de la Figure 11.

This method saves processing time because it is not necessary to test all the commands preceding the optimal command which is not fixed, in advance for classical methods.

4.3. *Some 3D benchmark models*

Surface representation and shape modeling play an increasingly important role in many computer vision and image processing applications such as astronomy, chemistry and medical imaging. Many techniques have been developed for the reconstruction of [24] surfaces. The SHs approach has been used mainly for the representation of zero-genus surfaces. This particular database makes it possible to perform an analysis of functional information or to classify different types of forms. In fact, the description of the SH shape makes it possible to efficiently approach complex 3D objects with minimal storage cost.

Figure 14 illustrates some examples of 3D surfaces with different complexities represented as 3D triangular meshes. The modeling process by SHs is performed on these surfaces to reconstruct the final models related to these surfaces.

The numerical results (see Table 2) show a considerable gain in computational complexity. For the recursive process, the number of SH coefficients to be calculated is always 2 regardless of the initial resolution, which also has an impact on the processing time. Table 2 illustrates for each surface the number of SH coefficients calculated for each modeling method.

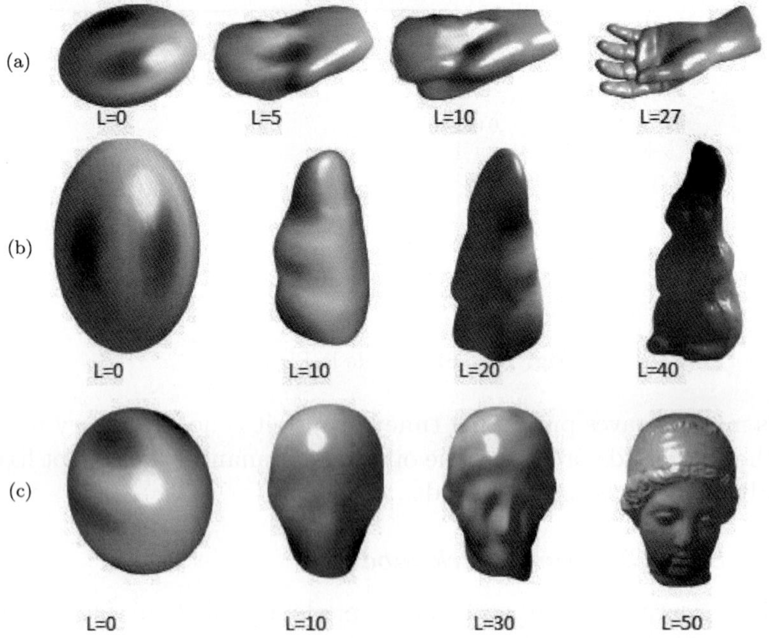

Fig. 14. Recursive SH representation of some 3D benchmark models: (a) The hand, (b) The rabbit, (c) Venus.

Table 2. Numerical result for the reconstruction of different 3D surfaces due to Figure 14.

3D Surface	Optimal Order	SH Coeff Number	Recursive SH Coeff Number
A	27	784	2
B	40	1681	2
C	50	2601	2

Now, as previously, the main problem of the reconstruction process by SHs resides in the definition of the optimal order representing the best model approximating at best the original object. Previous work Ref. [20] applied the mean squared error (MSE) to the SH coefficients via the formula

$$D_l^2 = \frac{1}{N}\sum_{i=0}^{N}(R_i(O_l) - R_i(O_{l-1}))^2, \tag{4.4}$$

where R_i is the ith radius calculated at the ith point of the SH model, O_l is the SH model reconstructed to the order of l, O_{l-1} is the SH model reconstructed to the order of $l-1$ and N is the number of points of the initial surface.

This distance measures the difference between two successive reconstructed SH models. This distance must be minimal when the optimal model is reached. In fact, when the optimal order l is reached, the reconstructed models at orders l and $l+1$ are almost the same.

Equation (4.4) is based on the difference between the radii calculated on two successive reconstructed models. This distance should not be efficient in our case, especially because the rays are information that does not define 3D surfaces, especially by not considering spherical angles.

Now, we show how the proposed SHsEnt should be representative to determine the optimal reconstruction order. Modeling by SH is a concept that is widely used in the literature. In this context, the objective of the research still remains the determination of the optimal reconstruction order which defines the model closest to the initial surface has been developed.

As a tool for determining the optimal reconstruction order of the model that best approximates the original object, we used the SHsEnt, a global measure of the invariance of the systems studied. The value of the found SHsEnt is constant unless we increase the level. This is shown in Figure 15.

These different curves show the variation of the SHsEnt during the process of modeling by SHs.

Note that for a precise value of l, the SHsEnt becomes stationary. Experimentation led us to conclude that this is the value of the

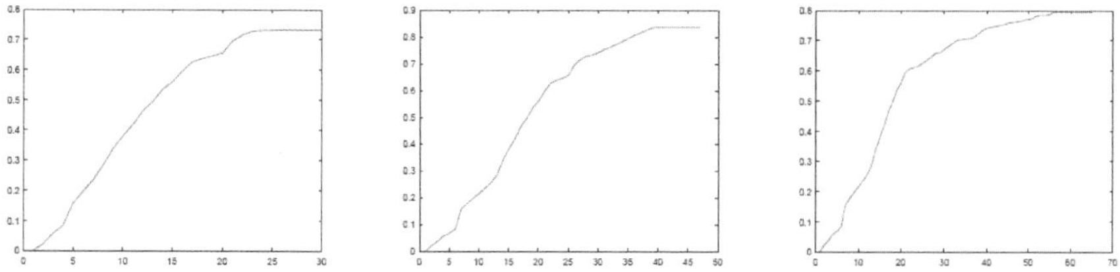

(a) (b) (c)

Fig. 15. SH entropy for 3D models: (a) The hand, (b) The rabbit, (c) Venus (due to Figure 14).

Table 3. Comparison between SHsEnt and MSE for 3D models due to Figure 14.

Surface	Level L	SHsEnt	MSE
Figure 14(a)	0..2	0.0000	0.0000
	5	0.1329	24.9390
	10	0.3829	0.0000
	20	0.6461	7.4459
	>= 27	0.79	0.0000
Figure 14(b)	0..2	0.0000	0.0000
	10	0.2086	10.0219
	20	0.4386	6.0000
	30	0.6960	8.0000
	>= 40	0.82	0.0000
Figure 14(c)	0..2	0.0000	0.0000
	20	0.4115	0.2510
	30	0.6370	3.0000
	40	0.6765	1.0600
	>= 50	0.79	0.0000

optimal reconstruction order. This is why we no longer have to test the entire order before the final rebuild order. This saves considerable processing time.

It is important to note that the experiment was performed for a large number of genus zero surfaces of varying complexity. In all cases, the results were confirmed.

Table 3 illustrates the evolution of the value of the spherical harmonic entropy during the reconstruction process. At each iteration, the root mean square error (MSE) between the SHs coefficients is calculated. It is clear that the more the order of reconstruction increases, the more the MSE decreases and the entropy stabilizes, until a complete stabilization of the value of the entropy and a cancellation of the MSE. This is explained by the fact that the following iterations will no longer show details at the level of the reconstructed model, and consequently, the optimal order of reconstruction is reached.

The variation of the value D_l is in particular due to the comparison between two successive reconstructed orders. Whereas the comparison must be made between the initial surface and the reconstructed model. That's why SHsEnt is a better and more accurate method to efficiently and automatically set the optimal order.

Table 4. Numerical result for the reconstruction of different 3D surfaces due to Figure 14.

3D Surface	Nb of Points	Optimal Level Due to SHsEnt	Optimal Level Due to MSE
Figure 14(a)	36619	27	31
Figure 14(b)	70658	40	45
Figure 14(c)	100759	50	59

Table 4 shows that for a simple surface, i.e. one that does not present many details, SHsEnt and MSE give the same results for the extraction of the final order. Whereas, when the surface is more complicated, SHsEnt detected the final order more easily and in a short time.

The comparison between the reconstructed model and the initial surface with SHsEnt is more relevant than the comparison of the radii of two successive reconstructed models, in particular because of the non-exhaustiveness of the information on the radii of a 3D surface.

4.4. *Heart left ventricle modeling*

The heart is an organ formed essentially of a muscle, the myocardium which is surrounded by a serious "bag", the pericardium which is lined with interior by a thin membrane and the endocardium. The heart has four cavities [17], two right cavities formed by the atrium and the right ventricle and two left cavities which are formed by the atrium and the left ventricle. The right and left cavities are totally separated by a partition (septum). These cavities are connected to veins and arteries that lead blood to or from the heart (see Figure 16).

The left ventricle (LV) is one of the two large lower chambers of the heart. Its walls are three times thicker than those of the right ventricle. It is the most powerful cavity of the heart muscle. She has 80% heart function. It consists of five parts; the apex or point of the left ventricle, the septum which constitutes the separation between the two ventricles, the lower wall on which the heart rests in the body, the lateral wall and the anterior wall upper located at the front next to the ribs (See Figure 17).

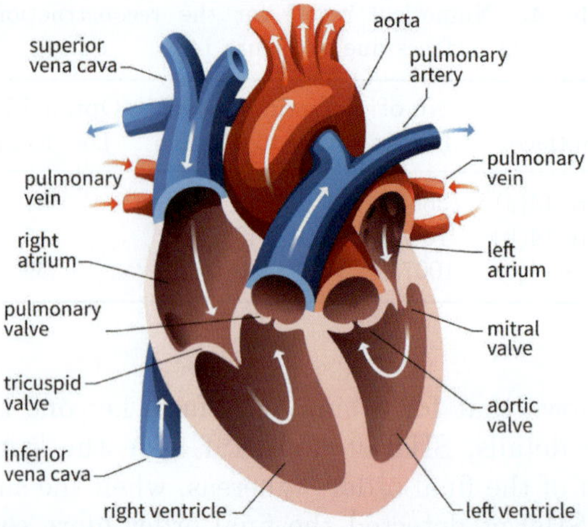

Fig. 16. The heart anatomy following a section where the four cavities are visible.

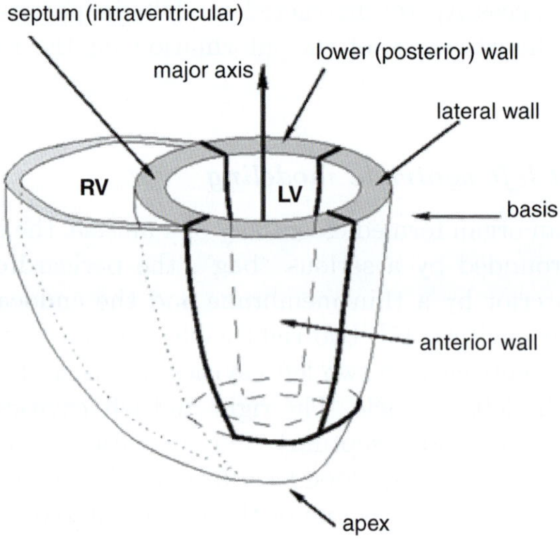

Fig. 17. Cutout of the left ventricle of the heart.

Cardiovascular disease is one of the leading causes of death among men worldwide. Functional and morphological analysis of the LV during the cardiac cycle provides insight into its functioning for early prevention and accurate treatment of associated diseases.

The LV is the engine of the heart pump. The analysis of LV is mainly between the phase of blood ejection called "systole" and the phase of dilation of the heart, when its cavities are again filled with blood, called "diastole". However, conventional diagnostic tests rely heavily on well-known parameters of global cardiac function, such as end diastolic volume (VDF), end stroke volume (VSF) or ejection fraction (FE) [1]. These parameters are based in particular on the ventricular volume and do not take into account its global deformation.

Assessment of cardiac function continues to be a problem due to the dynamic nature and non-rigid structure of the heart. In addition, acquiring data with good accuracy over time requires a considerable review period and storage capacity. For this, it remains interesting to seek to model the LV of the heart in order to extract quantitative parameters describing more effectively its function and its anatomy [19]. The model of the LV thus obtained will make it possible to recover the information necessary for understanding and interpreting the mechanical behavior at the over time [2] such as the creation of a dynamic cardiac model or the 2D/3D segmentation of one or more cavities of the heart [33].

The assessment of cardiac function has been greatly improved by the recent technical development of medical acquisition tools. Indeed, computer tomography makes it possible to provide, in a simple examination and in a reduced acquisition time, dynamic images of the walls and cardiac cavities. Functional LV parameters provide direct quantitative information about the patient's current cardiovascular health status and may be useful as a prognostic marker.

The most common parameters that qualify this function are essentially calculated between two specific instants of the cardiac cycle, namely the end-systolic and end-diastolic phases:

$$\begin{cases} E = \text{VDF} - \text{VSF}, \\ \text{FE} = \frac{\text{VE}}{\text{VDF}} * 100, \\ \text{FVG} = \text{VE} * \text{HR}, \end{cases} \tag{4.5}$$

where VE is Stroke Volume, VDF is Final Diastolic Volume, VSF is Final Systolic Volume, FE is Stroke Fraction, FVG is LV Flow, VE is Stroke Volume and finally, HR is heart rate.

These parameters offer important advantages for the study of cardiac function and provide quantitative information for diagnosing LV hypertrophy, which is generally associated with risk and increased cardiovascular mortality. Nevertheless, practitioners must establish a correlation between these functional parameters and those anatomical to obtain a more complete diagnosis.

The superquadrics have been applied to the modeling of the LV of the heart. They are defined by a small number of parameters and can reconstruct a surface model even from incomplete [5] data. This model was applied by Cohen *et al.* for the analysis of 3D cardiac data [11]. In Ref. [6], they have established a parametric superquadric model to describe the deformation of the LV over time.

Many researchers have also developed a local model of the heart in order to describe in a quantitative way the local cardiac deformation during the cardiac cycle. Amini and Duncan, in Ref. [12] were able to compare the LV at two successive instants of the cardiac cycle.

The harmonic analysis on the sphere is another variant of the global modeling of the heart [28]. This analysis extends the properties of the Fourier transform on zero surfaces of the genus. Computerized tomography allows providing dynamic 3D images leading to very large datasets from which cardiac structures must be extracted under strong constraints to cope with the clinical requirements. We concentrate in what follows on the modeling of the LV using SHs.

The 3D geometric modeling allows capturing an important series of parameters that identify the organ that is visualized (its size, its volume, ...). In addition, physical modeling is essential to help understand and interpret the mechanical and electrical behavior of the LV. In order to solve a complex medical imaging problem, skills related to acquisition, segmentation, modeling and visualization must be taken into consideration.

The processing approach for the analysis of scintigraphic and CT images of the heart is essentially composed of (see also Figure 18):

- A pre-processing phase of the acquired images (scintigraphy, scanner): segmentation of 2D slices, evaluation of some parameters of regional cardiac function such as parietal thickening.
- 3D reconstruction of the LV surface.

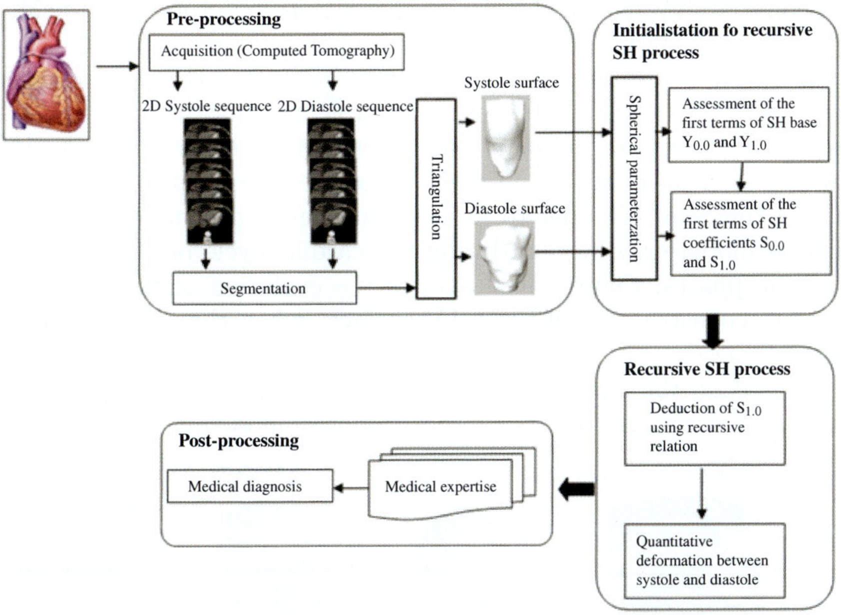

Fig. 18. Detailed LV global process.

- Modeling of the surface of the LV by the reconstruction of the SHs which consist in a parametrization on the sphere $\mathbb{S}^2$, the standardization of the triangulation, the calculation of the coefficients of SHs.
- Extraction of invariant descriptors by rotations of $\mathbb{R}^3$:
  - (a) Calculation of the deformation of the LV between the model reconstructed at effort and that reconstructed at rest for the scintigraphic images.
  - (b) Calculation of the deformation of the LV between the different phases of the cardiac cycle for CT images.
- Interpretation of results and assessment of the severity of any anomaly.

In what follows, we prove that SH can be a good modeler for LV surfaces. The 3-level recursion relation indeed allows the optimization of the computation time independently of the size and the complexity of the surface. The recursive SHs also make it possible to considerably reduce the size of the data to be processed, to simplify the calculation of the parameters of the global cardiac function and the LV deformation during different instants of the cardiac cycle and more particularly between the systole and diastole phases. The results obtained can help improve medical diagnosis based on precise and numerical data. The method proposed here aims to model the LV of the heart in order to extract anatomical parameters describing its global deformation during the cardiac cycle or between two specific phases. Figure 19 illustrates the evolution of the spherical model reconstruction of some LV surfaces when the reconstruction level L increases.

Based on the LV surfaces presented in Figure 19, a comparative study between the results of the proposed modeling process with

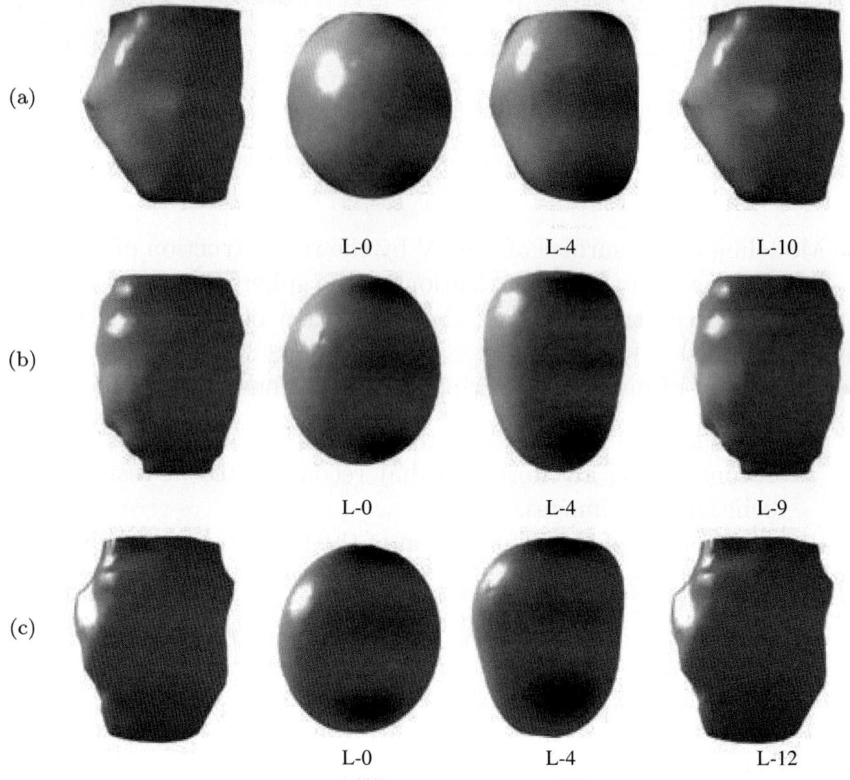

Fig. 19. SH Modeling of some LV surfaces at different reconstruction levels.

Table 5. Numerical result for the reconstruction of different surfaces.

LV Surface	Nb of Points	Optimal Level	SH Coeff Number	RSH Coeff Number
A	2601	10	120	2
B	2387	9	99	2
C	3150	12	168	2

Table 6. Comparison of algorithms running time.

LV Surface	Direct SH Methods (s)	RSH Methods (s)
A	6.65	2.10
B	3.72	1.95
C	9.98	3.09

the classic method of SHs has been carried out. Table 5 illustrates for each surface the number of SH coefficients calculated for each modeling method.

The numerical results of the array Table 5 show a considerable gain in computational complexity. In fact, this value is estimated at $O(L(L+2))$ for classical SH modeling. It therefore depends on the resolution of the initial surface. Whereas, for the recursive process, the number of SH coefficients to be calculated is always 2 regardless of the initial resolution, which also affects the processing time. Table 6 illustrates for each surface the processing time for the two modeling methods.

To complete this application, we return again to the entropy of the SH introduced previously to be used for fixing a priori the optimal reconstruction order which represents the model closest to the initial surface. The value of entropy must be maximum when the optimal model is reached. Figure 20 illustrates the variation of the entropy of the SHs of the different surfaces of the LV of Figure 19 during the modeling process.

For the surface Figure 19(a), we notice that for L greater than 10, the value of SHsEnt remains constant. For the surface Figure 19(b),

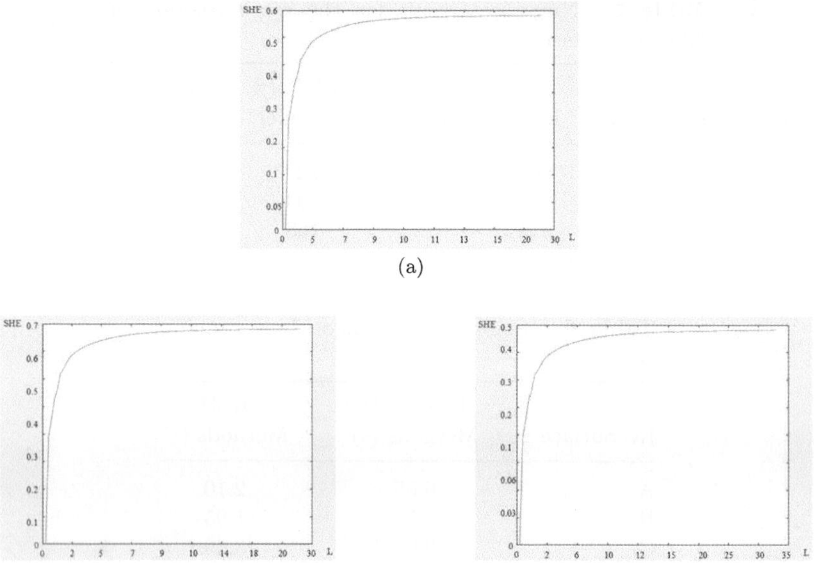

(a) (b) (c)

Fig. 20. SH entropy due to Figures 19(a), 19(b) and 19(c).

we find that when L is strictly less than 9, the value of SHsEnt increases and starts to be quasi-stationary when L equals 9 and almost constant when L is greater than 9. Whereas for the surface Figure 19(c), the value of SHsEnt remains constant when L is greater than 12. This is explained by the fact that the optimal order of reconstruction varies according to the complexity of the surface and it is impossible to define it before the end of the rebuilding process. From where the advantage of the entropy which determines, without having recourse to the 3D reconstruction of the model, to determine in a precise way the order of reconstruction.

Finally, an evaluation of the global deformation of the LV was carried out on a group of 50 patients and 10 volunteers from the LA RABTA university hospital in Tunisia. Cardiac computed tomography was obtained for the entire cardiac cycle. We are interested in the terminal systolic and final diastolic phases in order to calculate the global deformation of the LV and to establish a correlation between the global functional parameters defined at those two specific times.

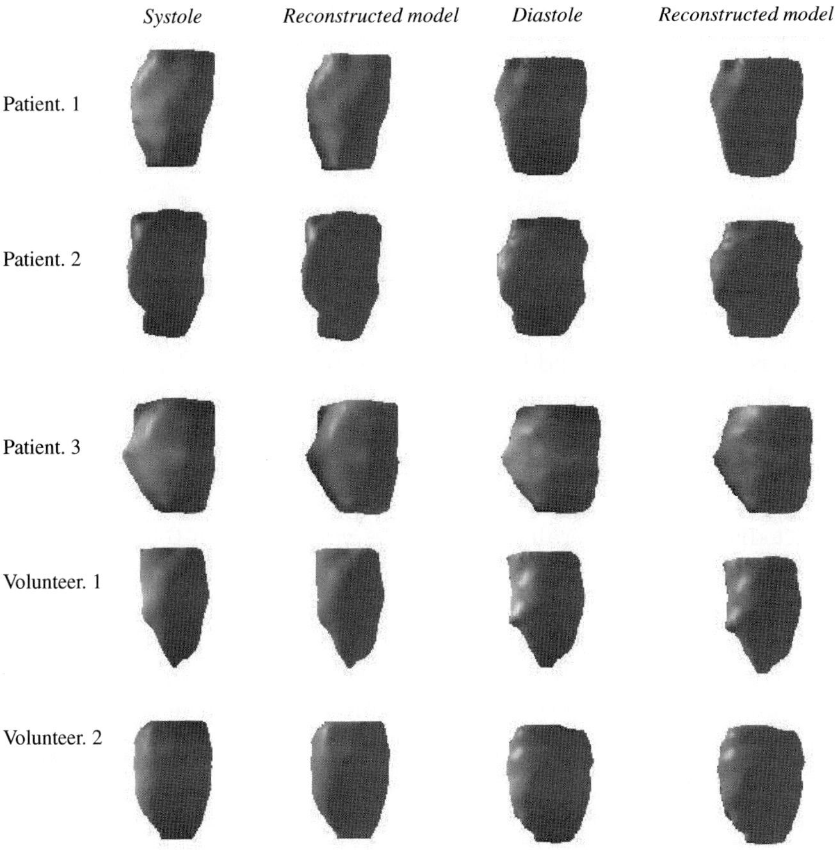

Fig. 21. RSH LV modeling for systolic and diastolic states.

Figure 21 illustrates the model of the recursive SHs of three patients reconstructed on systolic and diastolic phases. Table 7 illustrates the results obtained in some volunteers and patients.

The applied approach allowed the doctor to have both a functional and morphological description of the LV of the heart, which can contribute to a better clinical diagnosis. A normal left ventricular FE ranging from 50% to 70%, which means that, for example, 55% of the total amount of blood in the LV is pumped at every heartbeat. In case of heart failure (myocarditis or ischemic heart disease), the value of FE will decrease.

Table 7. Numerical results for the reconstruction of different 3D LV surfaces.

	VE (ml)	FE (%)	Global Deformation
Volunteer 1	70	51	1.4
Volunteer 2	65	48	0.98
Patient 1	47	32	0.15
Patient 2	36	26	0.11
Patient 3	29	18	0.09

It is important to add quantitative morphological parameters to the global function to allow a more precise diagnosis. For this, the global deformation is defined. Experiments have shown that a normal global deformation varying between 0.5 and 1.5. In the case of myocarditis, the strain is between 0.01 and 0.08 and for ischemic heart disease, the overall strain values vary between 0.08 and 0.2.

Experiments have been carried out on healthy volunteers, patients suffering from myocarditis (morphological pathology) and patients suffering from ischemic disease (functional pathology). The global strain is a new quantitative parameter that helps the practitioner to examine the abnormality through a simple examination of the nature of the cardiac pathology. Finally, the global deformation makes it possible to show the real nature of the cardiac pathology (functional or morphological) in 95% of the patients.

5. Conclusion

This chapter enters into the framework of the representation and the modeling of 3D data. Indeed, the volume of data currently provided by the acquisition hardware is generally large and unstructured. This shows that a modeling step turns out to be essential in certain applications, in particular medical applications. This is in order to have a compact and relevant representation, on the one hand, and to extract precise and quantitative parameters, on the other hand.

In this chapter, the SHs have been applied on synthetic images still medical imagery such as the human brain and medical cardiac images. The experimental results obtained showed that the recursive SHs are more efficient and precise than the classical SHs in terms of number of coefficients and execution time. As prospects for this work, we could consider the enrichment of the database in order to better test the elaborated algorithms. In addition, we will be able to deploy a statistical tool to be able to better judge the results.

References

[1] M. Alexa. Merging polyhedral shapes with scattered features. *Vis. Comput.*, 16(1), 2000, 26–37.

[2] A. Amini and J. Duncan. Pointwise tracking of left-ventricular motion in 3D. In *Proceedings of the IEEE Workshop on Visual Motion*, Princeton, New Jersey, 1991.

[3] S. Arfaoui, I. Rezgui and A. B. Mabrouk. *Wavelet Analysis on the Sphere, Spheroidal Wavelets.* Degruyter, Berlin/Boston, 2017.

[4] F. Aurenhammer. Voronoi diagrams-a survey of a fundamental geometric data structure. *ACM Comput. Surv.*, 23(3), 1991, 345–405.

[5] T. Bannour. Un nouveau pseudo reparamétrage numérique pour la représentation et description invariante de surfaces de $\mathbb{R}^3$. Application à l'analyse de structures osseuses. Thèse de doctorat, Ecoles Nationale des Sciences de l'Informatique, université de Manouba Tunis, Tunisie, 2003.

[6] A. H. Barr. Superquadrics and angle preserving transformations. *IEEE Comput. Graph. Appl.*, 1(1), 1981, 11–23.

[7] M. Bern, N. Amenta and M. Kamvysselis. A new Voronoi-based surface reconstruction algorithm. In *SIGGRAPH'98: Proceedings of the 25th Annual Conference on Computer Graphics and Interactive Techniques*, 1998, pp. 415–421.

[8] M. Bern and D. Eppstein. Mesh generation and optimal triangulation. In D.-Z. Du and F. Hwang (eds.), *Computing in Euclidean Geometry.* Lecture Notes Series on Computing, Vol. 1. World Scientific, Singapore, 1992.

[9] T. Bulow. Spherical diffusion for 3-D surface smoothing. *IEEE Trans. Pattern Anal.*, 26(12), 2004, 1650–1654.

[10] T. Bulow and K. Daniilidis. Surface Representations Using Spherical Harmonics and Gabor Wavelets on the Sphere. Technical Report, No. MS-CIS- 01-37, University of Pennsylvania, Department of Computer and Information Science, 2001.

[11] L. D. Cohen, E. Bardinet and N. Ayache. Tracking the cardiac ventricle with a parametric deformable model. *IEEE Trans. Med. Imaging*, 1(2), 1996, 129–149.

[12] K. Chatti, H. Essabbah, A. B. Abdallah, F. Ghorbel and M. H. Bedoui. A new uniform parameterization and invariant 3D spherical harmonic

shape descriptors for shape analysis of the heart's left ventricle—A pilot study. *Pattern Recognit. Lett.*, 31(13), 2010, 1981–1990.

[13] C. L. Bajaj, T. K. Dey and K. Sugihara. On good triangulations in three dimensions. In *Proceedings of the RST ACM Symposium on Solid Modeling Foundations and CAD/CAM Applications, SMA'91*, 1998, pp. 431–441.

[14] C. W. Chen, T. S. Huang and M. Arrott. Analysis and visualization of heart motion. In *Proceedings of SPIE. Biomedical Image Processing II*, Vol. 1450, 1 July 1991, pp. 231–242. https://doi.org/10.1117/12.44299. Event: *Electronic Imaging'91*, 1991, San Jose, CA, United States.

[15] S. Cheng and S. H. Poon. Three-dimensional Delaunay mesh generation. *Discrete Comput. Geom.*, 36, 2004, 419–456.

[16] L. P. Chew. Constrained delaunay triangulations. *Algorithmica*, 4(1), 1989, 97–108.

[17] I. Cohen and L. D. Cohen. A hybrid hyperquadric model for 2D and 3D data fitting. Rapp. Tech. 2188, INRIA, Rocquencourt, 1994.

[18] L. L. Chen, B. Wang and J. Cheng. New result on maximum entropy threshold image segmentation based on P system. *Optik*, 2018, 81–85.

[19] O. Devillers, B. Lévy, P. Alliez, D. Cohen-Steiner and M. Desbrun. Anisotropic polygonal remeshing. In *SIGGRAPH*, Vol. 22, No. 3, 2003.

[20] F. Ghorbel, W. B. Khelifa and A. B. Abdallah. Three dimensional modeling of the left ventricle of the heart using spherical harmonic analysis. In *ISBI*, 2008, pp. 1275–1278.

[21] G. Gerig, C. Brechbuhler and O. Kubler. Parametrization of closed surfaces for 3-D shape description. *Comput. Vision Image Underst.*, 16(2), 1995, 154–170.

[22] R. Green. Spherical harmonic lighting: The gritty details. In *Game Developers Conference*, 2003.

[23] P. J. Kostelec, D. M. Healy, D. N. Rockmore and S. Moore. FFTs for the 2-sphereimprovements and variations. In *Game Developers Conference*, Vol. 9, No. 4, 2003, pp. 341–385.

[24] C. Latxague, A. Boulenger de Hauteclocque, J. C. Bernhard, C. Michiels and E. Jambon. Nephrectomie partielle robotique pour tumeur renale assistée par modélisation 3D: série prospective de 100 cas et analyse comparative avec le registre national de néphrectomie partielle robot-assistée RoPaN (UroCCR study no 51). *Progrès en Urologie*, 28, 2018, 727–729.

[25] A. Meftah. Compression et codage scalable d'objets 3D pour leur visualisation et manipulation sur des systèmes heterogenes, 2011.

[26] S. A. Mitchell and S. A. Vavasis. Quality mesh generation in three dimensions, 1992, pp. 212–221.

[27] O. Nicolis and J. Mateu. 2D anisotropic wavelet entropy with an application to earthquakes in Chile. *Entropy*, 17, 2015, 4155–4172.

[28] V. Pascucci, J. Bennett and K. I. Joy. Genus oblivious approach to cross parameterization. *Comput. Aided Geom. Des.*, 25(8), 2008, 592–606.

[29] V. Rajan. Optimality of the Delaunay triangulation in R^d. *Discrete Comput. Geom.*, 12(1), 1994, 189–202.

[30] C. Shannon. A mathematical theory of communication. *Bell Syst. Tech. J.*, 27, 1948, 379–423.

[31] L. H. Staib and J. S. Duncan. Model-based deformable surface finding for medical imaging. *IEEE Trans. Med. Imaging*, 15(5), 1996, 720–731.

[32] M. S. Vasquez, M. Feixas and W. Heidrich. *Viewpoint Selection Using Viewpoint Entropy*. Stuttgart, Germany, 2001, pp. 21–23.

[33] B. Villegas, G. P. Aurigemma, W. H. Gaasch and T. E. Meyer. Noninvasive assessment of left ventricular mass, chamber volume, and contractile function. *Curr. Probl. Cardiol.*, 20(6), 1994, 361–440.

https://doi.org/10.1142/9789819820498_0009

Chapter 9

Strengthening Statistical Framework by Applied Mathematical Analysis

Christos P. Kitsos[*,‡] **and Filipe Amaral**[†,§]

[*] *University of West Attica, Athens, Greece*

[†] *Escola Superior de Educação Paula Frassinetti, Porto, Portugal*

[‡] *xkitsos@uniwa.gr*

[§] *fma@esepf.pt*

The target of this chapter is to discuss and extend the mathematical analysis support to the statistics theoretical framework. Among the number of applications that can be provided, we are focused on measure theory, linear transforms or Fourier transformation. The support of mathematics is strengthening the statistical framework, although statistics are restricted in $\mathbb{R}^n$ by definition.

Keywords: Measure theory, Fourier transformation, Michaelis–Menten.

1. Introduction

In 1901, in Paris at the 2nd International Conference of Mathematics, David Hilbert (1862–1943) presented the talk "The Problems of Mathematics", where he asks the Mathematical treatment of the axioms of Physics. Moreover, this formalism has to be extended to other fields as well so that mathematics and logic are the basis of their development. Indeed, as far as probability theory concerns, Andrey Kolmogorov (1903–1987) in 1931 introduced the probability space $(\Omega, \mathcal{A}, P)$ with Ω being the sample space, $\mathcal{A}$ the σ-Algebra (in statistics the Borel σ-algebra is needed) and the probability measure P. Note that due to the Carathéodory extension theorem, the probability P can be extended to $\sigma(\mathcal{A})$, the minimal σ-field containing $\mathcal{A}$, in a unique manner.

So, a nice theoretical development was building, due to Measure Theory, for probability and statistics. But statistics, since the time of Henry VIII, is collecting and analyzing real data. Indeed, in 1538, due to the plague in the English population, Henry VIII ordered parish priests to keep a record of christenings monthly data, provided a huge amount of data and the need to summarize these "mountains" of data were urgent. It was John Graunt, a London haberdasher, who presented to the Royal Society, in 1661, the paper on the "Natural and Political observations on the Bills of Mortality", published in 1662. He tried a near histogram to assign probabilities that a person in London would die in a particular age interval.

The required mathematical answer to undertake the improvement of Graunt's work took almost two hundred years with DeMoivre's work of the normal density function, Walker [53].

Indeed in his "The Doctrine of Changes" in 1738, he proved for the (Bernoulli trial, we would say in recent terminology) $B(n; k, p)$ that

$$\binom{n}{k} p^k q^{n-k} \cong \frac{1}{\sqrt{2\pi pq}} \exp\left[-\frac{(k-np)^2}{2\pi pq}\right], \quad q = 1 - p > 0. \tag{1.1}$$

Therefore, mathematical analysis was "unte portas", as well as the (Gaussian) Normal distribution, proposed by Gauss [13] though the development of the work on error. This work also resulted in the "Least Squares Method", Stigler [46]. Therefore, the mathematical strength was there but not related to statistics yet.

It was Karl Pearson (1857–1936) who in 1895 proposed a family of distribution; therefore, more mathematical analysis, of that time, oriented the founded work. He is also the one who established the discipline "Mathematical Statistics" and he founded the first ever Statistics Department at the University College London, in 1911, while Gosset [14] involved the Gamma function on his "Students t-distribution".

The Pearsonian methodology, and step-by-step mathematical improvement in statistics were covered by the work of Sir Ronald Fisher (1890–1962) on the Maximum Likelihood Estimator (MLE). The Fisherian line of thought was eventually victorious.

Let $x_1, x_2, \ldots, x_n$ be a sample from a distribution with probability density function (PDF) $f(x;\theta)$ continuous, with respect to its first θ derivative in parameter space $\Theta \subseteq \mathbb{R}^p$. Let $\frac{\partial}{\partial\theta}\log f(x;\theta)$ be continuous in θ for every $x \in \mathbb{R}$, except possibly for a set with probability zero. Let the MLE be

$$\mathcal{L}(f;n,\theta) \equiv f_n(x,\theta) := \prod_{i=1}^{n} f(x_i;\theta) \tag{1.2}$$

and the equation

$$\mathcal{U} = \frac{\partial}{\partial\theta}\log f_n(x;\theta) = 0 \tag{1.3}$$

has a unique solution, $\widehat{\theta}$ say. Then, the sequence $\widehat{\theta}_n\,(x_1, x_2, \ldots, x_n)$ converges almost surely (a.s.) to the true value of θ, θ_T. That is why, sometimes especially in nonlinear experiment designs, we are considering the parameter space $\Theta \subseteq \mathbb{R}^P$ to be compact: We need the convergence to be within Θ. An estimator $\widehat{\theta}_n$ for the true parameter θ_T is known as a consistent one if the sequence of estimators $\widehat{\theta}_n$ converges to θ_T in probability.

As the measure theory approach the a.s. convergence implies convergence in probability, the MLE is consistent and the sequence theory "hub" MLE. As consistency is rather weak property and it is not a necessary condition, it is evident why we need more "mathematical properties". Moreover, it is quite unlikely that one can define θ_T. So, we are looking for a much stronger asymptotic property. This is the Cramér–Rao lower bound, i.e. under some regularity conditions as the following, recall (1.3)

$$E\left[\mathcal{U}\right] = 0 \;\; B_n^2(\theta) = E\left[\mathcal{U}^2\right] = -E\left[\frac{\partial^2 \log f_n}{\partial\theta^2}\right] = -nE\left[\frac{\partial^2 \log f}{\partial\theta^2}\right]. \tag{1.4}$$

Then, the Cramér–Rao inequality holds:

$$\frac{1}{B_n^2(\theta)} \leq E(\widehat{\theta} - \theta)^2 \;\; \text{and MLE} \;\; \widehat{\theta} \sim N\left(\theta_T; B_n^2(\theta)\right). \tag{1.5}$$

For statistics, it is Fisher's (parameter) information which is a very important ingredient for the statistical framework and it is defined to be

$$I(\theta) = -E\left[\frac{\partial^2 \log f(x;\theta)}{\partial \theta^2}\right]. \tag{1.6}$$

Therefore, the Cramér–Rao inequality, for the unbiased estimation, Wilks [54], Kitsos [27], among others, is

$$\frac{1}{I(\theta)} \leq \operatorname{Var}(\widehat{\theta}) \Leftrightarrow \operatorname{Var}(\widehat{\theta}) I(\theta) \geq 1 \tag{1.7}$$

and for biased

$$\frac{[1+b'(\theta)]^2}{I(\theta)} \leq \operatorname{Var}(\widehat{\theta}), \quad b(\theta) = E(\widehat{\theta}) - \theta. \tag{1.8}$$

The entropy-type Fisher's information and the extension of Cramér–Rao inequality is discussed in Kitsos and Tavoularis [32], for the introduce γ-order generalized normal distribution that emerged from Logarithmic Sobolev Inequality (LSI).

Therefore, from simple curves (distributions), we moved to foundation (measure theory) to Gaussian and results (analysis) to the invariance, this treated the idea of applications from affine geometry, while the Fourier and Laplace transforms were there to examine the research on signal process, and now we are using LSI, to face probability and statistical problems.

2. Measure Theory

Kolmogorov also worked on the Strong Law of Large Numbers (SLLN) and extended Khinchin's Weak Law of Large Numbers (WLLN), by Aleksander Khinchin (1894–1954). The SLLN replaces the converges in probability (in p) and WLLN converges with probability one (w.p.1). As the convergence w.p.1 implies convergence (in p), it is indeed the SLLN, stronger than the WLLN. But the main difference, in statistical terms, is that the convergence w.p.1 applies to the individuals realization of a random sample/process, while convergence (in p) does not.

We would say that convergence w.p.1 is closer to the convergence of a sequence of numbers, while the interest is focused on random variables (r.v.'s)!

Recall that the Borel–Cantelli lemma, Borel [2], Cantelli [3], provides a condition ensuring convergence w.p.1.

All the measure theory proprieties (countable additivity, monotonicity, additivity of Lebesgue measure, etc.) were adopted and the development was brilliant.

The idea of measure $\mu(E)$ of a solid body E goes back to Archimedes, one could measure in one, two or three dimensions. In the classical approach to geometry, the measure of a body was computed by partitioning that body into a finite number of components. One also could obtain lower and upper bounds of the measure of the body. It was Archimedes who inscribed and circumscribed canonical polygons to a circle and obtained lower and upper bounds of the perimeter length — to approach the π value, $\pi = 3.14156$.

As it has been mentioned, it was Kolmogorov [36] who gave the axiomatic definition of probability, through the measure theory. Eventually, the probability function P in the σ-field A known as the probability distribution informs the experimentalist how the unit mass is distributed among different subsets A of Ω, with Ω the space of outcomes. Thus, the triplet (Ω, A, P) forms the probability space. Let $g : \Omega \to \Omega$ be $1-1$, onto and $gx := g(x)$. Recall onto means that $g\Omega = \Omega$. The collection of all sets $g\mathcal{A}$, with $\mathcal{A} \in A$ is a σ-field gA. Moreover, gP is the probability measure on $g\mathcal{A}$ "induced" by g in the sense that

$$gP(g\mathcal{A}) = P(\mathcal{A}), \quad \forall \mathcal{A} \in A.$$

Any function h on Ω generates a new function $gh : gh(gx) = h(x)$. Then, g creates an isomorphism

$$g(\Omega, A, P) \to (g\Omega, gA, gP).$$

Consider a group G, with elements g with properties

$$G = \{g : gA = \mathcal{A},\ gP \in \mathcal{P},\ P \in \mathcal{P}\}.$$

Then, the family

$$g(\Omega, A, P), \quad P \in \mathcal{P}$$

is said to be invariant under G. Fraser [10] worked his structure of inference adopting the sense of "orbit" with such an invariant family, through an affine geometry approach, with a prior distribution for the parameters, Kitos [23, 29].

One of the important theorems of measure theory is Fubini's theorem, mainly for product probability spaces, Fubini [11], which with Tonelli's theorem, Tonelli [49] form Fubini–Tonelli theorem. Practically, this theorem permits to "interchange" integration, in the sense that given (Ω_i, A_i, P_i) $i = 1,2$ and $(\Omega_1, A_1, P_1) \times (\Omega_2, A_2, P_2)|i = 1,2$ and w a non-negative measurable function on $\Omega_1 \times \Omega_2$, and λ the σ-finite product measure on $A_1 \times A_2$ ie $\lambda(A_1 \times A_2) = \mu_1(A_1)\mu_2(A_2)$, then

$$\int w d\lambda = \int\int w d\mu_1 d\mu_2 = \int\int w d\mu_2 d\mu_1. \tag{2.1}$$

Under this line of thought, a statistical tolerance region can be defined, and a relevant theory is develop, Kitsos [23], Wilks [54], Fraser [9], Guttman [15], we briefly refer to statistical tolerance regions, as where Fubini's theorem is applied, to a statistic provided a mathematical formulation to dined as

$$Q : \mathbb{R}^n \to B : (y_1, \dots, y_n) \to Q(y_1, y_2, \dots, y_n),$$

where B is the Borel σ-algebra in $\mathbb{R}$. Hence, there are functions $L(y_1, \dots, y_n)$, $U(y_1, \dots, y_n)$ such that

$$Q = [(L, U),\ L(\cdot) < U(\cdot)]\,. \tag{2.2}$$

The term statistical is needed to provide the special case of σ-algebra A to be the Borel σ-algebra and the $\mathbb{R}^n$ to be the special case of the set of all measurements, Wilks [54], sampled from a continuous cumulative distribution function F considered tolerance regions of the form

$$Q(\cdot) = \left[Y_{(k)},\ Y_{(k+m)}\right], \tag{2.3}$$

where $Y_{(l)}$ is the l-th-ordered statistic and proved that

$$\left[F(Y_{(k+m)}) - F(Y_{(k)})\right] \sim \text{Beta}(m, n - m + 1), \tag{2.4}$$

with Beta being the Beta distribution.

As $Q(\cdot)$ is not unique and we need to pay special attention to this characteristics, a criterion to what we choose is needed.

The "coverage" of the tolerance region is such one. It is based on the independent observations coming from distribution function P_θ, with parameters θ from the parameter space $\Theta \subseteq \mathbb{R}^p$ and denoting by $C(Q)$, it is

$$C(Q) = P_\theta\left[Q(y_1, \ldots, y_n)\right]. \tag{2.5}$$

The statistical tolerance region is a β-content tolerance region at confidence level γ if

$$P = \{P_\theta\left[Q(y_1, \ldots, y_n) \geq \beta\right]\} = \gamma, \quad \beta, \gamma \in [0, 1]. \tag{2.6}$$

From (2.4) with $k = r\ m = n - 2r + 1\ r < \frac{n+1}{2}$, the following can be proved, Kendall and Stuart [19]:

$$\gamma = \{P_\theta\left[Q(y_1, \ldots, y_n)\right] \geq \beta\} = 1 - I_\beta(n - 2r + 1, 2r), \tag{2.7}$$

with $I_\beta(p, q)$ the incomplete Beta distribution. With $r = 1$, equation (2.7) is reduced to

$$1 - \gamma = n\beta^{n-1}(1 - \beta). \tag{2.8}$$

From equation (2.8), given the probability levels β and γ, one can obtain, through Newton–Raphson iteration, from example, the appropriate sample size. As is stated in (2.8), it is a three parameter equation and not easy to be solved.

We need some thoughts to retrieve the number of tolerance regions. One idea is the statistical tolerance region $Q(\cdot)$ which is a β-expectation tolerance region if

$$E\left\{P_\theta\left[Q(y_1, \ldots, y_n)\right]\right\} = \beta. \tag{2.9}$$

The following proposition is a lemma by Paulson [41], who worked on the simplicity of proofs; it is an essential application of Fubini's theorem and the foundational definition in statistical theory, due to (2.9).

Lemma 1 (Paulson's Lemma). *If tolerance limits $L(y_1, \ldots, y_n)$ and $U(y_1, \ldots, y_n)$ of level β are determined and we let T to be a future sample of q new independent observations $(y_1^x, \ldots, y_n^x)$ say from the population (under investigation), then, with G being the distribution function of T,*

$$\text{if} \quad \mathcal{M} = \int_L^U dG(t), \quad \text{then} \quad E(\mathcal{M}) = \beta. \tag{2.10}$$

Indeed, let $h(l,u)dldu$ be the probability element of L, U, then the definition of expectation is as follows:

$$\begin{aligned} E(\mathcal{M}) &= \int_{-\infty}^{\infty}\int_{-\infty}^{\infty}\left\{\int_{L}^{U} dG(t)\right\} h(l,u)\,dl\,du, \\ \text{(Fubini's Theorem)} &= \int_{L}^{U} dG(t)\int_{-\infty}^{\infty}\int_{-\infty}^{\infty} h(l,u)\,dl\,du, \\ (L,U \text{ confidence limits}) &= P\left[L < T < U\right] = \beta. \end{aligned}$$

The extension of Paulson's lemma to tolerance regions is due to structural of inference approach of Statistics, Rinco [42], Kitsos [23], Fraser and Guttman [8], Ellerton *et al.* [6].

The structural tolerance region is statistic $Q = Q(y_1, \ldots, y_n)$ on $\mathbb{R}^n$, the space of the future responses, based on the data such that, recall (2.5),

$$\mathrm{C}\left[Q(\cdot)\right] = \int_{Q} P(y^*|\theta)dy^* \tag{2.11}$$

with $P(y^*|\theta)$ the probability element of y^*, the future response, which will depend on the parameter θ. Then, recall (2.9), the β-expectation tolerance region exists if

$$E_\theta\left\{\mathrm{C}\left[Q(\cdot)\right]\right\} = \beta, \tag{2.12}$$

where E_θ means the expectation on the parameters $\theta \in \Theta$ with Θ the parameter space. From (2.12), assuming Fubini's theorem holds, it is

$$\begin{aligned} \beta &= \int_{\Theta} \mathrm{C}\left[Q(\cdot)\right] g(\theta|y)d\theta, \\ \text{(due to (2.11))} &= \int_{\Theta}\left[\int_{Q} P(y^*|\theta)dy^*\right] g(\theta|y)d\theta, \\ \text{(due to Fubini's)} &= \int_{Q}\left[\int_{\Theta} P(y^*|\theta)g(\theta|y)d\theta\right] dy^*, \\ \text{(definition)} &= \int_{Q} h(y^*|y)dy^*, \end{aligned} \tag{2.13}$$

where the probability within in brackets was denoted by $h(y^*|y)$, i.e. by definition

$$h(y^*|y) = \int_{\Theta} P(y^*|y)g(\theta|y)d\theta. \tag{2.14}$$

Relation (2.14) it is known as the prediction distribution of y^* and thus, it is clear why (2.13) is Paulson's analogue for the structural β-expectation tolerance regions are applied in regression problems, Ellerton *et al.* [6] and in optimal design theory, Muller and Kitsos [39], in signal process, Kitsos and Toulias [21, 33], or for any general linear model, Kitos and Zarikas [35]. Note that these applications are based on a theoretical background, developed to analysis, and now are part of statistical theory.

3. Linear Transformations in Statistics

One of the earliest statistical-oriented transformation was introduced by Galton [12]. The idea behind was that non-normal data can be easier handled if it were transformed to normal, through a filter, i.e. through a transformation.

It was proposed that the non-normal $x_1, x_2, \ldots, x_n$ observations are transformed to a normal dataset $z_1, z_2, \ldots, z_n$ through

$$z = x + \beta \log\left[\frac{x-\kappa}{\lambda}\right]. \tag{3.1}$$

Then, $Z \sim \mathcal{N}(0,1)$ and $\mathcal{X}$ is said to be logNormal.

Under the γ-order generalized normal distribution, Kitos and Tavoularis [32], emerged from LSI, has been generalized to γ-order-generalized logNormal, Toulias and Kitsos [50], with PDF:

$$f_{LN}(x;\mu,\sigma^2,\gamma) = \wedge \exp\left\{-\frac{\gamma-1}{\gamma}\left(\frac{|\ln x-\mu|}{\sigma}\right)\frac{\gamma}{\gamma-1}\right\}$$
$$\wedge = \frac{1}{2x\sigma\Gamma\left(\frac{\gamma-1}{\gamma}\right)\left[\left(\frac{\gamma-1}{\gamma}\right)^{\frac{1}{\gamma}}\right]}. \tag{3.2}$$

Among the most popular transformations in mathematical analysis are the linear one, with the Jacobian of transformation being a link with the considered measure. Two examples are provided in the following in detail. But the linear transformation is related to "affine transformation" and Fraser [10] first worked on this affine framework, while Murray and Rice [40] clarified the influence of affine geometry through a group of transformations assuming a distribution for the involved parameters. This created criticism that the method was a particular Bayesian one and despite the number of applications created, Rinco [42], Kitsos [23] and Ellerton *et al.* [6] proposed an approach to fit the optimal polynomial for prediction of the future response. A new development was attempted by Murray and Rice [40], Kitsos [29], were the Differential Geometry and the relevant group of transformations were the fundamental ingredients of the analysis than the Bayesian underlying assumption, providing emphasis to invariance.

The location scale group, Fraser [10], Murray and Rice [40] is $G = \mathbb{R} \times \mathbb{R}^+$ defined by the action

$$G \times \mathbb{R} \to R : ((\mu, \sigma), x) \to \mu + \sigma x. \tag{3.3}$$

For the location scale action of densities, given by a function f of a single variable $x \in \mathbb{R}$, they are defined as

$$(\mu, \sigma) f(x) = f\left(\frac{x - \mu}{\sigma}\right). \tag{3.4}$$

Thus, for an action of the group G on probability densities $f(x)dx$, we define

$$(\mu, \sigma) f(x) dx = \frac{1}{\sigma} f\left(\frac{x - \mu}{\sigma}\right) dx. \tag{3.5}$$

Moreover, Kitsos [26, 29] proved that there is a "canonical form" invariant to affine transformation for the *Logit* model and the model we try to "solve"; its statistical binary problems are affine transformations of the canonical form. Note that the canonical form of the *Logit* model can be applied either to Ca problems, see also [31], or to engineering applications. Now, we are working with the Michaelis–Menten model which has a particular interest in biologic kinetics, Currie [4], Toulias and Kitsos [51, 52].

Proposition 1. *Consider the Michaelis–Menten model* $f(u, \theta) = \frac{\theta_1 u}{\theta_2 + u}$ *with all the information concerning the physical meaning as in references.*

Then, the group $SL(2, R)$ on the complex plane acts by a fractional linear transformations.

Proof. Recall that it is easy to see the essential decomposition

$$\nabla f(u, \theta) = \begin{pmatrix} 1 & 0 \\ 0 & \theta_1 \end{pmatrix} \begin{pmatrix} u(\theta_2 + u)^{-1} \\ u(\theta_2 + u)^{-2} \end{pmatrix} := Q\nu^T \tag{3.6}$$

with the definitions of Q and ν trivial.

This equation (3.6) provides a proof, Kitsos [24, 25], to confirm why the experiment design depends only on θ_2, i.e. needs prior information for θ_2. The essential is the decomposition $Q\nu^T$ which provides dependency on θ_2 only.

Now, recall $n \times n$ orthogonal matrices from the $O(n)$ orthogonal group, while the SO(n) has elements of the orthogonal matrices with determinant +1. So, $Q \notin O(n)$. The general linear group $GL_n(R)$ is defined by the $n \times n$ invertible matrices A, while when $|A| = 1$, the special linear group $SL_n(\mathbb{R})$ is defined.

Note that $SL_2(\mathbb{R})$ acts on $\mathbb{C}$-plane by fractional linear transformation:

$$\frac{pz + q}{rz + s} = \begin{pmatrix} p & q \\ r & s \end{pmatrix} z^* = Qz, \quad z^* = (z, 1)^T. \tag{3.7}$$

Note that only in spacial cases $Q \in SL_q$,

$$\begin{pmatrix} \theta_1 & 0 \\ 1 & \theta_2 \end{pmatrix} u^* = \frac{\theta_1 u}{u + \theta_2}, \quad u^* = (u, 1)^T. \tag{3.8}$$

For the case $\theta_1\theta_2 = 1$, $|Q_0| = +1$, so $Q_o \in O_2(\mathbb{R})$.

The affine transformations are not that popular in statistics, but in statistics, there was "more analysis" involved in the light of the Laplace and Fourier transformations. The characteristic function of a continuous random variable X with PDF f_X as well as the moment generating function (MGF) are very important in the statistical framework.

Here are the classic definition of the transformations:

$$\text{Fourier: } \mathcal{F}[f_X] = \widehat{f}_X(\xi) = \int f_x(x) \exp\{-i2\pi\xi x\}\, dx, \tag{3.9}$$

$$\text{Laplace: } \mathcal{L}[f_X] = \dot{f}_X(s) = \int f_X(x) \exp\{sx\}\, dx. \tag{3.10}$$

See also Section 4.

For continuous r.v. X with PDF, the characteristic function $\mathcal{M}_X$ of the r.v. (or for the PDF) is defined as

$$\mathcal{M}_X(\xi) = \int f_X(x) \exp\{i\xi x\}\, dx. \tag{3.11}$$

So, it is easy to see that

$$\mathcal{M}_X(\xi) = \widehat{f}_X\left(-\frac{\xi}{2\pi}\right) = \dot{f}_X(si). \tag{3.12}$$

The Jacobian determinant of a transformation is very essential in statistics. The following two examples, for two fundamental distributions, t_n and $F_{m,n}$, clarifies the application. □

Example 1. The t_n-distribution.
Let $Z \sim N(0,1)$ and $Y_n = \sum z_i^2 \sim \chi_n^2$, then we define

$$T = \frac{z}{\sqrt{\frac{Y_n}{n}}} \sim t_n. \tag{3.13}$$

The PDF of Z is

$$\varphi(z) = \frac{1}{\sqrt{2\pi}} e^{-\frac{z^2}{2}}. \tag{3.14}$$

The PDF of Y_n is

$$f_{Y_n}(y) = \frac{1}{2^{n/2}\Gamma(n/2)} y^{n/2-1} e^{-y/2}, \quad y > 0. \tag{3.15}$$

Let us define the transformation.

From equation (3.13) $\Rightarrow \begin{cases} T = Z\sqrt{n/Y_n} \\ S = Y_n \end{cases} \quad \therefore\ Z = T\sqrt{\frac{S}{n}}, \quad Y_n = S.$

Then, the Jacobian of the transformation is

$$J = \begin{pmatrix} \sqrt{S/n} & \frac{T}{2n\sqrt{S/n}} \\ 0 & 1 \end{pmatrix} \Rightarrow |J| = \sqrt{\frac{S}{n}}. \tag{3.16}$$

Since Z and Y_n are assumed independent, the joint density of Z and Y_n is

$$f_{Z,Y_n}(z,y) = f_Z(z) f_{Y_n}(y) \Rightarrow f_{T,S}(t,S) = f_Z\left(t\sqrt{S/n}\right) f_{Y_n}(S)\left|J\right|. \tag{3.17}$$

Thus, the joint density function becomes

$$f_{T,S}(t,s) = \frac{1}{\sqrt{2\pi n}\Gamma\left(\frac{n}{2}\right) 2^{n/2}} s^{\frac{n+1}{2}-1} \exp\left\{-\frac{s}{2}\left(\frac{t^2}{n}+1\right)\right\}. \tag{3.18}$$

The marginal density of T is required; therefore, an integration out of S is needed:

$$f_T(t) = \int_0^\infty f_{T,S}(t,s)\,ds = \frac{1}{\sqrt{2\pi n}\,\Gamma\left(\frac{n}{2}\right) 2^{n/2}} \int_0^\infty s^{\frac{n+1}{2}-1}$$
$$\times \exp\left\{-\frac{1}{2}\left(\frac{t^2}{n}+1\right)s\right\} ds,$$

$$f_T(t) = \text{const} \cdot \frac{\Gamma\left(\frac{n+1}{2}\right)}{\left[\frac{1}{2}\left(\frac{t^2}{n}+1\right)\right]^{\frac{n+1}{2}}} \int_0^\infty \frac{\left[\frac{1}{2}\left(\frac{t^2}{n}+1\right)\right]^{\frac{n+1}{2}}}{\Gamma\left(\frac{n+1}{2}\right)} s^{\frac{n+1}{2}-1}$$
$$\times \exp\left\{-\frac{1}{2}\left(\frac{t^2}{n}+1\right)s\right\} ds. \tag{3.19}$$

The integral is equal to the PDF of a Gamma distribution with parameters

$$\alpha = \frac{n+1}{2} \quad \beta = \frac{1}{2}\left(\frac{t^2}{n}+1\right)$$

and because the PDF integrates to one. Thus,

$$f_T(t) = \frac{\Gamma\left(\frac{n+1}{2}\right)}{\Gamma\left(\frac{n}{2}\right)\sqrt{n\pi}} \left(\frac{t^2}{n}+1\right)^{-\left(\frac{n+1}{2}\right)} \tag{3.20}$$

is the PDF of the t-distribution with n degrees of freedom.

Example 2. The $F_{m,n}$-distribution.

Let $X \sim \chi^2_m$ and $Y =\sim \chi^2_n$ two Chi-square distributions. Then, the ratio

$$F = \frac{\chi^2_m/m}{\chi^2_n/n} \sim F_{m,n} \tag{3.21}$$

follows the $F_{m,n}$ distribution with m and n degrees of freedom and with X, Y independent of each other.

The PDF of the Chi-Square is

$$f_\chi(x) = \frac{1}{\Gamma(m/2)2^{m/2}} x^{\frac{m}{2}-1} e^{-\frac{x}{2}}. \tag{3.22}$$

Define the r.v. F, S as a function of X and Y.

Let us define.

From equation (3.21) $\Rightarrow \begin{cases} F = \dfrac{\chi/m}{Y/n} \\ S = Y \end{cases} \therefore\ \chi = \frac{m}{n} FS,\ Y = S.$

The Jacobian of the transformation is

$$J = \begin{pmatrix} \frac{m}{n} S & \frac{m}{n} F \\ 0 & 1 \end{pmatrix} \Rightarrow |J| = \frac{m}{n} S. \tag{3.23}$$

Since X and Y are assumed independent, the join density equals the product of the marginals:

$$f_{X,Y}(x, y) = f_X(x) f_Y(y).$$

Thus,

$$f_{F,S}(f, s) = f_{X,Y}(x, y)\,|J| = f_X\left(\frac{m}{n} fs\right) f_Y(s)\,|J|,$$

$$f_{F,S}(f, s) = \frac{\left(\frac{m}{n}\right)^{\frac{m}{2}} f^{\frac{m}{2}-1}}{\Gamma\left(\frac{m}{2}\right)\Gamma\left(\frac{n}{2}\right) 2^{\frac{m+n}{2}}} s^{\frac{m+n}{2}-1} \exp\left\{-\frac{1}{2}\left(\frac{m}{n} f + 1\right) s\right\}. \tag{3.24}$$

The marginal density of F is obtained integrating over S, i.e.

$$\begin{aligned} f_F(f) &= \int_0^\infty f_{F,S}(f,s)\,ds = \text{const}\int_0^\infty s^{\frac{m+n}{2}-1} \\ &\quad\times \exp\left\{-\frac{1}{2}\left[\left(\frac{m}{n}f+1\right)s\right]\right\}ds, \\ f_F(f) &= \text{const}\cdot\frac{\Gamma(\alpha)}{\beta^\alpha}\int_0^\infty \frac{\beta^\alpha}{\Gamma(\alpha)}s^{\alpha-1}\exp\{-\beta s\}\,ds, \\ &\text{with } \alpha = \frac{m+n}{2},\quad \beta = \frac{1}{2}\left(\frac{m}{n}f+1\right). \end{aligned} \tag{3.25}$$

Thus, the integral equals to 1 as the PDF of Gamma distribution with parameter α, β, so

$$\begin{aligned} f_{F,S}(f,s) &= \text{const}\left(f^{\frac{m}{2}-1}\right)\left(s^{\frac{m+n}{2}-1}\right)\exp\left[-\frac{1}{2}s\left(\frac{m}{n}f+1\right)\right] \\ \text{with const} &= \frac{\frac{m}{n}}{\Gamma\left(\frac{m}{2}\right)\Gamma\left(\frac{n}{2}\right)2^{(m+n)/2}}. \end{aligned} \tag{3.26}$$

4. Fourier Transform

We now move to the Fourier transform, provided the mathematical background for it, Stein and Shakarchi [45], which is not discussed yet in statistical inferences. The Schwartz space on $\mathbb{R}^n$ is a function space of rapidly decreasing function on $\mathbb{R}^n$:

$$S(\mathbb{R}^n) = \left\{f \in C^\infty(\mathbb{R}^n,\mathbb{C}), \forall\alpha,\beta\in\mathbb{N}^n, \|f\|_{\alpha,\beta} < \infty\right\}, \tag{4.1}$$

where $C^\infty(\mathbb{R}^n,\mathbb{C})$ is the function space of "smooth functions" $f : \mathbb{R}^n \to \mathbb{C}$:

$$\|f\|_{\alpha,\beta} = \sup_{x\in\mathbb{R}^n}\left|x^\alpha\left(D^\beta f\right)(x)\right|, \tag{4.2}$$

with α, β used in a multi-index notation and D a linear partial differential operator. Then, it can be proved, Hormander [17], that Fourier transform is a linear isomorphism

$$F : S(\mathbb{R}^n) \to S(\mathbb{R}^n) : f \to Ff = \widehat{f} \tag{4.3}$$

with

$$\widehat{\xi} = \frac{1}{(2\pi)^{n/2}} \int_{\mathbb{R}^n} f(x) \exp\{-i\xi x\}\, dx, \quad \xi \in \mathbb{R}^n. \tag{4.4}$$

The transform is a decaying function; therefore, the Fourier transform maps the Schwartz space into itself.

Note that the reflection operator, $Rf(x) = f(-x)$, as well as the translation $T_h f(x) = f(x-h)$, can equip a Schwartz space. The convolution of a test function and a distribution can be defined, while the convolute of two test functions $\varphi, g \in S$ is defined as usual:

$$\begin{aligned} (\varphi * g)(x) &= \int_{\mathbb{R}^n} \varphi(x-y) g(y) dy, \\ (\varphi * g)(x) &= \int_{\mathbb{R}^n} (RT_x\varphi)(y) g(y) dy. \end{aligned} \tag{4.5}$$

The convolution of two functions equips statistic theory, especially in engineering application without, sometimes a rigorous mathematical form. It is easy to see that the convolution of a function φ with the δ function is

$$(\varphi * \delta)(x) = (RT_x\varphi)(0) = R\varphi(-x) = \varphi(x). \tag{4.6}$$

Thus, $\varphi * \delta = \varphi$, i.e. $\int \delta(x-y)\varphi(y)dy = \varphi(x)$. Moreover, the convolution with a derivative of δ is

$$(\varphi * \varrho^a \delta) = \varrho^a \varphi(x).$$

From a statistical point of view, it is interesting to see the delta sequence in $S(\mathbb{R})$ is given by

$$T_n(\varphi) = \frac{n}{2} \int_{-1/4}^{1/4} \varphi(x) dx. \tag{4.7}$$

For any continuous function φ, it is

$$T_n(\varphi) \to \varphi(0) = \delta(\varphi) \quad \text{as } n \to 0, \text{ i.e. } T_n \to \varphi. \tag{4.8}$$

That is, the approximate identity comes to a δ sequence with the Gaussian $\mathcal{N}(0, \tau^2)$ approximate identity.

Moreover, the δ-"distribution" can be defined with the support of $\mathcal{N}(0, \tau^2)$ as

$$\delta(\tau) = \lim \frac{1}{\sqrt{2\pi\tau}} \exp\left\{-\frac{1}{2}\frac{x^2}{\tau^2}\right\}. \tag{4.9}$$

It is also approached in limit as well by Kitsos and Toulias [34], through the γ-order generalized normal distribution, when $\gamma \to 0$.

Theorem 1. *Let $A \in \mathbb{R}^{n\times n}$ be a symmetric, positive definite matrix. Suppose that*

$$f(x) = \exp\left\{-\frac{1}{2}\langle x, Ax\rangle\right\} \tag{4.10}$$

is the n-dimensional Gaussian, then the Fourier transform of $f(x)$ is

$$\widehat{f}(\xi) = \frac{1}{|A|^{1/2}} \exp\left\{-\frac{1}{2}\langle \xi, A^{-1}\xi\rangle\right\}.$$

The covariance matrix A of the transform of a Gaussian is the inverse of the covariance matrix of the Gaussian.

Consider the Cauchy distribution

$$\mathrm{Cau}(x; x_0, \tau) = \frac{1}{\pi}\left[\frac{\tau}{(x - x_0)^2 + \tau^2}\right] \tag{4.11}$$

with τ the scale parameter and x_0 the location parameter.

The most well-known form is

$$\mathrm{Cau}(x; 0, 1) = \frac{1}{\pi(x^2 + 1)}. \tag{4.12}$$

For Cauchy distribution, despite that the mean is undefined, it is not differentiable at the origin of the Fourier transform (known also as characteristic function in Statistics):

$$\widehat{\mathrm{C}}\mathrm{au}(\xi) = E\left(e^{i\xi t}\right) = e^{-|\xi|}. \tag{4.13}$$

Therefore, one can see easy that for the sum of independent X_i for $i = 1, 2, \ldots, \sum_{i=1}^{n} X_i$, Cauchy distribution holds:

$$G_{\sum X_i} = e^{-n|t|}. \tag{4.14}$$

Therefore, for the Cauchy distribution, a kind of "central limit" theorem exists:

Theorem 2. *Let $X_1, X_2, \ldots, X_L$ be independent, identically distributed (i.i.d) random variables with PDF $f_C(x)$ such that*

$$\lim \frac{1}{c}\int_{-c}^{c} x^2 f_C(x)dx = \frac{2\delta}{\pi} \in (0, \infty).$$

Then,

$$\frac{1}{n}\sum_{i=1}^{L} X_i \xrightarrow{P} \mathrm{Cau}(0, \delta).$$

Such theorems provide elegant applications of mathematical analysis but not too much contribution to everyday statistical calculations. In principle, the theoretical results, witch are not applied too often, are very useful for the applications that were created. There is, we believe, enough cases to be strengthened by analysis in statistics. We are discussing some in the following section.

5. Discussion

Fourier transforms are adopting in the statistics analysis of input–output systems, Kitsos [27], among others; we would say in statistical signal process, while Kitsos and Toulias [21] worked only with statistics, comparing confidence and tolerance intervals. In engineering production, the tolerance regions, we believe, can be more useful, even in optimal experiment design theory, Muller and Kitsos [39]. In Section 4, it was very briefly discussed that the Dirac function is related to Fourier transform. This is due to the initial work of Dirac [5] on quantum mechanics and Sobolev [44] working on Cauchy problem. Actually, the Dirac function acted as the first step to tempered distributions, Jones [18]. This mathematical analysis work is related to the γ-order-generalized normal distribution, Kitsos and Tavouluris [32], while the δ distribution is within this family of distributions, when $\gamma \to 0$, Kitsos *et al.* [22]. Therefore, statistics can offer feedback to analysis. Even the hypergeometric functions can be applied to optimal design of experiments, Kitsos [28]. The Michaelis–Menten nonlinear model, discussed in Section 3, can be considered as a Mobius transformation in $\mathbb{C}$, Amaral [1], in the same terms that McCullagh [38] worked for Cauchy distributions, under this line of thought, strengthening statistical framework using results from the mathematical analysis, the statistical theory can be enlarged, despite the fact that applied statistics is working in a Euclidean space. Applied probability is still working on Euclidean spaces, but there are excellent results in Banach spaces, Schwartz [43], Ledoux and Talagrand [37].

As we noted in the introduction, the probability theory was equipped earlier than statistics with a solid mathematical background. Even recently, the LSI are, in principle, applied to probability theory and then to statistical data analysis, Kitsos and Nyamsi [20]. Although it would be useful to discuss isometric problem in statistics, comparing results for different areas in a large scale, this does not have, yet, the mathematical development, Tan and Xiang [48], it deserves.

Very briefly from Gauss [13], and the Gaussian as it is mentioned in engineering and the normal distribution as it is refereed to statistics, Walker [54], we moved to the Fisherian line of thought, Fisher [7], in statistics. The existent work was carefully displayed and analyzed by Kendall and Stuart [19] and a mathematical orientation was provided by Wilks [54], who traced the mathematical improvements. At the same time, from Apostol's line of development, we moved to Rudin's development of mathematical analysis and now we can work with Suzuki and Senba [47]. The question still remains the same in practice: mathematics or statistics, Halkos and Kitsos [16], deterministic or stochastic approach. The truth is that both approaches need, at the end, elegant calculations, either through numerical methods or facing big (and sensitive) data sets. In most cases, in practice, we cannot avoid the existence of the stochastic error to the data analysis we proceed. That is why an extended form of the normal distribution is needed, to cover "fat tail" distributions, the γ-order generalized normal covers such complicated cases, we believe, Kitsos and Nyamsi [20].

Statisticians always recall Sir David: "Keep it Simple", Kitsos [30]. But the notion of "simple" is a function of time. In the 21st century, the "simple" is different than 17th century or at the times of Pythagoras.

References

[1] F. Amaral. Weighted optimization criteria in mixtures DoE systems: Linear and non-linear models. PhD Thesis, Universidade Aberta, Lisboa, 2024. PhD Thesis (in preparation).

[2] E. M. Borel. Les probabilités dénombrables et leurs applications arithmétiques. *Rendiconti del Circolo Matematico di Palermo (1884–1940)*, 27(1), 1909, 247–271.

[3] F. P. Cantelli. Sulla probabilità come limite delta frequenza. *Rom. Acc. L. Rend., (5)*, 26(1), 1917, 39–45.

[4] D. J. Currie. Estimating Michaelis-Menten parameters: Bias, variance and experimental design. *Biometrics*, 1982, 907–919.

[5] P. A. M. Dirac. *The Principles of Quantum Mechanics.* Oxford University Press, Oxford, 1935.

[6] R. R. W. Ellerton, C. P. Kitsos and S. Rinco. Choosing the optimal order of a response polynomial-structural approach with minimax criterion. *Commun. Stat. Theory Methods*, 15, 1986, 129–136.

[7] R. A. Fisher. On the mathematical foundations of theoretical statistics. *Philos. Trans. R. Soc. Lond. Ser. A*, 222(594–604), 1922, 309–368.

[8] D. A. S. Fraser and I. Guttman. Tolerance regions. *Ann. Math. Stat.*, 27(1), 1956, 162–179.

[9] D. A. S. Fraser. *Nonparametric Methods in Statistics.* John Wiley & Sons Inc, London, 1956.

[10] D. A. S. Fraser. *The Structure of Inference.* John Wiley & Sons, London, 1968.

[11] G. Fubini. Sugli integrali multipli. *Rend. Acc. Naz. Lincei*, 16, 1907, 608–614.

[12] F. Galton. XII. The geometric mean, in vital and social statistics. *Proc. R. Soc. Lond.*, 29(196–199), 1879, 365–367.

[13] C. F. Gauss. *Theoria Combinationis Observationum: Erroribus Minimis Obnoxiae.* Dieterich, Gottingen, 1823.

[14] W. S. Gosset. The probable error of a mean. *Biometrika*, 1908, 1–25.

[15] I. Guttman. *Statistical Tolerance Regions: Classical and Bayesian.* Griffin Ltd, London, 1970.

[16] G. Halkos and C. Kitsos. Mathematics vs Statistics in Tackling Environmental Economics Uncertainty, 2018.

[17] L. Hörmander. *The Analysis of Linear Partial Differential Operators I: Distribution Theory and Fourier Analysis.* Springer, Berlin, 1990.

[18] D. S. Jones. The theory of generalised functions, Cambridge University Press, Cambridge, UK, 2009.

[19] M. G. Kendall and A. Stuart. *The Advanced Theory of Statistics*, Vol. 2. Hafner Publishing Company, London, 1968.

[20] C. P. Kitsos and U. E. Nyamsi. Applying the generalized normal distribution for modelling asset returns data. In *Proceedings of the Fifth International Congress of Applied Statistics (UYIK-2024)*, Istanbul, Turkey, 21–23 May 2024.

[21] C. P. Kitsos and L. T. Toulias. Confidence and tolerance regions for the signal process. *Recent Pat. Signal Process.*, 2(2), 2012, 149–155.

[22] C. Kitsos, V. Vassiliadis and T. Toulias. MLE for the γ-order generalized normal distribution. *Discuss. Math. Probab. Stat.*, 34(1–2), 2014, 143–158.

[23] C. P. Kitsos. *Choosing the Order of a Response Polynomial-Structural Approach with Minimax Criterion.* University of New Brunswick, Department of Mathematics, Canada, 1978.

[24] C. P. Kitsos. *Design and Inference in Nonlinear Problems*. University of Glasgow, United Kingdom, Glasgow, 1986.

[25] C. P. Kitsos. Design aspects for the Michaelis-Menten model. *Listy Biom.-Biom. Lett.*, 38(1), 2001, 53–66.

[26] C. P. Kitsos. On the logit methods for Ca problems (design and fit). *Commun. Depend. Qual. Manag.*, 10(2), 2007, 88–95.

[27] C. P. Kitsos. *Technological Mathematics and Statistics.* New Tech Pub. Co., Athens, 2009.

[28] C. P. Kitsos. Adopting hypergeometric functions for sequential statistical methods. *Bull. Greek Math. Soc.*, 57, 2010, 251–264.

[29] C. P. Kitsos. Invariant canonical form for the multiple logistic regression. *Math. Eng. Sci. Aerosp. (MESA)*, 2(3), 2011.

[30] C. P. Kitsos. Sir David Cox: A wise and noble statistician (1924–2022). *Eur. Math. Soc. Mag.*, 124, 2022, 27–32.

[31] C. P. Kitsos and C.-S. Nisiotis. Considering distance measures in statistics. *Biom. Lett.*, 59(1), 2022, 65–75.

[32] C. P. Kitsos and N. K. Tavoularis. Logarithmic Sobolev inequalities for information measures. *IEEE Trans. Inf. Theory*, 55(6), 2009, 2554–2561.

[33] C. P. Kitsos and T. L. Toulias. Adopting tolerance regions in environmental economics. *J. Adv. Math. Comput. Sci.*, 34(6), 2019, 1–12.

[34] C. P. Kitsos and T. L. Toulias. On the family of the γ-ordered Normal distributions. *Far East J. Theor. Stat.*, 35(2), 2011, 95–114.

[35] C. P. Kitsos and V. Zarikas. On the best predictive general linear model for data analysis: A tolerance region algorithm for prediction. *J. Appl. Sci.*, 13(4), 2013, 513–524.

[36] A. N. Kolmogorov. *Foundations of the Theory of Probability.* Julius Springer, Berlin, 1933.

[37] M. Ledoux and M. Talagrand. *Probability in Banach Spaces: Isoperimetry and Processes.* Springer Science & Business Media, London, 2013.
[38] P. McCullagh. Conditional inference and Cauchy models. *Biometrika*, 79(2), 1992, 247–259.
[39] C. H. Müller and C. P. Kitsos. Optimal design criteria based on tolerance regions, 2004, pp. 107–116.
[40] M. K. Murray and W. J. Rice. *Differential Geometry and Statistics.* Chapman & Hall, London, 1994.
[41] E. Paulson. A note on tolerance limits. *Ann. Math. Stat.*, 14(1), 1943, 90–93.
[42] S. Rinco. β-expectation tolerance regions based on the structural models, Ph. D. thesis, University of Western Ontario, London, Canada, 1973.
[43] L. Schwartz. *Geometry and Probability in Banach Spaces*, Vol. 852. Springer, London, 1982.
[44] S. L. Sobolev. Methode nouvelle a resoundre le probleme de Cauchy pour les equations lineaires hyperboliques normales. *Math. Sbornik*, 1, 1936, 43.
[45] E. M. Stein and R. Shakarchi. *Fourier Analysis: An Introduction*, Vol. 1. Princeton University Press, Princeton, NJ, 2011.
[46] S. M. Stigler. Gauss and the invention of least squares. *Ann. Stat.*, 1981, 465–474.
[47] T. Suzuki and T. Senba. *Applied Analysis: Mathematical Methods in Natural Science.* World Scientific Publishing Company, Singapore, 2011.
[48] L. Tan and S. Xiang. On the aleksandrov-rassias problem and the hyers-ulam-rassias stability problem. *Banach J. Math. Anal.*, 1(1), 2007, 11–22.
[49] L. Tonelli. Sull'integrazione per parti. *Rend. Acc. Naz. Lincei*, 5(18), 1909, 246–253.
[50] T. L. Toulias and C. P. Kitsos. On the generalized lognormal distribution. *J. Probab. Stat.*, 2013(1), 2013, 432642.
[51] T. L Toulias and C. P Kitsos. Estimation aspects of the Michaelis–Menten model. *REVSTAT-Stat. J.*, 14(2), 2016, 101–118.
[52] T. L. Toulias and C. P. Kitsos. Fitting the Michaelis-Menten model. *J. Comput. Appl. Math.*, 296, 2016, 303–319.
[53] H. M. Walker. *Studies in the History of Statistical Method: With Special Reference to Certain Educational Problems.* Williams & Wilkins Co, Baltimore, 1929.
[54] S. S. Wilks. *Mathematical Statistics.* J. Wiley & Sons Inc, New York, 1962.

https://doi.org/10.1142/9789819820498_0010

Chapter 10

New Fractional Hermite–Hadamard-Type Inequalities

Kuang Jichang[*,‡] **and Michael Th. Rassias**[†,§]

[*]*Department of Mathematics, Hunan Normal University, Changsha, Hunan, P.R. China*

[†]*Department of Mathematics and Engineering Sciences, Hellenic Military Academy, Athens, Greece*

[‡]*jckuang@163.com*

[§]*mrassias@sse.gr, mthrassias@yahoo.com*

We introduce new (β, λ, h)-exp convex functional, and by using fractional integral operators, some new fractional Hermite–Hadamard-type inequalities are established.

Keywords: Generalized convex functional, fractional Hermite–Hadamard inequality, fractional integral operators.

1. Introduction

Definition 1. A function $f : [a, b] \to R$ is called convex if

$$f(\lambda x_1 + (1 - \lambda)x_2) \leq \lambda f(x_1) + (1 - \lambda)f(x_2), \tag{1.1}$$

$\forall x_1, x_2 \in [a, b], \quad \forall \lambda \in [0, 1]$.

This classical inequality (1.1) plays an important role in analysis, optimization and the theory of inequalities, and it has a huge literature dealing with its applications, various generalizations and refinements. Further, convexity is one of the most fundamental and important notions in mathematics. Convexity theory and its inequalities are fields of interest of numerous mathematicians and there are many papers, books and monographs devoted to these fields and various applications (see, e.g. Refs. [1–6] and the references therein).

The Hermite–Hadamard inequality

$$f\left(\frac{x_1+x_2}{2}\right) \le \frac{1}{x_2-x_1}\int_{x_1}^{x_2} f(x)dx \le \frac{f(x_1)+f(x_2)}{2} \tag{1.2}$$

holds if and only if f is a convex function on $[a,b], [x_1,x_2] \subset [a,b]$.

The Hermite–Hadamard inequality is the most famous one in convex function theory; it gives both upper and lower bounds for the mean value of a convex function over a compact interval. This inequality finds many applications in various fields, including integral calculus, probability theory, statistics, optimization and number theory. Additionally, it is a useful tool for solving physical and engineering problems that require determining function mean. This inequality is characterized by the trapezoidal and midpoint inequalities on its right and left parts. Hence, it plays an important role in analysis, optimization and the theory of inequalities, and it has a huge literature dealing with its applications, various generalizations and refinements for different classes of convex functions (see, e.g. Refs. [1–12] and the references therein).

In the past decades, fractional calculus has gained remarkable popularity and importance due mainly to its application in diverse and widespread fields. As of late, it can be seen that popularize well-known integral inequalities using fractional integral operators have become an interesting topic. Using fractional integral operators, some authors obtained the fractional Hermite–Hadamard-type inequalities. For example, we have the following:

Definition 2 (cf. Ref. [1]). Let $f \in L[a,b]$, then Riemann–Liouville fractional integrals of f of order $\alpha > 0$ with $a \ge 0$ are defined by

$$T_{\mathbf{1},a}(f,x) = \frac{\mathbf{1}}{\Gamma(\alpha)} \int_a^x (x-t)^{\alpha-1} f(t)dt, \quad x > a, \tag{1.3}$$

and

$$T_{2,b}(f,x) = \frac{\mathbf{1}}{\Gamma(\alpha)} \int_x^b (t-x)^{\alpha-1} f(t)dt, \quad x < b, \tag{1.4}$$

respectively, where

$$\Gamma(\alpha) = \int_0^\infty t^{\alpha-1} e^{-t} dt \tag{1.5}$$

is the Gamma function and when $\alpha = 0$, $T_{\mathbf{1},a}(f,x) = T_{2,b}(f,x) = f(x)$.

In 2013–2014, Sarikaya [8, 9] first used Definition 2 to obtain the fractional Hermite–Hadamard type inequalities:

Theorem 1. *Let $f : [a,b] \to (\mathbf{0},\infty), f \in L[a,b]$. If f is a convex function on $[a,b]$, then*

$$(1) \quad f\left(\frac{a+b}{\mathbf{2}}\right) \le \frac{\Gamma(\alpha+\mathbf{1})}{\mathbf{2}(b-a)^\alpha} \{T_{\mathbf{1},a}(f,b) + T_{2,b}(f,a)\} \le \frac{f(a)+f(b)}{\mathbf{2}}, \tag{1.6}$$

$$(2) \quad f\left(\frac{a+b}{\mathbf{2}}\right) \le \frac{\mathbf{2}^{\alpha-\mathbf{1}}\Gamma(\alpha+\mathbf{1})}{(b-a)^\alpha} \{T_{\mathbf{1},c}(f,b) + T_{2,c}(f,a)\} \le \frac{f(a)+f(b)}{\mathbf{2}}, \tag{1.7}$$

where $c = \frac{a+b}{\mathbf{2}}$.

Theorem 2 ([10]). *Let $f\colon [a,b] \to (\mathbf{0},\infty), f \in L[a,b].a \ge \mathbf{0}, \mathbf{1} < p < \infty, \frac{\mathbf{1}}{p} + \frac{\mathbf{1}}{q} = \mathbf{1}$. If $|f'|^q$ is a convex function on $[a,b]$, then*

$$(1) \quad \left| \frac{f(a)+f(b)}{\mathbf{2}} - \frac{\Gamma(\alpha+\mathbf{1})}{\mathbf{2}(b-a)^\alpha} \{T_{\mathbf{1},a}(f,b) + T_{\mathbf{2},b}(f,a)\} \right| \le \frac{b-a}{(p\alpha+\mathbf{1})^{1/p}} \left\{ \frac{|f'(a)|^q + |f'(b)|^q}{\mathbf{2}} \right\}^{1/q}. \tag{1.8}$$

(2) If $q \geq \mathbf{1}$, then

$$\left|\frac{\mathbf{2}^{\alpha-\mathbf{1}}\Gamma(\alpha+\mathbf{1})}{(b-a)^{\alpha}}\left\{T_{\mathbf{1},c}(f,b)+T_{2,c}(f,a)\right\}-f\left(\frac{a+b}{\mathbf{2}}\right)\right|$$
$$\leq \frac{b-a}{\mathbf{4}(\alpha+\mathbf{1})}\left(\frac{\mathbf{1}}{\mathbf{2}(\alpha+\mathbf{2})}\right)^{\mathbf{1}/q}\Big\{[(\alpha+\mathbf{1})|f'(a)|^q$$
$$+(\alpha+\mathbf{3})|f'(b)|^q]^{1/q}+[(\alpha+\mathbf{3})|f'(a)|^q+(\alpha+\mathbf{1})|f'(b)|^q]^{\mathbf{1}/q}\Big\},$$

where $c = \frac{a+b}{\mathbf{2}}$.

(3) If $|f'|$ is convex on $[a, b]$, then

$$\left|\frac{f(a)+f(b)}{\mathbf{2}}-\frac{\Gamma(\alpha+\mathbf{1})}{\mathbf{2}(b-a)^{\alpha}}\left\{T_{\mathbf{1},a}(f,b)+T_{2,b}(f,a)\right\}\right|$$
$$\leq \frac{b-a}{\mathbf{2}(\alpha+\mathbf{1})}\left(\mathbf{1}-\frac{\mathbf{1}}{\mathbf{2}^{\alpha}}\right)\Big(|f'(a)|+|f'(b)|\Big).$$

Theorem 3 (Kuang [4]). *Let* $f \in L[a,b]$, $g\colon [a,b] \to (0,\infty)$ *be an increasing function, and* $g \in AC[a,b], k, c, \alpha > \mathbf{0},\ a \geq \mathbf{0}, \mathbf{1} \leq p < \infty, \frac{\mathbf{1}}{p}+\frac{\mathbf{1}}{q}=\mathbf{1}$, *and for* $p=\mathbf{1}$, *define* $q=\infty$, $\frac{\mathbf{1}}{\infty}=\mathbf{0}$. *If* $\mathbf{1}<p<\infty$, *then*

(1) $$\left|T_{\mathbf{1},a}(f,x)+T_{2,b}(f,x)-\frac{\mathbf{1}}{\alpha\Gamma(\alpha)}\left\{(b-x)^{\alpha}+(x-a)^{\alpha}\right\}f(x)\right|$$
$$\leq \frac{\{B(\mathbf{1}/\alpha,q+\mathbf{1})\}^{1/q}}{\alpha^{\mathbf{1}+(1/q)}\Gamma(\alpha)}\left\{(x-a)^{\alpha+(1/q)}+(b-a)^{\alpha+(1/q)}\right\}\|f'\|_p,$$

(2) $$\left|T_{\mathbf{1},a}\left(f,\frac{a+b}{\mathbf{2}}\right)+T_{2,b}\left(f,\frac{a+b}{\mathbf{2}}\right)-\frac{\mathbf{2}^{\mathbf{1}-\alpha}(b-a)^{\alpha}}{\alpha\Gamma(\alpha)}f\left(\frac{a+b}{\mathbf{2}}\right)\right|$$
$$\leq \frac{\mathbf{2}^{(1/p)-\alpha}\{B(\mathbf{1}/\alpha,q+\mathbf{1})\}^{1/q}(b-a)^{\alpha+(1/q)}}{\alpha^{1+(1/q)}\Gamma(\alpha)}\|f'\|_p,$$

where

$$B(\alpha, \beta) = \int_0^1 t^{\alpha-1}(\mathbf{1}-t)^{\beta-1} dt$$

is the Beta function.

If $p = \mathbf{1}$, then

$$(3) \quad \left| T_{\mathbf{1},a}(f,x) + T_{\mathbf{2},b}(f,x) - \frac{\mathbf{1}}{\alpha\Gamma(\alpha)} \{(b-x)^\alpha + (x-a)^\alpha\} f(x) \right| \leq \frac{\mathbf{1}}{\alpha\Gamma(\alpha)} \{(x-a)^\alpha + (b-x)^\alpha\} \|f'\|_{\mathbf{1}},$$

$$(4) \quad \left| T_{\mathbf{1},a}\left(f, \frac{a+b}{\mathbf{2}}\right) + T_{\mathbf{2},b}\left(f, \frac{a+b}{\mathbf{2}}\right) - \frac{\mathbf{2}^{1-\alpha}(b-a)^\alpha}{\alpha\Gamma(\alpha)} f\left(\frac{a+b}{\mathbf{2}}\right) \right| \leq \frac{\mathbf{2}^{1-\alpha}}{\alpha\Gamma(\alpha)}(b-a)^\alpha \|f'\|_{\mathbf{1}}.$$

Definition 3 (cf. Ref. [3]). Let $f \in L[a,b]$, then Riemann–Liouville k-fractional integrals of f of order $\alpha > 0$ with $a \geq 0$ are defined by

$$T_{\mathbf{3},a}(f,x) = \frac{1}{k\Gamma_k(\alpha)} \int_a^x (x-t)^{(\alpha/k)-1} f(t) dt, \quad x > a, \tag{1.9}$$

and

$$T_{\mathbf{4},b}(f,x) = \frac{1}{k\Gamma_k(\alpha)} \int_x^b (t-x)^{(\alpha/k)-1} f(t) dt, \quad x < b, \tag{1.10}$$

respectively, where

$$\Gamma_k(\alpha) = \int_0^\infty t^{\alpha-1} e^{-\frac{t^k}{k}} dt \quad (\alpha > 0) \tag{1.11}$$

is the k-Gamma function. Also, $\Gamma(x) = \lim_{k \to 1} \Gamma_k(x)$, $\Gamma_k(\alpha) = k^{(\alpha/k)-1}\Gamma(\frac{\alpha}{k})$, and $\Gamma_k(\alpha + k) = \alpha\Gamma_k(\alpha)$.

It is well known that the Mellin transform of f is defined by

$$M(f,\alpha) = \int_0^{\infty} t^{\alpha-1} f(t)dt, \quad (\alpha > 0).$$

Therefore, the Mellin transform of the exponential function exp $\left\{-\frac{t^k}{k}\right\}$ is the k-Gamma function.

Theorem 4 (Kuang [4]). *Let $f \in L[a,b], g : [a,b] \to (0,\infty)$ be an increasing function, and $g \in AC[a,b], k, c, \alpha > \mathbf{0},\ a \geq \mathbf{0}, \mathbf{1} \leq p < \infty, \frac{\mathbf{1}}{p} + \frac{\mathbf{1}}{q} = \mathbf{1}$, and for $p = \mathbf{1}$, define $q = \infty$, $\frac{\mathbf{1}}{\infty} = \mathbf{0}$. If $\mathbf{1} < p < \infty$, then*

$$(1)\ \left|T_{\mathbf{3},a}(f,x) + T_{4,b}(f,x) - \frac{\mathbf{1}}{\alpha\Gamma_k(\alpha)}\left\{(b-x)^{\alpha/k} + (x-a)^{\alpha/k}\right\} f(x)\right|$$

$$\leq \frac{\{kB(k/\alpha, q+\mathbf{1})\}^{\mathbf{1}/q}}{\alpha^{\mathbf{1}+\mathbf{1}/q}\Gamma_k(\alpha)}\{(x-a)^{(\alpha/k)+(\mathbf{1}/q)} + (b-a)^{(\alpha/k)+(\mathbf{1}/q)}\}\|f'\|_p,$$

$$(2)\ \left|T_{\mathbf{3},a}\left(f, \frac{a+b}{\mathbf{2}}\right) + T_{4,b}\left(f, \frac{a+b}{\mathbf{2}}\right) - \frac{\mathbf{2}^{\mathbf{1}-(\alpha/k)}(b-a)^{(\alpha/k)}}{\alpha\Gamma_k(\alpha)} f\left(\frac{a+b}{\mathbf{2}}\right)\right|$$

$$\leq \frac{\mathbf{2}^{(\mathbf{1}/p)-(\alpha/k)}\{kB(k/\alpha, q+\mathbf{1})\}^{\mathbf{1}/q}(b-a)^{(\alpha/k)+(\mathbf{1}/q)}}{\alpha^{\mathbf{1}+(\mathbf{1}/q)}\Gamma_k(\alpha)}\|f'\|_p,$$

where

$$B(\alpha,\beta) = \int_{\mathbf{0}}^{\mathbf{1}} t^{\alpha-\mathbf{1}}(\mathbf{1}-t)^{\beta-\mathbf{1}}dt$$

is the Beta function.

If $p = \mathbf{1}$, then

$$(3)\quad \left|T_{\mathbf{3},a}(f,x) + T_{\mathbf{4},b}(f,x) - \frac{\mathbf{1}}{\alpha\Gamma_k(\alpha)}\right.$$

$$\left.\times \left\{(b-x)^{\alpha/k} + (x-a)^{\alpha/k}\right\} f(x)\right|$$

$$\leq \frac{\mathbf{1}}{\alpha\Gamma_k(\alpha)}\left\{(x-a)^{(\alpha/k)} + (b-x)^{(\alpha/k)}\right\}\|f'\|_{\mathbf{1}},$$

$$(4) \quad \left| T_{\mathbf{3},a}\left(f, \frac{a+b}{\mathbf{2}}\right) + T_{\mathbf{4},b}\left(f, \frac{a+b}{\mathbf{2}}\right) - \frac{\mathbf{2}^{\mathbf{1}-(\alpha/k)}(b-a)^{(\alpha/k)}}{\alpha\Gamma_k(\alpha)} f\left(\frac{a+b}{\mathbf{2}}\right) \right| \leq \frac{\mathbf{1}}{\alpha\Gamma_k(\alpha)} \left\{ (x-a)^{(\alpha/k)} + (b-x)^{(\alpha/k)} \right\} \left\| f' \right\|_{\mathbf{1}}.$$

Definition 4 (cf. Refs. [4, 5]). Let $t^r f(t) \in L^1[a,b]$, $a \geq 0$, then generalized Riemann–Liouville fractional integrals of f of order $\alpha \geq 0$ and $r \geq 0$ are defined by

$$T_{\mathbf{5},a}(f,x) = \frac{(r+\mathbf{1})^{1-\alpha}}{\Gamma(\alpha)} \int_a^x \left(x^{r+1} - t^{r+1}\right)^{\alpha-1} t^r f(t)dt, \quad x > a, \tag{1.12}$$

$$T_{\mathbf{6},b}(f,x) = \frac{(r+\mathbf{1})^{1-\alpha}}{\Gamma(\alpha)} \int_x^b \left(t^{r+\mathbf{1}} - x^{r+\mathbf{1}}\right)^{\alpha-\mathbf{1}} t^r f(t)dt, \quad x < b, \tag{1.13}$$

and there are also (k, r) fractional of Riemann-type operators:

$$T_{\mathbf{7},a}(f,x) = \frac{(r+\mathbf{1})^{\mathbf{1}-(\alpha/k)}}{k\Gamma_k(\alpha)} \int_a^x \left(x^{r+\mathbf{1}} - t^{r+\mathbf{1}}\right)^{(\alpha/k)-\mathbf{1}} t^r f(t)dt, \; x > a, \tag{1.14}$$

$$T_{\mathbf{8},b}(f,x) = \frac{(r+\mathbf{1})^{\mathbf{1}-(\alpha/k)}}{k\Gamma_k(\alpha)} \int_x^b \left(t^{r+\mathbf{1}} - x^{r+\mathbf{1}}\right)^{(\alpha/k)-\mathbf{1}} t^r f(t)dt, \; x < b, \tag{1.15}$$

respectively, where $\alpha, k > 0, r \geq 0, x \in [a,b]$.

In particular, if $r = 0$, then Definition 4 reduces to Definitions 2 and 3.

Definition 5 (cf. Ref. [4]). Let f be a conformable integrable function on $[a,b] \subset [0,\infty)$. The left-sided and right-sided generalized conformable fractional integrals T_9 and T_{10} of f of order $\alpha > 0$ are defined by

$$T_{\mathbf{9},a}(f,x) = \frac{\mathbf{1}}{\Gamma(\alpha)}\int_a^x \left(\frac{x^{r+s}-t^{r+s}}{r+s}\right)^{\alpha-\mathbf{1}} t^{r+s-\mathbf{1}} f(t)dt, \quad x > a, \tag{1.16}$$

and

$$T_{\mathbf{10},b}(f,x) = \frac{\mathbf{1}}{\Gamma(\alpha)}\int_x^b \left(\frac{t^{r+s}-x^{r+s}}{r+s}\right)^{\alpha-\mathbf{1}} t^{r+s-\mathbf{1}} f(t)dt, \quad x < b, \tag{1.17}$$

respectively, where $r \in R^1, s \in (0,1], r+s \neq 0$.

In particular, if $s = 1$, then T_9, T_{10} reduce to T_5, T_6, respectively.

Definition 6 (cf. Ref. [4]). Let $f \in L[a,b]$, $g\colon [a,b] \to (0,\infty)$ be an increasing function, and $g' \in C[a,b], \alpha > 0$. Then, g-Riemann–Liouville fractional integrals of f with respect to the function g on $[a,b]$ are defined by

$$T_{\mathbf{11},a}(f,x) = \frac{\mathbf{1}}{\Gamma(\alpha)}\int_a^x g'(t)[g(x)-g(t)]^{\alpha-\mathbf{1}} f(t)dt, \; x > a, \tag{1.18}$$

and

$$T_{\mathbf{12},b}(f,x) = \frac{\mathbf{1}}{\Gamma(\alpha)}\int_x^b g'(t)[g(t)-g(x)]^{\alpha-\mathbf{1}} f(t)dt, \quad x < b, \tag{1.19}$$

respectively.

Theorem 5 ([4, 11]). *Let $f, g : [a,b] \to (\mathbf{0},\infty), \mathbf{0} \leq a < b, f \in L[a,b], \alpha \in (0,1)$, $g' \in C[a,b], g$ be an increasing function. If f is an s-convex function on $[a,b]$, that is,*

$$f(tx + (\mathbf{1}-t)y) \leq t^s f(x) + (\mathbf{1}-t)^s f(y), \mathbf{0} < s \leq \mathbf{1},$$

then

$$\begin{aligned} \mathbf{2}^{s-1} f\left(\frac{a+b}{\mathbf{2}}\right) &\leq \frac{\Gamma(\alpha+\mathbf{1})}{\mathbf{2}(b-a)^\alpha}\left\{T_{11,a}(f\mathbf{o}g)\left(g^{-\mathbf{1}}(b)\right)\right. \\ &\quad \left. + T_{\mathbf{12},b}(f\mathbf{o}g)\left(g^{-\mathbf{1}}(a)\right)\right\} \\ &\leq \frac{\alpha}{\alpha+s}\left(\mathbf{3}-\frac{\mathbf{1}}{\mathbf{2}^{\alpha+s}}\right)\frac{f(a)+f(b)}{\mathbf{2}}. \end{aligned}$$

Definition 7 (cf. Refs. [1, 4]). Let $f \in L[a,b]$, $g\colon [a,b] \to (0,\infty)$ be an increasing function, and $g \in AC[a,b], k, c, \alpha > \mathbf{0}, a \geq \mathbf{0}$. Then, the generalized fractional integrals of f with respect to the function g on $[a,b]$ are defined by

$$T_{\mathbf{13}}(f,x) = \frac{c}{k\Gamma_k(\alpha)} \int_a^b g^{'}(t)[g(x) - g(t)]^{(\alpha/k)-\mathbf{1}} f(t)dt, \quad x > a. \tag{1.20}$$

Let

$$T_{\mathbf{14},a}(f,x) = \frac{c}{k\Gamma_k(\alpha)} \int_a^x g^{'}(t)[g(x) - g(t)]^{(\alpha/k)-\mathbf{1}} f(t)dt, \quad x > a, \tag{1.21}$$

and

$$T_{\mathbf{15},b}(f,x) = \frac{c}{k\Gamma_k(\alpha)} \int_x^b g^{'}(t)[g(t) - g(x)]^{(\alpha/k)-\mathbf{1}} f(t)dt, \quad x < b, \tag{1.22}$$

Respectively, where $\Gamma_k(\alpha)$ is defined by (1.11).

Definition 8 (cf. Refs. [4, 11]). Let $f \in L[a,b]$, $a \geq \mathbf{0}$, $\alpha > \mathbf{0}$. The Hadamard fractional integrals of order α are defined by

$$T_{\mathbf{16},a}(f,x) = \frac{\mathbf{1}}{\Gamma(\alpha)} \int_a^x [\log x - \log t]^{\alpha-1} t^{-1} f(t)dt, \quad x > a, \tag{1.23}$$

and

$$T_{\mathbf{17},b}(f,x) = \frac{\mathbf{1}}{\Gamma(\alpha)} \int_x^b [\log t - \log x]^{\alpha-\mathbf{1}} t^{-\mathbf{1}} f(t)dt, \quad x < b, \tag{1.24}$$

respectively.

Theorem 6 ([11]). *Let $f \in L[a,b], a \geq \mathbf{0}$. If f is a GA-convex function, that is,*

$$f\left(x^t y^{\mathbf{1}-t}\right) \leq t f(x) + (\mathbf{1} - t) f(y)$$

then

$$f(\sqrt{ab}) \leq \frac{\Gamma(\alpha + \mathbf{1})}{\mathbf{2}(\log(b/a))^{\alpha}} \{T_{\mathbf{16},a}(f,b) + T_{\mathbf{17},b}(f,a)\} \leq \frac{f(a) + f(b)}{\mathbf{2}}.$$

Based on the above-mentioned Theorems 1–6, the subject of fractional Hermite–Hadamard-type inequalities has received considerable attention in recent years. For more details, see Refs. [1, 4, 8], *etc. It is of great significance to develop and explore Hermite–Hadamard inequality for several different types of convex functions by using various fractional integrals.*

2. New Fractional Hermite–Hadamard-Type Inequalities

In what follows, $(X, \|\cdot\|)$ denotes the real normed linear spaces, D is a convex subset of $X, h : (0,1) \to (0,\infty)$ is a given function, which h not identical to 0, we write

$$E_n(p) = \left\{ x = (x_1, x_2, \ldots, x_n) : x_k \in R^1, 1 \leq k \leq n, \right.$$

$$\left. \|x\|_p = \left(\sum_{k=1}^{n} |x_k|^p \right)^{1/p}, 1 \leq p < \infty \right\},$$

$B(0,r) = \{x \in E_n(p) : \|x\|_p \leq r\}$. In particular, $E_n(2)$ is an n-dimensional Euclidean space R^n.

Definition 9. A functional $f\colon D \to (0,\infty)$ is called (β, λ, h)-exp convex if

$$\begin{aligned} f(\lambda\|x_{\mathbf{1}}\| + (\mathbf{1}-\lambda)\|x_2\|) &\leq h(\lambda e^{\beta\lambda}) f(\|x_{\mathbf{1}}\|) \\ &\quad + h((\mathbf{1}-\lambda)e^{\beta(\mathbf{1}-\lambda)}) f(\|x_2\|) \end{aligned} \tag{2.1}$$

$\forall x_1, x_2 \in D, \quad \forall \lambda, t \in [\mathbf{0}, \mathbf{1}]$.

In particular, if $\beta = \mathbf{0}$, then (2.1) reduces to an h-convex functional:

$$f(\lambda\|x_1\| + (\mathbf{1}-\lambda)\|x_2\|) \leq h(\lambda) f(\|x_1\|) + h(\mathbf{1}-\lambda) f(\|x_2\|). \tag{2.2}$$

If $D = [a,b] \subset [\mathbf{0},\infty), x_1, x_2 \in [a,b]$, then (2.1), (2.2) reduces to

$$f(\lambda x_1 + (\mathbf{1}-\lambda)x_2) \leq h(\lambda e^{\beta\lambda}) f(x_1) + h((\mathbf{1}-\lambda)e^{\beta(\mathbf{1}-\lambda)}) f(x_2) \tag{2.1-1}$$

and

$$f(\lambda x_1 + (\mathbf{1}-\lambda)x_2) \leq h(\lambda) f(x_1) + h(\mathbf{1}-\lambda) f(x_2) \tag{2.2-1}$$

(see Ref. [13]). If

$$f(\lambda x_1 + (\mathbf{1} - \lambda)x_2) \leq h(e^{\lambda} - \mathbf{1})f(x_1) + h(e^{\mathbf{1}-\lambda} - \mathbf{1})f(x_2), \quad (2.2\text{-}2)$$

$\forall x_1, x_2 \in [a, b], \lambda \in [\mathbf{0}, \mathbf{1}]$, we say that f is a h-exp convex function.

In particular, if

$$f(\lambda x_1 + (\mathbf{1} - \lambda)x_2) \leq \left(e^{\lambda} - \mathbf{1}\right) f(x_1) + \left(e^{1-\lambda} - \mathbf{1}\right) f(x_2),$$

then f reduces to exp convex function in Ref. [17]. If

$$f(\lambda x_1 + (\mathbf{1} - \lambda)x_2) \leq \left(e^{\lambda} - \mathbf{1}\right)^{-1} f(x_1) + (e^{1-\lambda} - \mathbf{1})^{-1} f(x_2),$$

then f is called a harmonically exp convex function.

We require the following lemmas to prove our main results.

Lemma 1. *Let $[a, b] \subset [0, \infty), f \in L[a, b], g\colon [a, b] \to [0, \infty)$ be an increasing function, and $g \in AC[a, b], k, \alpha, c > 0$. Then,*

$$\begin{aligned} T_{\mathit{13}}(f, x) &= [G_{\mathit{14}}(\mathbf{1}, x) + G_{15}(\mathbf{1}, x)]\, f(x) \\ &\quad + \int_x^b G_{\mathbf{15}}(\mathbf{1}, t) f'(t)dt - \int_a^x G_{\mathit{14}}(\mathbf{1}, t) f'(t)dt, \end{aligned} \quad (2.3)$$

where $G_{\mathbf{14}}(\mathbf{1}, t)$ and $G_{15}(\mathbf{1}, t)$ are defined by

$$G_{\mathit{14}}(\mathbf{1}, t) = \frac{c}{k\Gamma_k(\alpha)} \int_a^t g'(u)[g(x) - g(u)]^{(\alpha/k)-1} du \quad (2.4)$$

and

$$G_{\mathit{15}}(\mathit{1}, t) = \frac{c}{k\Gamma_k(\alpha)} \int_t^b g'(u)[g(u) - g(x)]^{(\alpha/k)-1} du \quad (2.5)$$

and $T_{\mathbf{13}}, T_{\mathbf{14},a}, T_{\mathbf{15},b}$ and $\Gamma_k(\alpha)$ are defined by (1.20), (1.21), (1.22) *and* (1.11), *respectively.*

Proof. From (1.21) and (1.22), we have

$$T_{\mathbf{14},a}(\mathbf{1},t) = \frac{c}{k\Gamma_k(\alpha)} \int_a^t g^{'}(u)[g(t)-g(u)]^{(\alpha/k)-\mathbf{1}} du$$

and

$$T_{\mathbf{15},b}(\mathbf{1},t) = \frac{c}{k\Gamma_k(\alpha)} \int_t^b g^{'}(u)[g(u)-g(t)]^{(\alpha/k)-\mathbf{1}} du.$$

Thus,

$$G_{\mathbf{14}}(\mathbf{1},x) = T_{\mathbf{14},a}(\mathbf{1},x), \quad G_{\mathbf{15}}(\mathbf{1},x) = T_{\mathbf{15},b}(\mathbf{1},x).$$

Then, making use of integration by parts, we get

$$\begin{aligned}\int_a^x G_{\mathbf{14}}(\mathbf{1},t)f'(t)dt &= G_{\mathbf{14}}(\mathbf{1},t)f(t)|_a^x - \frac{c}{k\Gamma_k(\alpha)} \\ &\quad \times \int_a^x g'(t)[g(x)-g(t)]^{(\alpha/k)-1} f(t)dt \\ &= G_{\mathbf{14}}(\mathbf{1},x)f(x) - T_{\mathbf{14},a}(f,x),\end{aligned}$$

which leads to

$$T_{\mathbf{14},a}(f,x) = G_{\mathbf{14}}(\mathbf{1},x)f(x) - \int_a^x G_{\mathbf{14}}(\mathbf{1},t)f^{'}(t)dt. \tag{2.6}$$

Similarly, we have

$$T_{\mathbf{15},b}(f,x) = G_{\mathbf{15}}(\mathbf{1},x)f(x) + \int_x^b G_{\mathbf{15}}(\mathbf{1},t)f^{'}(t)dt. \tag{2.7}$$

Hence, (2.3) follows from (1.20), (2.6) and (2.7). The proof is completed. □

Theorem 7. *Let* $[a,b] \subset [0,\infty), f' \in L^p[a,b], g : [a,b] \to [0,\infty)$ *be an increasing function, and* $g \in AC[a,b], k, \alpha, c > 0$. *If* $\mathbf{1} < p < \infty$, *then*

$$|T_{13}(f,x) - [G_{\mathbf{14}}(\mathbf{1},x) + G_{\mathbf{15}}(\mathbf{1},x)]\, f(x)|$$
$$\leq \left\{ \left(\int_a^x |G_{\mathbf{14}}(\mathbf{1},t)|^q \, dt \right)^{\mathbf{1}/q} + \left(\int_x^b |G_{\mathbf{15}}(\mathbf{1},t)|^q \, dt \right)^{\mathbf{1}/q} \right\} \left\| f' \right\|_p . \tag{2.8}$$

If $p = \mathbf{1}$, *then*

$$|T_{13}(f,x) - [G_{\mathbf{14}}(\mathbf{1},x) + G_{15}(\mathbf{1},x)]\, f(x)| \leq \|G\|_\infty \left\| f' \right\|_1 , \tag{2.9}$$

where

$$G(t) = G_{\mathbf{15}}(\mathbf{1},t)\varphi_{D_2}(t) - G_{\mathbf{14}}(\mathbf{1}.t)\varphi_{D_1}(t), \tag{2.10}$$

$D_1 = [a,x], D_2 = [x,b], \varphi_D$ *is the characteristic function of the set* D, *that is,*

$$\varphi_D(t) = \begin{cases} \mathbf{1}, & t \in D, \\ \mathbf{0}, & t \in D^c. \end{cases}$$

Proof. If $\mathbf{1} < p < \infty$, by using (2.6), (2.7) and the Hölder inequality, we get

$$\begin{aligned} &|T_{13}(f,x) - [G_{\mathbf{14}}(\mathbf{1},x) + G_{\mathbf{15}}(\mathbf{1},x)]\, f(x)| \\ &\quad = |[T_{\mathbf{14},a}(f,x) - G_{\mathbf{14}}(\mathbf{1},x)f(x)] + [T_{\mathbf{15},b}(f,x) - G_{15}(\mathbf{1},x)f(x)]| \\ &\quad \leq \int_a^x \left| G_{\mathbf{14}}(\mathbf{1},t)f'(t) \right| dt + \int_x^b \left| G_{\mathbf{15}}(\mathbf{1},t)f'(t) \right| dt \\ &\quad \leq \left\{ \left(\int_a^x |G_{\mathbf{14}}(\mathbf{1},t)|^q \, dt \right)^{\mathbf{1}/q} + \left(\int_x^b |G_{\mathbf{15}}(\mathbf{1},t)|^q \, dt \right)^{\mathbf{1}/q} \right\} \left\| f' \right\|_p . \end{aligned}$$

If $p = \mathbf{1}$, then

$$\begin{aligned} &|T_{\mathbf{13}}(f,x) - [G_{\mathbf{14}}(\mathbf{1},x) + G_{\mathbf{15}}(\mathbf{1},x)]\, f(x)| \\ &\quad = |[T_{\mathbf{14},a}(f,x) - G_{\mathbf{14}}(\mathbf{1},x)f(x)] + [T_{\mathbf{15},b}(f,x) - G_{\mathbf{15}}(\mathbf{1},x)f(x)]| \\ &\quad \leq \int_a^x \left| G_{\mathbf{14}}(\mathbf{1},t)f'(t) \right| dt + \int_x^b \left| G_{\mathbf{15}}(\mathbf{1},t)f'(t) \right| dt \leq \|G\|_\infty \left\| f' \right\|_{\mathbf{1}} . \end{aligned}$$

The proof is completed. □

Theorem 8. *Under the assumptions of Lemma* 1, *if* $|f'|^p$ *is* (β,λ,h)-*exp convex on* $[a,b]$, *and* $1<p<\infty$, $\frac{1}{p}+\frac{1}{q}=1$, *then*

$$\begin{aligned}
&\left|T_{\mathbf{13}}(f,x)-\left[G_{\mathbf{14}}(\mathbf{1},x)+G_{\mathbf{15}}(\mathbf{1},x)\right]f(x)\right|\\
&\quad\le\left\{\left(\int_x^b |G_{\mathbf{15}}(\mathbf{1},t)|^q\,dt\right)^{\mathbf{1}/q}(b-x)\left(\left|f'(b)\right|^p+\left|f'(x)\right|^p\right)\right.\\
&\qquad\left.+\left(\int_a^x |G_{\mathbf{14}}(\mathbf{1},t)|^q\,dt\right)^{\mathbf{1}/q}(x-a)\left(\left|f'(x)\right|^p+\left|f'(a)\right|^p\right)\right\}\\
&\qquad\times\int_0^{\mathbf{1}} h(te^{\beta t})dt. \qquad (2.11)
\end{aligned}$$

If $p=1$, *then*

$$\begin{aligned}
&\left|T_{\mathbf{13}}(f,x)-\left[G_{\mathbf{14}}(\mathbf{1},x)+G_{\mathbf{15}}(\mathbf{1},x)\right]f(x)\right|\\
&\quad\le\left\{(b-x)\left(\left|f'(b)\right|+\left|f'(x)\right|\right)\|G_{\mathbf{15}}\|_\infty+(x-a)(\left|f'(x)\right|\right.\\
&\qquad\left.+\left|f'(a)\right|)\|G_{\mathbf{14}}\|_\infty\right\}\int_0^1 h(te^{\beta t})dt. \qquad (2.12)
\end{aligned}$$

Proof. For $1<p<\infty$, by using Lemma 1, and the Hölder inequality, we obtain

$$\begin{aligned}
&\left|T_{\mathbf{13}}(f,x)-\left[G_{\mathbf{14}}(\mathbf{1},x)+G_{\mathbf{15}}(\mathbf{1},x)\right]f(x)\right|\\
&\quad=\left|\left[T_{\mathbf{14},a}(f,x)-G_{\mathbf{14}}(\mathbf{1},x)f(x)\right]+\left[T_{\mathbf{15},b}(f,x)\right.\right.\\
&\qquad\left.\left.-G_{\mathbf{15}}(\mathbf{1},x)f(x)\right]\right|\\
&\quad\le\int_a^x\left|G_{\mathbf{14}}(\mathbf{1},t)f'(t)\right|dt+\int_x^b\left|G_{\mathbf{15}}(\mathbf{1},t)f'(t)\right|dt\\
&\quad\le\left(\int_a^x |G_{\mathbf{14}}(\mathbf{1},t)|^q\,dt\right)^{\mathbf{1}/q}\left(\int_a^x \left|f'(t)\right|^p dt\right)^{\mathbf{1}/p}\\
&\qquad+\left(\int_x^b |G_{\mathbf{15}}(\mathbf{1},t)|^q\,dt\right)^{\mathbf{1}/q}\left(\int_x^b \left|f'(t)\right|^p dt\right)^{\mathbf{1}/p}. \qquad (2.13)
\end{aligned}$$

Setting $u = x + (b - x)t$ and using (β, λ, h)-exp convexity of $|f'|^p$ on $[a, b]$, we have

$$\begin{aligned}
\int_x^b | f'(u)|^p \, du &= (b - x) \int_0^1 \left|f'(tb + (1 - t)x)\right|^p dt \\
&\leq (b - x) \int_{\mathbf{0}}^{\mathbf{1}} \left\{h\left(te^{\beta t}\right) \left|f'(b)\right|^p + h\left((\mathbf{1} - t)e^{\beta(\mathbf{1}-t)}\right) \left|f'(x)\right|^p\right\} dt \\
&= (b - x) \left(\left|f'(b)\right|^p + \left|f'(x)\right|^p\right) \int_0^1 h\left(te^{\beta t}\right) dt. \qquad (2.14)
\end{aligned}$$

By letting $u = a + (x - a)t$ and similar arguments, we get

$$\begin{aligned}
\int_a^x \left|f'(u)\right|^p du &= (x - a) \int_0^{\mathbf{1}} \left|f'(tx + (\mathbf{1} - t)a)\right|^p dt \\
&\leq (x - a) \int_0^1 \left\{h\left(te^{\beta t}\right) \left|f'(x)\right|^p + h\left((\mathbf{1} - t)e^{\beta(\mathbf{1}-t)}\right) \left|f'(a)\right|^p\right\} dt \\
&= (x - a) \left(\left|f'(x)\right|^p + \left|f'(a)\right|^p\right) \int_0^{\mathbf{1}} h(te^{\beta t}) dt. \qquad (2.15)
\end{aligned}$$

Hence, (2.11) follows from (2.13), (2.14) and (2.15).

The case $p = 1$ can be treated analogously. The proof is completed. □

By giving particular values to the parameters in Theorem 8, we get the corresponding inequalities for different fractional integral operators. Such as, taking $k = c = \mathbf{1}$ in Theorem 8, then T_{13} reduces to $T_{11} + T_{12}$, thus, we get the following corollary:

Corollary 1. *Under the assumptions of Theorem* 8, *if* $k = c = \mathbf{1}$, *and* $1 < p < \infty, \frac{1}{p} + \frac{1}{q} = 1$, *then*

$$\begin{aligned}
&\left|T_{\mathbf{11},a}(f, x) + T_{\mathbf{12},b}(f, x) - \left[G_{\mathbf{11}}(\mathbf{1}, x) + G_{\mathbf{12}}(\mathbf{1}, x)\right] f(x)\right| \\
&\quad \leq \left\{\left(\int_x^b |G_{\mathbf{12}}(\mathbf{1}, t)|^q \, dt\right)^{\mathbf{1}/q} (b - x) \left(\left|f'(b)\right|^p + \left|f'(x)\right|^p\right)\right.
\end{aligned}$$

$$+\left(\int_a^x |G_{\mathbf{11}}(\mathbf{1},t)|^q\,dt\right)^{\mathbf{1}/q}(x-a)\left(\left|f'(x)\right|^p+\left|f'(a)\right|^p\right)\Bigg\}$$

$$\times\int_0^{\mathbf{1}} h\left(te^{\beta t}\right)dt \tag{2.16}$$

where $G_{11}(1,t)$ *and* $G_{12}(1,t)$ *are defined by*

$$G_{11}(1,t)=\frac{1}{\Gamma(\alpha)}\int_a^t g'(u)[g(x)-g(u)]^{\alpha-1}du \tag{2.17}$$

and

$$G_{12}(1,t)=\frac{1}{\Gamma(\alpha)}\int_t^b g'(u)[g(u)-g(x)]^{\alpha-1}du, \tag{2.18}$$

respectively.

Corollary 2. *Under the assumptions of Theorem* 8*, if* $g(t)=t$*,* $c=1$*, then* $T_{\mathbf{13}}$ *reduces to* $T_{4,b}+T_{3,a}$*, thus, for* $1<p<\infty$*,* $\frac{1}{p}+\frac{1}{q}=1$*, we get*

$$|T_{\mathbf{3},a}(f,x)+T_{\mathbf{4},b}(f,x)-[G_{\mathbf{3}}(\mathbf{1},x)+G_{\mathbf{4}}(\mathbf{1},x)]\,f(x)|$$

$$\leq\Bigg\{\left(\int_x^b |G_{\mathbf{4}}(\mathbf{1},t)|^q\,dt\right)^{\mathbf{1}/q}(b-x)\left(\left|f'(b)\right|^p+\left|f'(x)\right|^p\right),$$

$$+\left(\int_a^x |G_3(\mathbf{1},t)|^q\,dt\right)^{\mathbf{1}/q}(x-a)\left(\left|f'(x)\right|^p+\left|f'(a)\right|^p\right)\Bigg\}$$

$$\times\int_0^{\mathbf{1}} h\left(te^{\beta t}\right)dt \tag{2.19}$$

where $G_3(1,t)$ *and* $G_4(1,t)$ *are defined by*

$$G_3(\mathbf{1},t)=\frac{\mathbf{1}}{k\Gamma_k(\alpha)}\int_a^t (t-u)^{(\alpha/k)-\mathbf{1}}du \tag{2.20}$$

and

$$G_{\mathbf{4}}(\mathbf{1},t)=\frac{\mathbf{1}}{k\Gamma_k(\alpha)}\int_t^b (u-t)^{(\alpha/k)-\mathbf{1}}du, \tag{2.21}$$

respectively.

If $k = \mathbf{1}$ in Corollary 2, then $T_{\mathbf{13}}$ reduces to $T_{2,b} + T_{\mathbf{1},a}$. Thus, we get the following corollary:

Corollary 3. *Under the assumptions of Theorem 8, if $k = \mathbf{1}$, then $T_{\mathbf{13}}$ reduces to $T_{2,b} + T_{\mathbf{1},a}$, thus, for $1 < p < \infty, \frac{1}{p} + \frac{1}{q} = 1$, we get*

$$
\begin{aligned}
&\left|T_{\mathbf{1},a}(f,x) + T_{2,b}(f,x) - \left[G_{\mathbf{1}}(\mathbf{1},x) + G_2(\mathbf{1},x)\right] f(x)\right| \\
&\leq \left\{ \left(\int_x^b |G_2(\mathbf{1},t)|^q \, dt \right)^{\mathbf{1}/q} (b-x) \left(|f'(b)|^p + |f'(x)|^p \right) \right. \\
&\quad \left. + \left(\int_a^x |G_1(\mathbf{1},t)|^q \, dt \right)^{\mathbf{1}/q} (x-a) \left(|f'(x)|^p + |f'(a)|^p \right) \right\} \\
&\quad \times \int_0^{\mathbf{1}} h(te^{\beta t}) dt, \qquad (2.22)
\end{aligned}
$$

where $G_1(\mathbf{1},t)$ and $G_2(\mathbf{1},t)$ are defined by

$$
G_{\mathbf{1}}(\mathbf{1},t) = \frac{\mathbf{1}}{\Gamma(\alpha)} \int_a^t (t-u)^{\alpha-1} du \qquad (2.23)
$$

and

$$
G_{\mathbf{2}}(\mathbf{1},t) = \frac{\mathbf{1}}{\Gamma(\alpha)} \int_t^b (u-t)^{\alpha-\mathbf{1}} du, \qquad (2.24)
$$

respectively.

Remark. Using (2.2-2), we can similarly prove the following:

Theorem 9. *Under the assumptions of Lemma 1, if $|f'|^p$ is h-exp convex on $[a,b]$, and $1 < p < \infty, \frac{1}{p} + \frac{1}{q} = 1$, then*

$$
\begin{aligned}
&\left|T_{\mathbf{13}}(f,x) - \left[G_{\mathbf{14}}(\mathbf{1},x) + G_{\mathbf{15}}(\mathbf{1},x)\right] f(x)\right| \\
&\leq \left\{ \left(\int_x^b |G_{\mathbf{15}}(\mathbf{1},t)|^q \, dt \right)^{1/q} (b-x) \left(|f'(b)|^p + |f'(x)|^p \right) \right.
\end{aligned}
$$

$$+\left(\int_a^x |G_{\mathbf{14}}(\mathbf{1},t)|^q\,dt\right)^{\mathbf{1}/q}(x-a)\left(\left|f'(x)\right|^p+\left|f'(a)\right|^p\right)\Bigg\}$$
$$\times\int_0^{\mathbf{1}} h\left(e^t-\mathbf{1}\right)dt. \tag{2.25}$$

If $p=1$, *then*

$$\begin{aligned}&|T_{\mathbf{13}}(f,x)-[G_{\mathbf{14}}(\mathbf{1},x)+G_{\mathbf{15}}(\mathbf{1},x)]\,f(x)|\\&\quad\le\{(b-x)(|f'(b)|+|f'(x)|)\|G_{\mathbf{15}}\|_\infty+(x-a)(|f'(x)|\\&\qquad+|f'(a)|)\|G_{\mathbf{14}}\|_\infty\}.\end{aligned}$$

References

[1] J. C. Kuang. *Applied Inequalities*. 5th edn. Shangdong Science and Technology Press, Jinan, 2021 (in Chinese).

[2] J. C. Kuang. *Real and Functional Analysis (Continuation)*, Vol. 2 Higher Education Press, Beijing, 2015.

[3] J. C. Kuang. Recent advances of convexity theory and its inequalities. In T. M. Rassias and P. M. Pardalos (eds.), *Mathematical Analysis and Applications*. Springer, 2019.

[4] J. C. Kuang. New generalized convexity and their applications. In N. J. Daras and T. M. Rassias (eds.), *Approximation and Computation in Science and Engineering*. Springer, 2021.

[5] J. C. Kuang. New trapezoid type inequalities for generalized exponentially strongly convex functions. In T. M. Rassias (ed.), *Approximation Theory and Analytic Inequalities*. Springer, 2021.

[6] S. S. Dragomir. On some new inequalities of Hermite–Hadamard type for $m-$convex functions. *Tamkang J. Math.*, 33(1), 2002, 55–65.

[7] S. S. Dragomir. Inequalities of Hermite–Hadamard type for $\lambda-$convex functions on linear spaces. Preprint *RGMIA Res. Rep. Coll.*, 17, 2014. Article 13.

[8] J. Ni, *et al.* Study on Hermite–Hadamard-type inequalities using a new generalized fractional integral operator. *J. Inequal. Appl.*, 2023, 59.

[9] W. L. Wang. *Approaches to Prove Inequalities*. Harbin Institute of Technology Press, 2011 (in Chinese).

[10] G. Abbas, *et al.* Generalizations of some fractional integral inequalities via generalized Mittag-Leffler function. *J. Inequal. Appl.*, 2017, 121.

[11] I. Iscan. Jensen-Mercer inequality for GA-convex functions and some related inequalities. *J. Inequal. Appl.*, 2020, 212.

[12] S. S. Dragomir. Bounding the Čebyšev functional for a differentiable function whose derivative is h or λ-convex in absolute value and applications. *Le Matematiche*, 71(1), 2016, 173–202.

[13] S. Varošanec. On h-convexity. *J. Math. Anal. Appl.*, 326, 2007, 303–311.

[14] M. A. Noor, K. I. Noor, and M. U. Awan. Some new estimates of Hermite–Hadamard inequalities via harmonically $r-$convex functions. *Le Matematiche*, 71(2), 2016, 117–127.

[15] P. O. Mohammed, *et al.* New fractional inequalities of Hermite–Hadamard type involving the incomplete gamma functions. *J. Inequal. Appl.*, 2020, 263.

[16] H. Kadakal and M. Kadakal. Some Hermite–Hadamard type integral inequalities for functions whose first derivatives are trigonometrically ρ-convex. *Creat. Math. Inform.*, 2024(33), 175–183.

[17] M. E. Özdemir, M. Avci and E. Set. On some inequalities of Hermite–Hadamard type via m-convexity. *Appl. Math. Lett.*, 23, 2010, 1065–1070.

[18] S. Ozcan and I. Iscan. Some new Hermite–Hadamard type inequalities for s-convex functions and their applications. *J. Inequal. Appl.*, 2019, 201.

[19] M. E. Ozdemir, C. Yildiz, A. C. Akdemir and E. Set. On some inequalities for s-convex functions and applications. *J. Inequal. Appl.*, 2013, 333.

[20] Y. C. Kwun, *et al.* Hermite–Hadamard type inequalities for functions whose derivatives are $\eta-$convex via fractional integrals. *J. Inequal. Appl.*, 2019, 44.

[21] D. Khan, *et al.* Analysis of (P, m)-superquadratic function and related fractional integral inequalities with applications. *J. Inequal. Appl.*, 2024, 137.

[22] P. Tiwari and C. R. Bhatta. Hermite–Hadamard integral inequality for differentiable m-convex function. *Nepali Math. Sci. Rep.*, 41(1), 2024, 68–78.

[23] J. Li, *et al.* A study of Hermite–Hadamard inequalities via Caputo-Fabrizio fractional integral operators using strongly (s, m)-convex functions in the second sense. *J. Inequal. Appl.*, 2025, 17.

https://doi.org/10.1142/9789819820498_0011

Chapter 11

Fluid–Structure Interaction in Arterial Hemodynamics: A Mathematical Approach

Konstantina C. Kyriakoudi[*,||], Grigorios T. Chrimatopoulos[†,], Anastasios C. Felias[§,††], Panagiotis D. Linardopoulos[*,‡‡], Michail Th. Rassias[¶,§§], Efstratios E. Tzirtzilakis[‡,¶¶] and Michail A. Xenos[*,|||]**

[*]*Department of Mathematics, University of Ioannina, Ioannina, 45110, Epirus, Greece*
[†]*Department of Mechanical Engineering, University of the Peloponnese, 26334, Patras, Greece*
[‡]*Department of Medicine, University of Ioannina, Ioannina, 45110, Epirus, Greece*
[§]*Department of Mathematics and Engineering Sciences, Hellenic Military Academy, 16673, Attiki, Greece*
[¶]*Department of Civil Engineering, University of the Peloponnese, 26334, Patras, Greece*
[||]*k.kyriakoudi@uoi.gr*
[**]*g.chrimatopoulos@go.uop.gr*
[††]*tsfelias1995@gmail.com*
[‡‡]*plinardopoulos@gmail.com*
[§§]*mrassias@sse.gr*
[¶¶]*etzirtzilakis@uop.gr*
[|||]*mxenos@uoi.gr*

This chapter focuses on a variety of pathological vessels that are of interest in hemodynamics. In fluid mechanics, partial differential equations are essential for describing physical phenomena, while Navier–Stokes equations are employed to describe the dynamics of fluid flow, providing a deeper comprehension of hemodynamics. The two-dimensional and three-dimensional formulations of the Navier–Stokes equations are introduced both aiming to simulate blood flow in such arteries. The numerical solution of the yielding systems requires the use of generalized curvilinear coordinates and the finite volume method. Finally, a comparative analysis is provided by the presentation of blood flows hydrodynamic cases through stenoses and aneurysms.

Keywords: Navier–Stokes, fluid structure interaction, finite volume method, stenotic arteries, aneurysms.

1. Introduction

The mathematical and numerical modeling of the cardiovascular and arterial systems is a research topic that has attracted remarkable interest from the mathematical community because of its intrinsic mathematical difficulty and the increasing impact of cardiovascular diseases worldwide. According to recent studies, cardiovascular diseases are the major cause of death worldwide, leading to more than 17.3 million deaths per year, a number that is expected to grow to more than 23.6 million by 2030. This corresponds to nearly half of the total amount of deaths in Europe [19].

More specifically, blood flow in arteries and the cardiovascular system may be studied quantitatively, which will offer scientists a set of tools to a better understanding of the underlying human physiology. The modeling can provide useful delineations of how the heart, vessels and blood itself interact with each other, aiding scientists to understand fundamental physiological processes and offer insights into the normal functioning of the cardiovascular system [9]. For instance, it can predict how pressure, flow rate, wall shear stress, cross-sectional area of a vessel and the compliance of its wall affect the system.

Moreover, the modeling process can contribute to various fields, such as diagnosing and treating various cardiovascular diseases by identifying regions of disturbed flow, potential blockages in the lumen or abnormal pressure gradients, while doctors will be able to use this information to make proper informed decisions about treatment options, as stent placement, bypass surgery or drug administration. The creation, optimization and further technological advancements of medical device such as stents, artificial heart valves and ventricular assist devices are significantly owed to the development of such models [23].

However, mathematical modeling itself imposes tremendous challenges due to the complexity of the cardiocirculatory system, the multiscale nature of the physiological processes involved and the necessity to devise computational methods that are stable, reliable and efficient. Despite that, with the continuous emergence of new diagnostic tools, the advancements in medical imaging and the technological advancements that keep on developing, the emergence of patient-specific computational fluid dynamics (CFD) has paved the way for the new field of computer-aided diagnostics [40].

Undoubtedly, computational fluid dynamics (CFD) is an uprising scientific field that utilizes the computational power of computers while using a conglomeration of software tools that calculate fluid-flow phenomena-related quantities numerically [12, 38] or even analytically. Those tools specifically include integrated solvers which use appropriate algorithms to solve large-scale, nonlinear and coupled systems of equations, which often occur from conservation laws, such as conservation of mass, momentum and energy equations.

In the final analysis, scientists are able to not only simulate more intricate physiological problems and diseases in significantly reduced computational time, predict and receive more accurate results of blood flow characteristics but also draw a path toward a patient-specific medical approach which would surpass various obstacles that the so-far invasive techniques set, as the mathematical models of the cardiovascular system evolves and the implementation of such models in CFD software advances.

2. Pathological Arteries

2.1. *Stenotic arteries*

Arterial stenosis is a common cardiovascular disease where a narrowing of the lumen disturbs and impedes blood flow in a local region of the blood vessel. Stenotic blood vessels create tangent stress pressure, which lessens the arterial side medium and thus it is prone to aneurysm creation [11, 30]. The arteries that are usually affected from stenosis that are large-to-mid sized arteries, such as the aorta.

This pathological condition can cause major health issues, such as thrombosis, sudden myocardial infarction and cerebral strokes; therefore, it is critical for the medical community to be able to investigate this condition. Fluid mechanics and, especially, hemodynamics can substantially contribute to this effort by modeling the blood flow process in stenotic arteries and extract significant results, for instance, by incorporating various degrees of occlusion in numerical simulations of a stenotic blood vessel and by studying the blood flow and pressure in pre-stenotic and post-stenotic areas.

Regarding the fundamental characteristics and the configuration of a stenosis, four are on the scope of interest:

- The location that it appears (in which segment of the arterial vessel);
- The extension it has (i.e. multiple occasions of stenosis in a vessel);
- The length of stenosis (with a threshold of characterization of small/big at 10 mm length);
- The shape and the geometric features of the diseased segment (including its diameter, the smoothness or lack of smoothness in contours, tapering, branchings, post-stenotic dilation and more) [30].

2.2. *Aneurysmatic arteries*

An aneurysm is a bulge in a blood vessel caused by a weakness in the wall of the vessel. The arterial wall can be weakened extensively by elevated blood pressure. There is a variety of arteries that can be affected by an aneurysm leading to pathological blood conditions. The most common locations for an aneurysm to be formed are the arteries supplying the brain and the heart, as well as the aorta. The aneurysm of the descending aorta is called an abdominal aortic aneurysm (AAA). An AAA does not usually threaten health, but there is a risk that a larger aneurysm could burst and rupture, which could be fatal in case of massive internal bleeding. Most unruptured aneurysms are asymptomatic and cannot be detected with the same ease; however, it could impede blood circulation to other tissues and organs. Suspicion of an unruptured aneurysm can be confirmed through diagnostic imaging, such as an magnetic resonance imaging (MRI) or a computed tomography (CT) scan [5, 22].

Abdominal aortic aneurysm (AAA) refers to the presence of an aneurysm in the lower part of the aorta and comprises the vast majority of aortic aneurysms, as the most common locations of aneurysm formation are the infrarenal abdominal aortic regions [14, 20]. The presence of an AAA can be detected either incidentally or after its rupture, while the enlargement of the AAA can be either observed as a stable increase of the diameter of the vessel or as a rapid enlargement in an unspecified period of time.

The significance of AAA in a clinical context stems from the elevated mortality risk linked to the rupture of the aneurysm as the constant progression of the disease ultimately leads to a rupture, since approximately 50–75% of individuals experiencing a ruptured AAA do not survive [36]. With the aid of simulation softwares where Fluid–Structure Interaction (FSI) is implemented, a detailed CFD model featuring

- parameters affecting the formation of AAAs,
- parameters affecting the weaking of the arterial wall,
- reconstructed aortic geometries and
- blood flow, pressure and wall-shear stresses, is achievable

can help medical experts monitor the gradual progression of a healthy aorta to a diseased one and identify a specific region of the arterial segment as a prone location to aneurysm formation, in an *a priori* fashion, which can be used as a medical predictive/prognostic tool.

2.3. *Other types of arteries*

The study of fluid motion in the maternal arterial system is a crucial aspect of maternal-fetal medicine and obstetrics. Understanding blood flow dynamics in the maternal arterial system is essential for monitoring and assessing the well-being of both the mother and the developing fetus during pregnancy. In this subsection, focus will be set on the applicability of CFD to blood flow in the uterine and umbilical arteries.

The uterine arteries play a vital role in supplying blood to the uterus, where the placenta develops. As vascular resistance in the uterine arterioles decreases, blood flow rises, causing the uterine artery diameter to expand throughout pregnancy. Doppler ultrasound is commonly used to assess blood flow, providing with waveforms, in the uterine arteries, giving useful insights about resistance and pulsatility indices. Abnormalities in these indices may indicate conditions such as preeclampsia or intrauterine growth restriction. Constructing a suitable CFD model usually relies on data obtained from ultrasound and/or CT-scan and magnetic resonance imaging (MRI).

In an equally important manner, the umbilical artery, being a paired artery, plays a crucial role in fetal circulation transporting oxygen-depleted blood from the fetus back to the placenta for oxygen and nutrient exchange. As its name suggests, it is part of the umbilical cord, which is the lifeline between the developing fetus and the mother. As the artery widens during development, there is a concurrent decrease in resistance to blood flow. Studying the umbilical arterial flow proves to be essential for assessing fetal well-being and detecting potential issues during pregnancy.

A mathematical modeling of these arteries usually centers around the Navier–Stokes equations, transformed to a generalized curvilinear domain accounting for the cylindrical and helical-like geometries of these vessels. Developing algorithms for simulating blood flow enables the ability to make observations for all the flow related parameters (pressure, velocity, wall shear stress, etc.), offering valuable insights into the impact of alterations in arterial geometry.

A CFD model was developed to extract key hemodynamic parameters from the maternal circulation, including the common iliac, internal and external iliac and uterine arteries [24]. The obtained results indicate that during pregnancy, the maternal arterial system undergoes adaptive changes to adequately supply blood to the fetus. The uterine artery's diameter and blood velocity increase as pregnancy progresses. While blood pressure remains constant, wall shear in uterine arteries is higher compared to other arteries during a pregnancy of no implications [24]. A three-dimensional CFD model of the coiled umbilical artery is also utilized in Ref. [25]. The purpose of this research is to investigate the individual and combined impacts of umbilical cord insertion, cord length and arterial diameter on the fluid dynamics of the umbilical artery. The study concludes that models of hypercoiled cords exhibit significant pressure gradients in blood flow. Additionally, hypercoiled cords with small diameters may not be viable, while longer hypercoiled cord models display reduced blood flow velocity [25]. In another novel study [10], a 3D computational model is developed to study blood flow in the umbilical arteries, evaluating the fetal-maternal heat exchange. Results showed a significant temperature drop in those arteries, where the helical geometry played a key role, transferring heat from the fetus to the placenta. The researchers concluded on fetal thermoregulation through the umbilical cord, preventing heat return to the mother [10].

3. Mathematical Model

This section outlines the mathematical modelling of blood flow within a domain, such as a vessel, in both two and three dimensions. Pathological vessels exhibit complex geometries, generating difficulties. To overcome these difficulties, the adoption of an alternative coordinates system is essential. In this chapter, the finite volume method is employed for discretization of the domain and solving the equations.

3.1. *The two-dimensional problem*

Blood flow in a pulsating artery is described by fluid dynamics principles. This part of this study focuses on a 2D formulation of a simplified abdominal aortic aneurysm (AAA). To convert the governing equations into a moving body-fitted coordinate system, generalized curvilinear coordinates (GCCs) were utilized [31]. Although density, ρ, remains constant in the equations, it is retained due to the adoption of specific methodologies that presume highly restricted compressibility in the fluid. This choice enables the integration of volume changes, as defined by the Jacobian of each control volume within this framework. This approach aligns with the work of various researchers [28, 38]. Continuity and momentum equations in Cartesian coordinates for two-dimensional, Newtonian fluid flow can be written in dimensional form as in (3.1)–(3.3) [28].

continuity equation

$$\frac{\partial \rho}{\partial \tilde{t}} + \frac{\partial (\rho \tilde{u})}{\partial \tilde{x}} + \frac{\partial (\rho \tilde{v})}{\partial \tilde{y}} = 0, \tag{3.1}$$

x-momentum equation

$$\frac{\partial(\rho\tilde{u})}{\partial\tilde{t}}+\frac{\partial(\rho\tilde{u}\tilde{u})}{\partial\tilde{x}}+\frac{\partial(\rho\tilde{u}\tilde{v})}{\partial\tilde{y}}=-\frac{\partial\tilde{p}}{\partial\tilde{x}}+\left[\frac{\partial}{\partial\tilde{x}}\left(\mu\frac{\partial\tilde{u}}{\partial\tilde{x}}\right)+\frac{\partial}{\partial\tilde{y}}\left(\mu\frac{\partial\tilde{u}}{\partial\tilde{y}}\right)\right], \tag{3.2}$$

y-momentum equation

$$\frac{\partial(\rho\tilde{v})}{\partial\tilde{t}}+\frac{\partial(\rho\tilde{u}\tilde{v})}{\partial\tilde{x}}+\frac{\partial(\rho\tilde{v}\tilde{v})}{\partial\tilde{y}}=-\frac{\partial\tilde{p}}{\partial\tilde{y}}+\left[\frac{\partial}{\partial\tilde{x}}\left(\mu\frac{\partial\tilde{v}}{\partial\tilde{x}}\right)+\frac{\partial}{\partial\tilde{y}}\left(\mu\frac{\partial\tilde{v}}{\partial\tilde{y}}\right)\right], \tag{3.3}$$

where $\tilde{u}$ and $\tilde{v}$ are the components of the velocity vector, $\bar{q}=(\tilde{u},\tilde{v})$, $\tilde{p}$ is the pressure, ρ is the density and μ is the viscosity of the fluid.

The system of equations (3.1)–(3.3) can be converted to a dimensionless form of equations (3.5)–(3.7), by introducing the parameters (3.4):

$$x=\frac{\tilde{x}}{R},\quad y=\frac{\tilde{y}}{R},\quad t=\frac{\tilde{t}}{R/u_0},\quad u=\frac{\tilde{u}}{u_0},$$
$$v=\frac{\tilde{v}}{u_0},\quad p=\frac{\tilde{p}}{\rho u_0^2},\quad c=\frac{\rho}{\rho_0}, \tag{3.4}$$

where u_0 is the characteristic inlet velocity and R is the inlet length of the aneurysmal geometry, ρu_0^2 is the dynamic pressure and c is a compressibility constant. The Reynolds number is introduced to the momentum equations in the dimensionless form, while the continuity equation remains the same.

continuity equation, dimensionless

$$\frac{\partial c}{\partial t}+\frac{\partial(cu)}{\partial x}+\frac{\partial(cv)}{\partial y}=0, \tag{3.5}$$

x-momentum equation, dimensionless

$$\frac{\partial u}{\partial t}+\frac{\partial(uu)}{\partial x}+\frac{\partial(uv)}{\partial y}=-\frac{\partial p}{\partial x}+\frac{\partial}{\partial x}\left(\frac{1}{Re}\frac{\partial u}{\partial x}\right)+\frac{\partial}{\partial y}\left(\frac{1}{Re}\frac{\partial u}{\partial y}\right), \tag{3.6}$$

y-momentum equation, dimensionless

$$\frac{\partial v}{\partial t}+\frac{\partial (uv)}{\partial x}+\frac{\partial (vv)}{\partial y}=-\frac{\partial p}{\partial y}+\frac{\partial}{\partial x}\left(\frac{1}{Re}\frac{\partial v}{\partial x}\right)+\frac{\partial}{\partial y}\left(\frac{1}{Re}\frac{\partial v}{\partial y}\right), \tag{3.7}$$

where u, v are the dimensionless components of the fluid velocity, p is the dimensionless fluid pressure and Re is the *Reynolds* number, defined as $Re=\frac{u_0 R}{\nu}$, where ν is the kinematic viscosity of the fluid, R is the inlet length of the geometry under consideration and u_0 is a characteristic flow velocity, e.g. the maximum inlet velocity of the parabolic flow.

Applying the generalized curvilinear coordinates (GCCs) transformation, the system of equations is written in a body-fitted approach. This is possible due to a local transformation from one domain, the physical domain, to a normalized one, as in Figure 1. The function of the transformation has to be a one-to-one thus ensuring the existence of the inverse. In this approach, the metrics of the transformation, $x_\xi, x_\eta, y_\xi, y_\eta$, can be computed locally, on each 2D finite volume. The determinant of the inverse Jacobian of the transformation for each volume is given by the expression, $J=x_\xi y_\eta - x_\eta y_\xi$. First-order derivatives with respect to x and y for a transport quantity, ϕ, e.g. the u, v velocity components, can be written as

$$\begin{aligned}\frac{\partial\phi}{\partial x}&=\frac{\partial\phi}{\partial\xi}\frac{\partial\xi}{\partial x}+\frac{\partial\phi}{\partial\eta}\frac{\partial\eta}{\partial x}=\xi_x\frac{\partial\phi}{\partial\xi}+\eta_x\frac{\partial\phi}{\partial\eta}=\frac{y_\eta}{J}\frac{\partial\phi}{\partial\xi}-\frac{y_\xi}{J}\frac{\partial\phi}{\partial\eta},\\ \frac{\partial\phi}{\partial y}&=\frac{\partial\phi}{\partial\xi}\frac{\partial\xi}{\partial y}+\frac{\partial\phi}{\partial\eta}\frac{\partial\eta}{\partial y}=\xi_y\frac{\partial\phi}{\partial\xi}+\eta_y\frac{\partial\phi}{\partial\eta}=-\frac{x_\eta}{J}\frac{\partial\phi}{\partial\xi}+\frac{x_\xi}{J}\frac{\partial\phi}{\partial\eta}.\end{aligned} \tag{3.8}$$

Due to equation (3.8), and since the fluid is considered incompressible, $\rho=$ constant, and $c\approx 1$ (almost incompressible) [12, 38], the equations (3.5)–(3.7) can be written in curvilinear coordinates, $\xi=\xi(x,y)$, $\eta=\eta(x,y)$, and in dimensionless form as in equations (3.9)–(3.11) [39].

continuity equation

$$\frac{\partial J}{\partial t}+\frac{\partial U}{\partial\xi}+\frac{\partial V}{\partial\eta}=0, \tag{3.9}$$

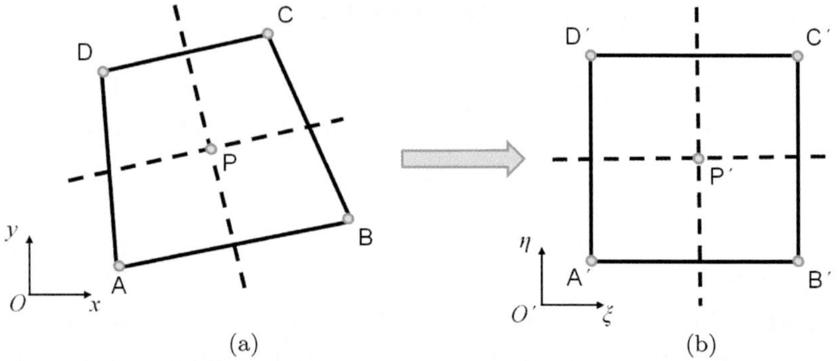

Fig. 1. The local transformation from the physical to the transformed domain. (a) Physical domain, x, y-coordinates, (b) Transformed domain, η, ξ-coordinates.

x-momentum equation

$$\frac{\partial (Ju)}{\partial t} + \frac{\partial (Uu)}{\partial \xi} + \frac{\partial (Vu)}{\partial \eta}$$
$$= -\left(y_\eta \frac{\partial p}{\partial \xi} - y_\xi \frac{\partial p}{\partial \eta}\right) + \frac{\partial}{\partial \xi}\left[\frac{1}{JRe}\left(q_1 \frac{\partial u}{\partial \xi} - q_2 \frac{\partial u}{\partial \eta}\right)\right]$$
$$+ \frac{\partial}{\partial \eta}\left[\frac{1}{JRe}\left(q_3 \frac{\partial u}{\partial \eta} - q_2 \frac{\partial u}{\partial \xi}\right)\right], \tag{3.10}$$

y-momentum equation

$$\frac{\partial (Jv)}{\partial t} + \frac{\partial (Uv)}{\partial \xi} + \frac{\partial (Vv)}{\partial \eta}$$
$$= -\left(x_\xi \frac{\partial p}{\partial \eta} - x_\eta \frac{\partial p}{\partial \xi}\right) + \frac{\partial}{\partial \xi}\left[\frac{1}{JRe}\left(q_1 \frac{\partial v}{\partial \xi} - q_2 \frac{\partial v}{\partial \eta}\right)\right]$$
$$+ \frac{\partial}{\partial \eta}\left[\frac{1}{JRe}\left(q_3 \frac{\partial v}{\partial \eta} - q_2 \frac{\partial v}{\partial \xi}\right)\right]. \tag{3.11}$$

Based on the Eulerian–Lagrangian approach, the dimensionless form of the continuity and momentum equations in Cartesian coordinates, equations (3.9)–(3.11), are transformed in a body-fitted approach [4, 27]. In equations (3.9)–(3.11), U and V are the contravariant velocity components defined in equation (3.12), taking under consideration the arbitrary motion of the domain; J is the determinant of the inverse Jacobian of the transformation from the physical moving domain to the normalized one:

$$U = (u - \dot{x})\, y_\eta - (v - \dot{y})\, x_\eta, \quad V = (v - \dot{y})\, x_\xi - (u - \dot{x})\, y_\xi, \tag{3.12}$$

where u, v are the fluid velocities and the arbitrary grid motion velocities and $\dot{x}$, $\dot{y}$ are given by a difference scheme [28]. Utilizing the Eulerian–Lagrangian approach, an interface tracking method is introduced, for moving objects or phase boundaries by expressing the motion of the interfaces and the computational domain. The quantities q_1, q_2, q_3, entering in equations (3.9)–(3.11), are defined as in equation (3.13):

$$q_1 = x_\eta^2 + y_\eta^2, \quad q_2 = x_\xi x_\eta + y_\xi y_\eta, \quad q_3 = x_\xi^2 + y_\xi^2, \tag{3.13}$$

where x_ξ, x_η, y_ξ and y_η are the metrics of the transformation that can be computed locally on each cell volume, Figure 1. In the arbitrary Euler–Lagrange grid motion, the inverse Jacobian of the transformation, J, is a function of time, and the transformed velocities are functions of the physical u, v fluid velocities and of the velocities of the grid motion, $\dot{x}$, $\dot{y}$ [28].

3.1.1. *Finite volume method*

The numerical method of the finite volume is applied in solving partial differential equations, where the values of the conserved variables are calculated, averaged across the volume. In this case, it would be natural to apply the collocated approach to place all state variables on the grid points.

The finite volume method provides second-order accuracy for spatial derivatives in the fluid domain and first-order accuracy for time derivative, leading to an implicit scheme, unconditionally stable [7, 8]. This formulation ensures that convective terms achieve first-order accuracy by minimizing their influence, while the diffusion terms remain unchanged (second-order accuracy), as the velocity increases in the flow field. The discretized form of equations (3.14)–(3.16) utilizing the finite volume approach is shown in the following for the continuity and momentum equations:

continuity equation

$$\frac{J - J_0}{\Delta t} + U_e - U_w + V_n - V_s = 0, \tag{3.14}$$

x-momentum equation

$$\begin{aligned}\frac{Ju - J_0 u_0}{\Delta t} &+ \left(Uu - \frac{1}{J\,\mathrm{Re}}\left(q_1\frac{\partial u}{\partial \xi} - q_2\frac{\partial u}{\partial \eta}\right) + \frac{\partial y}{\partial \eta}p\right)_e \\ &- \left(Uu - \frac{1}{J\,\mathrm{Re}}\left(q_1\frac{\partial u}{\partial \xi} - q_2\frac{\partial u}{\partial \eta}\right) + \frac{\partial y}{\partial \eta}p\right)_w \\ &+ \left(Vu - \frac{1}{J\,\mathrm{Re}}\left(-q_2\frac{\partial u}{\partial \xi} + q_3\frac{\partial u}{\partial \eta}\right) - \frac{\partial y}{\partial \xi}p\right)_n \\ &- \left(Vu - \frac{1}{J\,\mathrm{Re}}\left(-q_2\frac{\partial u}{\partial \xi} + q_3\frac{\partial u}{\partial \eta}\right) - \frac{\partial y}{\partial \xi}p\right)_s = 0, \end{aligned}\tag{3.15}$$

y-momentum equation

$$\begin{aligned}\frac{Jv - J_0 v_0}{\Delta t} &+ \left(Uv - \frac{1}{J\,\mathrm{Re}}\left(q_1\frac{\partial v}{\partial \xi} - q_2\frac{\partial v}{\partial \eta}\right) + \frac{\partial x}{\partial \eta}p\right)_e \\ &- \left(Uv - \frac{1}{J\,\mathrm{Re}}\left(q_1\frac{\partial v}{\partial \xi} - q_2\frac{\partial v}{\partial \eta}\right) + \frac{\partial x}{\partial \eta}p\right)_w \\ &+ \left(Vv - \frac{1}{J\,\mathrm{Re}}\left(-q_2\frac{\partial v}{\partial \xi} + q_3\frac{\partial v}{\partial \eta}\right) - \frac{\partial x}{\partial \xi}p\right)_n \\ &- \left(Vv - \frac{1}{J\,\mathrm{Re}}\left(-q_2\frac{\partial v}{\partial \xi} + q_3\frac{\partial v}{\partial \eta}\right) - \frac{\partial x}{\partial \xi}p\right)_s = 0, \end{aligned}\tag{3.16}$$

where the subscript 0 refers to the previous time level, and the subscripts e, w, n, s are referred to the four faces of each finite volume. The Cartesian components of the mesh velocity vector $(\dot{x}, \dot{y})$ are approximated by a first-order backward difference scheme, $\dot{x} = \frac{x - x_0}{\Delta t}$, $\dot{y} = \frac{y - y_0}{\Delta t}$, where x_0 and y_0 are the previous time step grid positions, x and y are the current time step grid positions, Δt is the time step [28].

3.2. *The three-dimensional mathematical formulation*

The three-dimensional mathematical model employs a similar approach, integrating generalized curvilinear coordinates and the finite volume method. To obtain the governing fluid flow equations in generalized curvilinear coordinates, allowing the study of fluid flow in complex domains such as aneurysms, cylinders or domains with a complicated geometry, the chain rule is applied to the unknown variables:

$$\begin{aligned}
\frac{\partial u}{\partial x} &= \xi_x \frac{\partial u}{\partial \xi} + \eta_x \frac{\partial u}{\partial \eta} + \zeta_x \frac{\partial \zeta}{\partial x}, \\
\frac{\partial v}{\partial y} &= \xi_y \frac{\partial v}{\partial \xi} + \eta_y \frac{\partial v}{\partial \eta} + \zeta_y \frac{\partial \zeta}{\partial y}, \\
\frac{\partial w}{\partial z} &= \xi_z \frac{\partial w}{\partial \xi} + \eta_z \frac{\partial w}{\partial \eta} + \zeta_z \frac{\partial \zeta}{\partial z}.
\end{aligned} \tag{3.17}$$

The Jacobian matrices are related as

$$\begin{pmatrix} \frac{\partial u}{\partial x} & \frac{\partial u}{\partial y} & \frac{\partial u}{\partial z} \\ \frac{\partial v}{\partial x} & \frac{\partial v}{\partial y} & \frac{\partial v}{\partial z} \\ \frac{\partial w}{\partial x} & \frac{\partial w}{\partial y} & \frac{\partial w}{\partial z} \end{pmatrix} = \begin{pmatrix} \frac{\partial u}{\partial \xi} & \frac{\partial u}{\partial \eta} & \frac{\partial u}{\partial \zeta} \\ \frac{\partial v}{\partial \xi} & \frac{\partial v}{\partial \eta} & \frac{\partial v}{\partial \zeta} \\ \frac{\partial w}{\partial \xi} & \frac{\partial w}{\partial \eta} & \frac{\partial w}{\partial \zeta} \end{pmatrix} \begin{pmatrix} \frac{\partial \xi}{\partial x} & \frac{\partial \xi}{\partial y} & \frac{\partial \xi}{\partial z} \\ \frac{\partial \eta}{\partial x} & \frac{\partial \eta}{\partial y} & \frac{\partial \eta}{\partial z} \\ \frac{\partial \zeta}{\partial x} & \frac{\partial \zeta}{\partial y} & \frac{\partial \zeta}{\partial z} \end{pmatrix}, \tag{3.18}$$

where the definition of the Jacobian matrix as well as the inverse Jacobian matrix are given in the following:

$$J = \begin{pmatrix} \frac{\partial \xi}{\partial x} & \frac{\partial \xi}{\partial y} & \frac{\partial \xi}{\partial z} \\ \frac{\partial \eta}{\partial x} & \frac{\partial \eta}{\partial y} & \frac{\partial \eta}{\partial z} \\ \frac{\partial \zeta}{\partial x} & \frac{\partial \zeta}{\partial y} & \frac{\partial \zeta}{\partial z} \end{pmatrix}, \quad J^{-1} = \begin{pmatrix} \frac{\partial x}{\partial \xi} & \frac{\partial x}{\partial \eta} & \frac{\partial x}{\partial \zeta} \\ \frac{\partial y}{\partial \xi} & \frac{\partial y}{\partial \eta} & \frac{\partial y}{\partial \zeta} \\ \frac{\partial z}{\partial \xi} & \frac{\partial z}{\partial \eta} & \frac{\partial z}{\partial \zeta} \end{pmatrix}. \tag{3.19}$$

Using relation between the Jacobian (J) matrix, the transpose comatrix of the inverse Jacobian (J^{-1}) and the determinant of the inverse Jacobian, the following relations hold true [1, 15, 29]:

$$\xi_x = \frac{y_\eta z_\zeta - y_\zeta z_\eta}{|J^{-1}|}, \quad \xi_y = \frac{x_\zeta z_\eta - x_\eta z_\zeta}{|J^{-1}|}, \quad \xi_z = \frac{x_\eta y_\zeta - x_\zeta y_\eta}{|J^{-1}|},$$

$$\eta_x = \frac{y_\zeta z_\xi - y_\xi z_\zeta}{|J^{-1}|}, \quad \eta_y = \frac{x_\xi z_\zeta - x_\zeta z_\xi}{|J^{-1}|}, \quad \eta_z = \frac{x_\zeta y_\xi - x_\xi y_\zeta}{|J^{-1}|},$$

$$\zeta_x = \frac{y_\xi z_\eta - y_\eta z_\xi}{|J^{-1}|}, \quad \zeta_y = \frac{x_\eta z_\xi - x_\xi z_\eta}{|J^{-1}|}, \quad \zeta_z = \frac{x_\xi y_\eta - x_\eta y_\xi}{|J^{-1}|}, \tag{3.20}$$

where the determinant of the inverse Jacobian was used at the denominator:

$$\begin{aligned}\left|J^{-1}\right| = \det\left(J^{-1}\right) &= x_\xi\left(y_\eta z_\zeta - y_\zeta z_\eta\right) - x_\eta\left(y_\xi z_\zeta - y_\zeta z_\zeta\right)\\ &\quad + x_\zeta\left(y_\xi z_\eta - y_\eta z_\xi\right).\end{aligned} \tag{3.21}$$

For simplicity from now on, the Jacobian will be written as $J^{-1} = \left|J^{-1}\right|$. For convenience setting,

$$\begin{aligned} c_1 &= y_\eta z_\zeta - y_\zeta z_\eta, & c_2 &= x_\zeta z_\eta - x_\eta z_\zeta, & c_3 &= x_\eta y_\zeta - x_\zeta y_\eta,\\ c_4 &= y_\zeta z_\xi - y_\xi z_\zeta, & c_5 &= x_\xi z_\zeta - x_\zeta z_\xi, & c_6 &= x_\zeta y_\xi - x_\xi y_\zeta,\\ c_7 &= y_\xi z_\eta - y_\eta z_\xi, & c_8 &= x_\eta z_\xi - x_\xi z_\eta, & c_9 &= x_\xi y_\eta - x_\eta y_\xi,\end{aligned} \tag{3.22}$$

a more compact form of the Navier–Stokes in curvilinear coordinates is obtained. An important observation is that the parameters c_i, where $i = 1, 2, \ldots, 9$, are functions of the new independent coordinates ξ, η and ζ. This significantly impacts the final result of the Navier–Stokes equations.

The continuity equation in three dimensions and cartesian coordinates is expressed as

$$\frac{\partial \rho}{\partial t} + \frac{\partial u}{\partial x} + \frac{\partial v}{\partial y} + \frac{\partial w}{\partial z} = 0. \tag{3.23}$$

Using the chain rule in (3.17), equation (3.23) divided by Jacobian yields

$$\begin{aligned}\frac{\partial J^{-1}}{\partial t} &+ \left[c_1(u - \dot{x}) + c_2(v - \dot{y}) + c_3(w - \dot{z})\right]_\xi\\ &+ \left[c_4(u - \dot{x}) + c_5(v - \dot{y}) + c_6(w - \dot{z})\right]_\eta\\ &+ \left[c_7(u - \dot{x}) + c_8(v - \dot{y}) + c_9(w - \dot{z})\right]_\zeta = 0,\end{aligned}$$

where $\dot{x}, \dot{y}, \dot{z}$ are the arbitrary grid motion velocities.

Setting the contravariant velocities,

$$U = c_1(u - \dot{x}) + c_2(v - \dot{y}) + c_3(w - \dot{z}),$$
$$V = c_4(u - \dot{x}) + c_5(v - \dot{y}) + c_6(w - \dot{z}),$$
$$W = c_7(u - \dot{x}) + c_8(v - \dot{y}) + c_9(w - \dot{z}),$$

the continuity equation transforms to

$$\frac{\partial J^{-1}}{\partial t} + \frac{\partial U}{\partial \xi} + \frac{\partial V}{\partial \eta} + \frac{\partial W}{\partial \zeta} = 0, \tag{3.24}$$

the non-dimensional Navier–Stokes in generalized curvilinear coordinates [8]. Equation (3.25) depict the momentum equation cartesian coordinates:

$$\frac{D\overline{q}}{Dt} + (\overline{q} \cdot \nabla)\overline{q} = -\nabla p + \frac{1}{Re}(\nabla^2 \overline{q}). \tag{3.25}$$

Introducing the GCC transformation, the following can be obtained:

x-momentum GCC equation

$$\begin{aligned}
\frac{\partial J^{-1}u}{\partial t} + \frac{\partial uU}{\partial \xi} + \frac{\partial uV}{\partial \eta} + \frac{\partial uW}{\partial \zeta} &= -\left(c_1\frac{\partial p}{\partial \xi} + c_4\frac{\partial p}{\partial \eta} + c_7\frac{\partial p}{\partial \zeta}\right) \\
&+ \frac{1}{Re}\frac{\partial}{\partial \xi}\left(\frac{1}{J^{-1}}\left(q_1\frac{\partial u}{\partial \xi} + q_2\frac{\partial u}{\partial \eta} + q_3\frac{\partial u}{\partial \zeta}\right)\right) \\
&+ \frac{1}{Re}\frac{\partial}{\partial \eta}\left(\frac{1}{J^{-1}}\left(q_2\frac{\partial u}{\partial \xi} + q_4\frac{\partial u}{\partial \eta} + q_5\frac{\partial u}{\partial \zeta}\right)\right) \\
&+ \frac{1}{Re}\frac{\partial}{\partial \zeta}\left(\frac{1}{J^{-1}}\left(q_3\frac{\partial u}{\partial \xi} + q_5\frac{\partial u}{\partial \eta} + q_6\frac{\partial u}{\partial \zeta}\right)\right),
\end{aligned} \tag{3.26}$$

where

$$q_1 = c_1^2 + c_2^2 + c_3^2, \quad q_2 = c_1c_4 + c_2c_5 + c_3c_6, \quad q_3 = c_1c_7 + c_2c_8 + c_3c_9$$
$$q_4 = c_4^2 + c_5^2 + c_6^2, \quad q_5 = c_4c_7 + c_5c_8 + c_6c_9, \quad q_6 = c_7^2 + c_8^2 + c_9^2.$$

Similarly, the y and z momentum of the three-dimensional Navier–Stokes for FSI applications are

y-momentum GCC equation

$$\frac{\partial J^{-1}v}{\partial t}+\frac{\partial vU}{\partial \xi}+\frac{\partial vV}{\partial \eta}+\frac{\partial vW}{\partial \zeta}=-\left(c_2\frac{\partial p}{\partial \xi}+c_5\frac{\partial p}{\partial \eta}+c_8\frac{\partial p}{\partial \zeta}\right)$$
$$+\frac{1}{Re}\frac{\partial}{\partial \xi}\left(\frac{1}{J^{-1}}\left(q_1\frac{\partial v}{\partial \xi}+q_2\frac{\partial v}{\partial \eta}+q_3\frac{\partial v}{\partial \zeta}\right)\right)$$
$$+\frac{1}{Re}\frac{\partial}{\partial \eta}\left(\frac{1}{J^{-1}}\left(q_2\frac{\partial v}{\partial \xi}+q_4\frac{\partial v}{\partial \eta}+q_5\frac{\partial v}{\partial \zeta}\right)\right) \tag{3.27}$$
$$+\frac{1}{Re}\frac{\partial}{\partial \zeta}\left(\frac{1}{J^{-1}}\left(q_3\frac{\partial v}{\partial \xi}+q_5\frac{\partial u}{\partial \eta}+q_6\frac{\partial u}{\partial \zeta}\right)\right)$$

and

z-momentum GCC equation

$$\frac{\partial J^{-1}w}{\partial t}+\frac{\partial wU}{\partial \xi}+\frac{\partial wV}{\partial \eta}+\frac{\partial wW}{\partial \zeta}=-\left(c_3\frac{\partial p}{\partial \xi}+c_6\frac{\partial p}{\partial \eta}+c_9\frac{\partial p}{\partial \zeta}\right)$$
$$+\frac{1}{Re}\frac{\partial}{\partial \xi}\left(\frac{1}{J^{-1}}\left(q_1\frac{\partial w}{\partial \xi}+q_2\frac{\partial w}{\partial \eta}+q_3\frac{\partial w}{\partial \zeta}\right)\right)$$
$$+\frac{1}{Re}\frac{\partial}{\partial \eta}\left(\frac{1}{J^{-1}}\left(q_2\frac{\partial w}{\partial \xi}+q_4\frac{\partial w}{\partial \eta}+q_5\frac{\partial w}{\partial \zeta}\right)\right) \tag{3.28}$$
$$+\frac{1}{Re}\frac{\partial}{\partial \zeta}\left(\frac{1}{J^{-1}}\left(q_3\frac{\partial w}{\partial \xi}+q_5\frac{\partial w}{\partial \eta}+q_6\frac{\partial w}{\partial \zeta}\right)\right).$$

3.2.1. *The finite volume method in the three-dimensional fluid flow equations*

Due to the nonlinearity of the Navier–Stokes, it is challenging to solve them analytically in three-dimensional form, although various methods have been developed to solve them numerically [6, 7, 26, 33, 34]

The finite volume method creates finite volumes inside the domain studied. Since the three-dimensional case is in consideration, the unknown variables are evaluated at the face of each volume by integrating over the control volume. The control volume faces are w, e, s, n, b and t, respectively, where the first faces are in the two-dimensional area of the volume and the last two are the new faces representing the top and bottom of the z-direction [18], as depicted

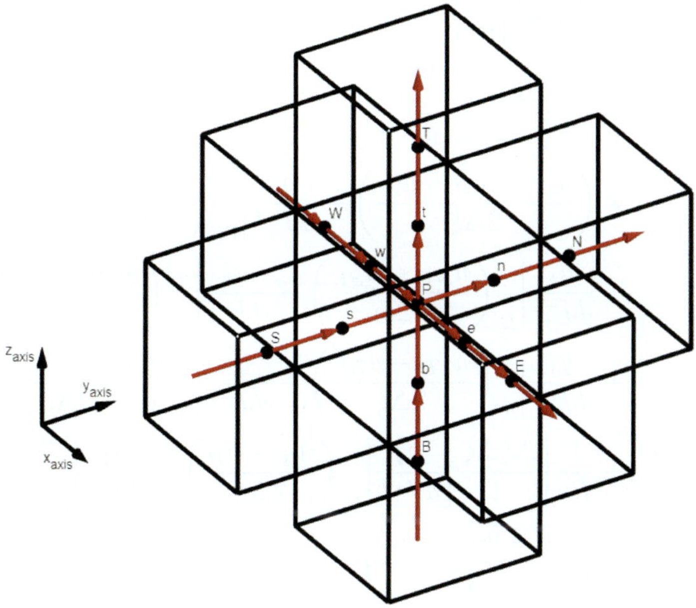

Fig. 2. The centroid cube of the discretization using the finite volume method. The schematic is the same as in the two-dimensional case if the image is observed from above.

in Figure 2. For convenience, equation (4.1) shows the discretized ξ-momentum of the Navier–Stokes (generalized curvilinear coordinates, corresponding x-momentum) using the finite volume method. As a result, the nonlinear PDE of the ξ-momentum is transformed into a coupled system of nonlinear algebraic equations.

Finite Volume discretized continuity equation

$$\frac{J^{-1} - J_0^{-1}}{\Delta t} + U_e - U_w + V_n - V_s + W_t - W_b = 0, \tag{3.29}$$

Finite Volume discretized ξ-momentum

$$\frac{J^{-1} - J_0^{-1}}{\Delta t} + \frac{\Delta\eta\Delta\zeta}{2}\left(uU\big|_E - uU\big|_W\right) + \frac{\Delta\xi\Delta\zeta}{2}\left(uV\big|_N - uV\big|_S\right)$$
$$+ \frac{\Delta\xi\Delta\eta}{2}\left(uW\big|_T - uW\big|_B\right) = -\frac{\Delta\eta\Delta\zeta}{2}\left(p\, c_1\big|_E - p\, c_1\big|_W\right)$$

$$
\begin{aligned}
&- \frac{\Delta\xi\Delta\zeta}{2}\left(p\, c_4\big|_N - p\, c_4\big|_S\right) - \frac{\Delta\xi\Delta\eta}{2}\left(p\, c_7\big|_T - p\, c_7\big|_B\right) \\
&+ \frac{1}{Re}\left[\frac{\Delta\eta\Delta\zeta}{2}\left[\frac{q_1}{dJ^{-1}}\bigg|_E\left(\frac{u_E - u_P}{2\Delta\xi}\right) + \frac{q_1}{dJ^{-1}}\bigg|_P\right.\right. \\
&\times\left(\frac{u_E - 2u_P + u_W}{2\Delta\xi}\right) - \frac{q_1}{dJ^{-1}}\bigg|_W\left(\frac{u_P - u_W}{2\Delta\xi}\right) \\
&+ \frac{q_2}{dJ^{-1}}\bigg|_E\left(\frac{u_{NE} - u_{SE}}{2\Delta\eta}\right) - \frac{q_2}{dJ^{-1}}\bigg|_W\left(\frac{u_{NW} - u_{SW}}{2\Delta\eta}\right) \\
&\left.\left.+ \frac{q_3}{dJ^{-1}}\bigg|_E\left(\frac{u_{TE} - u_{BE}}{2\Delta\zeta}\right) - \frac{q_3}{dJ^{-1}}\bigg|_W\left(\frac{u_{TW} - u_{BW}}{2\Delta\zeta}\right)\right]\right] \\
&+ \frac{1}{Re}\left[\frac{\Delta\xi\Delta\zeta}{2}\left[\frac{q_4}{dJ^{-1}}\bigg|_N\left(\frac{u_N - u_P}{2\Delta\eta}\right) + \frac{q_4}{dJ^{-1}}\bigg|_P\right.\right. \\
&\times\left(\frac{u_N - 2u_P + u_S}{2\Delta\xi}\right) - \frac{q_4}{dJ^{-1}}\bigg|_S\left(\frac{u_P - u_S}{2\Delta\eta}\right) \\
&+ \frac{q_2}{dJ^{-1}}\bigg|_N\left(\frac{u_{EN} - u_{WN}}{2\Delta\xi}\right) - \frac{q_2}{dJ^{-1}}\bigg|_S\left(\frac{u_{ES} - u_{WS}}{2\Delta\xi}\right) \\
&\left.\left.+ \frac{q_5}{dJ^{-1}}\bigg|_N\left(\frac{u_{TN} - u_{BN}}{2\Delta\zeta}\right) - \frac{q_5}{dJ^{-1}}\bigg|_S\left(\frac{u_{TS} - u_{BS}}{2\Delta\zeta}\right)\right]\right] \\
&+ \frac{1}{Re}\left[\frac{\Delta\xi\Delta\eta}{2}\left[\frac{q_6}{dJ^{-1}}\bigg|_T\left(\frac{u_T - u_P}{2\Delta\zeta}\right) + \frac{q_6}{dJ^{-1}}\bigg|_P\right.\right. \\
&\times\left(\frac{u_T - 2u_P + u_B}{2\Delta\zeta}\right) - \frac{q_6}{dJ^{-1}}\bigg|_B\left(\frac{u_P - u_B}{2\Delta\zeta}\right) \\
&+ \frac{q_3}{dJ^{-1}}\bigg|_T\left(\frac{u_{ET} - u_{WT}}{2\Delta\xi}\right) - \frac{q_3}{dJ^{-1}}\bigg|_B\left(\frac{u_{EB} - u_{WB}}{2\Delta\xi}\right) \\
&\left.\left.+ \frac{q_5}{dJ^{-1}}\bigg|_T\left(\frac{u_{NT} - u_{ST}}{2\Delta\eta}\right) - \frac{q_5}{dJ^{-1}}\bigg|_B\left(\frac{u_{NB} - u_{SB}}{2\Delta\eta}\right)\right]\right].
\end{aligned}
$$

In the same fashion, η and ζ momentum equations have similar forms.

In the following section, some applications in pathological arteries, in two and three dimensions, are studied based on the aforementioned approach.

4. Application to Vessels

4.1. *Stenosis*

Blood is considered a homogeneous Newtonian fluid, and blood flow is considered steady, incompressible, laminar and two-dimensional. In more detail, the mathematical model, as described in Ref. [37], is a flow between two impermeable plates of length $\bar{L}$ and distance between them $\bar{D}$, forming a constriction, as in Figure 3. The velocity profile at the entrance is parabolic and a fully developed flow boundary condition is considered at the outlet of the plates. The formulation of the position of the two plates is done by a function with respect to the longitudinal direction ($\bar{x}$-direction).

Under the aforementioned assumptions, the governing equations of blood flow in cartesian coordinates are

$$\frac{\partial \bar{u}}{\partial \bar{x}} + \frac{\partial \bar{v}}{\partial \bar{y}} = 0, \tag{4.1}$$

$$\bar{\rho}\left(\bar{u}\frac{\partial \bar{u}}{\partial \bar{x}} + \bar{v}\frac{\partial \bar{u}}{\partial \bar{y}}\right) = -\frac{\partial \bar{p}}{\partial \bar{x}} + \bar{\mu}\left(\frac{\partial^2 \bar{u}}{\partial \bar{x}^2} + \frac{\partial^2 \bar{u}}{\partial \bar{y}^2}\right), \tag{4.2}$$

$$\bar{\rho}\left(\bar{u}\frac{\partial \bar{v}}{\partial \bar{x}} + \bar{v}\frac{\partial \bar{v}}{\partial \bar{y}}\right) = -\frac{\partial \bar{p}}{\partial \bar{y}} + \bar{\mu}\left(\frac{\partial^2 \bar{v}}{\partial \bar{x}^2} + \frac{\partial^2 \bar{v}}{\partial \bar{y}^2}\right). \tag{4.3}$$

The respected boundary conditions are as follows:

- Inflow $(\bar{x} = 0, -\bar{D}/2 \leq \bar{y} \leq \bar{D}/2)$: $\bar{u} = \bar{u}(\bar{y}), \bar{v} = 0,$
- Outflow $(\bar{x} = \bar{L}, -\bar{D}/2 \leq \bar{y} \leq \bar{D}/2)$: $\partial \bar{R}/\partial \bar{x} = 0,$
- Upper plate $(\bar{y} = \bar{D}/2, 0 \leq \bar{x} \leq \bar{L})$: $\bar{u} = 0, \bar{v} = 0,$
- Lower plate $(\bar{y} = -\bar{D}/2, 0 \leq \bar{x} \leq \bar{L})$: $\bar{u} = 0, \bar{v} = 0.$

Here, $\vec{\boldsymbol{q}} = (\bar{u}, \bar{v})$ is the dimensional velocity, $\bar{p}$ is the dimensional pressure affecting fluid motion, $\boldsymbol{\mu}$ is the kinematic viscosity of the fluid, $\bar{\boldsymbol{R}}$ is either $\bar{u}$ or $\bar{v}$ and ρ is the blood density.

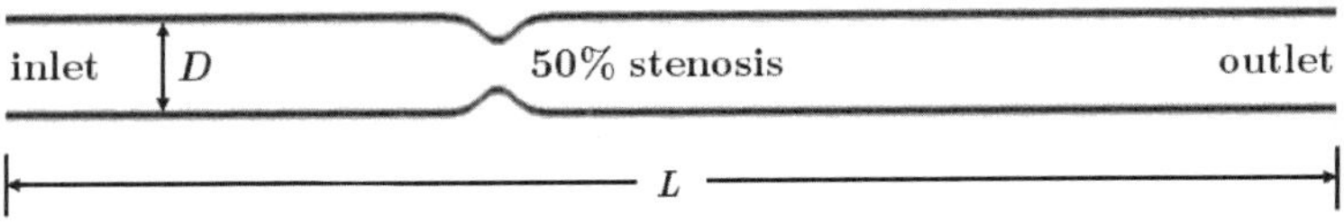

Fig. 3. An outline of the geometry and the boundary conditions applied on the stenotic model [37].

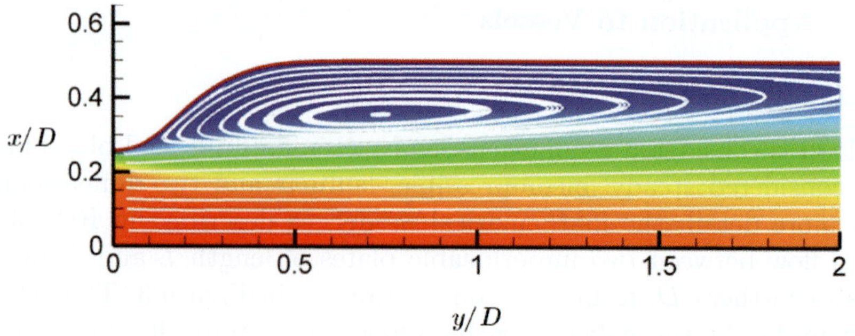

Fig. 4. Streamlines in stenotic area [37].

The results focus on half of the stenosis due to the symmetry of the domain as depicted in Figure 4. Vortices occur right after the stenotic region [3, 32]. Additionally, the pressure difference, close to the stenotic area, exhibits high values that are dependable to the percentage of the stenosis [21].

4.2. *Aneurysm*

This study focuses on a fusiform aneurysm, characterized by a symmetrical dilation across the entire circumference of the aorta. These aneurysms typically involve an elongated section of the aorta, resulting in a tubular or spindle-shaped enlargement. Fusiform aneurysms exhibit a simpler geometry compared to saccular ones and they are more common in pathology than their saccular counterparts. Furthermore, this form is less prone to rupture than saccular due to the even distribution of the stress on the vessel wall [17]. Specialized boundary conditions are necessary for the system of equations that represents blood flow in an aneurysmal geometry, when combined with a moving arterial wall. The pulsatility of the aneurysmatic vessel is described by a time-dependent condition for the inlet velocity during the cardiac cycle, which consists of the systolic and diastolic phases. Additionally, it is assumed that the flow reaches a state of full development at the outflow boundary. This results in a parabolic velocity profile for the fluid and the pressure distribution at the outlet is considered a known parameter. A kinematic boundary condition is imposed at the wall and symmetry boundary condition at the center of the fluid domain, as shown in Figure 5. Following previous studies

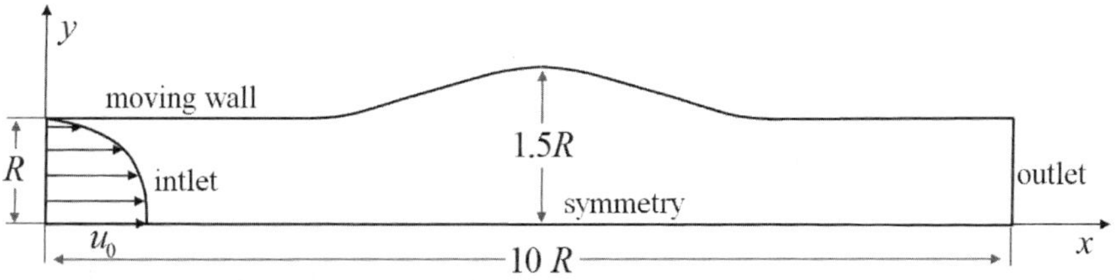

Fig. 5. An outline of the geometry and the boundary conditions applied on the aneurysmal model.

[38], the dimensionless boundary conditions of this problem for the system of equations (3.9)–(3.11) are as follows:

- at the inlet, for $t \geq 0$: $u(y,t) = \left[1 - \left(\frac{y}{R}\right)^2\right] \times$ velocity waveform(t), $v = 0$, $0 \leqslant y \leqslant R$;
- at the moving wall: $u = \dot{x}$, $v = \dot{y}$, kinematic boundary condition;
- at the symmetry: $\frac{\partial u}{\partial y} = 0$, $v = 0$, for $t \geq 0$;
- at the outlet: $p =$ pressure waveform(t), $\frac{\partial u}{\partial x} = 0$, $\frac{\partial v}{\partial x} = 0$, fully developed flow assumption.

Waveforms of velocity and pressure were extracted from the literature [16] and are fitted with Fourier series [13]. Recent studies employ the most advanced imaging techniques, such as intravascular ultrasonography, magnetic resonance angiography and computed tomographic angiography (CTA), to precisely quantify the pulsating motion of the aortic wall [2, 35]. In this particular study, the aortic motion pattern utilized represents a condition characterized by low-key pulsatility of the aortic wall.

The following results were obtained from a simulation of four cardiac cycles. Each cardiac cycle is divided to three intervals; the systolic acceleration, the systolic deceleration, and the diastole [38]. It is noted that during the systolic deceleration phase large vortices occur, as depicted in Figure 6. This happens due to an alteration of the initial velocity profile at the center of the aneurysm.

4.3. *The three-dimensional model*

The theory of the generalized curvilinear coordinates helps moving from complex geometries to less complicated geometries and vice versa. Such an example is the cylinder. Figure 7 shows the grid of

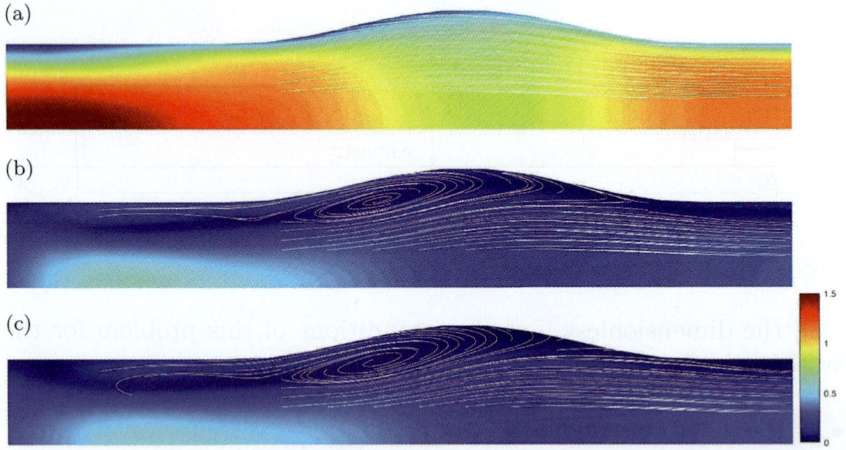

Fig. 6. Streamlines during one cardiac cycle. (a) Systolic acceleration, (b) Systolic deceleration, (c) Diastole.

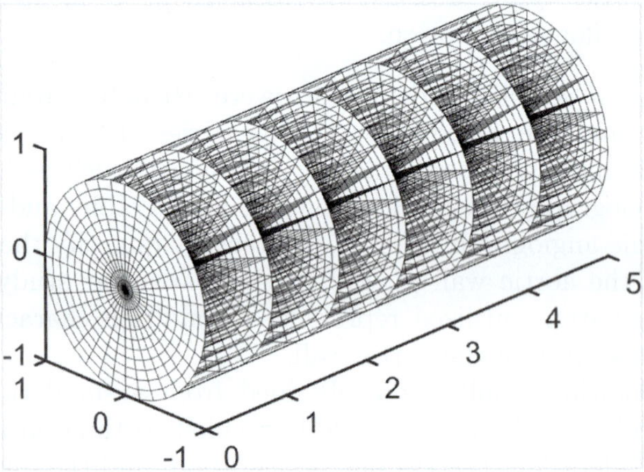

Fig. 7. Three-dimensional computational grid of a cylinder.

the cylinder yielding from the use of curvilinear coordinates. Instead of using the cylindrical coordinates, this scheme is utilized so that the descritized equations (momentum and continuity) can be solved in a three-dimensional cartesian grid and then depicted on a cylindrical one, as shown in Figure 8.

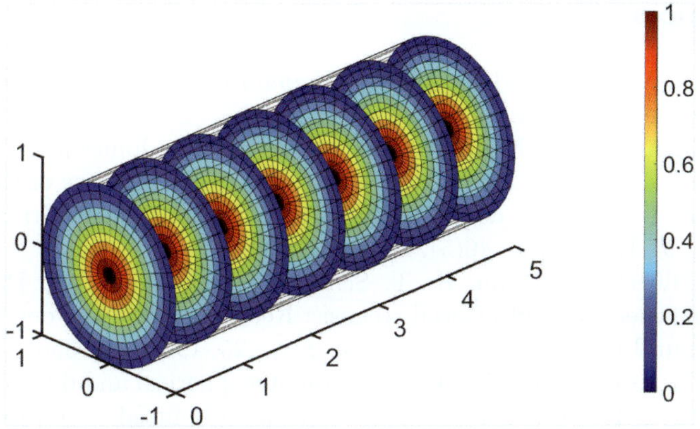

Fig. 8. The steady state parabolic flow through a cylinder for $Re = 300$.

5. Conclusions

Hemodynamics is a field of fluid mechanics. There is a great interest in studying blood flow through pathological vessels, especially in stenotic and aneurysmatic arteries, as such arterial disease have a significant impact on the flow, leading to undesired effects.

The simplest model that can simulate blood flow is the two-dimensional model of the Navier–Stokes equations. The results indicate that it can accurately predict blood flow in simple geometries such as stenosis or aneurysms. However, three-dimensional models can better describe the physical geometry of a vessel (physiological or pathological).

To accomplish both the two-dimensional and the three-dimensional models, the use of generalized curvilinear coordinates is essential to accurately represent the actual vessel geometry. Additionally, the finite volume method was utilized for the discretization of the domain.

The findings reveal recirculation zones downstream of the stenosis and in the centre of the aneurysm. Furthermore, high-pressure areas were observed, which could potentially exacerbate vascular conditions and cause complications such as internal bleeding in the event of rupture.

These challenges underscore the importance of investigating mathematical modeling for blood flow in pathological vessels.

References

[1] J. D. Anderson and J. Wendt. *Computational Fluid Dynamics*, Vol. 206. New York: McGraw-Hill, 1995.

[2] F. R. Arko, E. H. Murphy, C. M. Davis III, E. D. Johnson, S. T. Smith and C. K. Zarins. Dynamic geometry and wall thickness of the aortic neck of abdominal aortic aneurysms with intravascular ultrasonography. *J. Vasc. Surg.*, 46(5), 2007, 891–897.

[3] D. Bluestein, L. Niu, R. T. Schoephoerster and M. K. Dewanjee. Fluid mechanics of arterial stenosis: Relationship to the development of mural thrombus. *Ann. Biomed. Eng.*, 25, 1997, 344–356.

[4] M. E. Braaten and W. Shyy. Motion of a spherical and direct solution methods for viscous flow calculations in body-fitted co-ordinates. *Int. J. Numer. Methods Fluids*, 6(6), 1986, 325–349.

[5] M. J. Davies. Aortic aneurysm formation. *Circulation*, 98, July 1998, 193–195.

[6] J. H. Ferziger and M. Peric. *Computational Methods for Fluid Dynamics.* Springer, Berlin, 2012.

[7] C. A. J. Fletcher. *Computational Techniques for Fluid Dynamics 2: Specific Techniques for Different Flow Categories.* Computational Techniques for Fluid Dynamics. Springer-Verlag Berlin/Heidelberg, Germany, 1988.

[8] C. A. J. Fletcher. *Computational Techniques for Fluid Dynamics 1: Fundamental and General Techniques.* Springer Series in Computational Physics. Springer, Berlin, 2013.

[9] R. P. Hoskins, V. P. Lawford and J. B. Doyle. *Cardiovascular Biomechanics.* Springer International Publishing, Cham, Switzerland, 2017.

[10] D. Kasiteropoulou, A. Topalidou and S. Downe. A computational fluid dynamics modelling of maternal-fetal heat exchange and blood flow in the umbilical cord. *PLoS One*, 15(7), 2020, e0231997.

[11] R. Klabunde. *Cardiovascular Physiology Concepts.* Lippincott Williams & Wilkins, Philadelphia, PA, USA, 2011.

[12] K. C. Kyriakoudi and M. A. Xenos. Magnetohydrodynamic effects on a pathological vessel: An euler–lagrange approach. *Phys. Fluids*, 35(12), 2023.

[13] A. A. Linninger, M. Xenos, D. C. Zhu, M. R. Somayaji, S. Kondapalli, and R. D. Penn. Cerebrospinal fluid flow in the normal and hydrocephalic human brain. *IEEE Trans. Biomed. Eng.*, 54(2), 2007, 291–302.

[14] H. Lu and A. Daugherty. Aortic aneurysms. *Arterioscler. Thromb. Vasc. Biol.*, 37, June 2017, e59–e65.

[15] J. A. Michelsen. General curvilinear transformation of the Navier-Stokes equations in a 3d polar rotating frame, 1998.
[16] M. S. Olufsen, C. S. Peskin, W. Y. Kim, E. M. Pedersen, A. Nadim and J. Larsen. Numerical simulation and experimental validation of blood flow in arteries with structured-tree outflow conditions. *Ann. Biomed. Eng.*, 28(11), 2000, 1281–1299.
[17] S.-H. Park, M.-B. Yim, C.-Y. Lee, E. Kim and E.-I. Son. Intracranial fusiform aneurysms: It's pathogenesis, clinical characteristics and managements. *J. Korean Neurosurg. Soc.*, 44(3), 2008, 116.
[18] S. V. Patankar. *Numerical Heat Transfer and Fluid Flow*. Electro Skills Series. Hemisphere Publishing Corporation, Washington, DC, USA & New York, NY, USA, 1980.
[19] A. Quarteroni, A. Manzoni and C. Vergara. The cardiovascular system: Mathematical modelling, numerical algorithms and clinical applications. *Acta Numer.*, 26, 2017, 365–590.
[20] A. Raptis, M. Xenos, E. Georgakarakos, G. Kouvelos, A. Giannoukas, N. Labropoulos and M. Matsagkas. Comparison of physiological and post-endovascular aneurysm repair infrarenal blood flow. *Comput. Methods Biomech. Biomed. Eng.*, 20(3), 2017, 242–249.
[21] M. Roy, B. S. Sikarwar, M. Bhandwal and P. Ranjan. Modelling of blood flow in stenosed arteries. *Procedia Comput. Sci.*, 115, 2017, 821–830.
[22] N. Sakalihasan, R. Limet and O. D. Defawe. Abdominal aortic aneurysm. *Lancet*, 365(9470), 2005, 1577–1589.
[23] H. E. Salman and H. C. Yalcin. Computational modeling of blood flow hemodynamics for biomechanical investigation of cardiac development and disease. *J. Cardiovasc. Dev. Dis.*, 8(2), 2021, 14.
[24] A. Serrano, V. Cunha, J. P. Teixeira, M. B. Pires, J. G. O'Neill and V. Vassilenko. Hemodynamics in human uterine arteries: Modeling and computational fluid dynamics calculations. In *2019 IEEE 6th Portuguese Meeting on Bioengineering (ENBENG)*. IEEE, 2019, pp. 1–4.
[25] R. G. Shah, T. Girardi, G. Merz, P. Necaise and C. M. Salafia. Hemodynamic analysis of blood flow in umbilical artery using computational modeling. *Placenta*, 57, 2017, 9–12.
[26] A. Sharma. *Introduction to Computational Fluid Dynamics: Development, Application and Analysis*. Springer Nature, Cham, Switzerland, 2021.
[27] W. Shyy, S. S. Tong and S. M. Correa. Numerical recirculating flow calculation using a body-fitted coordinate system. *Numer. Heat Transf.*, 8(1), 1985, 99–113.
[28] W. Shyy, H. S. Udaykumar, M. M. Rao and R. W. Smith. *Computational Fluid Dynamics with Moving Boundaries*. Courier Corporation (Dover), Mineola, NY, USA, 2012.

[29] G. Strang. *Introduction to Linear Algebra.* Wellesley-Cambridge Press, Wellesley, MA, USA, 2016.

[30] M. Thiriet, M. Delfour and A. Garon. Vascular stenosis: An introduction. In P. Lanzer (eds.), *PanVascular Medicine*, Springer, Berlin, Heidelberg, 2015, pp. 781–868.

[31] J. F. Thompson, Z. U. A. Warsi and C. W. Mastin. *Numerical Grid Generation: Foundations and Applications.* Elsevier North-Holland, Inc., 1985.

[32] E. E. Tzirtzilakis. Biomagnetic fluid flow in a channel with stenosis. *Phys. D: Nonlinear Phenom.*, 237(1), 2008, 66–81.

[33] E. E. Tzirtzilakis. A simple numerical methodology for BFD problems using stream function vorticity formulation. *Commun. Numer. Methods Eng.*, 24(8), 2008, 683–700.

[34] E. E. Tzirtzilakis1, V. D. Sakalis, N. G. Kafoussias and P. M. Hatzikonstantinou. Biomagnetic fluid flow in a 3d rectangular duct. *Int. J. Numer. Methods Fluids*, 44(12), 2004, 1279–1298.

[35] J. W. Van Keulen, J. Van Prehn, M. Prokop, F. L. Moll and J. A. Van Herwaarden. Dynamics of the aorta before and after endovascular aneurysm repair: A systematic review. *Eur. J. Vasc. Endovasc. Surg.*, 38(5), 2009, 586–596.

[36] M. Xenos, N. Labropoulos, S. Rambhia, Y. Alemu, S. Einav, A. Tassiopoulos, N. Sakalihasan and D. Bluestein. Progression of abdominal aortic aneurysm towards rupture: Refining clinical risk assessment using a fully coupled fluid-structure interaction method. *Ann. Biomed. Eng.*, 43, January 2015, 139–153.

[37] M. Xenos and E. Tzirtzilakis. MHD effects on blood flow in a stenosis, 8, January 2013, 427–437.

[38] M. A. Xenos. An Euler–Lagrange approach for studying blood flow in an aneurysmal geometry. *Proc. R. Soc. A: Math. Phys. Eng. Sci.*, 473(2199), 2017, 20160774.

[39] L. Zhang, K. Kulkarni, M. R. Somayaji, M. Xenos and A. A. Linninger. Discovery of transport and reaction properties in distributed systems. *AIChE J.*, 53(2), 2007, 381–396.

[40] L. Zhong, J.-M. Zhang, B. Su, R. S. Tan, J. C. Allen and G. S. Kassab. Application of patient-specific computational fluid dynamics in coronary and intra-cardiac flow simulations: Challenges and opportunities. *Front. Physiol.*, 9, 2018, 742.

https://doi.org/10.1142/9789819820498_0012

Chapter 12

Ternary Derivation-Homomorphism Functional Inequalities

Jung Rye Lee[*,§], **Choonkil Park**[†,¶] **and Michael Th. Rassias**[‡,**]

[*] *Department of Data Science, Daejin University, Kyunggi 11159, Korea*
[†] *Department of Mathematics, Hanyang University, Seoul 04763, Korea*
[‡] *Department of Mathematics and Engineering Sciences, Hellenic Military Academy, 16673 Vari Attikis, Greece*
[§] *jrlee@daejin.ac.kr*
[¶] *baak@hanyang.ac.kr*
[**] *mthrassias@yahoo.com*

In this chapter, we introduce and solve the following additive-additive (s,t)-functional inequality:

$$\begin{aligned}&\left\|3g\left(\frac{x+y+z}{3}\right)-g(x)-g(y)-g(z)\right\|\\&\quad+\left\|3h\left(\frac{x+y+z}{3}\right)+h(x-2y+z)+h(x+y-2z)-3h(x)\right\|\\&\le \|s\left(g\left(x+y+z\right)-g(x)-g(y)-g(z)\right)\|\\&\quad+\|t\left(h(x+y+z)+h(x-2y+z)+h(x+y-2z)-3h(x)\right)\|, \quad (0.1)\end{aligned}$$

where s and t are fixed non-zero complex numbers with $|s|<1$ and $|t|<1$. Using the direct method and the fixed point method, we prove the Hyers–Ulam stability of ternary derivations and ternary homomorphisms in C^*-ternary algebras, associated with the additive-additive (s,t)-functional inequality (0.1) and the following functional inequality:

$$\begin{aligned}&\|g([x,y,z]) - [g(x),y,z] - [x,g(y),z] - [x,y,g(z)]\|\\&+\|h([x,y,z]) - [h(x),h(y),h(z)]\| \le \varphi(x,y,z).\end{aligned} \tag{0.2}$$

1. Introduction and Preliminaries

A C^*-ternary algebra is a complex Banach space A, equipped with a ternary product $(x,y,z) \mapsto [x,y,z]$ of A^3 into A, which is $\mathbb{C}$-linear in the outer variables, conjugate $\mathbb{C}$-linear in the middle variable, and associative in the sense that $[x,y,[z,w,v]] = [x,[w,z,y],v] = [[x,y,z],w,v]$, and satisfies $\|[x,y,z]\| \le \|x\|\cdot\|y\|\cdot\|z\|$ and $\|[x,x,x]\| = \|x\|^3$ [45].

Let A and B be C^*-ternary algebras. A $\mathbb{C}$-linear mapping $H : A \to B$ is called a *ternary homomorphism* if

$$H([x,y,z]) = [H(x),H(y),H(z)]$$

for all $x,y,z \in A$. A $\mathbb{C}$-linear mapping $\delta : A \to A$ is called a *ternary derivation* if

$$\delta([x,y,z]) = [\delta(x),y,z] + [x,\delta(y),z] + [x,y,\delta(z)]$$

for all $x,y,z \in A$ [4, 33].

The stability problem of functional equations originated from a question of Ulam [44] concerning the stability of group homomorphisms. Hyers [20] gave a first affirmative partial answer to the question of Ulam for Banach spaces. Hyers' theorem was generalized by Aoki [5] for additive mappings and by Rassias [42] for linear mappings by considering an unbounded Cauchy difference. A generalization of the Rassias theorem was obtained by Găvruta [17] by replacing the unbounded Cauchy difference by a general control function in the spirit of Rassias' approach.

Gilányi [18] showed that if f satisfies, the functional inequality

$$\|2f(x) + 2f(y) - f(x-y)\| \le \|f(x+y)\| \tag{1.1}$$

then f satisfies the Jordan-von Neumann functional equation

$$2f(x) + 2f(y) = f(x + y) + f(x - y).$$

See also [43]. Fechner [16] and Gilányi [19] proved the Hyers–Ulam stability of the functional inequality (1.1).

Park [36, 37, 39] defined additive ρ-functional inequalities and proved the Hyers–Ulam stability of the additive ρ-functional inequalities in Banach spaces and non-Archimedean Banach spaces. The stability problems of various functional equations and functional inequalities have been extensively investigated by a number of authors [1–3, 6, 10, 13–15, 21, 22, 24–30, 32, 34, 40].

We recall a fundamental result in fixed point theory.

Theorem 1 ([7, 11]). *Let (X, d) be a complete generalized metric space and let $J : X \to X$ be a strictly contractive mapping with Lipschitz constant $\alpha < 1$. Then, for each given element $x \in X$, either*

$$d(J^n x, J^{n+1} x) = \infty$$

for all non-negative integers n or there exists a positive integer n_0 such that

(1) $d(J^n x, J^{n+1} x) < \infty, \qquad \forall n \geq n_0;$
(2) *the sequence $\{J^n x\}$ converges to a fixed point y^* of J;*
(3) *y^* is the unique fixed point of J in the set $Y = \{y \in X \mid d(J^{n_0} x, y) < \infty\}$;*
(4) *$d(y, y^*) \leq \frac{1}{1-\alpha} d(y, Jy)$ for all $y \in Y$.*

In 1996, Isac and Rassias [23] were the first to provide applications of stability theory of functional equations for the proof of new fixed point theorems with applications. By using fixed point methods, the stability problems of several functional equations have been extensively investigated by a number of authors [8, 9, 12, 38, 41].

In this chapter, we solve the additive-additive (s, t)-functional inequality (0.1). Furthermore, we investigate ternary derivations and ternary homomorphisms in C^*-ternary algebras associated with the additive-additive (s, t)-functional inequality (0.1) and the functional inequality (0.2) by using the direct method and by the fixed point method.

Throughout this chapter, assume that A is a C^*-ternary algebra and that s and t are fixed non-zero complex numbers with $|s| < 1$ and $|t| < 1$.

2. Stability of Additive-Additive (s,t)-Functional Inequality (0.1): A Direct Method

In this section, we solve and investigate the additive-additive (s,t)-functional inequality (0.1) in C^*-ternary algebras.

Lemma 1. *If mappings $g, h : A \to A$ satisfy $g(0) = h(0) = 0$ and*

$$\begin{aligned}
&\left\|3g\left(\frac{x+y+z}{3}\right) - g(x) - g(y) - g(z)\right\| \\
&\quad + \left\|3h\left(\frac{x+y+z}{3}\right) + h(x-2y+z) + h(x+y-2z) - 3h(x)\right\| \\
&\leq \|s\left(g\left(x+y+z\right) - g(x) - g(y) - g(z)\right)\| \\
&\quad + \|t\left(h(x+y+z) + h(x-2y+z) + h(x+y-2z) - 3h(x)\right)\|,
\end{aligned} \tag{2.1}$$

for all $x, y, z \in A$, then the mappings $g, h : A \to A$ are additive.

Proof. Letting $y = z = 0$ in (2.1), we get

$$\left\|3g\left(\frac{x}{3}\right) - g(x)\right\| + \left\|3h\left(\frac{x}{3}\right) - h(x)\right\| \leq 0$$

for all $x \in A$. So, $g(3x) = 3g(x)$ and $h(3x) = 3h(x)$ for all $x \in A$. It follows from (2.1) that

$$\begin{aligned}
&\|g\left(x+y+z\right) - g(x) - g(y) - g(z)\| \\
&\quad + \|h(x+y+z) + h(x-2y+z) + h(x+y-2z) - 3h(x)\| \\
&\leq \|s\left(g\left(x+y+z\right) - g(x) - g(y) - g(z)\right)\| \\
&\quad + \|t\left(h(x+y+z) + h(x-2y+z) + h(x+y-2z) - 3h(x)\right)\|
\end{aligned}$$

for all $x, y, z \in A$. Thus,

$$\begin{aligned}
g\left(x+y+z\right) &= g(x) + g(y) + g(z), \\
h(x+y+z) + h(x-2y+z) + h(x+y-2z) &= 3h(x)
\end{aligned}$$

for all $x, y, z \in A$ since $|s| < 1$ and $|t| < 1$. So, the mappings $g, h : A \to A$ are additive. $\square$

Lemma 2 ([35, Theorem 2.1]). *Let $f : A \to A$ be an additive mapping such that*

$$f(\lambda a) = \lambda f(a)$$

for all $\lambda \in \mathbb{T}^1 := \{\xi \in \mathbb{C} : |\xi| = 1\}$ and all $a \in A$. Then, the mapping $f : A \to A$ is $\mathbb{C}$-linear.

Using the direct method, we prove the Hyers–Ulam stability of ternary derivations and ternary homomorphisms in C^*-ternary algebras associated with the additive-additive (s, t)-functional inequality (2.1).

Theorem 2. *Let $\varphi : A^3 \to [0, \infty)$ be a function such that*

$$\sum_{j=0}^{\infty} 27^j \varphi\left(\frac{x}{3^j}, \frac{y}{3^j}, \frac{z}{3^j}\right) < \infty \tag{2.2}$$

for all $x, y, z \in A$. Let $g, h : A \to A$ be mappings satisfying $g(0) = h(0) = 0$ and

$$\begin{aligned}
&\left\|3g\left(\lambda\frac{x+y+z}{3}\right) - \lambda(g(x) + g(y) + g(z))\right\| \\
&\quad + \left\|3h\left(\lambda\frac{x+y+z}{3}\right) + h(\lambda(x - 2y + z))\right. \\
&\qquad \left. + h(\lambda(x + y - 2z)) - 3\lambda h(x)\right\| \\
&\leq \|s\left(g\left(\lambda(x + y + z)\right) - \lambda(g(x) + g(y) + g(z))\right)\| \\
&\quad + \|t(h(\lambda(x + y + z)) + h(\lambda(x - 2y + z)) \\
&\quad + h(\lambda(x + y - 2z)) - 3\lambda h(x))\| + \varphi(x, y, z)
\end{aligned} \tag{2.3}$$

for all $\lambda \in \mathbb{T}^1$ and all $x, y, z \in A$. If the mappings $g, h : A \to A$ satisfy

$$\begin{aligned}
&\|g([x, y, z]) - [g(x), y, z] - [x, g(y), z] - [x, y, g(z)]\| \\
&\quad + \|h([x, y, z]) - [h(x), h(y), h(z)]\| \leq \varphi(x, y, z)
\end{aligned} \tag{2.4}$$

for all $x, y, z \in A$, then there exist a unique ternary derivation $D : A \to A$ and a unique ternary homomorphism $H : A \to A$ such that

$$\|g(x) - D(x)\| + \|h(x) - H(x)\| \leq \sum_{j=0}^{\infty} 3^j \varphi\left(\frac{x}{3^j}, 0, 0\right) \tag{2.5}$$

for all $x \in A$.

Proof. Letting $\lambda = 1$ and $y = z = 0$ in (2.3), we get

$$\left\|3g\left(\frac{x}{3}\right) - g(x)\right\| + \left\|3h\left(\frac{x}{3}\right) - h(x)\right\| \leq \varphi(x, 0, 0) \tag{2.6}$$

for all $x \in A$. Thus

$$\begin{aligned}
&\left\|3^l g\left(\frac{x}{3^l}\right) - 3^m g\left(\frac{x}{3^m}\right)\right\| + \left\|3^l h\left(\frac{x}{3^l}\right) - 3^m h\left(\frac{x}{3^m}\right)\right\| \\
&\leq \sum_{j=l}^{m-1} \left\|3^j g\left(\frac{x}{3^j}\right) - 3^{j+1} g\left(\frac{x}{3^{j+1}}\right)\right\| \\
&\quad + \sum_{j=l}^{m-1} \left\|3^j h\left(\frac{x}{3^j}\right) - 3^{j+1} h\left(\frac{x}{3^{j+1}}\right)\right\| \\
&\leq \sum_{j=l}^{m-1} 3^j \varphi\left(\frac{x}{3^j}, 0, 0\right)
\end{aligned} \tag{2.7}$$

for all non-negative integers m and l with $m > l$ and all $x \in A$. It follows from (2.7) that the sequences $\{3^k g(\frac{x}{3^k})\}$ and $\{3^k h(\frac{x}{3^k})\}$ are Cauchy for all $x \in A$. Since Y is a Banach space, the sequences $\{3^k g(\frac{x}{3^k})\}$ and $\{3^k h(\frac{x}{3^k})\}$ converge. So, one can define the mappings $D, H : A \to A$ by

$$D(x) := \lim_{k\to\infty} 3^k g\left(\frac{x}{3^k}\right), \qquad \& \qquad H(x) := \lim_{k\to\infty} 3^k h\left(\frac{x}{3^k}\right)$$

for all $x \in A$. Moreover, letting $l = 0$ and passing to the limit $m \to \infty$ in (2.7), we get (2.5).

It follows from (2.3) that

$$\begin{aligned}
&\left\|3D\left(\lambda\frac{x+y+z}{3}\right)-\lambda(D(x)+D(y)+D(z))\right\| \\
&\quad+\left\|3H\left(\lambda\frac{x+y+z}{3}\right)+H(\lambda(x-2y+z))\right. \\
&\quad\left.+H(\lambda(x+y-2z))-3\lambda H(x)\right\| \\
&=\lim_{n\to\infty}3^n\left\|3g\left(\lambda\frac{x+y+z}{3^{n+1}}\right)-\lambda\left(g\left(\frac{x}{3^n}\right)+g\left(\frac{y}{3^n}\right)+g\left(\frac{z}{3^n}\right)\right)\right\| \\
&\quad+\lim_{n\to\infty}3^n\left\|3h\left(\lambda\frac{x+y+z}{3^{n+1}}\right)+h\left(\lambda\frac{x-2y+z}{3^n}\right)\right. \\
&\quad\left.+h\left(\lambda\frac{x+y-2z}{3^n}\right)-3\lambda h\left(\frac{x}{3^n}\right)\right\| \\
&\le\lim_{n\to\infty}3^n\left\|s\left(g\left(\lambda\frac{x+y+z}{3^n}\right)-\lambda\left(g\left(\frac{x}{3^n}\right)\right.\right.\right. \\
&\quad\left.\left.\left.+g\left(\frac{y}{3^n}\right)+g\left(\frac{z}{3^n}\right)\right)\right)\right\| \\
&\quad+\lim_{n\to\infty}3^n\left\|t\left(h\left(\lambda\frac{x+y+z}{3^n}\right)+h\left(\lambda\frac{x-2y+z}{3^n}\right)\right.\right. \\
&\quad\left.\left.+h\left(\lambda\frac{x+y-2z}{3^n}\right)-3\lambda h\left(\frac{x}{3^n}\right)\right)\right\|+\lim_{n\to\infty}3^n\varphi\left(\frac{x}{3^n},\frac{y}{3^n},\frac{z}{3^n}\right) \\
&=\|s\left(D\left(\lambda(x+y+z)\right)-\lambda(D(x)+D(y)+D(z)\right))\| \\
&\quad+\|t(H(\lambda(x+y+z))+H(\lambda(x-2y+z)) \\
&\quad+H(\lambda(x+y-2z))-3\lambda H(x))\|
\end{aligned}$$

for all $\lambda\in\mathbb{T}^1$ and all $x,y,z\in A$. So,

$$\begin{aligned}
&\|3D\left(\lambda\frac{x+y+z}{3}\right)-\lambda(D(x)+D(y)+D(z))\| \\
&\quad+\|3H\left(\lambda\frac{x+y+z}{3}\right)+H(\lambda(x-2y+z))
\end{aligned}$$

$$
\begin{aligned}
&+H(\lambda(x+y-2z))-3\lambda H(x)\| \\
&\leq \|s\left(D\left(\lambda(x+y+z)\right)-\lambda(D(x)+D(y)+D(z))\right)\| \\
&\quad+\|t\left(H(\lambda(x+y+z))+H(\lambda(x-2y+z))\right. \\
&\quad\left.+H(\lambda(x+y-2z))-3\lambda H(x)\right)\| \qquad (2.8)
\end{aligned}
$$

for all $\lambda \in \mathbb{T}^1$ and all $x, y, z \in A$.

Let $\lambda = 1$ in (2.8). By Lemma 1, the mappings $D, H : A \to A$ are additive.

It follows from (2.8) and the additivity of D and H that

$$
\begin{aligned}
&\|D\left(\lambda(x+y+z)\right)-\lambda(D(x)+D(y)+D(z))\| \\
&\quad+\|H(\lambda(x+y+z))+H(\lambda(x-2y+z)) \\
&\quad+H(\lambda(x+y-2z))-3\lambda H(x)\| \\
&\leq \|s\left(D\left(\lambda(x+y+z)\right)-\lambda(D(x)+D(y)+D(z))\right)\| \\
&\quad+\|t\left(H(\lambda(x+y+z))+H(\lambda(x-2y+z))\right. \\
&\quad\left.+H(\lambda(x+y-2z))-3\lambda H(x)\right)\|
\end{aligned}
$$

for all $\lambda \in \mathbb{T}^1$ and all $x, y, z \in A$. Since $|s| < 1$ and $|t| < 1$,

$$
\begin{aligned}
D\left(\lambda(x+y+z)\right)-\lambda(D(x)+D(y)+D(z)) = 0, \\
H(\lambda(x+y+z))+H(\lambda(x-2y+z)) \\
+H(\lambda(x+y-2z))-3\lambda H(x) = 0
\end{aligned}
$$

and so $D(\lambda x) = \lambda D(x)$ and $H(\lambda x) = \lambda H(x)$ for all $\lambda \in \mathbb{T}^1$ and all $x, y, z \in A$. Thus, by Lemma 2, the additive mappings $D, H : A \to A$ are $\mathbb{C}$-linear.

It follows from (2.4) and the additivity of D, H that

$$
\begin{aligned}
&\|D([x,y,z])-[D(x),y,z]-[x,D(y),z]-[x,y,D(z)]\| \\
&\quad+\|H([x,y,z])-[H(x),H(y),H(z)]\| \\
&= 27^n \left\| g\left(\frac{[x,y,z]}{27^n}\right)-\left[g\left(\frac{x}{3^n}\right),\frac{y}{3^n},\frac{z}{3^n}\right]\right.
\end{aligned}
$$

$$
\begin{aligned}
&-\left[\frac{x}{3^n}, g\left(\frac{y}{3^n}\right), \frac{z}{3^n}\right]-\left[\frac{x}{3^n}, \frac{y}{3^n}, g\left(\frac{z}{3^n}\right)\right]\Bigg\| \\
&+27^n\left\|h\left(\frac{[x,y,z]}{27^n}\right)-\left[h\left(\frac{x}{3^n}\right), h\left(\frac{y}{3^n}\right), h\left(\frac{z}{3^n}\right)\right]\right\| \\
\leq{}& 27^n \varphi\left(\frac{x}{3^n}, \frac{y}{3^n}, \frac{z}{3^n}\right),
\end{aligned}
$$

which tends to zero as $n \to \infty$, by (2.2). So,

$$
\begin{aligned}
D([x,y,z])-[D(x),y,z]-[x,D(y),z]-[x,y,D(z)]=0,\\
H([x,y,z])-[H(x),H(y),H(z)]=0
\end{aligned}
$$

for all $x,y,z \in A$. Hence, the mapping $D : A \to A$ is a ternary derivation and the mapping $H : A \to A$ is a ternary homomorphism. □

Corollary 1. *Let $r > 3$ and θ be non-negative real numbers and $g,h : A \to A$ be mappings satisfying $g(0) = h(0) = 0$ and*

$$
\begin{aligned}
&\left\|3g\left(\lambda\frac{x+y+z}{3}\right)-\lambda(g(x)+g(y)+g(z))\right\| \\
&\quad+\Bigg\|3h\left(\lambda\frac{x+y+z}{3}\right)+h(\lambda(x-2y+z)) \\
&\quad+h(\lambda(x+y-2z))-3\lambda h(x)\Bigg\| \\
&\leq \|s\left(g\left(\lambda(x+y+z)\right)-\lambda(g(x)+g(y)+g(z))\right)\| \\
&\quad+\|t\left(h(\lambda(x+y+z))+h(\lambda(x-2y+z))\right. \\
&\quad\left.+h(\lambda(x+y-2z))-3\lambda h(x)\right)\|+\theta(\|x\|^r+\|y\|^r) \qquad (2.9)
\end{aligned}
$$

for all $\lambda \in \mathbb{T}^1$ and all $x,y,z \in A$. If the mappings $g,h : A \to A$ satisfy

$$
\begin{aligned}
&\|g([x,y,z])-[g(x),y,z]-[x,g(y),z]-[x,y,g(z)]\| \\
&\quad+\|h([x,y,z])-[h(x),h(y),h(z)]\| \leq \theta(\|x\|^r+\|y\|^r+\|z\|^r) \qquad (2.10)
\end{aligned}
$$

for all $x,y,z \in A$, then there exist a unique ternary derivation $D : A \to A$ and a unique ternary homomorphism $H : A \to A$ such that

$$
\|g(x)-D(x)\|+\|h(x)-H(x)\| \leq \frac{3^r\theta}{3^r-3}\|x\|^r
$$

for all $x \in A$.

Proof. The proof follows from Theorem 2 by $\varphi(x,y,z) = \theta(\|x\|^r + \|y\|^r + \|z\|^r)$ for all $x, y, z \in A$. □

Theorem 3. *Let $\varphi : A^3 \to [0,\infty)$ be a function and $g, h : A \to A$ be mappings satisfying $g(0) = h(0) = 0$, (2.3), (2.4) and*

$$\Phi(x,y,z) := \sum_{j=1}^{\infty} \frac{1}{3^j}\varphi(3^j x, 3^j y, 3^j z) < \infty \tag{2.11}$$

for all $x, y, z \in A$. Then, there exist a unique ternary derivation $D : A \to A$ and a unique ternary homomorphism $H : A \to A$ such that

$$\|g(x) - D(x)\| + \|h(x) - H(x)\| \leq \Phi(x,0,0) \tag{2.12}$$

for all $x \in A$.

Proof. It follows from (2.6) that

$$\left\|g(x) - \frac{1}{3}g(3x)\right\| + \left\|h(x) - \frac{1}{3}h(3x)\right\| \leq \frac{1}{3}\varphi(3x,0,0) \tag{2.13}$$

for all $x \in A$. Thus

$$\begin{aligned}
&\left\|\frac{1}{3^l}g\left(\frac{x}{3^l}\right) - \frac{1}{3^m}g\left(3^m x\right)\right\| + \left\|\frac{1}{3^l}h\left(3^l x\right) - \frac{1}{3^m}h\left(3^m x\right)\right\| \\
&\leq \sum_{j=l}^{m-1}\left\|\frac{1}{3^j}g\left(3^j x\right) - \frac{1}{3^{j+1}}g\left(3^{j+1}x\right)\right\| \\
&\quad + \sum_{j=l}^{m-1}\left\|\frac{1}{3^j}h\left(3^j x\right) - \frac{1}{3^{j+1}}h\left(3^{j+1}x\right)\right\| \\
&\leq \sum_{j=l+1}^{m}\frac{1}{3^j}\varphi\left(3^j x,0,0\right)
\end{aligned} \tag{2.14}$$

for all non-negative integers m and l with $m > l$ and all $x \in A$. It follows from (2.14) that the sequences $\{\frac{1}{3^k}g(3^k x)\}$ and $\{\frac{1}{3^k}h(3^k x)\}$ are Cauchy for all $x \in A$. Since Y is a Banach space, the sequences

$\{\frac{1}{3^k}g(3^kx)\}$ and $\{\frac{1}{3^k}h(3^kx)\}$ converge. So, one can define the mappings $D, H : A \to A$ by

$$D(x) := \lim_{k\to\infty} \frac{1}{3^k} g\left(3^k x\right),$$

$$H(x) := \lim_{k\to\infty} \frac{1}{3^k} h\left(3^k x\right)$$

for all $x \in A$. Moreover, letting $l = 0$ and passing to the limit $m \to \infty$ in (2.14), we get (2.12).

By the same reasoning as in the proof of Theorem 2, one can show that the mappings $D, H : A \to A$ are $\mathbb{C}$-linear.

It follows from (2.4) and the additivity of D and H that

$$\begin{aligned}
&\|D([x,y,z]) - [D(x),y,z] - [x,D(y),z] - [x,y,D(z)]\| \\
&\quad + \|H([x,y,z]) - [H(x),H(y),H(z)]\| \\
&= \frac{1}{27^n}\|g\left(27^n[x,y,z]\right) - [g\left(3^n x\right), 3^n y, 3^n z] - [3^n x, g(3^n y), 3^n z] \\
&\quad - [3^n x, 3^n y, g(3^n z)]\| + \frac{1}{27^n}\|h\left(27^n[x,y,z]\right) \\
&\quad - [h\left(3^n x\right), h\left(3^n y\right), h(3^n z)]\| \\
&\le \frac{1}{27^n}\varphi\left(3^n x, 3^n y, 3^n z\right) \le \frac{1}{3^n}\varphi\left(3^n x, 3^n y, 3^n z\right),
\end{aligned}$$

which tends to zero as $n \to \infty$, by (2.11). So

$$\begin{aligned}
D([x,y,z]) - [D(x),y,z] - [x,D(y),z] - [x,y,D(z)] &= 0, \\
H([x,y,z]) - [H(x),H(y),H(z)] &= 0
\end{aligned}$$

for all $x, y, z \in A$. Hence, the mapping $D : A \to A$ is a ternary derivation and the mapping $H : A \to A$ is a ternary homomorphism. □

Corollary 2. *Let $r < 1$ and θ be non-negative real numbers and $g, h : A \to A$ be mappings satisfying $g(0) = h(0) = 0$, (2.9) and (2.10). Then, there exist a unique ternary derivation $D : A \to A$ and a unique ternary homomorphism $H : A \to A$ such that*

$$\|g(x) - D(x)\| + \|h(x) - H(x)\| \le \frac{3^r\theta}{3 - 3^r}\|x\|^r$$

for all $x \in A$.

Proof. The proof follows from Theorem 3 by $\varphi(x,y,z) = \theta(\|x\|^r + \|y\|^r + \|z\|^r)$ for all $x, y, z \in A$. □

3. Stability of Additive-Additive (s,t)-Functional Inequality (0.1): A Fixed Point Method

Using the fixed point method, we prove the Hyers–Ulam stability of ternary derivations and ternary homomorphisms in C^*-ternary algebras associated with the additive-additive (s,t)-functional inequality (0.1).

Theorem 4. *Let $\varphi : A^3 \to [0,\infty)$ be a function such that there exists an $L < 1$ with*

$$\varphi\left(\frac{x}{3}, \frac{y}{3}, \frac{z}{3}\right) \le \frac{L}{27}\varphi(x,y,z) \le \frac{L}{3}\varphi(x,y,z) \tag{3.1}$$

for all $x, y, z \in A$. Let $g, h : A \to A$ be mappings satisfying $g(0) = h(0) = 0$, (2.3) and (2.4). Then, there exist a unique ternary derivation $D : A \to A$ and a unique ternary homomorphism $H : A \to A$ such that

$$\|g(x) - D(x)\| + \|h(x) - H(x)\| \le \frac{1}{1-L}\varphi(x,0,0) \tag{3.2}$$

for all $x \in A$.

Proof. It follows from (3.1) that

$$\sum_{j=1}^{\infty} 27^j \varphi\left(\frac{x}{3^j}, \frac{y}{3^j}, \frac{z}{3^j}\right) \le \sum_{j=1}^{\infty} 27^j \frac{L^j}{27^j}\varphi(x,y) = \frac{L}{1-L}\varphi(x,y,z) < \infty$$

for all $x, y, z \in A$. By Theorem 2, there exist a unique ternary derivation $D : A \to A$ and a unique ternary homomorphism $H : A \to A$ satisfying (2.5).

Letting $\lambda = 1$ and $y = z = 0$ in (2.3), we get

$$\left\|3g\left(\frac{x}{3}\right) - g(x)\right\| + \left\|3h\left(\frac{x}{3}\right) - h(x)\right\| \le \varphi(x,0,0) \tag{3.3}$$

for all $x \in A$.

Consider the set

$$S := \{(g, h) : (A, A) \to (A, A),\quad g(0) = h(0) = 0\}$$

and introduce the generalized metric on S:

$$d((g,h),(g_1,h_1)) = \inf\{\mu \in \mathbb{R}_+ : \|g(x) - g_1(x)\| + \|h(x) - h_1(x)\| \\ \leq \mu\varphi(x,0,0), \forall x \in A\},$$

where, as usual, $\inf \phi = +\infty$. It is easy to show that (S, d) is complete [31].

Now, we consider the linear mapping $J : S \to S$ such that

$$J(g,h)(x) := \left(3g\left(\frac{x}{3}\right), 3h\left(\frac{x}{3}\right)\right)$$

for all $x \in A$.

Let $(g, h), (g_1, h_1) \in S$ be given such that $d((g, h), (g_1, h_1)) = \varepsilon$. Then,

$$\|g(x) - g_1(x)\| + \|h(x) - h_1(x)\| \leq \varepsilon\varphi(x, 0, 0)$$

for all $x \in A$. Since

$$\left\|3g\left(\frac{x}{3}\right) - 3g_1\left(\frac{x}{3}\right)\right\| + \left\|3h\left(\frac{x}{3}\right) - 3h_1\left(\frac{x}{3}\right)\right\| \\ \leq \varepsilon\varphi\left(\frac{x}{3}, 0, 0\right) \leq 3\varepsilon\frac{L}{3}\varphi(x,0,0) = L\varepsilon\varphi(x,0,0)$$

for all $x \in A$, $d(J(g, h), J(g_1, h_1)) \leq L\varepsilon$. This means that

$$d(J(g,h), J(g_1,h_1)) \leq Ld((g,h),(g_1,h_1))$$

for all $(g, h), (g_1, h_1) \in S$.

It follows from (3.3) that $d((g, h), (Jg, Jh)) \leq 1$.

By Theorem 1, there exist mappings $D, H : A \to A$ satisfying the following:

(1) (D, H) is a fixed point of J, i.e.

$$D(x) = 3D\left(\frac{x}{3}\right), \qquad H(x) = 3H\left(\frac{x}{3}\right) \tag{3.4}$$

for all $x \in A$. The mapping (D, H) is a unique fixed point of J. This implies that (D, H) is a unique mapping satisfying (3.4) such that there exists a $\mu \in (0, \infty)$ satisfying

$$\|g(x) - D(x)\| + \|h(x) - H(x)\| \leq \mu\varphi(x,0,0)$$

for all $x \in A$;

(2) $d(J^l(g,h),(D,H)) \to 0$ as $l \to \infty$. This implies the equality

$$\lim_{l\to\infty} 3^l g\left(\frac{x}{3^l}\right) = D(x), \quad \lim_{l\to\infty} 3^l h\left(\frac{x}{3^l}\right) = H(x)$$

for all $x \in A$;

(3) $d((g,h),(D,H)) \le \frac{1}{1-L} d((g,h),J(g,h))$, which implies

$$\|g(x) - D(x)\| + \|h(x) - H(x)\| \le \frac{1}{1-L}\varphi(x,0,0)$$

for all $x \in A$. Thus, we get the inequality (3.2).

The rest of the proof is the same as in the proof of Theorem 2. □

Corollary 3. *Let $r > 3$ and θ be non-negative real numbers and $g, h : A \to A$ be mappings satisfying $g(0) = h(0) = 0$, (2.9) and (2.10). Then, there exist a unique ternary derivation $D : A \to A$ and a unique ternary homomorphism $H : A \to A$ such that*

$$\|g(x) - D(x)\| + \|h(x) - H(x)\| \le \frac{3^r\theta}{3^r - 3}\|x\|^r$$

for all $x \in A$.

Proof. The proof follows from Theorem 4 by taking $L = 3^{1-r}$ and $\varphi(x,y,z) = \theta(\|x\|^r + \|y\|^r + \|z\|^r)$ for all $x, y, z \in A$. □

Theorem 5. *Let $\varphi : A^3 \to [0,\infty)$ be a function such that there exists an $L < 1$ with*

$$\varphi(x,y,z) \le 27L\varphi\left(\frac{x}{2},\frac{y}{2},\frac{z}{2}\right) \tag{3.5}$$

for all $x, y, z \in A$. Let $g, h : A \to A$ be mappings satisfying $g(0) = h(0) = 0$, (2.3) and (2.4). Then, there exist a unique ternary derivation $D : A \to A$ and a unique ternary homomorphism $H : A \to A$ such that

$$\|g(x) - D(x)\| + \|h(x) - H(x)\| \le \frac{L}{1-L}\varphi(x,0,0) \tag{3.6}$$

for all $x \in A$.

Proof. It follows from (3.5) that

$$\sum_{j=1}^{\infty} \frac{1}{27^j} \varphi\left(3^j x, 3^j y, 3^j z\right) \leq \sum_{j=1}^{\infty} \frac{1}{27^j} (27L)^j \varphi(x, y, z)$$
$$= \frac{L}{1-L} \varphi(x, y, z) < \infty$$

for all $x, y, z \in A$. By Theorem 3, there exist a unique ternary derivation $D : A \to A$ and a unique ternary homomorphism $H : A \to A$ satisfying (2.12).

Let (S, d) be the generalized metric space defined in the proof of Theorem 4.

Now, we consider the linear mapping $J : S \to S$ such that

$$J(g, h)(x) := \left(\frac{1}{3} g\left(3x\right), \frac{1}{3} h\left(3x\right) \right)$$

for all $x \in A$.

It follows from (3.3) that

$$\left\| g(x) - \frac{1}{3} g(3x) \right\| + \left\| h(x) - \frac{1}{3} h(3x) \right\| \leq \frac{1}{3} \varphi(3x, 0, 0) \leq \frac{3L}{3} \varphi(x, 0, 0)$$

for all $x \in A$. Thus, we get the inequality (3.6).

The rest of the proof is similar to the proof of Theorem 4. □

Corollary 4. *Let $r < 1$ and θ be non-negative real numbers and $g, h : A \to A$ be mappings satisfying $g(0) = h(0) = 0$, (2.9) and (2.10). Then, there exist a unique ternary derivation $D : A \to A$ and a unique ternary homomorphism $H : A \to A$ such that*

$$\|g(x) - D(x)\| + \|h(x) - H(x)\| \leq \frac{3^r \theta}{3 - 3^r} \|x\|^r$$

for all $x \in A$.

Proof. The proof follows from Theorem 5 by taking $L = 3^{r-1}$ and $\varphi(x, y, z) = \theta(\|x\|^r + \|y\|^r + \|z\|^r)$ for all $x, y, z \in A$. □

References

[1] M. R. Abdollahpour, R. Aghayaria and M. T. Rassias. Hyers-Ulam stability of associated Laguerre differential equations in a subclass of analytic functions. *J. Math. Anal. Appl.*, 437, 2016, 605–612.

[2] M. R. Abdollahpour and M. T. Rassias. Hyers-Ulam stability of hypergeometric differential equations. *Aequ. Math.*, 93, 2019, 691–698.

[3] J. Aczel and J. Dhombres. *Functional Equations in Several Variables.* Cambridge University Press, Cambridge, 1989.

[4] M. Amyari and M. S. Moslehian. Approximate homomorphisms of ternary semigroups. *Lett. Math. Phys.*, 77, 2006, 1–9.

[5] T. Aoki. On the stability of the linear transformation in Banach spaces. *J. Math. Soc. Jpn.*, 2, 1950, 64–66.

[6] M. Amyari, C. Baak and M. Moslehian. Nearly ternary derivations. *Taiwan. J. Math.*, 11, 2007, 1417–1424.

[7] L. Cădariu and V. Radu. Fixed points and the stability of Jensen's functional equation. *J. Inequal. Pure Appl. Math.*, 4(1), 2003. Paper No. 4.

[8] L. Cădariu and V. Radu. On the stability of the Cauchy functional equation: A fixed point approach. *Grazer Math. Ber.*, 346, 2004, 43–52.

[9] L. Cădariu and V. Radu. Fixed point methods for the generalized stability of functional equations in a single variable. *Fixed Point Theory Appl.*, 2008, 2008. Article ID 749392.

[10] S. Czerwik. *Functional Equations and Inequalities in Several Variables.* World Scientific, Singapore, 2002.

[11] J. Diaz and B. Margolis. A fixed point theorem of the alternative for contractions on a generalized complete metric space. *Bull. Am. Math. Soc.*, 74, 1968, 305–309.

[12] I. EL-Fassi. Generalized hyperstability of a Drygas functional equation on a restricted domain using Brzdek's fixed point theorem. *J. Fixed Point Theory Appl.*, 19, 2017, 2529–2540.

[13] M. Eshaghi Gordji, A. Fazeli and C. Park. 3-Lie multipliers on Banach 3-Lie algebras. *Int. J. Geom. Meth. Mod. Phys.*, 9(7), 2012. Article ID 1250052.

[14] M. Eshaghi Gordji, M. B. Ghaemi and B. Alizadeh. A fixed point method for perturbation of higher ring derivations in non-Archimedean Banach algebras. *Int. J. Geom. Meth. Mod. Phys.*, 8(7), 2011, 1611–1625.

[15] G. Z. Eskandani and P. Găvruta. Hyers-Ulam-Rassias stability of pexiderized Cauchy functional equation in 2-Banach spaces. *J. Nonlinear Sci. Appl.*, 5, 2012, 459–465.

[16] W. Fechner. Stability of a functional inequalities associated with the Jordan-von Neumann functional equation. *Aequ. Math.*, 71, 2006, 149–161.

[17] P. Găvruta. A generalization of the Hyers-Ulam-Rassias stability of approximately additive mappings. *J. Math. Anal. Appl.*, 184, 1994, 431–436.

[18] A. Gilányi. Eine zur Parallelogrammgleichung äquivalente Ungleichung. *Aequ. Math.*, 62, 2001, 303–309.

[19] A. Gilányi. On a problem by K. Nikodem. *Math. Inequal. Appl.*, 5, 2002, 707–710.

[20] D. H. Hyers. On the stability of the linear functional equation. *Proc. Nat. Acad. Sci. USA*, 27, 1941, 222–224.

[21] D. H. Hyers, G. Isac and T. M. Rassias. *Stability of Functional Equations in Several Variables*. Birkhäuser, New York, 1998.

[22] D. H. Hyers and T. M. Rassias. Approximate homomorphisms. *Aequ. Math.*, 44, 1992, 125–153.

[23] G. Isac and T. M. Rassias. Stability of ψ-additive mappings: Applications to nonlinear analysis. *Int. J. Math. Math. Sci.*, 19, 1996, 219–228.

[24] S. Jung. Hyers–Ulam–Rassias. *Stability of Functional Equations in Nonlinear Analysis*. Springer, New York, 2011.

[25] S. Jung, C. Mortici and M. T. Rassias. On a functional equation of trigonometric type. *Appl. Math. Comput.*, 252, 2015, 294–303.

[26] S. Jung, D. Popa and M. T. Rassias. On the stability of the linear functional equation in a single variable on complete metric groups. *J. Global Optim.*, 59, 2014, 165–171.

[27] S. Jung and M. T. Rassias. A linear functional equation of third order associated to the Fibonacci numbers. *Abstr. Appl. Anal.*, 2014, 2014. Article ID 137468.

[28] Y. Lee, S. Jung and M. T. Rassias. On an n-dimensional mixed type additive and quadratic functional equation. *Appl. Math. Comput.*, 228, 2014, 13–16.

[29] Y. Lee, S. Jung and M. T. Rassias. Uniqueness theorems on functional inequalities concerning cubic-quadratic-additive equation. *J. Math. Inequal.*, 12, 2018, 43–61.

[30] P. Kannappan. *Functional Equations and Inequalities with Applications*. Springer, New York, 2009.

[31] D. Miheţ and V. Radu. On the stability of the additive Cauchy functional equation in random normed spaces. *J. Math. Anal. Appl.*, 343, 2008, 567–572.

[32] C. Mortici, S. Jung and M. T. Rassias. On the stability of a functional equation associated with the Fibonacci numbers. *Abstr. Appl. Anal.*, 2014, 2014. Article ID 546046.

[33] M. S. Moslehian. Almost derivations on C^*-ternary rings. *Bull. Belg. Math. Soc.-Simon Stevin*, 14, 2006, 135–142.

[34] I. Nikoufar. Jordan (θ, ϕ)-derivations on Hilbert C^*-modules. *Indag. Math.*, 26, 2015, 421–430.

[35] C. Park. Homomorphisms between Poisson JC^*-algebras. *Bull. Braz. Math. Soc.*, 36, 2005, 79–97.

[36] C. Park. Additive ρ-functional inequalities and equations. *J. Math. Inequal.*, 9, 2015, 17–26.

[37] C. Park. Additive ρ-functional inequalities in non-Archimedean normed spaces. *J. Math. Inequal.*, 9, 2015, 397–407.

[38] C. Park. Fixed point method for set-valued functional equations. *J. Fixed Point Theory Appl.*, 19, 2017, 2297–2308.

[39] C. Park. Biderivations and bihomomorphisms in Banach algebras. *Filomat*, 33, 2019, 2317–2328.

[40] C. Park and M. T. Rassias. Additive functional equations and partial multipliers in C^*-algebras. *Rev. R. Acad. Cien. Exactas Fís. Nat. Ser. A Mat. RACSAM*, 113, 2019, 2261–2275.

[41] V. Radu. The fixed point alternative and the stability of functional equations. *Fixed Point Theory*, 4, 2003, 91–96.

[42] T. M. Rassias. On the stability of the linear mapping in Banach spaces. *Proc. Am. Math. Soc.*, 72, 1978, 297–300.

[43] J. Rätz. On inequalities associated with the Jordan-von Neumann functional equation. *Aequ. Math.*, 66, 2003, 191–200.

[44] S. M. Ulam. *A Collection of the Mathematical Problems*. Interscience Publishers, New York, 1960.

[45] H. Zettl. A characterization of ternary rings of operators. *Adv. Math.*, 48, 1983, 117–143.

https://doi.org/10.1142/9789819820498_0013

Chapter 13

On Discrete Operator Linking Adjoint Bernoulli Polynomials

Deepak Malik[*,‡], **Vijay Gupta**[*,§] **and Michael Th. Rassias**[†,¶]

[*]*Department of Mathematics*
Netaji Subhas University of Technology
Sector 3 Dwarka, New Delhi 110078, India
[†]*Department of Mathematics and Engineering Sciences*
Hellenic Military Academy
Vari Attikis, Greece
[‡]*malikdeepak1510@gmail.com, deepak.malik.phd22@nsut.ac.in*
[§]*vijaygupta2001@hotmail.com, vijay@nsut.ac.in*
[¶]*mthrassias@yahoo.com*

In this chapter, we introduce new discrete operators defined on $[0, \infty)$, which are based on adjoint Bernoulli polynomials. We derive the moment generating function and establish several approximation properties for such operators including the rate of approximation in uniform norm, the quantitative Voronovskaja-type theorem in weighted space and for functions with the exponential growth. We also illustrate the convergence graphically and numerically. We continue by defining another new operators based on the composition with Szász–Mirakyan operators and study their convergence properties in terms of weighted modulus of continuity and also establish difference estimates.

1. Introduction

The theory of approximation by positive linear operators occupies a distinguished place in classical and modern analysis. It provides a robust and elegant framework for approximating functions by sequences of simpler, more tractable forms, typically polynomials.

Positive linear operators have been studied extensively for their ability to approximate continuous functions, with seminal contributions such as Bernstein polynomials laying the groundwork for this field. One of the most appealing aspects of this theory is that positivity preserves certain qualitative features of the original function, such as non-negativity and monotonicity, making these operators particularly suitable for applications in numerical analysis, computer-aided geometric design and statistical modeling.

In recent decades, the development of generalized positive operators and their convergence properties has drawn significant attention, especially in the context of functional spaces beyond $C[0,1]$ (class of all continuous functions on $[0,1]$). This ongoing research not only extends the classical results but also uncovers new connections with probability theory, sampling theory and even machine learning. We refer the reader to some key contributions in this field (see Refs. [1–6]).

Appell polynomials, named after the French mathematician Paul Émile Appell, are a significant class of polynomial sequences characterized by their unique derivative property. Specifically, for a sequence of polynomials $\{\tilde{A}_k(x)\}_{k=0}^{\infty}$, they satisfy the derivative rule

$$\frac{d}{dx}\tilde{A}_k(x) = k\tilde{A}_{k-1}(x), \quad k \geq 1.$$

This property indicates that the derivative of the kth polynomial is proportional to the $(k-1)$th polynomial, scaled by k. The polynomials are generated through an exponential generating function given by

$$A(t)e^{xt} = \sum_{k=0}^{\infty} \tilde{A}_k(x)\frac{t^k}{k!},$$

where $A(t)$ is an analytic function with $A(0) \neq 0$. Some of the most notable examples of Appell polynomials are the Bernoulli polynomials that arise when $A(t) = t(e^t - 1)^{-1}$, Hermite polynomials when

$A(t) = e^{-t^2/2}$, Gould–Hopper polynomials and Euler polynomials. These polynomials represent a rich area of study within mathematics, linking various branches such as analysis, probability and number theory. Their unique properties and applications make them a vital tool for mathematicians. Some notable operators based on these polynomials are introduced here (see Refs. [7, 8]).

Bernoulli polynomials play an important role in various areas of mathematics, including numerical analysis and approximation theory (see Refs. [9–11]). In numerical analysis, Bernoulli polynomials are used in the Euler–Maclaurin summation formula, which provides a powerful tool for approximating sums and integrals (see Ref. [12]). Bernoulli polynomials that involve in generalization of the Hurwitz–Lerch zeta function (see Ref. [13]).

Following Ref. [14, Definition 3.2], the adjoint Bernoulli polynomials are defined using a modified generating function, which involves changing the function $A(t)$ in the generating function of Bernoulli polynomials to $\frac{1}{A(t)}$.

In this chapter, we introduce new discrete operators, linking with adjoint Bernoulli polynomials. The adjoint Bernoulli polynomials $\tilde{B}_k(nx)$ are defined by means of the exponential generating function; the operators $\mathcal{H}_n$ for $f \in C[0,\infty), x \geq 0$ and $n \in \mathbb{N}$ are defined as follows:

$$(\mathcal{H}_n f)(x) := \frac{e^{-nx}}{(e-1)} \sum_{k=0}^{\infty} \tilde{B}_k(nx) \frac{f(kn^{-1})}{k!}, \tag{1.1}$$

where $\tilde{B}_k(nx)$ satisfy the generating function:

$$\sum_{k=0}^{\infty} \tilde{B}_k(nx) \frac{t^k}{k!} = \frac{e^{nxt}(e^t-1)}{t}. \tag{1.2}$$

As a special case, when $\tilde{B}_k(nx) = (nx)^k(e-1)$, then these operators $\mathcal{H}_n$ reduce to the well-known Szász–Mirakyan operators (see Refs. [15, 16]). However, unlike Szász–Mirakyan operators, these operators are not of the exponential type. Recently, Yilmaz [17] has introduced the Kantorovich variant of the operators $\mathcal{H}_n$ and studied some approximation properties on a compact set $[0,1]$. It should be noted that the operators are valid for the entire positive real axis. In Ref. [18], another linear positive operators based on adjoint

Bernoulli polynomials were defined on $[0, \infty)$. Throughout this chapter, we use the notations $\exp_\gamma(t) := e^{\gamma t}$ and $e_m(t) := t^m$.

2. Convergence Estimates

In this section, we derive the moment generating function (MGF) of the operators $\mathcal{H}_n$, and central moments, examine their convergence in the uniform norm and establish quantitative Voronovskaja-type results in various spaces.

Lemma 1. *The MGF of the operators $\mathcal{H}_n$ is given as*

$$(\mathcal{H}_n \exp_\gamma)(x) = \exp(nx(e^{\gamma/n} - 1)) \frac{\left(e^{e^{\frac{\gamma}{n}}} - 1\right)}{e^{\gamma/n}(e-1)}.$$

Suppose $\mathcal{H}_{n,e_m}$ be the mth moment of $\mathcal{H}_n$, i.e. $\mathcal{H}_{n,e_m} := (\mathcal{H}_n e_m)(x)$, then we have

$$\mathcal{H}_{n,e_0}(x) = 1, \quad \mathcal{H}_{n,e_1}(x) = x + \frac{1}{(e-1)n},$$

$$\mathcal{H}_{n,e_2}(x) = x^2 + \frac{(e-1) + n(e+1)x}{(e-1)n^2},$$

$$\mathcal{H}_{n,e_3}(x) = x^3 + \frac{3en^2x^2 + (4e-1)nx + e + 1}{(e-1)n^3},$$

$$\mathcal{H}_{n,e_4}(x) = x^4 + \frac{\begin{array}{c}(6e-2)n^3x^3 + (13e-1)n^2x^2 \\ + 11enx + nx + (4e-1)\end{array}}{(e-1)n^4},$$

$$\mathcal{H}_{n,e_5}(x) = x^5 + \frac{\begin{array}{c}(10e-5)n^4x^4 + (35e-5)n^3x^3 \\ + 55en^2x^2 + (41e-1)nx + 11e + 1\end{array}}{(e-1)n^5},$$

$$\mathcal{H}_{n,e_6}(x) = x^6 + \frac{\begin{array}{c}(15e-9)n^5x^5 + (80e-20)n^4x^4 \\ + (200e-10)n^3x^3 + (256e-1)n^2x^2\end{array}}{(e-1)n^6}$$

$$+ \frac{(162e+1)nx + 41e - 1}{(e-1)n^6}.$$

Proof. By using (1.2), we can write

$$(\mathcal{H}_n \exp_\gamma)(x) = \frac{e^{-nx}}{(e-1)} \sum_{k=0}^{\infty} \frac{B_k(nx)}{k!} e^{\frac{\gamma k}{n}}$$
$$= \exp\left(nx(e^{\gamma/n}-1) - \frac{\gamma}{n}\right) \frac{\left(e^{e^{\frac{\gamma}{n}}}-1\right)}{(e-1)}.$$

We have the following connection between the MGF and the mth-order moment

$$\mathcal{H}_{n,e_m}(x) = \left[\frac{\partial^m}{\partial \gamma^m}\left(\exp\left(nx(e^{\gamma/n}-1) - \frac{\gamma}{n}\right) \frac{\left(e^{e^{\frac{\gamma}{n}}}-1\right)}{(e-1)}\right)\right]_{\gamma=0}.$$

By doing some computations on the afore-mentioned relation, we derive the corresponding moments. □

Lemma 2. *The central moment of mth order of the operators $\mathcal{H}_n$ is denoted by $\mu_{n,m}(x) := (\mathcal{H}_{n,m}(e_1 - xe_0)^m)(x)$, where $m = 0, 1, 2, \ldots$, then we have*

$$\mu_{n,m}(x) = \left[\frac{\partial^m}{\partial \gamma^m}\left(\exp\left(nx(e^{\gamma/n}-1) - \frac{\gamma}{n}\right) \frac{\left(e^{e^{\frac{\gamma}{n}}}-1\right)}{(e-1)} - \gamma x\right)\right]_{\gamma=0}.$$

In particular,

$$\mu_{n,0}(x) = 1, \quad \mu_{n,1}(x) = \frac{1}{(e-1)n}, \quad \mu_{n,2}(x) = \frac{nx+1}{n^2},$$

$$\mu_{n,3}(x) = \frac{enx + 2nx + e + 1}{(e-1)n^3},$$

$$\mu_{n,4}(x) = \frac{(3e-3)n^2x^2 + 7enx - 3nx + 4e - 1}{(e-1)n^4},$$

$$\mu_{n,5}(x) = \frac{(10e+5)n^2x^2 + (21e+4)nx + 11e + 1}{(e-1)n^5},$$

$$\mu_{n,6}(x) = \frac{15(e-1)n^3x^3 + 10(7e-1)n^2x^2 + (96e-5)nx - 1}{(e-1)n^6},$$

$$\mu_{n,7}(x) = \frac{105en^3x^3 + (371e+14)n^2x^2 + (428e+6)nx + 162e + 1}{(e-1)n^7},$$

$$\mu_{n,8}(x) = \frac{\begin{array}{c}105(e-1)n^4x^4 + 70(13e-1)n^3x^3 \\ +\,7(317e-3)n^2x^2 + (2129e-7)nx\end{array}}{(e-1)n^8} + \frac{715e-1}{(e-1)n^8}.$$

Let us denote $C^*[0,\infty)$, as follows:

$$C^*[0,\infty) = \left\{ f \;:\; f \in C[0,\infty) \text{ and } \lim_{x\to\infty} |f(x)| < \infty \right\}.$$

Following Ref. [19], the weighted moduli of continuity $\omega^*(f,\xi)$ defined as follows:

$$\omega^*(f,\xi) = \sup_{\substack{|e^{-u}-e^{-v}|\le\xi \\ u,v\in[0,\infty)}} |f(u)-f(v)|, \quad \text{where } f \in C^*[0,\infty) \text{ and } \xi > 0.$$

Theorem 1. *Let $f \in C^*[0,\infty)$ and $\|.\|_\infty$ denote the uniform norm, then this holds*

$$\|(\mathcal{H}_n f)(x) - f(x)\|_\infty \le 2\omega^*(f, (\Theta_n + 2\Upsilon_n)^{1/2}),$$

where

$$\Upsilon_n = \|(\mathcal{H}_n \exp_{-1})(x) - \exp_{-1}(x)\|_\infty \to 0, \quad n \to \infty,$$

and

$$\Theta_n = \|(\mathcal{H}_n \exp_{-2})(x) - \exp_{-2}(x)\|_\infty \to 0, \quad n \to \infty.$$

Proof. Following Ref. [19], if the norms $\|(\mathcal{H}_n \exp_s)(x) - \exp_s(x)\|_\infty$, for $s = 0, -1, -2$, denote as Λ_n, Υ_n and Θ_n, respectively, which approach to zero as $n \to \infty$, then for $f \in C^*[0,\infty)$, we have

$$\begin{aligned}&\|(\mathcal{H}_n f)(x) - f(x)\|_\infty \\ &\quad \le \Lambda_n \|f\|_\infty + (2+\Lambda_n)\omega^*(f, (\Lambda_n + \Theta_n + 2\Upsilon_n)^{1/2}).\end{aligned}$$

These operators $\mathcal{H}_n$ preserve constant functions, i.e. $\Lambda_n = 0$. We have to compute Υ_n and Θ_n, we have

$$(\mathcal{H}_n \exp_{-1})(x) = \exp_{-1}(x) + \frac{e^{-x}}{n}\left(\frac{x}{2} - \frac{1}{(e-1)}\right) + \frac{e^{-x}}{24n^2}\left(3x^2 - \frac{4x(e+2)}{(e-1)} + 12\right) + \mathcal{O}(n^{-3}).$$

By observation for $a, r \in \mathbb{N}$, we have

$$\sup_{x\geq 0} x^r e^{-ax} = \left(\frac{r}{a}\right)^r e^{-r}. \tag{2.1}$$

By using (2.1), we obtain

$$\begin{aligned}\Upsilon_n &= \sup_{x\geq 0} |(\mathcal{H}_n \exp_{-1})(x) - \exp_{-1}(x)| \\ &\leq \frac{1}{n}\left(\frac{1}{(e-1)} + \frac{1}{2e} + 1\right) + \frac{1}{24n^2}\left(12 + \frac{4(e+2)}{e(e-1)} + \frac{4}{e^2}\right) \\ &\quad + \mathcal{O}(n^{-3}) \\ &\leq \mathcal{O}(n^{-1}).\end{aligned}$$

Similarly,

$$(\mathcal{H}_n \exp_{-2})(x) = \exp_{-2}(x) + \frac{2e^{-2x}}{n}\left(x - \frac{1}{(e-1)}\right) + \frac{2e^{-2x}}{n^2}\left(x^2 - \frac{2x(e+2)}{3(e-1)} + 1\right) + \mathcal{O}(n^{-3}).$$

Again by using (2.1), we obtain

$$\begin{aligned}\Theta_n &= \sup_{x\geq 0} |(\mathcal{H}_n \exp_{-2})(x) - \exp_{-2}(x)| \\ &\leq \frac{1}{n}\left(\frac{2}{(e-1)} + \frac{1}{e}\right) + \frac{2}{n^2}\left(\frac{1}{e^2} + \frac{e+2}{3(e-1)} + 1\right) + \mathcal{O}(n^{-3}) \\ &\leq \mathcal{O}(n^{-1}).\end{aligned}$$

□

Theorem 2. *Let $f, f'' \in C^*[0,\infty)$, then for any $x \in [0,\infty)$, we have*

$$\left| n[(\mathcal{H}_n f)(x) - f(x)] - \frac{f'(x)}{e-1} - \frac{x f''(x)}{2} \right|$$
$$\leq \frac{1}{2n} \left| f''(x) \right| + 2\omega^*(f'', n^{-1/2})$$
$$\times \left\{ \frac{nx+1}{n} + \sqrt{\frac{(3e-3)n^2x^2 + 7enx - 3nx + 4e - 1}{(e-1)n^2}} \right.$$
$$\left. \times \sqrt{n^2 \left(\mathcal{H}_n \left(\frac{1}{e^x} - \frac{1}{e^t} \right)^4 \right)(x)} \right\}.$$

Proof. Using Taylor's formula for $f \in C^*[0,\infty)$, with $x < \eta < t$, we have

$$f(t) = \sum_{k=0}^{2} \frac{(e_1 - xe_0)^k}{k!} f^{(k)}(x) + (e_1 - xe_0)^2 \frac{f''(\eta) - f''(x)}{2}.$$

Applying, the operators $\mathcal{H}_n$ on f and using Lemma 2, we get

$$\left| n[(\mathcal{H}_n f)(x) - f(x)] - \left(\frac{1}{e-1} \right) f'(x) - \left(\frac{x}{2} \right) f''(x) \right|$$
$$\leq \frac{1}{2n} \left| f''(x) \right| + \left| n \left(\mathcal{H}_n \left(\frac{f''(\eta) - f''(x)}{2} \right) (e_1 - xe_0)^2 \right)(x) \right|.$$

Next, by the property used in Ref. [20] and employing Cauchy–Schwarz inequality, we obtain

$$\left| n \left(\mathcal{H}_n \left(\frac{f''(\eta) - f''(x)}{2} \right) (e_1 - xe_0)^2 \right)(x) \right|$$
$$\leq 2n\omega^* \left(f'', \frac{1}{\sqrt{n}} \right) \mu_{n,2}(x)$$
$$+ 2n^2\omega^* \left(f'', \frac{1}{\sqrt{n}} \right) \mathcal{H}_n \left(\left(\frac{1}{e^x} - \frac{1}{e^t} \right)^2 (e_1 - xe_0)^2 \right)(x)$$

$$\leq 2\omega^*\left(f'', \frac{1}{\sqrt{n}}\right)\Bigg\{\frac{nx+1}{n}$$

$$+ n^2\sqrt{\frac{(3e-3)n^2x^2+7enx-3nx+4e-1}{(e-1)n^4}\left(\mathcal{H}_n\left(\frac{1}{e^x}-\frac{1}{e^t}\right)^4\right)(x)}\Bigg\}.$$

□

Remark 1. For the operators $\mathcal{H}_n$, the following limits hold:

- $\lim_{n\to\infty} n^2\mu_{n,4}(x) = 3x^2$,
- $\lim_{n\to\infty} n^2\left(\mathcal{H}_n\left(\frac{1}{e^x}-\frac{1}{e^t}\right)^4\right)(x) = 3x^2e^{-4x}$.

Corollary 1. *Let $f, f'' \in C^*[0,\infty)$, then for any $x \geq 0$, the following holds:*

$$\lim_{n\to\infty} n\left[(\mathcal{H}_nf)(x) - f(x)\right] = \frac{f'(x)}{(e-1)} + \frac{xf''(x)}{2}.$$

Proof. Apply the limit on the estimation given in Theorem 2 and using Remark 1, we get the desired result. □

$C_A[0,\infty)$ denotes the space of all functions $h(x) \in C[0,\infty)$ such that $|h(x)| \leq A(1+x^2)$ for any fixed $A \geq 0$ and $\lim_{x\to\infty}\frac{|h(x)|}{1+x^2}$ exists and is finite. Following Ref. [21], the weighted moduli of continuity $\Omega(h,\delta)$ is given by

$$\Omega(h,\delta) = \sup_{\substack{0\leq\xi\leq\delta \\ t\in[0,\infty)}} \frac{|h(t+\xi)-h(t)|}{(1+\xi^2)(1+t^2)}, \quad h \in C_A[0,\infty).$$

Theorem 3. *Let f be any function with polynomial growth, $f^{iv} \in C_A[0,\infty)$ and $x \in [0,\infty)$, then the following inequality holds:*

$$\Bigg|n\left[n[(\mathcal{H}_nf)(x)-f(x)] - \frac{f'(x)}{(e-1)} - \frac{xf''(x)}{2}\right]$$

$$-\frac{f''(x)}{2} - \frac{(e+2)xf'''(x)}{6(e-1)} - \frac{(3e-3)x^2f^{iv}(x)}{24(e-1)}\Bigg|$$

$$\leq \frac{(e+1)\,|f'''(x)|}{6(e-1)n} + \left(\frac{7ex-3x}{24n(e-1)} + \frac{4e-1}{24n^2(e-1)}\right)|f^{iv}(x)|$$
$$+ \frac{4}{3}(x^2+1)\left(\frac{(3e-3)n^2x^2+7enx-3nx+4e-1}{(e-1)n^2}\right)$$
$$\times\, \Omega\left(f^{iv}, \left(\frac{\mu_{n,8}(x)}{\mu_{n,4}(x)}\right)^{\frac{1}{4}}\right).$$

Proof. Using Taylor's formula for f, where $x < \eta < t$, we can write

$$f(t) = \sum_{k=0}^{4} \frac{(e_1 - xe_0)^k f^{(k)}(x)}{k!} + (e_1 - xe_0)^4 \frac{f^{iv}(\eta) - f^{iv}(x)}{4!}.$$

Applying the operators $\mathcal{H}_n$ and using Lemma 2, we immediately get

$$(\mathcal{H}_n f)(x) - f(x)$$
$$= \mu_{n,1}(x)f'(x) + 0.5\mu_{n,2}(x)f''(x) + \frac{\mu_{n,3}(x)f'''(x)}{3!}$$
$$+ \frac{\mu_{n,4}(x)f^{iv}(x)}{4!} + \left(\mathcal{H}_n \frac{(e_1 - xe_0)^4}{4!}(f^{iv}(\eta) - f^{iv}(x))\right)(x),$$

where $x < \eta < t$ and $f^{iv}(\eta) - f^{iv}(x) \to 0$ as $t \to x$. Now, again using Lemma 2, we get

$$\Bigg|(\mathcal{H}_n f)(x) - f(x) - \frac{f'(x)}{(e-1)n}$$
$$- \frac{(nx+1)f''(x)}{2n^2} - \frac{(e+2)nx+e+1}{(e-1)n^3}\frac{f'''(x)}{3!}$$
$$- \frac{(3e-3)n^2x^2+7enx-3nx+4e-1}{(e-1)n^4}\frac{f^{iv}(x)}{4!}\Bigg|$$
$$\leq \left|\left(\mathcal{H}_n\left(\frac{f^{iv}(\eta) - f^{iv}(x)}{4!}\right)(e_1 - xe_0)^4\right)(x)\right|. \qquad (2.2)$$

From Ref. [22], we have the following relation between function f and $\Omega(f, \delta)$, for $\delta > 0$:

$$\frac{|f^{iv}(\eta) - f^{iv}(x)|}{4} \leq \left(\frac{\delta^4 + (e_1 - xe_0)^4}{\delta^4}\right)(x^2+1)(1+\delta^2)^2\Omega(f^{iv}, \delta),$$

multiply this by $(e_1 - xe_0)^4$, then applying the operators $\mathcal{H}_n$, we get

$$\left|\left(\mathcal{H}_n\left(\frac{f^{iv}(\eta)-f^{iv}(x)}{4!}\right)(e_1-xe_0)^4\right)(x)\right|$$
$$\leq \frac{1}{6}\mu_{n,4}(x)(x^2+1)(1+\delta^2)^2\left(1+\frac{\mu_{n,8}(x)}{\delta^4\mu_{n,4}(x)}\right)\Omega(f^{iv},\delta),$$

multiply this by n^2, then choosing $\delta^4 := \frac{\mu_{n,8}(x)}{\mu_{n,4}(x)} \leq 1$ and using Lemma 2, we get

$$\left|n^2\left(\mathcal{H}_n\left(\frac{f^{iv}(\eta)-f^{iv}(x)}{4!}\right)(e_1-xe_0)^4\right)(x)\right|$$
$$\leq \frac{4}{3}\frac{(3e-3)n^2x^2+7enx-3nx+4e-1}{(e-1)n^2}(x^2+1)$$
$$\times\,\Omega\left(f^{iv},\frac{\mu_{n,8}(x)}{\mu_{n,4}(x)}\right),$$

rearranging the terms in (2.2) and multiplying it by n^2, then using above relation, we get the desired result. □

Corollary 2. *Under the assumptions of Theorem 3, we have*

$$\lim_{n\to\infty} n\left[n[(\mathcal{H}_n f)(x)-f(x)]-\frac{f'(x)}{(e-1)}-\frac{xf''(x)}{2}\right]$$
$$=\frac{f''(x)}{2}+\frac{(e+2)xf'''(x)}{6(e-1)}+\frac{(3e-3)x^2f^{iv}(x)}{24(e-1)}.$$

Theorem 4. *Let $f \in C_B[0,\infty)$ the class of all bounded continuous functions defined on $[0,\infty)$, then for each $m \geq 1$, $x \geq 0$ and $n \in \mathbb{N}$, this follows*

$$\lim_{n\to\infty}(\mathcal{H}_{mn}f(nt))\left(\frac{x}{n}\right)=(\mathcal{H}_m f)(x)$$

and

$$\lim_{n\to\infty}\left(\mathcal{H}_m f\left(\frac{t}{n}\right)\right)(nx)=f(x).$$

Proof. Following Lemma 1, the limit of characteristic function of the operators (1.1) for $s \in \mathbb{R}$ and $i = \sqrt{-1}$ can be written as

$$\begin{aligned}
&\lim_{n\to\infty} (\mathcal{H}_{mn} \exp_{isn}) \left(\frac{x}{n}\right) \\
&= \lim_{n\to\infty} \exp\left(mn(e^{isn/mn} - 1)\frac{x}{n} - \frac{isn}{mn}\right) \frac{\left(e^{e^{isn/mn}} - 1\right)}{(e-1)} \\
&= \exp\left(mx(e^{is/m} - 1) - \frac{is}{m}\right) \frac{\left(e^{e^{is/m}} - 1\right)}{(e-1)} \\
&= (\mathcal{H}_m \exp_{is})(x).
\end{aligned}$$

Similarly, we have

$$\begin{aligned}
&\lim_{n\to\infty} (\mathcal{H}_m \exp_{is/n})(nx) \\
&= \lim_{n\to\infty} \exp\left(mnx(e^{is/mn} - 1) - \frac{is}{mn}\right) \frac{\left(e^{e^{is/mn}} - 1\right)}{(e-1)} \\
&= \exp_{is}(x) = (Id \exp_{is})(x),
\end{aligned}$$

where Id is the identity operator. Thus, by using Ref. [23, Theorem 1], we get the required estimates. □

Here, we define the space G with exponential growth of order $\lambda > 0$, i.e.

$$G = \left\{ f \in C[0,\infty) \ : \ \sup_{x\in[0,\infty)} |f(x)e^{-\lambda x}| < \infty \right\}.$$

Following Ref. [24], ω_1^* is the modulus of continuity for exponential growth function f with exponential growth $\lambda > 0$ and $\delta > 0$ defined as follows:

$$\omega_1^*(f,\delta,\lambda) = \sup_{h\leq\delta, 0\leq y<\infty} |f(y) - f(y+h)|e^{-\lambda y},$$

and $\text{Lip}(\alpha,\lambda) = \{f \in C[0,\infty) \ : \ \omega_1^*(f,\delta,\lambda) < M\delta^\alpha$, for all $\delta < 1$, and some $0 < \alpha \leq 1\}$.

Theorem 5. *Let $f, f', f'' \in G$ and $f'' \in \mathrm{Lip}(\alpha, \lambda)$, then for $\lambda \geq \frac{nx(2nx+1)}{nx+1}$ and $x \geq \phi$, we have*

$$\left|(\mathcal{H}_n f)(x) - f(x) - \frac{f'(x)}{(e-1)n} - \frac{(nx+1)f''(x)}{2n^2}\right|$$

$$\leq \left[e^{2\lambda x} + \frac{C_x(\lambda)}{2} + \frac{\sqrt{C_x(2\lambda)}}{2}\right]\left(\frac{nx+1}{n^2}\right)\omega_1^*\left(f'', \sqrt{b_n(x)}, \lambda\right),$$

where ϕ is the golden ratio given by $\phi = \frac{1+\sqrt{5}}{2}$ and $b_n(x) = \frac{3x}{n} + \frac{4e}{(e-1)n^2} - \frac{1}{(e-1)(nx+1)n^2}$.

Proof. Using Taylor's formula and applying Lemma 2, we get

$$\left|(\mathcal{H}_n f)(x) - f(x) - \mu_{n,1}(x)f'(x) - 0.5\mu_{n,2}(x)f''(x)\right|$$

$$\leq \left|\left(\mathcal{H}_n(e_1 - xe_0)^2\left(\frac{f''(\eta) - f''(x)}{2!}\right)\right)(x)\right|,$$

where $t < \eta < x$. Following Ref. [24, (2.4)], for $h, \delta > 0$, we have

$$\left|\left(\mathcal{H}_n(e_1 - xe_0)^2\left(\frac{f''(\eta) - f''(x)}{2!}\right)\right)(x)\right|$$

$$\leq \frac{\omega_1^*(f'', h, \lambda)}{2}\left(\mathcal{H}_n\left(e^{2\lambda x} + e^{\lambda t}\right)\left((e_1 - xe_0)^2 + \frac{|e_1 - xe_0|^3}{h}\right)\right)(x)$$

$$\leq \frac{\omega_1^*(f'', h, \lambda)}{2}\left\{\left(\mathcal{H}_n(e^{2\lambda x}(e_1 - xe_0)^2)\right)(x)\right.$$

$$+\left(\mathcal{H}_n\left(e^{2\lambda x}\frac{|e_1 - xe_0|^3}{h}\right)\right)(x) + \left(\mathcal{H}_n(e^{\lambda t}(e_1 - xe_0)^2)\right)(x)$$

$$\left.+\left(\mathcal{H}_n\left(e^{\lambda t}\frac{|e_1 - xe_0|^3}{h}\right)\right)(x)\right\}.$$

For the operators $\mathcal{H}_n$, we have

$$\begin{aligned}
&\left(\mathcal{H}_n(e_1-e_0x)^2e^{\lambda t}\right)(x)\\
&= e^{nx(e^{\lambda/n}-1)-\lambda/n}\left\{\frac{\left(e^{e^{\lambda/n}}-1\right)(nx^2+1)}{(e-1)n^2}\right.\\
&\quad+\frac{nxe^{\lambda/n}(2nx+1)}{(e-1)n^2}-\frac{e^{\lambda/n}(2nx+1)e^{e^{\lambda/n}}(nx+1)}{(e-1)n^2}\\
&\quad\left.+\frac{e^{2\lambda/n}\left(e^{e^{\lambda/n}}(nx^2+1)-n^2x^2\right)}{(e-1)n^2}\right\}\\
&\le e^{nx(e^{\lambda/n}-1)-\lambda/n}\left\{\frac{\left(e^{e^{\lambda/n}}-1\right)(nx^2+1)}{(e-1)n^2}\right.\\
&\quad\left.+\frac{e^{\lambda/n}}{(e-1)}\left(2x^2+\frac{x}{n}\right)+\frac{e^{2\lambda/n}e^{e^{\lambda/n}}(nx+1)^2}{(e-1)n^2}\right\}.
\end{aligned}$$

For $x \ge \frac{1+\sqrt{5}}{2}$, $\lambda \ge \frac{nx(2nx+1)}{nx+1}$ and $n > 1$, then it is easy to verify that $2x^2+\frac{x}{n} \le \lambda\left(\frac{x}{n}+\frac{1}{n^2}\right)$ and $\left(x+\frac{1}{n}\right)^2 \le \lambda\left(\frac{x}{n}+\frac{1}{n^2}\right)$, then we get

$$\begin{aligned}
&\left(\mathcal{H}_n(e_1-xe_0)^2e^{\lambda t}\right)(x)\\
&\le e^{nx(e^{\lambda/n}-1)-\lambda}\left(\frac{e^{\lambda}}{e-1}+\frac{e^{e^{\lambda}}}{e-1}(e^{2\lambda}+1)\right)\mu_{n,2}(x).
\end{aligned}$$

By observation, we have $n(e^{\lambda/n}-1) < e^{\lambda}-1$; using this, we get

$$(\mathcal{H}_n(e_1-xe_0)^2e^{\lambda t})(x) \le C_x(\lambda)\mu_{n,2}(x), \tag{2.3}$$

where

$$C_x(\lambda)=\frac{1}{e^{\lambda}}\left(\frac{e^{\lambda}}{e-1}+\frac{e^{e^{\lambda}}}{e-1}(e^{2\lambda}+1)\right)e^{(e^{\lambda}-1)x},$$

by employing Cauchy–Schwarz inequality and using (2.3), we get

$$\begin{aligned}&\left(\mathcal{H}_n\left(e^{\lambda t}\left|e_1-e_0x\right|^3\right)\right)(x)\\&\le\left(\left(\mathcal{H}_n(e_1-e_0x)^2e^{2\lambda t}\right)(x)\times(\mathcal{H}_n(e_1-e_0x)^4)(x)\right)^{1/2}\\&\le(C_x(2\lambda)\times\mu_{n,2}(x)\times\mu_{n,4}(x))^{1/2},\end{aligned}$$

and $\mu_{n,3}(x) \le (\mu_{n,4}(x) \times \mu_{n,2}(x))^{1/2}$, by choosing $h := \sqrt{\mu_{n,4}(x)/\mu_{n,2}(x)}$ and using Lemma 2, we get

$$\begin{aligned}&\left|\left(\mathcal{H}_n(e_1-e_0x)^2\left(\frac{f''(\eta)-f''(x)}{2!}\right)\right)(x)\right|\\&\le\frac{1}{2}\left[2e^{2\lambda x}+C_x(\lambda)+\sqrt{C_x(2\lambda)}\right]\mu_{n,2}(x)\omega_1^*\left(f'',\sqrt{\frac{\mu_{n,4}(x)}{\mu_{n,2}(x)}},\lambda\right),\end{aligned}$$

finally, again using Lemma 2, we get the desired result. □

3. Graphical and Numerical Insights

The graphical convergence of the operators $\mathcal{H}_n$ in Figure 1 for the function $f(x) = e^{2x} - 7x^2 - x + 9.1$ with the following value of $n = 50, 100, 150, 300$ on the interval $[0, 1.3]$ and Figure 2 shows the

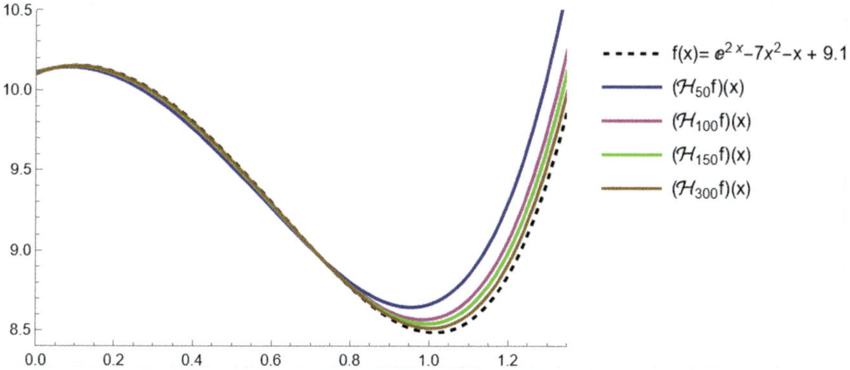

Fig. 1. Graphical representation of operators $\mathcal{H}_n$ for $f(x) = e^{2x} - 7x^2 - x + 9.1$.

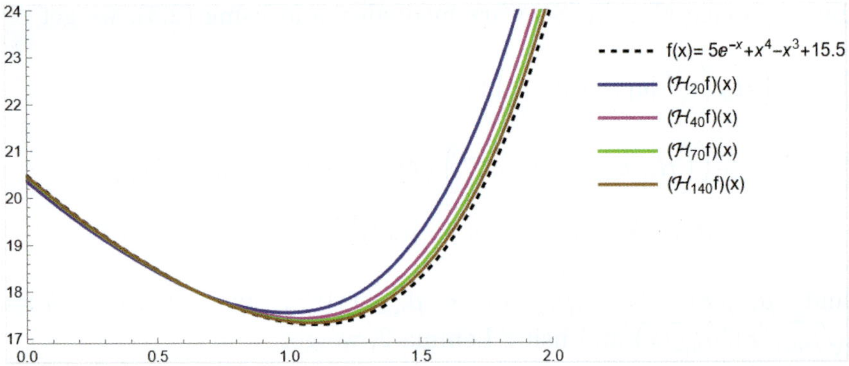

Fig. 2. Graphical representation of operators $\mathcal{H}_n$ for $f(x) = 5e^{-x} + x^4 - x^3 + 15.5$.

Table 1. Error estimates for $\mathcal{H}_n$, denoted by ϵ, where $|(\mathcal{H}_n f)(x) - f(x)| \leq \epsilon$.

	$f(x) = e^{2x} - 7x^2 - x + 9.1$			$f(x) = 5e^{-x} + x^4 - x^3 + 15.5$	
n	ϵ for $x = 1$	ϵ for $x = 0.8$	ϵ for $x = 0.6$	ϵ for $x = 0.4$	ϵ for $x = 0.2$
2000	0.00383691	0.000500863	0.000908655	0.000752790	0.001047790
4000	0.00191535	0.000248944	0.000454912	0.000376522	0.000523981
6000	0.00127621	0.000165633	0.000303404	0.000251043	0.000349339
8000	0.00095689	0.000124101	0.000227602	0.000188293	0.000262011
10000	0.00076539	0.000099221	0.000182105	0.000150639	0.000209613

convergence of $\mathcal{H}_n$ for the function $f(x) = 5e^{-x} + x^4 - x^3 + 15.5$ with the following values of $n = 20, 40, 70, 140$, on the interval $x \in [0, 2]$.

Moreover, Table 1 presents the values of ϵ, which represent the error estimates for $\mathcal{H}_n$ corresponding to various values of $n = 2000, 4000, 6000, 8000, 10000$. Specifically, ϵ is chosen such that the inequality $|(\mathcal{H}_n f)(x) - f(x)| \leq \epsilon$ holds for some fixed x. It can be seen from Figures 1, 2 and Table 1 that as n increases, the error decreases significantly, demonstrating the accelerated convergence of $\mathcal{H}_n$ to the given function.

4. Composition and Difference Estimates

In this section, we consider new operators $\mathcal{A}_n$ based on the composition of $\mathcal{H}_n$ with the well-known Szász–Mirakyan operators.

We continue to study the approximation properties of $\mathcal{A}_n$ and give two theorems based on difference of operators, asymptotic formula and convergence in terms of weighted moduli $\omega^*(f,\delta)$.

The Szász–Mirakyan operators for $f \in C[0,\infty)$ are defined by (see Ref. [15] and references therein)

$$(S_n f)(x) = \sum_{v=0}^{\infty} s_{n,v}(x) f(vn^{-1}), \quad \text{where } s_{n,v}(x) = e^{-nx}\frac{(nx)^v}{v!}.$$

Now, we consider the following composition operators $\mathcal{A}_n := \mathcal{H}_n \circ S_n$, defined for $x \geq 0$.

The operators $\mathcal{A}_n$ takes the following concise form:

$$(\mathcal{A}_n f)(x) = \sum_{j=0}^{\infty} \frac{e^{-nx}}{(e-1)} \sum_{k=0}^{\infty} \frac{\tilde{B}_k(nx)e^{-k}k^j}{k!} \frac{f(jn^{-1})}{j!}. \tag{4.1}$$

Lemma 3. *The MGF for the operators $\mathcal{A}_n$ is given by*

$$(\mathcal{A}_n \exp_\gamma)(x) = \exp\left(nx\left(e^{\left(e^{\frac{\gamma}{n}}-1\right)}-1\right)\right)\frac{\left(e^{e^{\left(e^{\frac{\gamma}{n}}-1\right)}}-1\right)}{e^{\left(e^{\frac{\gamma}{n}}-1\right)}(e-1)}.$$

Proof. Using definition of operators $\mathcal{A}_n$, we get

$$\begin{aligned}\left(\mathcal{A}_n \exp_\gamma\right)(x) &= \frac{e^{-nx}}{(e-1)}\sum_{k=0}^{\infty}\frac{\tilde{B}_k(nx)e^{-k}}{k!}\sum_{j=0}^{\infty}\frac{k^j e^{\frac{\gamma j}{n}}}{j!} \\ &= \frac{e^{-nx}}{(e-1)}\sum_{k=0}^{\infty}\frac{\tilde{B}_k(nx)}{k!}e^{k\left(e^{\frac{\gamma}{n}}-1\right)}.\end{aligned}$$

Then, using the generating function defined in (1.2), we get the desired result. □

Lemma 4. *The kth central moment for the operators $\mathcal{A}_n$ is denoted by $\zeta_{n,k}(x) := (\mathcal{A}_n(e_1 - xe_0)^k)(x)$, where $k = 0, 1, 2, \ldots,$*

then, we have

$$\zeta_{n,k}(x)=\left[\frac{\partial^k}{\partial\gamma^k}\left(\exp\left(nx\left(e^{\left(e^{\frac{\gamma}{n}}-1\right)}-1\right)-\gamma x\right)\frac{\left(e^{e^{\left(e^{\frac{\gamma}{n}}-1\right)}}-1\right)}{e^{\left(e^{\frac{\gamma}{n}}-1\right)}(e-1)}\right)\right]_{\gamma=0}.$$

In particular,

$$\zeta_{n,0}(x)=1,\quad \zeta_{n,1}(x)=\frac{1}{(e-1)n},\quad \zeta_{n,2}(x)=\frac{2x}{n}+\frac{e}{(e-1)n^2},$$

$$\zeta_{n,3}(x)=\frac{(5e+1)x}{(e-1)n^2}+\frac{(4e-1)}{(e-1)n^3},$$

$$\zeta_{n,4}(x)=\frac{12x^2}{n^2}+\frac{(27e+5)x}{(e-1)n^3}+\frac{(17e-1)}{(e-1)n^4},$$

$$\zeta_{n,5}(x)=\frac{(100e-40)n^2x^2+(182e+3)nx+91e+2}{(e-1)n^5},$$

$$\zeta_{n,6}(x)=\frac{120x^3}{n^3}+\frac{20(44e-5)n^2x^2+3(446e-7)nx+587e-9}{(e-1)n^6},$$

$$\zeta_{n,7}(x)=\frac{30(70e-43)n^3x^3+7(1229e-47)n^2x^2+6(1811e-12)nx+4327e+9}{(e-1)n^7},$$

$$\zeta_{n,8}(x)=\frac{1680x^4}{n^4}+\frac{280(107e-35)n^3x^3+7(12869e-235)n^2x^2}{(e-1)n^8}+\frac{2(48837e-11)nx+35604e-50}{(e-1)n^8}.$$

Theorem 6. *For $x\in[0,\infty)$, $f\in C_B[0,\infty)$ and ω the classical first-order moduli of continuity, the following inequality holds:*

$$|(\mathcal{A}_nf)(x)-(\mathcal{H}_nf)(x)|\le 2\omega\left(f,\sqrt{\frac{x}{n}+\frac{1}{(e-1)n^2}}\right).$$

Proof. By using the definition of operators $\mathcal{A}_n$, we have

$$|(\mathcal{A}_n f)(x) - (\mathcal{H}_n f)(x)| \leq \frac{e^{-nx}}{(e-1)} \sum_{k=0}^{\infty} \frac{B_k(nx)}{k!} |(S_n f - f)(t)|.$$

Following Ref. [25] and by using property of well-known Szász–Mirakyan operators, we have

$$|(S_n f)(x) - f(x)| \leq \left(1 + \frac{1}{\delta^2}\left(\frac{x}{n}\right)\right) \omega(f, \delta),$$

then we get

$$\begin{aligned} |(\mathcal{A}_n f)(x) - (\mathcal{H}_n f)(x)| \leq & \frac{e^{-nx}}{(e-1)} \sum_{k=0}^{\infty} \frac{\tilde{B}_k(nx)}{k!} \left(1 + \frac{1}{\delta^2}\left(\frac{t}{n}\right)\right) \omega(f, \delta) \\ \leq & \left(1 + \frac{1}{\delta^2}\left(\frac{x}{n} + \frac{1}{(e-1)n^2}\right)\right) \omega(f, \delta). \end{aligned}$$

Choosing $\delta = \sqrt{\frac{x}{n} + \frac{1}{(e-1)n^2}}$, we obtained the desired result. □

Theorem 7. *For $x \in [0, \infty)$, $f \in C_B[0, \infty)$ and ω the classical first-order moduli of continuity, this inequality holds*

$$|(\mathcal{A}_n f)(x) - (S_n f)(x)| \leq 2\left(2 - e^{-\sqrt{nx+1}}\right) \omega\left(f, \sqrt{\frac{nx+1}{n^2}}\right).$$

Proof. Let $S_n f := g$, then by using the definition of operators $\mathcal{A}_n$, we have

$$\begin{aligned} |(\mathcal{A}_n f)(x) - (S_n f)(x)| &= |(\mathcal{H}_n \circ S_n f)(x) - (S_n f)(x)| \\ &\leq |(\mathcal{H}_n g)(x) - g(x)| \\ &\leq \left(1 + \frac{\mathcal{H}_n(e_1 - x e_0)^2(x)}{\delta^2}\right) \omega(g, \delta) \\ &\leq \left(1 + \left(\frac{nx+1}{n^2}\right)\frac{1}{\delta^2}\right) \omega(g, \delta) \\ &\leq \left(1 + \left(\frac{nx+1}{n^2}\right)\frac{1}{\delta^2}\right) (2 - e^{-n\delta}) \omega(f, \delta), \end{aligned}$$

which follows from Ref. [26] and using Lemma 2, then by choosing $\delta = \sqrt{\frac{nx+1}{n^2}}$, we obtained the desired result. □

Theorem 8. *For $x \in [0, \infty)$, let f be any function with polynomial growth and $f'' \in C_A[0, \infty)$, then the following inequality holds:*

$$\left| n[(\mathcal{A}_n f)(x) - f(x)] - \frac{f'(x)}{(e-1)} - xf''(x) \right| \leq \frac{ef''(x)}{2(e-1)n} + 16(1+x^2)\left(\frac{e}{(e-1)n} - 2x \right) \times \Omega\left(f'', \left(\frac{\zeta_{n,6}(x)}{\zeta_{n,2}(x)} \right)^{\frac{1}{4}} \right).$$

Proof. The argument follows the same fundamental approach as the proof of Theorem 3, using Lemma 4. We omit the detailed computations. □

Corollary 3. *Let f be any function with polynomial growth and $f'' \in C_A[0, \infty)$, $x \in [0, \infty)$, then the following holds:*

$$\lim_{n \to \infty} n[(\mathcal{A}_n f)(x) - f(x)] = \frac{f'(x)}{(e-1)} + xf''(x).$$

By observations, we have

$$\lim_{n \to \infty} n[(\mathcal{A}_n f)(x) - f(x)] = \lim_{n \to \infty} n[(S_n f)(x) + (\mathcal{H}_n f)(x) - 2f(x)].$$

Theorem 9. *Let $f, f^{iv} \in C^*[0, \infty)$, then for any $x \geq 0$, this follows*

$$\left| n\left[n[(\mathcal{A}_n f)(x) - f(x)] - \frac{f'(x)}{(e-1)} - xf''(x) \right] - \frac{ef''(x)}{2(e-1)} - \frac{(5e+1)xf'''(x)}{6(e-1)} - \frac{x^2 f^{iv}(x)}{2} \right| \leq \frac{(4e-1)}{6(e-1)n}|f'''(x)| + \frac{(27e+5)x}{24(e-1)n}|f^{iv}(x)| + \frac{(17e-1)}{24(e-1)n^2}|f^{iv}(x)| + 2\omega^*\left(f^{iv}, \frac{1}{\sqrt{n}} \right)$$

$$\times \left\{ 12x^2 + \frac{(27e+5)x}{(e-1)n} + \frac{(17e-1)}{(e-1)n^2} \right.$$

$$\left. + n^3 \sqrt{\zeta_{n,8}(x) \left(\mathcal{A}_n \left(\frac{1}{e^x} - \frac{1}{e^t} \right)^8 \right)(x)} \right\}.$$

Proof. Using Taylor's formula for $f \in C^*[0,\infty)$ and $x < \eta < t$, we have

$$f(t) = \sum_{k=0}^{4} \frac{(e_1 - xe_0)^k}{k!} f^{(k)}(x) + (e_1 - xe_0)^4 \frac{f^{iv}(\eta) - f^{iv}(x)}{4!}.$$

Applying the operators A_n on f and using Lemma 4, we immediately have

$$\left| (\mathcal{A}_n f)(x) - f(x) - \frac{f'(x)}{(e-1)n} - \left(\frac{2xn(e-1)+e}{2(e-1)n^2} \right) f''(x) \right.$$

$$- \left(\frac{(5e+1)x}{(e-1)n^2} + \frac{(4e-1)}{6(e-1)n^3} \right) f'''(x)$$

$$\left. - \left(\frac{x^2}{2n^2} + \frac{(27e+5)x}{24(e-1)n^3} + \frac{(17e-1)}{24(e-1)n^4} \right) f^{iv}(x) \right|$$

$$\leq \left| \left(\mathcal{A}_n \left(\frac{f^{iv}(\eta) - f^{iv}(x)}{4!} \right) (e_1 - xe_0)^4 \right)(x) \right|. \tag{4.2}$$

Next, by the property used in Ref. [20] for $\delta > 0$, we can write

$$\left(\frac{f^{iv}(\eta) - f^{iv}(x)}{4!} \right) \leq 2 \left(1 + \frac{(\exp_{-1}(x) - \exp_{-1}(t))^4}{\delta^2} \right) \omega^*(f^{iv}, \delta).$$

Using the above relation, employing Cauchy–Schwarz inequality and selecting $\delta = n^{-1/2}$, we have

$$\left| n^2 \left(\mathcal{A}_n \left(\frac{f^{iv}(\eta) - f^{iv}(x)}{4!} \right) (e_1 - xe_0)^4 \right)(x) \right|$$

$$\leq 2n^2 \omega^* \left(f^{iv}, \frac{1}{\sqrt{n}} \right) (\mathcal{A}_n((e_1 - xe_0)^4)(x)$$

$$+ \mathcal{A}_n\left(n\left(\frac{1}{e^x}-\frac{1}{e^t}\right)^2(e_1-xe_0)^4\right)(x)\Bigg)$$

$$\leq 2\omega^*\left(f^{iv},\frac{1}{\sqrt{n}}\right)\Bigg\{12x^2+\frac{(27e+5)x}{(e-1)n}+\frac{(17e-1)}{(e-1)n^2}$$

$$+ n^3\sqrt{\zeta_{n,8}(x)\left(\mathcal{A}_n\left(\frac{1}{e^x}-\frac{1}{e^t}\right)^8\right)(x)}\Bigg\}.$$

Then, multiplying n^2 in the inequality (4.2) by rearranging the terms and using the above inequality, we get the desired result. □

Corollary 4. *Under to the assumptions of Theorem 9, we have*

$$\lim_{n\to\infty} n\left[n[(\mathcal{A}_n f)(x)-f(x)]-\frac{f'(x)}{(e-1)}-xf''(x)\right]$$
$$=\frac{ef''(x)}{2(e-1)}+\frac{(5e+1)xf'''(x)}{6(e-1)}+\frac{x^2 f^{iv}(x)}{2}.$$

5. Conclusion

In this chapter, we study the operators $\mathcal{H}_n$ and $\mathcal{A}_n$ linking adjoint Bernoulli polynomials. For the operators $\mathcal{H}_n$, our results include convergence in terms of quantitative Voronovskaja-type results in various weighted spaces, rate of approximation in uniform norm and Voronovskaja theorem for functions having exponential growth. In addition, we provide a graphical representation for the operators converging to the given functions, and numerical findings are given to support the convergence. Recently, the composition of operators has gained popularity in the theory of positive linear operators. We capture a sequence of operators $\mathcal{A}_n$ based on the composition method. Then, we provide results based on difference estimates and show convergence. The results of this chapter can be used to approximate functions on $[0,\infty)$ related to the sequence of operators linking to adjoint Bernoulli polynomials and may also be useful in establishing new difference estimation techniques for composition operators.

References

[1] V. Gupta, D. Agrawal and M. T. Rassias. Quantitative estimates for differences of Baskakov-type operators. *Complex Anal. Oper. Theory*, 13(361), 2019, 4045–4064.

[2] V. Gupta and M. T. Rassias. Asymptotic formula in simultaneous approximation for certain Ismail-May-Baskakov operators. *J. Numer. Anal. Approx. Theory*, 50(2), 2021, 153–163.

[3] V. Gupta and M. T. Rassias. Direct estimates for certain Szász type operators. *Appl. Math. Comput.*, 251, 2015, 469–474.

[4] V. Sharma and V. Gupta. Convergence properties of Durrmeyer-type sampling operators. *Comput. Appl. Math.*, 43(403), 2024, 1–23.

[5] V. Gupta and D. Malik. Semi Baskakov-type operators. *Bull. Iran. Math. Soc.*, 51(31), 2025, 1–17.

[6] V. Gupta and D. Malik. On convergence of certain Hermite-type operators. *Filomat*, 38(4), 2024, 1367–1374.

[7] V. Gupta and D. Malik. On a new Kantorovich operator based on Hermite polynomials. *J. Appl. Math. Comput.*, 70, 2024, 2587–2602.

[8] S. Sucu, G. İçoz and S. Varma. On some extensions of Szász operators including Boas Buck type polynomials. *Abstr. Appl. Anal.*, 2012, 2012, 1–15.

[9] S. Bazm. Bernoulli polynomials for the numerical solution of some classes of linear and nonlinear integral equations. *J. Comput. Appl. Math.*, 275, 2015, 44–60.

[10] E. Defez, J. A. Ibáñez, J. M. Alonso and P. Alonso-Jordá. On Bernoulli series approximation for the matrix cosine. *Math. Methods Appl. Sci.*, 45, 2022, 3239–3253.

[11] V. Gupta, M. T. Rassias, P. N. Agrawal and A. M. Acu. *Recent Advances in Constructive Approximation Theory.* Springer, Cham, 2018.

[12] L. Brandolini, L. Colzani and B. Gariboldi *et al.* Euler–Maclaurin summation formula on polytopes and expansions in multivariate Bernoulli polynomials. *J. Fourier Anal. Appl.*, 29(33), 2023, 1–49.

[13] S. Gaboury. Some relations involving generalized Hurwitz–Lerch zeta function obtained by means of fractional derivatives with applications to Apostol-type polynomials. *Adv. Differ. Equ.*, 2013(361), 2013, 1–13.

[14] P. Natalini and P. E. Ricci. Adjoint Appell-Euler and first kind Appell-Bernoulli polynomials. *Appl. Appl. Math.*, 14(2), 2019, 1112.

[15] R. Păltănea. Modified Szász-Mirakjan operators of integral form. *Carpathian J. Math.*, 24(3), 2008, 378–385.

[16] V. Gupta and V. Sharma. Durrmeyer variant of certain approximation operators. *J. Appl. Math. Comput.*, 70(4), 2024, 3717–3730.

[17] M. M. Yilmaz. Rate of convergence by Kantorovich type operators involving adjoint Bernoulli polynomials. *Publ. Inst. Math.*, 114, 2023, 51–62.

[18] N. Rao, M. Farid and R. A. Ali. Study of Szász–Durremeyer-type operators involving adjoint Bernoulli polynomials. *Mathematics*, 12(23), 2024, 1–15.

[19] A. Halhos. The rate of approximation of functions in an infinite interval by positive linear operators. *Stud. Univ. Babas-Bolyai Math.*, 55(2), 2010, 133–142.

[20] T. Acar, A. Aral and H. Gonska. On Szász-Mirakyan operators preserving $e^{2ax}, a > 0$. *Mediter. J. Math.*, 14(6), 2017, 1–14.

[21] N. Ispir. On modified Baskakov operators on weighted spaces. *Turk. J. Math.*, 25, 2001, 355–365.

[22] T. Acar, A. Aral and I. Raşa. The new forms of Voronovskayas theorem in weighted spaces. *Positivity*, 20, 2016, 25–40.

[23] A. Acu, V. Gupta, I. Raşa and F. Sofonea. Convergence of special sequences of semi exponential operators. *Mathematics*, 10, 2022, 1–13.

[24] G. Tachev, V. Gupta and A. Aral. Voronovskaja's theorem for functions with exponential growth. *Georgian Math. J.*, 27, 2018, 459–468.

[25] O. Shisha and B. Mond. The degree of convergence of linear positive operators. *Proc. Natl. Acad. Sci. USA*, 60(43), 1968, 1196–1200.

[26] J. A. Adell and J. de la Cal. Preservation of moduli of continuity for Bernstein-type operators. In G. A. Anastassiou and S. T. Rachev (eds.), *Proceedings of International Conference on Approximation, Probablity and Related Fields*, Santa Barbara, 1994.

https://doi.org/10.1142/9789819820498_0014

Chapter 14

Extended General Quasi-Variational Inclusions and Dynamical Systems

Muhammad Aslam Noor* **and Khalida Inayat Noor†**

Department of Mathematics, University of Wah,
Wah Cantt, Pakistan
**noormaslam@gmail.com*
†*khalidan@gmail.com*

Some new classes of extended general quasi-variational inclusions, which provide us with a unified, natural and simple framework to consider a wide class of unrelated problems arising in pure and applied sciences, are introduced and analyzed. It is proved that the general quasi-variational inclusions are equivalent to the fixed point problems. This alternative equivalence formulation is used to discuss the existence of a solution of the extended general quasi-variational inclusions. We also applied to suggest several inertial iterative schemes for solving the general quasi-variational inclusions. Convergence analysis is investigated under certain mild conditions. A new class of dynamical systems associated with extended general quasi-variational inclusions is considered. We have used the dynamical systems to suggest and analyze some implicit iterative methods for solving the extended general quasi-variational inclusions. Since the extended general quasi-variational inclusions include quasi-variational inclusions, complementarity problems, variational inequalities and related optimization problems as special cases, our results continue to hold for these problems. It is an interesting problem to implement these methods and compare with other techniques for solving extended general quasi-variational inclusions for further research activities. Some problems are indicated for future research.

1. Introduction

Variational inequalities theory, which was introduced by Stampacchia [55] in potential theory, has witnessed an explosive growth in theoretical advances, algorithmic development and applications across all disciplines of pure and applied sciences. This theory provides a unified and novel treatment of equilibrium problems arising in economics, finance, transportation, elasticity and structural analysis. It combines novel theoretical and algorithmic advances with new domain of applications. Variational inequalities theory contains a wealth of new ideas and techniques and can be viewed as novel extension of the variational principles. It is worth mentioning that the origin of variational principles and related optimization problems can be traced back to Euler, Lagrange, Newton and Bernoulli's brothers. As a result of interaction between different branches of mathematical and engineering sciences, we now have a variety of techniques to suggest and analyze various iterative algorithms for solving variational inequalities and related optimization problems. Though the problems in each of these areas may look completely different, the resulting algorithms can be very closely related. It is amazing that a wide class of unrelated problems such as potential theory, equilibrium problems, transport, structural analysis, quadratic programming and nonlinear optimization can be studied in a unified and general framework of variational inequalities. For more details, see, for example, Refs. [1, 2, 5, 6, 8–16, 19–22, 24–27, 27–51, 54–57] and the references therein.

It is well known that only even-order boundary value problems and initial value problems can be studied in the framework of variational inequalities. To overcome this drawback, Noor [28] introduced the class of variational inequality involving two operators, which is called the general variational inequalities. It have shown [32] that odd-order and non-symmetric boundary value problems can be studied in the unified framework of general variational inequalities. For the applications, numerical methods, sensitivity analysis, dynamical systems and other aspects of general variational inequalities, see Refs. [29, 30, 33–35, 45] and the references therein.

In order to develop the unified framework, Noor [36–38] considered another class of variational inequalities involving three operators, which is also called the extended general variational inequality. This class can be viewed as the unified one and includes as special cases general variational inequalities, generalized complementarity problems and related optimization problems.

Closely related to the variational inequalities, we have the complementarity problem, which was introduced by Lemake [18] in game theory. For the developments in the complementarity problems, see Refs. [6, 14, 18, 20, 26, 27, 41, 52] and the references therein. It is worth mentioning that if the underlying set in the variational inequalities and the complementarity problems is the convex cone, then both the problems are equivalent. This equivalent between these problems has played a very significant role in the studies of the qualitative and quantitative aspects.

One of the most difficult and important problems in variational inequality theory is the development of efficient numerical methods. Several numerical methods have been developed for solving the variational inclusions and their variant forms. Using the updating techniques of the solution, Noor [30, 33–35, 38, 45–47, 52, 53] suggested and analyzed some three-step forward-backward splitting algorithms for solving variational inequalities and quasi-variational inclusions. These forward-backward splitting algorithms are similar to those of Glowinski *et al.* [9–11] which they suggested by using the Lagrangian technique. These three-step schemes are a natural generalization of the splitting methods for solving partial differential equations. For applications of the splitting techniques to partial differential equations, see Ames [3] and the references therein. For novel applications of the three-step methods, see Ashish *et al.* [4]. These methods include the Mann and Ishikawa iterative schemes and modified forward-backward splitting methods of Tseng [56], Noor [30, 33] and Noor *et al.* [45–47, 52, 53] as special cases, Alvarez *et al.* [1, 2] used the inertial type projection methods for solving variational inequalities, the origin of which can be traced back to Polyak [53]. Noor [33] suggested and investigated inertial-type projection methods for solving general variational inequalities. These inertial-type methods have been modified in various directions for solving variational inequalities and related optimization problems. Recently, Shehu *et al.* [54], Noor *et al.* [44, 45] and Jabeen *et al.* [12, 13] analyzed some inertial projection methods for some classes of general quasi-variational inequalities. For more details and applications of the inertial methods, see Refs. [1, 2, 13, 39, 40, 45, 46, 49] and reference therein.

Another important and significant generalization of the variational inequalities is called the extended general variational inequality, which was introduced and analyzed by Noor [36]. It have been shown [36] that several classes of variational inequalities and optimization are important special cases of the extended general variational inequalities. Motivated and inspired by the recent research activities, we consider and study some new classes of variational inclusions involving five arbitrary operators, which are called the extended general quasi-variational inclusions. Some special important cases are also discussed. We have shown that the general quasi-variational inclusions are equivalent to the fixed point problems. This alternative formulation is used to study the existence of a solution of the extended general quasi-variational inequalities. Inspired and motivated by the usefulness and applications of the splitting-type methods, we suggest and analyze a new class of three-step approximation schemes for solving extended general quasi-variational inclusions and related problems. The convergence of the proposed three-step iterative methods is analyzed under some suitable conditions. Also, some inertial iterative are proposed. Our results represent an improvement and refinement of the previously known results in these fields. Dynamical systems associated with extended general quasi-variational are also introduced. Using the dynamical systems technique coupled with forward finite differences schemes, some new iterative methods are proposed and analyzed. We have only considered theoretical aspects of the suggested methods. It is an interesting problem to implement these methods and to illustrate the efficiency. Comparison with other methods need further research efforts. The ideas and techniques of this chapter may be extended for other classes of quasi-variational inclusions and related optimization problems. Some open problems are also suggested for future research.

2. Basic Definitions and Results

Let $\mathcal{H}$ be a real Hilbert space whose inner product and norm are denoted by $\langle \cdot, \cdot \rangle$ and $\|\cdot\|$, respectively. Let $\mathcal{A}(\cdot,\cdot) : \mathcal{H} \times \mathcal{H} \longrightarrow \mathcal{H}$ be a maximal monotone operator with respect to the first argument. For given nonlinear operators $\mathcal{T}, g, h : \mathcal{H} \longrightarrow \mathcal{H}$, consider the problem of finding $\mu \in \mathcal{H}$ such that

$$0 \in \mathcal{T}\mu + h(\mu) - g(\mu) + \mathcal{A}(h(\mu), h(\mu)), \tag{2.1}$$

which is called the extended general quasi-variational inclusion. Here, the operators $\mathcal{T}, g, h$ are strongly monotone and Lipschitz continuous and the bifunction $\mathcal{A}(\cdot, \cdot)$ is maximal monotone operator with respect to first argument. A number of problems arising in structural analysis, mechanics and economics can be studied in the framework of the general quasi-variational inclusions in a unified manner.

We now discuss some interesting and important problems, which can obtained from the problem (2.1).

Special Cases.

(1) For $g = I$ and $h = I$, the identity operator, the problem (2.1) reduces to finding $\mu \in \mathcal{H}$ such that

$$0 \in \mathcal{T}\mu + \mathcal{A}(\mu, \mu), \tag{2.2}$$

which is called the quasi-variational inclusion, introduced and studied by Noor *et al.* [39].

(2) For $\mathcal{A}(\cdot, \mu) = \partial\varphi(\cdot, \mu) : \mathcal{H} \times \mathcal{H} \longrightarrow R \cup \{+\infty\}$, the subdifferential of a convex, proper and lower semi-continuous function $\varphi(\cdot, \mu)$ with respect to the first argument, the problem (2.1) is equivalent to finding $\mu \in \mathcal{H}$ such that

$$\begin{aligned} &\langle \mathcal{T}\mu, g(\nu) - h(\mu)\rangle + \varphi(h(\nu), g(\mu)) \\ &\quad - \varphi(h(\mu), h(\mu)) \geq 0, \quad \forall \nu \in \mathcal{H}, \end{aligned} \tag{2.3}$$

which is called the mixed extended general quasi-variational inequality.

(3) For $\mathcal{A}(h(\mu), h(\nu)) = \mathcal{A}(h(\mu))$, for all $\nu \in \mathcal{H}$, the problem (2.1) is equivalent to finding $\mu \in \mathcal{H}$ such that

$$0 \in \mathcal{T}\mu + h(\mu) - g(\mu) + \mathcal{A}(h(\mu)), \tag{2.4}$$

a problem considered and studied using the resolvent equation technique.

(4) For $\mathcal{A}(h(\mu)) = \partial\varphi(h(\mu))$ the subdifferential of a proper, convex and lower, semicontinuous function $\varphi : \mathcal{H} \longrightarrow R \cup \{+\infty\}$, the problem (2.4) reduces to the following: find $\mu \in \mathcal{H}$ such that

$$\langle \mathcal{T}\mu, g(\nu) - h(\mu)\rangle + \varphi(g(\nu)) - \varphi(h(\mu)) \geq 0, \quad \nu \in \mathcal{H}. \tag{2.5}$$

Problem (2.5) is known as the mixed extended general variational inequality.

(5) If the function $\varphi(\cdot,\cdot)$ is the indicator function of a closed convex-valued set $\mathcal{K}(\mu)$ in $\mathcal{H}$, that is,

$$\varphi(\mu,\mu) = \mathcal{K}_{(\mu)}(\mu) = \begin{cases} 0, & \text{if } \mu \in \mathcal{K}(\mu) \\ +\infty, & \text{otherwise}\ , \end{cases}$$

then problem (2.3) is equivalent to finding $\mu \in \mathcal{H}$, $g(\mu) \in \mathcal{K}(u)$ such that

$$\langle \mathcal{T}\mu, g(\nu) - h(\mu)\rangle \geq 0, \quad \forall \nu \in \mathcal{K}(u), \tag{2.6}$$

a problem considered and studied by using the projection method and the implicit Wiener–Hopf equations technique.

(6) If $\mathcal{K}(u) = \mathcal{K}$, then problem (2.6) reduces to finding $\mu \in \mathcal{K}$ such that

$$\langle \mathcal{T}\mu, g(\nu) - h(\mu)\rangle \geq 0, \quad \forall \nu \in \mathcal{K}, \tag{2.7}$$

which is called the extended general variational inequality, introduced and studied by Noor [36] in 2009. For the applications, numerical methods, generalizations and other aspects of the extended general variational inequalities, see Refs. [22, 36–38, 42, 46, 50].

(7) If $h = g$, then the problem (2.7) collapses to finding $\mu \in \mathcal{K}$ such that

$$\langle \mathcal{T}\mu, g(\nu) - g(\mu)\rangle \geq 0, \quad \forall \nu \in \mathcal{K}, \tag{2.8}$$

is called the general variational inequalities. For the applications, formulation, numerical analysis, merit functions, dynamical systems and generalizations, see Refs. [29, 31, 32, 44–48].

(8) *For* $h = g = I$, the problem (2.8) collapses to finding $\mu \in \mathcal{K}$ such that

$$\langle \mathcal{T}\mu, \nu - \mu \rangle \geq 0, \quad \forall \nu \in \mathcal{K}, \tag{2.9}$$

is called the classical variational inequality, introduced and studied by Stampacchia [55] in 1964. Variational inequalities can be viewed as novel extensions of the variational principles. Variational inequality has influenced several branches of mathematical, engineering, economics, transformation, regional and medical sciences and continues to inspire researchers to find its applications. For more details, see Refs. [5, 7–16, 19, 21–23, 25, 26, 28–32, 34–36, 38–42, 44–46, 49, 50, 52, 53, 56] and the references therein.

(9) If $\mathcal{K}^*(\mu) = \{\mu \in \mathcal{H} : \langle \mu, \nu \rangle \geq 0, \ \forall \nu \in \mathcal{K}(\mu)\}$ is a polar cone of the convex-valued cone $\mathcal{K}(\mu)$ in $\mathcal{H}$, then problem (2.6) is equivalent to finding $\mu \in \mathcal{H}$, such that

$$h(\mu) \in \mathcal{K}(\mu), \quad \mathcal{T}(\mu) \in \mathcal{K}^*(\mu), \quad \langle \mathcal{T}(\mu), h(\mu) \rangle = 0, \tag{2.10}$$

which is called the generalized quasi-complementarity problem.

(10) If $\mathcal{K}^*(\mu) = \mathcal{K}^*$, then problem (2.10) reduces to finding $\mu \in \mathcal{H}$ such that

$$h(\mu) \in \mathcal{K}, \quad \mathcal{T}\mu \in \mathcal{K}^*, \quad \langle \mathcal{T}\mu, h(\mu) \rangle = 0, \tag{2.11}$$

which is known as the general complementarity problem. For $h = I$, the identity operator, problem (2.11) was introduced by Karamardian [14] and is known as nonlinear complementarity problem. Karamardian [14] proved that the nonlinear complementarity problems are equivalent to the variational inequalities. This result enabled us to use the techniques of complementarity problems to solve the variational inequalities and vice versa. For the applications, formulations and generalizations of the complementarity problems, see, for example, Refs. [6, 14, 18, 20, 26–28, 30, 33, 43, 51, 52].

Remark 1. For special choices of the operators $\mathcal{T}$, g, h, $\mathcal{A}(.,.)$ and the convex-valued set $\mathcal{K}(\mu)$, one can obtain a large number of implicit (quasi)-complementarity problems and variational inequality problems, which are very special cases of problem (2.1). Thus, it is clear that problem (2.1) is a general and unifying one and has numerous applications in pure and applied sciences.

We now recall some well-known results and notions.

Definition 1. If $\mathcal{T}$ is a maximal monotone operator on H, then, for a constant $\rho > 0$, the resolvent operator associated with $\mathcal{T}$ is defined by

$$\mathcal{J}_{\mathcal{T}}(\mu) = (I + \rho\mathcal{T})^{-1}(\mu), \quad \forall \mu \in \mathcal{H},$$

where I is the identity operator. Also, the resolvent operator $\mathcal{J}_{\mathcal{T}}$ is single-valued and non-expansive.

Remark 2 ([31]). Since the operator $\mathcal{A}(.,.)$ is a maximal monotone operator with respect to the first argument, we denote by

$$\mathcal{J}_{\mathcal{A}(\mu)} \equiv (I + \rho\mathcal{A}(\mu))^{-1}(\mu), \quad \forall \mu \in \mathcal{H},$$

the resolvent operator associated with $\mathcal{A}(.,\mu) \equiv \mathcal{A}(\mu)$. For example, if $\mathcal{A}(.,\mu) = \partial\varphi(.,\mu)$, for all $\mu \in \mathcal{H}$, and $\varphi(.,.) : \mathcal{H} \times \mathcal{H} \longrightarrow R \cup \{\infty\}$ is a proper, convex and lower semicontinuous with respect to the first argument, then it is well known that $\partial\varphi(.,\mu)$ is a maximal monotone operator with respect to the first argument. In this case, the resolvent operator $\mathcal{J}_{\mathcal{A}(\mu)} = \mathcal{J}_{\varphi(\mu)}$ is defined as

$$\mathcal{J}_{\varphi(\mu)} = (I + \rho\partial\varphi(.,\mu))^{-1}(\mu) = (I + \rho\partial\varphi(\mu))^{-1}, \quad \forall \mu \in \mathcal{H},$$

which is defined everywhere on the whole space $\mathcal{H}$, where $\partial\varphi(\mu) \equiv \partial\varphi(.,\mu)$.

We need the following well-known definitions and results in obtaining our results.

Definition 2.

(1) The mapping $\mathtt{T}$ is called $\alpha \geq 0$-strongly monotone ($\alpha \geq 0$), if

$$\left\langle \mathtt{T}\mu - \mathtt{T}\nu,\, \mu - \nu \right\rangle \geq \alpha \left\|\mu - \nu\right\|^2, \quad \forall\, \mu,\, \nu \,\in\, \mathcal{H}.$$

(2) The mapping T is called $\xi-$ cocoercive ($\xi > 0$), if

$$\langle \mathtt{T}\mu - \mathtt{T}\nu,\, \mu - \nu \rangle \geq \xi \,\| \mathtt{T}\mu - \mathtt{T}\nu \|^2, \quad \forall\, \mu,\, \nu \in \mathcal{H}.$$

(3) The mapping T is called $\beta-$ Lipschitz continuous ($\beta > 0$), if

$$\| \mathtt{T}\mu - \mathtt{T}\nu \| \leq \beta \,\|\mu - \nu\|, \quad \forall\, \mu,\, \nu \in \mathcal{H}.$$

The implicit resolvent operator $\mathcal{J}_{\mathtt{A}}(\mu)$ is non-expansive and has the following characterization.

Assumption 1 ([31]). The implicit resolvent operator $\mathcal{J}_{\mathtt{A}(\mu)}$, satisfies the condition

$$\| \mathcal{J}_{\mathtt{A}(\mu)}[\omega] - \mathcal{J}_{\mathtt{A}(\nu)}[\omega] \| \leq \eta \;\| \mu - \nu \|, \quad \forall \mu, \nu, \omega \in \mathtt{H}, \tag{2.12}$$

where $\eta > 0$ is a constant.

3. Main Results

In this section, we prove the equivalence between the problem (2.1) and the fixed point problems. This alternative formulation is used to discuss the existence of a solution as well as to suggest some new inertial-type approximation schemes for solving the general quasi-variational inclusion (2.1).

Lemma 1. *The function $\mu \in \mathcal{H}$ is a solution of the general quasi-variational inclusion* (2.1)*, if and only if, $\mu \in \mathcal{H}$ satisfies the relation*

$$h(\mu) = \mathcal{J}_{\mathcal{A}(\mu)}[g(\mu) - \rho \mathcal{T}\mu], \tag{3.1}$$

where $\mathcal{J}_{\mathcal{A}(\mu)}$ is the resolvent operator and $\rho > 0$ is a constant.

Proof. Let $\mu \in \mathcal{H}$ be a solution of (2.1), then, for a constant ρ,

$$\rho\mathcal{T}\mu \quad + \quad \rho\mathcal{A}(h(\mu), h(\mu)) \ni 0,$$
$$\Longleftrightarrow$$
$$-g(\mu) + \rho\mathcal{T}\mu \quad + \quad h(\mu) + \rho\mathcal{A}(h(\mu), h(\mu)) \ni 0$$
$$\Longleftrightarrow$$
$$h(\mu) = \mathcal{J}_{\mathcal{A}(\mu)}[g(\mu) - \rho\mathcal{T}\mu],$$

the required (3.1). □

Lemma 1 implies that the general quasi-variational inclusion (2.1) is equivalent to the fixed point problem (3.1).

Clearly, from (3.1), we have

$$\mu = \mu - h(\mu) + \mathcal{J}_{\mathcal{A}(\mu)}[g(\mu) - \rho\mathcal{T}\mu], \tag{3.2}$$

which play an important part to derive the main results.

Using (3.2), we define the mapping Φ associated with (3.1) as

$$\Phi(\mu) = \mu - h(\mu) + \mathcal{J}_{\mathcal{A}(\mu)}[g(\mu) - \rho\mathcal{T}\mu]. \tag{3.3}$$

To prove the existence of the solution of problem (2.1), it is enough that the mapping Φ defined by (3.3) is a contraction mapping.

Theorem 1. *Let the operators $\mathcal{T}, g, h$ be strongly monotone with constants $\alpha > 0, \delta > 0, \delta_1 > 0$ and Lipschitz continuous with constants $\beta > 0, \sigma > 0, \sigma_1 > 0$, respectively. Let Assumption 1 hold. If there exists a constant $\rho > 0$, such that*

$$\left|\rho - \frac{\alpha}{\beta^2}\right| < \frac{\sqrt{\alpha)^2 - \beta^2 k(2-k)}}{\beta^2}, \quad \alpha > \beta\sqrt{\beta^2 k(2-k)} \quad k < 1, \tag{3.4}$$

where

$$k = \sqrt{(1 - 2\delta + \sigma^2)} + \sqrt{(1 - 2\delta_1 + \sigma_1^2)} + \eta, \tag{3.5}$$

then there exists a solution $\mu \in \mathcal{H}$ satisfying problem (2.1).

Proof. Let $\mu \neq \nu \in \mathcal{H}$ be two solutions of problem (2.1). Then, from (3.3) and Assumption 1, we have

$$\begin{aligned}
\|\Phi(\nu) - \Phi(\mu)\| &\leq \|\nu - \mu - (h(\nu) - h(\mu))\| \\
&\quad + \|\mathcal{J}_{\mathcal{A}(\nu)}[g(\nu) - \rho\mathcal{T}\nu] - \mathcal{J}_{\mathcal{A}(\mu)}[g(\mu) - \rho\mathcal{T}\mu]\| \\
&= \|\nu - \mu - (h(\nu) - h(\mu))\| \\
&\quad + \|\mathcal{J}_{\mathcal{A}(\nu)}[g(\nu) - \rho\mathcal{T}\nu] - \mathcal{J}_{\mathcal{A}(\mu)}[g(\nu) - \rho\mathcal{T}\nu]\| \\
&\quad + \|\mathcal{J}_{\mathcal{A}(\mu)}[g(\nu) - \rho\mathcal{T}\nu] - \mathcal{J}_{\mathcal{A}(\mu)}[g(\mu) - \rho(\mathcal{T}\mu]\| \\
&\leq \|\nu - \mu - (h(\nu) - h(\mu))\| + \|g(\nu) - g(\mu) \\
&\quad - \rho(\mathcal{T}\mu - \mathcal{T}\nu)\| + \eta\|\nu - \mu\| \\
&\leq \|\nu - \mu - (h(\nu) - h(\mu))\| + \|\nu - \mu - (g(\nu) - g(\mu))\| \\
&\quad + \|\nu - \mu - \rho(\mathcal{T}\nu - \mathcal{T}\mu)\| + \eta\|\nu - \mu\|.
\end{aligned} \tag{3.6}$$

Since the operator $\mathcal{T}$ is strongly monotonicity with constant $\alpha > 0$ and Lipschitz continuous with constant $\beta > 0$, so

$$\begin{aligned}\|\mu - \nu - \rho(\mathcal{T} - \mathcal{T}\nu)\|^2 &= \|\mu - \nu\|^2 - \rho\langle \mathcal{T}\mu - \mathcal{T}\nu, \mu - \nu\rangle \\ &\quad + \rho^2\|\mathcal{T}\mu - \mathcal{T}\nu\|^2, \\ &\leq (1 - 2\alpha\rho + \beta^2\rho^2)\|\mu - \nu\|^2. \end{aligned} \tag{3.7}$$

Similarly, using the strongly monotonicity with constants δ, δ_1 and Lipschitz continuity with constants σ, σ_1 of the operator g, h, we have

$$\begin{aligned}\|\mu - \nu - (g(\mu) - g(\nu))\|^2 &= \|\mu - \nu\|^2 - \langle g(\mu) - g(\nu), \mu - \nu\rangle \\ &\quad + \|g(\mu) - g(\nu)\|^2, \\ &\leq (1 - 2\delta + \sigma^2)\|\mu - \nu\|^2 \end{aligned} \tag{3.8}$$

and

$$\begin{aligned}\|\mu - \nu - (h(\mu) - h(\nu))\|^2 &= \|\mu - \nu\|^2 - \langle h(\mu) - h(\nu), \mu - \nu\rangle \\ &\quad + \|h(\mu) - h(\nu)\|^2, \\ &\leq (1 - 2\delta_1 + \sigma_1^2)\|\mu - \nu\|^2. \end{aligned} \tag{3.9}$$

Combining (3.8), (3.7), (3.6) and (3.9), we have

$$\begin{aligned}\|\Phi(\nu) - \Phi(\mu)\| &\leq \Big\{\sqrt{(1 - 2\alpha\rho + \beta^2\rho^2)} \\ &\quad + \sqrt{(1 - 2\delta + \sigma^2)} + \sqrt{(1 - 2\delta_1 + \sigma_1^2)} + \eta\Big\}\|\mu - \nu\| \\ &= \theta\|\mu - \nu\|, \end{aligned} \tag{3.10}$$

where

$$\begin{aligned}\theta &= \Big\{\sqrt{(1 - 2\alpha\rho + \beta^2\rho^2)} + \sqrt{(1 - 2\delta + \sigma^2)} + \sqrt{(1 - 2\delta_1 + \sigma_1^2)} + \eta\Big\} \\ &= \Big\{\sqrt{(1 - 2\alpha\rho + \beta^2\rho^2)} + k\Big\}, \end{aligned} \tag{3.11}$$

where

$$k = \sqrt{(1 - 2\delta + \sigma^2)} + \sqrt{(1 - 2\delta_1 + \sigma_1^2)} + \eta.$$

From (3.5), it follows that $\theta < 1$. Thus, it follows that the mapping $\Phi(\mu)$ defined by (3.3) is a contraction mapping and consequently, the mapping $\Phi(\mu)$ has a fixed point $\Phi(\mu) = \mu \in \mathcal{H}$ satisfying (2.1), the required result. □

This alternative equivalent formulation (3.1) is used to suggest the following iterative methods for solving problem (2.1).

Algorithm 1. For a given $\mu_0 \in \mathcal{H}$, compute the approximate solutions $\{\mu_n\}$, $\{w_n\}$ and $\{y_n\}$ by the iterative schemes

$$h(y_n) = \mathcal{J}_{\mathcal{A}(\mu_n)}[g(\mu_n) - \rho\mathcal{T}\mu_n],$$
$$h(w_n) = \mathcal{J}_{\mathcal{A}(y_n)}[g(y_n) - \rho\mathcal{T}y_n],$$
$$h(\mu_{n+1}) = \mathcal{J}_{\mathcal{A}(w_n)}[g(w_n) - \rho\mathcal{T}w_n].$$

Algorithm 7 is a three-step forward-backward splitting algorithm for solving general variational inclusions (2.1). This method is very much similar to that of Glowinski and Le Tallec's [10] for variational inequalities, which they suggested by using the Lagrangian technique.

We now suggested and analyzed the three-step scheme for solving the general quasi-variational inclusion (2.1).

Algorithm 2. For a given $\mu_0 \in \mathcal{H}$, compute the approximate solution $\{\mu_{n+1}\}$ by the iterative schemes

$$y_n = (1-\gamma_n)\mu_n + \gamma_n\{\mu_n - h(\mu_n) + \mathcal{J}_{\mathcal{A}(\mu)}[g(\mu_n) - \rho\mathcal{T}\mu_n]\}, \tag{3.12}$$

$$w_n = (1-\beta_n)\mu_n + \beta_n\{y_n - h(y_n) + \mathcal{J}_{\mathcal{A}(y_n)}[g(y_n) - \rho\mathcal{T}y_n]\}, \tag{3.13}$$

$$\mu_{n+1} = (1-\alpha_n)\mu_n + \alpha_n\{w_n - h(w_n) + \mathcal{J}_{\mathcal{A}(w_n)}[g(w_n) - \rho\mathcal{T}w_n]\}. \tag{3.14}$$

For $\gamma_n = 0$, Algorithm 2 reduces to the following:

Algorithm 3. For a given $\mu_0 \in \mathcal{H}$, compute $\{\mu_{n+1}\}$ by the iterative schemes

$$w_n = (1-\beta_n)\mu_n + \beta_n\{\mu_n - h(\mu_n) + \mathcal{J}_{\mathcal{A}(\mu_n)}[g(\mu_n) - \rho\mathcal{T}\mu_n]\},$$
$$\mu_{n+1} = (1-\alpha_n)\mu_n + \alpha_n\{w_n - h(w_n) + \mathcal{J}_{\mathcal{A}(w_n)}[g(w_n) - \rho\mathcal{T}w_n]\},$$

which is known as the Ishikawa iterative scheme for the general quasi variational inclusion (2.1).

Note that for $\gamma_n = 0$ *and* $\beta_n = 0$, Algorithm 7 is called the Mann iterative method.

Algorithm 4. For a given $\mu_0 \in \mathcal{H}$, compute $\{\mu_{n+1}\}$ by the iterative schemes

$$\mu_{n+1} = (1-\beta_n)\mu_n + \beta_n\{\mu_n - h(\mu_n) + \mathcal{J}_{\mathcal{A}(\mu_n)}[g(\mu_n) - \rho\mathcal{T}\mu_n]\}.$$

Now, we suggest a perturbed iterative scheme for solving the extended general quasi-variational inclusion (2.1).

Algorithm 5. For a given $\mu_o \in \mathcal{H}$, compute the approximate solution $\{\mu_n\}$ by the iterative schemes

$$\begin{aligned} y_n &= (1-\gamma_n)\mu_n + \gamma_n\{\mu_n - h(\mu_n) + \mathcal{J}_{\mathcal{A}(\mu_n)}[g(\mu_n) - \rho T\mu_n]\} + \gamma_n h_n, \\ w_n &= (1-\beta_n)\mu_n + \beta_n\{y_n - h(y_n) + \mathcal{J}_{\mathcal{A}(y_n)}[g(y_n) - \rho\mathcal{T}y_n]\} + \beta_n f_n, \\ \mu_{n+1} &= (1-\alpha_n)\mu_n + \alpha_n\{w_n - h(w_n) + \mathcal{J}_{\mathcal{A}(w_n)} \\ &\qquad \times [g(w_n) - \rho\mathcal{T}w_n\} + \alpha_n e_n, \end{aligned}$$

where $\{e_n\}$, $\{f_n\}$ and $\{h_n\}$ are the sequences of the elements of $\mathcal{H}$ introduced to take into account possible inexact computations and $\mathcal{J}_{A_n}$ is the corresponding perturbed resolvent operator and the sequences $\{\alpha_n\}$, $\{\beta_n\}$ and $\{\gamma_n\}$ satisfy

$$0 \le \alpha_n, \beta_n, \gamma_n \le 1; \quad \forall n \ge 0, \quad \sum_{n=0}^{\infty} \alpha_n = \infty.$$

For $\gamma_n = 0$, we obtain the perturbed Ishikawa iterative method, and for $\gamma_n = 0$ and $\beta_n = 0$, we obtain the perturbed Mann iterative schemes for solving general variational inclusion (2.1).

We now study the convergence analysis of Algorithm 2, which is the main motivation of our following result.

Theorem 2. *Let the operators $\mathcal{T}, g, h$ satisfy all the assumptions of Theorem* 1. *If the condition* (3.23) *holds, then the approximate solution $\{u_n\}$ obtained from Algorithm* 2 *converges to the exact solution $\mu \in \mathcal{H}$ of the general quasi variational inclusion* (2.1) *strongly in $\mathcal{H}$.*

Proof. From Theorem 1, we see that there exists a unique solution $\mu \in \mathcal{H}$ of the general quasi variational inclusion (2.1). Let $\mu \in \mathcal{H}$ be the unique solution of (2.1). Then, using Lemma 1, we have

$$\mu = (1-\alpha_n)\mu + \alpha_n\{\mu - h(\mu) + \mathcal{J}_{\mathcal{A}(\mu)}[g(\mu) - \rho\mathcal{T}\mu]\} \quad (3.15)$$

$$= (1-\beta_n)\mu + \beta_n\{\mu - h(\mu) + \mathcal{J}_{\mathcal{A}(\mu)}[g(\mu) - \rho\mathcal{T}\mu]\} \quad (3.16)$$
$$= (1-\gamma_n)\mu + \gamma_n\{\mu - h(\mu) + \mathcal{J}_{\mathcal{A}(\mu)}[g(\mu) - \rho\mathcal{T}\mu]\}. \quad (3.17)$$

From (3.14), (3.15) and (3.5) and Assumption 1, we have

$$\begin{aligned}
\|\mu_{n+1} - \mu\| &= \|(1-\alpha_n)(\mu_n - \mu) + \alpha_n(w_n - \mu - (h(w_n) - h(\mu))) \\
&\quad + \alpha_n\{\mathcal{J}_{\mathcal{A}(w_n)}[g(w_n) - \rho\mathcal{T}w_n] - \mathcal{J}_{\mathcal{A}(\mu)}[g(\mu) - \rho\mathcal{T}\mu]\}\| \\
&\leq (1-\alpha_n)\|\mu_n - \mu\| + \alpha_n\|w_n - \mu - (h(w_n) - h(\mu))\| \\
&\quad + \alpha_n\{\mathcal{J}_{\mathcal{A}(w_n)}[g(w_n) - \rho\mathcal{T}w_n] - \mathcal{J}_{\mathcal{A}(w_n)}[g(\mu_n) - \rho\mathcal{T}\mu\}\| \\
&\quad + \alpha_n\{\mathcal{J}_{\mathcal{A}(w_n)}[g(\mu_n) - \rho\mathcal{T}\mu] + \mathcal{J}_{\mathcal{A}(\mu)}[g(\mu) - \rho\mathcal{T}\mu\}\| \\
&\leq (1-\alpha_n)\|\mu_n - \mu\| + \alpha_n\|w_n - \mu - (g(w_n) - g(\mu))\| \\
&\quad + \alpha_n\|w_n - \mu - (h(w_n) - h(\mu))\| \\
&= \alpha_n\{\|w_n - \mu - (g(w_n) - g(\mu))\|\} \\
&\quad + \alpha_n\|w_n - \mu - \rho(\mathcal{T}w_n - \mathcal{T}\mu)\| + \alpha_n\eta\|w_n - \mu\| \\
&\leq (1-\alpha_n)\|\mu_n - \mu\| + \alpha_n(k + t(\rho))\|w_n - \mu\| \\
&= (1-\alpha_n)\|u_n - \mu\| + \alpha_n\theta\|w_n - \mu\|, \qquad (3.18)
\end{aligned}$$

where θ is defined by (3.2).

In a similar way, from (3.5), (3.12) and (3.16), we have

$$\begin{aligned}
\|w_n - \mu\| &\leq (1-\beta_n)\|\mu_n - \mu\| + \beta_n\theta\|y_n - \mu - (h(y_n) - h(\mu))\| \\
&\quad + \beta_n\theta\|y_n - \mu - (g(y_n) - g(\mu))\| \\
&\quad + \beta_n\|y_n - \mu - \rho(Ty_n - T\mu)\| + \beta_n\eta\|y_n - \mu\| \\
&\leq (1-\beta_n)\|\mu_n - \mu\| + \beta_n(k + t(\rho))\|y_n - \mu\|, \\
&\leq (1-\beta_n)\|\mu_n - \mu\| + \beta_n\theta\|y_n - \mu\|, \qquad (3.19)
\end{aligned}$$

where θ is defined by (3.2).

From (3.5), (3.12) and (3.17), we obtain

$$\begin{aligned}
\|y_n - \mu\| &\leq (1-\gamma_n)\|\mu_n - \mu\| + \gamma_n\theta\|\mu_n - \mu\| \\
&\leq (1 - (1-\theta)\gamma_n)\|\mu_n - \mu\| \\
&\leq \|\mu_n - \mu\|, \qquad (3.20)
\end{aligned}$$

where θ is defined by (3.2).

From (3.19) and (3.20), we obtain

$$\begin{aligned}\|w_n - \mu\| &\leq (1-\beta_n)\|\mu_n - \mu\| + \beta_n\theta\|\mu_n - \mu\| \\ &= (1-(1-\theta)\beta_n)\|\mu_n - \mu\| \\ &\leq \|\mu_n - \mu\|. \end{aligned} \tag{3.21}$$

Form the above equations, we have

$$\begin{aligned}\|\mu_{n+1} - \mu\| &\leq (1-\alpha_n)\|\mu_n - \mu\| + \alpha_n\theta\|\mu_n - \mu\| \\ &= [1-(1-\theta)\alpha_n]\|\mu_n - \mu\| \\ &\leq \prod_{i=0}^{n}[1-(1-\theta)\alpha_1]\|\mu_0 - \mu\|. \end{aligned}$$

Since $\sum_{n=0}^{\infty}\alpha_n$ diverges and $1-\theta > 0$, we have $\prod_{i=0}^{n}[1-(1-\theta)\alpha_i] = 0$. Consequently, the sequence $\{u_n\}$ convergences strongly to μ. From (3.20) and (3.21), it follows that the sequences $\{y_n\}$ and $\{w_n\}$ also converge to μ strongly in $\mathcal{H}$. This completes the proof. □

We can propose some iterative approximation schemes for solving the extended general quasi-variational inclusion (2.1).

Algorithm 6. For given $\mu_0 \in \mathcal{H}$, compute μ_{n+1} by the recurrence relation

$$\begin{aligned}\mu_{n+1} &= (1-\alpha_n)\mu_n + \alpha_n\{h(\mu_n) - \mathcal{J}_{\mathtt{A}(\mu_n)}[\mathtt{g}(\mu_n) - \rho\mathtt{T}\mu_{n+1}]\}, \\ &\alpha_n \in [0,1],\end{aligned}$$

which is known as the extraresolvent method in the sense of Koperlevich [17] and is equivalent to the following two-step method.

Algorithm 7. For given $\mu_0 \in \mathcal{H}$, compute μ_{n+1} by the recurrence relation

$$\begin{aligned}\mathtt{h}(\omega_n) &= \mathcal{J}_{\mathtt{A}(\mu_n)}[\mathtt{g}(\mu_n) - \rho\mathtt{T}\mu_n], \\ \mu_{n+1} &= (1-\alpha_n)\mu_n + \alpha_n\{h(\mu_n) - \mathcal{J}_{\mathtt{A}(\mu_n)}[\mathtt{g}(\mu_n) - \rho\mathtt{T}\omega_n]\}, \\ &\alpha_n \in [0,1],\end{aligned}$$

which is called the predictor–corrector resolvent method.

We can rewrite (3.1) as

$$\mathtt{h}(\mu) = \mathcal{J}_{\mathtt{A}(\mu)}\{\mathtt{g}((1-\Theta_n)\mu + \Theta_n\mu) - \rho\mathtt{T}\mu\},$$

where $\Theta_n \in [0,1]$, $\forall n \geq 1$.

This fixed point formulation is used to suggest the following two-step method for solving general quasi-variational inclusion (2.1) using the ideas of Alvarez [1], Alvarez *et al.* [2] and Noor *et al.* [33, 44, 46, 51].

Algorithm 8. For given $\mu_0, \mu_1 \in \mathcal{H}$, compute μ_{n+1} by the recurrence relation

$$\begin{aligned} \omega_n &= \mu_n - \Theta_n \left(\mu_n - \mu_{n-1}\right), \\ \mathtt{g}(\mu_{n+1}) &= (1-\alpha_n)\mu_n + \alpha_n\{h(\mu_n) - \mathcal{J}_{\mathtt{A}(\mu_n)}\left[\mathtt{g}(\omega_n) - \rho\mathtt{T}\mu_n - \rho\mathtt{B}\mu_n\right]\}, \\ & n = 1, 2, \ldots, \end{aligned}$$

where $\Theta_n \in [0,1]$, $\forall\, n \geq 1$.

Such types of inertial projection methods for solving general quasi-variational inequalities have been considered by Jabeen *et al.* [12, 13], Noor [33] and Noor *et al.* [33, 44–46, 51].

We can rewrite (3.1) as

$$\begin{aligned} \mathtt{h}(\mu) = \mathcal{J}_{\mathtt{A}((1-\Theta_n)\mu+\Theta_n\mu))}\{&\mathtt{g}((1-\Theta_n)\mu + \Theta_n\mu)) \\ &- \rho\mathtt{T}((1-\Theta_n)\mu + \Theta_n\mu)))\}, \end{aligned}$$

where $\Theta_n \in [0,1]$, $\forall n \geq 1$.

Using this fixed point formulation, we can suggest the following inertial-type methods for solving general quasi-variational inclusion (2.1).

Algorithm 9. For given $\mu_0, \mu_1 \in \mathcal{H}$, compute μ_{n+1} by the recurrence relation

$$\begin{aligned} \omega_n &= \mu_n - \Theta_n \left(\mu_n - \mu_{n-1}\right), \\ \mathtt{h}(\mu_{n+1}) &= 1 - \alpha_n)\mu_n + \alpha_n\{h(\mu_n) - \mathcal{J}_{\mathtt{A}(\omega_n)}\left[\mathtt{g}(\omega_n) - \rho\mathtt{T}\omega_n)\right], \end{aligned}$$

where $\Theta_n \in [0,1]$, $\forall n \geq 1$.

Algorithm 9 is known as the modified inertial resolvent method for solving general quasi-variational inclusion (2.1).

Algorithm 10. For given $\mu_0, \mu_1 \in \mathcal{H}$, compute μ_{n+1} by the recurrence relation

$$\begin{aligned}\omega_n &= \mu_n - \Theta_n\left(\mu_n - \mu_{n-1}\right), \\ \mathtt{h}(\mathtt{y}_n) &= \mathcal{J}_{\mathtt{A}(\omega_n)}\left[\mathtt{g}(\omega_n) - \rho\mathrm{T}\omega_n\right], \\ \mathtt{h}(\mu_{n+1}) &= \mathcal{J}_{\mathtt{A}(\mathtt{y}_n)}\left[\mathtt{g}(\mathtt{y}_n) - \rho\mathrm{T}\mathtt{y}_n)\right],\end{aligned}$$

where $\Theta_n \in [0,1], \forall n \geq 1$.

Algorithm 10 is a three-step modified inertial method for solving general quasi-variational inclusion (2.1).

We now suggest a four-step inertial method for solving the general quasi-variational inclusion (2.1).

Algorithm 11. For given $\mu_0, \mu_1 \in \mathcal{H}$, compute μ_{n+1} by the recurrence relation

$$\begin{aligned}\omega_n &= \mu_n - \Theta_n\left(\mu_n - \mu_{n-1}\right), \\ \mathtt{x}_n &= (1-\gamma_n)\mu_n + \gamma_n\big\{\omega_n - \mathtt{h}(\omega_n) + \mathcal{J}_{\mathtt{A}(\omega_n)}\left[\mathtt{g}(\omega_n) - \rho\mathrm{T}\omega_n\right]\big\}, \\ \mathtt{y}_n &= (1-\beta_n)\mu_n + \beta_n\big\{\mathtt{x}_n - \mathtt{h}(\mathtt{x}_n) + \mathcal{J}_{\mathtt{A}(\mathtt{x}_n)}\left[\mathtt{g}(\mathtt{x}_n) - \rho\mathrm{T}\mathtt{x}_n\right]\big\}, \\ \mu_{n+1} &= (1-\alpha_n)\mu_n + \alpha_n\big\{\mathtt{y}_n - \mathtt{h}(\mathtt{y}_n) + \mathcal{J}_{\mathtt{A}(\mathtt{y}_n)}\left[\mathtt{g}(\mathtt{y}_n) - \rho\mathrm{T}\mathtt{y}_n\right]\big\},\end{aligned}$$

where $\alpha_n, \beta_n, \gamma_n, \Theta_n \in [0,1], \quad \forall n \geq 1$.

We now suggest and propose some new iterative forward-backward for solving problem (2.1).

Let $\mu \in \mathcal{H}$ satisfy (2.1). Then, $\mu \in \mathcal{H}$ satisfies (3.1), if and only if, $\mu \in \mathcal{H}$ such that

$$h(\mu) = \mathcal{J}_{\mathcal{A}(\mu)}z, \tag{3.22}$$

where

$$\begin{aligned}z &= g(\mu) - \rho\mathcal{T}\mu && (3.23)\\ &= g(\mu) - \rho\mathcal{T}h^{-1}\mathcal{J}_{\mathcal{A}(\mu)}z && (3.24)\\ &= g(h^{-1}\mathcal{J}_{\mathcal{A}(\mu)}z) - \rho\mathcal{T}h^{-1}\mathcal{J}_{\mathcal{A}(\mu)}z. && (3.25)\end{aligned}$$

From (3.25) and (3.22), it follows that

$$h(\mu) = \mathcal{J}_{\mathcal{A}(\mu)}\{g(h^{-1}\mathcal{J}_{\mathcal{A}(\mu)}z) - \rho\mathcal{T}h^{-1}\mathcal{J}_{\mathcal{A}(\mu)}z\}$$

which can be written as

$$\mu = \mu - \Theta_n\{h(\mu) - \mathcal{J}_{\mathcal{A}(\mu)}\{g(h^{-1}\mathcal{J}_{\mathcal{A}(\mu)}z) - \rho\mathcal{T}h^{-1}\mathcal{J}_{\mathcal{A}(\mu)}z\}\}, \tag{3.26}$$

where $\Theta_n \in [0,1], \quad \forall n \geq 1$.

The fixed point formulation (3.26) is used to suggest the forward-backward predictor–corrector iterative methods.

Algorithm 12. For given $\mu_0 \in \mathcal{H}$, compute μ_{n+1} by the recurrence relation

$$\begin{aligned} \mathtt{h}(\omega_n) &= \mathcal{J}_{\mathtt{A}(\mu_n)}\left[\mathtt{g}(\mu_n) - \rho\mathtt{T}\mu_n\right], \\ \mu_{n+1} &= \mu_n - \Theta_n\{h(\mu_n) - \mathcal{J}_{\mathcal{A}(\mu)}\{g(\omega_n) - \rho\mathcal{T}\omega_n\}\}, \end{aligned}$$

which is also called the two-step iterative method. Here, $\Theta_n \in [0,1], \quad \forall n \geq 1$.

Algorithm 13. For given $\mu_0 \in \mathcal{H}$, compute μ_{n+1} by the recurrence relation

$$\begin{aligned} y_n &= \mu_n - \Theta_n\left(\mu_n - \mu_{n-1}\right), \\ \mathtt{h}(\omega_n) &= \mathcal{J}_{\mathtt{A}(\mu_n)}\left[\mathtt{g}(y_n) - \rho\mathtt{T}y_n\right], \\ \mu_{n+1} &= \mu_n - \Theta_n\{h(\mu_n) - \mathcal{J}_{\mathcal{A}(\mu)}\{g(\omega_n) - \rho\mathcal{T}\omega_n\}\}, \end{aligned}$$

which is also called the three-step inertial forward-backward iterative method. Here, $\Theta_n \in [0,1], \ \forall n \geq 1$.

Remark 3. In a similar way, one can suggest and investigate several inertial iterative methods by appropriate, suitable choices of the operators and combining equations (3.22), (3.23), (3.24) and (3.25). We have only given the glimpse of the techniques of the resolvent methods and its variations. It is an interesting problem to consider the implementation of these various iterative methods and comparison with other methods. For a different and suitable choice of operators and spaces in Algorithm 13, one can obtain numerous new and previous iterative schemes for solving general quasi-variational inclusion (2.1) and related problems. This shows that Algorithm 11 is quite flexible and unifying ones. Convergence criteria of Algorithm 11 can be considered using the technique of Noor *et al.* [39].

4. Dynamical Systems Technique

Though the problems in each of these areas may look completely different, the resulting algorithms can be very closely related. Using the projection technique, one can show that the variational inequalities are equivalent to the fixed point problem. This equivalence has been used [8, 12, 21, 31, 33, 40, 45, 46, 51] to suggest and analyze some dynamical systems, in which the right-hand side of the ordinary differential equation is a projection operator. Projected dynamical systems are characterized by a discontinuous right-hand side. The discontinuity arises from the constraints governing the applications in question. The novel feature of the projected dynamical system is that the set of the stationary points of the dynamical systems corresponds to the set of the solution of the variational inequalities. Consequently, the equilibrium problems which can be formulated in the setting of variational inequalities can now be studied in the more general setting of the dynamical systems. They are intrinsically as framework for creating behavioral model for describing disequilibrium trajectories of real economics and physical process prior to reaching steady states. The projected dynamical systems can be used effectively in designing neural network for solving variational inequalities. The neural networks are computing systems composed of highly interconnected simple information processing units and thus can solve variational inequalities and optimization problems in execution times at the order of magnitude mush faster than most iterative algorithms for general purpose digital computers.

In this section, we consider the projected dynamical systems associated with variational inequalities, which were introduced and studied by Dupuis *et al.* [8]. For the motivation, applications, formulation and numerical methods of dynamical systems, see Nagurney and Zhang [21] and Noor [31–33].

We now define the residue vector $R(\mu)$ by the relation

$$R(u) = h(\mu) - J_{A(\mu)}[g(\mu) - \rho \mathcal{T}\mu]. \tag{4.1}$$

Invoking Lemma 1, one can easily conclude that $\mu \in \mathcal{H}$ is a solution of (2.1), if and only if, $\mu \in \mathcal{H}$ is a zero of the equation

$$R(\mu) = 0. \tag{4.2}$$

We now consider a projected dynamical system associated with the variational inequalities. Using the equivalent formulation (4.2), we suggest a class of projected dynamical systems as

$$\frac{dh(\mu)}{dt} = \lambda J_{A(\mu)}[g(\mu) - \rho \mathcal{T}\mu] - h(\mu)\}, \quad \mu(t_0) = u_0 \in \mathcal{H}, \quad (4.3)$$

where λ is a parameter. The system of type (4.3) is called the resolvent dynamical system associated with extended general quasi-variational inclusion (2.6). Here, the right hand is related to the resolvent and is discontinuous on the boundary. From the definition, it is clear that the solution of the dynamical system always stays in $\mathcal{H}$. This implies that the qualitative results such as the existence, uniqueness and continuous dependence of the solution of (4.3) can be studied. These resolvent dynamical systems are associated with the extended general quasi-variational inclusions (2.6), which have been studied extensively.

We use the resolvent dynamical system (4.3) to suggest some iterative for solving quasi-variational inclusions (2.6). These methods can be viewed in the sense of Koperlevich [17] and Noor [31, 32] involving the double resolvent operator.

For simplicity, we consider the dynamical system

$$\frac{dh(\mu)}{dt} + h(\mu) = J_{A(\mu)}[g(\mu) - \rho \mathcal{T}\mu], \quad u(t_0) = \alpha. \quad (4.4)$$

We construct the implicit iterative method using the forward difference scheme.

Discretizing equation (4.4), we have

$$\frac{h(\mu_{n+1}) - h(\mu_n)}{h_1} + h(\mu_n) = J_{A(\mu_n)}[g(\mu_n) - \rho T\mu_{n+1})], \quad (4.5)$$

where h_1 is the step size. Now, for $h_1 = 1$, we can suggest the following implicit iterative method for solving the variational inequality (2.6).

Algorithm 14. For a given $\mu_0 \in H$, compute μ_{n+1} by the iterative scheme

$$h(\mu_{n+1}) = J_{A(\mu_n)}\left[g(\mu_n) - \rho T\mu_{n+1}\right], \quad n = 0, 1, 2, \ldots.$$

This is an implicit method in the sense of Koperlevich [17]. Using the predictor-corrector technique, Algorithm 14 can be rewritten in the equivalent form as follows:

Algorithm 15. For a given $\mu_0 \in H$, compute μ_{n+1} by the iterative scheme

$$y(\mu_n) = J_{A(\mu_n)}\left[g(\mu_n) - \rho T\mu_n\right],$$
$$h(\mu_{n+1}) = J_{A(\mu_n)}\left[g(\mu_n) - \rho T y_n\right],$$
$$n = 0, 1, 2, \ldots,$$

where $\alpha_n, \beta_n, \gamma_n, \Theta_n \in [0,1], \ \forall n \geq 1$.

Using the inertial scheme idea, Algorithm 14 can be rewritten in the equivalent form as follows:

Algorithm 16. For a given $\mu_0 \in H$, compute μ_{n+1} by the iterative scheme

$$\omega_n = \mu_n - \Theta_n\left(\mu_n - \mu_{n-1}\right),$$
$$h(\mu_{n+1}) = J_{A(\mu_n)}[g(\mu_n) - \rho T\omega_n], \quad n = 0, 1, 2, \ldots,$$

where $\alpha_n, \beta_n, \gamma_n, \Theta_n \in [0,1], \quad \forall n \geq 1$.

We now suggest an other implicit iterative method for solving (2.6).

Discretizing (4.6), we have

$$\frac{h(u_{n+1}) - h(u_n)}{h_1} + h(u_n) = J_{A(\mu_{n+1})}[g(u_{n+1}) - \rho T u_{n+1}], \quad (4.6)$$

where h is the step size.

For $h = 1$, this formulation enables us to suggest the following iterative method.

Algorithm 17. For a given $\mu_0 \in \mathcal{H}$, compute μ_{n+1} by the iterative scheme

$$\omega_n = \mu_n - \Theta_n\left(\mu_n - \mu_{n-1}\right),$$
$$h(\mu_{n+1}) = J_{A(\mu_n)}[g(\omega_n) - \rho T\omega_n], \quad n = 0, 1, 2, \ldots,$$

which is two-step resolvent method in the sense of Noor [31, 32].

Note that Algorithms 16 and 17 are quite different methods for solving the extended general quasi-variational inclusions.

Again, using the dynamical systems, we can suggested some iterative methods for solving the extended general quasi-variational inclusions and related optimization problems.

Algorithm 18. For a given $\mu_0 \in \mathcal{H}$, compute μ_{n+1} by the iterative scheme

$$h(\mu_{n+1}) = J_{A(\mu_n)}[(1-\Theta_n)g(\mu_n) + \Theta_n g(\mu_{n+1}) - \rho\mathcal{T}\mu_{n+1}],$$
$$n = 0, 1, 2, \ldots.$$

Using the predictor–corrector technique, Algorithm 18 can be written as follows:

Algorithm 19. For a given $\mu_0 \in \mathcal{H}$, compute μ_{n+1} by the iterative scheme

$$\omega_n = \mu_n - \Theta_n (\mu_n - \mu_{n-1}),$$
$$h(\mu_{n+1}) = J_{A(\mu_n)}[(1-\Theta_n)g(\mu_n) + \Theta_n g(\omega_n) - \rho\mathcal{T}\omega_n].$$

In a similar way, one can suggest a wide class of implicit iterative methods for solving variational inequalities and related optimization problems using the dynamical system technique.

5. Future Research

We would like to mention that the ideas and techniques developed in this paper can be extended for suggesting the resolvent dynamical systems for multivalued quasi-variational inclusions. To be more precise, let $C(H)$ be a family of nonempty compact subsets of H. Let $T, V : H \longrightarrow C(H)$ be the multivalued operators. For a given nonlinear operator $N(.,.) : H \times H \longrightarrow H$, consider the problem of finding $\mu \in \mathcal{H}, w \in T(\mu), y \in V(\mu)$ such that

$$0 \in \rho N(w, y) + h(\mu) - g(\mu) + \rho A(h(\mu), h(\mu)), \tag{5.1}$$

which are called the multivalued extended general quasi-variational inclusions.

Using the resolvent operator technique, it can be shown that the multivalued extended general quasi-variational inclusions (5.1) are equivalent to finding $\mu \in \mathcal{H}, w \in T(\mu), y \in V(\mu)$ such that

$$h(\mu) = J_{A(\mu)}[g(\mu) - \rho N(w, y)], \tag{5.2}$$

which is a multivalued fixed point problem. Using this, the alternative equivalence between these problems can be used to suggest and analyze several iterative methods for solving multivalued extended general quasi-variational inclusions using the technique of Section 3.

We use this equivalence to suggest an multivalued extended dynamical system of the type

$$a\frac{d^2h(\mu)}{d^2t} + b\frac{dh(\mu)}{dt} = \lambda\{J_{A(\mu)}[g(\mu) - \rho N(w, y)] - h(\mu)\},$$
$$\mu(t_0) = \alpha, \quad \mu'(t_0] = \beta, \tag{5.3}$$

associated with the multivalued extended general quasi-variational inclusions (5.1), where a, b, λ, α and β are constants.

Using essentially techniques and ideas of Noor [34, 35] and Noor *et al.* [33, 45, 46, 51], one can consider the properties of the multivalued extended general dynamical system (5.3).

Remark 4. The theory of the multivalued extended general dynamical systems associated with multivalued extended general quasi-variational inclusions is a relatively new one and does not appear to have developed to an extent that it provides a complete framework for studying these problems. There are still many open problems involving the globally asymptotic stability of the multivalued extended general dynamical systems. Despite the recent activities, much more work is needed in all these areas to develop a sound basis for applications and numerical methods. It is hoped that the results obtained in this paper may inspire and motivate the readers to discover new, innovative and novel applications of the multivalued extended general quasi-variational inclusions in various branches of pure and applied sciences.

6. Conclusion

In this chapter, we have introduced and studied some new classes of general quasi-variational inclusions. Several important new and previous special cases are discussed. We have proved that the extended general quasi-variational inclusions are equivalent to the implicit fixed point problems. This alternative formulation is used to discuss the existence of a solution of the extended general quasi-variational inclusions. Several new three-step iterative and inertial methods for solving extended general quasi-variational inclusions are proposed and analyzed. Convergence analysis of these three step resolvent methods is investigated. Dynamical systems associated with extended general quasi-variational inclusions are also introduced. Using the finite difference schemes and discretizing the dynamical systems, some iterative schemes are suggested for solving the general quasi-variational inequalities. Since the dynamical systems associated with variational inclusions are basically initial value problems, one may use the Euler method, Runge–Kutta method and shooting methods to find the approximate solutions of the variational inclusions and related optimization problems. This is a challenging research problems for future research. We would like to point out that very few numerical examples are available for classical extended general quasi-variational inclusions due to their complex and completed structure. In spite of these activities, further efforts are needed to develop numerical implementable methods. It is an interesting problem to compare the efficiency of the proposed methods with other known methods. We expect that the ideas and techniques of this chapter will motivate and inspire the interested readers to explore its applications in various fields.

Acknowledgments

The authors wish to express their deepest gratitude to our teachers, students, colleagues, collaborators, reviewers, referees and friends, who have directly or indirectly contributed in the preparation of this chapter.

References

[1] F. Alvarez. Weak convergence of a relaxed and inertial hybrid projection-proximal point algorithm for maximal monotone operators in Hilbert space. *SIAM J. Optim.*, 14, 2003, 773–782.

[2] F. Alvarez and H. Attouch. An inertial proximal method for maximal monotone operators via discretization of a nonlinear oscillator with damping. *Set-Valued Anal.*, 9, 2001, 3–11.

[3] W. F. Ames. *Numerical Methods for Partial Differential Equations*, 3rd edn. Academic Press, New York, 1992.

[4] K. Ashish, M. Rani and R. Chugh. Julia sets and Mandelbrot sets in Noor orbit. *Appl. Math. Comput.*, 228(1), 2014, 615–631.

[5] A. Bensoussan and J. L. Lions. *Application des Inequalities Variationnelles en Control eten Stochastique.* Dunod, Paris, 1978.

[6] R. W. Cottle, J.-S. Pang and R. E. Stone. *The Linear Complementarity Problem.* SIAM Publication, Philadelphia, USA, 1992.

[7] G. Cristescu and L. Lupsa. *Non-Connected Convexities and Applications.* Kluwer Academic Publishers, Dordrecht, Holland, 2002.

[8] P. Dupius and A. Nagurney. Dynamical systems and variational inequalities. *Ann. Oper. Res.*, 44, 1993, 19–42.

[9] R. Glowinski, J. L. Lions and R. Tremolieres. *Numerical Analysis of Variational Inequalities.* North-Holland, Amsterdam, 1981.

[10] R. Glowinski and P. Le Tallec. *Augmented Lagrangian and Operator Splitting Methods in Nonlinear Mechanics.* SIAM, Philadelphia, Pennsylvania, 1989.

[11] S. Haubruge, V. H. Nguyen and J. J. Strodiot. Convergence analysis and applications of the Glowinski-Le Tallec splitting method for finding a zero of the sum of two maximal monotone operators. *J. Optim. Theory Appl.*, 97, 1998, 645–673.

[12] S. Jabeen, M. A. Noor and K. I. Noor. Inertial iterative methods for general quasi variational inequalities and dynamical systems. *J. Math. Anal.*, 11(3), 2020, 14–29.

[13] S. Jabeen, B. Bin- Mohsen, M. A. Noor and K. I. Noor. Inertial projection methods for solving general quasi-variational inequalities. *AIMS Math.*, 6(2), 2021, 1075–1086. DOI: 10.3934/math.2021064.

[14] S. Karamardian. Generalized complementarity problem. *J. Optim. Theory Appl.*, 8, 1971, 161–168.

[15] N. Kikuchi and J. T. Oden. *Contact Problems in Elasticity.* SAIM Publishing Co., Philadelphia, 1988.

[16] D. Kinderlehrer and G. Stampacchia. *An Introduction to Variational Inequalities and Their Applications.* SIAM, Philadelphia, 2000.

[17] G. M. Koperlevich. The extragradient method for finding saddle points and other problems. *Matecom*, 12, 1976, 203–213.

[18] C. E. Lemke. Bimatrix equilibrium points and mathematical programming. *Manag. Sci.*, 11, 1965, 681–689.

[19] N. Mijajlovic, J. Milojica and M. A. Noor. Gradient-type projection methods for quasi variational inequalities. *Optim. Lett.*, 13, 2019, 1885–1896.

[20] K. G. Murty. *Linear Complementarity: Linear and Nonlinear Programming.* Heldermann, Berlin, 1988.

[21] A. Nagurney and D. Zhang. *Projected Dynamical Systems and Variational Inequalities With Applications.* Kluwer Academic Publishers, Boston, 1996.

[22] T. V. Nghi and N. N. Tam. General variational inequalities: Existence of solutions, Tikhonov-type regularization and well-posedness. *Acta Math. Vietnam*, 47, 2022, 539–552. https://doi.org/10.1007/s40306-021-00435-0.

[23] C. P. Niculescu and L. E. Persson. *Convex Functions and Their Applications.* Springer Verlag, New York, 2018.

[24] M. A. Noor. On Variational Inequalities. PhD Thesis, Brunel University, London, UK, 1975.

[25] M. A. Noor. An iterative scheme for class of quasi variational inequalities. *J. Math. Anal. Appl.*, 110, 1985, 463–468.

[26] M. A. Noor. The quasi-complementarity problem. *J. Math. Anal. Appl.*, 130, 1988, 344–353.

[27] M. A. Noor. Fixed point approach for complementarity problems. *J. Math. Anal. Appl.*, 133, 1988, 437–448.

[28] M. A. Noor. General variational inequalities. *Appl. Math. Lett.*, 1, 1988, 119–121.

[29] M. A. Noor. Quasi variational inequalities. *Appl. Math. Lett.*, 1(4), 1988, 367–370.

[30] M. A. Noor. New approximation schemes for general variational inequalities. *J. Math. Anal. Appl.*, 251, 2000, 217–230.

[31] M. A. Noor. Stability of the modified projected dynamical systems. *Comput. Math. Appl.*, 44, 2002, 1–5.

[32] M. A. Noor. Implicit dynamical systems and quasi variational ineualities. *Appl. Math. Comput.*, 134, 2003, 69–81.

[33] M. A. Noor. Some developments in general variational inequalities. *Appl. Math. Comput.*, 152, 2004, 199–277.

[34] M. A. Noor. Generalized set-valued variational inclusions and resolvent equations. *J. Math. Anal. Appl.*, 228, 1998, 206–220.

[35] M. A. Noor. Three-step iterative algorithms for multivalued quasi variational inclusions. *J. Math. Anal. Appl.*, 255, 2001, 589–604.

[36] M. A. Noor. Extended general variational inequalities. *Appl. Math, Lett.*, 22(2), 2009, 186–192.

[37] M. A. Noor. Projection iterative methods for extended general variational inequalities. *J. Appl. Math. Comput.*, 32, 2010, 83–95.

[38] M. A. Noor. Some aspects of extended general variational inequalities. *Abstr. Appl. Anal.*, 2012, 2012, 16 p. Article ID 303569. DOI: 10.1155/2012/303569.

[39] M. A. Noor and K. I. Noor. Sensitivity analysis of quasi variational inclusions. *J. Math. Anal. Appl.*, 236, 1999, 290–299.

[40] M. A. Noor and K. I. Noor. Some novel aspects of quasi variational inequalities. *Earthline J. Math. Sci.*, 10(1), 2022. https://doi.org/10.34198/ejms.10122.166.

[41] M. A. Noor and K. I. Noor. From representation theorems to variational inequalities. In N. J. Daras and T. M. Rassias (eds.), *Computational Mathematics and Variational Analysis.* Springer Optimization and Its Applications, Vol. 159, 2020, pp. 261–277. https://doi.org/10.1007/978-3-030-44625-3-15261.

[42] M. A. Noor, K. I. Noor and A. G. Khan. Parallel schemes for solving a system of extended general quasi variational inequalities. *Appl. Math. Comput.*, 245, 2014, 566–574.

[43] M. A. Noor and W. Oettli. On general nonlinear complementarity problems and quasi equilibria. *Le Mathematiche* (Catania), 49, 1994, 313–331.

[44] M. A. Noor, K. I. Noor and B. B. Mohsen. Some new classes of general quasi variational inequalities. *AIMS Math.*, 6(6), 2021, 6406–6421.

[45] M. A. Noor, K. I. Noor and M. T. Rassias. New trends in general variational inequalities. *Acta Appl. Math.*, 170(1), 2020, 981–1046.

[46] M. A. Noor, K. I. Noor and M. T. Rassias. A unified approach to extended general quasi variational inclusions. In T. M. Rassias (ed.), *Exploring Mathematical Analysis, Approximation Theory and Optimization: 270 Years, Since A. M. Legendre's Birth.* Springer, Berlin, Germany, 2023.

[47] M. A. Noor, K. I. Noor and T. M. Rassias. Some aspects of variational inequalities. *J. Comput. Appl. Math.*, 47, 1993, 285–312.

[48] M. A. Noor, K. I. Noor, A. Hamdi and E. H. El-Shemas. On difference of two monotone operators. *Optim. Lett.*, 3, 2009, 329–335.

[49] M. A. Noor, K. I. Noor, M. U. Awan and A. G. Khan. Quasi variational inclusions involving three operators. *Inf. Sci. Lett.*, 11(2), 2022, 589–399.

[50] M. A. Noor, S. Ullah, K. I. Noor and E. Al-Said. Iterative methods for solving extended general mixed variational inequalities. *Comput. Math. Appl.*, 62(2), 2011, 804–813.

[51] M. A. Noor, K. I. Noor, S. Treanta and K. Nonlaopon. On three-step iterative schemes associated with general quasi-variational inclusions. *Alex. Eng. J.*, 61(12), 2022, 12051–12059.

[52] M. Patriksson. *Nonlinear Programming and Variational Inequalities: A Unified Approach.* Kluwer Academic Publishers, Drodrecht, 1998.

[53] B. T. Polyak. Some methods of speeding up the convergence of iterative methods. *Zh. Pychisl. Mat. Mat. Fiz.*, 4, 1964, 791–803.

[54] Y. Shehu, A. Gibali and S. Sagratella. Inertial projection-type method for solving quasi variational inequalities in real Hilbert space. *J. Optim. Theory Appl.*, 184, 2020, 877–894.

[55] G. Stampacchia. Formes bilineaires coercivites sur les ensembles convexes. *C. R. Acad. Sci.* (Paris), 258, 1964, 4413–4416.

[56] P. Tseng. A modified forward-backward splitting method for maximal monotone mappings. *SIAM J. Control Optim.*, 38, 2000, 431–446.

[57] D. L. Zhu and P. Marcotte. Cocoercivity and its role in the convergence of iterative schemes for solving variational inequalities. *SIAM J. Optim.*, 6, 1996, 714–726.

https://doi.org/10.1142/9789819820498_0015

Chapter 15

Two Kinds of Hardy-Type Integral Inequalities and Their Equivalent Forms

Michael Th. Rassias[*,¶] **and Bicheng Yang**[§,‖]

[*]*Department of Mathematics and Engineering Sciences, Hellenic Military Academy, 16673 Vari Attikis, Greece*

[§]*School of Mathematics, Guangdong University of Education, Guangzhou, Guangdong 510303, P. R. China*

[¶]*mthrassias@yahoo.com*

[‖]*bcyang@gdei.edu.cn, bcyang818@163.com*

By using the way of real analysis, we obtain an example and two lemmas, and then deduce two kinds of Hardy-type integral inequalities as well as their equivalent forms with multi-parameters. We also consider their operator expressions.

Keywords: Hardy-type integral inequality, weight function, equivalent form, operator expression, norm.

1. Introduction

Assuming that $0 < \int_0^\infty f^2(x)dx < \infty$ and $0 < \int_0^\infty g^2(y)dy < \infty$, we have the following well-known Hilbert's integral inequality (cf. Ref. [1]):

$$\int_0^\infty \int_0^\infty \frac{f(x)g(y)}{x+y} dxdy < \pi \left(\int_0^\infty f^2(x)dx \int_0^\infty g^2(y)dy \right)^{\frac{1}{2}}, \quad (1.1)$$

where the constant factor π is the best possible. In 1925, by introducing one pair of conjugate exponents $(p,q)(\frac{1}{p}+\frac{1}{q}=1)$, Hardy [2] gave an extension of (1.1) as follows: For $p > 1, f(x), g(y) \geq 0$, such that $0 < \int_0^\infty f^p(x)dx < \infty$ and $0 < \int_0^\infty g^q(y)dy < \infty$, we have

$$\int_0^\infty \int_0^\infty \frac{f(x)g(y)}{x+y} dxdy < \frac{\pi}{\sin(\pi/p)} \left(\int_0^\infty f^p(x)dx \right)^{\frac{1}{p}} \left(\int_0^\infty g^q(y)dy \right)^{\frac{1}{q}}, \quad (1.2)$$

where, the constant factor $\frac{\pi}{\sin(\pi/p)}$ is the best possible. Inequalities (1.1) and (1.2) are important in analysis and its applications (cf. Refs. [3, 4]).

In 1934, Hardy *et al.* gave an extension of (1.2) as follows: If $k_1(x,y)$ is a non-negative homogeneous function of degree -1,

$$k_p = \int_0^\infty k_1(u,1)u^{\frac{-1}{p}} du \in \mathbf{R}_+ = (0,\infty),$$

then we have the following integral inequality:

$$\int_0^\infty \int_0^\infty k_1(x,y)f(x)g(y)dxdy < k_p \left(\int_0^\infty f^p(x)dx \right)^{\frac{1}{p}} \left(\int_0^\infty g^q(y)dy \right)^{\frac{1}{q}}, \quad (1.3)$$

where the constant factor k_p is the best possible (cf. Ref. [3], Theorem 319). Also, a Hilbert-type integral inequality with the non-homogeneous kernel is provided as follows: If $h(u) > 0$, $\phi(\sigma) =$

$\int_0^\infty h(u)u^{\sigma-1}du \in \mathbf{R}_+$, then

$$\int_0^\infty \int_0^\infty h(xy)f(x)g(y)dxdy < \phi\left(\frac{1}{p}\right)\left(\int_0^\infty x^{p-2}f^p(x)dx\right)^{\frac{1}{p}}\left(\int_0^\infty g^q(y)dy\right)^{\frac{1}{q}}, \tag{1.4}$$

where the constant factor $\phi(\frac{1}{p})$ is the best possible (cf. Ref. [3], Theorem 350).

In 1998, by introducing an independent parameter $\lambda > 0$, Yang gave an extension of (1.1) with the kernel $\frac{1}{(x+y)^\lambda}$ (cf. Refs. [5, 6]). In 2004, by introducing another pair of conjugate exponents (r, s), Yang [7] gave an extension of (1.2) as follows: If $\lambda > 0, r > 1, \frac{1}{r} + \frac{1}{s} = 1$, $f(x), g(y) \geq 0$, such that $0 < \int_0^\infty x^{p(1-\frac{\lambda}{r})-1}f^p(x)dx < \infty$ and $0 < \int_0^\infty y^{q(1-\frac{\lambda}{s})-1}g^q(y)dy < \infty$, then we have

$$\int_0^\infty \int_0^\infty \frac{f(x)g(y)}{x^\lambda + y^\lambda}dxdy < \frac{\pi}{\lambda\sin(\pi/r)}\left[\int_0^\infty x^{p(1-\frac{\lambda}{r})-1}f^p(x)dx\right]^{\frac{1}{p}} \times \left[\int_0^\infty y^{q(1-\frac{\lambda}{s})-1}g^q(y)dy\right]^{\frac{1}{q}}, \tag{1.5}$$

where the constant factor $\frac{\pi}{\lambda\sin(\pi/r)}$ is the best possible. For $\lambda = 1$, $r = q, s = p$, (1.5) reduces to (1.2). For $\lambda = 1, r = p, s = q$, (1.5) reduces to the dual form of (1.2) as follows:

$$\int_0^\infty \int_0^\infty \frac{f(x)g(y)}{x + y}dxdy < \frac{\pi}{\sin(\pi/p)}\left(\int_0^\infty x^{p-2}f^p(x)dx\right)^{\frac{1}{p}} \times \left(\int_0^\infty y^{q-2}g^q(y)dy\right)^{\frac{1}{q}}. \tag{1.6}$$

For $p = q = 2$, both (1.2) and (1.6) reduce to (1.1).

In 2005, Yang *et al.* [8] also gave an extension of (1.2) and (1.5) with the kernel $\frac{1}{(x+y)^\lambda}$. Krnić *et al.* [9–14] provided some extensions and particular cases of (1.2), (1.3) and (1.4) with parameters.

In 2009, Yang gave an extension of (1.3), (1.5) as follows (cf. Refs. [15, 17]): If $\lambda_1 + \lambda_2 = \lambda \in \mathbf{R} = (-\infty, \infty)$, $k_\lambda(x, y)$ is a non-negative homogeneous function of degree $-\lambda$ satisfying

$$k_\lambda(ux, uy) = u^{-\lambda} k_\lambda(x, y)(u, x, y > 0),$$

$k(\lambda_1) = \int_0^\infty k_\lambda(u, 1) u^{\lambda_1 - 1} du \in \mathbf{R}_+$, then we have

$$\int_0^\infty \int_0^\infty k_\lambda(x, y) f(x) g(y) dx dy < k(\lambda_1) \left(\int_0^\infty x^{p(1-\lambda_1)-1} f^p(x) dx \right)^{\frac{1}{p}} \times \left(\int_0^\infty y^{q(1-\lambda_2)-1} g^q(y) dy \right)^{\frac{1}{q}}, \tag{1.7}$$

where the constant factor $k(\lambda_1)$ is the best possible. For $\lambda = 1, \lambda_1 = \frac{1}{q}, \lambda_2 = \frac{1}{p}$, (1.7) reduces to (1.3). Also, an extension of (1.4) was given as follows:

$$\int_0^\infty \int_0^\infty h(xy) f(x) g(y) dx dy < \phi(\sigma) \left(\int_0^\infty x^{p(1-\sigma)-1} f^p(x) dx \right)^{\frac{1}{p}} \times \left(\int_0^\infty y^{q(1-\sigma)-1} g^q(y) dy \right)^{\frac{1}{q}}, \tag{1.8}$$

where the constant factor $\phi(\sigma)$ is the best possible (cf. Ref. [16]). For $\sigma = \frac{1}{p}$, (1.8) reduces to (1.4).

Some equivalent inequalities of (1.7) and (1.8) are considered by Ref. [17]. In 2013, Yang [16] also studied the equivalency between (1.7) and (1.8) by adding a condition. In 2017, Hong [18] studied a equivalent condition between (1.7) and a few parameters. The other further results were provided by Refs. [19–28].

Remark 1 (cf. Ref. [16]). If $h(xy) = 0$ $(xy > 1)$, then $\phi(\sigma) = \int_0^1 h(u) u^{\sigma - 1} du = \phi_1(\sigma) \in \mathbf{R}_+$, and then (1.8) reduces to the following first kind of Hardy-type integral inequality with the non-homogeneous kernel:

$$\int_0^\infty g(y)\left(\int_0^{\frac{1}{y}} h(xy)f(x)dx\right)dy < \phi_1(\sigma)\left(\int_0^\infty x^{p(1-\sigma)-1}f^p(x)dx\right)^{\frac{1}{p}} \times \left(\int_0^\infty y^{q(1-\sigma)-1}g^q(y)dy\right)^{\frac{1}{q}}, \tag{1.9}$$

where the constant factor $\phi_1(\sigma)$ is the best possible. If $h(xy) = 0$ ($xy < 1$), then $\phi(\sigma) = \int_1^\infty h(u)u^{\sigma-1}du = \phi_2(\sigma) \in \mathbf{R}_+$, and (1.8) reduces to the following second kind of Hardy-type integral inequality with the non-homogeneous kernel:

$$\int_0^\infty g(y)\left(\int_{\frac{1}{y}}^\infty h(xy)f(x)dx\right)dy < \phi_2(\sigma)\left(\int_0^\infty x^{p(1-\sigma)-1}f^p(x)dx\right)^{\frac{1}{p}} \times \left(\int_0^\infty y^{q(1-\sigma)-1}g^q(y)dy\right)^{\frac{1}{q}}, \tag{1.10}$$

where the constant factor $\phi_2(\sigma)$ is the best possible.

In this chapter, by using the way of real analysis, we obtain an example and two lemmas, and then deduce two kinds of Hardy-type integral inequalities with the non-homogeneous kernel as $\prod_{k=1}^s \frac{(\min\{xy,c_k\})^{\alpha/s}}{(\max\{xy,c_k\})^{(\lambda+\alpha)/s}}$ and the homogeneous kernel as $\prod_{k=1}^s \frac{(\min\{x,c_ky\})^{\alpha/s}}{(\max\{x,c_ky\})^{(\lambda+\alpha)/s}}$, as well as their equivalent forms. We also consider the operator expressions.

2. An Example

Example 1. For $s \in \mathbf{N} = \{1, 2, \ldots\}$, $0 = c_0 < c_1 \leq \cdots \leq c_s < c_{s+1} = \infty$, $\lambda_1, \lambda_2 \in \mathbf{R}$, $\lambda_1 + \lambda_2 = \lambda$, we set a function as follows:

$$h(u) := \prod_{k=1}^s \frac{(\min\{u, c_k\})^{\frac{\alpha}{s}}}{(\max\{u, c_k\})^{\frac{\lambda+\alpha}{s}}} (u \in \mathbf{R}_+). \tag{2.1}$$

(a) For $\lambda_1 > -\alpha$, we define

$$k_1(\lambda_1) := \int_0^1 h(u)u^{\lambda_1-1}du = \int_0^1 \left[\prod_{k=1}^{s} \frac{(\min\{u,c_k\})^{\frac{\alpha}{s}}}{(\max\{u,c_k\})^{\frac{\lambda+\alpha}{s}}}\right] u^{\lambda_1-1}du \tag{2.2}$$

and set $A := \{i \in \{1,\ldots,s\}; c_i < 1\}$.

(i) If $A = \Phi$, namely, $1 \leq c_1 \leq \cdots \leq c_s$, then by (2.2), we obtain

$$\begin{aligned} k_1(\lambda_1) &= \int_0^1 \left(\prod_{k=1}^{s} \frac{u^{\frac{\alpha}{s}}}{c_k^{\frac{\lambda+\alpha}{s}}}\right) u^{\lambda_1-1}du = \frac{1}{\prod_{k=1}^{s} c_k^{\frac{\lambda+\alpha}{s}}} \int_0^1 u^{\lambda_1+\alpha-1}du \\ &= \frac{1}{\lambda_1+\alpha}\frac{1}{\prod_{k=1}^{s} c_k^{\frac{\lambda+\alpha}{s}}} \in \mathbf{R}_+. \end{aligned} \tag{2.3}$$

(ii) If $A \neq \Phi$, $s_0 = \max A$, namely, $c_1 \leq \cdots c_{s_0} < 1$, and $1 \leq c_{s_0+1} \cdots \leq c_s$ (if there exists), then for $\lambda_1 + \alpha - \frac{\lambda+2\alpha}{s} i \neq 0$ $(i \in \{1,\ldots,s_0\})$, we obtain

$$\begin{aligned} k_1(\lambda_1) &= \int_0^{c_1} \left[\prod_{k=1}^{s} \frac{(\min\{u,c_k\})^{\frac{\alpha}{s}}}{(\max\{u,c_k\})^{\frac{\lambda+\alpha}{s}}}\right] u^{\lambda_1-1}du \\ &\quad + \int_{c_{s_0}}^1 \left[\prod_{k=1}^{s} \frac{(\min\{u,c_k\})^{\frac{\alpha}{s}}}{(\max\{u,c_k\})^{\frac{\lambda+\alpha}{s}}}\right] u^{\lambda_1-1}du \\ &\quad + \sum_{i=1}^{s_0-1} \int_{c_i}^{c_{i+1}} \left[\prod_{k=1}^{i} \frac{(\min\{u,c_k\})^{\frac{\alpha}{s}}}{(\max\{u,c_k\})^{\frac{\lambda+\alpha}{s}}}\right. \\ &\quad \left. \times \prod_{k=i+1}^{s} \frac{(\min\{u,c_k\})^{\frac{\alpha}{s}}}{(\max\{u,c_k\})^{\frac{\lambda+\alpha}{s}}}\right] u^{\lambda_1-1}du \\ &= \int_0^{c_1} \left(\prod_{k=1}^{s} \frac{u^{\frac{\alpha}{s}}}{c_k^{\frac{\lambda+\alpha}{s}}}\right) u^{\lambda_1-1}du \\ &\quad + \int_{c_{s_0}}^1 \left(\prod_{k=1}^{s_0} \frac{c_k^{\frac{\alpha}{s}}}{u^{\frac{\lambda+\alpha}{s}}}\right)\left(\prod_{k=s_0+1}^{s} \frac{u^{\frac{\alpha}{s}}}{c_k^{\frac{\lambda+\alpha}{s}}}\right) u^{\lambda_1-1}du \end{aligned}$$

$$+\sum_{i=1}^{s_0-1}\left(\frac{\prod_{k=1}^{i} c_k^{\frac{\alpha}{s}}}{\prod_{k=i+1}^{s} c_k^{\frac{\lambda+\alpha}{s}}}\right)\int_{c_i}^{c_{i+1}} u^{\lambda_1+\alpha-\frac{\lambda+2\alpha}{s}i-1}du$$

$$=\frac{c_1^{\lambda_1+\alpha}}{\lambda_1+\alpha}\frac{1}{\prod_{k=1}^{s} c_k^{\frac{\lambda+\alpha}{s}}}+\frac{1-c_{s_0}^{\lambda_1+\alpha-\frac{\lambda+2\alpha}{s}s_0}}{\lambda_1+\alpha-\frac{\lambda+2\alpha}{s}s_0}\frac{\prod_{k=1}^{s_0} c_k^{\frac{\alpha}{s}}}{\prod_{k=s_0+1}^{s} c_k^{\frac{\lambda+\alpha}{s}}}$$

$$+\sum_{i=1}^{s_0-1}\left[\frac{c_{i+1}^{\lambda_1+\alpha-\frac{\lambda+2\alpha}{s}i}-c_i^{\lambda_1+\alpha-\frac{\lambda+2\alpha}{s}i}}{\lambda_1+\alpha-\frac{\lambda+2\alpha}{s}i}\frac{\prod_{k=1}^{i} c_k^{\frac{\alpha}{s}}}{\prod_{k=i+1}^{s} c_k^{\frac{\lambda+\alpha}{s}}}\right]\in \mathbf{R}_+. \tag{2.4}$$

Note. If there exists a $i_0 \in \{1,\ldots,s_0\}$, such that $\lambda_1+\alpha-\frac{\lambda+2\alpha}{s}i_0=0$, and

$$\ln\left(\frac{c_{i_0+1}}{c_{i_0}}\right)=\int_{c_{i_0}}^{c_{i_0+1}} t^{\lambda_1+\alpha-\frac{\lambda+2\alpha}{s}i_0-1}dt=\lim_{i\to i_0}\int_{c_i}^{c_{i+1}} t^{\lambda_1+\alpha-\frac{\lambda+2\alpha}{s}i-1}dt$$

$$=\lim_{i\to i_0}\frac{c_{i+1}^{\lambda_1+\alpha-\frac{\lambda+2\alpha}{s}i}-c_i^{\lambda_1+\alpha-\frac{\lambda+2\alpha}{s}i}}{\lambda_1+\alpha-\frac{\lambda+2\alpha}{s}i}$$

(we assume $c_{s_0+1}=1$ in the above), then we indicate $\ln(\frac{c_{i_0+1}}{c_{i_0}})$ by the following formal expression:

$$\frac{c_{i_0+1}^{\lambda_1+\alpha-\frac{\lambda+2\alpha}{s}i_0}-c_{i_0}^{\lambda_1+\alpha-\frac{\lambda+2\alpha}{s}i_0}}{\lambda_1+\alpha-\frac{\lambda+2\alpha}{s}i_0}.$$

In this case, we still have (2.4). (In (b), we still use this viewpoint.)

We indicate $\prod_{k=1}^{0} c_k^{\frac{\alpha}{s}}=1$. In view of (2.3) and (2.4) with Note, we have

$$k_1(\lambda_1)=\begin{cases}\frac{1}{(\lambda_1+\alpha)\prod_{k=1}^{s} c_k^{\frac{\lambda+\alpha}{s}}}, & A=\Phi\\ \frac{(1-c_{s_0}^{\lambda_1+\alpha-\frac{\lambda+2\alpha}{s}s_0})\prod_{k=1}^{s_0} c_k^{\frac{\alpha}{s}}}{(1+\alpha-\frac{\lambda+2\alpha}{s}s_0)\prod_{k=s_0+1}^{s} c_k^{\frac{\lambda+\alpha}{s}}} & \\ \quad+\sum_{i=0}^{s_0-1}\frac{(c_{i+1}^{\lambda_1+\alpha-\frac{\lambda+2\alpha}{s}i}-c_i^{\lambda_1+\alpha-\frac{\lambda+2\alpha}{s}i})\prod_{k=1}^{i} c_k^{\frac{\alpha}{s}}}{(\lambda_1+\alpha-\frac{\lambda+2\alpha}{s}i)\prod_{k=i+1}^{s} c_k^{\frac{\lambda+\alpha}{s}}}, & A\neq\Phi\ (s_0=\max A)\end{cases}. \tag{2.5}$$

(b) For $\lambda_2 > -\alpha$, we define

$$k_2(\lambda_1) := \int_1^\infty h(u)u^{\lambda_1-1}du = \int_1^\infty \left[\prod_{k=1}^{s} \frac{(\min\{u,c_k\})^{\frac{\alpha}{s}}}{(\max\{u,c_k\})^{\frac{\lambda+\alpha}{s}}}\right] u^{\lambda_1-1}du \tag{2.6}$$

and then set $B := \{i \in \{1,\ldots,s\}; c_i > 1\}$.

(i) If $B = \Phi$, namely, $c_1 \le \cdots \le c_s \le 1$, then we obtain

$$\begin{aligned} k_2(\lambda_1) &= \int_1^\infty \left(\prod_{k=1}^{s} \frac{c_k^{\frac{\alpha}{s}}}{u^{\frac{\lambda+\alpha}{s}}}\right) u^{\lambda_1-1}du = \left(\prod_{k=1}^{s} c_k^{\frac{\alpha}{s}}\right)\int_1^\infty u^{-\lambda_2-\alpha-1}du \\ &= \frac{1}{\lambda_2+\alpha}\prod_{k=1}^{s} c_k^{\frac{\alpha}{s}} \in \mathbf{R}_+. \end{aligned} \tag{2.7}$$

(ii) If $B \ne \Phi$, $s_1 = \min B$, namely, $1 < c_{s_1} \le \cdots \le c_s$ and $c_1 \le \cdots \le c_{s_1-1} \le 1$ (if there exists), then we obtain

$$\begin{aligned} k_2(\lambda_1) &= \int_1^{c_{s_1}} \left[\prod_{k=1}^{s} \frac{(\min\{u,c_k\})^{\frac{\alpha}{s}}}{(\max\{u,c_k\})^{\frac{\lambda+\alpha}{s}}}\right] u^{\lambda_1-1}du \\ &\quad + \int_{c_s}^\infty \left[\prod_{k=1}^{s} \frac{(\min\{u,c_k\})^{\frac{\alpha}{s}}}{(\max\{u,c_k\})^{\frac{\lambda+\alpha}{s}}}\right] u^{\lambda_1-1}du \\ &\quad + \sum_{i=s_1}^{s-1}\int_{c_i}^{c_{i+1}} \left[\prod_{k=1}^{i} \frac{(\min\{u,c_k\})^{\frac{\alpha}{s}}}{(\max\{u,c_k\})^{\frac{\lambda+\alpha}{s}}}\right. \\ &\quad \left.\times \prod_{k=i+1}^{s} \frac{(\min\{u,c_k\})^{\frac{\alpha}{s}}}{(\max\{u,c_k\})^{\frac{\lambda+\alpha}{s}}}\right] u^{\lambda_1-1}du \\ &= \int_1^{c_{s_1}} \left(\prod_{k=1}^{s_1-1} \frac{c_k^{\frac{\alpha}{s}}}{u^{\frac{\lambda+\alpha}{s}}} \prod_{k=s_1}^{s} \frac{u^{\frac{\alpha}{s}}}{c_k^{\frac{\lambda+\alpha}{s}}}\right) u^{\lambda_1-1}du \\ &\quad + \int_{c_s}^\infty \left(\prod_{k=1}^{s} \frac{c_k^{\frac{\alpha}{s}}}{u^{\frac{\lambda+\alpha}{s}}}\right) u^{\lambda_1-1}du \end{aligned}$$

$$
\begin{aligned}
&+\sum_{i=s_1}^{s-1}\left(\frac{\prod_{k=1}^{i} c_k^{\frac{\alpha}{s}}}{\prod_{k=i+1}^{s} c_k^{\frac{\lambda+\alpha}{s}}}\right)\int_{c_i}^{c_{i+1}} u^{\lambda_1+\alpha-\frac{\lambda+2\alpha}{s}i-1}du\\
&=\frac{c_{s_1}^{\lambda_1+\alpha-\frac{\lambda+2\alpha}{s}(s_1-1)}-1}{\lambda_1+\alpha-\frac{\lambda+2\alpha}{s}(s_1-1)}\frac{\prod_{k=1}^{s_1-1} c_k^{\frac{\alpha}{s}}}{\prod_{k=s_1}^{s} c_k^{\frac{\lambda+\alpha}{s}}}+\frac{c_s^{-\lambda_2-\alpha}}{\lambda_2+\alpha}\prod_{k=1}^{s} c_k^{\frac{\alpha}{s}}\\
&+\sum_{i=s_1}^{s-1}\frac{c_{i+1}^{\lambda_1+\alpha-\frac{\lambda+2\alpha}{s}i}-c_i^{\lambda_1+\alpha-\frac{\lambda+2\alpha}{s}i}}{\lambda_1+\alpha-\frac{\lambda+2\alpha}{s}i}\frac{\prod_{k=1}^{i} c_k^{\frac{\alpha}{s}}}{\prod_{k=i+1}^{s} c_k^{\frac{\lambda+\alpha}{s}}}\in\mathbf{R}_+.
\end{aligned}
\tag{2.8}
$$

We indicate $\prod_{k=s+1}^{s} c_k^{\frac{\lambda+\alpha}{s}}=1$. In view of (2.7) and (2.8), it follows that

$$
k_2(\lambda_1)=\begin{cases}\frac{1}{\lambda_2+\alpha}\prod_{k=1}^{s} c_k^{\frac{\alpha}{s}}, & B=\Phi\\[2mm] \frac{[c_{s_1}^{\lambda_1+\alpha-\frac{\lambda+2\alpha}{s}(s_1-1)}-1]\prod_{k=1}^{s_1-1} c_k^{\frac{\alpha}{s}}}{[\lambda_1+\alpha-\frac{\lambda+2\alpha}{s}(s_1-1)]\prod_{k=s_1}^{s} c_k^{\frac{\lambda+\alpha}{s}}} & \\ \quad+\sum_{i=s_1}^{s}\frac{(c_{i+1}^{\lambda_1+\alpha-\frac{\lambda+2\alpha}{s}i}-c_i^{\lambda_1+\alpha-\frac{\lambda+2\alpha}{s}i})\prod_{k=1}^{i} c_k^{\frac{\alpha}{s}}}{(\lambda_1+\alpha-\frac{\lambda+2\alpha}{s}i)\prod_{k=i+1}^{s} c_k^{\frac{\lambda+\alpha}{s}}}, & \begin{array}{c}B\neq\Phi\\(s_1=\min B).\end{array}\end{cases}
\tag{2.9}
$$

In particular, (1) if $s=1$ (or $c_s=\cdots=c_1$), $h(u)=\frac{(\min\{u,c_1\})^{\alpha}}{(\max\{u,c_1\})^{\lambda+\alpha}}$, then we find for $\lambda_1>-\alpha$,

$$
\begin{aligned}
k_1(\lambda_1)&=\int_0^1\frac{(\min\{u,c_1\})^{\alpha}}{(\max\{u,c_1\})^{\lambda+\alpha}}u^{\lambda_1-1}du\\
&=\begin{cases}\frac{1}{(\lambda_1+\alpha)c_1^{\lambda+\alpha}}, c_1\geq 1\\ \left(\frac{1}{\lambda_1+\alpha}+\frac{1-c_1^{\lambda_2+\alpha}}{\lambda_2+\alpha}\right)\frac{1}{c_1^{\lambda_2}}, 0<c_1<1\end{cases};
\end{aligned}
$$

for $\lambda_2>-\alpha$,

$$
\begin{aligned}
k_2(\lambda_1)&=\int_1^{\infty}\frac{(\min\{u,c_1\})^{\alpha}}{(\max\{u,c_1\})^{\lambda+\alpha}}u^{\lambda_1-1}du\\
&=\begin{cases}\frac{c_1{}^{\alpha}}{\lambda_2+\alpha}, 0<c_1\leq 1\\ \left(\frac{1}{\lambda_2+\alpha}+\frac{1-c_1^{-\lambda_1-\alpha}}{\lambda_1+\alpha}\right)\frac{1}{c_1{}^{\lambda_2}}, c_1>1\end{cases};
\end{aligned}
$$

(2) if $\alpha = 0, h(u) = \frac{1}{\prod_{k=1}^{s}(\max\{u,c_k\})^{\frac{\lambda}{s}}}$, then we find for $\lambda_1 > 0$,

$$k_1(\lambda_1) = \begin{cases} \frac{1}{\lambda_1 \prod_{k=1}^{s} c_k^{\frac{\lambda}{s}}}, & A = \Phi \\ \frac{1-c_{s_0}^{\lambda_1 - \frac{s_0\lambda}{s}}}{(\lambda_1 - \frac{s_0\lambda}{s})\prod_{k=s_0+1}^{s} c_k^{\frac{\lambda}{s}}} + \sum_{i=0}^{s_0-1} \frac{c_{i+1}^{\lambda_1 - \frac{i\lambda}{s}} - c_i^{\lambda_1 - \frac{i\lambda}{s}}}{(\lambda_1 - \frac{i\lambda}{s})\prod_{k=i+1}^{s} c_k^{\frac{\lambda}{s}}}, & ; \\ A \neq \Phi\,(s_0 = \max A) \end{cases}$$

for $\lambda_2 > 0$,

$$k_2(\lambda_1) = \begin{cases} \frac{1}{\lambda_2}, & B = \Phi \\ \frac{c_{s_1}^{\lambda_1 - \frac{\lambda}{s}(s_1-1)} - 1}{[\lambda_1 - \frac{\lambda}{s}(s_1-1)]\prod_{k=s_1}^{s} c_k^{\frac{\lambda}{s}}} + \sum_{i=s_1}^{s} \frac{c_{i+1}^{\lambda_1 - \frac{i\lambda}{s}} - c_i^{\lambda_1 - \frac{i\lambda}{s}}}{(\lambda_1 - \frac{i\lambda}{s})\prod_{k=i+1}^{s} c_k^{\frac{\lambda}{s}}}, & ; \\ B \neq \Phi\,(s_1 = \min B) \end{cases}$$

(3) if $\lambda = 0, h(u) = \prod_{k=1}^{s}\left(\frac{\min\{u,c_k\}}{\max\{u,c_k\}}\right)^{\frac{\alpha}{s}}$, then we find for $\lambda_1 > -\alpha$,

$$k_1(\lambda_1) = \begin{cases} \frac{1}{(\lambda_1+\alpha)\prod_{k=1}^{s} c_k^{\frac{\alpha}{s}}}, & A = \Phi \\ \frac{[1-c_{s_0}^{\lambda_1+\alpha(1-\frac{2s_0}{s})}]\prod_{k=1}^{s_0} c_k^{\frac{\alpha}{s}}}{[\lambda_1+\alpha(1-\frac{2s_0}{s})]\prod_{k=s_0+1}^{s} c_k^{\frac{\alpha}{s}}} & ; \\ + \sum_{i=0}^{s_0-1} \frac{[c_{i+1}^{\lambda_1+\alpha(1-\frac{2i}{s})} - c_i^{\lambda_1+\alpha(1-\frac{2i}{s})}]\prod_{k=1}^{i} c_k^{\frac{\alpha}{s}}}{[\lambda_1+\alpha(1-\frac{2i}{s})]\prod_{k=i+1}^{s} c_k^{\frac{\alpha}{s}}}, \\ A \neq \Phi\,(s_0 = \max A) \end{cases}$$

for $\lambda_2 > -\alpha$,

$$k_2(\lambda_1) = \begin{cases} \frac{1}{\lambda_2+\alpha}\prod_{k=1}^{s} c_k^{\frac{\alpha}{s}}, & B = \Phi \\ \frac{[c_{s_1}^{\lambda_1+\alpha-\frac{2\alpha}{s}(s_1-1)} - 1]\prod_{k=1}^{s_1-1} c_k^{\frac{\alpha}{s}}}{[\lambda_1+\alpha-\frac{2\alpha}{s}(s_1-1)]\prod_{k=s_1}^{s} c_k^{\frac{\alpha}{s}}} & ; \\ + \sum_{i=s_1}^{s} \frac{[c_{i+1}^{\lambda_1+\alpha(1-\frac{2i}{s})} - c_i^{\lambda_1+\alpha(1-\frac{2i}{s})}]\prod_{k=1}^{i} c_k^{\frac{\alpha}{s}}}{[\lambda_1+\alpha(1-\frac{2i}{s}]\prod_{k=i+1}^{s} c_k^{\frac{\alpha}{s}}}, \\ B \neq \Phi\,(s_1 = \min B) \end{cases}$$

(4) if $\lambda = -\alpha, h(u) = \prod_{k=1}^{s}(\min\{u, c_k\})^{\frac{\alpha}{s}}$, then we find for $\lambda_1 > -\alpha$,

$$k_1(\lambda_1) = \begin{cases} \frac{1}{\lambda_1+\alpha}, \qquad A = \Phi \\ \frac{1-c_{s_0}^{\lambda_1+\alpha(1-\frac{s_0}{s})}}{\lambda_1+\alpha(1-\frac{s_0}{s})} \prod_{k=1}^{s_0} c_k^{\frac{\alpha}{s}} \\ +\sum_{i=0}^{s_0-1} \frac{c_{i+1}^{\lambda_1+\alpha(1-\frac{i}{s})}-c_i^{\lambda_1+\alpha(1-\frac{i}{s})}}{\lambda_1+\alpha(1-\frac{i}{s})} \prod_{k=1}^{i} c_k^{\frac{\alpha}{s}}, \\ A \neq \Phi\,(s_0 = \max A) \end{cases};$$

for $\lambda_2 > -\alpha$,

$$k_2(\lambda_1) = \begin{cases} \frac{1}{\lambda_2+\alpha} \prod_{k=1}^{s} c_k^{\frac{\alpha}{s}}, \qquad B = \Phi \\ \frac{c_{s_1}^{\lambda_1+\alpha(1-\frac{s_1-1}{s})}-1}{\lambda_1+\alpha(1-\frac{s_1-1}{s})} \prod_{k=1}^{s_1-1} c_k^{\frac{\alpha}{s}} \\ +\sum_{i=s_1}^{s} \frac{c_{i+1}^{\lambda_1+\alpha(1-\frac{i}{s})}-c_i^{\lambda_1+\alpha(1-\frac{i}{s})}}{\lambda_1+\alpha(1-\frac{i}{s})} \prod_{k=1}^{i} c_k^{\frac{\alpha}{s}}, \\ B \neq \Phi\,(s_1 = \min B) \end{cases}.$$

3. Two Lemmas

In the following, we make appointment that $p > 1, \frac{1}{p} + \frac{1}{q} = 1, s \in \mathbf{N}$, $0 = c_0 < c_1 \leq \cdots \leq c_s < c_{s+1} = \infty$, $\lambda_1, \lambda_2 \in \mathbf{R}, \lambda_1 + \lambda_2 = \lambda, \sigma_1$, $\sigma_2 \in \mathbf{R}$.

Lemma 1. *If $\lambda_1 > -\alpha$, there exists a constant M_1, such that for any non-negative measurable functions $f(x)$ and $g(y)$ in $(0, \infty)$, the following inequality*

$$\int_0^\infty g(y)\left[\int_0^{\frac{1}{y}} \prod_{k=1}^{s} \frac{(\min\{xy, c_k\})^{\frac{\alpha}{s}}}{(\max\{xy, c_k\})^{\frac{\lambda+\alpha}{s}}} f(x)dx\right] dy$$

$$\leq M_1 \left[\int_0^\infty x^{p(1-\lambda_1)-1} f^p(x)dx\right]^{\frac{1}{p}} \left[\int_0^\infty y^{q(1-\sigma_1)-1} g^q(y)dy\right]^{\frac{1}{q}} \tag{3.1}$$

holds, then we have $\sigma_1 = \lambda_1$. In this case, we deduce $M_1 \geq k_1(\lambda_1)$ ($k_1(\lambda_1)$ is indicated by (2.5)).

Proof. If $\sigma_1 > \lambda_1$, then for $n \geq \frac{1}{\sigma_1 - \lambda_1}$ ($n \in \mathbf{N}$), we set the following two functions:

$$f_n(x) := \begin{cases} x^{\lambda_1 + \frac{1}{pn} - 1}, 0 < x \leq 1 \\ 0, x > 1 \end{cases}, \quad g_n(y) := \begin{cases} 0, 0 < y < 1 \\ y^{\sigma_1 - \frac{1}{qn} - 1}, y \geq 1 \end{cases}.$$

We find

$$J_1 := \left[\int_0^\infty x^{p(1-\lambda_1)-1} f_n^p(x) dx\right]^{\frac{1}{p}} \left[\int_0^\infty y^{q(1-\sigma_1)-1} g_n^q(y) dy\right]^{\frac{1}{q}}$$
$$= \left(\int_0^1 x^{\frac{1}{n}-1} dx\right)^{\frac{1}{p}} \left(\int_1^\infty y^{-\frac{1}{n}-1} dy\right)^{\frac{1}{q}} = n.$$

Setting $u = xy$, we obtain

$$I_1 := \int_0^\infty g_n(y) \left(\int_0^{\frac{1}{y}} \prod_{k=1}^s \frac{(\min\{xy, c_k\})^{\frac{\alpha}{s}}}{(\max\{xy, c_k\})^{\frac{\lambda+\alpha}{s}}} f_n(x) dx\right) dy$$
$$= \int_1^\infty \left(\int_0^{\frac{1}{y}} \prod_{k=1}^s \frac{(\min\{xy, c_k\})^{\frac{\alpha}{s}}}{(\max\{xy, c_k\})^{\frac{\lambda+\alpha}{s}}} x^{\lambda_1 + \frac{1}{pn} - 1} dx\right) y^{\sigma_1 - \frac{1}{qn} - 1} dy$$
$$= \int_1^\infty y^{(\sigma_1 - \lambda_1) - \frac{1}{n} - 1} dy \int_0^1 \prod_{k=1}^s \frac{(\min\{u, c_k\})^{\frac{\alpha}{s}}}{(\max\{u, c_k\})^{\frac{\lambda+\alpha}{s}}} u^{\lambda_1 + \frac{1}{pn} - 1} du,$$

and then by (3.1), we have

$$\int_1^\infty y^{(\sigma_1 - \lambda_1) - \frac{1}{n} - 1} dy \int_0^1 \prod_{k=1}^s \frac{(\min\{u, c_k\})^{\frac{\alpha}{s}}}{(\max\{u, c_k\})^{\frac{\lambda+\alpha}{s}}} u^{\lambda_1 + \frac{1}{pn} - 1} du$$
$$= I_1 \leq M_1 J_1 = M_1 n. \tag{3.2}$$

Since $(\sigma_1 - \lambda_1) - \frac{1}{n} \geq 0$, it follows that $\int_1^\infty y^{(\sigma_1 - \lambda_1) - \frac{1}{n} - 1} dy = \infty$. By (3.2), in view of

$$\int_0^1 \prod_{k=1}^s \frac{(\min\{u, c_k\})^{\frac{\alpha}{s}}}{(\max\{u, c_k\})^{\frac{\lambda+\alpha}{s}}} u^{\lambda_1 + \frac{1}{pn} - 1} du > 0,$$

we find that $\infty \leq M_1 n < \infty$, which is a contradiction.

If $\sigma_1 < \lambda_1$, then for $n \geq \frac{1}{\lambda_1 - \sigma_1}$ ($n \in \mathbf{N}$), we set the following two functions:

$$\widetilde{f}_n(x) := \begin{cases} 0, 0 < x < 1 \\ x^{\lambda_1 - \frac{1}{pn} - 1}, \ x \geq 1 \end{cases}, \quad \widetilde{g}_n(y) := \begin{cases} y^{\sigma_1 + \frac{1}{qn} - 1}, \ 0 < y \leq 1 \\ 0, y > 1 \end{cases}$$

and find

$$\begin{aligned} \widetilde{J}_1 &:= \left[\int_0^\infty x^{p(1-\lambda_1)-1} \widetilde{f}_n^p(x) dx\right]^{\frac{1}{p}} \left[\int_0^\infty y^{q(1-\sigma_1)-1} \widetilde{g}_n^q(y) dy\right]^{\frac{1}{q}} \\ &= \left(\int_1^\infty x^{-\frac{1}{n}-1} dx\right)^{\frac{1}{p}} \left(\int_0^1 y^{\frac{1}{n}-1} dy\right)^{\frac{1}{q}} = n. \end{aligned}$$

Setting $u = xy$, we obtain

$$\begin{aligned} \widetilde{I}_1 &:= \int_0^\infty \widetilde{f}_n(x) \left[\int_0^{\frac{1}{x}} \prod_{k=1}^s \frac{(\min\{xy, c_k\})^{\frac{\alpha}{s}}}{(\max\{xy, c_k\})^{\frac{\lambda+\alpha}{s}}} \widetilde{g}_n(y) dy\right] dx \\ &= \int_1^\infty \left[\int_0^{\frac{1}{x}} \prod_{k=1}^s \frac{(\min\{xy, c_k\})^{\frac{\alpha}{s}}}{(\max\{xy, c_k\})^{\frac{\lambda+\alpha}{s}}} y^{\sigma_1 + \frac{1}{qn} - 1} dy\right] x^{\lambda_1 - \frac{1}{pn} - 1} dx \\ &= \int_1^\infty x^{(\lambda_1 - \sigma_1) - \frac{1}{n} - 1} dx \int_0^1 \prod_{k=1}^s \frac{(\min\{u, c_k\})^{\frac{\alpha}{s}}}{(\max\{u, c_k\})^{\frac{\lambda+\alpha}{s}}} u^{\sigma_1 + \frac{1}{qn} - 1} du, \end{aligned}$$

and then by Fubini theorem (cf. Ref. [29]) and (3.1), we have

$$\begin{aligned} &\int_1^\infty x^{(\lambda_1 - \sigma_1) - \frac{1}{n} - 1} dx \int_0^1 \prod_{k=1}^s \frac{(\min\{u, c_k\})^{\frac{\alpha}{s}}}{(\max\{u, c_k\})^{\frac{\lambda+\alpha}{s}}} u^{\sigma_1 + \frac{1}{qn} - 1} du \\ &\quad = \widetilde{I}_1 = \int_0^\infty \widetilde{g}_n(y) \left[\int_0^{\frac{1}{y}} \prod_{k=1}^s \frac{(\min\{xy, c_k\})^{\frac{\alpha}{s}}}{(\max\{xy, c_k\})^{\frac{\lambda+\alpha}{s}}} \widetilde{f}_n(x) dx\right] dy \\ &\quad \leq M_1 \widetilde{J}_1 = M_1 n. \end{aligned} \tag{3.3}$$

Since $(\lambda_1 - \sigma_1) - \frac{1}{n} \geq 0$, it follows that $\int_1^\infty x^{(\lambda_1 - \sigma_1) - \frac{1}{n} - 1} dx = \infty$. By (3.3), in view of

$$\int_0^1 \prod_{k=1}^s \frac{(\min\{u, c_k\})^{\frac{\alpha}{s}}}{(\max\{u, c_k\})^{\frac{\lambda+\alpha}{s}}} u^{\sigma_1 + \frac{1}{qn} - 1} du > 0,$$

we find that $\infty \leq M_1 n < \infty$, which is a contradiction.

Hence, we conclude that $\sigma_1 = \lambda_1$.
For $\sigma_1 = \lambda_1$, we reduce (3.3) as follows:

$$M_1 \geq \int_0^1 \prod_{k=1}^{s} \frac{(\min\{u, c_k\})^{\frac{\alpha}{s}}}{(\max\{u, c_k\})^{\frac{\lambda+\alpha}{s}}} u^{\lambda_1 + \frac{1}{qn} - 1} du. \tag{3.4}$$

Since $\{\prod_{k=1}^{s} \frac{(\min\{u,c_k\})^{\alpha/s}}{(\max\{u,c_k\})^{(\lambda+\alpha)/s}} u^{\lambda_1+\frac{1}{qn}-1}\}_{n=1}^{\infty}$ is non-negative and increasing in $(0, 1]$, by Levi theorem (cf. Ref. [29]), we find

$$\begin{aligned} M_1 &\geq \lim_{n\to\infty} \int_0^1 \prod_{k=1}^{s} \frac{(\min\{u, c_k\})^{\frac{\alpha}{s}}}{(\max\{u, c_k\})^{\frac{\lambda+\alpha}{s}}} u^{\lambda_1 + \frac{1}{qn} - 1} du \\ &= \int_0^1 \lim_{n\to\infty} \prod_{k=1}^{s} \frac{(\min\{u, c_k\})^{\frac{\alpha}{s}}}{(\max\{u, c_k\})^{\frac{\lambda+\alpha}{s}}} u^{\lambda_1 + \frac{1}{qn} - 1} du = k_1(\lambda_1). \end{aligned}$$

The lemma is proved. □

Lemma 2. *If $\lambda_2 > -\alpha$, there exists a constant M_2, such that for any non-negative measurable functions $f(x)$ and $g(y)$ in $(0, \infty)$, the following inequality*

$$\begin{aligned} &\int_0^{\infty} g(y) \left[\int_{\frac{1}{y}}^{\infty} \prod_{k=1}^{s} \frac{(\min\{xy, c_k\})^{\frac{\alpha}{s}}}{(\max\{xy, c_k\})^{\frac{\lambda+\alpha}{s}}} f(x) dx \right] dy \\ &\quad \leq M_2 \left[\int_0^{\infty} x^{p(1-\lambda_1)-1} f^p(x) dx \right]^{\frac{1}{p}} \left[\int_0^{\infty} y^{q(1-\sigma_1)-1} g^q(y) dy \right]^{\frac{1}{q}} \end{aligned} \tag{3.5}$$

holds, then we have $\sigma_1 = \lambda_1$. In this case, we deduce $M_2 \geq k_2(\lambda_1)$ ($k_2(\lambda_1)$ is indicated by (2.9)).

Proof. If $\sigma_1 < \lambda_1$, then for $n \geq \frac{1}{\lambda_1 - \sigma_1}$ ($n \in \mathbf{N}$), we set two functions $\widetilde{f}_n(x)$ and $\widetilde{g}_n(y)$ as in Lemma 1 and find

$$\widetilde{J}_1 = \left[\int_0^{\infty} x^{p(1-\lambda_1)-1} \widetilde{f}_n^p(x) dx \right]^{\frac{1}{p}} \left[\int_0^{\infty} y^{q(1-\sigma_1)-1} \widetilde{g}_n^q(y) dy \right]^{\frac{1}{q}} = n.$$

Setting $u = xy$, we obtain

$$\widetilde{I}_2 := \int_0^\infty \widetilde{g}_n(y)\left[\int_{\frac{1}{y}}^\infty \prod_{k=1}^s \frac{(\min\{xy,c_k\})^{\frac{\alpha}{s}}}{(\max\{xy,c_k\})^{\frac{\lambda+\alpha}{s}}}\widetilde{f}_n(x)dx\right]dy$$

$$= \int_0^1 \left[\int_{\frac{1}{y}}^\infty \prod_{k=1}^s \frac{(\min\{xy,c_k\})^{\frac{\alpha}{s}}}{(\max\{xy,c_k\})^{\frac{\lambda+\alpha}{s}}}x^{\lambda_1-\frac{1}{pn}-1}dx\right]y^{\sigma_1+\frac{1}{qn}-1}dy$$

$$= \int_0^1 y^{(\sigma_1-\lambda_1)+\frac{1}{n}-1}dy\int_1^\infty \prod_{k=1}^s \frac{(\min\{u,c_k\})^{\frac{\alpha}{s}}}{(\max\{u,c_k\})^{\frac{\lambda+\alpha}{s}}}u^{\lambda_1-\frac{1}{pn}-1}du,$$

and then by (3.5), we obtain

$$\int_0^1 y^{(\sigma_1-\lambda_1)+\frac{1}{n}-1}dy\int_1^\infty \prod_{k=1}^s \frac{(\min\{u,c_k\})^{\frac{\alpha}{s}}}{(\max\{u,c_k\})^{\frac{\lambda+\alpha}{s}}}u^{\lambda_1-\frac{1}{pn}-1}du$$

$$= \widetilde{I}_2 \le M_2\widetilde{J}_1 = M_2 n. \tag{3.6}$$

Since $(\sigma_1-\lambda_1)+\frac{1}{n}\le 0$, it follows that $\int_0^1 y^{(\sigma_1-\lambda_1)+\frac{1}{n}-1}dy = \infty$. By (3.6), in view of

$$\int_1^\infty \prod_{k=1}^s \frac{(\min\{u,c_k\})^{\frac{\alpha}{s}}}{(\max\{u,c_k\})^{\frac{\lambda+\alpha}{s}}}u^{\lambda_1-\frac{1}{pn}-1}du > 0,$$

we have $\infty \le M_2 n < \infty$, which is a contradiction.

If $\sigma_1 > \lambda_1$, then for $n \ge \frac{1}{\sigma_1-\lambda_1}$ $(n \in \mathbf{N})$, we set two functions $f_n(x)$ and $g_n(y)$ as in Lemma 1 and find

$$J_1 = \left[\int_0^\infty x^{p(1-\lambda_1)-1}f_n^p(x)dx\right]^{\frac{1}{p}}\left[\int_0^\infty y^{q(1-\sigma_1)-1}g_n^q(y)dy\right]^{\frac{1}{q}} = n.$$

Setting $u = xy$, we obtain

$$I_2 := \int_0^\infty f_n(x)\left[\int_{\frac{1}{x}}^\infty \prod_{k=1}^s \frac{(\min\{xy,c_k\})^{\frac{\alpha}{s}}}{(\max\{xy,c_k\})^{\frac{\lambda+\alpha}{s}}}g_n(y)dy\right]dx$$

$$= \int_0^1 \left[\int_{\frac{1}{x}}^\infty \prod_{k=1}^s \frac{(\min\{xy,c_k\})^{\frac{\alpha}{s}}}{(\max\{xy,c_k\})^{\frac{\lambda+\alpha}{s}}}y^{\sigma_1-\frac{1}{qn}-1}dy\right]x^{\lambda_1+\frac{1}{pn}-1}dx$$

$$= \int_0^1 x^{(\lambda_1-\sigma_1)+\frac{1}{n}-1}dx\int_1^\infty \prod_{k=1}^s \frac{(\min\{u,c_k\})^{\frac{\alpha}{s}}}{(\max\{u,c_k\})^{\frac{\lambda+\alpha}{s}}}u^{\sigma_1-\frac{1}{qn}-1}du,$$

and then by Fubini theorem (cf. Ref. [29]) and (3.5), we have

$$\int_0^1 x^{(\lambda_1-\sigma_1)+\frac{1}{n}-1}dx \int_1^\infty \prod_{k=1}^s \frac{(\min\{u,c_k\})^{\frac{\alpha}{s}}}{(\max\{u,c_k\})^{\frac{\lambda+\alpha}{s}}} u^{\sigma_1-\frac{1}{qn}-1}du$$

$$= I_2 = \int_0^\infty g_n(y)\left[\int_{\frac{1}{y}}^\infty \prod_{k=1}^s \frac{(\min\{xy,c_k\})^{\frac{\alpha}{s}}}{(\max\{xy,c_k\})^{\frac{\lambda+\alpha}{s}}} f_n(x)dx\right]dy$$

$$\leq M_2 J_1 = M_2 n. \tag{3.7}$$

Since $(\lambda_1-\sigma_1)+\frac{1}{n}\leq 0$, it follows that $\int_0^1 x^{(\lambda_1-\sigma_1)+\frac{1}{n}-1}dx = \infty$. By (3.7), in view of

$$\int_1^\infty \prod_{k=1}^s \frac{(\min\{u,c_k\})^{\frac{\alpha}{s}}}{(\max\{u,c_k\})^{\frac{\lambda+\alpha}{s}}} u^{\sigma_1-\frac{1}{qn}-1}du > 0,$$

we have $\infty \leq M_2 n < \infty$, which is a contradiction.

Hence, we conclude the fact that $\sigma_1 = \lambda_1$.

For $\sigma_1=\lambda_1$, we reduce (3.7) as follows:

$$M_2 \geq \int_1^\infty \prod_{k=1}^s \frac{(\min\{u,c_k\})^{\frac{\alpha}{s}}}{(\max\{u,c_k\})^{\frac{\lambda+\alpha}{s}}} u^{\lambda_1-\frac{1}{qn}-1}du. \tag{3.8}$$

Since $\{\prod_{k=1}^s \frac{(\min\{u,c_k\})^{\alpha/s}}{(\max\{u,c_k\})^{(\lambda+\alpha)/s}} u^{\lambda_1-\frac{1}{qn}-1}\}_{n=1}^\infty$ is non-negative and increasing in $[1,\infty)$, still by Levi theorem (cf. Ref. [29]), we have

$$M_2 \geq \lim_{n\to\infty}\int_1^\infty \prod_{k=1}^s \frac{(\min\{u,c_k\})^{\frac{\alpha}{s}}}{(\max\{u,c_k\})^{\frac{\lambda+\alpha}{s}}} u^{\lambda_1-\frac{1}{qn}-1}du$$

$$= \int_1^\infty \lim_{n\to\infty}\prod_{k=1}^s \frac{(\min\{u,c_k\})^{\frac{\alpha}{s}}}{(\max\{u,c_k\})^{\frac{\lambda+\alpha}{s}}} u^{\lambda_1-\frac{1}{qn}-1}du = k_2(\lambda_1).$$

The lemma is proved. □

4. Main Results

Theorem 1. *If* $\lambda_1 > -\alpha$, *then the following conditions are equivalent:*

(i) There exists a constant M_1*, such that for any* $f(x) \geq 0$*, satisfying*

$$0 < \int_0^\infty x^{p(1-\lambda_1)-1} f^p(x)dx < \infty,$$

we have the following first kind of Hardy-type integral inequality with the non-homogeneous kernel:

$$J := \left\{ \int_0^\infty y^{p\sigma_1 - 1} \left[\int_0^{\frac{1}{y}} \prod_{k=1}^{s} \frac{(\min\{xy, c_k\})^{\frac{\alpha}{s}}}{(\max\{xy, c_k\})^{\frac{\lambda+\alpha}{s}}} f(x)dx \right]^p dy \right\}^{\frac{1}{p}}$$
$$< M_1 \left[\int_0^\infty x^{p(1-\lambda_1)-1} f^p(x)dx \right]^{\frac{1}{p}}. \quad (4.1)$$

(ii) There exists a constant M_1*, such that for any* $f(x), g(y) \geq 0$*, satisfying* $0 < \int_0^\infty x^{p(1-\lambda_1)-1} f^p(x)dx < \infty$ *and* $0 < \int_0^\infty y^{q(1-\sigma_1)-1} g^q(y)dy < \infty$*, we have the following inequality:*

$$I := \int_0^\infty g(y) \left[\int_0^{\frac{1}{y}} \prod_{k=1}^{s} \frac{(\min\{xy, c_k\})^{\frac{\alpha}{s}}}{(\max\{xy, c_k\})^{\frac{\lambda+\alpha}{s}}} f(x)dx \right] dy$$
$$< M_1 \left[\int_0^\infty x^{p(1-\lambda_1)-1} f^p(x)dx \right]^{\frac{1}{p}} \left[\int_0^\infty y^{q(1-\sigma_1)-1} g^q(y)dy \right]^{\frac{1}{q}}. \quad (4.2)$$

(iii) $\sigma_1 = \lambda_1$.

If Condition (iii) holds, then $M_1 \geq k_1(\lambda_1)$ *and the constant factor* $M_1 = k_1(\lambda_1)$ *in* (4.1) *and* (4.2) *is the best possible.*

Proof. "*(i)* => *(ii)*". By Hölder's inequality (cf. Ref. [30]), we have

$$I = \int_0^\infty \left[y^{\sigma_1 - \frac{1}{p}} \int_0^{\frac{1}{y}} \prod_{k=1}^{s} \frac{(\min\{xy, c_k\})^{\frac{\alpha}{s}}}{(\max\{xy, c_k\})^{\frac{\lambda+\alpha}{s}}} f(x)dx \right] \left(y^{\frac{1}{p} - \sigma_1} g(y) \right) dy$$
$$\leq J \left[\int_0^\infty y^{q(1-\sigma_1)-1} g^q(y)dy \right]^{\frac{1}{q}}. \quad (4.3)$$

Then, by (3.1), we have (3.2).

"*(ii)* => *(iii)*". By Lemma 1, we have $\sigma_1 = \lambda_1$.

"(iii) => (i)". Setting $u = xy$, we obtain the following weight function:

$$\omega_1(\lambda_1, y) := y^{\lambda_1} \int_0^{\frac{1}{y}} \prod_{k=1}^{s} \frac{(\min\{xy, c_k\})^{\frac{\alpha}{s}}}{(\max\{xy, c_k\})^{\frac{\lambda+\alpha}{s}}} x^{\lambda_1 - 1} dx$$

$$= \int_0^1 \prod_{k=1}^{s} \frac{(\min\{u, c_k\})^{\frac{\alpha}{s}}}{(\max\{u, c_k\})^{\frac{\lambda+\alpha}{s}}} u^{\lambda_1 - 1} du = k_1(\lambda_1)(y > 0). \tag{4.4}$$

By Hölder's inequality with weight and (4.4), for $y \in (0, \infty)$, we have

$$\left[\int_0^{\frac{1}{y}} \prod_{k=1}^{s} \frac{(\min\{xy, c_k\})^{\frac{\alpha}{s}}}{(\max\{xy, c_k\})^{\frac{\lambda+\alpha}{s}}} f(x) dx\right]^p$$

$$= \left\{\int_0^{\frac{1}{y}} \prod_{k=1}^{s} \frac{(\min\{xy, c_k\})^{\frac{\alpha}{s}}}{(\max\{xy, c_k\})^{\frac{\lambda+\alpha}{s}}} \left[\frac{y^{(\lambda_1-1)/p}}{x^{(\lambda_1-1)/q}} f(x)\right] \left[\frac{x^{(\lambda_1-1)/q}}{y^{(\lambda_1-1)/p}}\right] dx\right\}^p$$

$$\leq \int_0^{\frac{1}{y}} \prod_{k=1}^{s} \frac{(\min\{xy, c_k\})^{\frac{\alpha}{s}}}{(\max\{xy, c_k\})^{\frac{\lambda+\alpha}{s}}} \frac{y^{\lambda_1-1} f^p(x)}{x^{(\lambda_1-1)p/q}} dx$$

$$\times \left[\int_0^{\frac{1}{y}} \prod_{k=1}^{s} \frac{(\min\{xy, c_k\})^{\frac{\alpha}{s}}}{(\max\{xy, c_k\})^{\frac{\lambda+\alpha}{s}}} \frac{x^{\lambda_1-1} dx}{y^{(\lambda_1-1)q/p}}\right]^{p-1}$$

$$= \left[\frac{\omega_1(\lambda_1, y)}{y^{q(\lambda_1-1)+1}}\right]^{p-1} \int_0^{\frac{1}{y}} \prod_{k=1}^{s} \frac{(\min\{xy, c_k\})^{\frac{\alpha}{s}}}{(\max\{xy, c_k\})^{\frac{\lambda+\alpha}{s}}} \frac{y^{\lambda_1-1}}{x^{(\lambda_1-1)p/q}} f^p(x) dx$$

$$= \frac{(k_1(\lambda_1))^{p-1}}{y^{p\lambda_1-1}} \int_0^{\frac{1}{y}} \prod_{k=1}^{s} \frac{(\min\{xy, c_k\})^{\frac{\alpha}{s}}}{(\max\{xy, c_k\})^{\frac{\lambda+\alpha}{s}}} \frac{y^{\lambda_1-1}}{x^{(\lambda_1-1)p/q}} f^p(x) dx. \tag{4.5}$$

If (4.5) takes the form of equality for a $y \in (0, \infty)$, then (cf. Ref. [30]) there exists constants A and B, such that they are not all zero,

and

$$A\frac{y^{\lambda_1-1}}{x^{(\lambda_1-1)p/q}}f^p(x)=B\frac{x^{\lambda_1-1}}{y^{(\lambda_1-1)q/p}}\text{ a.e. in }\mathbf{R}_+.$$

We suppose that $A\neq 0$ (otherwise $B=A=0$). It follows that

$$x^{p(1-\lambda_1)-1}f^p(x)=y^{q(1-\lambda_1)}\frac{B}{Ax}\text{ a.e. in }\mathbf{R}_+,$$

which contradicts the fact that $0<\int_0^\infty x^{p(1-\lambda_1)-1}f^p(x)dx<\infty$. Hence, (4.5) takes the form of strict inequality.

For $\sigma_1=\lambda_1$, by (4.5) and Fubini theorem (cf. Ref. [29]), we obtain

$$\begin{aligned}J&<(k_1(\lambda_1))^{\frac{1}{q}}\left\{\int_0^\infty\left[\int_0^{\frac{1}{y}}\prod_{k=1}^s\frac{(\min\{xy,c_k\})^{\frac{\alpha}{s}}}{(\max\{xy,c_k\})^{\frac{\lambda+\alpha}{s}}}\right.\right.\\&\qquad\left.\left.\times\frac{y^{\lambda_1-1}}{x^{(\lambda_1-1)p/q}}f^p(x)dx\right]dy\right\}^{\frac{1}{p}}\\&=(k_1(\lambda_1))^{\frac{1}{q}}\left\{\int_0^\infty\left[\int_0^{\frac{1}{x}}\prod_{k=1}^s\frac{(\min\{xy,c_k\})^{\frac{\alpha}{s}}}{(\max\{xy,c_k\})^{\frac{\lambda+\alpha}{s}}}\right.\right.\\&\qquad\left.\left.\times\frac{y^{\lambda_1-1}}{x^{(\lambda_1-1)(p-1)}}dy\right]f^p(x)dx\right\}^{\frac{1}{p}}\\&=(k_1(\lambda_1))^{\frac{1}{q}}\left[\int_0^\infty\omega_1(\lambda_1,x)x^{p(1-\lambda_1)-1}f^p(x)dx\right]^{\frac{1}{p}}\\&=k_1(\lambda_1)\left[\int_0^\infty x^{p(1-\lambda_1)-1}f^p(x)dx\right]^{\frac{1}{p}}.\end{aligned}$$

Setting $M_1\geq k_1(\lambda_1)$, (4.1) follows.

Therefore, conditions (i), (ii) and (iii) are equivalent.

While Condition (iii) follows, if there exists a constant factor $M_1\leq k_1(\lambda_1)$, such that (4.2) is valid, then by Lemma 1, we have $M_1\geq k_1(\lambda_1)$, and the constant factor $M_1=k_1(\lambda_1)$ in (4.2) is the best possible. The constant factor $M_1=k_1(\lambda_1)$ in (4.1) is still the best possible. Otherwise, by (4.3) (for $\sigma_1=\lambda_1$), we can conclude that the constant factor $M_1=k_1(\lambda_1)$ in (4.2) is not the best possible. □

Setting $y = \frac{1}{Y}$, $G(Y) = Y^{\lambda-2}g(\frac{1}{Y}), \sigma_2 = \lambda - \sigma_1$ in Theorem 1, then replacing Y $(G(Y))$ by y $(g(y))$, we have the following:

Corollary 1. *If $\lambda_1 > -\alpha$, then the following conditions are equivalent:*

(I) There exists a constant M_1, such that for any $f(x) \geq 0$, satisfying

$$0 < \int_0^\infty x^{p(1-\lambda_1)-1} f^p(x)dx < \infty,$$

we have the following first kind of Hardy-type inequality with the homogeneous kernel:

$$\left\{\int_0^\infty y^{p\sigma_2 - 1}\left[\int_0^y \prod_{k=1}^s \frac{(\min\{x, c_k y\})^{\frac{\alpha}{s}}}{(\max\{x, c_k y\})^{\frac{\lambda+\alpha}{s}}} f(x)dx\right]^p dy\right\}^{\frac{1}{p}}$$
$$< M_1 \left[\int_0^\infty x^{p(1-\lambda_1)-1} f^p(x)dx\right]^{\frac{1}{p}}. \tag{4.6}$$

(II) There exists a constant M_1, such that for any $f(x), g(y) \geq 0$, satisfying $0 < \int_0^\infty x^{p(1-\lambda_1)-1} f^p(x)dx < \infty$ and $0 < \int_0^\infty y^{q(1-\sigma_2)-1}$ $g^q(y)dy < \infty$, we have the following inequality:

$$\int_0^\infty g(y)\left[\int_0^y \prod_{k=1}^s \frac{(\min\{x, c_k y\})^{\frac{\alpha}{s}}}{(\max\{x, c_k y\})^{\frac{\lambda+\alpha}{s}}} f(x)dx\right] dy$$
$$< M_1 \left[\int_0^\infty x^{p(1-\lambda_1)-1} f^p(x)dx\right]^{\frac{1}{p}} \left[\int_0^\infty y^{q(1-\sigma_2)-1} g^q(y)dy\right]^{\frac{1}{q}}. \tag{4.7}$$

(III) $\sigma_2 = \lambda_2$.

If Condition (III) holds, then we have $M_1 \geq k_1(\lambda_1)$, and the constant $M_1 = k_1(\lambda_1)$ in (4.6) and (4.7) is the best possible.

Remark 2. On the other hand, setting $y = \frac{1}{Y}$, $G(Y) = Y^{\lambda-2}$ $g(\frac{1}{Y})$, $\sigma_2 = \lambda - \sigma_1$ in Corollary 1, then replacing Y $(G(Y))$ by

$y(g(y))$, we have Theorem 1. Hence, Theorem 1 and Corollary 1 are equivalent.

We set Condition (0): $\sigma_2 = \lambda - \sigma_1 = \lambda_2$. We may unify Theorem 1 and Corollary 1 as follows: For $\lambda_1 > -\alpha$, the conditions (0), (i), (ii), (I) and (II) are equivalent. If Condition (0) holds, then we have $M_1 \geq k_1(\lambda_1)$, and the constant $M_1 = k_1(\lambda_1)$ in (4.1), (4.2), (4.6) and (4.7) is the best possible.

Similarly, we obtain the following weight function:

$$\omega_2(\lambda_1, y) := y^{\lambda_1} \int_{\frac{1}{y}}^{\infty} \prod_{k=1}^{s} \frac{(\min\{xy, c_k\})^{\frac{\alpha}{s}}}{(\max\{xy, c_k\})^{\frac{\lambda+\alpha}{s}}} x^{\lambda_1 - 1} dx$$

$$= \int_{1}^{\infty} \prod_{k=1}^{s} \frac{(\min\{u, c_k\})^{\frac{\alpha}{s}}}{(\max\{u, c_k\})^{\frac{\lambda+\alpha}{s}}} u^{\lambda_1 - 1} du = k_2(\lambda_1)(y > 0),$$

and then in view of Lemma 2 and in the same way, we have the following:

Theorem 2. *If* $\lambda_2 > -\alpha$, *then the following conditions are equivalent:*

(i)' There exists a constant M_2, *such that for any* $f(x) \geq 0$, *satisfying*

$$0 < \int_{0}^{\infty} x^{p(1-\lambda_1)-1} f^p(x) dx < \infty,$$

we have the following second kind of Hardy-type inequality with the non-homogeneous kernel:

$$\left\{ \int_{0}^{\infty} y^{p\sigma_1 - 1} \left[\int_{\frac{1}{y}}^{\infty} \prod_{k=1}^{s} \frac{(\min\{xy, c_k\})^{\frac{\alpha}{s}}}{(\max\{xy, c_k\})^{\frac{\lambda+\alpha}{s}}} f(x) dx \right]^p dy \right\}^{\frac{1}{p}}$$

$$< M_2 \left[\int_{0}^{\infty} x^{p(1-\lambda_1)-1} f^p(x) dx \right]^{\frac{1}{p}}. \tag{4.8}$$

(ii)' There exists a constant M_2, *such that for any* $f(x), g(y) \geq 0$, *satisfying* $0 < \int_0^\infty x^{p(1-\lambda_1)-1} f^p(x) dx < \infty$, *and* $0 < \int_0^\infty y^{q(1-\sigma_1)-1}$

$g^q(y)dy < \infty$, *we have the following inequality:*

$$\int_0^\infty g(y)\left[\int_{\frac{1}{y}}^\infty \prod_{k=1}^s \frac{(\min\{xy, c_k\})^{\frac{\alpha}{s}}}{(\max\{xy, c_k\})^{\frac{\lambda+\alpha}{s}}} f(x)dx\right] dy$$

$$< M_2\left[\int_0^\infty x^{p(1-\lambda_1)-1} f^p(x)dx\right]^{\frac{1}{p}}\left[\int_0^\infty y^{q(1-\sigma_1)-1} g^q(y)dy\right]. \tag{4.9}$$

(iii)' $\sigma_1 = \lambda_1$.

If Condition (iii)' holds, then we have $M_2 \geq k_2(\lambda_1)$, *and the constant factor* $M_2 = k_2(\lambda_1)$ *in* (4.8) *and* (4.9) *is the best possible.*

Setting $y = \frac{1}{Y}$, $G(Y) = Y^{\lambda-2}g(\frac{1}{Y}), \sigma_2 = \lambda - \sigma_1$ in Theorem 2, then replacing Y $(G(Y))$ by y $(g(y))$, we have the following:

Corollary 2. *If* $\lambda_2 > -\alpha$, *then the following conditions are equivalent:*

(I)' There exists a constant M_2, *such that for any* $f(x) \geq 0$, *satisfying*

$$0 < \int_0^\infty x^{p(1-\lambda_1)-1} f^p(x)dx < \infty,$$

we have the following second kind of Hardy-type inequality with the homogeneous kernel:

$$\left\{\int_0^\infty y^{p\sigma_2-1}\left[\int_y^\infty \prod_{k=1}^s \frac{(\min\{x, c_k y\})^{\frac{\alpha}{s}}}{(\max\{x, c_k y\})^{\frac{\lambda+\alpha}{s}}} f(x)dx\right]^p dy\right\}^{\frac{1}{p}}$$

$$< M_2\left[\int_0^\infty x^{p(1-\lambda_1)-1} f^p(x)dx\right]^{\frac{1}{p}}. \tag{4.10}$$

(II)' There exists a constant M_2, *such that for any* $f(x), g(y) \geq 0$, *satisfying* $0 < \int_0^\infty x^{p(1-\lambda_1)-1} f^p(x)dx < \infty$ *and* $0 < \int_0^\infty y^{q(1-\sigma_2)-1}$

$g^q(y)dy < \infty$, *we have the following inequality:*

$$\int_0^\infty g(y)\left[\int_y^\infty \prod_{k=1}^{s} \frac{(\min\{x, c_k y\})^{\frac{\alpha}{s}}}{(\max\{x, c_k y\})^{\frac{\lambda+\alpha}{s}}} f(x)dx\right] dy$$

$$< M_2 \left[\int_0^\infty x^{p(1-\lambda_1)-1} f^p(x)dx\right]^{\frac{1}{p}} \left[\int_0^\infty y^{q(1-\sigma_2)-1} g^q(y)dy\right]^{\frac{1}{q}}. \tag{4.11}$$

(III)' $\sigma_2 = \lambda_2$.

If Condition (III)' holds, then we have $M_2 \geq k_2(\lambda_1)$, *and the constant* $M_2 = k_2(\lambda_1)$ *in* (4.10) *and* (4.11) *is the best possible.*

Remark 3. Similarly, Theorem 2 and Corollary 2 are equivalent. We may unify Theorem 2 and Corollary 2 as follows: For $\lambda_2 > -\alpha$, the conditions (0), (i)', (ii)', (I)' and (II)' are equivalent. If Condition (0) holds, then we have $M_1 \geq k_2(\lambda_1)$, and the constant $M_1 = k_2(\lambda_1)$ in (4.8), (4.9), (4.10) and (4.11) is the best possible.

5. Operator Expressions

We set the following functions: $\varphi(x) := x^{p(1-\lambda_1)-1}, \psi(y) := y^{q(1-\lambda_1)-1}, \phi(y) := y^{q(1-\lambda_2)-1}$, wherefrom $\psi^{1-p}(y) = y^{p\lambda_1-1}, \phi^{1-p}(y) = y^{p\lambda_2-1}$ $(x, y \in \mathbf{R}_+)$. Define the following real normed linear spaces:

$$L_{p,\varphi}(\mathbf{R}_+) := \left\{f : ||f||_{p,\varphi} := \left(\int_0^\infty \varphi(x)|f(x)|^p dx\right)^{\frac{1}{p}} < \infty\right\},$$

$$L_{q,\psi}(\mathbf{R}_+) := \left\{g : ||g||_{q,\psi} := \left(\int_0^\infty \psi(y)|g(y)|^q dy\right)^{\frac{1}{q}} < \infty\right\},$$

$$L_{q,\phi}(\mathbf{R}_+) := \left\{g : ||g||_{q,\phi} := \left(\int_0^\infty \phi(y)|g(y)|^q dy\right)^{\frac{1}{q}} < \infty\right\},$$

$$L_{p,\psi^{1-p}}(\mathbf{R}_+) := \left\{ h : ||h||_{p,\psi^{1-p}} = \left(\int_0^\infty \psi^{1-p}(y)|h(y)|^p dy \right)^{\frac{1}{p}} < \infty \right\},$$

$$L_{q,\phi^{1-p}}(\mathbf{R}_+) := \left\{ h : ||h||_{p,\phi^{1-p}} = \left(\int_0^\infty \phi^{1-p}(y)|h(y)|^p dy \right)^{\frac{1}{p}} < \infty \right\}.$$

(a) In view of Theorem 1 (putting $\sigma_1 = \lambda_1$), for $f \in L_{p,\varphi}(\mathbf{R}_+)$, setting

$$h_1(y) := \int_0^{\frac{1}{y}} \prod_{k=1}^{s} \frac{(\min\{xy, c_k\})^{\frac{\alpha}{s}}}{(\max\{xy, c_k\})^{\frac{\lambda+\alpha}{s}}} f(x)dx \ (y \in \mathbf{R}_+),$$

by (4.1), we have

$$||h_1||_{p,\psi^{1-p}} = \left[\int_0^\infty \psi^{1-p}(y) h_1^p(y) dy \right]^{\frac{1}{p}} < M_1 ||f||_{p,\varphi} < \infty. \quad (5.1)$$

Definition 1. Define a first kind of Hardy-type integral operator with the non-homogeneous kernel $T_1^{(1)} : L_{p,\varphi}(\mathbf{R}_+) \to L_{p,\psi^{1-p}}(\mathbf{R}_+)$ as follows: For any $f \in L_{p,\varphi}(\mathbf{R}_+)$, there exists a unique representation $T_1^{(1)} f = h_1 \in L_{p,\psi^{1-p}}(\mathbf{R}_+)$, satisfying for any $y \in \mathbf{R}_+, T_1^{(1)} f(y) = h_1(y)$.

In view of (5.1), it follows that $||T_1^{(1)} f||_{p,\psi^{1-p}} = ||h_1||_{p,\psi^{1-p}} \leq M_1 ||f||_{p,\varphi}$, and then the operator $T_1^{(1)}$ is bounded satisfying

$$||T_1^{(1)}|| = \sup_{f(\neq\theta)\in L_{p,\varphi}(\mathbf{R}_+)} \frac{||T_1^{(1)} f||_{p,\psi^{1-p}}}{||f||_{p,\varphi}} \leq M_1.$$

If we define the formal inner product of $T_1^{(1)} f$ and g as follows:

$$(T_1^{(1)} f, g) := \int_0^\infty \left[\int_0^{\frac{1}{y}} \prod_{k=1}^{s} \frac{(\min\{xy, c_k\})^{\frac{\alpha}{s}}}{(\max\{xy, c_k\})^{\frac{\lambda+\alpha}{s}}} f(x)dx \right] g(y)dy,$$

then we can rewrite Theorem 1 (for $\sigma_1 = \lambda_1$) as follows:

Theorem 3. *For $\lambda_1 > -\alpha$, the following conditions are equivalent:*

(i) There exists a constant M_1, such that for any $f(x) \geq 0, f \in L_{p,\varphi}(\mathbf{R}_+)$, $||f||_{p,\varphi} > 0$, we have the following inequality:

$$||T_1^{(1)} f||_{p,\psi^{1-p}} < M_1 ||f||_{p,\varphi}. \tag{5.2}$$

(ii) There exists a constant M_1, such that for any $f(x), g(y) \geq 0, f \in L_{p,\varphi}(\mathbf{R}_+)$, $g \in L_{q,\psi}(\mathbf{R}_+)$, $||f||_{p,\varphi}, ||g||_{q,\psi} > 0$, we have the following inequality:

$$(T_1^{(1)} f, g) < M_1 ||f||_{p,\varphi} ||g||_{q,\psi}. \tag{5.3}$$

We still have $||T_1^{(1)}|| = k_1(\lambda_1) \leq M_1$.

(b) In view of Corollary 1 (putting $\sigma_2 = \lambda_2$), for $f \in L_{p,\varphi}(\mathbf{R}_+)$, setting

$$h_2(y) := \int_0^y \prod_{k=1}^{s} \frac{(\min\{x, c_k y\})^{\frac{\alpha}{s}}}{(\max\{x, c_k y\})^{\frac{\lambda+\alpha}{s}}} f(x) dx \ (y \in \mathbf{R}_+),$$

by (4.6), we have

$$||h_2||_{p,\phi^{1-p}} = \left[\int_0^\infty \phi^{1-p}(y) h_2^p(y) dy\right]^{\frac{1}{p}} < M_1 ||f||_{p,\varphi} < \infty. \tag{5.4}$$

Definition 2. Define a first kind of Hardy-type integral operator with the homogeneous kernel $T_1^{(2)} : L_{p,\varphi}(\mathbf{R}_+) \to L_{p,\phi^{1-p}}(\mathbf{R}_+)$ as follows: For any $f \in L_{p,\varphi}(\mathbf{R})$, there exists a unique representation $T_1^{(2)} f = h_2 \in L_{p,\phi^{1-p}}(\mathbf{R}_+)$, satisfying for any $y \in \mathbf{R}_+, T_1^{(2)} f(y) = h_2(y)$.

In view of (5.4), it follows that $||T_1^{(2)} f||_{p,\phi^{1-p}} = ||h_2||_{p,\phi^{1-p}} \leq M_1 ||f||_{p,\varphi}$, and then the operator $T_1^{(2)}$ is bounded satisfying

$$||T_1^{(2)}|| = \sup_{f(\neq\theta) \in L_{p,\varphi}(\mathbf{R}_+)} \frac{||T_1^{(2)} f||_{p,\phi^{1-p}}}{||f||_{p,\varphi}} \leq M_1.$$

If we define the formal inner product of $T_1^{(2)}f$ and g as follows:

$$(T_1^{(2)}f,g):=\int_0^\infty\left[\int_0^y\prod_{k=1}^s\frac{(\min\{x,c_ky\})^{\frac{\alpha}{s}}}{(\max\{x,c_ky\})^{\frac{\lambda+\alpha}{s}}}f(x)dx\right]g(y)dy,$$

then we can rewrite Corollary 1 (for $\sigma_2=\lambda_2$) as follows:

Corollary 3. *For $\lambda_2>-\alpha$, the following conditions are equivalent:*

(I) There exists a constant M_1, such that for any $f(x)\geq 0, f\in L_{p,\varphi}(\mathbf{R}_+)$, $||f||_{p,\varphi}>0$, we have the following inequality:

$$||T_1^{(2)}f||_{p,\phi^{1-p}}<M_1||f||_{p,\varphi}. \tag{5.5}$$

(II) There exists a constant M_1, such that for any $f(x)$, $g(y)\geq 0, f\in L_{p,\varphi}(\mathbf{R}_+), g\in L_{q,\phi}(\mathbf{R}_+), ||f||_{p,\varphi}, ||g||_{q,\phi}>0$, we have the following inequality:

$$(T_1^{(2)}f,g)<M_1||f||_{p,\varphi}||g||_{q,\phi}. \tag{5.6}$$

We still have $||T_1^{(2)}||=k_1(\lambda_1)\leq M_1$.

Remark 4. Similarly, Theorem 3 and Corollary 3 are equivalent.

(c) In view of Theorem 2 (putting $\sigma_1=\lambda_1$), for $f\in L_{p,\varphi}(\mathbf{R}_+)$, setting

$$H_1(y):=\int_{\frac{1}{y}}^\infty\prod_{k=1}^s\frac{(\min\{xy,c_k\})^{\frac{\alpha}{s}}}{(\max\{xy,c_k\})^{\frac{\lambda+\alpha}{s}}}f(x)dx\ (y\in\mathbf{R}_+),$$

by (4.8), we have

$$||H_1||_{p,\psi^{1-p}}=\left[\int_0^\infty\psi^{1-p}(y)H_1^p(y)dy\right]^{\frac{1}{p}}<M_2||f||_{p,\varphi}<\infty. \tag{5.7}$$

Definition 3. Define a second kind of Hardy-type integral operator with the non-homogeneous kernel $T_2^{(1)}:L_{p,\varphi}(\mathbf{R}_+)\to L_{p,\psi^{1-p}}(\mathbf{R}_+)$

as follows: For any $f \in L_{p,\varphi}(\mathbf{R}_+)$, there exists a unique representation $T_2^{(1)} f = H_1 \in L_{p,\psi^{1-p}}(\mathbf{R}_+)$, satisfying for any $y \in \mathbf{R}_+, T_2^{(1)} f(y) = H_1(y)$.

In view of (5.7), it follows that $||T_2^{(1)} f||_{p,\psi^{1-p}} = ||H_1||_{p,\psi^{1-p}} \le M_2||f||_{p,\varphi}$, and then the operator $T_2^{(1)}$ is bounded satisfying

$$||T_2^{(1)}|| = \sup_{f(\neq\theta)\in L_{p,\varphi}(\mathbf{R}_+)} \frac{||T_2^{(1)} f||_{p,\psi^{1-p}}}{||f||_{p,\varphi}} \le M_2.$$

If we define the formal inner product of $T_2^{(1)} f$ and g as follows:

$$(T_2^{(1)} f, g) := \int_0^\infty \left[\int_{\frac{1}{y}}^\infty \prod_{k=1}^s \frac{(\min\{xy, c_k\})^{\frac{\alpha}{s}}}{(\max\{xy, c_k\})^{\frac{\lambda+\alpha}{s}}} f(x)dx \right] g(y)dy,$$

then we can rewrite Theorem 2 (for $\sigma_1 = \lambda_1$) as follows:

Theorem 4. *For $\lambda_2 > -\alpha$, the following conditions are equivalent:*

(i)' There exists a constant M_2, such that for any $f(x) \ge 0, f \in L_{p,\varphi}(\mathbf{R}_+)$, $||f||_{p,\varphi} > 0$, we have the following inequality:

$$||T_2^{(1)} f||_{p,\psi^{1-p}} < M_2||f||_{p,\varphi}. \tag{5.8}$$

(ii)' There exists a constant M_2, such that for any $f(x), g(y) \ge 0, f \in L_{p,\varphi}(\mathbf{R}_+), g \in L_{q,\psi}(\mathbf{R}_+), ||f||_{p,\varphi}, ||g||_{q,\psi} > 0$, we have the following inequality:

$$(T_2^{(1)} f, g) < M_2||f||_{p,\varphi}||g||_{q,\psi}. \tag{5.9}$$

We still have $||T_2^{(1)}|| = k_2(\lambda_1) \le M_2$.

(d) In view of Corollary 2 (putting $\sigma_2 = \lambda_2$), for $f \in L_{p,\varphi}(\mathbf{R}_+)$, setting

$$H_2(y) := \int_y^\infty \prod_{k=1}^s \frac{(\min\{x, c_k y\})^{\frac{\alpha}{s}}}{(\max\{x, c_k y\})^{\frac{\lambda+\alpha}{s}}} f(x)dx \ (y \in \mathbf{R}_+),$$

by (4.10), we have

$$||H_2||_{p,\phi^{1-p}} = \left[\int_0^\infty \phi^{1-p}(y) H_2^p(y)dy \right]^{\frac{1}{p}} < M_2||f||_{p,\varphi} < \infty. \tag{5.10}$$

Definition 4. Define a second kind of Hardy-type integral operator with the homogeneous kernel $T_2^{(2)} : L_{p,\varphi}(\mathbf{R}_+) \to L_{p,\phi^{1-p}}(\mathbf{R}_+)$ as follows: For any $f \in L_{p,\varphi}(\mathbf{R})$, there exists a unique representation $T_2^{(2)} f = H_2 \in L_{p,\phi^{1-p}}(\mathbf{R}_+)$, satisfying for any $y \in \mathbf{R}_+, T_2^{(2)} f(y) = H_2(y)$.

In view of (5.10), it follows that $||T_2^{(2)} f||_{p,\phi^{1-p}} = ||H_2||_{p,\phi^{1-p}} \leq M_2||f||_{p,\varphi}$, and then the operator $T_2^{(2)}$ is bounded satisfying

$$||T_2^{(2)}|| = \sup_{f(\neq\theta)\in L_{p,\varphi}(\mathbf{R}_+)} \frac{||T_2^{(2)} f||_{p,\phi^{1-p}}}{||f||_{p,\varphi}} \leq M_2.$$

If we define the formal inner product of $T_1^{(2)} f$ and g as follows:

$$(T_2^{(2)} f, g) := \int_0^\infty \left[\int_y^\infty \prod_{k=1}^s \frac{(\min\{x, c_k y\})^{\frac{\alpha}{s}}}{(\max\{x, c_k y\})^{\frac{\lambda+\alpha}{s}}} f(x)dx \right] g(y)dy,$$

then we can rewrite Corollary 2 (for $\sigma_2 = \lambda_2$) as follows:

Corollary 4. *For $\lambda_2 > -\alpha$, the following conditions are equivalent:*

(I)' There exists a constant M_2, such that for any $f(x) \geq 0, f \in L_{p,\varphi}(\mathbf{R}_+)$, $||f||_{p,\varphi} > 0$, we have the following inequality:

$$||T_2^{(2)} f||_{p,\phi^{1-p}} < M_2||f||_{p,\varphi}. \tag{5.11}$$

(II)' There exists a constant M_2, such that for any $f(x)$, $g(y) \geq 0, f \in L_{p,\varphi}(\mathbf{R}_+), g \in L_{q,\phi}(\mathbf{R}_+)$, $||f||_{p,\varphi}, ||g||_{q,\phi} > 0$, we have the following inequality:

$$(T_2^{(2)} f, g) < M_2||f||_{p,\varphi}||g||_{q,\phi}. \tag{5.12}$$

We still have $||T_2^{(2)}|| = k_2(\lambda_1) \leq M_2$.

Remark 5. Similarly, Theorem 4 and Corollary 4 are equivalent.

6. Conclusions

In this chapter, by using the way of real analysis, we obtain an example and two lemmas and then deduce two kinds of Hardy-type integral inequalities with the non-homogeneous kernel as $\prod_{k=1}^{s} \frac{(\min\{xy,c_k\})^{\alpha/s}}{(\max\{xy,c_k\})^{(\lambda+\alpha)/s}}$ and the homogeneous kernel as $\prod_{k=1}^{s} \frac{(\min\{x,c_k y\})^{\alpha/s}}{(\max\{x,c_k y\})^{(\lambda+\alpha)/s}}$, as well as their equivalent forms in Theorems 1 and 2, Corollaries 1 and 2. We also consider the operator expressions in Theorems 3 and 4, Corollaries 3 and 4. The lemmas and theorems provide an extensive account of these types of inequalities.

Acknowledgments

This work is supported by the National Natural Science Foundation of China (61772140). We are grateful for this help.

References

[1] I. Schur. Bernerkungen sur Theorie der beschrankten Billnearformen mit unendlich vielen Veranderlichen. *J. Math.*, 140, 1911, 1–28.
[2] G. H. Hardy. Note on a theorem of Hilbert concerning series of positive terms. *Proc. Lond. Math. Soc.*, 23(2), 1925. Records of Proc. xlv–xlvi.
[3] G. H. Hardy, J. E. Littlewood and G. Pólya. *Inequalities.* Cambridge University Press, Cambridge, USA, 1934.
[4] D. S. Mitrinović, J. E. Pečarić and A. M. Fink. *Inequalities Involving Functions and Their Integrals and Derivatives.* Kluwer Academic, Boston, USA, 1991.
[5] B. C. Yang. On Hilbert's integral inequality. *J. Math. Anal. Appl.*, 220, 1998, 778–785.
[6] B. C. Yang. A note on Hilbert's integral inequality. *Chin. Q. J. Math.*, 13(4), 1998, 83–86.
[7] B. C. Yang. On an extension of Hilbert's integral inequality with some parameters. *Aust. J. Math. Anal. Appl.*, 1(1), 2004, 1–8. Article 11.
[8] B. C. Yang, I. Brnetić, M. Krnić and J. E. Pečarić. Generalization of Hilbert and Hardy-Hilbert integral inequalities. *Math. Inequal. and Appl.*, 8(2), 2005, 259–272.

[9] M. Krnić and J. E. Pečarić. Hilbert's inequalities and their reverses. *Publ. Math. Debrecen*, 67(3–4), 2005, 315–331.

[10] Y. Hong. On Hardy-Hilbert integral inequalities with some parameters. *J. Inequal. Pure Appl. Math.*, 6(4), 2005, 1–10. Article 92.

[11] B. Arpad and O. Choonghong. Best constant for certain multi linear integral operator. *J. Inequal. Appl.*, 28582, 2006.

[12] Y. J. Li and B. He. On inequalities of Hilbert's type. *Bull. Aust. Math. Soc.*, 76(1), 2007, 1–13.

[13] W. Y. Zhong and B. C. Yang. On multiple Hardy-Hilbert's integral inequality with kernel. *J. Inequal. Appl.*, 2007, 2007, 17 p. Article ID 27962.

[14] J. S. Xu. Hardy-Hilbert's inequalities with two parameters. *Adv. Math.*, 36(2), 2007, 63–76.

[15] B. C. Yang. *The Norm of Operator and Hilbert-Type Inequalities.* Science Press, Beijing, China, 2009.

[16] B. C. Yang. On Hilbert-type integral inequalities and their operator expressions. *J. Guangaong Univ. Educ.*, 33(5), 2013, 1–17.

[17] B. C. Yang. *Hilbert-Type Integral Inequalities.* Bentham Science Publishers Ltd., The United Emirates, 2009.

[18] Y. Hong. On the structure character of Hilbert's type integral inequality with homogeneous kernal and applications. *J. Jilin Univ. (Sci. Ed.)*, 55, 2017 (In press).

[19] B. He, Y. Hong and Z. Li. Conditions for the validity of a class of optimal Hilbert type multiple integral inequalities with non-homogeneous. *J. Inequal. Appl.*, 2021, 2021, 64

[20] Q. Chen, B. He, Y. Hong and Z. Li. Equivalent parameter conditions for the validity of half-discrete Hilbert-type multiple integral inequality with generalized homogeneous kernel. *J. Funct. Spaces*, 2020, 2020, 6 p. Article ID 7414861.

[21] B. He, Y. Hong and Q. Chen. The equivalent parameter conditions for constructing multiple integral half-discrete Hilbert-type inequalities with a class of non-homogeneous kernels and their applications. *Open Math.*, 19, 2021, 400–411.

[22] Y. Hong, Q. L. Huang and Q. Chen. The parameter conditions for the existence of the Hilberttype multiple integral inequality and its best constant factor. *Ann. Funct. Anal.* https://doi.org/10.1007/s43034-020-00087-5.

[23] Y. Hong and Q. Chen. Equivalent parameter conditions for the construction of Hilbert–type integral inequalities with a class of non–homogeneous kernels. *J. South China Normal Univ. (Nat. Sci. Ed.)*, 52(5), 2020, 124–128.

[24] Y. Hong and Q. Chen. Research progress and applications of Hilbert-type series inequalities. *J. Jilin Univ. (Sci. Ed.)*, 59(5), 2021, 1131–1140.

[25] Y. Hong, Q. Chen and C. Y. Wu. The best matching parameters for semi-discrete Hilbert-type inequality with quasi-homogeneous kernel. *Math. Appl.*, 34(3), 2021, 779–785.

[26] Y. Hong and B. He. The optimal matching parameter of half-discrete Hilbert-type multiple integral inequalities with non-homogeneous kernels and applications. *Chin. Q. J. Math.*, 36(3), 2021, 252–262.

[27] Y. Hong, J. Q. Liao, B. C. Yang and Q. Chen. A class of Hilbert-type multiple integral inequalities with the kernel of generalized homogeneous function and its applications. *J. Inequal. Appl.*, 2020, 2020, 140.

[28] M. T. Rassias, B. C. Yang and A. Raigorodskii. An equivalent form related to a Hilbert-type integral inequality. *Axioms*, 12, 2023, 677.

[29] J. C. Kuang. *Real and Functional Analysis (Continuation)*, Vol. 2. Higher Education Press, Beijing, 2015.

[30] J. C. Kuang. *Applied Inequalities.* Shangdong Science and Technology Press, Jinan, China, 2004.

https://doi.org/10.1142/9789819820498_0016

Chapter 16

A Multidimensional Reverse Hardy–Hilbert's Integral Inequality Involving One Derivative Function of Higher Order

Michael Th. Rassias[*,§] **and Bicheng Yang**[‡,¶]

[*] *Department of Mathematics and Engineering Sciences, Hellenic Military Academy, 16673 Vari Attikis, Greece*

[‡] *School of Mathematics, Guangdong University of Education, Guangzhou, Guangdong 510303, P. R. China*

[§] *mthrassias@yahoo.com*

[¶] *bcyang@gdei.edu.cn, bcyang818@163.com*

By means of the weight functions, the idea of introduced parameters and the transfer formula, a multidimensional reverse Hardy–Hilbert's integral inequality involving one derivative function of higher order is obtained. A few equivalent statements of the best possible constant factor in the new inequality related to several parameters are considered. Some corollaries are obtained.

Keywords: Weight function, derivative function of higher order, parameter, Beta function, multidimensional reverse Hardy–Hilbert's integral inequality.

1. Introduction

Assuming that $p > 1, \frac{1}{p} + \frac{1}{q} = 1, a_m, b_n \geq 0, 0 < \sum_{m=1}^{\infty} a_m^p < \infty$ and $0 < \sum_{n=1}^{\infty} b_n^q < \infty$, we have the following well-known Hardy–Hilbert's inequality with the best possible constant factor $\frac{\pi}{\sin(\pi/p)}$ (cf. Ref. [1], Theorem 315):

$$\sum_{m=1}^{\infty}\sum_{n=1}^{\infty}\frac{a_m b_n}{m+n} < \frac{\pi}{\sin(\pi/p)}\left(\sum_{m=1}^{\infty} a_m^p\right)^{\frac{1}{p}}\left(\sum_{n=1}^{\infty} b_n^q\right)^{\frac{1}{q}}. \tag{1.1}$$

Setting $f(x), g(y) \geq 0$, $0 < \int_0^{\infty} f^p(x)dx < \infty$ and $0 < \int_0^{\infty} g^q(y)dy < \infty$, we have the following integral analogue of (1.1) with the same best possible constant factor (cf. Ref. [1], Theorem 316):

$$\int_0^{\infty}\int_0^{\infty}\frac{f(x)g(y)}{x+y}dxdy < \frac{\pi}{\sin(\pi/p)}\left(\int_0^{\infty} f^p(x)dx\right)^{\frac{1}{p}}\left(\int_0^{\infty} g^q(y)dy\right)^{\frac{1}{q}}. \tag{1.2}$$

In 2006, by means of the Euler–Maclaurin summation formula and the techniques of real analysis, Krnić *et al.* [2] gave the following extension of (1.1):

$$\sum_{m=1}^{\infty}\sum_{n=1}^{\infty}\frac{a_m b_n}{(m+n)^{\lambda}} < B(\lambda_1, \lambda_2)\left[\sum_{m=1}^{\infty} m^{p(1-\lambda_1)-1} a_m^p\right]^{\frac{1}{p}}\left[\sum_{n=1}^{\infty} n^{q(1-\lambda_2)-1} b_n^q\right]^{\frac{1}{q}}, \tag{1.3}$$

where, $\lambda_1, \lambda_2 \in (0, 2]$, $\lambda_1 + \lambda_2 = \lambda \in (0, 4]$, the constant factor $B(\lambda_1, \lambda_2)$ is the best possible and

$$B(u, v) = \int_0^{\infty}\frac{t^{u-1}}{(1+t)^{u+v}}dt, \quad u, v > 0 \tag{1.4}$$

is the Beta function.

In 2019, by means of inequality (1.3) and Abel's partial summation formula, Adiyasuren *et al.* [3] obtained an extension of (1.3) involving two partial sums. In 2020, Mo *et al.* [4] gave an extension of (1.2) involving two upper limit functions. Inequalities (1.1)–(1.2) with their extensions played an important role in analysis and its applications (cf. Refs. [5–15]).

In 2016–2017, Hong *et al.* [16, 17] considered several equivalent conditions of the extensions of (1.1) and (1.2) with a few parameters related to the best possible constant factors. Some further results were provided by Refs. [18–20, 20]. In 2023, Hong *et al.* [21] gave a more accurate half-discrete multidimensional Hilbert-type inequality involving one derivative function of m-order. Some other results were obtained by Refs. [22, 26–30].

In this chapter, following the way of Ref. [21], by means of the weight functions, the idea of introduced parameters and the transfer formula, a multidimensional reverse Hardy–Hilbert's integral inequality with the kernel as $\frac{1}{(x+||y||_\beta^\alpha)^\lambda}$ $(\alpha, \lambda > 0)$ involving one derivative function of m-order is obtained. The equivalent statements of the best possible constant factor related to several parameters in the new inequality are considered. Some corollaries are provided.

2. Some Lemmas

In what follows, we suppose that (H1): $m \in \mathbf{N}_0 = \{0, 1, \ldots\}$, $n \in \mathbf{N} = \{1, 2, \ldots\}$, $p < 0$ $(0 < q < 1)$, $\frac{1}{p} + \frac{1}{q} = 1$, $\alpha, \beta \in \mathbf{R}_+ = (0, \infty)$, $\lambda > 0$, $\lambda_1, \lambda_2 \in (0, \lambda)$, $\widehat{\lambda}_1 = \frac{\lambda - \lambda_2}{p} + \frac{\lambda_1}{q}$, $\widehat{\lambda}_2 = \frac{\lambda - \lambda_1}{q} + \frac{\lambda_2}{p}$, $f(x)$ (≥ 0) is a differentiable function of m-order, $f^{(m)}(x)$ (≥ 0) is a continuous function unless at finite points in $\mathbf{R}_+$, satisfying for $m \in \mathbf{N}$, $k = 1, \ldots, m$,

$$f^{(k-1)}(x) = o(e^{tx}) \ (t > 0; x \to \infty), f^{(k-1)}(0^+) = 0,$$

and for $f^{(m)}(x)$, $g(y) \geq 0$ $(x \in \mathbf{R}_+,\ y \in \mathbf{R}_+^n,\ m \in \mathbf{N}_0)$, $||y||_\beta := (\sum_{i=1}^n y_i^\beta)^{\frac{1}{\beta}}$, we have

$$0 < \int_0^\infty x^{p(1-\widehat{\lambda}_1)-1} (f^{(m)}(x))^p dx < \infty \quad \text{and}$$

$$0 < \int_{\{y \in \mathbf{R}_+^n\}} ||y||_\beta^{q(n-\alpha\widehat{\lambda}_2)-n} g^q(y) dy < \infty.$$

If $M > 0, \psi(u)$ $(u > 0)$ is a non-negative measurable function, then we have the following transfer formula (cf. Ref. [5], (9.1.5)):

$$\int \cdots \int_{\{y\in\mathbf{R}_+^n;0<\sum_{i=1}^n\left(\frac{y_i}{M}\right)^\beta\leq 1\}} \psi\left(\sum_{i=1}^n\left(\frac{y_i}{M}\right)^\beta\right) dy_1\cdots dy_n$$
$$= \frac{M^n\Gamma^n(\frac{1}{\beta})}{\beta^n\Gamma(\frac{n}{\beta})}\int_0^1 \psi(u)u^{\frac{n}{\beta}-1}du. \tag{2.1}$$

In particular, (i) for $||y||_\beta = M\left[\sum_{i=1}^n(\frac{y_i}{M})^\beta\right]^{\frac{1}{\beta}}$, by (2.1), we have

$$\int_{\{y\in\mathbf{R}_+^n\}}\varphi(||y||_\beta)dy$$
$$= \lim_{M\to\infty}\int\cdots\int_{\{y\in\mathbf{R}_+^n;0<\sum_{i=1}^n(\frac{y_i}{M})^\beta\leq 1\}}\varphi\left(M\left[\sum_{i=1}^n\left(\frac{y_i}{M}\right)^\beta\right]^{\frac{1}{\beta}}\right)dy_1\cdots dy_n$$
$$= \lim_{M\to\infty}\frac{M^n\Gamma^n(\frac{1}{\beta})}{\beta^n\Gamma(\frac{n}{\beta})}\int_0^1\varphi(Mu^{\frac{1}{\beta}})u^{\frac{n}{\beta}-1}du$$
$$= \frac{\Gamma^n(\frac{1}{\beta})}{\beta^{n-1}\Gamma(\frac{n}{\beta})}\int_0^\infty\varphi(v)v^{n-1}dv\ (v = Mu^{\frac{1}{\beta}}); \tag{2.2}$$

(ii) for $\psi(u) = \varphi(Mu^{\frac{1}{\beta}}) = 0, u < (\frac{b}{M})^\beta$ $(b > 0)$, by (5), we have

$$\int_{\{y\in\mathbf{R}_+^n;||y||_\beta\geq b\}}\varphi(||y||_\beta)dy = \lim_{M\to\infty}\frac{M^n\Gamma^n(\frac{1}{\beta})}{\beta^n\Gamma(\frac{n}{\beta})}\int_{(\frac{b}{M})^\beta}^1\varphi(Mu^{\frac{1}{\beta}})u^{\frac{n}{\beta}-1}du$$
$$= \frac{\Gamma^n(\frac{1}{\beta})}{\beta^{n-1}\Gamma(\frac{n}{\beta})}\int_b^\infty\varphi(v)v^{n-1}dv. \tag{2.3}$$

Remark 1. For $b = 1, c \in \mathbf{R}_+$, by (2.3), we have

$$\int_{\{y\in\mathbf{R}_+^n;||y||_\beta\geq 1\}}||y||_\beta^{-\alpha c-n}dy = \frac{\Gamma^n(\frac{1}{\beta})}{\alpha c\beta^{n-1}\Gamma(\frac{n}{\beta})}. \tag{2.4}$$

Lemma 1. *With regard to the assumption H1, for* $s \in (0, \infty), s_1, s_2 \in (0, s)$, *we define the following weight functions:*

$$\omega_s(s_1, y) := ||y||_\beta^{\alpha(s-s_1)} \int_0^\infty \frac{x^{s_1-1}}{(x + ||y||_\beta^\alpha)^s} dx \quad (y \in \mathbf{R}_+^n), \quad (2.5)$$

$$\varpi_s(s_2, x) := x^{s-s_2} \int_{\{y \in \mathbf{R}_+^n\}} \frac{||y||_\beta^{\alpha s_2 - n}}{(x + ||y||_\beta^\alpha)^s} dy \quad (x \in \mathbf{R}_+). \quad (2.6)$$

The following expressions are value:

$$\omega_s(s_1, y) = B(s_1, s - s_1) \quad (y \in \mathbf{R}_+^n), \quad (2.7)$$

$$\varpi_s(s_2, x) = \frac{\Gamma^n(\frac{1}{\beta})}{\alpha\beta^{n-1}\Gamma(\frac{n}{\beta})} B(s_2, s - s_2) \quad (x \in \mathbf{R}_+). \quad (2.8)$$

Proof. In (2.5), setting $u = \frac{x}{||y||_\beta^\alpha}$, we find

$$\omega_s(s_1, y) = ||y||_\beta^{\alpha(s-s_1)} \int_0^\infty \frac{(u||y||_\beta^\alpha)^{s_1-1}||y||_\beta^\alpha}{(u||y||_\beta^\alpha + ||y||_\beta^\alpha)^s} du$$

$$= \int_0^\infty \frac{u^{s_1-1}}{(u+1)^s} du = B(s_1, s - s_1),$$

and then we have (2.7).

In (2.2), for $\varphi(v) = \frac{v^{\alpha s_2 - n}}{(x+v^\alpha)^s}$, we have

$$\varpi_s(s_2, x) = \frac{\Gamma^n(\frac{1}{\beta})}{\beta^{n-1}\Gamma(\frac{n}{\beta})} x^{s-s_2} \int_0^\infty \frac{v^{\alpha s_2 - n}}{(x + v^\alpha)^s} v^{n-1} dv$$

$$= \frac{\Gamma^n(\frac{1}{\beta})}{\beta^{n-1}\Gamma(\frac{n}{\beta})} x^{s-s_2} \int_0^\infty \frac{v^{\alpha s_2 - 1}}{(x + v^\alpha)^s} dv$$

$$= \frac{\Gamma^n(\frac{1}{\beta})}{\alpha\beta^{n-1}\Gamma(\frac{n}{\beta})} \int_0^\infty \frac{t^{s_2-1}}{(1+t)^s} dt$$

$$= \frac{\Gamma^n(\frac{1}{\beta})}{\alpha\beta^{n-1}\Gamma(\frac{n}{\beta})} B(s_2, s - s_2),$$

and then we have (2.8).

The lemma is proved. □

Lemma 2. *With regard to the assumption H1, for* $t > 0, m \in \mathbf{N}_0$, *we have the following expression:*

$$\int_0^\infty e^{-tx} f(x)dx = t^{-m} \int_0^\infty e^{-tx} f^{(m)}(x)dx. \tag{2.9}$$

Proof. For $m = 0$, (2.9) is naturally valid; for $m \in \mathbf{N}$, $k = 1, \ldots, m$, since $f^{(k-1)}(0^+) = 0, f^{(k-1)}(x) = o(e^{tx})$ $(t > 0; x \to \infty)$, we find

$$\int_0^\infty e^{-tx} f^{(k)}(x)dx = \int_0^\infty e^{-tx} df^{(k-1)}(x) = e^{-tx} f^{(k-1)}(x)|_0^\infty$$
$$- \int_0^\infty f^{(k-1)}(x)de^{-tx} = t \int_0^\infty e^{-tx} f^{(k-1)}(x)dx.$$

Substitution of the above expression in $k = 1, \ldots m$, by simplification, we have (2.9).

The lemma is proved. □

Lemma 3. *With regard to the assumption H1, we have the following reverse inequality:*

$$I_\lambda := \int_{\{y \in \mathbf{R}_+^n\}} \int_0^\infty \frac{f^{(m)}(x)g(y)}{(x + ||y||_\beta^\alpha)^\lambda} dxdy$$
$$> \left(\frac{\Gamma^n(\frac{1}{\beta})}{\alpha\beta^{n-1}\Gamma(\frac{n}{\beta})} \right)^{\frac{1}{p}} B^{\frac{1}{p}}(\lambda_2, \lambda - \lambda_2) B^{\frac{1}{q}}(\lambda_1, \lambda - \lambda_1)$$
$$\times \left[\int_0^\infty x^{p(1-\widehat{\lambda}_1)-1} (f^{(m)}(x))^p dx \right]^{\frac{1}{p}}$$
$$\times \left[\int_{\{y \in \mathbf{R}_+^n\}} ||y||_\beta^{q(n-\alpha\widehat{\lambda}_2)-n} g^q(y)dy \right]^{\frac{1}{q}}. \tag{2.10}$$

Proof. By the reverse Hölder's inequality (cf. Ref. [31]), we have

$$I_\lambda = \int_{\{y \in \mathbf{R}_+^n\}} \int_0^\infty \frac{1}{(x + ||y||_\beta^\alpha)^\lambda} \left[\frac{x^{(1-\lambda_1)/q} f^{(m)}(x)}{||y||_\beta^{(n-\alpha\lambda_2)/p}} \right]$$
$$\times \left[\frac{||y||_\beta^{(n-\alpha\lambda_2)/p} g(y)}{x^{(1-\lambda_1)/q}} \right] dxdy$$

$$\geq \left[\int_0^\infty \int_{\{y\in\mathbf{R}_+^n\}} \frac{1}{(x+||y||_\beta^\alpha)^\lambda} \frac{x^{(1-\lambda_1)(p-1)}(f^{(m)}(x))^p}{||y||_\beta^{n-\alpha\lambda_2}} dydx\right]^{\frac{1}{p}}$$

$$\times \left[\int_{\{y\in\mathbf{R}_+^n\}} \int_0^\infty \frac{1}{(x+||y||_\beta^\alpha)^\lambda} \frac{||y||_\beta^{(n-\alpha\lambda_2)(q-1)} g^q(y)}{x^{1-\lambda_1}} dxdy\right]^{\frac{1}{q}}$$

$$= \left\{\int_0^\infty \left[x^{\lambda-\lambda_2} \int_{\{y\in\mathbf{R}_+^n\}} \frac{||y||_\beta^{\alpha\lambda_2-n} dy}{(x+||y||_\beta^\alpha)^\lambda}\right] x^{p(1-\widehat{\lambda}_1)-1}(f^{(m)}(x))^p dx\right\}^{\frac{1}{p}}$$

$$\times \left\{\int_{\{y\in\mathbf{R}_+^n\}} \left[||y||_\beta^{\alpha(\lambda-\lambda_2)} \int_0^\infty \frac{x^{\lambda_1-1}dx}{(x+||y||_\beta^\alpha)^\lambda}\right]\right.$$

$$\left. \times \, ||y||_\beta^{q(n-\alpha\widehat{\lambda}_2)-n} g^q(y)dy\right\}^{\frac{1}{q}}$$

$$= \left[\int_0^\infty \varpi_\lambda(\lambda_2,x) x^{p(1-\widehat{\lambda}_1)-1}(f^{(m)}(x))^p dx\right]^{\frac{1}{p}}$$

$$\times \left[\int_{\{y\in\mathbf{R}_+^n\}} \omega_\lambda(\lambda_1,y) ||y||_\beta^{q(n-\alpha\widehat{\lambda}_2)-n} g^q(y)dy\right]^{\frac{1}{q}}. \tag{2.11}$$

If (2.11) keeps the form of equality, then there exist constants A and B, such that they are not both zero, satisfying

$$A\frac{x^{(1-\lambda_1)(p-1)}(f^{(m)}(x))^p}{||y||_\beta^{n-\alpha\lambda_2}} = B\frac{||y||_\beta^{(n-\alpha\lambda_2)(q-1)} g^q(y)}{x^{1-\lambda_1}} \text{ a.e. in } \mathbf{R}_+ \times \mathbf{R}_+^n. \tag{2.12}$$

Assuming that $A \neq 0$, by (2.12), there exists a $y \in \mathbf{R}_+^n$, such that

$$x^{p(1-\widehat{\lambda}_1)-1}(f^{(m)}(x))^p = \left[\frac{B}{A}||y||_\beta^{q(n-\alpha\lambda_2)} g^q(y)\right] x^{-1-(\lambda-\lambda_1-\lambda_2)} \text{ a.e. in } \mathbf{R}_+.$$

Since for any $\lambda - \lambda_1 - \lambda_2 \in \mathbf{R}$, $\int_0^\infty x^{-1-(\lambda-\lambda_1-\lambda_2)}dx = \infty$, the above expression contradicts the fact that $0 < \int_0^\infty x^{p(1-\widehat{\lambda}_1)-1} (f^{(m)}(x))^p dx < \infty$.

Therefore, by (2.7) and (2.8) (for $s = \lambda, s_1 = \lambda_1, s_2 = \lambda_2$), we have (2.10).

The lemma is proved. □

3. Main Results

Theorem 1. *With regard to the assumption H1, we have the following multidimensional reverse Hardy–Hilbert's integral inequality involving one derivative function of m-order:*

$$
\begin{aligned}
I &:= \int_{\{y \in \mathbf{R}_+^n\}} \int_0^\infty \frac{f(x)g(y)}{(x + ||y||_\beta^\alpha)^{\lambda+m}} dx dy \\
&> \frac{\Gamma(\lambda)}{\Gamma(\lambda+m)} \left(\frac{\Gamma^n(\frac{1}{\beta})}{\alpha\beta^{n-1}\Gamma(\frac{n}{\beta})} \right)^{\frac{1}{p}} B^{\frac{1}{p}}(\lambda_2, \lambda - \lambda_2) B^{\frac{1}{q}}(\lambda_1, \lambda - \lambda_1) \\
&\times \left[\int_0^\infty x^{p(1-\widehat{\lambda}_1)-1} (f^{(m)}(x))^p dx \right]^{\frac{1}{p}} \\
&\times \left[\int_{\{y \in \mathbf{R}_+^n\}} ||y||_\beta^{q(n-\alpha\widehat{\lambda}_2)-n} g^q(y) dy \right]^{\frac{1}{q}}. \qquad (3.1)
\end{aligned}
$$

In particular, for $\lambda = \lambda_1 + \lambda_2$ in (3.1), we have

$$
0 < \int_0^\infty x^{p(1-\lambda_1)-1} (f^{(m)}(x))^p dx < \infty,
$$

$$
0 < \int_{\{y \in \mathbf{R}_+^n\}} ||y||_\beta^{q(n-\alpha\lambda_2)-n} g^q(y) dy < \infty,
$$

and the following reverse inequality:

$$
\begin{aligned}
&\int_{\{y \in \mathbf{R}_+^n\}} \int_0^\infty \frac{f(x)g(y)}{(x + ||y||_\beta^\alpha)^{\lambda+m}} dx dy \\
&\quad > \frac{\Gamma(\lambda)}{\Gamma(\lambda+m)} \left(\frac{\Gamma^n(\frac{1}{\beta})}{\alpha\beta^{n-1}\Gamma(\frac{n}{\beta})} \right)^{\frac{1}{p}} B(\lambda_1, \lambda_2)
\end{aligned}
$$

$$\times\left[\int_0^\infty x^{p(1-\lambda_1)-1}(f^{(m)}(x))^p dx\right]^{\frac{1}{p}}$$

$$\times\left[\int_{\{y\in\mathbf{R}_+^n\}} ||y||_\beta^{q(n-\alpha\lambda_2)-n} g^q(y)dy\right]^{\frac{1}{q}}. \tag{3.2}$$

Proof. By the following expression of the Gamma function:

$$\frac{1}{(x+||y||_\beta^\alpha)^{\lambda+m}} = \frac{1}{\Gamma(\lambda+m)}\int_0^\infty t^{\lambda+m-1}e^{-(x+||y||_\beta^\alpha)t}dt \ (x>0), \tag{3.3}$$

equation (2.9) and Fubini theorem (cf. Ref. [32]), we have

$$\begin{aligned}
I &= \frac{1}{\Gamma(\lambda+m)}\int_{\{y\in\mathbf{R}_+^n\}}\int_0^\infty f(x)g(y)\left[\int_0^\infty t^{\lambda+m-1}e^{-(x+||y||_\beta^\alpha)t}dt\right]dxdy \\
&= \frac{1}{\Gamma(\lambda+m)}\int_0^\infty t^{\lambda+m-1}\left(\int_0^\infty e^{-xt}f(x)dx\right) \\
&\quad\times\left(\int_{\{y\in\mathbf{R}_+^n\}} e^{-||y||_\beta^\alpha t}g(y)dy\right)dt \\
&= \frac{1}{\Gamma(\lambda+m)}\int_0^\infty t^{\lambda+m-1}\left(t^{-m}\int_0^\infty e^{-xt}f^{(m)}(x)dx\right) \\
&\quad\times\left(\int_{\{y\in\mathbf{R}_+^n\}} e^{-||y||_\beta^\alpha t}g(y)dy\right)dt \\
&= \frac{1}{\Gamma(\lambda+m)}\int_{\{y\in\mathbf{R}_+^n\}}\int_0^\infty f^{(m)}(x)g(y)\left[\int_0^\infty t^{\lambda-1}e^{-(x+||y||_\beta^\alpha)t}dt\right]dxdy \\
&= \frac{\Gamma(\lambda)}{\Gamma(\lambda+m)}\int_{\{y\in\mathbf{R}_+^n\}}\int_0^\infty \frac{f^{(m)}(x)g(y)}{(x+||y||_\beta^\alpha)^\lambda}dxdy.
\end{aligned}$$

Then, by (2.10), we have (3.1). For $\lambda = \lambda_1 + \lambda_2$ in (3.1), we have (3.2).

The theorem is proved. □

Theorem 2. *With regard to the assumption H1, if* $\lambda_1+\lambda_2=\lambda$, *then the constant factor*

$$\frac{\Gamma(\lambda)}{\Gamma(\lambda+m)}\left(\frac{\Gamma^n(\frac{1}{\beta})}{\alpha\beta^{n-1}\Gamma(\frac{n}{\beta})}\right)^{\frac{1}{p}} B^{\frac{1}{p}}(\lambda_2,\lambda-\lambda_2)B^{\frac{1}{q}}(\lambda_1,\lambda-\lambda_1)$$

in (3.1) *is the best possible.*

Proof. We need to prove that the constant factor in (3.2) is the best possible. For any $0<\varepsilon<|p|\lambda_2$, and $m\in\mathbf{N}$, we set

$$\widetilde{f}^{(m)}(x):=\begin{cases}0, & 0<x<1,\\ \prod_{i=0}^{m-1}(\lambda_1+i-\frac{\varepsilon}{p})x^{\lambda_1-\frac{\varepsilon}{p}-1}, & x\ge 1\end{cases},$$

$$\widetilde{g}(y):=\begin{cases}0, & \|y\|_\beta<1,\\ \|y\|_\beta^{\alpha(\lambda_2-\frac{\varepsilon}{q})-n}, & \|y\|_\beta\ge 1\end{cases},$$

$$\widetilde{f}(x):=\begin{cases}0, & 0<x<1,\\ \prod_{i=0}^{m-1}(\lambda_1+i-\frac{\varepsilon}{p}) & \\ \int_1^x(\int_1^{t_m}\cdots\int_1^{t_2}t_1^{\lambda_1-\frac{\varepsilon}{p}-1}dt_1\cdots dt_{m-1})dt_m, & x\ge 1\end{cases},$$

namely,

$$\widetilde{f}(x)=\begin{cases}0, & 0<x<1,\\ x^{\lambda_1+m-\frac{\varepsilon}{p}-1}-p_{m-1}(x), & x\ge 1\end{cases},$$

where $p_{m-1}(x)$ is indicated as a polynomial of $(m-1)$-order with all the positive coefficients, satisfying

$$p_{m-1}^{(i)}(1)=(x^{\lambda_1+m-\frac{\varepsilon}{p}-1})^{(i)}|_{x=1}(i=0,\ldots,m-1).$$

We denote $p_{-1}(1):=0,\prod_{i=0}^{-1}(a+i):=1$ $(a>0)$. Then, the above expressions satisfy for $m\in\mathbf{N}_0$.

If there exists a constant M, with

$$M\ge\frac{\Gamma(\lambda)}{\Gamma(\lambda+m)}\left(\frac{\Gamma^n(\frac{1}{\beta})}{\alpha\beta^{n-1}\Gamma(\frac{n}{\beta})}\right)^{\frac{1}{p}}B(\lambda_1,\lambda_2) \tag{3.4}$$

such that (3.2) is valid when we replace the constant factor by M, then in particular, by (2.4) (for $c = \varepsilon$), we have

$$\begin{aligned}
\widetilde{I} &:= \int_{\{y\in\mathbf{R}_+^n\}}\int_0^\infty \frac{\widetilde{f}(x)\widetilde{g}(y)}{(x+||y||_\beta^\alpha)^{\lambda+m}}dxdy \\
&> M\left[\int_0^\infty x^{p(1-\lambda_1)-1}(\widetilde{f}^{(m)}(x))^p dx\right]^{\frac{1}{p}} \\
&\quad\times\left[\int_{\{y\in\mathbf{R}_+^n\}} ||y||_\beta^{q(n-\alpha\lambda_2)-n}\widetilde{g}^q(y)dy\right]^{\frac{1}{q}} \\
&= M\prod_{i=0}^{m-1}\left(\lambda_1+i-\frac{\varepsilon}{p}\right)\left(\int_1^\infty x^{-1-\varepsilon}dx\right)^{\frac{1}{p}} \\
&\quad\times\left(\int_{\{y\in\mathbf{R}_+^n;||y||_\beta\geq 1\}} ||y||_\beta^{-\alpha\varepsilon-n}dy\right)^{\frac{1}{q}} \\
&= \frac{M}{\varepsilon}\prod_{i=0}^{m-1}\left(\lambda_1+i-\frac{\varepsilon}{p}\right)\left(\frac{\Gamma^n(\frac{1}{\beta})}{\alpha\beta^{n-1}\Gamma(\frac{n}{\beta})}\right)^{\frac{1}{q}}.
\end{aligned}$$

By (2.4), (2.5) and (2.7) (for $s = \lambda + m > 0, s_1 = \lambda_1 - \frac{\varepsilon}{p} + m \in (0, \lambda + m)$), we have

$$\begin{aligned}
\widetilde{I} &= \int_{\{y\in\mathbf{R}_+^n;||y||_\beta\geq 1\}}\int_1^\infty \frac{(x^{\lambda_1+m-\frac{\varepsilon}{p}-1}-p_{m-1}(x))||y||_\beta^{\alpha(\lambda_2-\frac{\varepsilon}{q})-n}}{(x+||y||_\beta^\alpha)^{\lambda+m}}dxdy \\
&\leq \int_{\{y\in\mathbf{R}_+^n;||y||_\beta\geq 1\}} ||y||_\beta^{-\alpha\varepsilon-n}\left[||y||_\beta^{\alpha(\lambda_2+\frac{\varepsilon}{p})}\int_0^\infty \frac{x^{\lambda_1+m-\frac{\varepsilon}{p}-1}}{(x+||y||_\beta^\alpha)^{\lambda+m}}dx\right]dy \\
&= \int_{\{y\in\mathbf{R}_+^n;||y||_\beta\geq 1\}} ||y||_\beta^{-\alpha\varepsilon-n}\omega_{\lambda+m}\left(\lambda_1-\frac{\varepsilon}{p}+m,y\right)dy \\
&= B\left(\lambda_1-\frac{\varepsilon}{p}+m,\lambda_2+\frac{\varepsilon}{p}\right)\int_{\{y\in\mathbf{R}_+^n;||y||_\beta\geq 1\}} ||y||_\beta^{-\alpha\varepsilon-n}dy \\
&= B\left(\lambda_1-\frac{\varepsilon}{p}+m,\lambda_2+\frac{\varepsilon}{p}\right)\frac{\Gamma^n(\frac{1}{\beta})}{\varepsilon\alpha\beta^{n-1}\Gamma(\frac{n}{\beta})}.
\end{aligned}$$

Based on the above results, we have the following inequality:

$$B\left(\lambda_1 - \frac{\varepsilon}{p} + m, \lambda_2 + \frac{\varepsilon}{p}\right) \frac{\Gamma^n(\frac{1}{\beta})}{\alpha\beta^{n-1}\Gamma(\frac{n}{\beta})}$$

$$\geq \varepsilon\widetilde{I} > M \prod_{i=0}^{m-1} \left(\lambda_1 + i - \frac{\varepsilon}{p}\right) \left(\frac{\Gamma^n(\frac{1}{\beta})}{\alpha\beta^{n-1}\Gamma(\frac{n}{\beta})}\right)^{\frac{1}{q}}.$$

For $\varepsilon \to 0^+$, in view of the continuity of the Beta function, we have

$$\frac{B(\lambda_1 + m, \lambda_2)\Gamma^n(\frac{1}{\beta})}{\alpha\beta^{n-1}\Gamma(\frac{n}{\beta})} \geq M \prod_{i=0}^{m-1} (\lambda_1 + i) \left(\frac{\Gamma^n(\frac{1}{\beta})}{\alpha\beta^{n-1}\Gamma(\frac{n}{\beta})}\right)^{\frac{1}{q}},$$

namely,

$$\frac{\Gamma(\lambda)}{\Gamma(\lambda + m)} \left(\frac{\Gamma^n(\frac{1}{\beta})}{\alpha\beta^{n-1}\Gamma(\frac{n}{\beta})}\right)^{\frac{1}{p}} B(\lambda_1, \lambda_2)$$

$$= \frac{B(\lambda_1 + m, \lambda_2)}{\prod_{i=0}^{m-1} (\lambda_1 + i)} \left(\frac{\Gamma^n(\frac{1}{\beta})}{\alpha\beta^{n-1}\Gamma(\frac{n}{\beta})}\right)^{\frac{1}{p}} \geq M.$$

By (3.4), it follows that

$$M = \frac{\Gamma(\lambda)}{\Gamma(\lambda + m)} \left(\frac{\Gamma^n(\frac{1}{\beta})}{\alpha\beta^{n-1}\Gamma(\frac{n}{\beta})}\right)^{\frac{1}{p}} B(\lambda_1, \lambda_2)$$

is the best possible constant factor of (3.2) (namely, for $\lambda_1 + \lambda_2 = \lambda$ in (3.1)).

The theorem is proved. □

Theorem 3. *With regard to the assumption H1, if the constant factor in* (3.1) *is the best possible, then for*

$$-q\lambda_2 < \lambda - \lambda_1 - \lambda_2 < q(\lambda - \lambda_2),$$

we have $\lambda_1 + \lambda_2 = \lambda$.

Proof. For $\widehat{\lambda}_1 = \frac{\lambda-\lambda_2}{p} + \frac{\lambda_1}{q}, \widehat{\lambda}_2 = \frac{\lambda-\lambda_1-\lambda_2}{q} + \lambda_2$, we find $\widehat{\lambda}_1 + \widehat{\lambda}_2 = \lambda$. For $-q\lambda_2 < \lambda - \lambda_1 - \lambda_2 < q(\lambda - \lambda_2)$, we obtain $0 < \widehat{\lambda}_2 < \lambda$, and then $0 < \widehat{\lambda}_1 = \lambda - \widehat{\lambda}_2 < \lambda$.

By the reverse Hölder's inequality (cf. Ref. [28]), we obtain

$$\begin{aligned}
B(\widehat{\lambda}_1, \widehat{\lambda}_2) &= \int_0^\infty \frac{u^{\widehat{\lambda}_1 - 1}}{(1+u)^\lambda} du = \int_0^\infty \frac{u^{\frac{\lambda-\lambda_2}{p} + \frac{\lambda_1}{q} - 1}}{(1+u)^\lambda} du \\
&= \int_0^\infty \frac{1}{(1+u)^\lambda} (u^{\frac{\lambda-\lambda_2-1}{p}})(u^{\frac{\lambda_1-1}{q}}) du \\
&\geq \left[\int_0^\infty \frac{u^{\lambda-\lambda_2-1}}{(1+u)^\lambda} du\right]^{\frac{1}{p}} \left[\int_0^\infty \frac{u^{\lambda_1-1}}{(1+u)^\lambda} du\right]^{\frac{1}{q}} \\
&= B^{\frac{1}{p}}(\lambda_2, \lambda - \lambda_2) B^{\frac{1}{q}}(\lambda_1, \lambda - \lambda_1). \qquad (3.5)
\end{aligned}$$

Since the constant factor

$$\frac{\Gamma(\lambda)}{\Gamma(\lambda+m)} \left(\frac{\Gamma^n(\frac{1}{\beta})}{\alpha\beta^{n-1}\Gamma(\frac{n}{\beta})}\right)^{\frac{1}{p}} B^{\frac{1}{p}}(\lambda_2, \lambda - \lambda_2) B^{\frac{1}{q}}(\lambda_1, \lambda - \lambda_1)$$

in (3.1) is the best possible, comparing with the constant factors in (3.1) and (3.2) (for $\lambda_1 = \widehat{\lambda}_1, \lambda_2 = \widehat{\lambda}_2$), we have

$$\begin{aligned}
&\frac{\Gamma(\lambda)}{\Gamma(\lambda+m)} \left(\frac{\Gamma^n(\frac{1}{\beta})}{\alpha\beta^{n-1}\Gamma(\frac{n}{\beta})}\right)^{\frac{1}{p}} B^{\frac{1}{p}}(\lambda_2, \lambda - \lambda_2) B^{\frac{1}{q}}(\lambda_1, \lambda - \lambda_1) \\
&\geq \frac{\Gamma(\lambda)}{\Gamma(\lambda+m)} \left(\frac{\Gamma^n(\frac{1}{\beta})}{\alpha\beta^{n-1}\Gamma(\frac{n}{\beta})}\right)^{\frac{1}{p}} B(\widehat{\lambda}_1, \widehat{\lambda}_2),
\end{aligned}$$

namely, $B(\widehat{\lambda}_1, \widehat{\lambda}_2) \leq B^{\frac{1}{p}}(\lambda_2, \lambda - \lambda_2) B^{\frac{1}{q}}(\lambda_1, \lambda - \lambda_1)$. Hence, (3.5) keeps the form of equality. The necessary and sufficient condition for taking an equal sign is that there exist constants A and B, such that they are not both zero, and (cf. Ref. [28]) $Au^{\lambda-\lambda_2-1} = Bu^{\lambda_1-1}$ a.e. in $\mathbf{R}_+$. Assuming that $A \neq 0$, we have $u^{\lambda-\lambda_2-\lambda_1} = \frac{B}{A}$ a.e. in $\mathbf{R}_+$. It follows that $\lambda - \lambda_2 - \lambda_1 = 0$, and then $\lambda_1 + \lambda_2 = \lambda$.

The theorem is proved. □

Corollary 1. *For $n = 1$ in (3.1), we have the following reverse inequality:*

$$\int_0^\infty \int_0^\infty \frac{f(x)g(y)}{(x+y^\alpha)^{\lambda+m}}dxdy$$
$$> \frac{\Gamma(\lambda)}{\Gamma(\lambda+m)}\left(\frac{1}{\alpha}\right)^{\frac{1}{p}} B^{\frac{1}{p}}(\lambda_2, \lambda-\lambda_2)B^{\frac{1}{q}}(\lambda_1, \lambda-\lambda_1)$$
$$\times \left[\int_0^\infty x^{p(1-\widehat{\lambda}_1)-1}(f^{(m)}(x))^p dx\right]^{\frac{1}{p}} \left[\int_0^\infty y^{q(1-\alpha\widehat{\lambda}_2)-1} g^q(y)dy\right]^{\frac{1}{q}}. \tag{3.6}$$

In particular, for $\lambda_1 + \lambda_2 = \lambda$, we have

$$\int_0^\infty \int_0^\infty \frac{f(x)g(y)}{(x+y^\alpha)^{\lambda+m}}dxdy > \frac{\Gamma(\lambda)}{\Gamma(\lambda+m)}\left(\frac{1}{\alpha}\right)^{\frac{1}{p}} B(\lambda_1, \lambda_2)$$
$$\times \left[\int_0^\infty x^{p(1-\lambda_1)-1}(f^{(m)}(x))^p dx\right]^{\frac{1}{p}} \left[\int_0^\infty y^{q(1-\alpha\lambda_2)-1} g^q(y)dy\right]^{\frac{1}{q}}. \tag{3.7}$$

Corollary 2. *If $\lambda_1 + \lambda_2 = \lambda$, then the constant factor*

$$\frac{\Gamma(\lambda)}{\Gamma(\lambda+m)}\left(\frac{1}{\alpha}\right)^{\frac{1}{p}} B^{\frac{1}{p}}(\lambda_2, \lambda-\lambda_2)B^{\frac{1}{q}}(\lambda_1, \lambda-\lambda_1)$$

in (3.6) is the best possible. On the other hand, if the same constant factor in (3.6) is the best possible, then for $-q\lambda_2 < \lambda - \lambda_1 - \lambda_2 < q(\lambda-\lambda_2)$, we have $\lambda_1 + \lambda_2 = \lambda$.

Acknowledgment

This work is supported by the National Science Foundation of China (No. 61772140).

References

[1] G. H. Hardy, J. E. Littlewood and G. Polya. *Inequalities.* Cambridge University Press, Cambridge, 1934.

[2] M. Krnić and J. Pečarić. Extension of Hilbert's inequality. *J. Math. Anal. Appl.*, 324(1), 2006, 150–160.

[3] V. Adiyasuren, T. Batbold and E. A. Azar. A new discrete Hilbert-type inequality involving partial sums. *J. Inequal. Appl.*, 2019, 2019, 127.

[4] H. M. Mo and B. C. Yang. On a new Hilbert-type integral involving the upper limit functions. *J. Inequal. Appl.*, 2020, 2020, 5.

[5] B. C. Yang. *The Norm of Operator and Hilbert-Type Inequality.* Science Press, Beijing, 2009.

[6] M. Krnić and J. Pečarić. General Hilbert's and Hardy's inequalities. *Math. Inequal. Appl.*, 8(1), 2005, 29–51.

[7] I. Perić and P. Vuković. Multiple Hilbert's type inequalities with a homogeneous kernel. *Banach J. Math. Anal.*, 5(2), 2011, 33–43.

[8] Q. L. Huang. A new extension of Hardy-Hilbert-type inequality. *J. Inequal. Appl.*, 2015, 2015, 397.

[9] B. He. A multiple Hilbert-type discrete inequality with a new kernel and best possible constant factor. *J. Math. Anal. Appl.*, 431, 2015, 902–990.

[10] J. S. Xu. Hardy-Hilbert's inequalities with two parameters. *Adv. Math.*, 36(2), 2007, 63–76.

[11] Z. T. Xie, Z. Zeng and Y. F. Sun. A new Hilbert-type inequality with the homogeneous kernel of degree-2. *Adv. Appl. Math. Sci.*, 12(7), 2013, 391–401.

[12] Z. Zeng, K. R. R. Gandhi and Z. T. Xie. A new Hilbert-type inequality with the homogeneous kernel of degree-2 and with the integral. *Bull. Math. Sci. Appl.*, 3(1), 2014, 11–20.

[13] D. M. Xin. A Hilbert-type integral inequality with the homogeneous kernel of zero degree. *Math. Theory Appl.*, 30(2), 2010, 70–74.

[14] L. E. Azar. The connection between Hilbert and Hardy inequalities. *J. Inequal. Appl.*, 2013, 2013, 452.

[15] V. Adiyasuren, T. Batbold and M. Krnić. Hilbert–type inequalities involving differential operators, the best constants and applications. *Math. Inequal. Appl.*, 18, 2015, 111–124.

[16] Y. Hong and Y. M. Wen. A necessary and sufficient condition of that Hilbert type series inequality with homogeneous kernel has the best constant factor. *Ann. Math.*, 37A(3), 2016, 329–336.

[17] Y. Hong. On the structure character of Hilbert's type integral inequality with homogeneous kernel and applications. *J. Jilin Univ. Sci. Ed.*, 55(2), 2017, 189–194.

[18] Y. Hong, Q. L. Huang, B. C. Yang and J. Q. Liao. The necessary and sufficient conditions for the existence of a kind of Hilbert-type multiple integral inequality with the non-homogeneous kernel and its applications. *J. Inequal. Appl.*, 2017, 2017, 316.

[19] Q. Chen, B. He, Y. Hong and Z. Li. Equivalent parameter conditions for the validity of half-discrete Hilbert-type multiple integral inequality with generalized homogeneous kernel. *J. Funct. Spaces*, 2020, 2020, 7414861.

[20] B. He, Y. Hong and Z. Li. Conditions for the validity of a class of optimal Hilbert type multiple integral inequalities with non-homogeneous. *J. Inequal. Appl.*, 2021, 2021, 64.

[21] Y. Hong, Y. R. Zhong and B. C. Yang. On a more accurate half-discrete multidimensional Hilbert-type inequality involving one derivative function of m-order. *J. Inequal. Appl.*, 2023, 2023, 74.

[22] R. C. Luo, B. C. Yang and L. P. He. A Hardy-Hilbert-type integral inequality involving two multiple upper-limit functions. *J. Inequal. Appl.*, 2023, 2023, 19.

[23] M. T. Rassias and B. C. Yang. A multidimensional half-discrete Hilbert-type inequality and the Riemann zeta function. *Appl. Math. Comput.*, 225, 2013, 263–277.

[24] M. T. Rassias and B. C. Yang. On a multidimensional half C discrete Hilbert C type inequality related to the hyperbolic cotangent function. *Appl. Math. Comput.*, 242, 2014, 800–813.

[25] M. T. Rassias and B. C. Yang. On a multidimensional Hilbert-type integral inequality associated to the gamma function. *Appl. Math. Comput.*, 249, 2014, 408–418.

[26] Y. Hong, Y. R. Zhong and B. C. Yang. A more accurate halif-discrete multidimensional Hilbert-type inequality involving one multiple upper limit function. *J. Inequal. Appl.*, 2023, 2023, 74.

[27] Y. Hong, Y. R. Zhong and B. C. Yang. A more accurate half-discrete multidimensional Hilbert-type inequality involving one multiple upper limit function. *Axioms*, 12, 2023, 211. https://doi.org/10.3390/axioms12020211.

[28] J. Q. Liao and B. C. Yang. A new reverse half-discrete Hilbert-type inequality with one partial sum involving one derivative function of higher order. *Open Math.*, 21, 2023, 20230139.

[29] A. Z. Wang and B. C. Yang. A new half-discrete Hilbert-type inequality involving one multiple upper limit function and one partial sums. *Ann. Math.*, 45A(1), 2024, 25–38.

[30] Z. H. Zeng and B. C. Yang. A Hardy-Hilbert-type integral inequality involving the derivative functions of n-order. *J. South China Normal Univ. (Nat. Sci. Ed.)*, 56(4), 2024, 123–128.

[31] J. C. Kuang. *Applied Inequalities.* Shandong Science and Technology Press, Jinan, 2004.

[32] J. C. Kuang. *Real and Functional Analysis (Continuation)*, Vol. 2. Higher Education Press, Beijing, 2015.

https://doi.org/10.1142/9789819820498_0017

Chapter 17

Oscillation and Non-Oscillation Criteria of Second-Order Functional Differential Equations

Abhay Kumar Sethi[*,§], **Choonkil Park**[†,¶] **and Michael Th. Rassias**[‡,‖]

[*] *Department of Mathematics, Sambalpur University, Sambalpur-768019, India*

[†] *Department of Mathematics, Hanyang University, Seoul 04763, Korea*

[‡] *Department of Mathematics and Engineering Sciences, Hellenic Military Academy, 16673 Vari Attikis, Greece*

[§] *sethiabhaykumar100@gmail.com*

[¶] *baak@hanyang.ac.kr*

[‖] *mthrassias@yahoo.com*

In this work, we establish the necessary conditions for oscillation of the second-order neutral delay differential equations of the form

$$(r(t)((x(t)+p(t)x(\tau(t)))')^{\gamma})' + q(t)x^{\alpha}(\sigma(t)) + v(t)x^{\beta}(\eta(t)) = 0$$

under the assumption that

$$\int_0^{\infty} \left(\frac{1}{r(t)}\right)^{\frac{1}{\gamma}} dt = \infty$$

or

$$\int_0^{\infty} \left(\frac{1}{r(t)}\right)^{\frac{1}{\gamma}} dt < \infty,$$

for various ranges of $p(t)$, where γ is a quotient of odd positive integers.

1. Introduction

Consider the nonlinear neutral delay differential equation of the form

$$(r(t)((x(t)+p(t)x(\tau(t)))')^{\gamma})'+q(t)x^{\alpha}(\sigma(t))+v(t)x^{\beta}(\eta(t)) = 0, \quad (1.1)$$

where γ, α, β are quotient of odd positive integers and r, q, v, τ, $\sigma, \eta \in C(\mathbb{R}_+, \mathbb{R}_+)$, $p \in C(\mathbb{R}_+, \mathbb{R})$, $\tau(t) \leq t$, $\sigma(t) \leq t$, $\eta(t) \leq t$ with $\lim_{t\to\infty} \tau(t) = \infty = \lim_{t\to\infty} \sigma(t) = \infty = \lim_{t\to\infty} \eta(t)$.

In this work, our objective is to establish the sufficient condition results for oscillation of all solutions of (1.1) without the comparison results as well as without Riccati transformation technique under the assumption

$$(H_0) \int_0^{\infty} \frac{ds}{r(s)} = \infty$$

or

$$(H_{00}) \int_0^{\infty} \frac{ds}{r(s)} < \infty,$$

for all ranges of $p(t)$ with $|p(t)| < \infty$.

Li [15] established the oscillation of second-order neutral delay differential equations with integrable coefficients of the form

$$[a(t)(x(t) + p(t)x(t - \tau))']' + q(t)f(x(t - \sigma)) = 0,$$

where $t \geq t_0$, τ and σ are nonnegative constants, $a, p, q \in C([t_0, \infty), \mathbb{R})$, $f \in C(\mathbb{R}, \mathbb{R})$, under the following assumptions:

(a) $0 \leq p(t) \leq 1,\ q(t) > 0,\ a(t) > 0$;
(b) $\int^{\infty} \frac{dt}{a(t)} = \infty$;
(c) $\frac{f(x)}{x} \geq \gamma > 0,$ for $x \neq 0,$ γ is a constant.

Li [15] studied some necessary conditions for the non-oscillation of solutions of the second-order delay differential equations.

Baculikova and Dzurina [6] studied the oscillation the oscillatory behavior of solutions of

$$(a(t)[z'(t)]^{\gamma} + q(t)x^{\beta}(\sigma(t))) = 0,$$

where $z(t) = x(t) + p(t)x(\tau(t))$:

(a) γ, β are the ratio of two odd positive integers;
(b) $\sigma(t) \leq t$, $\sigma(t)$ non-decreasing;
(c) a, p, q, are positive $o \leq p(t) \leq p_0 < \infty$, $\lim_{t\to\infty} \tau(t) = \infty = \lim_{t\to\infty} \sigma(t)$, $\tau'(t) \geq \tau_0 > 0$ and $\tau o \sigma = \sigma o \tau$;

and they established the sufficient conditions for oscillation by means of comparison results.

Li *et al.* [17] obtained a class of nonlinear second-order neutral differential equations:

$$(r(t)((x(t) + p(t)x(t-\tau))')^{\gamma})' + q(t)f(x(t), x(\sigma(t)))) = 0,$$

where $t \geq t_0 > 0$, $\tau \geq 0$ and $\gamma > 1$ is a quotient of two odd positive integers, under the following assumptions:

(a) $r \in C^1([t_0, +\infty), (0, +\infty))$;
(b) $p, q \in C([t_0, +\infty), [0, +\infty))$ and $q(t)$ is not identically zero for large t;
(c) $f \in C(\mathbb{R}^2, \mathbb{R})$ and $\frac{f(x,y)}{y^{\gamma}} \geq K$ for all $y \neq 0$ and for some $K > 0$;
(d) $\sigma \in C'(t_0, +\infty), \mathbb{R})$, $\sigma(t) \leq t$, $\sigma'(t) > 0$ and $\lim_{t\to\infty} \sigma(t) = +\infty$.

They obtained the sufficient conditions for oscillation by means of Riccati transformation technique.

Tripathy *et al.* [22] studied the nonlinear neutral delay differential equations of the form

$$((r(t)((y(t) + p(t)y(\delta(t)))')^{\alpha})' + q(t)(y^{\beta}(\tau(t)) = 0, \tag{1.2}$$

where α, β are quotients of odd positive integers and r, q, δ, τ, $\in C(\mathbb{R}_+, \mathbb{R}_+)$, $p \in C(\mathbb{R}_+, \mathbb{R})$, $\tau(t) \leq t$, $\delta(t) \leq t$, with $\lim_{t\to\infty} \delta(t) = \infty = \lim_{t\to\infty} \tau(t) = \infty$, and they established the sufficient conditions for oscillation without comparison results. Clearly, (1.2) is a special case of (1.1). However, there are few results to study the oscillation of (1.1). The purpose of this work is to obtain some sufficient condition results for oscillation of (1.1). This work would be interesting than the works of Refs. [14, 16, 20] as long as (1.1) is concerned.

Neutral delay differential equations find numerous applications in electric network. For example, they are frequently used for the study of distributed networks containing lossless transmission lines which arises in high-speed computers where the lossless transmission lines are used to interconnect switching circuits (see Ref. [13]). The problem of obtaining sufficient conditions to ensure the second-order differential equations which are special cases of (1.1) are oscillatory has received a great attention. Since the second-order equations have the applied applications, there is permanent interest in obtaining new sufficient conditions for oscillation or non-oscillation of solutions of varietal type of the second-order equations (see Refs. [1–5, 7–12, 18, 19, 21, 23]).

Definition 1. By a solution of (1.1), we mean a continuously differentiable function $x(t)$, which is defined for $t \geq T^* = \min\{\tau(t_0),\ \sigma(t_0),\ \eta(t_0)\}$ such that $x(t)$ satisfies (1.1) for all $t \geq t_0$. In the sequel, it will always be assumed that the solutions of (1.1) exist on some half line $[t_1, \infty)$, $t_1 \geq t_0$. A solution of (1.1) is said to be oscillatory if it has arbitrarily large zeros, otherwise, it is called non-oscillatory. Equation (1.1) is called oscillatory if all its solutions are oscillatory.

2. Main Results

In this section, we deal with the sufficient conditions for oscillation of all solutions of (1.1). Throughout our discussion, we use the notation

$$z(t) = x(t) + p(t)x(\tau(t)). \tag{2.1}$$

Lemma 1 ([6]). *Let $p(t) \geq 0$. Assume that (H_0) holds. If $x(t)$ is an eventually positive solution of (1.1), then the corresponding function $z(t) = x(t) + p(t)x(\tau(t))$ satisfies $z(t) > 0$, $z'(t) > 0$, $(r(t)(z'(t))^\gamma)' \leq 0$, for any large value of t, where γ is a quotient of odd positive integers.*

Lemma 2 ([6]). *Assume that $A \geq 0, B \geq 0$ and $\lambda \geq 1$. Then,*

$$(A + B)^\lambda \leq 2^{\lambda-1}(A^\lambda + B^\lambda). \tag{2.2}$$

If $0 < \lambda \leq 1$, *then*

$$(A+B)^{\lambda} \leq (A^{\lambda} + B^{\lambda}).$$

Theorem 1. *Let* $0 \leq p(t) \leq a < \infty$, $\tau(\sigma(t)) = \sigma(\tau(t))$ *and* $\tau(\eta(t)) = \eta(\tau(t))$ *for all* $t > 0$ *and* $\gamma = \alpha = \beta$ *and* (H_0) *hold. Assume that*

$$(H_{1\beta}) \quad \int_{t_0}^{\infty} 2^{1-\alpha} C^{\beta}[Q(t) + V(t)]dt = \infty,$$

where $Q(t) = \min\{q(t), q(\tau(t))\}, V(t) = \min\{v(t), v(\tau(t))\}$. *Then, every solution of* (1.1) *oscillates.*

Proof. Suppose on the contrary that $x(t)$ is a non-oscillatory solution of (1.1). Without loss of generality, we may assume that $x(t) > 0$ for $t \geq t_0$. Hence, there exists $t_1 > t_0$ such that $x(t) > 0$, $x(\tau(t)) > 0$, $x(\sigma(t)) > 0$ and $x(\eta(t)) > 0$ for $t \geq t_1$. By (2.1), (1.1) becomes

$$(r(t)(z'(t))^{\gamma})' = -q(t)x^{\alpha}(\sigma(t)) - v(t)x^{\beta}(\eta(t)) \leq 0, \not\equiv 0 \quad \text{for } t \geq t_1. \tag{2.3}$$

So, $r(t)(z'(t))^{\gamma}$ is non-increasing on $[t_1, \infty)$, that is, either $z'(t) > 0$ or $z'(t) < 0$. From Lemma 1, it follows that $z'(t) > 0$ for $t \geq t_2$. By (1.1), it is easy to see that

$$\begin{aligned}
&(r(t)(z'(t))^{\gamma})' + a^{\alpha}(r(\tau(t))(z'(\tau(t))^{\gamma})' + q(t)x^{\alpha}(\sigma(t)) \\
&\quad + a^{\alpha}q(\tau(t))x^{\alpha}(\sigma(\tau(t)) + v(t)x^{\beta}(\eta(t)) \\
&\quad + a^{\alpha}v(\tau(t))x^{\beta}(\eta(\tau(t)) = 0, \quad \text{for } t \geq t_1. \ (\because \gamma = \alpha = \beta).
\end{aligned} \tag{2.4}$$

By Lemma 2, (2.4) implies that

$$\begin{aligned}
&(r(t)(z'(t))^{\gamma})' + a^{\alpha}(r(\tau(t))(z'(\tau(t))^{\gamma})' + 2^{1-\alpha}Q(t)z^{\beta}(\sigma(t)) \\
&\quad + 2^{1-\alpha}V(t)z^{\beta}(\eta(t)) \leq 0.
\end{aligned} \tag{2.5}$$

From Lemma 1, it follows that $z(t)$ is non-decreasing on $[t_1, \infty)$, and hence, $z(t) \geq C$ for $t \geq t_2 > t_1$. Hence, (2.5) implies that

$$\begin{aligned}
&(r(t)(z'(t))^{\gamma})' + a^{\alpha}(r(\tau(t))(z'(\tau(t))^{\gamma})' \\
&\quad + 2^{1-\alpha}Q(t)C^{\beta} + 2^{1-\alpha}V(t)C^{\beta} \leq 0.
\end{aligned}$$

Hence, there exists $t_3 > t_2$ such that

$$2^{1-\alpha}Q(t)C^{\beta} + 2^{1-\alpha}V(t)C^{\beta} \le -(r(t)(z'(t))^{\gamma})' - a^{\alpha}(r(\tau(t))(z'(\tau(t))^{\gamma})', \tag{2.6}$$

for $t \ge t_3$. Integrating (2.6) from t_3 to $+\infty$, we get a contradiction to $(H_{1\beta})$. This completes the proof of the theorem. □

Theorem 2. *Let* $0 \le p(t) \le a < \infty$, $\tau(\sigma(t)) = \sigma(\tau(t))$ *and* $\tau(\eta(t)) = \eta(\tau(t))$ *for all* $t > 0$ *and* $\gamma > \alpha > \beta$. *If* (H_0) *and* $(H_{1\beta})$ *hold, then every solution of* (1.1) *oscillates.*

Proof. Suppose on the contrary that $x(t)$ is a non-oscillatory solution of (1.1). Without loss of generality, we may assume that $x(t) > 0$ for $t \ge t_0$. Hence, there exists $t_1 > t_0$ such that $x(t) > 0, x(\tau(t)) > 0$, $x(\sigma(t)) > 0$ and $x(\eta(t)) > 0$ for $t \ge t_1$. Using (2.1), (1.1) becomes

$$(r(t)(z'(t))^{\gamma})' = -q(t)x^{\alpha}(\sigma(t)) - v(t)x^{\beta}(\eta(t)) \le 0, \not\equiv 0 \quad \text{for } t \ge t_1.$$

So, $r(t)(z'(t))^{\gamma}$ is non-increasing on $[t_1, \infty)$, that is, either $z'(t) > 0$ or $z'(t) < 0$. From Lemma 1, it follows that $z'(t) > 0$ for $t \ge t_2$. By (1.1), it is easy to see that

$$\begin{aligned}&(r(t)(z'(t))^{\alpha})' + a^{\alpha}(r(\tau(t))(z'(\tau(t))^{\alpha})' + q(t)x^{\alpha}(\sigma(t))\\&+ a^{\alpha}q(\tau(t))x^{\alpha}(\sigma(\tau(t)) + v(t)x^{\beta}(\eta(t))\\&+ a^{\alpha}v(\tau(t))x^{\beta}(\eta(\tau(t)) = 0, \quad \text{for } t \ge t_1.\ (\because \gamma > \alpha > \beta).\end{aligned} \tag{2.7}$$

By Lemma 2, (2.7) implies that

$$\begin{aligned}&(r(t)(z'(t))^{\alpha})' + a^{\beta}(r(\tau(t))(z'(\tau(t))^{\alpha})' + 2^{1-\alpha}Q(t)z^{\beta}(\sigma(t))\\&+ 2^{1-\alpha}V(t)z^{\beta}(\eta(t)) \le 0.\end{aligned}$$

The rest of the proof follows from Theorem 1. Hence, the details will be omitted. □

Theorem 3. *Let* $0 \le p(t) \le a < \infty$, $\tau(\sigma(t)) = \sigma(\tau(t))$ *and* $\tau(\eta(t)) = \eta(\tau(t))$ *for all* $t > 0$ *and* $\gamma < \alpha < \beta$. *Assume that* (H_0) *holds and*

$$(H_{1\alpha}) \int_{t_0}^{\infty} 2^{1-\beta}C^{\alpha}[Q(t) + V(t)]dt = \infty$$

holds. Then, every solution of (1.1) *oscillates.*

Proof. Suppose on the contrary that $x(t)$ is a non-oscillatory solution of (1.1). Without loss of generality, we may assume that $x(t) > 0$ for $t \geq t_0$. Hence, there exists $t_1 > t_0$ such that $x(t) > 0$, $x(\tau(t)) > 0$, $x(\sigma(t)) > 0$ and $x(\eta(t)) > 0$ for $t \geq t_1$. Using (2.1), (1.1) becomes

$$(r(t)(z'(t))^{\gamma})' = -q(t)x^{\alpha}(\sigma(t)) - v(t)x^{\beta}(\eta(t)) \leq 0, \not\equiv 0 \quad \text{for } t \geq t_1.$$

So, $r(t)(z'(t))^{\gamma}$ is non-increasing on $[t_1, \infty)$, that is, either $z'(t) > 0$ or $z'(t) < 0$. From Lemma 1, it follows that $z'(t) > 0$ for $t \geq t_2$. By (1.1), it is easy to see that

$$\begin{aligned}(r(t)(z'(t))^{\alpha})' &+ a^{\beta}(r(\tau(t))(z'(\tau(t))^{\alpha})' + q(t)x^{\alpha}(\sigma(t)) \\ &+ a^{\beta}q(\tau(t))x^{\alpha}(\sigma(\tau(t)) + v(t)x^{\beta}(\eta(t)) + a^{\beta}v(\tau(t))x^{\beta}(\eta(\tau(t)) = 0\end{aligned} \tag{2.8}$$

for all $t \geq t_1$, since $\gamma < \alpha < \beta$. By Lemma 2, (2.8) yields that

$$\begin{aligned}(r(t)(z'(t))^{\alpha})' &+ a^{\beta}(r(\tau(t))(z'(\tau(t))^{\alpha})' + 2^{1-\alpha}Q(t)z^{\alpha}(\sigma(t)) \\ &+ 2^{1-\beta}V(t)z^{\alpha}(\eta(t)) \leq 0.\end{aligned}$$

The rest of the proof follows from Theorem 2. Hence, the details will be omitted. □

Theorem 4. *Let* $0 \leq p(t) \leq a < \infty$, $\tau(\sigma(t)) = \sigma(\tau(t))$ *and* $\tau(\eta(t)) = \eta(\tau(t))$ *for all* $t > 0$ *and* $\gamma < \beta < \alpha$ *and* (H_0) *hold. If* $(H_{1\beta})$ *holds, then every solution of* (1.1) *oscillates.*

Proof. The proof is similar to the proof of Theorem 2. Hence, the details will be omitted. □

Theorem 5. *Let* $0 \leq p(t) \leq a < \infty$, $\tau(\sigma(t)) = \sigma(\tau(t))$ *and* $\tau(\eta(t)) = \eta(\tau(t))$ *for all* $t > 0$ *and* $\alpha > \gamma > \beta$. *If* (H_0) *and* $(H_{1\beta})$ *hold, then every solution of* (1.1) *oscillates.*

Proof. The proof is similar to the proof of Theorem 2. Hence, the details will be omitted. □

Theorem 6. *Let* $0 \leq p(t) \leq a < \infty$, $\tau(\sigma(t)) = \sigma(\tau(t))$ *and* $\tau(\eta(t)) = \eta(\tau(t))$ *for all* $t > 0$ *and* $\alpha < \beta < \gamma$. *If* (H_0) *and* $(H_{1\alpha})$ *hold, then every solution of* (1.1) *oscillates.*

Proof. The proof is similar to the proof of Theorem 3. Hence, the details will be omitted. □

Theorem 7. *Let* $0 \le p(t) \le a < \infty$ $\tau(\sigma(t)) = \sigma(\tau(t))$ *and* $\tau(\eta(t)) = \eta(\tau(t))$ *for all* $t > 0$ *and* $\alpha < \gamma < \beta$. *If* (H_0) *and* $(H_{1\alpha})$ *hold, then every solution of* (1.1) *oscillates.*

Proof. The proof is similar to the proof of Theorem 3. Hence, the details will be omitted. □

Theorem 8. *Let* $-1 < a \le p(t) \le 0,\ t \in [t_0, \infty)$ *and* $\gamma > \alpha > \beta$. *Assume that* (H_0) *holds,*

$$(H_{2\beta})\ \int_{t_0}^{\infty} [q(s) + v(s)]\, ds = \infty$$

and

$$(H_3)\ \int_{t_0}^{\infty} [q(\tau(s)) + v(\tau(s))] ds = \infty.$$

Then, every solution of (1.1) *oscillates or tends to zero as* $t \to \infty$.

Proof. By the same method as in the proof of Theorem 1, we get (2.3) for $t \in [t_2, \infty)$. Thus, $z(t)$ and $z'(t)$ are monotonic functions on $[t_2, \infty)$. If $z'(t) < 0$ for $t \ge t_2$, then $(r(t)(z'(t))^{\gamma})' \le 0$ implies that

$$r(t)(z'(t))^{\gamma} \le r(t_2)(z'(t_2))^{\gamma},$$

that is,

$$(z'(t))^{\gamma} \le \frac{r(t_2)(z'(t_2))^{\gamma}}{r(t)},$$

and hence,

$$z(t) \le z(t_2) + r(t_2)^{\frac{1}{\gamma}}(z'(t_2)) \int_{t_1}^{t} \frac{1}{r(s)^{\frac{1}{\gamma}}} ds \to -\infty \text{ as } t \to \infty,$$

which contradicts to the fact that $z(t) > 0$. Also, the above contradiction holds true if $z(t) < 0$. Indeed, $x(t)$ is bounded when $z(t) < 0$. Suppose this does not hold. Then, there exists a sequence of points

$\{l_n\}$ such that $l_n \to \infty$ as $n \to \infty$ and $x(l_n) = max\{x(t) : t_3 \leq t \leq l_n\}$. Since $\sigma(l_n) \leq l_n$, $x(\sigma(l_n)) \leq x(l_n)$ implies that

$$\begin{aligned} z(l_n) &\geq x(l_n) + ax(\sigma(l_n)) \\ &= (1+a)x(l_n) \ (\because 1 + a > 0) \\ &\to +\infty \quad as\ n \to \infty, \end{aligned}$$

which is absurd. Therefore, $x(t)$ is bounded, that is, $\lim_{t\to\infty} z(t)$ exists. Consequently, we consider the following two cases, viz.,

$$(i)\ \ z(t) > 0, \quad z'(t) > 0,$$
$$(ii)\ \ z(t) < 0, \quad z'(t) > 0.$$

Case (i). In this case, $z(t) \leq x(t)$ and $\lim_{t\to\infty} r(t)z'(t)$ exists. Therefore, (1.1) reduces to

$$(r(t)(z'(t))^{\gamma})' + q(t)z^{\beta}(\sigma(t)) + v(t)z^{\beta}(\eta(t)) \leq 0 \tag{2.9}$$

for $t \geq t_3 > t_2$ since $\gamma > \alpha > \beta$. By Lemma 2, we get $z(t) \geq C$. Hence, (2.9) yields that

$$\begin{aligned} &(r(t)(z'(t))^{\gamma})' + q(t)C^{\beta} + v(t)C^{\beta} \leq 0, \\ &C^{\beta}\left[q(t) + v(t)\right] \leq -(r(t)(z'(t))^{\gamma})'. \end{aligned} \tag{2.10}$$

Integrating (2.10), from t_3 to $+\infty$, we get a contradiction to $(H_{2\beta})$.

Case (ii). Let $\lim_{t\to\infty} z(t)$ exist. Let $\lim_{t\to\infty} z(t) = \rho$, $\rho \in (-\infty, 0]$. We assert that $\rho = 0$. If not, then there exist $l < 0$ and $t_3 > t_2$ such that $z(\sigma(t)) \leq z(t) < l$, $z(\eta(t)) \leq z(t) < l$ for $t \geq t_3$. From (2.1), it follows that $z(t) > ax(\tau(t))$ and hence $x(\tau(\sigma(t)) > \frac{1}{a}z(\sigma(t))$, that is, $x(\sigma(\tau(t)) > \left(\frac{l}{a}\right)$ for $t \geq t_3$. Also, $x(\eta(\tau(t)) > \left(\frac{l}{a}\right)$ for $t \geq t_3$. Since (1.1) can be written as

$$(r(\tau(t))(z'(\tau(t))^{\gamma})' + q(\tau(t))x^{\alpha}(\sigma(\tau(t)) + v(\tau(t))x^{\beta}(\eta(\tau(t)) = 0,$$

for $t \geq t_3$, we get

$$\begin{aligned} &(r(\tau(t))(z'(\tau(t))^{\gamma})' + \left(\frac{l}{a}\right)^{\beta} q(\tau(t)) \\ &\quad + \left(\frac{l}{a}\right)^{\beta} v(\tau(t)) \leq 0. \ (\because \gamma > \alpha > \beta). \end{aligned}$$

Consequently,

$$\begin{aligned}
&\left(\frac{l}{a}\right)^{\beta}\left[\int_{t_3}^{t} q(\tau(s)) + \int_{t_3}^{t} v(\tau(s))\right] ds \\
&\quad \le -\left[(r(\tau(t))(z'(\tau(t))^{\gamma})'\right]_{t_3}^{t} \\
&\quad < -r(\tau(t))(z'(\tau(t)))^{\gamma} < \infty \quad as\ t \to \infty,
\end{aligned}$$

which is a contradicts to (H_3). Hence, $\rho = 0$. By the same reasoning as above, it is easy to show that $x(t)$ is bounded. Therefore,

$$\begin{aligned}
0 \ge \lim_{t\to\infty} z(t) &= \limsup_{t\to\infty} z(t) \\
&\ge \limsup_{t\to\infty}(x(t) + ax(\tau(t))) \\
&\ge \limsup_{t\to\infty} x(t) + \liminf_{t\to\infty}(ax(\tau(t))) \\
&= \limsup_{t\to\infty} x(t) + a \limsup_{t\to\infty} x(\tau(t)) \\
&= (1+a) \limsup_{t\to\infty} x(t),
\end{aligned}$$

which implies that $\limsup_{t\to\infty} x(t) = 0$, that is, $\lim_{t\to\infty} x(t) = 0$.
This completes the proof of the theorem. □

Theorem 9. *Let $-1 < a \le p(t) \le 0$, $t \in [t_0, \infty)$. Assume that $\gamma < \alpha < \beta$. If (H_0), $(H_{2\alpha})$ and (H_3) hold, then every solution of* (1.1) *oscillates or tends to zero as $t \to \infty$.*

Proof. By the same reasoning as in the proof of Theorem 1, we get (2.3) for $t \in [t_2, \infty)$. Thus, $z(t)$ and $z'(t)$ are monotonic functions on $[t_2, \infty)$. In what follows, we consider the following four possible cases:

$$\begin{aligned}
&(i)\ \ z(t) > 0, \quad z'(t) > 0, \\
&(ii)\ \ z(t) < 0, \quad z'(t) > 0, \\
&(iii)\ \ z(t) > 0, \quad z'(t) < 0, \\
&(iv)\ \ z(t) < 0, \quad z'(t) < 0.
\end{aligned}$$

Case (i). In this case, $z(t) \leq x(t)$ and $\lim_{t\to\infty} r(t)z'(t)$ exists. Therefore, (1.1) reduces to

$$(r(t)(z'(t))^{\gamma})' + q(t)z^{\alpha}(\sigma(t)) + v(t)z^{\alpha}(\eta(t)) \leq 0. \ (\because \gamma < \alpha < \beta) \tag{2.11}$$

for $t \geq t_3 > t_2$. By Lemma 2, we get $z(t) \geq C$. Hence, (2.11) yields that

$$(r(t)(z'(t))^{\gamma})' + q(t)C^{\alpha} + v(t)C^{\alpha} \leq 0,$$

$$C^{\alpha}\left[q(t) + v(t)\right] \leq -(r(t)(z'(t))^{\gamma})'. \tag{2.12}$$

Integrating (2.12) from t_3 to $+\infty$, we get a contradiction to $(H_{2\alpha})$.

The rest of the proof is similar to the proof of Theorem 8. Hence, the details will be omitted. □

Theorem 10. *Let* $-1 < a \leq p(t) \leq 0$, $t \in [t_0, \infty)$. *Assume that* $\gamma < \beta < \alpha$. *If* (H_0), $(H_{2\beta})$ *and* (H_3) *hold, then every solution of* (1.1) *oscillates or tends to zero as* $t \to \infty$.

Proof. The proof is similar to the proof of Theorem 8. Hence, the details will be omitted. □

Theorem 11. *Let* $-1 < a \leq p(t) \leq 0$, $t \in [t_0, \infty)$. *Assume that* $\alpha > \gamma > \beta$. *If* (H_0), $(H_{2\beta})$ *and* (H_3) *hold, then every solution of* (1.1) *oscillates or tends to zero as* $t \to \infty$.

Proof. The proof is similar to the proof of Theorem 8. Hence, the details will be omitted. □

Theorem 12. *Let* $-1 < a \leq p(t) \leq 0$, $t \in [t_0, \infty)$. *Assume that that* $\alpha < \beta < \gamma$. *If* (H_0), $(H_{2\alpha})$ *and* (H_3) *hold, then every solution of* (1.1) *oscillates or tends to zero as* $t \to \infty$.

Proof. The proof is similar to the proof of Theorem 9. Hence, the details will be omitted. □

Theorem 13. *Let* $-1 < a \leq p(t) \leq 0$, $t \in [t_0, \infty)$. *Assume that that* $\alpha < \gamma < \beta$. *If* (H_0), $(H_{2\alpha})$ *and* (H_3) *hold, then every solution of* (1.1) *oscillates or tends to zero as* $t \to \infty$.

Proof. The proof is similar to the proof of Theorem 9. Hence, the details will be omitted. □

In Theorem 8, it was shown that $x(t)$ is bounded ultimately for the case (ii). Keeping this fact in view, $x(t)$ converges to zero as $t \to \infty$. Hence, for unbounded $x(t)$, Case (*ii*) does not exist ultimately. Therefore, we have proved the following result.

Theorem 14. *Let* $-1 < a \le p(t) \le 0$, $\gamma > \alpha > \beta$ *for* $t \in [t_0, \infty)$. *If* (H_0), $(H_{2\beta})$ *and* (H_3) *hold, then every unbounded solution of* (1.1) *oscillates.*

Theorem 15. *Let* $-\infty < a \le p(t) \le d < -1$, $\tau(\sigma(t)) = \sigma(\tau(t))$ *and* $\tau(\eta(t)) = \eta(\tau(t))$ *holds for all* $t \in [t_0, \infty)$. *Assume that all the conditions of Theorem* 8 *hold. Then, every bounded solution of* (1.1) *oscillates or tends to zero as* $t \to \infty$.

Proof. The proof of the theorem follows from Theorem 8. In case $\lim_{t\to\infty} z(t) = 0$, we use the fact that $x(t)$ is bounded and it happens that

$$\begin{aligned}0 = \lim_{t\to\infty} z(t) &= \liminf_{t\to\infty} z(t)\\ &\le \liminf_{t\to\infty}(x(t) + dx(\tau(t)))\\ &\le \limsup_{t\to\infty} x(t) + \liminf_{t\to\infty}(dx(\tau(t)))\\ &= \limsup_{t\to\infty} x(t) + d\limsup_{t\to\infty} x(\tau(t))\\ &= (1+d)\limsup_{t\to\infty} x(t),\end{aligned}$$

which implies that $\limsup_{t\to\infty} x(t) = 0$, that is, $\lim_{t\to\infty} x(t) = 0$. Hence, the proof of the theorem is complete. □

Theorem 16. *Let* $-\infty < a \le p(t) \le d < -1$, $\tau(\sigma(t)) = \sigma(\tau(t))$ *and* $\tau(\eta(t)) = \eta(\tau(t))$ *holds for all* $t \in [t_0, \infty)$. *If all the conditions of Theorem* 9, *then every bounded solution of* (1.1) *oscillates or tends to zero as* $t \to \infty$.

Proof. The proof is similar to the proof of Theorem 9. Hence, the details will be omitted. □

Theorem 17. *Let* $-\infty < a \leq p(t) \leq d < -1$, $\tau(\sigma(t)) = \sigma(\tau(t))$ *and* $\tau(\eta(t)) = \eta(\tau(t))$ *holds for all* $t \in [t_0, \infty)$*. If all the conditions of Theorem* 10, *then every bounded solution of (1.1) oscillates or tends to zero as* $t \to \infty$.

Proof. The proof follows from Theorem 10. Hence, the details will be omitted. □

Theorem 18. *Let* $-\infty < a \leq p(t) \leq d < -1$, $\tau(\sigma(t)) = \sigma(\tau(t))$ *and* $\tau(\eta(t)) = \eta(\tau(t))$ *holds for all* $t \in [t_0, \infty)$*. Assume that all the conditions of Theorem* 11*. Then, every bounded solution of* (1.1) *oscillates or tends to zero as* $t \to \infty$.

Proof. The proof follows from Theorem 11. Hence, the details will be omitted. □

Theorem 19. *Let* $-\infty < a \leq p(t) \leq d < -1$, $\tau(\sigma(t)) = \sigma(\tau(t))$ *and* $\tau(\eta(t)) = \eta(\tau(t))$ *holds for all* $t \in [t_0, \infty)$*. If all the conditions of Theorem* 12, *then every bounded solution of* (1.1) *oscillates or tends to zero as* $t \to \infty$.

Proof. The proof follows from Theorem 12. Hence, the details will be omitted. □

Theorem 20. *Let* $-\infty < a \leq p(t) \leq d < -1$, $\tau(\sigma(t)) = \sigma(\tau(t))$ *and* $\tau(\eta(t)) = \eta(\tau(t))$ *holds for all* $t \in [t_0, \infty)$*. If all the conditions of Theorem* 13, *then every bounded solution of* (1.1) *oscillates or tends to zero as* $t \to \infty$.

Proof. The proof follows from Theorem 13. Hence, the details will be omitted. □

3. Oscillation Criteria with (H_{00})

Lemma 3 ([6]). *Assume that* (H_{00}) *holds. Let* $u(t)$ *be an eventually positive continuous function on* $[t_0, \infty)$, $t_0 \geq 0$ *such that* $r(t)u'(t)$ *is a continuously differentiable function with* $((r(t)u'(t))^{\gamma})' \leq 0, \not\equiv 0$

for large $t \in [t_0, \infty)$, where $r(t)$ is a continuous function defined on $[t_0, \infty)$. Then, the following statements hold:

(i) If $u'(t) > 0$, then there exists a constant $C^\gamma > 0$ such that $u(t) > C^\gamma R_\gamma(t)$ for large t.

(ii) If $u'(t) < 0$, then $u(t) \geq -(r(t)(u'(t))^\gamma)^{\frac{1}{\gamma}} R_\gamma(t)$, where $R_\gamma(t) = \int_t^\infty \left(\frac{1}{r(s)}\right)^{\frac{1}{\gamma}} ds$.

Theorem 21. *Let $0 \leq p(t) \leq a < \infty$, $\tau(\sigma(t)) = \sigma(\tau(t))$ and $\tau(\eta(t)) = \eta(\tau(t))$ for all $t > 0$ and $\gamma = \alpha = \beta$ and (H_{00}) hold. Assume that*

$$(H_{3\beta}) \int_{t_0}^\infty \left[R_\gamma^\beta(\sigma(t))Q(t) + R_\gamma^\beta(\eta(t))V(t)\right] dt = \infty$$

and

$$(H_4) \int_0^\infty \left[\frac{1}{r(t)} \int_T^t [R_\gamma^\beta(\sigma(t))Q(t) + R_\gamma^\beta(\eta(t))V(t)]ds\right]^{\frac{1}{\beta}} dt = \infty, \; T > 0,$$

where $Q(t) = \min\{q(t), q(\tau(t))\}, V(t) = \min\{v(t), v(\tau(t))\}$. Then, every solution of (1.1) oscillates.

Proof. Suppose on the contrary that $x(t)$ is a non-oscillatory solution of (1.1). Without loss of generality, we may assume that $x(t) > 0$ for $t \geq t_0$. Hence, there exists $t_1 > t_0$ such that $x(t) > 0$, $x(\tau(t)) > 0$, $x(\sigma(t)) > 0$ and $x(\eta(t)) > 0$ for $t \geq t_1$. By (2.1), (1.1) becomes

$$(r(t)(z'(t))^\gamma)' = -q(t)x^\alpha(\sigma(t)) - v(t)x^\beta(\eta(t)) \leq 0, \not\equiv 0 \quad \text{for } t \geq t_1.$$

So, $r(t)(z'(t))^\gamma$ is non-increasing on $[t_1, \infty)$, that is, either $z'(t) > 0$ or $z'(t) < 0$. From Lemma 1, it follows that $z'(t) > 0$ for $t \geq t_2$. By (1.1), it is easy to see that

$$\begin{aligned}(r(t)(z'(t))^\gamma)' &+ a^\alpha(r(\tau(t))(z'(\tau(t))^\gamma)' + q(t)x^\alpha(\sigma(t)) \\ &+ a^\alpha q(\tau(t))x^\alpha(\sigma(\tau(t)) + v(t)x^\beta(\eta(t)) + a^\alpha v(\tau(t))x^\beta(\eta(\tau(t)) = 0\end{aligned} \tag{3.1}$$

for $t \geq t_1$ since $\gamma = \alpha = \beta$. By Lemma 2, (3.1) yields that

$$\begin{aligned}(r(t)(z'(t))^\gamma)' &+ a^\alpha(r(\tau(t))(z'(\tau(t))^\gamma)' + 2^{1-\alpha}Q(t)z^\beta(\sigma(t)) \\ &+ 2^{1-\alpha}V(t)z^\beta(\eta(t)) \leq 0.\end{aligned} \tag{3.2}$$

From Lemma 3, it follows that $z(t)$ is non-decreasing on $[t_1, \infty)$ and hence, $z(t) \geq C^\gamma R_\gamma(t)$ for $t \geq t_2 > t_1$. Hence, (3.2) yields that

$$(r(t)(z'(t))^\gamma)' + a^\alpha(r(\tau(t))(z'(\tau(t))^\gamma)' + 2^{1-\alpha}Q(t)C^\beta R_\gamma^\beta(\sigma(t))$$
$$+ 2^{1-\alpha}V(t)C^\beta R_\gamma^\beta(\eta(t)) \leq 0.$$

Hence, there exists $t_3 > t_2$ such that

$$\begin{aligned} &2^{1-\alpha}Q(t)C^\beta R_\gamma^\beta(\sigma(t)) + 2^{1-\alpha}V(t)C^\beta R_\gamma^\beta(\eta(t)) \\ &\quad \leq -(r(t)(z'(t))^\gamma)' - a^\alpha(r(\tau(t))(z'(\tau(t))^\gamma)', \end{aligned} \tag{3.3}$$

for $t \geq t_3$. Integrating (3.3) from t_3 to $+\infty$, we get a contradiction to $(H_{3\beta})$. By Lemma 3, we have that $z(t) \geq -R_\gamma(t)r^{\frac{1}{\gamma}}(t)z'(t)$ for $t \geq t_2 > t_1$. Since $r(t)(z'(t))^\gamma$ is non-increasing, we can find a constant $C^\gamma > 0$ and $t_3 > t_2$ such that $r(t)(z'(t))^\gamma \leq -C^\gamma$ and $z(t) \geq CR_\gamma(t)$ for $t \geq t_3$. Integrating (3.3) from t_3 to t, we obtain that

$$\begin{aligned} &\frac{2^{1-\alpha}C^\beta}{r(t)} \int_{t_3}^t [R_\gamma^\beta(\sigma(s))Q(s) + R_\gamma^\beta(\eta(s))V(s)]ds \\ &\quad \leq -(r(t)(z'(t))^\gamma)' - a^\alpha(r(\tau(t))(z'(\tau(t))^\gamma)', \end{aligned}$$

due to non-increasing $r(t)(z'(t))^\gamma$. Hence,

$$\frac{2^{1-\alpha}C^\beta}{r(t)} \int_{t_3}^t [R_\gamma^\beta(\sigma(s))Q(s) + R_\gamma^\beta(\eta(s))V(s)]ds \leq -(1 + a^\alpha)(z'(t))^\gamma.$$

Since $\lim_{t\to\infty} z(t)$ exists, it follows from the above inequality that

$$\left[\frac{2^{1-\alpha}C^\beta}{(1 + a^\alpha)}\right]^{\frac{1}{\beta}} \int_{t_3}^\infty \left[\frac{1}{r(t)} \int_{t_3}^t [R_\gamma^\beta(\sigma(s))Q(s) + R_\gamma^\beta(\eta(s))V(s)]ds\right]^{\frac{1}{\beta}} dt < \infty,$$

which is a contradiction to (H_4). □

Theorem 22. *Let* $0 \leq p(t) \leq a < \infty$, $\gamma > \alpha > \beta$, $\tau(\sigma(t)) = \sigma(\tau(t))$ *and* $\tau(\eta(t)) = \eta(\tau(t))$ *for all* $t > 0$. *If* (H_{00}), $(H_{3\beta})$ *and* (H_4) *hold, then every solution of* (1.1) *oscillates.*

Proof. The proof follows from Theorem 21. Hence, the details will be omitted. □

Theorem 23. *Let $0 \leq p(t) \leq a < \infty$, $\gamma < \alpha < \beta$, $\tau(\sigma(t)) = \sigma(\tau(t))$ and $\tau(\eta(t)) = \eta(\tau(t))$ for all $t > 0$. Assume that (H_{00}) holds,*

$$(H_{3\alpha}) \int_{t_0}^{\infty} \left[R_\gamma^\alpha(\sigma(t))Q(t) + R_\gamma^\alpha(\eta(t))V(t)\right] dt = \infty$$

and

$$(H_5) \quad \int_0^\infty \left[\frac{1}{r(t)} \int_T^t R_\gamma^\alpha(\sigma(s))Q(s) + R_\gamma^\alpha(\eta(s))V(s)ds\right]^{\frac{1}{\alpha}} dt = \infty, \quad T > 0.$$

Then every solution of (1.1) *oscillates.*

Proof. Suppose on the contrary that $x(t)$ is a non-oscillatory solution of (1.1). Without loss of generality, we may assume that $x(t) > 0$ for $t \geq t_0$. Hence, there exists $t_1 > t_0$ such that $x(t) > 0$, $x(\tau(t)) > 0$, $x(\sigma(t)) > 0$ and $x(\eta(t)) > 0$ for $t \geq t_1$. By (2.1), (1.1) becomes

$$(r(t)(z'(t))^\gamma)' = -q(t)x^\alpha(\sigma(t)) - v(t)x^\beta(\eta(t)) \leq 0, \not\equiv 0 \text{ for } t \geq t_1.$$

So, $r(t)(z'(t))^\gamma$ is non-increasing on $[t_1, \infty)$, that is, either $z'(t) > 0$ or $z'(t) < 0$. From Lemma 1, it follows that $z'(t) > 0$ for $t \geq t_2$. By (1.1), it is easy to see that

$$\begin{aligned}(r(t)(z'(t))^\gamma)' &+ a^\beta(r(\tau(t))(z'(\tau(t))^\gamma)' + q(t)x^\alpha(\sigma(t)) \\ &+ a^\beta q(\tau(t))x^\alpha(\sigma(\tau(t)) + v(t)x^\beta(\eta(t)) + a^\beta v(\tau(t))x^\beta(\eta(\tau(t)) = 0\end{aligned} \tag{3.4}$$

for $t \geq t_1$ since $\gamma < \alpha < \beta$. By Lemma 2, (3.4) yields that

$$\begin{aligned}(r(t)(z'(t))^\gamma)' &+ a^\alpha(r(\tau(t))(z'(\tau(t))^\gamma)' + 2^{1-\beta}Q(t)z^\alpha(\sigma(t)) \\ &+ 2^{1-\beta}V(t)z^\alpha(\eta(t)) \leq 0.\end{aligned}$$

From Lemma 3, it follows that $z(t)$ is non-decreasing on $[t_1, \infty)$ and hence, $z(t) \geq C^\gamma R_\gamma(t)$ for $t \geq t_2 > t_1$. Hence, (3.2) yields that

$$\begin{aligned}(r(t)(z'(t))^\gamma)' &+ a^\alpha(r(\tau(t))(z'(\tau(t))^\gamma)' + 2^{1-\beta}Q(t)C^\alpha R_\gamma^\alpha(\sigma(t)) \\ &+ 2^{1-\beta}V(t)C^\alpha R_\gamma^\alpha(\eta(t)) \leq 0.\end{aligned}$$

Hence, there exists $t_3 > t_2$ such that

$$2^{1-\beta}Q(t)C^{\alpha}R_{\gamma}^{\alpha}(\sigma(t)) + 2^{1-\beta}V(t)C^{\alpha}R_{\gamma}^{\alpha}(\eta(t)) \\ \leq -(r(t)(z'(t))^{\gamma})' - a^{\alpha}(r(\tau(t))(z'(\tau(t))^{\gamma})' \tag{3.5}$$

for $t \geq t_3$. Integrating (3.5) from t_3 to $+\infty$, we get a contradiction to $(H_{3\alpha})$. By Lemma 3, we have that $z(t) \geq -R_{\gamma}(t)r^{\frac{1}{\gamma}}(t)z'(t)$ for $t \geq t_2 > t_1$. Since $r(t)(z'(t))^{\gamma}$ is non-increasing, we can find a constant $C^{\gamma} > 0$ and $t_3 > t_2$ such that $r(t)(z'(t))^{\gamma} \leq -C^{\gamma}$ and $z(t) \geq CR_{\gamma}(t)$ for $t \geq t_3$. Integrating (3.5) from t_3 to t, we obtain that

$$\frac{2^{1-\alpha}C^{\alpha}}{r(t)}\int_{t_3}^{t}\left[R_{\gamma}^{\alpha}(\sigma(s))Q(s) + R_{\gamma}^{\alpha}(\eta(s))V(s)\right]ds \\ \leq -(r(t)(z'(t))^{\gamma})' - a^{\alpha}(r(\tau(t))(z'(\tau(t))^{\gamma})',$$

due to non-increasing $r(t)(z'(t))^{\gamma}$. Hence,

$$\frac{2^{1-\alpha}C^{\alpha}}{r(t)}\int_{t_3}^{t}\left[R_{\gamma}^{\alpha}(\sigma(s))Q(s) + R_{\gamma}^{\alpha}(\eta(s))V(s)\right]ds \leq -(1+a^{\alpha})(z'(t))^{\gamma}.$$

Since $\lim_{t\to\infty} z(t)$ exists, it follows from the above inequality that

$$\left[\frac{2^{1-\alpha}C^{\alpha}}{(1+a^{\alpha})}\right]^{\frac{1}{\alpha}}\int_{t_3}^{\infty}\left[\frac{1}{r(t)}\int_{t_3}^{t}R_{\gamma}^{\alpha}(\sigma(s))Q(s) + R_{\gamma}^{\alpha}(\eta(s))V(s)ds\right]^{\frac{1}{\alpha}}dt < \infty,$$

which is a contradiction to (H_5). □

Theorem 24. *Let* $0 \leq p(t) \leq a < \infty$, $\gamma < \beta < \alpha$, $\tau(\sigma(t)) = \sigma(\tau(t))$ *and* $\tau(\eta(t)) = \eta(\tau(t))$ *for all* $t > 0$. *If* (H_{00}), (H_5) *and* $(H_{3\beta})$ *hold, then every solution of* (1.1) *oscillates.*

Proof. The proof follows from Theorem 22. Hence, the details will be omitted. □

Theorem 25. *Let* $0 \leq p(t) \leq a < \infty$, $\alpha > \gamma > \beta$, $\tau(\sigma(t)) = \sigma(\tau(t))$ *and* $\tau(\eta(t)) = \eta(\tau(t))$ *for all* $t > 0$. *If* (H_{00}), (H_5) *and* $(H_{3\beta})$ *hold, then every solution of* (1.1) *oscillates.*

Proof. Th proof follows from the proof of Theorem 22. Hence, the details will be omitted. □

Theorem 26. *Let $0 \le p(t) \le a < \infty$, $\alpha < \beta < \gamma$, $\tau(\sigma(t)) = \sigma(\tau(t))$ and $\tau(\eta(t)) = \eta(\tau(t))$ for all $t > 0$. If (H_{00}), (H_5) and $(H_{3\alpha})$ hold, then every solution of* (1.1) *oscillates.*

Proof. The proof follows from Theorem 23. Hence, the details will be omitted. □

Theorem 27. *Let $0 \le p(t) \le a < \infty$, $\alpha < \gamma < \beta$, $\tau(\sigma(t)) = \sigma(\tau(t))$ and $\tau(\eta(t)) = \eta(\tau(t))$ for all $t > 0$. If (H_{00}), (H_5) and $(H_{3\alpha})$ hold, then every solution of* (1.1) *oscillates.*

Proof. The proof follows from Theorem 23. Hence, the details will be omitted. □

Theorem 28. *Let $-1 < a \le p(t) \le 0$, $t \in [t_0, \infty)$. Assume that $\gamma > \alpha > \beta$ and (H_{00}) holds. If*

$$(H_{4\beta}) \int_{t_0}^{\infty} [q(s)R_{\gamma}^{\beta}\sigma(s) + v(s)R_{\gamma}^{\beta}\eta(s)]ds = \infty$$

and

$$(H_6) \int_{T_1}^{\infty} \left[\frac{1}{r(\theta)} \int_{t_0}^{\theta} [q(s)R_{\gamma}^{\beta}\sigma(s) + v(s)R_{\gamma}^{\beta}\eta(s)]ds\right]^{\frac{1}{\beta}} d\theta = \infty, T_1 > 0,$$

then every solution of (1.1) *oscillates or tends to zero as $t \to \infty$.*

Proof. By the same reasoning as in the proof of Theorem 1, we get (2.3) for $t \in [t_2, \infty)$. Thus, $z(t)$ and $z'(t)$ are monotonic functions on $[t_2, \infty)$. In what follows, we consider the following four possible cases:

$$\begin{aligned} &(i)\ \ z(t) > 0, \quad z'(t) > 0,\\ &(ii)\ \ z(t) < 0, \quad z'(t) > 0,\\ &(iii)\ \ z(t) > 0, \quad z'(t) < 0,\\ &(iv)\ \ z(t) < 0, \quad z'(t) < 0. \end{aligned}$$

Case (i). In this case, $z(t) \le x(t)$ and $\lim_{t\to\infty} r(t)z'(t)$ exists. Therefore, (1.1) reduces to

$$(r(t)(z'(t))^{\gamma})' + q(t)z^{\beta}(\sigma(t)) + v(t)z^{\beta}(\eta(t)) \le 0. \ (\because \gamma > \alpha > \beta) \tag{3.6}$$

for $t \geq t_3 > t_2$. Using Lemma 3, we get $z(t) \geq C^\gamma R_\gamma(t)$. Hence, (3.6) yields that

$$(r(t)(z'(t))^\gamma)' + +q(t)C^\beta R_\gamma^\beta(\sigma(t)) + v(t)C^\beta R_\gamma^\beta(\eta(t)) \leq 0,$$

$$q(t)C^\beta R_\gamma^\beta(\sigma(t)) + v(t)C^\beta R_\gamma^\beta(\eta(t)) \leq -(r(t)(z'(t))^\gamma)'. \quad (3.7)$$

Integrating (3.7) from t_3 to $+\infty$, we get a contradiction to $(H_{4\beta})$.

The rest of the proof follows from Theorem 22. Hence, the details will be omitted. □

Theorem 29. *Let $-1 < a \leq p(t) \leq 0$ and $\gamma < \alpha < \beta$ for all $t \in [t_0, \infty)$. Assume that (H_{00}), $(H_{4\alpha})$ and*

$$(H_7)\quad \int_{T_1}^{\infty} \left[\frac{1}{r(\theta)} \int_{t_0}^{\theta} [q(s)R_\gamma^\alpha \sigma(s) + v(s)R_\gamma^\alpha \eta(s)]ds\right]^{\frac{1}{\alpha}} d\theta = \infty, T_1 > 0$$

hold. Then, every solution of (1.1) *oscillates or tends to zero as $t \to \infty$.*

Proof. By the same reasoning as in the proof of Theorem 1, we get (2.3) for $t \in [t_2, \infty)$. Thus, $z(t)$ and $z'(t)$ are monotonic functions on $[t_2, \infty)$. In what follows, we consider the following four possible cases:

$$(i)\ \ z(t) > 0, \quad z'(t) > 0,$$
$$(ii)\ \ z(t) < 0, \quad z'(t) > 0,$$
$$(iii)\ \ z(t) > 0, \quad z'(t) < 0,$$
$$(iv)\ \ z(t) < 0, \quad z'(t) < 0.$$

Case (i). In this case, $z(t) \leq x(t)$ and $\lim_{t\to\infty} r(t)z'(t)$ exists. Therefore, (1.1) reduces to

$$(r(t)(z'(t))^\gamma)' + q(t)z^\alpha(\sigma(t)) + v(t)z^\alpha(\eta(t)) \leq 0$$

for $t \geq t_3 > t_2$ since $\gamma < \alpha < \beta$. Using Lemma 3, we get $z(t) \geq C^\gamma R_\gamma(t)$. Hence, (3.7) yields that

$$(r(t)(z'(t))^\gamma)' + +q(t)C^\alpha R_\gamma^\alpha(\sigma(t)) + v(t)C^\alpha R_\gamma^\alpha(\eta(t)) \leq 0,$$

$$q(t)C^\alpha R_\gamma^\alpha(\sigma(t)) + v(t)C^\alpha R_\gamma^\alpha(\eta(t)) \leq -(r(t)(z'(t))^\gamma)'. \quad (3.8)$$

Integrating (3.8) from t_3 to $+\infty$, we get a contradiction to $(H_{4\alpha})$.

The rest of the proof follows from Theorem 28. Hence, the details will be omitted. □

Theorem 30. *Let* $-1 < a \leq p(t) \leq 0$ *and* $\gamma < \beta < \alpha$ *for* $t \in [t_0, \infty)$. *If* (H_{00}), $(H_{4\beta})$ *and* (H_7) *hold, then every solution of* (1.1) *oscillates or tends to zero as* $t \to \infty$.

Proof. The proof follows from Theorem 28. Hence, the details will be omitted. □

Theorem 31. *Let* $-1 < a \leq p(t) \leq 0$ *and* $\alpha > \gamma > \beta$ *for* $t \in [t_0, \infty)$. *If* (H_{00}), $(H_{4\beta})$ *and* (H_7) *hold, then every solution of* (1.1) *oscillates or tends to zero as* $t \to \infty$.

Proof. The proof follows from Theorem 28. Hence, the details will be omitted. □

Theorem 32. *Let* $-1 < a \leq p(t) \leq 0$ *and* $\alpha < \beta < \gamma$ *for* $t \in [t_0, \infty)$. *If* (H_{00}) $(H_{4\alpha})$ *and* (H_7) *hold, then every solution of* (1.1) *oscillates or tends to zero as* $t \to \infty$.

Proof. The proof of follows from Theorem 29. Hence, the details will be omitted. □

Theorem 33. *Let* $-1 < a \leq p(t) \leq 0$ *and* $\alpha < \gamma < \beta$ *for* $t \in [t_0, \infty)$. *If* (H_{00}), $(H_{4\alpha})$ *and* (H_7) *hold, then every solution of* (1.1) *oscillates or tends to zero as* $t \to \infty$.

Proof. The proof follows from Theorem 29. Hence, the details will be omitted. □

Theorem 34. *Let* $-\infty < a \leq p(t) \leq d < -1$, $\tau(\sigma(t)) = \sigma(\tau(t))$ *and* $\tau(\eta(t)) = \eta(\tau(t))$ *for all* $t \in [t_0, \infty)$. *If* (H_3) *and all the conditions of Theorem* 28 *hold, then every bounded solution of* (1.1) *oscillates or tends to zero as* $t \to \infty$.

Proof. The proof follows from Theorems 15 and 28. Hence, the details will be omitted. □

Theorem 35. *Let* $-\infty < a \leq p(t) \leq d < -1$, $\tau(\sigma(t)) = \sigma(\tau(t))$ *and* $\tau(\eta(t)) = \eta(\tau(t))$ *for all* $t \in [t_0, \infty)$. *If all the conditions of Theorem* 29 *hold, then every bounded solution of* (1.1) *oscillates or tends to zero as* $t \to \infty$.

Proof. The proof follows from Theorems 16 and 29. Hence, the details will be omitted. □

Theorem 36. *Let* $-\infty < a \le p(t) \le d < -1$, $\tau(\sigma(t)) = \sigma(\tau(t))$ *and* $\tau(\eta(t)) = \eta(\tau(t))$ *for all* $t \in [t_0, \infty)$. *If all the conditions of Theorem* 30 *hold, then every bounded solution of* (1.1) *oscillates or tends to zero as* $t \to \infty$.

Proof. The proof follows from Theorems 17 and 30. Hence, the details will be omitted. □

Theorem 37. *Let* $-\infty < a \le p(t) \le d < -1$, $\tau(\sigma(t)) = \sigma(\tau(t))$ *and* $\tau(\eta(t)) = \eta(\tau(t))$ *for all* $t \in [t_0, \infty)$. *If all the conditions of Theorem* 31 *hold, then every bounded solution of* (1.1) *oscillates or tends to zero as* $t \to \infty$.

Proof. The proof follows from Theorems 18 and 31. Hence, the details will be omitted. □

Theorem 38. *Let* $-\infty < a \le p(t) \le d < -1$, $\tau(\sigma(t)) = \sigma(\tau(t))$ *and* $\tau(\eta(t)) = \eta(\tau(t))$ *for all* $t \in [t_0, \infty)$. *If all the conditions of Theorem* 32 *hold, then every bounded solution of* (1.1) *oscillates or tends to zero as* $t \to \infty$.

Proof. The proof follows from Theorems 19 and 33. Hence, the details will be omitted. □

Theorem 39. *Let* $-\infty < a \le p(t) \le d < -1$, $\tau(\sigma(t)) = \sigma(\tau(t))$ *and* $\tau(\eta(t)) = \eta(\tau(t))$ *for all* $t \in [t_0, \infty)$. *If all the conditions of Theorem* 33 *hold, then every bounded solution of* (1.1) *oscillates or tends to zero as* $t \to \infty$.

Proof. The proof follows from Theorems 20 and 33. Hence, the details will be omitted. □

Example 1. Let

$$((t^{\gamma}((\sin(t) + \sin(t-\pi)))')^{\frac{5}{3}})' + e^t \sin^{\frac{5}{3}}(t - 2\pi) + e^t \sin^{\frac{5}{3}}(t-\pi) = 0,$$

where $r(t) = t^{\gamma}$, $p(t) = 1$ and $\gamma = \frac{5}{3} = \alpha = \beta$. Then, all the conditions of Theorem 1 are satisfied for (1.1). Hence, every solution of (1.1) oscillates. In particular, $x(t) = \sin t$ is one of such solutions of (1.1).

Example 2. Let

$$((e^{\gamma t}((\sin(t)+\sin(t-\pi)))')^{\gamma})'+e^{t}\sin^{\alpha}(t-2\pi)+e^{t}\sin^{\beta}(t-3\pi)=0,$$

where $r(t) = e^{\gamma t}$, $p(t) = 1$ and $\gamma = 1 = \alpha = \beta$. Then, all the conditions of Theorem 21 are satisfied for (1.1). Hence, every solution of (1.1) oscillates. In particular, $x(t) = \sin t$ is one of such solutions of (1.1).

4. Conclusion

In this work, we established the necessary conditions for oscillation of the second-order neutral delay differential equations of the form

$$(r(t)((x(t)+p(t)x(\tau(t)))')^{\gamma})'+q(t)x^{\alpha}(\sigma(t))+v(t)x^{\beta}(\eta(t))=0$$

under the assumption that

$$\int_0^{\infty}\left(\frac{1}{r(t)}\right)^{\frac{1}{\gamma}}dt=\infty$$

or

$$\int_0^{\infty}\left(\frac{1}{r(t)}\right)^{\frac{1}{\gamma}}dt<\infty,$$

for various ranges of $p(t)$, where γ is a quotient of odd positive integers.

Acknowledgments

This work was supported by Rajiv Gandhi National Fellowship (UGC), New Delhi, India, through the letter No. F1-17.1/2013-14/ RGNF-2013-14-SC-ORI-42425, dated February 6, 2014.

References

[1] R. P. Agarwal, M. Bohner, T. Li and C. Zhang. Oscillation of second order differential equations with a sublinear neutral term. *Carpathian J. Math.*, 30, 2014, 1–6.

[2] R. P. Agarwal and S. R. Grace. Oscillation theorem for certain neutral functional differential equations. *Comput. Math. Appl.*, 38(11–12), 1999, 1–11.

[3] R. P. Agarwal, S. L. Shieh and C. C. Yeh. Oscillation criteria for second order retarded differential equations. *Math. Comput. Modell.*, 26(4), 1997, 1–11.

[4] B. Baculikova. Oscillation criteria for second order nonlinear differential equations. *Arch. Math.*, 42, 2006, 141–149.

[5] B. Baculikova and J. Dzurina. Oscillation theorems for second order neutral differential equations. *Comput. Math. Appl.*, 61, 2011, 94–99.

[6] B. Baculikova and J. Dzurina. Oscillation theorems for second order nonlinear neutral differential equations. *Comput. Math. Appl.*, 62, 2011, 4472–4478.

[7] B. Baculikova, T. Li and J. Dzurina. Oscillation theorems for second order neutral differential equations. *Elect. J. Qual. Theory Differ. Equ.*, 2011, 2011. Paper No. 74.

[8] D. Bainov and E. Minchev. Forced oscillations of solutions of implusive nonlinear hyperbolic differential-difference equations. *Note Mat.*, 19(2), 2001, 173–181.

[9] P. Boulanger and M. Hayes. The effect of a linear constraint on the small oscillations of a dynamical system with three degrees of freedom. *Note Mat.*, 27(2), 2007, 13–26.

[10] J. Dzurina. Oscillation theorems for second order advanced neuteal differential equations. *Tatramt. Math. Publ.*, 48, 2011, 61–71.

[11] L. H. Erbe, Q. Kong and B. G. Zhang. *Oscillation Theory for Functional Differential Equations*. Marcel Dekker, Inc., New York, 1995.

[12] I. Gyori and G. Ladas. *Oscillation Theory of Delay Differential Equations with Applications*. Clarendon Press, Oxford, 1991.

[13] J. K. Hale. *Theory of Functional Differential Equations*. Springer-Verlag, New York, 1977.

[14] Z. Han, T. Li, C. Zhang and Y. Sun. Oscillation criteria for certain second-order nonlinear neutral differential equations of mixed type. *Abstr. Appl. Anal.*, 2011, 2011. Article ID 387483.

[15] J. H. Li. Oscillation of solutions of second order neutral delay differential equations with integrable coefficients. *Math. Comput. Modell.*, 25, 1997, 69–79.

[16] Q. Li, R. Wang, F. Chen and T. Li. Oscillation of second order nonlinear delay differential equations with nonpositive neutral coefficients. *Adv. Differ. Equ.*, 2015, 2015. Paper No. 35.

[17] T. Li, Y. V. Rogovchenko and C. Zhang. Oscillation results for second order nonlinear neutral differential equations. *Adv. Differ. Equ.*, 2013, 2013. Paper No. 336.

[18] A. K. Sethi. Oscillation of a functional differential equations of second order. *Global J. Pure Appl. Math.*, 13, 2017, 5625–5634.

[19] A. K. Sethi and A. K. Tripathy. Riccati transformation and oscillation of superlinear second order functional differential equations. In *Proceedings of Mathematical and Numerical Aspects of Dynamical System Analysis*, 2017, pp. 479–490.

[20] A. A. Soliman and A. M. Hassan. Oscillation criteria of second order nonlinear neutral differential equations. *Int. J. Appl. Math. Res.*, 1(3), 2012, 314–322.

[21] S. Tamilvanan, E. Thandapani and J. Dzurina. Oscillation of second order nonlinear differential equations with sublinear neutral term. *Differ. Equ. Appl.*, 9, 2017, 29–35.

[22] A. K. Tripathy and A. K. Sethi. On oscillatory nonlinear second order neutral delay differential equations. *Differ. Equ. Appl.*, 8, 2016, 247–258.

[23] A. K. Tripathy and A. K. Sethi. Sufficient condition for oscillation and nonoscillation of a class of second-order neutral differential equations. *Adv. Dyn. Syst. Appl.*, 13(1), 2018, 73–90.

https://doi.org/10.1142/9789819820498_0018

Chapter 18

Use of Moments Formulas for Geometric and Negative Binomial Distributions with Special Numbers Polynomials in Data Sciences Applications

Buket Simsek

Department of Interdisciplinary Regenerative Medicine,
Prof. Dr. Tuncer Karpuzoğlu Organ Transplantation and
Advanced Health Research Institute,
Akdeniz University, Antalya, Türkiye
buketsimsek@akdeniz.edu.tr; simsekbukett@gmail.com

With the aid of moment formulas with their applications, data science, which is a fundamental modern technique skill to discover trends and patterns with them, has been used in interdisciplinary areas. Calculating moments based on both higher-order derivative operators and infinite series formulas and convergence conditions is quite challenging. For example, the geometric distribution and negative binomial distribution contain such moment formulas. Our moment calculation formulas are based on new methods and formulas we have discovered in recent years. In this chapter, an effort is made to present the known results about moment formulas by analyzing them with new results. The aim of this chapter is to give and survey some applications of moments discrete probability distributions including geometric distribution and negative binomial distribution with aid of special numbers and polynomials involving the Bernoulli numbers and polynomials (of higher order), the Apostol–Bernoulli numbers and polynomials (of higher order), the Stirling numbers and the Eulerian numbers. This chapter also covers the following results: Refs. [49–52, 54] and also [53].

By applying moment formulas, many values of moment fare given and new number series produced by moments are mentioned.

Keywords: Data science, probability distribution, geometric distribution, moment generating function, Polygamma function, characteristic function, Apostol–Bernoulli numbers and polynomials, Stirling numbers, Eulerian numbers.

1. Introduction, Definitions and Notations

Data science is interested in all aspects of data science and its application in the field of statistics, mathematical statistic and probability, business, economics, finance, operations, engineering, healthcare, transportation, agriculture, energy, environment, sports and social management.

Such current areas of data science can be given as follows: statistics, mathematical statistics and probability, distribution functions, moments and their applications, management, machine learning and intelligent management, data mining and business analytics, data statistics and decision-making, intelligent computing and algorithms, data-driven management decisions, data-based policy evaluation, business intelligence, business data science, digital economy, data quality and data privacy, data and knowledge management, enterprise digital management, cyberspace management, data security management, smart city management, smart society management, digital engineering management, data visualization, data-driven intelligent system management, data science applications, etc. The moment formulas given in the following are resources for researchers in investigating the relationships among the above topics and data science, especially statistics, mathematical statistics and probability, distribution functions, moments and their applications, and in applying them to real-world problems. Therefore, there are two types of Probability Distributions (PDs). PDs are divided into two main types based on the nature of the random variables: discrete random variables and continuous random variables. In this chapter, we focus on the discrete data distributions with discrete random variables. It known that the discrete distribution can be used when the random variable can take on countable, specific values (cf. Refs. [6, 16, 18, 24, 40, 41, 43, 58, 64]).

With the help of the Bernoulli distribution which is a special case of the binomial distribution and the geometric distribution which is a special case of the negative binomial distribution, Sugiyama [58] gave interesting results for the probabilities form a geometric progression.

Margolin and Winokur [41] gave the exact moments of the order statistics of the geometric distribution. They also gave their relation to inverse sampling and reliability of redundant systems with statistics applications. They showed that formulas for the first two moments of the order statistics from a geometric distribution are given in closed form. They gave related tables of the first two moments of the order statistics from a geometric distribution, which have two related dice games. They also showed that not only the distribution of the range of a set of independent observations from a geometric distribution is presented in closed form, but also a system of redundant components which performs a given task repetitively at discrete times and its statistical equivalent, an inverse sampling scheme. Finally, the concluded that the distribution of the number of stages until the scheme terminates or the system fails is shown to be identical to the distribution of a specific order statistic from a geometric distribution.

Howell [24] studied the moments and the moment generating functions of statistical distributions with explanation into statistical moments and how to derive them and how to use them. By the aid of his physics background, Howell discussed the idea of moments in statistics and why they are important and also what moments are?

Due to work of Howell [24], moments describe the location, shape and size of a probability distribution. That is, the following list of the first four moments can be easily deduced: mean (central tendency), variance (spread), skewness (asymmetry) and also kurtosis (outlier prone).

The applications of the moments describe as follows:

The zeroth moment is considered as the area under any probability distribution is 1. The first moment (mean) is considered as an uncentered moment, whereas the remaining three are centered moments as they relate to the mean. The higher-order moments, which are given in the following sections, exist and just provide more information about the skewness and kurtosis of the distribution. Moreover, the odd-powered moments infer information about skewness and even powered moments infer information about the kurtosis (cf. Refs. [6, 8, 24, 40, 45, 58]).

Let $M_X(w)$ denote the moment generating function (MGF) with respect to random variable X and w some dummy variable that makes it easy to calculate the moments and w a real variable. By taking derivatives of the function $M_X(w)$ with respect to w at $w = 0$, all moments can be computed. Thus, moments describe the location (mean), size (variance) and shape (skewness and kurtosis) of a probability density function. Consequently, MGFs are used to calculate these moments with the aid their derivatives.

In Ref. [64], we have many useful information about probability distribution functions (PDFs). That is, mathematical PDF describes the probability of obtaining different values of a random variable in a particular probability distribution. PDF is related to probability statistics and Probability Data Distributions (PDDs) in data science. PDFs and Probability Distributions (PDs) help us to understand how data behaves, which is one of the first steps in data science. When we study with complex data such as customer purchases, stock prices and weather. PD helps us to not only find following questions answers but also make better predictions and understand uncertainty:

What is most likely to happen? What are rare or unusual outcomes? Are values close together or spread out?

Consequently, due to "Explain how data behaves (clustered or spread)", "Form the basis of machine learning models", "Used in statistical tests (e.g. p-value)", "Help identify outliers and make predictions", etc., PD and Probability Mass Function (PMF), which is used for discrete variables, are important tools and topics to evaluate predictions and understand uncertainty, random variables, which assign numbers to outcomes of random events, etc. (cf. Ref. [64]).

Moments are specific quantitative measurements of the graph of a probability distribution function. If the function is representative of mass density, it is well known that the zeroth moment corresponds to the total mass. The first moment corresponds to the center of mass. The second moment is known as the moment of inertia.

Additionally, the first moment is also called the expected value. Variance is defined by the first and second moments. The third standardized moment is known as skewness, and the fourth standardized moment is known as kurtosis. Moments have inspired the work of many researchers. Probability theory was one of the scientific developments that reached their peak especially in the 1700s. For example, with the development of applied statistics, the theory of errors and the method of least squares, probability and its applications to the statistical theory of thermodynamics, applications to genetics, etc. developed rapidly after 1700. Considering the historical developments, it is known that by the middle of the 19th century, Russian mathematician Pafnuty Lvovich Chebyshev (16 May 1821–1828 December 1894) was the first researcher to conduct systematic research in terms of moments of random variables. As a result of Chebyshev's studies and experimental studies, moment calculation problems in the theory of probability distribution functions have many important applications (cf. Refs. [6, 16, 40]). Similarly, generating functions and moment generating functions have a variety of applications in many areas, ranging from probability and statistics, many branches of mathematics, engineering and computer science. Likewise, characteristic functions associated with moment generating functions have also many applications in mathematics, especially probability theory, analysis, combinatorics and other applied sciences. Recently, by using moment generating function and characteristic function of the geometric distribution, Simsek [50] gave new formulas for moments of the geometric distribution in terms of the Apostol–Bernoulli polynomials and numbers. The motivation of this chapter is to construct a new generating function for the moment of the geometric distribution. By using this generating function and its derivative and functional equations, we give new formulas and recurrence relations for the moments involving the Apostol–Bernoulli numbers and polynomials, the Bernstein polynomials, the Apostol–Euler numbers and polynomials, the Fubini numbers, and the Eulerian numbers and polynomials. These new results are related to special numbers with their generating functions, which are used in combinatorics to solve problems involving partitions, permutations, combinations and recurrence relations. For example, the Eulerian numbers and the Fubini numbers are used to enumerate many different combinatorial situations.

The following notations, relations and formulas associated with the results of this chapter are inspired by the references (see Refs. [1–62]).

Let

$$\mathbb{N} = \{1, 2, 3, \ldots\}, \quad \mathbb{N}_0 := \mathbb{N} \cup \{0\},$$

and also $\mathbb{Z}$, $\mathbb{R}$, $\mathbb{C}$ denote the set of integers, the set of real numbers, the set of complex numbers, respectively. Moreover, we assume that $\exp(t) := e^t$, $i^2 = -1$,

$$0^r = \begin{cases} 1, & r = 0 \\ 0, & r \in \mathbb{N} \end{cases}$$

and

$$(y)_r = \prod_{s=0}^{r-1} (y - s)$$

with $(y)_0 = 1$ and $r \in \mathbb{N}$ (cf. Refs. [1–65]).

1.1. *The Apostol–Bernoulli numbers and polynomials*

The classical method, which involves quite complex operations to calculate each moment of the geometric distribution, is generally based on derivatives of the geometric series. In order to solve these difficulties, Simsek [50] gave novel relations between the $M_X(u)$ and the $K_X(u)$ in terms of the Apostol–Bernoulli numbers and polynomials. These numbers and polynomials are defined in the following.

The Apostol–Bernoulli numbers $\mathcal{B}_n(\beta)$ are given by the following generating function:

$$\frac{t}{\beta \exp(t) - 1} = \sum_{n=0}^{\infty} \mathcal{B}_n(\beta) \frac{t^n}{n!} \tag{1.1}$$

(cf. Ref. [5]).

From (1.1), we have

$$\beta \exp(t) \sum_{n=0}^{\infty} \mathcal{B}_n(\beta) \frac{t^n}{n!} - \sum_{n=0}^{\infty} \mathcal{B}_n(\beta) \frac{t^n}{n!} = t.$$

Putting

$$\exp(t) = \sum_{n=0}^{\infty} \frac{t^n}{n!}$$

in the LHS of the previous equation, we have

$$\beta \sum_{n=0}^{\infty} \frac{t^n}{n!} \sum_{n=0}^{\infty} \mathcal{B}_n(\beta) \frac{t^n}{n!} - \sum_{n=0}^{\infty} \mathcal{B}_n(\beta) \frac{t^n}{n!} = t.$$

Using multiplying series condition by the Cauchy product formula in the previous equation, we have

$$\beta \sum_{n=0}^{\infty} \sum_{b=0}^{n} \mathcal{B}_b(\beta) \frac{t^n}{b!(n-b)!} - \sum_{n=0}^{\infty} \mathcal{B}_n(\beta) \frac{t^n}{n!}$$

$$= 0 + t + 0t^2 + \cdots + 0t^n + \cdots.$$

It is time to comparing coefficients t^n on the both sides of the previous equation; after that the following equation are obtained:

For $n = 0$,

$$\beta \sum_{b=0}^{0} \mathcal{B}_b(\beta) \frac{1}{b!(0-b)!} - \frac{1}{0!} \mathcal{B}_0(\beta) = 0.$$

Since $0! = 1$, we have

$$(1-\beta)\mathcal{B}_0(\beta) = 0.$$

If $\beta \neq 1$, we have

$$\mathcal{B}_0(\beta) = 0.$$

For $n = 1$,

$$\beta \sum_{b=0}^{1} \mathcal{B}_b(\beta) \frac{1}{b!(1-b)!} - \frac{1}{1!} \mathcal{B}_1(\beta) = 1.$$

Thus,

$$\beta \mathcal{B}_0(\beta) + \beta \mathcal{B}_1(\beta) - \mathcal{B}_1(\beta) = 1.$$

From this equation, we have

$$\mathcal{B}_1(\beta) = \frac{1}{\beta - 1},$$

where $\beta \neq 1$.

For $n > 1$, we also have

$$\beta \sum_{b=0}^{n} \mathcal{B}_b(\beta) \frac{t^n}{b!(n-b)!} - \frac{1}{n!} \mathcal{B}_n(\beta) = 0.$$

Since

$$\binom{n}{b} = \frac{n!}{b!(n-b)!},$$

we have the following well-known formula:

$$\mathcal{B}_n(\beta) = \beta \sum_{b=0}^{n} \binom{n}{b} \mathcal{B}_b(\beta),$$

where $n \in \mathbb{N}$ with $n > 1$ (cf. Ref. [5]). We modify the above formulas as follows:

$$\begin{aligned}\mathcal{B}_n(\beta) - \beta \mathcal{B}_n(\beta) &= \beta \sum_{b=0}^{n-1} \binom{n}{b} \mathcal{B}_b(\beta) \\ &= \beta \mathcal{B}_0(\beta) + \beta \sum_{b=1}^{n-1} \binom{n}{b} \mathcal{B}_b(\beta).\end{aligned}$$

Therefore, the Apostol–Bernoulli numbers can be given by the following computational formula:

$$\mathcal{B}_n(\beta) = \frac{\beta}{1-\beta} \sum_{b=1}^{n-1} \binom{n}{b} \mathcal{B}_b(\beta), \tag{1.2}$$

where $n \in \mathbb{N}$ with $n > 1$.

Putting $n = 2$ in (1.2), we have

$$\begin{aligned}\mathcal{B}_2(\beta) &= \frac{\beta}{1-\beta} \sum_{b=1}^{1} \binom{2}{b} \mathcal{B}_b(\beta) \\ &= \frac{\beta}{1-\beta} 2\mathcal{B}_1(\beta).\end{aligned}$$

Thus,

$$\mathcal{B}_2(\beta) = -\frac{2\beta}{(1-\beta)^2}.$$

Putting $n = 3$ in (1.2), we have

$$\begin{aligned}\mathcal{B}_3(\beta) &= \frac{\beta}{1-\beta}\sum_{b=1}^{2}\binom{3}{b}\mathcal{B}_b(\beta)\\ &= \frac{\beta}{1-\beta}(3\mathcal{B}_1(\beta) + 3\mathcal{B}_2(\beta))\\ &= \frac{3\beta}{1-\beta}\left(\frac{1}{\beta-1} - \frac{2\beta}{(1-\beta)^2}\right).\end{aligned}$$

Thus,

$$\mathcal{B}_4(\beta) = \frac{3\beta(\beta+1)}{(\beta-1)^3}.$$

Putting $n = 4$ in (1.2), we have

$$\begin{aligned}\mathcal{B}_4(\beta) &= \frac{\beta}{1-\beta}\sum_{b=1}^{3}\binom{4}{b}\mathcal{B}_b(\beta)\\ &= \frac{\beta}{1-\beta}\left(4\mathcal{B}_1(\beta) + 6\mathcal{B}_2(\beta) + 4\mathcal{B}_3(\beta)\right)\\ &= \frac{3\beta}{1-\beta}\left(\frac{4}{\beta-1} - \frac{12\beta}{(1-\beta)^2} + \frac{12\beta\,(\beta+1)}{(\beta-1)^3}\right).\end{aligned}$$

Thus,

$$\mathcal{B}_4(\beta) = \frac{-4\beta\left(\beta^2 + 4\beta + 1\right)}{(\beta-1)^4}.$$

Putting $n = 5$ in (1.2), we have

$$\begin{aligned}\mathcal{B}_5(\beta) &= \frac{\beta}{1-\beta}\sum_{b=1}^{4}\binom{5}{b}\mathcal{B}_b(\beta)\\ &= \frac{\beta}{1-\beta}\left(\binom{5}{1}\mathcal{B}_1\,(\beta) + \binom{5}{2}\mathcal{B}_2(\beta) + \binom{5}{3}\mathcal{B}_3(\beta) + \binom{5}{4}\mathcal{B}_4\,(\beta)\right)\\ &= \frac{\beta}{1-\beta}\left(\frac{\binom{5}{1}}{\beta-1} - \frac{2\binom{5}{2}\beta}{(1-\beta)^2} + \frac{3\binom{5}{3}\beta\,(\beta+1)}{(\beta-1)^3} - \frac{4\binom{5}{4}\beta\left(\beta^2+4\beta+1\right)}{(\beta-1)^4}\right).\end{aligned}$$

Thus,

$$\mathcal{B}_5(\beta) = \frac{5\beta\left(\beta^3 + 11\beta^2 + 11\beta + 1\right)}{(\beta - 1)^5}.$$

Putting $n = 6$ in (1.2), we have

$$\begin{aligned}
\frac{1-\beta}{\beta}\mathcal{B}_6(\beta) &= \sum_{b=1}^{5}\binom{6}{b}\mathcal{B}_b(\beta) \\
&= \binom{6}{1}\mathcal{B}_1(\beta) + \binom{6}{2}\mathcal{B}_2(\beta) + \binom{6}{3}\mathcal{B}_3(\beta) \\
&\quad + \binom{6}{4}\mathcal{B}_4(\beta) + \binom{6}{5}\mathcal{B}_5(\beta). \\
\mathcal{B}_6(\beta) &= \frac{-6\beta\left(\beta^4 + 26\beta^3 + 66\beta^2 + 26\beta + 1\right)}{(\beta - 1)^6}.
\end{aligned}$$

Putting $n = 7$ in (1.2), we have

$$\begin{aligned}
\frac{1-\beta}{\beta}\mathcal{B}_7(\beta) &= \sum_{b=1}^{6}\binom{7}{b}\mathcal{B}_b(\beta) \\
&= \binom{7}{1}\mathcal{B}_1(\beta) + \binom{7}{2}\mathcal{B}_2(\beta) + \binom{7}{3}\mathcal{B}_3(\beta) + \binom{7}{4}\mathcal{B}_4(\beta) \\
&\quad + \binom{7}{5}\mathcal{B}_5(\beta) + \binom{7}{6}\mathcal{B}_6(\beta).
\end{aligned}$$

Thus,

$$\mathcal{B}_7(\beta) = \frac{7\beta\left(\beta^5 + 57\beta^4 + 302\beta^3 + 302\beta^2 + 57\beta + 1\right)}{(\beta - 1)^7}.$$

The Apostol–Bernoulli polynomials are defined by

$$\frac{t\exp(ty)}{\beta\exp(t) - 1} = \sum_{n=0}^{\infty}\mathcal{B}_n(y;\beta)\frac{t^n}{n!} \tag{1.3}$$

(cf. Ref. [5]).

By using (1.3), we get

$$\beta \exp(t) \sum_{n=0}^{\infty} \mathcal{B}_n(y;\beta)\frac{t^n}{n!} - \sum_{n=0}^{\infty} \mathcal{B}_n(y;\beta)\frac{t^n}{n!} = t\exp(ty).$$

Putting

$$\exp(yt) = \sum_{n=0}^{\infty} y^n \frac{t^n}{n!}$$

in the LHS of the previous equation, we have

$$\beta \sum_{n=0}^{\infty} \frac{t^n}{n!} \sum_{n=0}^{\infty} \mathcal{B}_n(y;\beta)\frac{t^n}{n!} - \sum_{n=0}^{\infty} \mathcal{B}_n(y;\beta)\frac{t^n}{n!} = \sum_{n=0}^{\infty} y^n \frac{t^n}{n!}.$$

Using multiplying series condition by the Cauchy product formula in the previous equation, we have

$$\beta \sum_{n=0}^{\infty}\sum_{b=0}^{n} \mathcal{B}_b(y;\beta)\frac{t^n}{b!(n-b)!} - \sum_{n=0}^{\infty} \mathcal{B}_n(y;\beta)\frac{t^n}{n!} = \sum_{n=0}^{\infty} y^n \frac{t^n}{n!}.$$

Comparing coefficients $\frac{t^n}{n!}$ on both sides of the previous equation, after that the following known formula for the Apostol–Bernoulli polynomials $\mathcal{B}_n(y;\beta)$ is obtained:

$$\beta \sum_{b=0}^{n} \binom{n}{b} \mathcal{B}_b(y;\beta) - \mathcal{B}_n(y;\beta) = y^n. \tag{1.4}$$

Some properties of the Apostol–Bernoulli polynomials $\mathcal{B}_n(y;\beta)$ are given as follows:

$$\mathcal{B}_n(y;\beta) = \sum_{j=0}^{n} \binom{n}{j} y^{n-j} \mathcal{B}_n(\beta) \tag{1.5}$$

(cf. Ref. [5]),

$$\frac{d}{dy}\{\mathcal{B}_n(y;\beta)\} = n\mathcal{B}_{n-1}(y;\beta)$$

(cf. Refs. [5, 60]),

$$\mathcal{B}_n(0) = \mathcal{B}_n(0;\beta)$$

(cf. Refs. [5, 60]).

For $\beta = 1$ into (1.3) and (1.1), we have the Bernoulli polynomials $B_n(x)$ and the Bernoulli numbers B_n, respectively.

The following recurrence relations are very important to evaluate the Apostol–Bernoulli polynomials:

$$\frac{\beta\mathcal{B}_n(y+1;\beta) - \mathcal{B}_n(y;\beta)}{n} = y^{n-1}, \tag{1.6}$$

where $n \in \mathbb{N}$ (cf. Ref. [5]).

Putting $y = 0$ into (1.6), we have

$$\mathcal{B}_n(1;\beta) = \begin{cases} \frac{1+\mathcal{B}_1(\beta)}{\beta} & \text{if } n = 1, \\ \frac{\mathcal{B}_n(\beta)}{\beta} & \text{if } n \geq 2 \end{cases} \tag{1.7}$$

(cf. Ref. [5]).

By using (1.5) and (1.2), we have the following values for the polynomials $\mathcal{B}_n(x;\lambda)$:

$$\mathcal{B}_0(x;\lambda) = 0,$$

$$\mathcal{B}_1(x;\lambda) = \frac{1}{\lambda-1},$$

$$\mathcal{B}_2(x;\lambda) = \frac{1}{\lambda-1}x - \frac{2\lambda}{(\lambda-1)^2},$$

$$\mathcal{B}_3(x;\lambda) = \frac{3}{\lambda-1}x^2 - \frac{6\lambda}{(\lambda-1)^2}x + \frac{3\lambda(\lambda+1)}{(\lambda-1)^3},$$

$$\mathcal{B}_4(x;\lambda) = \frac{4}{\lambda-1}x^3 - \frac{12\lambda}{(\lambda-1)^2}x^2 + \frac{12\lambda(\lambda+1)}{(\lambda-1)^3}x - \frac{4\lambda(\lambda^2+4\lambda+1)}{(\lambda-1)^4},$$

$$\mathcal{B}_5(x;\lambda) = \frac{5}{\lambda-1}x^4 - \frac{20\lambda}{(\lambda-1)^2}x^3 + \frac{30\lambda(\lambda+1)}{(\lambda-1)^3}x^2 - \frac{20t(\lambda^2+4\lambda+1)}{(\lambda-1)^4}x + \frac{5\lambda(\lambda^3+11\lambda^2+11\lambda+1)}{(\lambda-1)^5}.$$

In Ref. [50], we observed that when the first coefficients of each numerator of the Apostol–Bernoulli numbers are examined, we gave the following sequence $\{b_n\}_{n=1}^{\infty}$:

$$\{1, -2, 3, -4, \ldots\} = \left\{(-1)^{n+1}n\right\}_{n=1}^{\infty}.$$

That is,

$$b_n = (-1)^{n+1} n,$$

where a positive integer n is an index of the Apostol–Bernoulli numbers.

Generating function for the sequence $\{b_n\}_{n=1}^{\infty}$ is given by

$$\frac{t}{(1+t)^2} = \sum_{n=0}^{\infty} b_n t^n. \tag{1.8}$$

Applying binomial theorem to the above equation yields

$$\sum_{n=0}^{\infty} b_n t^n = \sum_{n=1}^{\infty} \binom{-2}{n-1} t^n.$$

We now compare the coefficients of $\frac{t^n}{n!}$ on both sides of the previous equation; for $n \in \mathbb{N}$, we have

$$b_n = \frac{(-2)_{n-1}}{(n-1)!}$$

(cf. Ref. [50]).

The Bernoulli polynomials and numbers of order d are defined by

$$\frac{t^d \exp(yt)}{(\exp(t)-1)^d} = \sum_{n=0}^{\infty} B_n^{(d)}(y) \frac{t^n}{n!} \tag{1.9}$$

and

$$\frac{t^d}{(\exp(t)-1)^d} = \sum_{n=0}^{\infty} B_n^{(d)} \frac{t^n}{n!} \tag{1.10}$$

(cf. Refs. [44, 61]).

By combining (1.9) with (1.10), we have

$$B_n^{(d)}(y) = \sum_{j=0}^{n} \binom{n}{j} y^{n-j} B_j^{(d)} \tag{1.11}$$

(cf. Refs. [44, 61]).

When $d=1$ in (1.9) and (1.10), we have

$$B_n^{(1)}(y)=B_n(y)$$

and

$$B_n^{(1)}=B_n.$$

By using (1.10), we get

$$\begin{aligned} t^d &= \sum_{j=0}^{d}(-1)^{d-j}\binom{d}{j}\exp(jt)\sum_{n=0}^{\infty}B_n^{(d)}\frac{t^n}{n!} \\ &= \sum_{j=0}^{d}(-1)^{d-j}\binom{d}{j}\sum_{n=0}^{\infty}\sum_{k=0}^{n}\binom{n}{k}j^{n-k}B_k^{(d)}\frac{t^n}{n!}. \end{aligned}$$

If $n=d$, we have

$$\sum_{j=0}^{d}(-1)^{d-j}\binom{d}{j}\sum_{k=0}^{d}\binom{d}{k}j^{d-k}B_k^{(d)}=d!$$

If $n\neq d$, we have

$$\sum_{j=0}^{d}(-1)^{d-j}\binom{d}{j}\sum_{k=0}^{n}\binom{n}{k}j^{n-k}B_k^{(d)}=0.$$

For $n,k\in\mathbb{N}$, Carlitz *et al.* [11] gave the following novel formulas:

$$B_n^{(d)}=\left(1-\frac{n}{d-1}\right)B_n^{(d-1)}-nB_{n-1}^{(d-1)} \tag{1.12}$$

and

$$B_d^{(d+1)}=(-1)^d d! \tag{1.13}$$

(cf. Refs. [10, 11, 23, 60]).

$$B_n^{(k+1)}=(k+1)\binom{n}{p+1}\sum_{j=0}^{k}(-1)^{k-j}\binom{k}{j}\frac{B_{n-j}B_j^{(k+1)}}{n-j}, \tag{1.14}$$

$$(x-1)(x-2)\cdots(x-n)=\sum_{j=0}^{n}\binom{n}{j}B_j^{(n+1)}x^{n-j}, \tag{1.15}$$

$$B_p^{(p)}=p\sum_{j=0}^{p-1}(-1)^j\binom{p-1}{j}\frac{B_{p-j}B_j^{(p)}}{p-j}$$

(cf. Refs. [10, 11], [44, Chapter 6]).

Carlitz [10] also defined the numbers $C_n^{(m)}$ as follows:

$$(x-1)(x-2)\cdots(x-n)=\sum_{j=0}^{n} C_j^{(n)} x^{n-j-1}$$

which gives us

$$C_j^{(n+1)}=(-1)^j\binom{n}{j}B_j^{(n+1)}$$

with

$$C_j^{(n+1)}=C_j^{(n)}+nC_{j-1}^{(n+1)}.$$

By combining (1.12) with (1.14), one has

$$\begin{aligned}
B_1^{(k)} &= -\frac{k}{2},\\
B_2^{(k)} &= \frac{3k^2-k}{12},\\
B_3^{(k)} &= \frac{-k^3+k^2}{8},\\
B_1^{(1)} &:= B_1=-\frac{1}{2},\\
B_2^{(1)} &:= B_2=\frac{1}{6},\\
B_3^{(1)} &:= B_3=0,\\
B_1^{(2)} &= -1,\\
B_2^{(2)} &= \frac{5}{6},\\
B_3^{(2)} &= \frac{1}{2},\\
B_1^{(3)} &= -\frac{3}{2},\\
B_2^{(3)} &= \frac{13}{12},
\end{aligned}$$

$$B_3^{(3)} = -\frac{9}{4},$$

$$B_1^{(4)} = -2,$$

$$B_2^{(4)} = \frac{11}{3},$$

$$B_3^{(4)} = -6$$

and so on.

By using (1.11), some values of the polynomials $B_n^{(d)}(y)$:

$$\begin{aligned} B_1^{(k)}(y) &= \sum_{j=0}^{1} \binom{1}{j} y^{n-j} B_j^{(k)} \\ &= B_0^{(k)} y + B_1^{(k)} \\ &= y - \frac{k}{2}, \\ B_2^{(k)}(y) &= \sum_{j=0}^{2} \binom{2}{j} y^{2-j} B_j^{(k)} \\ &= B_0^{(k)} y^2 + 2B_1^{(k)} y + B_2^{(k)} \\ &= B_0^{(k)} y^2 - \frac{k}{2} y + \frac{3k^2-k}{12}, \\ B_2^{(k)}(y) &= \sum_{j=0}^{2} \binom{2}{j} y^{2-j} B_j^{(k)} \\ &= B_0^{(k)} y^2 + 2B_1^{(k)} y + B_2^{(k)} \\ &= B_0^{(k)} y^3 - \frac{3k}{2} y^2 + \frac{3k^2-k}{12} y - \frac{k^3-k^2}{8} \end{aligned}$$

and so on.

Let $r \in \mathbb{N}_0$. The higher-order generalized Apostol–Bernoulli polynomials and numbers are defined by means of the following generating functions, respectively:

$$\sum_{n=0}^{\infty} \mathcal{B}_n^{(r)}(y;\beta) \frac{w^n}{n!} = \frac{w^r \exp(wy)}{(\beta \exp(w) - 1)^r} \tag{1.16}$$

and

$$\sum_{n=0}^{\infty} \mathcal{B}_n^{(r)}(\beta)\frac{w^n}{n!} = \frac{w^r}{(\beta\exp(w)-1)^r}, \tag{1.17}$$

where $|w| < 2\pi$ when $\beta = 1$ and $|w| < |\log(\beta)|$ when $\beta \neq 1$, with

$$\mathcal{B}_n^{(0)}(y;\beta) = y^n$$

and $1^r := 1$ (cf. Refs. [38, 38, 39, 61]).

Using (1.16) and (1.17), we have the some following known formula:

$$\mathcal{B}_n^{(r)}(y;\beta) = \sum_{k=0}^{n} \binom{n}{k} y^{n-k}\mathcal{B}_k^{(r)}(\beta) \tag{1.18}$$

(cf. Refs. [38, 38, 39, 61]).

By applying derivative operator with respect to t to equation (1.16), we have the some following known formula:

$$\mathcal{B}_n^{(r)}(s+1;\beta) = \frac{n\mathcal{B}_{n-1}^{(r-1)}(s;\beta) + \mathcal{B}_n^{(r)}(s;\beta)}{\beta}$$

(cf. Refs. [38, 38, 39, 61]).

Replacing r with $r+f$ in equation (1.16), we also have

$$\mathcal{B}_n^{(r+f)}(\beta) = \sum_{k=0}^{n} \binom{n}{k} \mathcal{B}_k^{(r)}(\beta)\mathcal{B}_{n-k}^{(f)}(\beta)$$

(cf. Refs. [38, 39, 61]).

With the aid of equation (1.17), Srivastava and Luo gave the following formula for the numbers $\mathcal{B}_n^{(r)}(\beta)$:

$$\mathcal{B}_n^{(r)}(\beta) = \binom{n}{r} r! \sum_{v=0}^{n-r} (-1)^v \binom{v+r-1}{v} \frac{v!\beta^v}{(\beta-1)^{v+r}} S_2(n-r,v), \tag{1.19}$$

where $n, r \in \mathbb{N}_0$ and $\beta \neq 1$ (cf. Refs. [38, 39, 61]).

Replacing n with $n+r$ and β with q in equation (1.19), we have

$$\mathcal{B}_{n+r}^{(r)}(q) = (-1)^r \frac{(n+r)!}{n!} \sum_{v=0}^{n} \frac{(v+r-1)!}{(r-1)!} \frac{q^v}{(1-q)^{v+r}} S_2(n,v). \quad (1.20)$$

Combining (1.19) with (1.18), we also get

$$\mathcal{B}_n^{(r)}(y;\beta) = \sum_{k=0}^{n} \sum_{v=0}^{k-r} (-1)^r \frac{\binom{v+r-1}{v} n!}{(n-k)!(k-r)!} \frac{v!\beta^v y^{n-k}}{(1-\beta)^{v+r}} S_2(k-r,v),$$

where $\sum_{v=0}^{k-r} = 0$ if $k-r<0$.

The Apostol–Euler polynomials are defined by means of the following generating function:

$$\frac{2\exp(ty)}{\beta\exp(t)+1} = \sum_{n=0}^{\infty} \mathcal{E}_n(y;\beta) \frac{t^n}{n!} \quad (1.21)$$

(cf. Refs. [38, 39, 61]).

The Apostol–Euler polynomials and numbers of order d are defined by

$$\frac{2^d \exp(yt)}{(\beta\exp(t)+1)^d} = \sum_{n=0}^{\infty} \mathcal{E}_n^{(d)}(y;\beta) \frac{t^n}{n!} \quad (1.22)$$

and

$$\frac{2^d}{(\beta\exp(t)+1)^d} = \sum_{n=0}^{\infty} \mathcal{E}_n^{(d)} \frac{t^n}{n!}, \quad (1.23)$$

where $|t| < \pi$ when $\beta = 1$; $|t| < |\log(-\beta)|$ when $\beta \neq 1$; $1^d := 1$ (cf. Refs. [26, 38, 39, 61]).

When $y=0$ and $\beta=1$ in (1.21), one obtains the Apostol–Euler numbers and the Euler polynomials, respectively:

$$\mathcal{E}_n(0;\beta) = \mathcal{E}_n(\beta)$$

and

$$\mathcal{E}_n\left(y;1\right)=E_n(y).$$

Thus, Stirling numbers of the first kind are defined by means of the following formula:

$$(z)_v=\sum_{d=0}^{v}S_1(v,d)z^k \tag{1.24}$$

(cf. Refs. [14, 15, 61]; see also the references cited in each of these earlier works).

Few values of Stirling numbers of the first kind $S_1(m,v)$ are given by

$m\backslash v$	0	1	2	3	4	5
0	1	0	0	0	0	0
1	0	1	0	0	0	0
2	0	−1	1	0	0	0.
3	0	2	−3	1	0	0
4	0	−6	11	−6	1	0
5	0	24	−50	35	−10	1

Stirling numbers of the second kind are defined by

$$\frac{\left(\exp(t)-1\right)^r}{r!}=\sum_{n=0}^{\infty}S_2\left(n,r\right)\frac{t^n}{n!} \tag{1.25}$$

and

$$z^v=\sum_{d=0}^{v}S_2(v,d)\left(z\right)_d \tag{1.26}$$

and their computational formula

$$S_2\left(n,r\right)=\frac{1}{r!}\sum_{v=0}^{r}(-1)^{r-v}\binom{r}{v}v^n \tag{1.27}$$

with $S_2\left(0,0\right)=1$, $S_2\left(n,r\right)=0$ when $r>n$ (cf. Refs. [17, 30, 50, 55, 61]).

By using (1.27), few values of Stirling numbers of the second kind $S_2(m, v)$ are given by

$m\backslash v$	0	1	2	3	4	5
0	1	0	0	0	0	0
1	0	1	0	0	0	0
2	0	1	1	0	0	0.
3	0	1	3	1	0	0
4	0	1	7	6	1	0
5	0	1	15	25	10	1

Using (1.1) and (1.25), one has the following relation:

$$\mathcal{B}_n(\beta) = \frac{n\beta}{(\beta-1)^n} \sum_{s=1}^{n-1} (-1)^s \, s! \beta^{s-1} (\beta-1)^{n-1-s} \, S_2(n-1, s), \tag{1.28}$$

where $n \in \mathbb{N}$ and $\beta \neq 1$ (cf. Ref. [5]).

By using (1.10), Gun and Simsek showed that

$$\sum_{k=0}^{n} \binom{n}{k} S_2(n-k, d) B_k^{(d)} = 0.$$

From the above equation, we also get

$$\frac{1}{d!} \sum_{k=0}^{n} \binom{n}{k} \sum_{v=0}^{d} (-1)^{d-v} \binom{d}{v} v^{n-k} B_k^{(d)} = 0.$$

The array polynomials are defined by

$$\frac{(\exp(t)-1)^r \exp(xt)}{r!} = \sum_{n=0}^{\infty} S_r^n(x) \frac{t^n}{n!} \tag{1.29}$$

(cf. Ref. [9, 56]).

When $x = 0$ in (1.29), we derive

$$S_r^n(0) = S_2(n, r).$$

The Eulerian polynomials are defined by

$$\frac{k-1}{k - \exp(t(k-1))} = \sum_{n=0}^{\infty} A_n(k) \frac{t^n}{n!} \tag{1.30}$$

(cf. Refs. [17, 47]).

Using (1.30), we have

$$A_n(k) = \sum_{s=0}^{n} A_{n,s} k^s, \tag{1.31}$$

where $A_{n,s}$ denotes the Eulerian numbers, the number of permutations of the numbers 1 to n in which exactly s elements, are given by the following alternating sum formula

$$A_{n,s} = \sum_{v=0}^{s} (-1)^v \binom{n+1}{v} (s-v+1)^n \tag{1.32}$$

(cf. Refs. [17, 47]).

The Bernstein basis functions $B_r^n(w)$ are defined by

$$B_r^n(w) = \binom{n}{r} w^r (1-w)^{n-r},$$

where $0 \leq r \leq n$, and $n, r \in \mathbb{N}_0$ and also for $r < 0$ or $r > n$,

$$B_r^n(w) = 0$$

(cf. Refs. [22, 52]).

Fubini numbers are defined by

$$\frac{1}{2-\exp(t)} = \sum_{n=0}^{\infty} w_g(n) \frac{t^n}{n!}$$

(cf. Ref. [17]).

Fubini numbers, also known as the ordered Bell numbers, count the number of weak ordering of a set of n elements. Thereby, these numbers have many important applications in number theory, probability theory and combinatorics. The numbers $w_g(n)$ are also related to the many special numbers. Some of them are given as follows:

$$\mathcal{B}_n\left(\frac{1}{2}\right) = -2nw_g(n-1), \tag{1.33}$$

$$w_g(n) = \sum_{r=0}^{n} r! S_2(n,r),$$

$$w_g(n) = \sum_{r=0}^{n-1} 2^r A_{n,r}$$

and

$$w_g(n) = A_n(2)$$

(cf. Refs. [17, 26, 30]).

The Fubini-type polynomials and numbers of order d are defined as follows:

$$\frac{2^d}{(2-\exp(t))^{2d}}\exp(xt) = \sum_{n=0}^{\infty} a_n^{(d)}(x)\frac{t^n}{n!} \tag{1.34}$$

and

$$\frac{2^d}{(2-\exp(t))^{2d}} = \sum_{n=0}^{\infty} a_n^{(d)}\frac{t^n}{n!} \tag{1.35}$$

(cf. Ref. [30]).

The generalized Fubini polynomials of order r are defined as follows:

$$\left(\frac{\exp(t)-1}{k+1-k\exp(t)}\right)^d \exp(xt) = \sum_{n=0}^{\infty} f_{n,k}^{(d)}(x)\frac{t^n}{n!}, \tag{1.36}$$

where $k, d \in \mathbb{N}_0$ (cf. Ref. [26]).

For $x = 0$ in (1.36), the generalized Fubini numbers of order r are given by

$$\left(\frac{\exp(t)-1}{k+1-k\exp(t)}\right)^d = \sum_{n=1}^{\infty} \frac{f_{n,k}^{(d)}}{n!}t^n \tag{1.37}$$

(cf. Ref. [30]).

1.2. *Geometric distributions*

First, we state that the geometric distribution models the number of trials needed to get the first success in repeated independent attempts. This distribution has many applications in the theory of probability and statistics, and also other sciences.

It is well known that discrete probability distributions, which have a very important place in the theory of probability and statistics, have been applied in many different areas in recent years. Some of these are given as follows: in probability and statistics, cost–benefit analysis, sports applications, tossing a coin, feedback from customers, number of supporters of a law, number of faulty products manufactured at an industry, number of bugs in a code, a teacher examining test records, playing a game, throwing darts at a dartboard, number of network failures, biology, etc. (cf. Refs. [6, 13–15, 18, 19, 64]; see also the references cited in each of these earlier works).

Geometric distributions helps us to find the number of trials until the first success occurs. In order to understand how many attempts it takes before achieving the first success, e.g. how many times you need to flip a coin before getting heads, geometric distributions easily help us in these topics (cf. Ref. [64]). This distribution can be briefly described as follows:

The probability distribution of the number X of the Bernoulli trials needed to get one success, supported on the set $\mathbb{N} = \{1, 2, \ldots\}$, which of these is called the geometric distribution (cf. Refs. [6, 13–15, 18, 19]). That is, X is a geometric random variable with parameter p. Thus, the geometric distribution gives the probability that the first occurrence of success requires r independent trials, each with success probability p.

If the probability of success on each trial is parameter p, then the probability that the r-th trial is the first success is

$$P(X = r) = pq^{r-1},$$

where $r \in \mathbb{N}$ and

$$p + q = 1$$

with $0 < p < 1$ (cf. Refs. [6, 16, 18, 19]).

For example, see for details Ref. [64], how many emails you must send before a customer makes a purchase. It helps predict the chance of success happening at each trial. Consequently, count trials until first success; each trial is independent with fixed success probability; useful for questions such as "How many emails until first purchase?"; graph shows a decreasing curve–fewer trial is more likely.

In this distribution, the mean represents the average value that can be expected as a result of an experiment repeated several times. Variance is a measure of the spread of data.

The moment generating function and the characteristic function of the geometric distribution with success probability p are defined as follows, respectively:

$$M_x(w;p,q) := M_X(w) = \mathbf{E}(\exp(Xw))$$

and

$$K_X(w) = \mathbf{E}(\exp(iXw)),$$

where $i^2 = -1$ (cf. Refs. [6, 18, 37]).

For $n \in \mathbb{N}$, taking n times derivative of the moment generating function $M_X(w)$ for $w = 0$, one has the following moments of geometric random variable with parameter p:

$$\begin{aligned}\mu_n &:= \mu_n(p,q) \\ &:= \mathbf{E}(X^n;p,q) \\ &= \frac{d^n}{dw^n}\{M_X(w;p,q)\}\,|_{w=0}\end{aligned}$$

and

$$\mu_n(p,q) = \frac{d^n}{dw^n}\{\mathbf{E}(\exp(xw))\}\,|_{w=0} \tag{1.38}$$

(cf. Refs. [6, 18, 37, 52]).

Suppose that an infinite sequence of Bernoulli trials with probability of success p are available. The number X is a negative binomial variable with parameters p and r ($r = 1, 2, 3, \ldots,$ and, $0 < p < 1$). The number X of failures that occur before the rth success has the following probability distribution functions known as the *negative binomial distribution*:

$$f_N(k;p,q;r) = \binom{k+r-1}{k} p^r (1-p)^k, \tag{1.39}$$

where $k \in \mathbb{N}_0$, r is the number of successes, k is the number of failures and p is the probability of success on each trial, $p + q = 1$:

$$f_N(k;p,q;r) = 0$$

if $k \notin \mathbb{N}_0$ (cf. Ref. [18, p. 297]; see also Refs. [6, 8, 13, 20], and see also the references cited in each of these earlier works).

Observe that negative binomial distribution is also expressed as follows: A random variable X has the negative binomial distribution with parameters r and p ($r = 1, 2, \ldots$ and $0 < p < 1$) if X has a discrete distribution for which the probability function $f_N(k; p, q; r)$ is as specified by equation (1.39), see (cf. Ref. [18]).

This distribution is a discrete distribution and is also known as the Pascal distribution, named after French mathematician Blaise Pascal (19 June 1623–19 August 1662) (cf. Ref. [18, p. 298], see also Refs. [6, 8, 13, 20]).

Observe that the most proverbial special case of a Pascal distribution random variable is one for which $r = 1$. That is, putting $r = 1$ in equation (1.39), the variable X becomes the number of trials before the first failure. Then, the Pascal distribution reduces to the geometric distribution.

Some special values of the function $f_{G,r}(k; p, q)$ are given by

$$f_N(1; p, q; 1) = p(1-p) = pq,$$
$$f_N(0; p, q; 1) = 1,$$
$$f_N(0; p, q; r) = (-1)^r \binom{-r}{k} p^r$$

and

$$f_N(k; p, q; 0) = (1-p)^k = q^k.$$

This distribution gives us that not only the mean represents the average value that can be expected as a result of an experiment repeated several times, but also the variance is a measure of the spread of data (cf. Refs. [6, 8, 13, 18, 20]).

1.3. *Moment generating function of the geometric distribution in terms of the Apostol–Bernoulli numbers and polynomials*

Here, many new formulas for the moment-generating function of the geometric distribution in terms of the Apostol–Bernoulli numbers and polynomials and other special functions are given. Using this function, computational formulas for nth moment of the geometric random variable with parameter p are also given. These formulas give us direct computation formula for these moments nth moment of the geometric random variable with parameter p without any infinite series formulas involving geometric series, etc.

Theorem 1. *Let $p+q=1$. Let $M_X(w) := M_X(w;p,q)$ be the moment-generating function of the X geometric random variable with parameter p. Then, we have*

$$M_X(w) = -p\sum_{n=0}^{\infty} \frac{\mathcal{B}_{n+1}(1;q)}{n+1}\frac{w^n}{n}. \tag{1.40}$$

Proof. Proof of (1.40) was given by the author [52]. We give briefly this proof methods with their related equations as follows. Moment-generating function for the geometric distribution is given by

$$\begin{aligned} M_X(w) &= \mathbf{E}(\exp(Xw)) \\ &= p\exp(w)\sum_{k=1}^{\infty} q^{k-1}\exp((k-1)w) \end{aligned} \tag{1.41}$$

(cf. Refs. [6, 18, 37]). By using geometric series expansion for $|q\exp(w)| < 1$, we have

$$M_X(w) = p\exp(w)\frac{1}{1-q\exp(w)}.$$

Therefore,

$$M_X(w) = -\frac{p\exp(w)}{q\exp(w)-1} \tag{1.42}$$

(cf. Refs. [6, 18]). Thus, we have

$$M_X(w) = -\frac{p}{w}\left(\frac{w\exp(w)}{q\exp(w)-1}\right).$$

Now, combining (1.42) with (1.3), we have

$$M_X(w) = -\frac{p}{w}\sum_{n=0}^{\infty}\mathcal{B}_n(1;q)\frac{w^n}{n!}.$$

Therefore,

$$M_X(w) = -p\sum_{n=0}^{\infty} \frac{\mathcal{B}_{n+1}(1;q)}{(n+1)!} w^n. \tag{1.43}$$

Due to equation (1.43), we get proof of the theorem. □

By using (1.41), we have

$$\mathbf{E}(\exp(Xw)) = \mathbf{E}\left(\sum_{n=0}^{\infty} \frac{X^n}{n!} w^n\right).$$

With the aid of the above equation, equation (1.40) reduces to the following equation:

$$\mathbf{E}\left(\sum_{n=0}^{\infty} \frac{X^n}{n!} w^n\right) = -p\sum_{n=0}^{\infty} \frac{\mathcal{B}_{n+1}(1;q)}{(n+1)!} w^n.$$

Since

$$\mathbf{E}\,(aX+b) = a\mathbf{E}\,(X) + b,$$

we have

$$\sum_{n=0}^{\infty} \frac{\mathbf{E}(X^n)}{n!} w^n = -p\sum_{n=0}^{\infty} \frac{\mathcal{B}_{n+1}(1;q)}{(n+1)!} w^n.$$

It is well known that

$$\mu_n(p,q) := \mu_n = \mathbf{E}(X^n),$$

which denotes moments of geometric random variable with parameter p; we also have

$$\sum_{n=0}^{\infty} \mathbf{E}(X^n)\frac{w^n}{n!} = \sum_{n=0}^{\infty} \left(\frac{-p\mathcal{B}_{n+1}(1;q)}{n+1}\right) \frac{w^n}{n!}.$$

We now compare the coefficients of $\frac{w^n}{n!}$ on both sides of the previous equation; we have

$$\mathbf{E}(X^n) = \frac{-p\mathcal{B}_{n+1}(1;q)}{n+1}$$

or

$$\mu_n(p,q) = \frac{-p\mathcal{B}_{n+1}(1;q)}{n+1}.$$

By the above equation, we arrive at the following theorem:

Theorem 2 (cf. Ref. [52, Theorem 2.2]). *Let $p+q=1$ and $n \in \mathbb{N}_0$. Let $\mu_n(p,q)$ be nth moment of the X geometric random variable with parameter p. Then, we have*

$$\mu_n := \mu_n(p,q) = -\frac{p\mathcal{B}_{n+1}(1;q)}{n+1}. \tag{1.44}$$

By combining (1.44) with (1.7), we get

$$\mu_n(p,q) = -\frac{p\mathcal{B}_{n+1}(1;q)}{n+1} = \begin{cases} -p\frac{1+\mathcal{B}_1(q)}{q} & \text{if } n=0, \\ -p\frac{\mathcal{B}_{n+1}(q)}{q(n+1)} & \text{if } n \geq 1. \end{cases}$$

By combining (1.44) with (1.38) and the above formula, for $n \geq 1$, we have

$$\begin{aligned} \mu_n(p,q) &= -\frac{p\mathcal{B}_{n+1}(1;q)}{n+1} \\ &= -p\frac{\mathcal{B}_{n+1}(q)}{q(n+1)} \\ &= \frac{d^n}{dw^n}\{\mathbf{E}(\exp(Xw))\}\,|_{w=0}. \end{aligned}$$

By using the above formula, the Apostol–Bernoulli numbers can be computed by the following formula:

Theorem 3. *Let $p+q=1$ and $n \in \mathbb{N}_0$. Let $\mu_n(p,q)$ be nth moment of the X geometric random variable with parameter p. Then, we have*

$$\mathcal{B}_{n+1}(q) = -\frac{q}{p}(n+1)\frac{d^n}{dw^n}\{\mathbf{E}(\exp(Xw))\}\,|_{w=0}. \tag{1.45}$$

We [50] also gave decomposition formulas for the nth moment of the geometric random variable (with parameter p) in terms of the

Apostol–Bernoulli polynomials:

$$\mu_{2n}(p,q) = -\frac{p\mathcal{B}_{2n+1}(1;q)}{2n+1} = \begin{cases} \frac{q-1}{q}\left(1+\mathcal{B}_1(q)\right) & \text{if } n=0, \\ \frac{q-1}{q}\frac{\mathcal{B}_{2n+1}(q)}{2n+1} & \text{if } n \geq 1 \end{cases} \tag{1.46}$$

and

$$\mu_{2n+1}(p,q) = -\frac{p\mathcal{B}_{2n+2}(1;q)}{2n+2} \tag{1.47}$$

or

$$\mu_{2n+1}(p,q) = \frac{q-1}{2q}\frac{\mathcal{B}_{2n+2}(q)}{n+1},$$

where $p+q=1$ with $0<p<1$ and $n \in \mathbb{N}_0$.

By substituting (1.28) into (1.44), for $n \geq 2$, we have

$$\begin{aligned} \mu_n(p,q) &= -\frac{p\mathcal{B}_{n+1}(1;q)}{n+1} \\ &= -p\frac{\mathcal{B}_{n+1}(q)}{q(n+1)} \\ &= -\frac{p}{(q-1)^{n+1}}\sum_{s=1}^{n}(-1)^s s! q^{s-1}(q-1)^{n-s} S_2(n,s). \end{aligned}$$

Since

$$p = 1-q,$$

from the previous equation, we have

$$\mu_n(p,q) = \sum_{s=1}^{n} s!\frac{q^{s-1}}{p^s} S_2(n,s).$$

After some calculations in the above equation, we deduce the following theorem:

Theorem 4 (cf. Ref. [52, Theorem 2.2]). *Let $p+q=1$ and $n \in \mathbb{N}$. Let $\mu_n(p,q)$ be nth moment of the X geometric random variable with parameter p. Then, we have*

$$\mu_n(p,q) = \frac{1}{q}\sum_{s=1}^{n} s!\left(\frac{q}{p}\right)^s S_2(n,s) \tag{1.48}$$

or

$$\mu_n(p,q) = p\sum_{s=1}^{n}\frac{q^{s-1}}{(1-q)^{s+1}} s! S_2(n,s). \tag{1.49}$$

Due to $0 < q < 1$, it is possible to apply binomial theorem to (1.49); we have

$$\mu_n(p,q) = p\sum_{s=1}^{n} q^{s-1} s! S_2(n,s) \sum_{m=0}^{\infty} \binom{s+m}{m} q^m$$
$$= p\sum_{s=1}^{n} q^{s-1} s! S_2(n,s) \sum_{m=0}^{\infty} \binom{-s}{m} (-1)^m q^m,$$

where

$$\frac{1}{(1-q)^{s+1}} = \sum_{m=0}^{\infty} \binom{s+m}{m} q^m$$
$$= \sum_{m=0}^{\infty} \binom{-s}{m} (-1)^m q^m,$$

thus,

$$\binom{s+m}{m} = (-1)^m \binom{-s}{m},$$

where

$$\binom{-s}{m} = \frac{-s(-s-1)(-s-2)\cdots(-s-m+1)}{m!}.$$

Therefore,

$$\mu_n(p,q) = p\sum_{s=1}^{n} s! S_2(n,s) \sum_{m=0}^{\infty} \binom{-s}{m} (-1)^m q^{m+s-1}.$$

Substituting the following well-known formula

$$s! = (-1)^s (s+1) \sum_{j=1}^{s} S_1(s,j) B_j$$

(cf. Refs. [32, 48]), we get the following theorem involving relation among the Bernoulli numbers B_n, the Stirling numbers of the first kind $S_1(n,s)$, the Stirling numbers of the second kind $S_2(n,s)$ and nth moment of the X geometric random variable with parameter p:

Theorem 5. *Let $p + q = 1$ and $n \in \mathbb{N}$. Let $\mu_n(p, q)$ be nth moment of the X geometric random variable with parameter p. Then, we have*

$$\mu_n(p,q) = \sum_{s=1}^{n} \sum_{j=0}^{s} (-1)^s (s+1) \frac{q^{s-1}}{p^s} S_2(n,s) S_1(s,j) B_j. \quad (1.50)$$

Problem 1. *By using the orthogonality relation of the Stirling numbers in equation* (1.50), *find other new formulas among the Bernoulli numbers, finite sums and nth moment of the X geometric random variable with parameter p.*

Problem 2. *Combining equation* (1.13) *with* (1.50), *find relations among the Bernoulli numbers, the Bernoulli numbers of higher order, the Stirling numbers and nth moment of the X geometric random variable with parameter p.*

Substituting explicit formula for the Stirling numbers of the second kind $S_2(n, s)$ in equation (1.27) into (1.49), we obtain the following formula for the moments of geometric random variable with parameter p:

$$\mu_n(p,q) = \frac{p}{q} \sum_{s=1}^{n} \sum_{j=0}^{s} (-1)^{s-j} \binom{s}{j} \frac{q^s}{(1-q)^{s+1}} j^n. \quad (1.51)$$

In Ref. [52], We gave modification of the formula (1.51) by the following theorem, which has the same proof of that of the formula (1.51), so we omit it.

Theorem 6. *Let $p + q = 1$ and $j \in \mathbb{N}_0$. Let $\mu_n(p, q)$ be nth moment of the X geometric random variable with parameter p. Then, we have*

$$\mu_n(p,q) = \frac{1}{q} \sum_{v=1}^{n} \sum_{j=0}^{v} (-1)^{v-j} \binom{v}{j} \left(\frac{q}{p}\right)^v j^n. \quad (1.52)$$

1.4. *n*th moment of the *X* geometric random variable with parameter *p* in terms of the polylogarithm function

The polygamma function of order n, $\mathrm{Li}_n(w)$, is a meromorphic function on the set of complex numbers. This function is related to derivative of the logarithm of the Euler gamma function. The polygamma function of order n is defined by the following infinite series:

$$\mathrm{Li}_n(w) = \sum_{v=1}^{\infty} \frac{w^v}{v^n},$$

where $w \in \mathbb{C}$ with $|w| \leq 1$ and $n \in \mathbb{Z}$ with $n \neq 1$ (cf. Refs. [60, 61]).

When $s = 2$, we have the polygamma function of order 2 which is known as the dilogarithm function

$$\mathrm{Li}_2(w) = \sum_{v=1}^{\infty} \frac{w^v}{v^2}$$

with $|w| \leq 1$ (cf. Refs. [60, 61]).

Relations among $\mathrm{Li}_n(w)$, the Apostol–Bernoulli numbers and polynomials, and the nth moment of the X geometric random variable with parameter p are given as follows:

$$\mu_n(p,q) = \frac{p}{q} \sum_{v=1}^{\infty} v^n q^v \tag{1.53}$$

(cf. Ref. [19]). Since $0 < q < 1$, we have the following known series in terms of the Apostol–Bernoulli polynomials:

$$\sum_{v=1}^{\infty} v^j q^v = -\frac{q\mathcal{B}_{j+1}(1;q)}{j+1}, \tag{1.54}$$

$$\sum_{v=1}^{\infty} v^j q^v = \frac{q}{p} \mathrm{Li}_{-j}(q).$$

In Ref. [52], we studied on the following known formula:

$$\mu_n(p,q) = \frac{q}{p} \mathrm{Li}_{-n}(q) \tag{1.55}$$

(cf. Refs. [19, 60, 61]). Hence, one has

$$\mathrm{Li}_{-n}(q) = -\frac{p\mathcal{B}_{n+1}(1;q)}{n+1}$$

(cf. Refs. [19, 60, 61]). For $n \geq 1$, we have

$$\mathrm{Li}_{-n}(q) = \frac{(q-1)\mathcal{B}_{n+1}(q)}{q(n+1)}.$$

1.5. *Characteristic function of the geometric distribution*

The characteristic functions were used to solve certain problems in probability theory. Characteristic functions for random variables of any probability distribution are defined on more general cases. These functions are also used to examine the limit distributions of sums of independent random variables. It is used in statistics, probability and mathematical statistics in applications of a wide variety of problems including variance, expected value and other statistical parameters that can be approached by the method of characteristic functions. The characteristic functions are an important and useful branch for mathematical statisticians, as they provide various different and important applications involving real-world problems in applied sciences (see Ref. [37]). Unlike the moment-generating functions of any probability distribution, the characteristic functions always exist when treated as a function of a real-valued argument. It is well known that interesting, applicable relations exist between the behavior of the characteristic functions and properties of the distribution functions, such as the existence of moments and the density functions. These important features and extended properties give us the existence of relationships between the behavior of a distribution's characteristic function and its properties, such as the existence of moments and the existence of the probability functions (cf. Refs. [6, 12, 16, 19, 20, 36, 37, 40–42, 50, 54, 58, 59]; see also the references cited in each of these earlier works).

The characteristic function of the geometric distribution is given as follows:

$$K_X(-iw; p, q) := K_X(-iw) = \mathbf{E}\left(\exp(Xiw)\right) \tag{1.56}$$

(cf. Refs. [6, 37, 54]).

A characteristic function of the geometric random variable with parameter p in terms of the Apostol–Bernoulli numbers is given by the following theorem:

Theorem 7. *Let $p + q = 1$ and $0 < q < 1$. Then, we have*

$$K_X(w) = 1 - \frac{p}{q}\sum_{n=1}^{\infty} i^n \frac{\mathcal{B}_{n+1}(q)}{n+1}\frac{w^n}{n!}. \tag{1.57}$$

Proof of (1.57) was given in Ref. [50]. For details about the function $K_X(w)$, see also Ref. [50].

1.6. *Moment generating function for the negative binomial random in terms of the Apostol–Bernoulli numbers of order r*

In Ref. [51], we gave the relation among the moment generating function for the negative binomial random variable with parameters p and r, the Apostol–Bernoulli numbers of higher order and the Stirling numbers. Here, we give these relations by the following theorems and some novel formulas for the negative binomial distribution with parameters r and p.

Theorem 8. *If X has the negative binomial distribution with parameters r and p, then the moment-generating function of X is given by the following formula, which is related to the generating function for the Apostol–Bernoulli numbers:*

$$M_X(t;p,q,r) = \left(\frac{p}{1-q\exp(t)}\right)^r, \tag{1.58}$$

where $|t| < \log\left(\frac{1}{q}\right)$ and $p+q=1$ with $0<p<1$. Here, r is not necessarily an integer.

Proofs of this theorem were given by many different method, see Ref. [18, Theorem 5.5.3, p. 299] and see also (cf. Refs. [6, 8, 13, 20]):

$$M_M(t;p,q,r) = \mathbf{E}(\exp(xt);p,q,r) := \mathbf{E}(\exp(xt)).$$

From Equation (1.58), we have

$$M_X(t;p,q,r) = p^r \sum_{k=0}^{\infty} \exp(kt) \binom{k+r-1}{k} q^k, \tag{1.59}$$

the series on the right-hand side of the above equation is convergence for

$$0 < q\exp(kw) < 1,$$

that is,

$$t < \log\left(\frac{1}{q}\right)$$

or

$$t < \log\left(\frac{1}{1-p}\right).$$

With the aid of equations (1.17) and (1.59), we have

$$M_x(w;p,q,r) = (-p)^r \sum_{n=0}^{\infty} \mathcal{B}_n^{(r)}(q)\frac{w^{n-r}}{n!}.$$

By the above relation, we derive the following theorem:

Theorem 9 (cf. Ref. [51]). *Let $p+q=1$ with $0<p<1$. Then, we have*

$$M_x(w;p,q,r) = (-p)^r \sum_{n=0}^{\infty} \mathcal{B}_{n+r}^{(r)}(q)\frac{w^n}{\binom{n+r}{r}r!n!}. \tag{1.60}$$

The jth moments of geometric random variable with parameter p are given by

$$\mu_j(p,q,r) = \frac{d^j}{du^j}\left\{\mathbf{E}(\exp(xw,;p,q,r))\right\}|_{w=0}. \tag{1.61}$$

By combining (1.60) with (1.61), we have the following theorems:

Theorem 10 (cf. Ref. [51]). *Let $p+q=1$ and $n\in\mathbb{N}_0$. Let $\mu_n(p,q,r)$ be nth moment of the X negative binomial random variable with parameter p and r. Then,*

$$\mu_n(p,q,r) = (-p)^r\frac{n!\mathcal{B}_{n+r}^{(r)}(q)}{(r+n)!}. \tag{1.62}$$

By combining (1.62) with (1.20), we get

$$\mu_n(p,q,r) = p^r\sum_{v=0}^{n}(-1)^v\binom{v+r-1}{v}\frac{v!q^v}{(q-1)^{v+r}}S_2(n,v).$$

Therefore, we arrive at the following theorem:

Theorem 11. *Let $p+q=1$ and $n\in\mathbb{N}_0$. Let $\mu_n(p,q,r)$ be nth moment of the X negative binomial random variable with parameter p and r. Then,*

$$\mu_n(p,q,r) = \sum_{v=0}^{n}\frac{(v+r-1)!}{(r-1)!}\frac{q^v}{p^v}S_2(n,v), \tag{1.63}$$

where $r\geq 1$.

Substituting (1.27) into (1.63), we have the following theorem:

Theorem 12 (cf. Ref. [51]). *Let* $p + q = 1$ *and* $n \in \mathbb{N}_0$. *Let* $\mu_n(p, q, r)$ *be nth moment of the X negative binomial random variable with parameter p and r. Then,*

$$\mu_n(p, q, r) = \sum_{v=0}^{n} \sum_{k=0}^{v} (-1)^{v-k} \binom{v + r - 1}{v} \binom{v}{k} \left(\frac{q}{p}\right)^v k^n \tag{1.64}$$

By the aid of formula in (1.64), the following moments of the X negative binomial random variable with parameter p and r can be given:

$$\mu_1(p, q, r) = \frac{rq}{p}, \tag{1.65}$$

$$\mu_2(p, q, r) = \frac{rqp + \left(r^2 + r\right) q^2}{p^2},$$

$$\mu_3(p, q, r) = \frac{rqp^2 + 3(r^2 + r)q^2 p + 4(r^3 + 3r^2 + 2r)q^3}{p^3}$$

and so on (cf. Ref. [51]).

Using the above moments, the mean and the variance of the negative binomial distribution with parameters r and p are given as follows:

The mean of X is

$$\mathbf{E}(X; p, q, r) = \mu_1(p, q, r) = \frac{rq}{p}.$$

The variance of X is

$$\begin{aligned} Var(X, p, q, r) &= \mu_2(p, q, r) - (\mu_1(p, q, r))^2 \\ &= \frac{rqp + \left(r^2 + r\right) q^2}{p^2} - \left(\frac{rq}{p}\right)^2 \\ &= \frac{rq(1 - q) + rq^2}{p^2} \\ &= \frac{rq}{p^2}. \end{aligned}$$

Thus,

$$Var(X,p,q,r) = \mu_2(p,q,r) - (\mu_1(p,q,r))^2 = \frac{rq}{p^2}.$$

The above calculations were given by Refs. [18, p. 300] and [51].

The other jth moment of the X negative binomial random variable with parameter p and r are also given as follows:

Theorem 13 (cf. Ref. [51]). *Let $p+q=1$ with $0<p<1$ and $j \in \mathbb{N}_0$. Let $\mu_j(p,q,r)$ be jth moment of the X negative binomial random variable with parameter p and r. Then, we have*

$$\mu_j(p,q,r) = p^r \sum_{k=0}^{\infty} \binom{k+r-1}{k} k^j q^k. \tag{1.66}$$

1.7. *Factorial moments derived from z-transform related to random variable*

Here, we survey the z-transform, which is called the probability generating function, is associated with a random variable. The z-transform is denoted by $G_X(z)$. This transform is an alternative representation of a probability law. It is associated with not only a random variable X but also moment-generating function (cf. Refs. [6, 45]).

In order to give definition of the z-transform, setting

$$f(x) = P(X = x)$$

is the probability distribution of the non-negative discrete random variable X. This function satisfies the following relations:

If $x < 0$, then

$$f(x) = 0.$$

The z-transform of the probability distribution $f(x)$ is given by

$$\begin{aligned} G_X(z) &= \mathbf{E}(z^X) \\ &= \sum_{x=0}^{\infty} z^x f(x) \end{aligned}$$

(cf. Ref. [6]),

$$\begin{aligned} G_x(1) &= \mathbf{E}(1) \\ &= \sum_{x=0}^{\infty} f(x) \\ &= 1. \end{aligned}$$

It is easy to see that using the higher-order derivatives of the probability generating function evaluated at $z = 1$, the factorial moments

$$\mu_k\left((x)_k\right) := \mu_k\left((x)_k; p, q\right)$$

can be computed by the following derivative equation:

$$\begin{aligned} \mu_k\left((x)_k; p, q\right) &= \frac{d^k}{dz^k}\left\{G_x(z)\right\}|_{z=1} \\ &= \sum_{x=0}^{\infty} (x)_k f(x). \end{aligned}$$

Substituting (1.24) into the above equation, we have

$$\mu_k\left((x)_{(k)}; p, q\right) = \sum_{j=0}^{k} S_1(k,j) \sum_{x=0}^{\infty} x^j f(x) \tag{1.67}$$

(cf. Refs. [45, 52]).

Combining (1.67), (1.53) and (1.24) with the geometric random variable X with parameter p, we have the following factorial moments:

$$\mu_k\left((x)_k; p, q\right) = \frac{1-q}{q} \sum_{j=0}^{k} S_1(k,j) \sum_{x=0}^{\infty} x^j q^x.$$

Combining the above equation with the following definition

$$E(X^j) := \mu_j(p,q) = \sum_{x=0}^{\infty} x^j q^x,$$

we have

$$\mu_k\left((x)_k; p, q\right) = \sum_{j=0}^{k} S_1(k,j) \mu_j(p,q). \tag{1.68}$$

Combining (1.68) with equations (1.53) and (1.44), we [52] found the following novel relation between the kth moments and the kth factorial moments of the geometric random variable X:

Theorem 14 (cf. Ref. [52]). *Let $p+q=1$ and $k \in \mathbb{N}_0$. Let $\mu_k(p,q)$ be kth moment of the X geometric random variable with parameter p. Then, we have*

$$\mu_k\left((x)_k;p,q\right)=(q-1)\sum_{j=0}^{k}\frac{S_1(k,j)\mathcal{B}_{j+1}(1;q)}{j+1}.$$

1.8. *Open problems*

We [50] also gave the following open problems:

Problem 3. *Is there any relationship between moments and the triangle of the Eulerian numbers?*

Problem 4. *How can the polynomial of q in the numerators of the moments be expressed in terms of the triangle of the Eulerian numbers?*

1.9. *Partial answer of the problems*

Simsek and Kilar [53] gave partial answers to these open problems with the aid of relations relations between moment of the geometric random variable with parameter p and the Eulerian numbers and polynomials.

Combining (1.46) and (1.47) with the following identity

$$A_{m-1}\left(q\right)=-\frac{\left(1-q\right)^{m}}{mq}\mathcal{B}_{m}\left(q\right), \tag{1.69}$$

where $m \in \mathbb{N}$ (cf. Refs. [7], [55, Corollary 4]), Simsek and Kilar [53] gave the following formula:

$$\mu_{2m}\left(p,q\right)=p^{-2m}A_{2m}\left(q\right) \tag{1.70}$$

and

$$\mu_{2m+1}\left(p,q\right)=p^{-2m-1}A_{2m+1}\left(q\right). \tag{1.71}$$

Therefore, we have

$$\mu_m(p,q) = p^{-m} A_m(q) \tag{1.72}$$

(cf. Ref. [53]).

Combining (1.72) with (1.32) and (1.31), Simsek and Kilar [53] proved the following theorem:

Theorem 15. *Let $p + q = 1$ with $0 < p < 1$, $n \in \mathbb{N}_0$. Let $\mu_n(p,q)$ be nth moment of the geometric random variable with parameter p. Then, we have*

$$\mu_n(p,q) = \sum_{r=0}^{n} \sum_{v=0}^{r} (-1)^v \binom{n+1}{v} q^r \left(\frac{r-v+1}{1-q}\right)^n. \tag{1.73}$$

Here, we note that using the following infinite series representation for $A_n(q)$, proved in Ref. [47],

$$A_n(q) = (1-q)^{n+1} \sum_{k=0}^{\infty} (k+1)^n q^k,$$

Simsek and Kilar [53] also found the formula given in (1.55).

By the aid of moment formulas with their applications, we obtain some data, given in the following.

Simsek and Kilar [53] calculated the following list of many values of the moment $\mu_n(p,q)$ in terms of the Eulerian numbers via implementation of the equation (1.73) in the Wolfram language:

$$\mu_0(p,q) = 1,$$

$$\mu_1(p,q) = \frac{1}{p},$$

$$\mu_2(p,q) = \frac{1+q}{p^2},$$

$$\mu_3(p,q) = \frac{1+4q+q^2}{p^3},$$

$$\mu_4(p,q) = \frac{1+11q+11q^2+q^3}{p^4},$$

$$\mu_5(p,q) = \frac{1+26q+66q^2+26q^3+q^4}{p^5},$$

$$\mu_6(p,q) = \frac{1+57q+302q^2+302q^3+57q^4+q^5}{p^6},$$

$$\mu_7(p,q) = \frac{1+120q+1191q^2+2416q^3+1191q^4+120q^5+q^6}{p^7},$$

$$\mu_8(p,q) = \frac{1+247q+4293q^2+15619q^3+15619q^4+4293q^5+247q^6+q^7}{p^8},$$

$$\mu_9(p,q) = \frac{1+502q+14608q^2+88243q^3+156190q^4+88263q^5}{p^9} + \frac{14608q^6+502q^7+q^8}{p^9}$$

and

$$\mu_{10}(p,q) = \frac{1+1013q+47840q^2+455192q^3+1310354q^4+1310354q^5}{p^{10}} + \frac{455192q^6+47840q^7+1013q^8+q^9}{p^{10}},$$

$$\mu_{11}(p,q) = \frac{1+2036q+152637q^2+2203488q^3+9738114q^4+15724248q^5}{p^{11}} + \frac{9738114q^6+2203488q^7+152637q^8+2036q^9+q^{10}}{p^{11}},$$

$$\mu_{12}(p,q) = \frac{1+4083q+478271q^2+10187685q^3+66318474q^4+162512286q^5}{p^{12}} + \frac{162512286q^6+66318474q^7+10187685q^8+478271q^9+4083q^{10}+q^{11}}{p^{12}},$$

$$\mu_{13}(p,q) = \frac{1+8178q+1479726q^2+45533450q^3+423281535q^4+1505621508q^5}{p^{13}} + \frac{2275172004q^6+1505621508q^7+423281535q^8+45533450q^9}{p^{13}} + \frac{1479726q^{10}+8178q^{11}+q^{12}}{p^{13}},$$

$$\mu_{14}(p,q) = \frac{1+16369q+4537314q^2+198410786q^3+2571742175q^4+12843262863q^5}{p^{14}} + \frac{27971176092q^6+27971176092q^7+12843262863q^8+2571742175q^9}{p^{14}} + \frac{198410786q^{10}+4537314q^{11}+16369q^{12}+q^{13}}{p^{14}},$$

$$\mu_{15}(p,q) = \frac{1+32752q+13824739q^2+848090912q^3+15041229521q^4 +102776998928q^5}{p^{15}} + \frac{311387598411q^6+447538817472q^7+311387598411q^8+102776998928q^9}{p^{15}} + \frac{15041229521q^{10}+848090912q^{11}+13824739q^{12}+32752q^{13}+q^{14}}{p^{15}},$$

$$\mu_{16}(p,q) = \frac{1+65519q+41932745q^2+3572085255q^3+85383238549q^4 +782115518299q^5}{p^{16}} + \frac{3207483178157q^6+6382798925475q^7+6382798925475q^8 +3207483178157q^9}{p^{16}} + \frac{782115518299q^{10}+85383238549q^{11}+3572085255q^{12}+41932745q^{13}}{p^{16}} + \frac{65519q^{14}+q^{15}}{p^{16}},$$

$$\mu_{17}(p,q) = \frac{1+131054q+126781020q^2+14875399450q^3+473353301060q^4 +5717291972382q^5}{p^{17}} + \frac{31055652948388q^6+83137223185370q^7+114890380658550q^8 +83137223185370q^9}{p^{17}}$$

$$+\frac{31055652948388q^{10}+5717291972382q^{11}+473353301060q^{12}}{p^{17}}$$

$$+\frac{14875399450q^{13}126781020q^{14}+131054q^{15}+q^{16}}{p^{17}},$$

$\mu_{18}(p,q)$

$$=\frac{1+262125q+382439924q^{2}+61403313100q^{3}+2575022097600q^{4}}{p^{18}}$$

$$+\frac{40457344748072q^{5}+285997074307300q^{6}+1006709967915228q^{7}}{p^{18}}$$

$$+\frac{1865385657780650q^{8}+1865385657780650q^{9}+1006709967915228q^{10}}{p^{18}}$$

$$+\frac{285997074307300q^{11}+40457344748072q^{12}+2575022097600q^{13}}{p^{18}}$$

$$+\frac{61403313100q^{14}+382439924q^{15}+262125q^{16}+q^{17}}{p^{18}},$$

$\mu_{19}(p,q)$

$$=\frac{1+524268q+1151775897q^{2}+251732291184q^{3}+13796160184500q^{4}}{p^{19}}$$

$$+\frac{278794377854832q^{5}+2527925001876036q^{6}+11485644635009424q^{7}}{p^{19}}$$

$$+\frac{27862280567093358q^{8}+37307713155613000q^{9}+27862280567093358q^{10}}{p^{19}}$$

$$+\frac{11485644635009424q^{11}+2527925001876036q^{12}+278794377854832q^{13}}{p^{19}}$$

$$+\frac{\begin{array}{l}13796160184500q^{14}+251732291184q^{15}+1151775897q^{16}\\ \quad +524268q^{17}+q^{18}\end{array}}{p^{19}},$$

$\mu_{20}(p,q)$

$$=\frac{1+1048555q+3464764515q^{2}+1026509354985q^{3}}{p^{20}}$$

$$+\frac{73008517581444q^{4}+1879708669896492q^{5}+21598596303099900q^{6}}{p^{20}}$$

$$+\frac{\begin{array}{l}124748182104463860q^{7}+388588260723953310q^{8}\\ \quad +679562217794156938q^{9}\end{array}}{p^{20}}$$

$$+\frac{679562217794156938q^{10}+388588260723953310q^{11} +124748182104463860q^{12}}{p^{20}}$$

$$+\frac{21598596303099900q^{13}+1879708669896492q^{14}+73008517581444q^{15}}{p^{20}}$$

$$+\frac{1026509354985q^{16}+q3464764515q^{17}+1048555q^{18}+q^{19}}{p^{20}}.$$

We now give some special values of the formula given by (1.72).

In order to give these values, we assume that p and q can be members of nonnegative rational numbers on the interval $[0,1]$. For $k \in \mathbb{N}$, assume that

$$p=\frac{1}{k}$$

and

$$q=\frac{k-1}{k}.$$

Thus, equation (1.72) reduces to the following formula:

$$\mu_n\left(\frac{1}{k},\frac{k-1}{k}\right)=k^n A_n\left(\frac{k-1}{k}\right), \tag{1.74}$$

where $k \in \mathbb{N}$ and $n \in \mathbb{N}_0$ (*cf.* Ref. [53]).

Combining equation (1.74) with the following formula

$$A_n\left(\frac{k-1}{k}\right)=\sum_{r=1}^{n}\frac{r!}{\binom{n}{r}}S_2(n,r)\,B_r^n\left(\frac{k-1}{k}\right),$$

where $B_r^n(x)$ denotes the Bernstein polynomials (cf. Ref. [55, Theorem 4]), we have

$$\mu_n\left(\frac{1}{k},\frac{k-1}{k}\right)=k^n\sum_{r=1}^{n}\frac{r!}{\binom{n}{r}}S_2(n,r)\,B_r^n\left(\frac{k-1}{k}\right)$$

(*cf.* Ref. [53]).

In order to find new sequence via terms of the moments $\mu_n\left(\frac{1}{k},\frac{k-1}{k}\right)$, using the same implementation method that of Ref. [53] with the aid of Wolfram language in equation (1.74), we have the following polynomials with the variable k:

$$\mu_0\left(\frac{1}{k},\frac{k-1}{k}\right)=1,$$

$$\mu_1\left(\frac{1}{k},\frac{k-1}{k}\right)=k,$$

$$\mu_2\left(\frac{1}{k},\frac{k-1}{k}\right)=k(-1+2k),$$

$$\mu_3\left(\frac{1}{k},\frac{k-1}{k}\right)=k\left(1-6k+6k^2\right),$$

$$\mu_4\left(\frac{1}{k},\frac{k-1}{k}\right)=k\left(-1+14k-36k^2+24k^3\right),$$

$$\mu_5\left(\frac{1}{k},\frac{k-1}{k}\right)=k\left(1-30k+150k^2-240k^3+120k^4\right),$$

$$\mu_6\left(\frac{1}{k},\frac{k-1}{k}\right)=k\left(-1+62k-540k^2+1560k^3-1800k^4+720k^5\right),$$

$$\mu_7\left(\frac{1}{k},\frac{k-1}{k}\right)=k\left(1-126k+1806k^2-8400k^3+16800k^4\right)$$
$$+k\left(-15120k^5+5040k^6\right),$$

$$\mu_8\left(\frac{1}{k},\frac{k-1}{k}\right)=k\left(-1+254k-5796k^2+40824k^3-126000k^4\right)$$
$$+k\left(191520k^5-141120k^6+40320k^7\right),$$

$$\mu_9\left(\frac{1}{k},\frac{k-1}{k}\right)=k\left(1-510k+18150k^2-186480k^3+834120k^4\right)$$
$$+k(-1905120k^5+2328480k^6-1451520k^7$$
$$+362880k^8)$$

and

$$
\begin{aligned}
&\mu_{10}\left(\frac{1}{k},\frac{k-1}{k}\right)\\
&\quad = k\left(-1+1022k-55980k^2+818520k^3-5103000k^4\right)\\
&\qquad + k\left(16435440k^5-29635200k^6+30240000k^7\right)\\
&\qquad + k\left(-16329600k^8+3628800k^9\right),\\
&\mu_{11}\left(\frac{1}{k},\frac{k-1}{k}\right)\\
&\quad = k\left(1-2046k+171006k^2-3498000k^3+29607600k^4\right)\\
&\qquad + k\left(-129230640k^5+322494480k^6-479001600k^7+419126400k^8\right)\\
&\qquad + k\left(-199584000k^9+39916800k^{10}\right),\\
&\mu_{12}\left(\frac{1}{k},\frac{k-1}{k}\right)\\
&\quad = k\left(-1+4094k-519156k^2+14676024k^3-165528000k^4\right)\\
&\qquad + k\left(953029440k^5-3162075840k^6+6411968640k^7-8083152000k^8\right)\\
&\qquad + k\left(6187104000k^9-2634508800k^{10}+479001600k^{11}\right),\\
&\mu_{13}\left(\frac{1}{k},\frac{k-1}{k}\right)\\
&\quad = k\left(1-8190k+1569750k^2-60780720k^3+901020120k^4\right)\\
&\qquad + k\left(-6711344640k^5+28805736960k^6-76592355840k^7\right)\\
&\qquad + k\left(130456085760k^8-142702560000k^9+97037740800k^{10}\right)\\
&\qquad + k\left(-37362124800k^{11}+6227020800k^{12}\right),\\
&\mu_{14}\left(\frac{1}{k},\frac{k-1}{k}\right)\\
&\quad = k\left(-1+16382k-4733820k^2+249401880k^3-4809004200k^4\right)\\
&\qquad + k\left(45674188560k^5-248619571200k^6+843184742400k^7\right)\\
&\qquad + k\left(-1863435974400k^8+2731586457600k^9-2637143308800k^{10}\right)\\
&\qquad + k\left(-1612798387200k^{11}+566658892800k^{12}+87178291200k^{13}\right),
\end{aligned}
$$

$$\mu_{15}\left(\frac{1}{k},\frac{k-1}{k}\right)$$
$$= k\left(1 - 32766k + 14250606k^2 - 1016542800k^3 + 25292030400k^4\right)$$
$$+ k\left(302899156560k^5 + 2060056318320k^6 - 8734434508800k^7\right)$$
$$+ k\left(24359586451200k^8 - 45950224320000k^9 + 59056027430400k^{10}\right)$$
$$+ k\left(50999300352000k^{11} + 28332944640000k^{12} - 9153720576000k^{13}\right)$$
$$+ k\left(1307674368000k^{14}\right),$$

$$\mu_{16}\left(\frac{1}{k},\frac{k-1}{k}\right)$$
$$= k\left(-1 + 65534k - 42850116k^2 + 4123173624k^3 - 131542866000k^4\right)$$
$$+ k\left(1969147121760k^5 - 16540688324160k^6 + 86355926616960k^7\right)$$
$$+ k(-297846188640000k^8 + 703098107712000k^9$$
$$- 1155068769254400k^{10}) + k(1320663933388800k^{11}$$
$$- 1031319184896000k^{12} + 524813313024000k^{13})$$
$$+ k\left(-156920924160000k^{14} + 20922789888000k^{15}\right),$$

$$\mu_{17}\left(\frac{1}{k},\frac{k-1}{k}\right)$$
$$= k\left(1 - 131070k + 128746950k^2 - 16664094960k^3 + 678330198120k^4\right)$$
$$+ k(-12604139926560k^5 + 129568848121440k^6$$
$$- 823172919528960k^7) + k(3457819037312640k^8$$
$$- 10009442963520000k^9 + 20439835646630400k^{10})$$
$$+ k(-29708792431718400k^{11} + 30575780537702400k^{12}$$
$$- 21785854970880000k^{13}) + k(10226013557760000k^{14}$$
$$- 2845499424768000k^{15} + 355687428096000k^{16}),$$

$$\mu_{18}\left(\frac{1}{k},\frac{k-1}{k}\right)$$
$$= k(-1 + 262142k - 386634060k^2 + 67171367640k^3$$
$$- 3474971465400k^4) + k(79694820748080k^5 - 995210916336000k^6$$

$$+ 7621934141203200k^{7}) + k(-38528927611574400k^{8}$$
$$+ 134672620008326400k^{9} - 334942064711654400k^{10})$$
$$+ k(601783536940185600k^{11} - 783699448602470400k^{12}$$
$$+ 733062897120153600k^{13}) + k(-480178027929600000k^{14}$$
$$+ 209144207720448000k^{15} - 54420176498688000k^{16})$$
$$+ k\left(6402373705728000k^{17}\right),$$

$$\mu_{19}\left(\frac{1}{k}, \frac{k-1}{k}\right)$$
$$= k(1 - 524286k + 1160688606k^{2} - 270232006800k^{3}$$
$$+ 17710714165200k^{4}) + k(-499018753280880k^{5}$$
$$+ 7524340159588560k^{6} - 68937160460313600k^{7})$$
$$+ k(415357755774998400k^{8} - 1732015476199008000k^{9}$$
$$+ 5165761531919788800k^{10}) + k(-11240707219822080000k^{11}$$
$$+ 18011278812054528000k^{12}) + k(-21234672840116736000k^{13}$$
$$+ 18198613875746304000k^{14}) + k(-11029155770400768000k^{15}$$
$$+ 4480594531725312000k^{16}) + k(-1094805903679488000k^{17}$$
$$+ 121645100408832000k^{18}),$$

$$\mu_{20}\left(\frac{1}{k}, \frac{k-1}{k}\right)$$
$$= k(-1 + 1048574k - 3483638676k^{2} + 1085570781624k^{3}$$
$$- 89904730860000k^{4}) + k(3100376804676480k^{5}$$
$$- 56163512390086080k^{6} + 611692004959217280k^{7})$$
$$+ k\left(-4358654246117808000k^{8} + 21473732319740064000k^{9}\right)$$
$$+ k\left(-75875547089306764800k^{10} + 196877625020902425600k^{11}\right)$$
$$+ k\left(-380275818414395904000k^{12} + 549443323130397696000k^{13}\right)$$
$$+ k\left(-591499300737945600000k^{14} + 467644314338353152000k^{15}\right)$$
$$+ k\left(-263665755136143360000k^{16} + 100357207837286400000k^{17}\right)$$
$$+ k(-23112569077678080000k^{18} + 2432902008176640000k^{19}).$$

Problem 5. *It is time to investigate exponential and ordinary generating functions of the following polynomials:*

$$P_n(k) = \sum_{v=0}^{n} a_v k^v = \mu_n\left(\frac{1}{k}, \frac{k-1}{k}\right).$$

Problem 6. *If*

$$F(w,k) = \sum_{n=0}^{\infty} P_n(k)\frac{w^n}{n!},$$

then find $F(w,k)$.

Problem 7. *If*

$$H(w,k) = \sum_{n=0}^{\infty} P_n(k)w^n,$$

then find $H(w,k)$.

Here, study on the construction of the function $F(w,k)$. But the construction of the function $H(w,k)$ is not given here. That is the solution of Problem 7 left to the readers.

Recently, the solution of Problem 6 was given by Simsek and Kilar [53]. Some parts of their solutions are mentioned here.

Simsek and Kilar [53] constructed the following generating functions for the polynomial

$$P_n(k) = \mu_n\left(\frac{1}{k}, \frac{k-1}{k}\right)$$

as follows:

$$\sum_{n=0}^{\infty} \mu_n\left(\frac{1}{k}, \frac{k-1}{k}\right)\frac{w^n}{n!} = \frac{\frac{\exp(w)}{k-1}}{\frac{k}{k-1} - \exp(w)}, \tag{1.75}$$

which implies that $k \in \mathbb{N}$ with $k \neq 1$.

Note that there is one generating function for each value of k.

Using (1.75), (1.46) and (1.47), for $n \in \mathbb{N}_0$ and $k \in \mathbb{N}$, Simsek and Kilar [53] gave the following novel recurrence relations:

$$\mu_n\left(\frac{1}{k}, \frac{k-1}{k}\right) = 1 + (k-1)\sum_{r=0}^{n-1}\binom{n}{r}\mu_r\left(\frac{1}{k}, \frac{k-1}{k}\right), \tag{1.76}$$

$$\mu_n\left(\frac{1}{k},\frac{k-1}{k}\right) = -k\sum_{r=1}^{n}(-1)^r\binom{n}{r}\mu_{n-r}\left(\frac{1}{k},\frac{k-1}{k}\right) \tag{1.77}$$

and

$$\mu_{n+1}\left(\frac{1}{k},\frac{k-1}{k}\right) = -\frac{1}{n+1}\sum_{r=0}^{n+1}\binom{n+1}{r}\mu_r\left(\frac{1}{k},\frac{k-1}{k}\right) \times \mathcal{B}_{n+1-r}\left(\frac{k-1}{k}\right).$$

By applying the Herschel theorem, used also in Ref. [5], to equation (1.75), we obtained the following functional equation:

$$\sum_{n=0}^{\infty}\mu_n\left(\frac{1}{k},\frac{k-1}{k}\right)\frac{w^n}{n!} = e^w\left(1+\sum_{m=1}^{\infty}\left(\left(e^w-1\right)(k-1)\right)^m\right).$$

Combining this equation with (1.25), after some elementary calculations, Simsek and Kilar [53] showed that the following identity holds true:

$$\mu_j\left(\frac{1}{k},\frac{k-1}{k}\right) = 1+\sum_{r=1}^{j}\binom{j}{r}\sum_{n=1}^{r}(k-1)^n\, n!S_2(r,n). \tag{1.78}$$

Using (1.75), we have

$$\mu_n\left(\frac{1}{k},\frac{k-1}{k}\right) = \frac{1}{2k}\sum_{v=0}^{n}\binom{n}{v}\mathcal{E}_v\left(\frac{1-k}{k}\right)$$

(cf. Ref. [53]).

Substituting $k=2$ into (1.74) with the aid of equation (1.69), one has

$$\mu_j\left(\frac{1}{2},\frac{1}{2}\right) = -\frac{\mathcal{B}_{j+1}\left(\frac{1}{2}\right)}{j+1} = \sum_{r=1}^{j}(-1)^{j-r}r!2^r S_2(j,r),$$

$$\mu_j\left(\frac{1}{2},\frac{1}{2}\right) = 2w_g(j) \tag{1.79}$$

(cf. Ref. [53]).

Few values of the $\mu_j\left(\frac{1}{2},\frac{1}{2}\right)$ are given as follows:

$$\mu_0\left(\frac{1}{2},\frac{1}{2}\right) = 1,$$

$$\mu_1\left(\frac{1}{2},\frac{1}{2}\right) = 2,$$

$$\mu_2\left(\frac{1}{2},\frac{1}{2}\right) = 6,$$

$$\mu_3\left(\frac{1}{2},\frac{1}{2}\right) = 26,$$

$$\mu_4\left(\frac{1}{2},\frac{1}{2}\right) = 150,$$

$$\mu_5\left(\frac{1}{2},\frac{1}{2}\right) = 1082,$$

$$\mu_6\left(\frac{1}{2},\frac{1}{2}\right) = 9366,$$

$$\mu_7\left(\frac{1}{2},\frac{1}{2}\right) = 94586,$$

$$\mu_8\left(\frac{1}{2},\frac{1}{2}\right) = 1091670,$$

$$\mu_9\left(\frac{1}{2},\frac{1}{2}\right) = 14174522,$$

$$\mu_{10}\left(\frac{1}{2},\frac{1}{2}\right) = 204495126,$$

$$\mu_{11}\left(\frac{1}{2},\frac{1}{2}\right) = 3245265146,$$

$$\mu_{12}\left(\frac{1}{2},\frac{1}{2}\right) = 56183135190,$$

$$\mu_{13}\left(\frac{1}{2},\frac{1}{2}\right) = 1053716696762,$$

$$\mu_{14}\left(\frac{1}{2},\frac{1}{2}\right) = 21282685940886,$$

$$\mu_{15}\left(\frac{1}{2},\frac{1}{2}\right) = 460566381955706,$$

$$\mu_{16}\left(\frac{1}{2},\frac{1}{2}\right) = 10631309363962710$$

and

$$\mu_{17}\left(\frac{1}{2},\frac{1}{2}\right) = 260741534058271802,$$

$$\mu_{18}\left(\frac{1}{2},\frac{1}{2}\right) = 6771069326513690646,$$

$$\mu_{19}\left(\frac{1}{2},\frac{1}{2}\right) = 185603174638656822266,$$

$$\mu_{20}\left(\frac{1}{2},\frac{1}{2}\right) = 5355375592488768406230,$$

see for details A000629 in the On-Line Encyclopedia of Integer Sequences [46, 53].

When $k = 3$, we also have

$$\mu_j\left(\frac{1}{3},\frac{2}{3}\right) = \sum_{r=0}^{j}(-1)^{j-r}3^r S_2(j,r)r!,$$

$$\sum_{j=0}^{\infty}\mu_j\left(\frac{1}{3},\frac{2}{3}\right)\frac{t^j}{j!} = \frac{\exp(t)}{3-2\exp(t)}$$

and

$$\mu_j\left(\frac{1}{3},\frac{2}{3}\right) = 1+2\sum_{r=0}^{j-1}\binom{j}{r}\mu_r\left(\frac{1}{3},\frac{2}{3}\right),$$

for details, see A201339 [46, 53].

Few values of the $\mu_j\left(\frac{1}{3},\frac{2}{3}\right)$ are given as follows:

$$\mu_0\left(\frac{1}{3},\frac{2}{3}\right)=1,$$

$$\mu_1\left(\frac{1}{3},\frac{2}{3}\right)=3,$$

$$\mu_2\left(\frac{1}{3},\frac{2}{3}\right)=15,$$

$$\mu_3\left(\frac{1}{3},\frac{2}{3}\right)=111,$$

$$\mu_4\left(\frac{1}{3},\frac{2}{3}\right)=1095,$$

$$\mu_5\left(\frac{1}{3},\frac{2}{3}\right)=13503,$$

$$\mu_6\left(\frac{1}{3},\frac{2}{3}\right)=199815,$$

$$\mu_7\left(\frac{1}{3},\frac{2}{3}\right)=3449631,$$

$$\mu_8\left(\frac{1}{3},\frac{2}{3}\right)=68062695,$$

$$\mu_9\left(\frac{1}{3},\frac{2}{3}\right)=1510769343,$$

$$\mu_{10}\left(\frac{1}{3},\frac{2}{3}\right)=37260156615,$$

$$\mu_{11}\left(\frac{1}{3},\frac{2}{3}\right)=1010843385951,$$

$$\mu_{12}\left(\frac{1}{3},\frac{2}{3}\right)=29916558512295,$$

$$\mu_{13}\left(\frac{1}{3},\frac{2}{3}\right)=959183053936383,$$

$$\mu_{14}\left(\frac{1}{3},\frac{2}{3}\right) = 33118910817665415,$$

$$\mu_{15}\left(\frac{1}{3},\frac{2}{3}\right) = 122521926629616707,$$

$$\mu_{16}\left(\frac{1}{3},\frac{2}{3}\right) = 4834820029818476989 5$$

and

$$\mu_{17}\left(\frac{1}{3},\frac{2}{3}\right) = 20271026745163995226 23,$$

$$\mu_{18}\left(\frac{1}{3},\frac{2}{3}\right) = 899901062055417779262 15,$$

$$\mu_{19}\left(\frac{1}{3},\frac{2}{3}\right) = 42169152997726594598729 91,$$

$$\mu_{20}\left(\frac{1}{3},\frac{2}{3}\right) = 208003856087549167497555495,$$

for details, see A201339 [46, 53].

For $k = 4$, Simsek and Kilar [53] showed that

$$\mu_j\left(\frac{1}{4},\frac{3}{4}\right) = \sum_{r=0}^{j}(-1)^{j-r}4^r S_2(j,r)r!,$$

$$\sum_{j=0}^{\infty}\mu_j\left(\frac{1}{4},\frac{3}{4}\right)\frac{t^j}{j!} = \frac{\exp(t)}{4-3\exp(t)},$$

see for details A201354 [46].

When $k = 4$ in (1.76) and (1.77), we have the recurrence relations of the $\mu_j\left(\frac{1}{4},\frac{3}{4}\right)$ which are given by

$$\mu_j\left(\frac{1}{4},\frac{3}{4}\right) = -4\sum_{r=1}^{j}(-1)^r\binom{j}{r}\mu_{j-r}\left(\frac{1}{4},\frac{3}{4}\right)$$

and

$$\mu_j\left(\frac{1}{4},\frac{3}{4}\right) = 1+3\sum_{r=0}^{j-1}\binom{j}{r}\mu_r\left(\frac{1}{4},\frac{3}{4}\right),$$

see also A201354 [46].

Few values of the $\mu_j\left(\frac{1}{4},\frac{3}{4}\right)$ are given by

$$\mu_0\left(\frac{1}{4},\frac{3}{4}\right)=1,$$

$$\mu_1\left(\frac{1}{4},\frac{3}{4}\right)=4,$$

$$\mu_2\left(\frac{1}{4},\frac{3}{4}\right)=28,$$

$$\mu_3\left(\frac{1}{4},\frac{3}{4}\right)=292,$$

$$\mu_4\left(\frac{1}{4},\frac{3}{4}\right)=4060,$$

$$\mu_5\left(\frac{1}{4},\frac{3}{4}\right)=70564,$$

$$\mu_6\left(\frac{1}{4},\frac{3}{4}\right)=1471708,$$

$$\mu_7\left(\frac{1}{4},\frac{3}{4}\right)=35810212,$$

$$\mu_8\left(\frac{1}{4},\frac{3}{4}\right)=995827420,$$

$$\mu_9\left(\frac{1}{4},\frac{3}{4}\right)=31153998244,$$

$$\mu_{10}\left(\frac{1}{4},\frac{3}{4}\right)=1082931514588,$$

$$\mu_{11}\left(\frac{1}{4},\frac{3}{4}\right)=41407678132132,$$

$$\mu_{12}\left(\frac{1}{4},\frac{3}{4}\right)=1727226633730780,$$

$$\mu_{13}\left(\frac{1}{4},\frac{3}{4}\right)=78051253062575524,$$

$$\mu_{14}\left(\frac{1}{4},\frac{3}{4}\right)=379835119221483 7468,$$

$$\mu_{15}\left(\frac{1}{4},\frac{3}{4}\right)=19804942100718605 4052,$$

$$\mu_{16}\left(\frac{1}{4},\frac{3}{4}\right)=1101490513158794549 0140,$$

$$\mu_{17}\left(\frac{1}{4},\frac{3}{4}\right)=65090391500979282065 0404$$

and

$$\mu_{18}\left(\frac{1}{4},\frac{3}{4}\right)=4072645323472515853547 2348,$$

$$\mu_{19}\left(\frac{1}{4},\frac{3}{4}\right)=26897839161996335312205 67972,$$

$$\mu_{20}\left(\frac{1}{4},\frac{3}{4}\right)=1869969785239554386918036 81500,$$

for details, see also A201354 [46, 53].

Few values of the $\mu_j\left(\frac{1}{5},\frac{4}{5}\right)$ are given as follows:

$$\mu_{0}\left(\frac{1}{5},\frac{4}{5}\right)=1,$$

$$\mu_{1}\left(\frac{1}{5},\frac{4}{5}\right)=5,$$

$$\mu_{2}\left(\frac{1}{5},\frac{4}{5}\right)=45,$$

$$\mu_{3}\left(\frac{1}{5},\frac{4}{5}\right)=605,$$

$$\mu_{4}\left(\frac{1}{5},\frac{4}{5}\right)=10845,$$

$$\mu_{5}\left(\frac{1}{5},\frac{4}{5}\right)=243005,$$

$$\mu_{6}\left(\frac{1}{5},\frac{4}{5}\right)=6534045,$$

$$\mu_7\left(\frac{1}{5},\frac{4}{5}\right) = 204972605,$$

$$\mu_8\left(\frac{1}{5},\frac{4}{5}\right) = 7348546845,$$

$$\mu_9\left(\frac{1}{5},\frac{4}{5}\right) = 296387331005,$$

$$\mu_{10}\left(\frac{1}{5},\frac{4}{5}\right) = 13282361478045,$$

$$\mu_{11}\left(\frac{1}{5},\frac{4}{5}\right) = 654762261324605,$$

$$\mu_{12}\left(\frac{1}{5},\frac{4}{5}\right) = 35211177242722845,$$

$$\mu_{13}\left(\frac{1}{5},\frac{4}{5}\right) = 2051349014835939005,$$

$$\mu_{14}\left(\frac{1}{5},\frac{4}{5}\right) = 128701394409842982045,$$

$$\mu_{15}\left(\frac{1}{5},\frac{4}{5}\right) = 8651475271312083756605,$$

$$\mu_{16}\left(\frac{1}{5},\frac{4}{5}\right) = 620334325261670875138845$$

and

$$\mu_{17}\left(\frac{1}{5},\frac{4}{5}\right) = 47259638324026516284867005,$$

$$\mu_{18}\left(\frac{1}{5},\frac{4}{5}\right) = 3812225290950213997141446045,$$

$$\mu_{19}\left(\frac{1}{5},\frac{4}{5}\right) = 324599479131089667413427468605,$$

$$\mu_{20}\left(\frac{1}{5},\frac{4}{5}\right) = 29093332719619507982147967394845.$$

Few values of the $\mu_j\left(\frac{1}{6},\frac{5}{6}\right)$ are given by

$$\mu_0\left(\frac{1}{6},\frac{5}{6}\right)=1,$$

$$\mu_1\left(\frac{1}{6},\frac{5}{6}\right)=6,$$

$$\mu_2\left(\frac{1}{6},\frac{5}{6}\right)=66,$$

$$\mu_3\left(\frac{1}{6},\frac{5}{6}\right)=1086,$$

$$\mu_4\left(\frac{1}{6},\frac{5}{6}\right)=23826,$$

$$\mu_5\left(\frac{1}{6},\frac{5}{6}\right)=653406,$$

$$\mu_6\left(\frac{1}{6},\frac{5}{6}\right)=21502866,$$

$$\mu_7\left(\frac{1}{6},\frac{5}{6}\right)=825574686,$$

$$\mu_8\left(\frac{1}{6},\frac{5}{6}\right)=36224995026,$$

$$\mu_9\left(\frac{1}{6},\frac{5}{6}\right)=1788186547806,$$

$$\mu_{10}\left(\frac{1}{6},\frac{5}{6}\right)=98078723067666,$$

$$\mu_{11}\left(\frac{1}{6},\frac{5}{6}\right)=5917380109712286,$$

$$\mu_{12}\left(\frac{1}{6},\frac{5}{6}\right)=389468818527014226,$$

$$\mu_{13}\left(\frac{1}{6},\frac{5}{6}\right)=27770137167998698206,$$

$$\mu_{14}\left(\frac{1}{6},\frac{5}{6}\right) = 2132396888160363224466,$$

$$\mu_{15}\left(\frac{1}{6},\frac{5}{6}\right) = 175437034900669687673886,$$

$$\mu_{16}\left(\frac{1}{6},\frac{5}{6}\right) = 15395834742585900180201426$$

and

$$\mu_{17}\left(\frac{1}{6},\frac{5}{6}\right) = 1435536176996041621338144606,$$

$$\mu_{18}\left(\frac{1}{6},\frac{5}{6}\right) = 141725705069152492266550453266,$$

$$\mu_{19}\left(\frac{1}{6},\frac{5}{6}\right) = 14769446047222344555286187619486,$$

$$\mu_{20}\left(\frac{1}{6},\frac{5}{6}\right) = 1620153568994981974751970242476626.$$

Few values of the $\mu_j\left(\frac{1}{7},\frac{6}{7}\right)$ are given by

$$\mu_{0}\left(\frac{1}{7},\frac{6}{7}\right) = 1,$$

$$\mu_{1}\left(\frac{1}{7},\frac{6}{7}\right) = 7,$$

$$\mu_{2}\left(\frac{1}{7},\frac{6}{7}\right) = 91,$$

$$\mu_{3}\left(\frac{1}{7},\frac{6}{7}\right) = 1771,$$

$$\mu_{4}\left(\frac{1}{7},\frac{6}{7}\right) = 45955,$$

$$\mu_{5}\left(\frac{1}{7},\frac{6}{7}\right) = 1490587,$$

$$\mu_{6}\left(\frac{1}{7},\frac{6}{7}\right) = 58018051,$$

$$\mu_7\left(\frac{1}{7},\frac{6}{7}\right) = 2634606331,$$

$$\mu_8\left(\frac{1}{7},\frac{6}{7}\right) = 136728885475,$$

$$\mu_9\left(\frac{1}{7},\frac{6}{7}\right) = 7982838419227,$$

$$\mu_{10}\left(\frac{1}{7},\frac{6}{7}\right) = 517859436505411,$$

$$\mu_{11}\left(\frac{1}{7},\frac{6}{7}\right) = 36953802655576891,$$

$$\mu_{12}\left(\frac{1}{7},\frac{6}{7}\right) = 2876702408084408995,$$

$$\mu_{13}\left(\frac{1}{7},\frac{6}{7}\right) = 242601144198680472667,$$

$$\mu_{14}\left(\frac{1}{7},\frac{6}{7}\right) = 22033091405030196352771,$$

$$\mu_{15}\left(\frac{1}{7},\frac{6}{7}\right) = 2143982572414264604067451,$$

$$\mu_{16}\left(\frac{1}{7},\frac{6}{7}\right) = 222533700124249094305148515$$

and

$$\mu_{17}\left(\frac{1}{7},\frac{6}{7}\right) = 24541396160896319049913194907,$$

$$\mu_{18}\left(\frac{1}{7},\frac{6}{7}\right) = 2865670987576276722432064424131,$$

$$\mu_{19}\left(\frac{1}{7},\frac{6}{7}\right) = 353211214021005692475504205934011,$$

$$\mu_{20}\left(\frac{1}{7},\frac{6}{7}\right) = 45826747493662070905606729593536035.$$

Few values of the $\mu_j\left(\frac{1}{8},\frac{7}{8}\right)$ are given by

$$\mu_0\left(\frac{1}{8},\frac{7}{8}\right)=1,$$

$$\mu_1\left(\frac{1}{8},\frac{7}{8}\right)=8,$$

$$\mu_2\left(\frac{1}{8},\frac{7}{8}\right)=120,$$

$$\mu_3\left(\frac{1}{8},\frac{7}{8}\right)=2696,$$

$$\mu_4\left(\frac{1}{8},\frac{7}{8}\right)=80760,$$

$$\mu_5\left(\frac{1}{8},\frac{7}{8}\right)=3024008,$$

$$\mu_6\left(\frac{1}{8},\frac{7}{8}\right)=135878520,$$

$$\mu_7\left(\frac{1}{8},\frac{7}{8}\right)=7123041416,$$

$$\mu_8\left(\frac{1}{8},\frac{7}{8}\right)=426748573560,$$

$$\mu_9\left(\frac{1}{8},\frac{7}{8}\right)=28762803162248,$$

$$\mu_{10}\left(\frac{1}{8},\frac{7}{8}\right)=2154010573612920,$$

$$\mu_{11}\left(\frac{1}{8},\frac{7}{8}\right)=177442291614284936,$$

$$\mu_{12}\left(\frac{1}{8},\frac{7}{8}\right)=15946119167339370360,$$

$$\mu_{13}\left(\frac{1}{8},\frac{7}{8}\right)=1552440554247180823688,$$

$$\mu_{14}\left(\frac{1}{8},\frac{7}{8}\right) = 1627644805635698845633320,$$

$$\mu_{15}\left(\frac{1}{8},\frac{7}{8}\right) = 18283844423900514626821256,$$

$$\mu_{16}\left(\frac{1}{8},\frac{7}{8}\right) = 219080700824417148615583116 0$$

and

$$\mu_{17}\left(\frac{1}{8},\frac{7}{8}\right) = 2789135828522110581356122963 28,$$

$$\mu_{18}\left(\frac{1}{8},\frac{7}{8}\right) = 37597484701270239713364238329720,$$

$$\mu_{19}\left(\frac{1}{8},\frac{7}{8}\right) = 534969489408999096080896433473 0376,$$

$$\mu_{20}\left(\frac{1}{8},\frac{7}{8}\right) = 8012640007631488154249430092997 15960.$$

Few values of the $\mu_j\left(\frac{1}{9},\frac{8}{9}\right)$ are given by

$$\mu_{0}\left(\frac{1}{9},\frac{8}{9}\right) = 1,$$

$$\mu_{1}\left(\frac{1}{9},\frac{8}{9}\right) = 9,$$

$$\mu_{2}\left(\frac{1}{9},\frac{8}{9}\right) = 153,$$

$$\mu_{3}\left(\frac{1}{9},\frac{8}{9}\right) = 3897,$$

$$\mu_{4}\left(\frac{1}{9},\frac{8}{9}\right) = 132345,$$

$$\mu_{5}\left(\frac{1}{9},\frac{8}{9}\right) = 5618169,$$

$$\mu_{6}\left(\frac{1}{9},\frac{8}{9}\right) = 286195833,$$

$$\mu_7\left(\frac{1}{9},\frac{8}{9}\right) = 17008993017,$$

$$\mu_8\left(\frac{1}{9},\frac{8}{9}\right) = 1155276253305,$$

$$\mu_9\left(\frac{1}{9},\frac{8}{9}\right) = 88276603008249,$$

$$\mu_{10}\left(\frac{1}{9},\frac{8}{9}\right) = 7494848686510713,$$

$$\mu_{11}\left(\frac{1}{9},\frac{8}{9}\right) = 699959337031654137,$$

$$\mu_{12}\left(\frac{1}{9},\frac{8}{9}\right) = 71313428097441158265,$$

$$\mu_{13}\left(\frac{1}{9},\frac{8}{9}\right) = 7871036436616843591929,$$

$$\mu_{14}\left(\frac{1}{9},\frac{8}{9}\right) = 935571998960137422505593,$$

$$\mu_{15}\left(\frac{1}{9},\frac{8}{9}\right) = 119147718567410580173035257,$$

$$\mu_{16}\left(\frac{1}{9},\frac{8}{9}\right) = 16185382610108085023566383225$$

and

$$\mu_{17}\left(\frac{1}{9},\frac{8}{9}\right) = 2336087729769131108135202217209,$$

$$\mu_{18}\left(\frac{1}{9},\frac{8}{9}\right) = 357008790794953590249430769572473,$$

$$\mu_{19}\left(\frac{1}{9},\frac{8}{9}\right) = 57590356601888896820075172145648377,$$

$$\mu_{20}\left(\frac{1}{9},\frac{8}{9}\right) = 9779057957022130499831618769039064185$$

and so on. Observe that the moments generated from each of the above sequences provide a numerical framework for characterizing the mathematical structure. Since moment-based sequences are widely used in probability theory, statistical inference, distribution functions, and data-driven modeling, they enable further research in applied mathematics, in data science, and in related interdisciplinary areas.

Here we note that it is clear on the above all related sequences of moments are increases sequence. For positive integer k, the above sequence is satisfies the following relation:

$$\mu_k\left(\frac{1}{9}, \frac{8}{9}\right) < \mu_{k+1}\left(\frac{1}{9}, \frac{8}{9}\right).$$

Recently, Kilar [28] constructed the following generating functions for the moments $\mu_n\left(\frac{1}{k}, \frac{k-1}{k}\right)$ of order d:

$$\frac{\exp(wd)}{(k-(k-1)\exp(w))^d} = \sum_{n=0}^{\infty} \mu_n^{(d)}\left(\frac{1}{k}, \frac{k-1}{k}\right)\frac{w^n}{n!}, \tag{1.80}$$

where $d \in \mathbb{N}$ and $k \in \mathbb{N}$.

Note that there is one generating function for each value of k.

Kilar [28] gave many properties of (1.80). With the aid of functional equations and derivative equations, many identities were also given by Kilar [28]

Observe that for $d = 1$, equation (1.80) reduces to equation (1.75).

By using (1.80), we have

$$\sum_{j=0}^{d}(-1)^{d-j}\binom{d}{j}k^j\,(k-1)^{d-j}\exp(w(d-j))$$
$$\times\sum_{n=0}^{\infty}\mu_n^{(d)}\left(\frac{1}{k}, \frac{k-1}{k}\right)\frac{w^n}{n!} = \exp(wd).$$

Therefore,

$$\sum_{j=0}^{d}(-1)^{d-j}\binom{d}{j}k^j\,(k-1)^{d-j}\sum_{n=0}^{\infty}\sum_{v=0}^{n}\binom{n}{v}\mu_v^{(d)}$$
$$\times\left(\frac{1}{k}, \frac{k-1}{k}\right)\frac{(d-j)^{n-v}w^n}{n!}$$
$$\sum_{n=0}^{\infty}\frac{d^n w^n}{n!}.$$

We now compare the coefficients of $\frac{w^n}{n!}$ on both sides of the previous equation; we arrive at the following theorem:

Theorem 16. *Let $n, d \in \mathbb{N}_0$, and $k \in \mathbb{N}$ with $k \neq 1$. Then, we have*

$$1 = \sum_{j=0}^{d} \sum_{v=0}^{n} \sum_{c=0}^{n-v} \binom{n-v-c}{c} (-1)^{n+d-j-v-c} \binom{d}{j} \binom{n}{v} k^j (k-1)^{d-j}$$

$$d^{j-n} j^{n-v-c} \mu_v^{(d)} \left(\frac{1}{k}, \frac{k-1}{k} \right).$$

References

[1] J. A. Adell, Probabilistic Stirling numbers of the second kind and applications, *J. Theor. Prob.*, 35(1), 2022, 636–652.

[2] J. A. Adell and A. Lekuona, A probabilistic generalization of the Stirling numbers of the second kind, *J. Number Theory*, 194, 2019, 335–355.

[3] J. A. Adell and A. Lekuona, Explicit expressions for a certain subset of Appell polynomials: A probabilistic perspective, *Integers*, 21, 2021, Article ID: A60.

[4] E. Agyuz, Identities derived from a particular class of generating functions for Frobenius–Euler type Simsek numbers and polynomials, *Filomat*, 38(5), 2024, 1531–1545.

[5] T. M. Apostol, On the Lerch zeta function, *Pacific J. Math.*, 1, 1951, 161–167.

[6] D. P. Bertsekas and S. N. Tsitsiklis, *Introduction to Probability*, Second Edition. (Belmont Massachusetles, Athena Scientific, 2008).

[7] K.N. Boyadzhiev, Apostol–Bernoulli functions, derivative polynomials and Eulerian polynomials. https://arxiv.org/ftp/arxiv/423papers/0710/0710.1124.pdf.

[8] K. Bury, *Statistical Distributions in Engineering.* (Cambridge University Press, Cambridge, 1999).

[9] N. P. Cakic and G. V. Milovanovic, On generalized Stirling numbers and polynomials, *Math. Balk.*, 18, 2004, 241–248.

[10] L. Carlitz, Some theorems on Bernoulli numbers of higher order, *Pac. Math.*, 2(2), 1952, 127–139.

[11] L. Carlitz, D.P. Roselle and R. Scoville, Permutations and sequences with repetitions by number of increase, *J. Comb. Theory*, 1, 1966, 350–374.

[12] G. Casella and R. L Berger, *Statistical Inference*, 2nd Edition. (Duxbury Press, Pacific Grove, 2002).

[13] R. Chattamvelli and R. Shanmugam, *Discrete Distributions in Engineering and the Applied Sciences.* (Morgan and Claypool Publishers Series, 2020).

[14] C. A. Charalambides, *Combinatorial Methods in Discrete Distributions.* (A John Wiley and Sons Inc. Publication, New Jersey, 2015).

[15] C. A. Charalambides, *Enumerative Combinatorics.* (Chapman and Hall/Crc Press Company, London, New York, 2002).

[16] R. Chattamvelli and R. Shanmugam, *Discrete Distributions in Engineering and the Applied Sciences.* Springer Nature, Switzerland, 2020).

[17] L. Comtet, *Advanced Combinatorics: The Art of Finite and Infinite Expansions.* (Reidel Publishing Company, Dordrecht-Holland, Boston, Translated from the French by J. W. Nienhuys, 1974).

[18] M.H. Degroot and M.J. Schervish, *Probability and Statistics* 4. addition. (Addison-Wesley, New York, London, Tokyo, 2012).

[19] Dekking FM, Kraaikamp C, Lopuhaa HP, Meester LE, A Modern *Introduction to Probability and Statistics*: *Understanding Why and How.* (Springer Nature, Switzerland, 2005).

[20] M. Evans and N. Hastings and B. Peacock, *Statistical Distributions.* (Wiley-Interscience, USA, 2000).

[21] G. Grimmett and D. Stirzaker, *Probability and Random Processes*, 3rd Edition. (Oxford University Press, Oxford, UK, 2001).

[22] R. Goldman, *Pyramid Algorithms, A Dynamic Programming Approach to Curves and Surfaces for Geometric Modeling.* (Morgan Kaufmann, Burlington, MA, 2003).

[23] D. Gun and Y. Simsek, Some new identities and inequalities for Bernoulli polynomials and numbers of higher order related to the Stirling and Catalan numbers, *RACSAM* 114, 2020, 167, 1–12. https://doi.org/10.1007/s13398-020-00899-z.

[24] E. Howell, *Moments and Moment Generating Functions of Statistical Distributions, An explanation into Statistical Moments and how to derive them using*, Mar 28, 2022, https://towardsdatascience.com/moments-and-moment-generating-functions-of-statistical-distributions-e216cb11a079/.

[25] Y. He, A new formula of products of the Apostol–Bernoulli and Apostol–Euler polynomials, *Bull. Malays. Math. Sci. Soc.*, 39, 2016, 1307–1318.

[26] N. Kilar, A new class of generalized Fubini polynomials and their computational algorithms, *Appl. Anal. Discrete Math.*, 17(2), 2023, 496–524.

[27] N. Kilar, Formulas for Fubini type numbers and polynomials of negative higher order, *Montes Taurus J. Pure Appl. Math.*, 5(3), 2023, 23–36.

[28] N. Kilar, New identities for generalized Fubini polynomials and moments of geometric distribution, *Montes Taurus J. Pure Appl. Math.*, 7(4), 2025.

[29] N. Kilar, On a family of probabilistic Frobenius–Euler type Simsek numbers and polynomials with an analysis of their generating functions, to appear in *Montes Taurus J. Pure Appl. Math.*, 8, 2026.

[30] N. Kilar and Y. Simsek, A new family of Fubini numbers and polynomials associated with Apostol–Bernoulli numbers and polynomials, *J. Korean Math. Soc.*, 54(5), 2017, 1605–1621.

[31] N. Kilar and Y. Simsek, Identities and relations for Fubini type numbers and polynomials via generating functions and p-adic integral approach, *Publ. Inst. Math., Nouv. Ser.*, 106(120), 2019, 113–123.

[32] D. S. Kim and T. Kim, Daehee numbers and polynomials. *Appl. Math. Sci.* (Ruse) 1167(120), 2013, 5969–5976.

[33] T. Kim and D. S. Kim, Probabilistic degenerate Bell polynomials associated with random variables. *Russ. J. Math. Phys.*, 30(4), 2023, 528–542.

[34] T. Kim, D. S. Kim and J. Kwon, Probabilistic degenerate Stirling polynomials of the second kind and their applications, *Math. Comput. Model Dyn. Syst.*, 30(1), 2024, 16–30.

[35] V. Kotesovec, *Non-attacking Chess Pieces*, Sixth Edition. (Chess and Mathematics, 2013).

[36] J. Kotz and A. W. Kemp, *Univariate Discrete Distributions*, 2nd Edition. (Wiley-Interscience, 1992).

[37] E. Lukacs, *Characteristic Function*, Second Edition. (Charles griffin and Company Limited, London, 1970).

[38] Q.M. Luo and H.M. Srivastava, Some generalizations of the Apostol–Bernoulli and Apostol–Euler polynomials, *J. Math. Anal. Appl.*, 308, 2005, 290–302.

[39] Q.M. Luo and H.M. Srivastava, Some generalizations of the Apostol–Genocchi polynomials and the Stirling numbers of the second kind, *Appl. Math. Comput.*, 217(12), 2011, 5702–5728.

[40] G. Mackey, Harmonic analysis as the exploitation of symmetry — A historical survey, *Bull. Am. Math. Soc. New Ser.*, 3(1), 1980, 543–698.

[41] B. H. Margolin and H. S. Winokur, Exact moments of the order statistics of the geometric distribution and their relation to inverse sampling and reliability of redundant systems, *J. Amer. Statist. Associat.*, 62(319), 1967, 915–925. https://doi.org/10.1080/01621459.1967.10500902.

[42] G. Mastroianni and G. Milovanovic, *Interpolation Processes: Basic Theory and Applications.* (Springer Monographs in Mathematics, Heidelberg, 2008).

[43] B. Mishra, *Data Science and Interdisciplinary Research: Recent Trends and Applications (Advances in Computing Communications and Informatics).* Kindle Edition; Editor B. K. Mishra, Bentham Book, 2023.

[44] N. E. Norlund, *Vorlesungen uber Differenzenrechnung.* (Verlag Von Julius Springer, Berlin, 1924).

[45] R. W. Oldford and P. Shuldiner, Bernoulli sums: The only random variables that count. https://arxiv.org/pdf/2110.02363.

[46] OEIS Foundation Inc.: The On-Line Encyclopedia of Integer Sequences. Published electronically at https://oeis.org, 2024.

[47] T. K. Petersen, *Eulerian Numbers.* (Birkhauser, Basel, 2015).

[48] J. Riordan, *Introduction to Combinatorial Analysis.* (Princeton University Press, Princeton, 1958).

[49] B. Simsek, Formulas derived from moment generating functions and Bernstein polynomials, *Appl. Anal. Discrete Math.*, 13(6), 2019, 839–848.

[50] B. Simsek, New moment formulas for moments and characteristic function of the geometric distribution in terms of Apostol–Bernoulli polynomials and numbers, *Math. Meth. Appl. Sci.*, 47(11), 2024, 9169–9179. https://doi.org/10.1002/mma.10066.

[51] B. Simsek, Novel formulas of moments of Negative Binomial distribution connected with Apostol–Bernoulli numbers of higher order and Stirling numbers, *Rev. Real Acad. Cienc. Exactas Fis. Nat. Ser. A-Mat.*, 118, 2024, 142, 1–12. https://doi.org/10.1007/s13398-024-01640-w.

[52] B. Simsek, Certain mathematical formulas for moments of Geometric distribution by means of the Apostol–Bernoulli polynomials, *Filomat*, 39(6), 2025, 1843–1853.

[53] B. Simsek and N. Kilar, By analysis of moments of geometric distribution: New formulas involving Eulerian and Fubini numbers. *Appl. Anal. Discrete Math.*, 19, 2025, 233–252.

[54] B. Simsek and B. Simsek, The computation of expected values and moments of special polynomials via characteristic and generating functions, *AIP Conf. Proc.*, 2017, 1863, 300012–1–300012-5. http://www.fixedpointtheoryandapplications.com/content/2013/1/87.

[55] Y. Simsek, Novel formulas for B-splines, Bernstein basis functions and special numbers: Approach to derivative and functional equations of generating functions, *Mathematics*, 12(1), 2024, 1–20. https://doi.org/10.3390/math12010065.

[56] Y. Simsek, Generating functions for generalized Stirling type numbers, array type polynomials, Eulerian type polynomials and their applications, *Fixed Point Theory Appl.*, 87, 2013, 1–28.

[57] Y. Simsek and N. Kilar, Generating Functions for the Fubini Type Polynomials and Their Applications. In N. J. Daras, M. T. Rassias, N. B. Zographopoulos (eds.), *Exploring Mathematical Analysis, Approximation Theory, and Optimization* (Springer Optimization and Its Applications, Springer, Cham., Volume 207, 2023, pp. 279–380), https://doi.org/10.1007/978-3-031-46487-4_15.

[58] M. Sugiyama, Chapter 3 - Examples of discrete probability distributions, In Masashi Sugiyama (ed.), *Introduction to Statistical Machine Learning*, Morgan Kaufmann, 2016, pp. 25–36, https://doi.org/10.1016/B978-0-12-802121-7.00014-5.

[59] P. Sun, Product of uniform distribution and Stirling numbers of the first kind, *Acta Math. Sin.* (Engl. Ser.), 21(6), 2005, 1435–1442.

[60] H. M. Srivastava and J. Choi, *Series Associated with the Zeta and Related Functions.* (Kluwer Academic Publishers, Dordrecht, Boston, 2001).

[61] H. M. Srivastava and J. Choi, *Zeta and q-Zeta Functions and Associated Series and Integrals.* (Elsevier Science Publishers, Amsterdam, 2012).

[62] A. Xu, Computing a family of probabilistic numbers in terms of probabilistic Stirling numbers of the second kind, *Appl. Math. Sci. Eng.*, 33(1), 2025, Article ID: 2485250.

[63] Wolfram Research Inc.: Mathematica Online (Wolfram Cloud). Champaign, IL, USA, 2020.

[64] https://www.geeksforgeeks.org/data-science/probability-data-distributions-in-data-science/ "Probability Data Distributions in Data Science, Last Updated: 24 May, 2025; arrival date 10 July, 2025.

[65] https://www Wikipedia – Negative Binomial Distribution, https://en.wikipedia.org/wiki/Negative_binomial_distribution, arrival date 29.07.2025.

https://doi.org/10.1142/9789819820498_0019

Chapter 19

Applications of Frobenius–Euler-Type Polynomials and Numbers with Their Generating Functions

Yilmaz Simsek
Department of Mathematics Faculty of Science Akdeniz University
Antalya, Turkey
ysimsek@akdeniz.edu.tr, ysimsek63@gmail.com

The inspiration for this chapter comes from the (*twisted*) (q-) Frobenius–Euler- type polynomials and numbers, and their generating and interpolation functions. Starting from Euler, one of the world's most famous mathematicians, to the present day, numerous books, thousands of articles and monographs have been written about these numbers and polynomials and their applications, and they can be easily found in the literature. This section aims to bring to light the hidden parts of these numbers and polynomials, as well as to present interesting exercises to researchers and readers, by blending my articles, we wrote years ago, with the work of other researchers. Therefore, we should emphasize that this section includes both my own articles and articles written with co-authors: Refs. [1, 2, 29, 54, 57–59, 62–66, 74, 94–120, 127–129] and others. Finally, we give many interesting exercises for the readers.

1. Introduction

The history of Euler numbers (and polynomials) goes back to 1600. Since then, these numbers were studied by Leonard Euler. After Euler, many researchers have investigated generalizations of these numbers and polynomials, such as Frobenius, Norlund, Schoenberg, Rademacher, Shiratani, Gould, Carlitz, Apostol, T. Kim, D.S. Kim, Srivatava, Choi, Luo, Putzer, Tsumura, Satoh, Schempp, Yamamoto, Srivatava, Kim, Simsek, Bayad, Kurt, Feng, Kozak, Cenkci, Can, Kucukoglu Kilar, Gun and Sukruoglu, see the literature. These numbers and polynomials recently have many kinds of generalizations, such as Eulerian numbers, Euler–Frobenius numbers, generalized Euler–Frobenius numbers attached to character(s) with conductor, twisted Euler–Frobenius numbers, q-Euler–Frobenius numbers, Euler–Frobenius numbers of higher order, Barnes-type Euler–Frobenius numbers and Hermite-base Euler–Frobenius numbers. These kinds of numbers are also related to the same kind of special polynomials. Many kinds of splines are defined by means of these numbers and polynomials. For instance, exponential splines, Eulerian splines, etc. There are many applications of these numbers in analytic number theory, complex functions theory, algebra, geometry, interpolation functions involving zeta and Dirichlet L-type functions, Euler and Gnocchi l-type functions, and other applied sciences.

The B-spline and asymptotic behavior of zeroes of the generalized Frobenius–Euler polynomials have been studied by Feng and Kozak [25]. In order to give applications of the cardinal spline functions, Schempp [88, 89] gave a contour integral representation of the Euler–Frobenius polynomials.

$S_{m,r}^N$ denotes the space of all N-periodic splines of degree m and defect r subordinate to the equidistant lattice $\mathbb{Z}$ on the real line, with the aid of the generalized Frobenius–Euler polynomials; Reimer [84] gave the following interpolation problem:

$$s^{(k)}(j-\lambda) = y^{(k)},$$

where $j \in \mathbb{Z}$, set of integers, $k = 0, \ldots, r-1$, $s \in S_{m,r}^N$, for $\lambda \in [0,1)$ is a fixed shift parameter and $y^{(k)}$ are given N-periodic data.

Notations and Definitions:

$$[x] = [x:q] = \begin{cases} \dfrac{1-q^x}{1-q}, & \text{if } q \nrightarrow 1 \\ x, & \text{if } q \to 1. \end{cases} \tag{1.1}$$

Using the above definition, one has the following well-known identities:

$$[x+y] = [x] + q^x[y],$$
$$[xy] = [x : q][y : q^x].$$

For $x = n \in \mathbb{N}$, a set of positive integer and $\mathbb{N} = \{0, 1, 2, 3, \ldots\}$, with the aid of (1.1), we have

$$[n] = \prod_{j=0}^{n-1} q^j$$

(cf. Ref. [30]).

The q-Frobenius–Euler numbers $H_m(u, q)$ are given by means of the following generating function:

$$F_{u,q}(t) = \frac{u-1}{u} \sum_{n=0}^{\infty} \frac{e^{[n]t}}{u^n} = \sum_{m=0}^{\infty} H_m(u, q) \frac{t^m}{m!}, \tag{1.2}$$

where $| u |> 1$ and $q \in \mathbb{C}$, the set of complex numbers, with $|q| < 1$ and $u \neq q$ (cf. Ref. [87], see also Ref. [98]).

When $q \to 1$, (1.2) reduces to generating function for the Frobenius–Euler numbers:

$$\sum_{m=0}^{\infty} H_m(u) \frac{t^m}{m!} = \frac{1-u}{e^t - u} \tag{1.3}$$

(cf. Refs. [1, 2, 32–41, 43–45, 79, 80, 91–93, 97, 119, 120, 123]).

Schoenberg [91, Eq. (3.11), p. 96] denoted these numbers by a_m:

Using (1.2), we have

$$\sum_{m=0}^{\infty} H_m(u, q) \frac{t^m}{m!} = \frac{u-1}{u} \sum_{n=0}^{\infty} \frac{e^{[n]t}}{u^n} \sum_{m=0}^{\infty} [n]^m \frac{t^m}{m!}.$$

By comparing coefficients $\frac{t^m}{m!}$ in the above equation, we also have the following computational formula for the numbers $H_m(u, q)$:

$$H_m(u, q) = \frac{u-1}{u} \sum_{n=0}^{\infty} \frac{1}{u^n} [n]^m, \tag{1.4}$$

where $| u |> 1$ and $m \in \mathbb{N}_0$.

Putting $m = 0$ in (1.4), we get

$$H_0(u,q) = \frac{u-1}{u}\sum_{n=0}^{\infty}\frac{1}{u^n}$$
$$= \frac{u-1}{u}\frac{1}{1-\frac{1}{u}}.$$

Thus,

$$H_0(u,q) = 1.$$

Putting $m = 1$ in (1.4), we get

$$H_1(u,q) = (u-1)\sum_{n=0}^{\infty}\frac{[n]}{u^{n+1}}.$$

Combining the above equation with (1.1) yields

$$H_1(u,q) = (u-1)\sum_{n=0}^{\infty}\frac{1}{u^{n+1}}\left(\frac{1-q^n}{1-q}\right).$$

Hence,

$$H_1(u,q) = \frac{u-1}{u(1-q)}\left(\sum_{n=0}^{\infty}\frac{1}{u^n} - \sum_{n=0}^{\infty}\left(\frac{q}{u}\right)^n\right).$$

Assuming that $|\ u\ | > 1$ and $|\ \frac{q}{u}\ | < 1$, then the above equation reduces to

$$H_1(u,q) = \frac{u-1}{u(1-q)}\left(\frac{1}{1-\frac{1}{u}} - \frac{1}{1-\frac{q}{u}}\right)$$
$$= \frac{1}{u-q},$$

hence, $u \neq q$ and $q \neq 1$.

Putting $m = 2$ in (1.4), we get not only

$$H_2(u,q) = (u-1)\sum_{n=0}^{\infty}\frac{\left(1+q+\cdots+q^{n-1}\right)^2}{u^{n+1}}$$

but also with the aid of equation (1.1)

$$H_2(u,q) = \frac{u-1}{u}\sum_{n=0}^{\infty}\frac{1}{u^n}\left(\frac{1-q^n}{1-q}\right)^2$$

$$= \frac{u-1}{u\,(1-q)^2}\left(\sum_{n=0}^{\infty}\frac{1}{u^n} - 2\sum_{n=0}^{\infty}\left(\frac{q}{u}\right)^n + \sum_{n=0}^{\infty}\left(\frac{q^2}{u}\right)^n\right).$$

Therefore,

$$H_2(u,q) = \frac{u-1}{u\,(1-q)^2}\left(\frac{1}{1-\frac{1}{u}} - \frac{2}{1-\frac{q}{u}} + \frac{1}{1-\frac{q^2}{u}}\right).$$

Thus,

$$H_2(u,q) = \frac{q^3+(u-2)\,q^2+(1-2u)\,q+u}{(1-q)^2\,(u-q)\,(u-q^2)}.$$

By a similar method, using binomial expansion, we get the following general formula for (1.4):

$$H_m(u,q) = \frac{1-u}{u\,(1-q)^m}\sum_{k=0}^{m}\binom{m}{k}(-1)^{m(k-1)+1}\sum_{n=0}^{\infty}\left(\frac{q^k}{u}\right)^n,$$

where $\left|\frac{q^k}{u}\right| < 1$. Therefore, using geometric series in the above series, we obtain the following formula:

$$H_m(u,q) = \frac{u-1}{u\,(q-1)^m}\sum_{k=0}^{m}\binom{m}{k}(-1)^{m-k}\frac{1}{1-\frac{q^k}{u}}.$$

After some calculation, we arrive at the following elegant computational formula for the numbers $H_m(u,q)$:

Theorem 1. *Let $m \in \mathbb{N}_0$. Then,*

$$H_m(u,q) = \frac{1-u}{(1-q)^m}\sum_{k=0}^{m}(-1)^{k+1}\binom{m}{k}\frac{1}{u-q^k}. \qquad (1.5)$$

The generating function for q-generalized Frobenius–Euler numbers $F_{u,q,\chi}(t)$ is given by means of the following functional equation:

$$F_{u,q,\chi}(t) = \sum_{a=1}^{f} \chi(a) u^{f-a} e^{[a]t} F_{u^f,q^f}(t[f]q^a)$$

(cf. Ref. [98]). By using this equation, we have

$$\begin{aligned} F_{u,q,\chi}(t) &= \left(1 - \frac{1}{u^f}\right) \sum_{a=1}^{f} \chi(a) u^{f-a} \sum_{n=0}^{\infty} \frac{e^{[a+fn]t}}{u^{fn}} \\ &= \sum_{n=0}^{\infty} H_{n,\chi}(u,q) \frac{t^n}{n!}, \end{aligned} \tag{1.6}$$

where $H_{n,\chi}(u,q)$ denotes the nth q-generalized Frobenius–Euler number (cf. Ref. [98]).

If $q \to 1$, then

$$F_u(t) = \frac{1-u}{e^t - u} = \sum_{n=0}^{\infty} H_n(u) \frac{t^n}{n!}, \tag{1.7}$$

which is given by (cf. Refs. [32–41, 43–45, 79, 80, 92, 93, 97, 119, 120, 123]).

Carlitz's q-Frobenius–Euler numbers $H_k(u,q)$ can be determined inductively by (cf. Refs. [14, 33, 61, 67, 86, 119, 120, 123])

$$H_0(u,q) = 1$$

and for $k \geq 1$,

$$(qH+1)^k - uH_k(u,q) = 0,$$

where u is a complex number with $|\, u \,| > 1$.

For $k \geq 0$,

$$H_k(x,u,q) = (q^x H + [x])^k, \tag{1.8}$$

with the usual convention of replacing H^j with H_j.

When

$$q \to 1,$$

we have

$$H_k(u,q) \to H_k(u)$$

and

$$H_k(x,u,q) \to H_k(x,u),$$

where $H_k(u)$ and $H_k(x,u)$ are the Frobenius–Euler numbers and the Frobenius–Euler polynomials, respectively (cf. Refs. [98, 119, 123]).

Using binomial expansion in equation (1.8), we also have

$$H_k(x,u,q) = \sum_{j=0}^{k} \binom{k}{j} q^{xj} H_j(u,q)[x]^{k-j}.$$

The generating function of q-Frobenius–Euler polynomials is given by means of the following generating function:

Setting

$$F_{u,q}(t,x) = F_{u,q}(q^x t)e^{[x]t},$$

therefore,

$$F_{u,q}(t,x) = (u-1)\sum_{n=0}^{\infty} \frac{e^{(q^x[n]+[x])t}}{u^{n+1}},$$

which gives us the following equation:

$$F_{u,q}(t,x) = (u-1)\sum_{n=0}^{\infty} \frac{e^{[n+x]t}}{u^{n+1}} = \sum_{n=0}^{\infty} H_n(x,u,q)\frac{t^n}{n!} \tag{1.9}$$

(cf. Refs. [86, 98, 119, 123]).

When

$$q \to 1,$$

$$H_n(u,x,q) \to H_n(u,x),$$

which is called the exponential Euler polynomials by Schoenberg [91, Eq. (3.13), p. 390]. Therefore, these polynomials are defined by means of the following generating function:

$$\frac{e^{tx}}{e^t - u} = \sum_{n=0}^{\infty} \frac{H_n(x,u)}{1-u}\frac{t^n}{n!} \tag{1.10}$$

(cf. Refs. [14, 17, 18, 29, 91, 104, 113, 122]).

These polynomials are members of the Appell sequence of polynomials because

$$H_0(x,u) = 1 \quad \text{and}$$
$$\frac{d}{dx}\{H_n(x,u)\} = nH_{n-1}(x,u),$$

which is derived from (1.10) with the aid of partial derivative operator $\frac{\partial}{\partial x}$. Carlitz [17, 18] called $H_k(x,u)$ Eulerian polynomials and showed that $H_k(x|u)$.

By using (1.10), we deduce that

$$\sum_{n=0}^{\infty} \frac{u}{1-u} H_n(x,u)\frac{t^n}{n!} = \frac{te^{tx}}{\frac{1}{u}e^t - 1}.$$

Since LHS of the above equation gives generating function for the Apostol–Bernoulli polynomials, that is,

$$P(t,\beta) = \frac{t}{\beta e^t - 1}e^{tx} = \sum_{n=0}^{\infty} \mathcal{B}_n(x;\beta)\frac{t^n}{n!} \tag{1.11}$$

(cf. Ref. [6]), we get

$$\sum_{n=0}^{\infty} H_n(x,u)\frac{t^n}{n!} = \sum_{n=0}^{\infty} \frac{1-u}{u}\mathcal{B}_n\left(x;\frac{1}{u}\right)\frac{t^n}{n!}.$$

By comparing coefficients $\frac{t^n}{n!}$ in the above equation, we have the following relation between the polynomials $H_n(x,u)$ and $\mathcal{B}_n\left(x;\frac{1}{u}\right)$:

$$uH_n(x,u) + (u-1)\mathcal{B}_n\left(x;\frac{1}{u}\right) = 0$$

(cf. Ref. [113]).

When $u = -1$, from (1.10), we have

$$E_n(x) := H_n(-1,x),$$

which denotes the Euler polynomials (of the first kind) (cf. Refs. [73, 123]).

That is, for $u = -1$, (1.10) reduces to the generating function for these polynomials which are given by

$$\frac{e^{tx}}{e^t+1} = \sum_{n=0}^{\infty} \frac{E_n(x)}{2}\frac{t^n}{n!}$$

(cf. Refs. [3–132]).

Schoenberg [91, Eq. (3.13), p. 390] defined the polynomials $\eta_n(u)$ by means of the following generating function:

$$\frac{1}{u-e^t}=\sum_{n=0}^{\infty}\frac{\eta_n(u)}{(u-1)^{n+1}}\frac{t^n}{n!} \tag{1.12}$$

with

$$\eta_0(u)=1.$$

Combining (1.3) with (1.12), we deduce that

$$\eta_n(u)=(u-1)^n\,H_n(u)$$

(cf. Ref. [91, Eq. (3.15), p. 390]).

The polynomial $\eta_n(u)$ was defined by Ref. [24]. The polynomial $\eta_n(u)$ is a reciprocal monic polynomial of degree $n-1$ having integer coefficients. Frobenius [26] showed that the polynomial $\eta_n(u)$ has simple negative zeros. Due to these observations, Schoenberg [91, p. 391] call them *Euler–Frobenius polynomials.*

Schoenberg [91, LEMMA 7. 1] proved the following theorem which gives us fundamental properties of the polynomial $\eta_n(u)$:

Theorem 2. *The Euler–Frobenius polynomial $\eta_n(u)$, defined by the generating function in equation (1.12), is a monic reciprocal polynomial of degree $n-1$ having integer coefficients. It is related to following the B-spline member of the exponential splines:*

$$n!\ell_{n+1}(x)=x_+^n-\binom{n+1}{1}(x-1)_+^n+\cdots+(-1)^{n+1}(x-1-n)_+^n,$$

where

$$x_+=\max\{0,x\},$$

by the relation

$$\eta_n(u)=n!\sum_{j=0}^{n}\ell_{n+1}(j+1)u^j.$$

$\eta_n(u)$ can be independently defined by the expansion

$$\frac{\eta_n(u)}{(u-1)^{n+1}} = \sum_{k=0}^{\infty}(k+1)^n u^k. \tag{1.13}$$

It satisfies the recurrence relation

$$\eta_{n+1}(u) = (1+nu)\eta_n(u) + u(u-1) + \frac{d}{du}\left\{\eta_n(u)\right\}$$

with

$$\eta_1(u) = 1.$$

The zeros λ_k of $\eta_n(u)$ are simple and negative. We label them so that

$$\lambda_{n-1} < \lambda_{n-2} < \cdots < \lambda_2 < \lambda_1 < 0.$$

Recently, the Euler–Frobenius polynomial $\eta_n(u)$ (or the Eulerian polynomials) has many applications in many different areas. These polynomials are also defined by means of the following generating function:

$$F_a(t,u) = \frac{1}{1-ue^{t(1-u)}} = \sum_{n=0}^{\infty} \frac{\eta_n(u)}{1-u}\frac{t^n}{n!}, \tag{1.14}$$

where $u \neq 1$, $\eta_n(u)$ is a polynomial in u of degree $n-1$ for $n > 0$:

$$\eta_n(u) = \sum_{j=0}^{n} A_{n,j}u^j, \tag{1.15}$$

where $A_{n,j}$ is an integer numbers, known as the Eulerian numbers:

$$A_{n,j} = \sum_{v=0}^{j}(-1)^v\binom{n+1}{v}(j-v)^n, \tag{1.16}$$

$j = 1, 2, \ldots, n$, $0 \leq j < n$, $n \in \mathbb{N}$ (cf. Refs. [13, 20, 82]). The Worpitzky identity for the Eulerian numbers is given as follows:

$$\alpha^n = \sum_{v=0}^{n}\binom{\alpha+v-1}{n}A_{n,v}, \tag{1.17}$$

where $\alpha \in \mathbb{R}$, $n \in \mathbb{N}_0$ (cf. Refs. [13, 20, 82]).

In (cf. Ref. [113]), we gave the following remarks and observation on the numbers $A_{n,v}$:

The numbers $A_{n,v}$ have many useful combinatoric applications. That is, let $A = \{1, 2, 3, \ldots, n\}$ and $P(n, j)$ denote the number of permutation of the elements of the set A that show exactly j increases between adjacent elements, the first element always being counted as a jump. Consequently,

$$P(n, j) = A_{n,j}$$

(cf. Refs. [78, 113]). It is known that in the literature, the numbers $A_{n,j}$ are represented by many different notations. Some of these are presented as follows: $A_{n,j}$, $A(n, k)$, $\left\langle {n \atop j} \right\rangle$, $E(n, k)$, $W_{n,k}$.

By using umbral calculus method in (1.14), we get

$$\eta_n(u) = u \sum_{j=0}^{n} \binom{n}{j} (1-u)^{n-j} \, \eta_j(u),$$

where $\eta_0(u) = 1$ (cf. Ref. [113]).

Theorem 3 (cf. Ref. [113]). *Let $m \in \mathbb{N}_0$. Then, we have*

$$\eta_m(u) = \sum_{n=0}^{m} \left(\frac{u}{1-u} \right)^n n!(1-u)^m S_2(m, n). \tag{1.18}$$

Proof. From (1.14), we get the following functional equation:

$$\eta_0(u) + \sum_{m=1}^{\infty} \eta_m(u) \frac{t^m}{m!} = 1 + \sum_{n=1}^{\infty} \left(\frac{u}{1-u} \right)^n \left(e^{t(1-u)} - 1 \right)^n. \tag{1.19}$$

By combining (1.19) with following generating functions for the Stirling numbers of the second kind $S_2(m, n)$:

$$\left(e^t - 1 \right)^n = \sum_{m=0}^{\infty} n! S_2(m, n) \frac{t^m}{m!} \tag{1.20}$$

(cf. Refs. [3–132]), we get

$$\sum_{m=1}^{\infty} \eta_m(u) \frac{t^m}{m!} = \sum_{n=1}^{\infty} \left(\frac{u}{1-u} \right)^n \sum_{m=0}^{\infty} n! S_2(m, n) \frac{(1-u)^m t^m}{m!}.$$

Since

$$S_2(m,n) = 0,$$

if $n > m$, then we have

$$\sum_{m=1}^{\infty} \eta_m(u) \frac{t^m}{m!} = \sum_{m=0}^{\infty} \sum_{n=1}^{m} \left(\frac{u}{1-u} \right)^n n!(1-u)^m S_2(m,n) \frac{t^m}{m!}.$$

After comparing the coefficients of $\frac{t^m}{m!}$ on both sides of the above equation, we arrive at the desired result. □

Corollary 1 (cf. Ref. [113]). *Let $m \in \mathbb{N}_0$. Then, we have*

$$\eta_m(u) = \sum_{n=1}^{m} \frac{n!}{\binom{m}{n}} B_n^m(u) S_2(m,n),$$

where $B_n^m(u)$ denotes the Bernstein base functions, defined by

$$B_n^m(u) = \binom{m}{n} u^n (1-u)^{m-n}.$$

By combining (1.7) with (1.14), we have the following functional equation:

$$F_a(t,u) = F_{\frac{1}{u}}((1-u)t)$$

(cf. Ref. [113]). Using this equation, we get

$$\sum_{m=0}^{\infty} \eta_m(u) \frac{t^m}{m!} = \sum_{m=0}^{\infty} (1-u)^m H_m \left(\frac{1}{u} \right) \frac{t^m}{m!}.$$

After comparing the coefficients of $\frac{t^m}{m!}$ on both sides of the above equation, we arrive at the following identity:

$$\eta_m(u) = (1-u)^m H_m \left(\frac{1}{u} \right)$$

(cf. Refs. [90, 91, 113]).

By combining (1.11) with (1.14), we have the following functional equation:

$$-tF_a(t,u) = P((1-u)t, 0, u)$$

(cf. Ref. [113]). Using this equation, we get

$$-t\sum_{m=0}^{\infty}\eta_m(u)\frac{t^m}{m!}=\sum_{m=0}^{\infty}(1-u)^m\mathcal{B}_m\left(u\right)\frac{t^m}{m!}.$$

Therefore,

$$\sum_{m=0}^{\infty}m\eta_{m-1}(u)\frac{t^m}{m!}=\sum_{m=0}^{\infty}(-1)^{m+1}(u-1)^m\mathcal{B}_m\left(u\right)\frac{t^m}{m!}.$$

After comparing the coefficients of $\frac{t^m}{m!}$ on both sides of the above equation, we arrive at the following identity:

$$A_{m-1}(u)=(-1)^{m+1}\frac{(u-1)^m}{m}\mathcal{B}_m(u), \tag{1.21}$$

where $m\in\mathbb{N}$ (cf. Ref. [113]).

By using (1.21), a few values of the polynomials $A_n(u)$ are given as follows:

$$\begin{aligned}
A_1(u) &= \frac{(u-1)^2}{-2u}\mathcal{B}_2\left(u\right)\\
&= A_{1,0}=1,\\
A_2(u) &= \frac{(u-1)^3}{3u}\mathcal{B}_3\left(u\right)\\
&= A_{2,0}\rho+A_{2,1}=u+1,\\
A_3(u) &= \frac{(u-1)^4}{-4u}\mathcal{B}_4\left(u\right)\\
&= A_{3,0}u^2+A_{3,1}u+A_{3,2}=u^2+4u+1,\\
A_4(u) &= \frac{(u-1)^5}{5u}\mathcal{B}_5\left(u\right)=A_{4,0}u^3+A_{4,1}u^2+A_{4,2}u+A_{4,3}\\
&= u^3+11u^2+11u+1,
\end{aligned}$$

$$\begin{aligned}
A_5(u) &= \frac{(u-1)^6}{-6u}\mathcal{B}_6(u) \\
&= A_{5,0}u^4 + A_{5,1}u^3 + A_{5,2}u^2 + A_{5,3}u + A_{5,4} \\
&= u^4 + 26u^3 + 66u^2 + 26u + 1, \\
A_6(u) &= \frac{(u-1)^7}{7u}\mathcal{B}_7(u) \\
&= A_{6,0}u^5 + A_{6,1}u^4 + A_{6,2}u^3 + A_{6,3}u^2 + A_{5,4}u + A_{6,5} \\
&= u^5 + 57u^4 + 302u^3 + 302u^2 + 57u + 1
\end{aligned}$$

and so on.

Boyadzhiev [9] also gave a relation between the Apostol–Bernoulli polynomials and the geometric polynomials $W_n(u)$, which is given as follows:

$$\mathcal{B}_n(u) = \frac{n}{u-1}W_{n-1}\left(\frac{u}{1-u}\right) \tag{1.22}$$

and

$$\mathcal{B}_n\left(\frac{u}{1+u}\right) = -n(u+1)W_{n-1}(u),$$

where $n \in \mathbb{N}$ (cf. Ref. [113]) and

$$W_n(u) = \sum_{j=0}^{n} j!S_2(n,j)u^j$$

(cf. Refs. [9, 113]).

With the Euler (derivative) operator $u\frac{d}{du}$, Boyadzhiev [9] also showed that

$$\begin{aligned}
\left(u\frac{d}{du}\right)^k \left\{\frac{1}{1-u}\right\} &= \frac{1}{1-u}W_k\left(\frac{u}{1-u}\right) \\
&= \sum_{v=0}^{\infty} v^k u^v,
\end{aligned} \tag{1.23}$$

where $|u| < 1$ (cf. Refs. [9, 113]).

We [113] showed that

$$\eta_k(u) = (1-u)^{k+1}\left(u\frac{d}{du}\right)^k\left\{\frac{1}{1-u}\right\},$$

see also Ref. [9].

Combining the above equation with (1.21) yields

$$-\frac{\mathcal{B}_{k+1}(u)}{n+1} = \left(u\frac{d}{du}\right)^k\left\{\frac{1}{1-u}\right\} \tag{1.24}$$

(cf. Refs. [9, 113]).

Since

$$(n+1)\,\mathcal{E}_n\,(x;u) = -2\mathcal{B}_{n+1}\,(x;-u) \tag{1.25}$$

(cf. Ref. [122]), we also get

$$\mathcal{E}_n\,(u) = \left(u\frac{d}{du}\right)^k\left\{\frac{1}{1-u}\right\} \tag{1.26}$$

(cf. Ref. [113]).

Here, we note that there are many different methods for the proof of (1.24) and (1.26).

By combining (1.13) with (1.23), we get

$$\begin{aligned}\frac{u\eta_n(u)}{(u-1)^{n+1}} &= \sum_{k=1}^{\infty} k^n u^k \\ &= \left(u\frac{d}{du}\right)^k\left\{\frac{1}{1-u}\right\} \\ &= \frac{1}{1-u}W_k\left(\frac{u}{1-u}\right).\end{aligned}$$

Let λ be a complex numbers with $\lambda \neq 1$. The Frobenius–Euler numbers $H_n^{(\alpha)}\,(\lambda)$ are defined by means of the following generating function:

$$F_H\,(t;\lambda,\alpha) = \left(\frac{1-\lambda}{e^t-\lambda}\right)^\alpha = \sum_{n=0}^{\infty} H_n^{(\alpha)}\,(\lambda)\,\frac{t^n}{n!}$$

(cf. [7, 55, 64, 104, 119, 122]).

Let $\alpha, \lambda \in \mathbb{C}$. The Apostol–Euler numbers and polynomials of order α are respectively defined by

$$F_{\alpha}(t;\lambda) = \frac{2^{\alpha}}{(\lambda e^{t}+1)^{\alpha}} = \sum_{n=0}^{\infty} \mathcal{E}_{n}^{(\alpha)}(\lambda)\frac{t^{n}}{n!} \tag{1.27}$$

and

$$F_{\alpha}(t,x;\lambda) = e^{xt}F_{\alpha}(t;\lambda) = \sum_{n=0}^{\infty} \mathcal{E}_{n}^{(\alpha)}(x;\lambda)\frac{t^{n}}{n!}, \tag{1.28}$$

where $|t| < \pi$ when $\lambda = 1$; $|t| < |\log(-\lambda)|$ when $\lambda \neq 1$ and also $1^{\alpha} = 1$ (cf. Refs. [23, p. 253], [68–70, 122]).

By applying binomial series to (1.28), assuming that $\left|\lambda e^{t}\right| < 1$, we have

$$\sum_{n=0}^{\infty} \mathcal{E}_{n}^{(\alpha)}(x;\lambda)\frac{t^{n}}{n!} = 2^{\alpha}\sum_{v=0}^{\infty}\binom{\alpha}{v}\lambda^{v}e^{(v+x)t}.$$

Thus, we have

$$\sum_{n=0}^{\infty} \mathcal{E}_{n}^{(\alpha)}(x;\lambda)\frac{t^{n}}{n!} = 2^{\alpha}\sum_{v=0}^{\infty}\binom{\alpha}{v}\lambda^{v}\sum_{n=0}^{\infty}\frac{(v+x)^{n}t^{n}}{n!}.$$

Comparing the coefficient of $\frac{t^{v}}{v!}$ in both sides of the above equation, we get the following formula:

$$\mathcal{E}_{n}^{(\alpha)}(x;\lambda) = 2^{\alpha}\sum_{v=0}^{\infty}\binom{\alpha}{v}\lambda^{v}(v+x)^{n}, \tag{1.29}$$

where $|\lambda| < 1$ and

$$\binom{\alpha}{v} = \begin{cases} \alpha(\alpha-1)\ldots(\alpha-v+1) & \text{if } v \in \mathbb{N}, \\ 1 & v = 0. \end{cases}$$

Due to the values of α, $\binom{\alpha}{v}$ is related to the rising factorial (the Pochhammer symbol) $(\alpha)_{v}$, given by

$$(\alpha)_{v} = \begin{cases} \alpha(\alpha+1)\ldots(\alpha+v-1), & v \in \mathbb{N}, \\ 1, & v = 0. \end{cases}$$

Some values of (1.29) are given as follows:

$$\mathcal{E}_n^{(0)}(x;\lambda) = \sum_{v=0}^{\infty} \binom{0}{v} \lambda^v (v+x)^n = x^n,$$

$$\mathcal{E}_n^{(1)}(x;\lambda) = 2(v+x)^n (1+\lambda),$$

$$\begin{aligned} \mathcal{E}_n^{(m)}(x;\lambda) &= \sum_{v=0}^{m} \binom{m}{v} \lambda^v (v+x)^n \\ &= \sum_{j=0}^{n} \binom{n}{j} x^{n-j} \sum_{v=0}^{m} \binom{m}{v} v^j \lambda^v \\ &= m! \sum_{j=0}^{n} \binom{n}{j} x^{n-j} y_1(j,m;\lambda), \end{aligned}$$

where

$$\sum_{v=0}^{m} \binom{m}{v} v^j \lambda^v = m! y_1(j,m;\lambda),$$

where the numbers $y_1(j,m;\lambda)$ are defined by the author [108].

Generating function for the numbers $y_1(j,m;\lambda)$ is given by

$$\frac{\left(\lambda e^t + 1\right)^n}{n!} = \sum_{n=0}^{\infty} y_1(m,n;\lambda) \frac{t^n}{n!}$$

(cf. Ref. [108]).

The numbers $y_1(m,n;\lambda)$ have been called Simsek numbers by Goubi [27, 28]. Recently, in Kucukoglu [62, 63], called them the combinatorial Simsek numbers. Oussi [75] gave degenerate Simsek numbers, which generalized and modified the numbers $y_1(m,n;\lambda)$. Xu [132] studied open problem for the numbers.

In Refs. [62, 63], Kucukoglu constructed the q-combinatorial Simsek numbers and polynomials via analyzing with computational implementations. Kucukoglu gave relations between the numbers $y_1(m,n;\lambda)$ and new number families related to finite sums, Thacker's (totient) function, divisors, totatives and higher powers of binomial coefficients using the decomposition of the generating functions and symbolic computation of the number families and their generating functions in the Wolfram language.

Exercise 1. Find relations between the polynomials $\mathcal{E}_n^{(\alpha)}(x;\lambda)$, the polynomials $H_n(x;\lambda)$, arithmetic functions, the Stirling numbers of the first kind and the Lah numbers.

Hint: Combine definition $\binom{\alpha}{v}$ with the following well-known formulas:

$$(\alpha)_v = \sum_{k=0}^{v} |S_1(v,k)| \alpha^k,$$

where $S_1(v,k)$ denotes the Stirling numbers of the first kind and

$$(\alpha)_v = \sum_{k=1}^{v} L(v,k) \binom{\alpha}{k} k! \tag{1.30}$$

where $L(n,k)$ denotes the Lah numbers which are defined by

$$L(n,k) = (-1)^n \frac{n!}{k!} \binom{n-1}{k-1},$$

$$|L(n,k)| = \frac{n!}{k!} \binom{n-1}{k-1}$$

(cf. Refs. [8, 21, 85, 110, 122]).

It is easy to see that

$$\mathcal{E}_n^{(\alpha)}(\lambda) = \mathcal{E}_n^{(\alpha)}(0;\lambda).$$

Putting $\lambda = 1$ in (1.28), one has the Euler polynomials $E_n^{(\alpha)}(x)$ of order α.

Some properties of the polynomials $\mathcal{E}_n^{(\alpha)}(x;\lambda)$ and the numbers $\mathcal{E}_n^{(\alpha)}(\lambda)$ are given as follows:

$$\mathcal{E}_n^{(\alpha)}(x;\lambda) = \sum_{k=0}^{n} \binom{n}{k} x^{n-k} \mathcal{E}_k^{(\alpha)}(\lambda) \tag{1.31}$$

and

$$\mathcal{E}_n^{(\beta+\gamma)}(\lambda) = \sum_{k=0}^{n} \binom{n}{k} \mathcal{E}_k^{(\beta)}(\lambda) \mathcal{E}_{n-k}^{(\gamma)}(\lambda) \tag{1.32}$$

(cf. Refs. [29, 69, 70, 122]).

The relation between the numbers $\mathcal{E}_n^{(\alpha)}(\lambda)$ and $H_n^{(\alpha)}(\lambda)$ is given by

$$\mathcal{E}_n^{(\alpha)}(\lambda) = \frac{2^\alpha}{(\lambda+1)^\alpha} H_n^{(\alpha)}\left(-\frac{1}{\lambda}\right)$$

(cf. Refs. [8, 9, 29, 102, 122]).

In Ref. [29], $d \in \mathbb{N}$, Gun and Simsek gave the following PDE for the function $\mathcal{F}_\alpha(t;\lambda)$:

$$\frac{\partial^d}{\partial t^d}\{\mathcal{F}_\alpha(t;\lambda)\} = \sum_{j=1}^{d}(-1)^j \frac{\lambda^j}{2^j}(\alpha)^{(j)} S_2(d,j) F_{\alpha+j}(t,j;\lambda). \quad (1.33)$$

Using (1.33), Gun and Simsek [29] showed that the numbers $\mathcal{E}_{n+d}^{(\alpha)}(\lambda)$ have the following recurrence relations:

Theorem 4. *Let $d \in \mathbb{N}$ and $n \in \mathbb{N}_0$. Then, we have*

$$\mathcal{E}_{n+d}^{(\alpha)}(\lambda) = \sum_{j=0}^{d}(-1)^{j+1}\left(\frac{\lambda}{2}\right)^{j+1}(\alpha)^{(j+1)} S_2(d,j+1) \times \mathcal{E}_n^{(\alpha+j+1)}(j+1;\lambda) \quad (1.34)$$

and

$$\mathcal{E}_d^{(\alpha)}(\lambda) = 2^\alpha \sum_{j=1}^{d}(-1)^j \lambda^j S_2(d,j)\frac{(\alpha)^{(j)}}{(\lambda+1)^{\alpha+j}}. \quad (1.35)$$

When $\alpha = \beta+\gamma$, equation (1.35) reduces to

$$\mathcal{E}_d^{(\beta+\gamma)}(\lambda) = 2^{\beta+\gamma}\sum_{j=1}^{d}(-1)^j \lambda^j S_2(d,j)\frac{(\beta+\gamma)^{(j)}}{(\lambda+1)^{\beta+\gamma+j}}.$$

Combining the above equation with the following well-known Vandermonde's convolution formula

$$(\beta+\gamma)^{(j)} = \sum_{v=0}^{j}\binom{j}{v}(\beta)^{(v)}(\gamma)^{(j-v)}$$

into the above equation, we also have

$$\mathcal{E}_{d}^{(\beta+\gamma)}(\lambda)=2^{\beta+\gamma}\sum_{j=1}^{d}\sum_{v=0}^{j}(-1)^{j}\binom{j}{v}\frac{(\beta)^{(v)}(\gamma)^{(j-v)}\lambda^{j}S_{2}(d,j)}{(\lambda+1)^{\beta+\gamma+j}}, \tag{1.36}$$

where $\beta,\gamma\in\mathbb{C}$, $d\in\mathbb{N}$ and $n\in\mathbb{N}_0$ (cf. Ref. [29]). If $|\lambda|<1$ and applying binomial series to equation (1.36), we deduce the following novel formula for the numbers $\mathcal{E}_{d}^{(\beta+\gamma)}(\lambda)$:

Theorem 5. *Let $\beta,\gamma\in\mathbb{C}$, $d\in\mathbb{N}$ and $n\in\mathbb{N}_0$. If $|\lambda|<1$, then we have*

$$\mathcal{E}_{d}^{(\beta+\gamma)}(\lambda)=2^{\beta+\gamma}\sum_{j=1}^{d}\sum_{v=0}^{j}(-1)^{j}\binom{j}{v}(\beta)^{(v)}(\gamma)^{(j-v)}S_{2}(d,j)\times\sum_{n=0}^{\infty}\binom{-\beta-\gamma-j}{n}\lambda^{n+j}.$$

In Ref. [29], Gun and Simsek gave many examples. For instance, for $n=0,1,2,3,4,5$, and $\alpha\in\mathbb{C}$, using (1.35), we give few values of the numbers $\mathcal{E}_{n}^{(\alpha)}(\lambda)$ as follows:

$$\mathcal{E}_{0}^{(\alpha)}(\lambda)=\frac{2^{\alpha}}{(\lambda+1)^{\alpha}},$$

$$\mathcal{E}_{0}^{(5+i)}(\lambda)=\frac{2^{5+i}}{(\lambda+1)^{5+i}},$$

$$\mathcal{E}_{0}^{(1+i)}(\lambda)=\frac{2^{1+i}}{(\lambda+1)^{1+i}},$$

$$\mathcal{E}_{1}^{(\alpha)}(\lambda)=-\frac{\alpha\lambda 2^{\alpha}}{(\lambda+1)^{\alpha+1}},$$

$$\mathcal{E}_{1}^{(-2i)}(\lambda)=\frac{i\lambda 2^{-2i+1}}{(\lambda+1)^{-2i+1}},$$

$$\mathcal{E}_{1}^{(1+i)}(\lambda)=-\frac{(1+i)\lambda 2^{1+i}}{(\lambda+1)^{2+i}},$$

$$\mathcal{E}_2^{(\alpha)}(\lambda) = -\frac{\alpha\lambda 2^{\alpha}}{(\lambda+1)^{\alpha+1}} + \frac{\alpha(\alpha+1)\lambda^2 2^{\alpha}}{(\lambda+1)^{\alpha+2}},$$

$$\mathcal{E}_2^{(1+i)}(\lambda) = -\frac{(1+i)\lambda 2^{1+i}}{(\lambda+1)^{2+i}} + \frac{(1+i)(2+i)\lambda^2 2^{1+i}}{(\lambda+1)^{3+i}},$$

$$\mathcal{E}_3^{(\alpha)}(\lambda) = \frac{-\alpha\lambda 2^{\alpha}}{(\lambda+1)^{\alpha+1}} + \frac{3\alpha(\alpha+1)\lambda^2 2^{\alpha}}{(\lambda+1)^{\alpha+2}} - \frac{\alpha(\alpha+1)(\alpha+2)\lambda^3 2^{\alpha}}{(\lambda+1)^{\alpha+3}},$$

$$\mathcal{E}_3^{(1+i)}(\lambda) = \frac{-(1+i)\lambda 2^{1+i}}{(\lambda+1)^{2+i}} + \frac{3(1+i)(2+i)\lambda^2 2^{1+i}}{(\lambda+1)^{3+i}} - \frac{(1+i)(2+i)(3+i)\lambda^3 2^{1+i}}{(\lambda+1)^{4+i}},$$

$$\mathcal{E}_4^{(\alpha)}(\lambda) = \frac{-\alpha\lambda 2^{\alpha}}{(\lambda+1)^{\alpha+1}} + \frac{7\alpha(\alpha+1)\lambda^2 2^{\alpha}}{(\lambda+1)^{\alpha+2}} - \frac{6\alpha(\alpha+1)(\alpha+2)\lambda^3 2^{\alpha}}{(\lambda+1)^{\alpha+3}} + \frac{\alpha(\alpha+1)(\alpha+2)(\alpha+3)\lambda^4 2^{\alpha}}{(\lambda+1)^{\alpha+4}},$$

$$\mathcal{E}_4^{(1+i)}(\lambda) = -\frac{(1+i)\lambda 2^{\alpha}}{(\lambda+1)^{2+i}} + \frac{7(1+i)(2+i)\lambda^2 2^{\alpha}}{(\lambda+1)^{3+i}} - \frac{6(1+i)(2+i)(3+i)\lambda^3 2^{\alpha}}{(\lambda+1)^{4+i}} + \frac{(1+i)(2+i)(3+i)(4+i)\lambda^4 2^{\alpha}}{(\lambda+1)^{5+i}},$$

$$\mathcal{E}_5^{(\alpha)}(\lambda) = -\frac{\alpha\lambda 2^{\alpha}}{(\lambda+1)^{\alpha+1}} + \frac{15\alpha(\alpha+1)\lambda^2 2^{\alpha}}{(\lambda+1)^{\alpha+2}} - \frac{25\alpha(\alpha+1)(\alpha+2)\lambda^3 2^{\alpha}}{(\lambda+1)^{\alpha+3}} + \frac{10\alpha(\alpha+1)(\alpha+2)(\alpha+3)\lambda^4 2^{\alpha}}{(\lambda+1)^{\alpha+4}} - \frac{\alpha(\alpha+1)(\alpha+2)(\alpha+3)(\alpha+4)\lambda^5 2^{\alpha}}{(\lambda+1)^{\alpha+5}},$$

$$\mathcal{E}_5^{(1+i)}(\lambda) = -\frac{(1+i)\lambda 2^{1+i}}{(\lambda+1)^{2+i}} + \frac{15(1+i)(2+i)\lambda^2 2^{1+i}}{(\lambda+1)^{3+i}}$$

$$-\frac{25(1+i)(2+i)(3+i)\lambda^3 2^{1+i}}{(\lambda+1)^{4+i}}$$

$$+\frac{10(1+i)(2+i)(3+i)(4+i)\lambda^4 2^{1+i}}{(\lambda+1)^{5+i}}$$

$$-\frac{(1+i)(2+i)(3+i)(4+i)(5+i)\lambda^5 2^{1+i}}{(\lambda+1)^{6+i}}$$

and also

$$\mathcal{E}_d^{(i)}(i-1) = \sum_{j=1}^{d}\sum_{k=0}^{j}(-1)^j S_2(d,j)\left|S_1(j,k)\right| e^{j\ln\sqrt{2}+\frac{\pi}{2}+i\frac{(2k+j)\pi}{4}},$$

where $S_1(j,k)$ denotes the Stirling numbers of the first kind (cf. Ref. [29]).

Putting $\alpha = -m$ $(m \in \mathbb{N})$ in (1.35), we have

$$\mathcal{E}_d^{(-m)}(\lambda) = 2^{-m}\sum_{j=1}^{d}(-1)^j \lambda^j S_2(d,j)\frac{(-m)^{(j)}}{(\lambda+1)^{-m+j}} \tag{1.37}$$

(cf. Ref. [29]). Substituting the following beta-type rational functions

$$\mathfrak{M}_{j,m}(\lambda) = \lambda^j(\lambda+1)^{m-j} \tag{1.38}$$

(cf. Ref. [105, Definition 1.1]) into (1.37), one has

$$\mathcal{E}_d^{(-m)}(\lambda) = 2^{-m}\sum_{j=1}^{d}\binom{m}{j} j! S_2(d,j)\,\mathfrak{M}_{j,m}(\lambda), \tag{1.39}$$

where $d, m \in \mathbb{N}$ (cf. Ref. [29]).

By integrating equation (1.39) from −1 to 0, and using the following integral formulas of the function $\mathfrak{M}_{j,m}(\lambda)$:

$$\int_{-1}^{0} \mathfrak{M}_{j,m}(\lambda)d\lambda = \sum_{k=0}^{m-j}(-1)^{m-k}\binom{m-j}{k}\frac{1}{m+1-k},$$

$$\int_{-1}^{0} \mathfrak{M}_{j,m}(\lambda)d\lambda = (-1)^{j}\frac{1}{(m+1)\binom{m}{j}}$$

(cf. Ref. [105, Eqs. (18)–(19)]) and using the following known result:

$$\sum_{j=1}^{d}(-1)^{j-d}j!S_2(d,j) = 1,$$

we have

$$\int_{-1}^{0} \mathcal{E}_d^{(-m)}(\lambda)\,d\lambda = \sum_{j=1}^{d}\sum_{k=0}^{m-j}(-1)^{m-k}\binom{m-j}{k}\binom{m}{j}\frac{j!S_2(d,j)}{2^m(m+1-k)} \tag{1.40}$$

and

$$\int_{-1}^{0} \mathcal{E}_d^{(-m)}(\lambda)\,d\lambda = \frac{1}{2^m(m+1)}\sum_{j=1}^{d}(-1)^{j}j!S_2(d,j) \tag{1.41}$$

$$= \frac{(-1)^d}{2^m(m+1)}$$

(cf. Ref. [29]).

Combining (1.40) with (1.41), we have the following theorem:

Theorem 6 (cf. Ref. [29]). *Let $d, m \in \mathbb{N}$. Then,*

$$\sum_{j=1}^{d}\sum_{k=0}^{m-j}(-1)^{m-k-d}\binom{m}{j}\binom{m-j}{k}\frac{j!S_2(d,j)}{m+1-k} = \frac{1}{m+1}.$$

1.1. *The polynomials* $\mathfrak{u}_n^{(\alpha)}(x;\lambda)$

Let α and λ be arbitrary real or complex parameters. Gun and Simsek [29] defined the following generating function for the polynomials $\mathfrak{u}_n^{(\alpha)}(x;\lambda)$ with real variable x:

$$h(t,x;\lambda,\alpha)=\frac{e^{tx}}{(\lambda e^t+1)^\alpha}=\sum_{n=0}^{\infty}\frac{\mathfrak{u}_n^{(\alpha)}(x;\lambda)}{(1+\lambda)^\alpha}\frac{t^n}{n!}, \tag{1.42}$$

where $|t|<\pi$ when $\lambda=1$; $|t|<|\log(-\lambda)|$ when $\lambda\neq 1$, and also $1^\alpha=1$. When $\alpha=0$, we have

$$\mathfrak{u}_n^{(0)}(x;\lambda)=x^n.$$

Motivations of (1.42) are given as follows:

When $\lambda=1$, generating function in (1.42) reduces to the generating function for the Euler polynomials (of the first kind) of order α. That is,

$$\begin{aligned}\frac{e^{tx}}{(e^t+1)^\alpha}&=\sum_{n=0}^{\infty}2^{-\alpha}\mathfrak{u}_n^{(\alpha)}(x;1)\frac{t^n}{n!}\\&=\sum_{n=0}^{\infty}E_n^{(\alpha)}(x)\frac{t^n}{n!}.\end{aligned}$$

For details about the Euler polynomials (of the first kind) of order α, cf. Refs. [73, 85, 102, 122, 123].

By using the following functional equations

$$2^\alpha h(t,x;\lambda,\alpha)-(1+\lambda)^\alpha F_\alpha(t,x;\lambda)=0$$

and

$$2^\alpha h(t,0;\lambda,\alpha)-(1+\lambda)^\alpha F_\alpha(t;\lambda)=0,$$

Gune and Simsek [29] showed that $\mathfrak{u}_n^{(\alpha)}(x;\lambda)$ is related to the polynomials $\mathcal{E}_n^{(\alpha)}(x;\lambda)$ and $H_n^{(\alpha)}(x;\lambda)$, respectively:

$$2^\alpha\mathfrak{u}_n^{(\alpha)}(x;\lambda)-(1+\lambda)^\alpha\mathcal{E}_n^{(\alpha)}(x;\lambda)=0 \tag{1.43}$$

and

$$\mathfrak{u}_n^{(\alpha)}(x;\lambda) - H_n^{(\alpha)}\left(x;-\frac{1}{\lambda}\right) = 0.$$

Gune and Simsek [29] defined the following polynomials $\mathfrak{q}_n(\lambda;\alpha)$ in terms of the polynomials $\mathfrak{u}_n^{(\alpha)}(\lambda)$:

$$\mathfrak{q}_n(\lambda;\alpha) = (1+\lambda)^n \mathfrak{u}_n^{(\alpha)}(\lambda), \tag{1.44}$$

which satisfies the following computational formula:

$$\mathfrak{q}_n(\lambda;\alpha) = \sum_{j=1}^{n} (-1)^j S_2(n,j)(\alpha)^{(n)} \mathfrak{M}_{j,n}(\lambda),$$

where $n \in \mathbb{N}_0$.

Exercise 2. Find the twisted version of the polynomials $\mathfrak{q}_n(\lambda;\alpha)$ and of the function $\mathfrak{M}_{j,n}(\lambda)$ with their generating and interpolation functions.

2. *q*-Apostol-Type Frobenius–Euler Polynomials and Numbers

The q-Apostol-type Frobenius–Euler polynomials $\mathcal{H}_n(x;u;a,b;\lambda;q)$ and the q-Apostol-type Frobenius–Euler numbers $\mathcal{H}_n(u;a,b;\lambda;q)$ are defined by means of the following generating functions, respectively:

$$\begin{aligned} F_{\lambda,q}(x,t;u,a,b) &= \left(1-\frac{a^t}{u}\right)\sum_{n=0}^{\infty}\left(\frac{\lambda}{u}\right)^n b^{[n+x]t} \\ &= \sum_{n=0}^{\infty} \mathcal{H}_n(x;u;a,b;\lambda;q)\frac{t^n}{n!} \end{aligned} \tag{2.1}$$

and

$$\begin{aligned} F_{\lambda,q}(t;u,a,b) &= \left(1-\frac{a^t}{u}\right)\sum_{n=0}^{\infty}\left(\frac{\lambda}{u}\right)^n b^{[n]t} \\ &= \sum_{n=0}^{\infty} \mathcal{H}_n(u;a,b;\lambda;q)\frac{t^n}{n!}, \end{aligned} \tag{2.2}$$

where $\lambda, q \in \mathbb{C}$, $a,b \in \mathbb{R}^+$ $(a \neq b)$ and $u \in \mathbb{C}\setminus\{1\}$ with $\left|\frac{\lambda}{u}\right| < 1$ (cf. Ref. [103]).

For $x = 0$, we have

$$\mathcal{H}_n(u;a,b;\lambda;q) := \mathcal{H}_n(0;u;a,b;\lambda;q)$$

(cf. Ref. [103]).

With the aid of (2.1) and (2.2), it is easy to obtain the following functional equation:

$$F_{\lambda,q}(x,t;u,a,b) = b^{t[x]} F_{\lambda,q}(q^x t;u,a,b)$$

(cf. Ref. [103]).

When $q \to 1$ in (2.1), we have

$$\frac{a^t - u}{\lambda b^t - u} b^{tx} = \sum_{n=0}^{\infty} \mathcal{H}_n(x;u;a,b;\lambda;1) \frac{t^n}{n!}$$

(cf. Refs. [104, 106]). Putting $a = 1$ and $b = e$, the above equation reduces to following the generating functions for the Apostol-type Frobenius–Euler numbers and polynomials, respectively:

$$\frac{1-u}{\lambda e^t - u} = \sum_{n=0}^{\infty} H_n(\lambda|u) \frac{t^n}{n!} \tag{2.3}$$

and

$$\frac{1-u}{\lambda e^t - u} e^{tx} = \sum_{n=0}^{\infty} H_n(x;\lambda|u) \frac{t^n}{n!} \tag{2.4}$$

(cf. [8, 104]).

When $a = 1$, $\lambda = 1$ and $b = e$, the functions $F_{\lambda,q}(t;u,a,b)$ and the functions $F_{\lambda,q}(x,t;u,a,b)$ reduce to equations (1.9) and (1.2), which are the generating functions for the q-Frobenius–Euler polynomials and numbers.

Generating functions, related to generating functions for the q-Frobenius–Euler polynomials and numbers, for the numbers $E_{n,q}$ and the polynomials $E_{n,q}(x)$, respectively:

$$2\sum_{n=0}^{\infty} (-1)^n e^{[n]t} = \sum_{n=0}^{\infty} E_{n,q} \frac{t^n}{n!},$$

and

$$2\sum_{n=0}^{\infty}(-1)^n e^{([n+x]t)} = \sum_{n=0}^{\infty} E_{n,q}(x)\frac{t^n}{n!}$$

(cf. Refs. [74, 103, 117, 123]).

Theorem 7 (cf. Ref. [64]). *Then, we have*

$$\mathcal{H}_m(x;u;a,b;\lambda;q) = \sum_{n=0}^{\infty}\frac{\lambda^n}{u^{n+1}}(u([n+x]\ln b)^m - ([n+x]\ln b + \ln a)^m),$$

when $\lambda \neq 1$, $\left|\frac{\lambda}{u}\right| < 1$ *and when* $\lambda = 1$, $|u| > 1$.

When $a = 1$ and $b = e$, we have

$$\mathcal{H}_m(u;1,e;\lambda;q) = \frac{u-1}{u(1-q)^m}\sum_{j=0}^{m}(-1)^{m-j}\binom{m}{j}\sum_{n=0}^{\infty}\left(\frac{\lambda}{u}\right)^n q^{n(m-j)}$$

or

$$\mathcal{H}_m(u;1,e;\lambda;q) = \frac{u-1}{u(1-q)^m}\sum_{j=0}^{m}(-1)^{j}\binom{m}{j}\sum_{n=0}^{\infty}\left(\frac{\lambda}{u}\right)^n q^{nj}.$$

Assuming that $\left|\frac{q^j\lambda}{u}\right| < 1$, using geometric series, the above equation reduces to

$$\mathcal{H}_m(u;1,e;\lambda;q) = \frac{u-1}{u(1-q)^m}\sum_{j=0}^{m}(-1)^{j}\binom{m}{j}\frac{1}{1-\frac{\lambda q^j}{u}}.$$

After some calculations, we arrive at the following theorem:

Theorem 8.

$$\mathcal{H}_m(u;1,e;\lambda;q) = \frac{u-1}{(1-q)^m}\sum_{j=0}^{m}(-1)^{j}\binom{m}{j}\frac{1}{u-\lambda q^j}. \tag{2.5}$$

Substituting $\lambda = 1$ into (2.5), we arrive at (1.5).

When we modify (2.2), we have

$$\sum_{n=0}^{\infty} \left(\frac{\lambda}{u}\right)^{n} b^{([n]t)} = \frac{u}{u - a^{t}} \sum_{m=0}^{\infty} \mathcal{H}_{m}\left(u; a, b; \lambda; q\right) \frac{t^{m}}{m!}. \tag{2.6}$$

By using (2.6), Kucukoglu and Simsek [64] proved the following theorem:

Theorem 9. *Let* $\left|\frac{\lambda}{u}\right| < 1$ *and* $a \neq 1$*. Then, we have*

$$(u-1) \sum_{n=0}^{\infty} \left(\frac{\lambda}{u}\right)^{n} \left([n] \frac{\ln b}{\ln a}\right)^{m} = \sum_{k=0}^{m} \binom{m}{k} \frac{u \mathcal{H}_{k}\left(u; a, b; \lambda; q\right) H_{m-k}\left(u\right)}{\left(\ln a\right)^{k}}. \tag{2.7}$$

Substituting $a = b = e$ into (2.7), we get the following corollary:

Corollary 2. *Let* $\left|\frac{\lambda}{u}\right| < 1$*. Then, we have*

$$\sum_{n=0}^{\infty} \frac{\lambda^{n} \left(1 + q + q^{2} + \cdots + q^{n-1}\right)^{m}}{u^{n}} = \sum_{k=0}^{m} \binom{m}{k} \frac{u \mathcal{H}_{k}\left(u; e, e; \lambda; q\right) H_{m-k}\left(u\right)}{u-1}.$$

Kucukoglu and Simsek [64] defined generating functions for the numbers $\mathcal{I}_{m}\left(u; a, b; \lambda; q\right)$, related to the Frobenius–Euler numbers and the Apostol–Bernoulli numbers:

$$F_{\mathcal{I}}\left(t, u; a, b; \lambda; q\right) = \sum_{m=0}^{\infty} \mathcal{I}_{m}\left(u; a, b; \lambda; q\right) \frac{t^{m}}{m!} = \left(1 - \frac{1}{u}\right) \sum_{n=0}^{\infty} \left(\frac{\lambda}{u}\right)^{n} e^{\left(t[n] \frac{\ln b}{\ln a}\right)}, \tag{2.8}$$

where $a \neq 1$.

From the above function, we have

$$\mathcal{I}_m\left(u;a,b;\lambda;1\right)=\left(\frac{\ln b}{\ln a}\right)^m H_m\left(\lambda|u\right)$$

and

$$\lim_{q\to 1}F_{\mathcal{I}}\left(t,u;a,b;\lambda;q\right)=\frac{1-u}{\lambda e^{\left(t\left(\frac{\ln b}{\ln a}\right)\right)}-u}.$$

It easy to see that the above function is a generating function for the Apostol–Bernoulli numbers (cf. Ref. [64]).

Theorem 10 (cf. Ref. [64]). (*Recurrence relation*) *Let* $a\neq 1$. *Then, we have*

$$\mathcal{I}_{m+v}\left(u;a,b;\lambda;q\right)=\left(\frac{\ln b}{(1-q)\ln a}\right)^v\sum_{j=0}^{v}(-1)^{v-j}\times\binom{v}{j}\mathcal{I}_m\left(u;a,b;\lambda q^{v-j};q\right).$$

Taking partial derivative v times both side of (2.8), with respect to t, we also have

$$\left.\frac{\partial^v}{\partial t^v}F_{\mathcal{I}}\left(t,u;a,b;\lambda;q\right)\right|_{t=0}=\mathcal{I}_v\left(u;a,b;\lambda;q\right)$$

(cf. Ref. [64]). When $q\to 1$, the above equation reduces to the Lerch zeta function at negative integers:

$$\lim_{q\to 1}\left.\frac{\partial^v}{\partial t^v}F_{\mathcal{I}}\left(t,u;a,b;\lambda;q\right)\right|_{t=0}=\left(\frac{\ln b}{\ln a}\right)^v\left(1-\frac{1}{u}\right)\Phi\left(\frac{\lambda}{u},-v,0\right) \tag{2.9}$$

and

$$\lim_{q\to 1}\left.\frac{\partial^v}{\partial t^v}F_{\mathcal{I}}\left(t,u;a,b;\lambda;q\right)\right|_{t=0}=\left(\frac{\ln b}{\ln a}\right)^v\left(\frac{1}{u}-1\right)\frac{B_{v+1}\left(\frac{\lambda}{u}\right)}{v+1} \tag{2.10}$$

(cf. Ref. [64]).

Putting $a = b = e$ and $u = -1$ in (2.10), we also have

$$\lim_{q\to 1} \frac{\partial^v}{\partial t^v} F_{\mathcal{I}}(t, -1; e, e; \lambda; q)\bigg|_{t=0} = \mathcal{E}_v(\lambda)$$

(cf. Ref. [64]).

By using (2.8), the polynomials $\mathcal{I}_m(x; u; a, b; \lambda; q)$ defined by means of the following generating function:

$$G_{\mathcal{I}}(x, t; u; a, b; \lambda; q) = \left(1 - \frac{1}{u}\right) \sum_{n=0}^{\infty} \left(\frac{\lambda}{u}\right)^n e^{\left(t[n+x]\frac{\ln b}{\ln a}\right)}$$
$$= \sum_{m=0}^{\infty} \mathcal{I}_m(x; u; a, b; \lambda; q) \frac{t^m}{m!}, \tag{2.11}$$

where $a \neq 1$ (cf. Ref. [64]).

For $x = 0$, we have

$$\mathcal{I}_m(u; a, b; \lambda; q) = \mathcal{I}_m(0; u; a, b; \lambda; q).$$

With the aid of (2.8), we have the following functional equation:

$$G_{\mathcal{I}}(x, t; u; a, b; \lambda; q) = e^{\left(t[x]\frac{\ln b}{\ln a}\right)} F_{\mathcal{I}}(q^x t, u; a, b; \lambda; q). \tag{2.12}$$

Using the above functional equation, we have the following recurrence relation:

Theorem 11 (cf. Ref. [64]). *Let $a \neq 1$. Then, we have*

$$\mathcal{I}_m(x; u; a, b; \lambda; q) = \sum_{j=0}^{m} \binom{m}{j} \left([x]\frac{\ln b}{\ln a}\right)^{m-j} q^{xj} \mathcal{I}_j(u; a, b; \lambda; q)$$
$$\times \sum_{m=0}^{n-1} \left(\frac{\lambda}{u}\right)^m \left(\left(1 + q + q^2 + \cdots + q^{m-1}\right) \ln b\right)^v$$
$$= \frac{u\mathcal{I}_v(u; e, b; \lambda; q) - \lambda^n u^{n-1} \mathcal{I}_v(n; u; e, b; \lambda; q)}{u - 1} \tag{2.13}$$

and

$$\mathcal{I}_m(u; a, b; \lambda; q) = \left(\frac{\ln b}{(q-1)\ln a}\right)^m \sum_{j=0}^{m} (-1)^{m-j} \binom{m}{j} \frac{u-1}{u - \lambda q^j}. \tag{2.14}$$

Remark 1. For $\lambda = 1$, $a = b = e$ and replace u by u^{-1}; one has the following equation:

$$H\left(u^{-1} : q|1\right) = \frac{1-u}{(1-q)^m} \sum_{j=0}^{m} (-1)^j \binom{m}{j} \frac{1}{1-q^j u},$$

where $H\left(u^{-1} : q|1\right)$ denotes the Euler–Barnes-type Daehee q-Euler numbers (cf. Ref. [123]).

Multiplication formula for the polynomials $\mathcal{I}_m(x; u; a, b; \lambda; q)$ is given by the following theorem:

Theorem 12 (cf. Ref. [64]). (*Multiplication formula*) *Let* $m \in \mathbb{N}_0$ *and* $d \in \mathbb{N}$. *Then,*

$$\begin{aligned} &\mathcal{I}_m(xd; u; a, b; \lambda; q) \\ &= [d]^m \sum_{k=0}^{d-1} \frac{\lambda^k \left(u^{d+1} - u^d\right)}{u^{d+k+1} - u^{k+1}} \mathcal{I}_m\left(x + \frac{k}{d}; u^d; a, b; \lambda^d; q^d\right). \end{aligned} \tag{2.15}$$

Remark 2. When $a = b = e$, $\lambda = 1$ and $q \to 1$, (2.15) reduces to

$$\mathcal{I}_m(xd; u; e, e; 1; 1) = d^m \sum_{k=0}^{d-1} \frac{u^{d+1} - u^d}{u^{d+k+1} - u^{k+1}} \mathcal{I}_m\left(x + \frac{k}{d}; u^d; e, e; 1; 1\right)$$

(cf. Ref. [104]). Putting $u = -1$ (with odd d integer) in the above equation, multiplication formula for the Euler polynomials can be found. The multiplication formula of the polynomials $\mathcal{I}_m(x; u; a, b; \lambda; q)$ is related to the normalized polynomials, which are satisfy

$$f_n(xd) = d^{n-1} \sum_{k=0}^{d-1} f_n\left(x + \frac{k}{d}\right),$$

where f_n is a polynomial (on finite fields) (cf. Refs. [22, 64, 104]).

Exercise 3. Find the twisted version of the polynomials $\mathcal{I}_m(u; a, b; \lambda; q)$ with their generating functions and their interpolation functions.

2.1. *Formulas for sum of powers of consecutive integers related to the q-Apostol-type Frobenius–Euler numbers and polynomials and other special polynomials*

Alternative summation formulas and some of powers of consecutive q-integers have been studied by many different methods. Using the same method of Kim *et al.* [50, 101], Kucukoglu and Simsek [64] gave some novel formulas for sum of powers of consecutive integers involving the q-Apostol-type Frobenius–Euler numbers and polynomials and other special polynomials.

From (2.2) and (2.1), one has

$$u^n F_{\lambda,q}(t;u,a,b) - \lambda^n F_{\lambda,q}(n,t;u,a,b) = \left(u - a^t\right) \sum_{m=0}^{n-1} \frac{\lambda^m b^{[m]t}}{u^{m+1-n}}.$$

Combining the above equation with the following relation:

$$a^t = \sum_{v=0}^{\infty} (\ln a)^v \frac{t^v}{v!}$$

and

$$b^{[m]t} = \sum_{v=0}^{\infty} \left(\left(1 + q + q^2 + \cdots + q^{m-1}\right) \ln b\right)^v \frac{t^v}{v!},$$

one has

$$\sum_{v=0}^{\infty} \left(\mathcal{H}_v(u;a,b;\lambda;q) - \left(\frac{\lambda}{u}\right)^n \mathcal{H}_v(n;u;a,b;\lambda;q)\right) \frac{t^v}{v!}$$

$$= \sum_{v=0}^{\infty} \sum_{m=0}^{n-1} \left(\frac{\lambda}{u}\right)^m ([m] \ln b)^v \frac{t^v}{v!}$$

$$- \sum_{v=0}^{\infty} \sum_{k=0}^{v} \sum_{m=0}^{n-1} \binom{v}{k} \frac{\lambda^m (\ln a)^{v-k} ([m] \ln b)^k}{u^{m+1}} \frac{t^v}{v!}.$$

Comparing the coefficient of $\frac{t^v}{v!}$ in both sides of the above equation, one has

$$u^n\mathcal{H}_v(u;a,b;\lambda;q) - \lambda^n\mathcal{H}_v(n;u;a,b;\lambda;q) = \sum_{m=0}^{n-1}\frac{\lambda^m([m]\ln b)^v}{u^{m-n}} - \sum_{k=0}^{v}\sum_{m=0}^{n-1}\binom{v}{k}\frac{\lambda^m(\ln a)^{v-k}([m]\ln b)^k}{u^{m+1-n}}. \tag{2.16}$$

After some calculations in (2.16), the following theorem is given:

Theorem 13 (cf. Ref. [64]). *Let $u \neq 0$. Let $n \in \mathbb{N}$ and $v \in \mathbb{N}_0$. Then, we have*

$$u^n\mathcal{H}_v(u;a,b;\lambda;q) - \lambda^n\mathcal{H}_v(n;u;a,b;\lambda;q) = \sum_{m=0}^{n-1}\frac{\lambda^m}{u^{m-n+1}}\begin{pmatrix} u\left(\left(1+q+q^2+\cdots+q^{m-1}\right)\ln b\right)^v \\ -\left(\left(1+q+q^2+\cdots+q^{m-1}\right)\ln b+\ln a\right)^v\end{pmatrix}. \tag{2.17}$$

Putting $a = 1$ in (2.17), we have

$$\sum_{m=0}^{n-1}\frac{\lambda^m\left(1+q+q^2+\cdots+q^{m-1}\right)^v}{u^m} = \frac{u\mathcal{H}_v(u;1,b;\lambda;q) - \lambda^n u^{n-1}\mathcal{H}_v(n;u;1,b;\lambda;q)}{(u-1)\ln^v b} \tag{2.18}$$

(cf. Ref. [64]).

Putting $\lambda = 1$ and $b = e$ in (2.18), we have the following summation formula in terms of the q-Frobenius–Euler numbers and polynomials:

$$\sum_{m=0}^{n-1}\left(\frac{1}{u^{m-1}} - 1\right)[m]^v = uH_v(u,q) - u^{n-1}H_v(n,u,q) \tag{2.19}$$

(cf. Refs. [50, 52, 54–56, 64, 102, 123]).

Putting $\lambda = 1$, $b = e$ and $u = -1$ in (2.18) and when $q \to 1$, one has the following summation formula involving the q-Euler numbers and polynomials:

$$2\sum_{m=0}^{n-1} (-1)^m \left(1 + q + q^2 + \cdots + q^{m-1}\right)^v = E_{v,q} + (-1)^{n-1} E_{v,q}(n)$$

(cf. Refs. [50, 54, 64]). In the above equation, when $q \to 1$, we have

$$2\sum_{m=0}^{n-1} (-1)^m m^v = E_v + (-1)^{n-1} E_v(n)$$

(cf. Refs. [50, 54, 64]).

Putting $\lambda = 1$, $b = e$ and $q \to 1$ in (2.18), we have the following summation formula involving the Frobenius–Euler numbers and polynomials:

$$\sum_{m=0}^{n-1} \frac{m^v}{u^m} = \frac{uH_v(u) - u^{n-1} H_v(n; u)}{u - 1},$$

where $u \in \mathbb{C} \setminus \{1\}$ (cf. Refs. [50, 54, 55, 64, 123]).

Substituting $\lambda = 1$, $b = e$, $q \to 1$ and $u = -\alpha^{-h}$, with $\alpha \in \mathbb{C}$ and $|\alpha| < 1$, into (2.18), one has the following summation formula involving the (h, q)-extension of the Euler polynomials $E_{v,\alpha}^{(h)}(x)$ and the (h, q)-extension of Euler numbers $E_{v,\alpha}^{(h)}$:

$$\sum_{m=0}^{n-1} (-1)^m \alpha^{hm} m^v = \frac{E_{v,\alpha}^{(h)} + (-1)^{n-1} \alpha^{-hn} E_{v,\alpha}^{(h)}(n)}{2}$$

(cf. Ref. [102]).

The well-known Faulhaber formula is also given as follows:

$$\sum_{m=0}^{n-1} m^v = \frac{B_{v+1}(n) - B_{v+1}}{v + 1}$$

(cf. Refs. [50–55, 59, 64, 114–116, 122, 123]).

Exercise 4. Find the applications of the above formulas for the sum of powers of consecutive integers related to the Apostol-type Frobenius–Euler numbers and polynomials in theory diophantine equations.

2.2. Two parametric kinds of Eulerian-Type polynomials associated with Euler's formula

In order to give definitions and some properties of the two parametric kinds of Frobenius–Eulerian polynomials, we start this section with some definitions, formulas and notations.

We can use the following well-known Euler's formula:

$$e^{iz} = \cos(z) + i\sin(z),$$

where $i^2 = -1$.

The Apostol–Bernoulli polynomials of order m are defined by means of the following generating function:

$$F_{AB}^{(m)}(t, x; \lambda) = \left(\frac{t}{\lambda e^t - 1}\right)^m e^{xt} = \sum_{n=0}^{\infty} \mathcal{B}_n^{(m)}(x; \lambda)\frac{t^n}{n!}, \tag{2.20}$$

where $|t| < 2\pi$ when $\lambda = 1$; $|t| < |\log(\lambda)|$ when $\lambda \neq 1$ and

$$\mathcal{B}_n^{(m)}(\lambda) = \mathcal{B}_n^{(m)}(0; \lambda),$$

where $\mathcal{B}_n^{(m)}(\lambda)$ denotes the Apostol–Bernoulli number of order m and $\mathcal{B}_n(\lambda)$ denotes the Apostol–Bernoulli numbers (cf. Refs. [6, 8, 55, 66, 69, 74, 114–116, 122]).

The Apostol–Euler polynomials $\mathcal{E}_n^{(m)}(x; \lambda)$ of order m are defined by means of the following generating function:

$$F_{AE}^{(m)}(t, x; \lambda) = \left(\frac{2}{\lambda e^t + 1}\right)^m e^{xt} = \sum_{n=0}^{\infty} \mathcal{E}_n^{(m)}(x; \lambda)\frac{t^n}{n!}, \tag{2.21}$$

where $|t| < \pi$ when $\lambda = 1$; $|t| < |\log(-\lambda)|$ when $\lambda \neq 1$ and

$$\mathcal{E}_n^{(m)}(\lambda) = \mathcal{E}_n^{(m)}(0; \lambda),$$

where $\mathcal{E}_n^{(m)}(\lambda)$ denotes the Apostol–Euler numbers of order m and $\mathcal{E}_n(\lambda)$ denotes the Apostol–Euler numbers (cf. Refs. [6, 8, 55, 66, 69, 74, 114–116, 122]).

The λ-Stirling numbers of the second kind are defined by means of the following generating function:

$$F_{S2}(t, m; \lambda) = \frac{(\lambda e^t - 1)^m}{m!} = \sum_{n=0}^{\infty} S_2(n, m; \lambda)\frac{t^n}{n!}, \tag{2.22}$$

where $m \in \mathbb{N}_0$ and $\lambda \in \mathbb{C}$ (cf. Refs. [69, 104, 114–116, 122, 123]).

Putting $\lambda = 1$ in (2.22), the numbers $S_2(n, m; \lambda)$ reduce to the Stirling numbers of the second kind:

$$S_2(n, m) = S_2(n, m; 1).$$

Using (2.21) and (2.22), computation formula for the Apostol–Euler polynomials of order m is given by

$$\mathcal{E}_n^{(m)}(x; \lambda) = 2^m \sum_{k=0}^{n} \binom{n}{k} x^{n-k} \sum_{j=0}^{k} \binom{m+j-1}{j} \frac{j!(-\lambda)^j}{(\lambda+1)^{j+m}} S_2(k, j) \tag{2.23}$$

(cf. Refs. [29, 68, 69, 122]).

We now show that the computation formula (2.23) can also be written in terms of the beta function:

$$\mathcal{E}_n^{(m)}(x; \lambda) = 2^m \sum_{k=0}^{n} \binom{n}{k} x^{n-k} \sum_{j=0}^{k} (-1)^j \times \frac{(m+j-1)!\mathfrak{M}_{j,-m}(\lambda) S_2(k, j)}{(m-1)!}.$$

The Apostol–Genocchi polynomials $\mathcal{G}_n^{(m)}(x; \lambda)$ of order m are defined by

$$F_{AG}^{(m)}(t, x; \lambda) = \left(\frac{2t}{\lambda e^t + 1}\right)^m e^{xt} = \sum_{n=0}^{\infty} \mathcal{G}_n^{(m)}(x; \lambda) \frac{t^n}{n!}, \tag{2.24}$$

where $|t| < \pi$ when $\lambda = 1$; $|t| < |\log(-\lambda)|$ when $\lambda \neq 1$ and

$$\mathcal{G}_n^{(m)}(\lambda) = \mathcal{G}_n^{(m)}(0; \lambda),$$

where $\mathcal{G}_n^{(m)}(\lambda)$ denotes the so-called Apostol–Genocchi numbers of order m (cf. [68, 69, 122]).

The Apostol-type Frobenius–Euler polynomials $\mathcal{H}_n^{(m)}(x;\lambda,u)$ of order m are defined by means of the following generating function:

$$F_{AH}^{(m)}(t,x;\lambda,u)=\left(\frac{1-u}{\lambda e^t-u}\right)^m e^{xt}=\sum_{n=0}^{\infty}\mathcal{H}_n^{(m)}(x;\lambda,u)\frac{t^n}{n!}, \quad (2.25)$$

where $u\in\mathbb{C}$ with $u\neq\lambda$, $\lambda\neq 1$ and

$$\mathcal{H}_n^{(m)}(\lambda,u)=\mathcal{H}_n^{(m)}(0;\lambda,u),$$

where $\mathcal{H}_n^{(m)}(\lambda,u)$ denotes the so-called Apostol-type Frobenius–Euler numbers of order m (cf. Refs. [7, 8, 58, 64, 65, 104, 114–116, 122]).

Putting $m=1$ in (2.25), we arrive at (2.4).

Substituting $u=-1$ into (2.25), we have

$$\mathcal{H}_n^{(m)}(x;\lambda,-1)=\mathcal{E}_n^{(m)}(x;\lambda)$$

(cf. Refs. [7, 8, 58, 64, 65, 104, 114–116, 122]).

Substituting $\lambda=1$ into (2.25), we have

$$H_n^{(m)}(x;u)=\mathcal{H}_n^{(m)}(x;1,u), \quad (2.26)$$

where $H_n^{(m)}(x;u)$ denotes the so-called Frobenius–Euler polynomials of order m.

Substituting $x=0$ into (2.26), we have

$$H_n^{(m)}(0;u)=H_n^{(m)}(u)$$

(cf. Refs. [7, 8, 58, 64, 65, 104, 114–116, 122])

The polynomials $C_n(x,y)$ are defined by means of the following generating function:

$$F_C(t,x,y)=e^{xt}\cos(yt)=\sum_{n=0}^{\infty}C_n(x,y)\frac{t^n}{n!} \quad (2.27)$$

(cf. Refs. [53, 72, 125]).

By combining (2.27) with

$$\cos(yt) = \sum_{n=0}^{\infty} (-1)^n y^{2n} \frac{t^{2n}}{(2n)!}$$

and

$$e^{xt} = \sum_{n=0}^{\infty} x^n \frac{t^n}{n!},$$

we have

$$\sum_{n=0}^{\infty} C_n(x,y) \frac{t^n}{n!} = \sum_{n=0}^{\infty} x^n \frac{t^n}{n!} \sum_{n=0}^{\infty} (-1)^n y^{2n} \frac{t^{2n}}{(2n)!}.$$

Therefore,

$$\sum_{n=0}^{\infty} C_n(x,y) \frac{t^n}{n!} = \sum_{n=0}^{\infty} \sum_{k=0}^{\left[\frac{n}{2}\right]} (-1)^k \binom{n}{2k} x^{n-2k} y^{2k} \frac{t^n}{n!}.$$

After comparing the coefficients of $\frac{t^n}{n!}$ on both sides of the above equation, we arrive at the following known formula for the polynomials $C_n(x,y)$:

$$C_n(x,y) = \sum_{k=0}^{\left[\frac{n}{2}\right]} (-1)^k \binom{n}{2k} x^{n-2k} y^{2k}$$

(cf. Refs. [53, 72, 125]).

The polynomials $S_n(x,y)$ are defined by means of the following generating function:

$$F_S(t,x,y) = e^{xt} \sin(yt) = \sum_{n=0}^{\infty} S_n(x,y) \frac{t^n}{n!} \tag{2.28}$$

(cf. Refs. [53, 72, 125]).

By combining (2.28) with

$$\sin(yt) = \sum_{n=0}^{\infty} (-1)^n y^{2n+1} \frac{t^{2n+1}}{(2n+1)!}$$

and

$$e^{xt} = \sum_{n=0}^{\infty} x^n \frac{t^n}{n!},$$

we have

$$\sum_{n=0}^{\infty} S_n(x,y) \frac{t^n}{n!} = \sum_{n=0}^{\infty} x^n \frac{t^n}{n!} \sum_{n=0}^{\infty} (-1)^n y^{2n+1} \frac{t^{2n+1}}{(2n+1)!}.$$

Therefore,

$$\sum_{n=0}^{\infty} S_n(x,y) \frac{t^n}{n!} = \sum_{n=0}^{\infty} \sum_{k=0}^{\left[\frac{n-1}{2}\right]} (-1)^k \binom{n}{2k+1} x^{n-2k-1} y^{2k+1} \frac{t^n}{n!}.$$

After comparing the coefficients of $\frac{t^n}{n!}$ on both sides of the above equation, we arrive at the following known formula for the polynomials $S_n(x,y)$:

$$S_n(x,y) = \sum_{k=0}^{\left[\frac{n-1}{2}\right]} (-1)^k \binom{n}{2k+1} x^{n-2k-1} y^{2k+1}$$

(cf. Refs. [53, 72, 125]).

The Fubini-type polynomials $a_n^{(m)}(x)$ of order m are defined by means of following generating function:

$$F_a^{(m)}(t,x) = \frac{2^m}{(2-e^t)^{2m}} e^{xt} = \sum_{n=0}^{\infty} a_n^{(m)}(x) \frac{t^n}{n!} \tag{2.29}$$

(cf. Ref. [57]).

For $x = 0$, from (2.29), one has

$$a_n^{(m)}(0) = a_n^{(m)},$$

where $a_n^{(m)}$ denotes the Fubini-type numbers of order m (cf. Ref. [57]).

By using (2.29) with $\left|e^t - 1\right| < 1$, we get

$$2^m \sum_{n=0}^{\infty} x^n \frac{t^n}{n!} = \sum_{j=0}^{\infty} \binom{-2m}{j} \left(e^t - 1\right)^j \sum_{n=0}^{\infty} a_n^{(m)}(x) \frac{t^n}{n!}.$$

Combining the above equation with (1.20), we obtain

$$\sum_{n=0}^{\infty} x^n \frac{t^n}{n!} = \sum_{j=0}^{\infty} (-1)^j 2^{m+1} m(2m+1) \cdots (2m+j-1)$$
$$\times \sum_{n=0}^{\infty} S_2(n,j) \frac{t^m}{m!} \sum_{n=0}^{\infty} a_n^{(m)}(x) \frac{t^n}{n!}.$$

Since

$$S_2(n,j) = 0$$

when $j > n$, we get

$$\sum_{n=0}^{\infty} x^n \frac{t^n}{n!} = \sum_{n=0}^{\infty} \sum_{j=0}^{n} (-1)^j 2^{m+1} \left(m(2m+1) \cdots (2m+j-1)\right)$$
$$\times S_2(n,j) \frac{t^m}{m!} \sum_{n=0}^{\infty} a_n^{(m)}(x) \frac{t^n}{n!}.$$

Therefore,

$$\sum_{n=0}^{\infty} x^n \frac{t^n}{n!} = \sum_{n=0}^{\infty} \sum_{v=0}^{n} \sum_{j=0}^{v} (-1)^j \binom{n}{v} 2^{m+1} \left(m(2m+1) \cdots (2m+j-1)\right)$$
$$\times S_2(v,j) a_{n-v}^{(m)}(x) \frac{t^n}{n!}.$$

After comparing the coefficients of $\frac{t^n}{n!}$ on both sides of the above equation, we arrive at the following formula:

Theorem 14. *Let $n, m \in \mathbb{N}_0$. Then,*

$$\sum_{v=0}^{n} \sum_{j=0}^{v} (-1)^j \binom{n}{v} 2^{m+1} \left(m(2m+1) \cdots (2m+j-1)\right)$$
$$\times S_2(v,j) a_{n-v}^{(m)}(x) = x^n. \tag{2.30}$$

Note the different version of equation (2.30) given by Kilar and Simsek [57] via Theorem 4.8.

By using (2.25), (2.20), (2.29) and (2.21), find the relations among $\mathcal{H}_n^{(m)}(x;\lambda,u)$, $a_n^{(m)}(x)$, $\mathcal{B}_n^{(m)}(x;\lambda)$, $\mathcal{E}_n^{(m)}(x;\lambda)$ and $\mathcal{G}_n^{(m)}(x;\lambda)$.

Let r-tuples $\overrightarrow{v}=(v_1,v_2,\ldots,v_r)$. The two kinds of Hermite-based r-parametric Milne–Thomson-type polynomials are defined by means of the following generating functions:

$$\begin{aligned} B\left(t,x,y,z,\overrightarrow{v},r,a,b\right) &= 2\left(b+f\left(t,a\right)\right)^z \exp\left(xt\right) M_4\left(t,y,\overrightarrow{v},r\right) \\ &= \sum_{n=0}^{\infty} \mathfrak{h}_1\left(n,x,y,z;\overrightarrow{v},r,a,b\right)\frac{t^n}{n!}, \end{aligned} \tag{2.31}$$

where

$$M_4\left(t,y,\overrightarrow{v},r\right)=\exp\left(\sum_{j=1}^{r} v_j t^j\right)\cos\left(yt\right)=\sum_{n=0}^{\infty} C_n\left(\overrightarrow{v},y;r\right)\frac{t^n}{n!}$$

and

$$\begin{aligned} B_1\left(t,x,y,z,\overrightarrow{v},r,a,b\right) &= 2\left(b+f\left(t,a\right)\right)^z \exp(xt) M_5\left(t,y,\overrightarrow{v},r\right) \\ &= \sum_{n=0}^{\infty} \mathfrak{h}_2\left(n,x,y,z;\overrightarrow{v},r,a,b\right)\frac{t^n}{n!}, \end{aligned} \tag{2.32}$$

where

$$M_5\left(t,y,\overrightarrow{v},r\right)=\exp\left(\sum_{j=1}^{r} v_j t^j\right)\sin\left(yt\right)=\sum_{n=0}^{\infty} S_n\left(\overrightarrow{v},y;r\right)\frac{t^n}{n!}$$

(cf. Ref. [58]).

The two parametric kinds of Eulerian-type polynomials of order k_1 and k_2 are denoted by $\mathcal{H}_n^{(C,k_1)}(x,y;\lambda,u)$ and $\mathcal{H}_n^{(S,k_2)}(x,y;\lambda,u)$, respectively, by means of the following generating functions:

$$F_{HC}^{(k_1)}(t,x,y;\lambda,u) = \left(\frac{1-u}{\lambda e^t - u}\right)^{k_1} e^{xt}\cos(yt)$$

with

$$F_{HC}^{(k_1)}(t,x,y;\lambda,u) = \sum_{n=0}^{\infty} \mathcal{H}_n^{(C,k_1)}(x,y;\lambda,u)\frac{t^n}{n!}, \tag{2.33}$$

and

$$F_{HS}^{(k_2)}(t,x,y;\lambda,u) = \left(\frac{1-u}{\lambda e^t - u}\right)^{k_2} e^{xt}\sin(yt)$$

with

$$F_{HS}^{(k_2)}(t,x,y;\lambda,u) = \sum_{n=0}^{\infty} \mathcal{H}_n^{(S,k_2)}(x,y;\lambda,u)\frac{t^n}{n!}, \tag{2.34}$$

where k_1, $k_2 \in \mathbb{N}_0$ (cf. Ref. [58]).

Substituting $z = x + iy$ into (2.25), we get

$$F_{AH}^{(m)}(t,x+iy;\lambda,u) = \sum_{n=0}^{\infty} \mathcal{H}_n^{(m)}(x+iy;\lambda,u)\frac{t^n}{n!}.$$

Combining the above equation with

$$e^{(x+iy)t} = e^{xt}(\cos(yt) + i\sin(yt)),$$

we get the following functional equations:

$$\begin{aligned}
F_{AH}^{(m)}(t,x+iy;\lambda,u) &= F_{AH}^{(m)}(t,x;\lambda,u)\cos(yt) \\
&\quad + iF_{AH}^{(m)}(t,x;\lambda,u)\sin(yt), \\
F_{AH}^{(m)}(t,x+iy;\lambda,u) &= F_{HC}^{(m)}(t,x,y;\lambda,u) + iF_{HS}^{(m)}(t,x,y;\lambda,u), \\
F_{AH}^{(m)}(t,x+iy;\lambda,u) &= F_{AH}^{(m)}(t,0;\lambda,u)(F_C(t,x,y) + iF_S(t,x,y)), \\
F_{AH}^{(m)}(t,x+iy;\lambda,u) &= \frac{1}{(ut)^m}F_{AB}^{(m)}\left(t,0;\frac{\lambda}{u}\right)(F_C(t,x,y) \\
&\quad + iF_S(t,x,y))
\end{aligned} \tag{2.35}$$

and

$$F_{AH}^{(m)}(t,x;-\lambda,u)=\left(\frac{u-1}{2u}\right)^m F_{AE}^{(m)}\left(t,0;\frac{\lambda}{u}\right)(F_C(t,x,y)+iF_S(t,x,y)).$$

The above functions can also be written as follows:

$$F_{AH}^{(m)}(t,x;\lambda,u)\cos(yt)+iF_{AH}^{(m)}(t,x;\lambda,u)\sin(yt)=\sum_{n=0}^{\infty}\mathcal{H}_n^{(m)}(x+iy;\lambda,u)\frac{t^n}{n!},\tag{2.36}$$

$$F_{HC}^{(m)}(t,x,y;\lambda,u)+iF_{HS}^{(m)}(t,x,y;\lambda,u)=\sum_{n=0}^{\infty}\mathcal{H}_n^{(m)}(x+iy;\lambda,u)\frac{t^n}{n!},\tag{2.37}$$

$$F_{AH}^{(m)}(t,0;\lambda,u)(F_C(t,x,y)+iF_S(t,x,y))=\sum_{n=0}^{\infty}\mathcal{H}_n^{(m)}(x+iy;\lambda,u)\frac{t^n}{n!},\tag{2.38}$$

$$\frac{1}{u^m}F_{AB}^{(m)}\left(t,0;\frac{\lambda}{u}\right)(F_C(t,x,y)+iF_S(t,x,y))=\sum_{n=0}^{\infty}\mathcal{H}_n^{(m)}(x+iy;\lambda,u)\frac{t^{n+m}}{n!}\tag{2.39}$$

and

$$\left(\frac{u-1}{2u}\right)^m F_{AE}^{(m)}\left(t,0;\frac{\lambda}{u}\right)(F_C(t,x,y)+iF_S(t,x,y))=\sum_{n=0}^{\infty}\mathcal{H}_n^{(m)}(x+iy;\lambda,u)\frac{t^n}{n!}.\tag{2.40}$$

By using (2.36), we get

$$\sum_{n=0}^{\infty} \mathcal{H}_n^{(m)} (x+iy;\lambda,u) \frac{t^n}{n!}$$
$$= \sum_{n=0}^{\infty} \mathcal{H}_n^{(m)} (x;\lambda,u) \frac{t^n}{n!} \sum_{n=0}^{\infty} \frac{(-1)^m y^{2m} t^{2m}}{(2m)!}$$
$$+ i \sum_{n=0}^{\infty} \mathcal{H}_n^{(m)} (x;\lambda,u) \frac{t^n}{n!} \sum_{n=0}^{\infty} \frac{(-1)^n y^{2n+1} t^{2n+1}}{(2n+1)!}.$$

Therefore,

$$\sum_{n=0}^{\infty} \mathcal{H}_n^{(m)} (x+iy;\lambda,u) \frac{t^n}{n!}$$
$$= \sum_{n=0}^{\infty} \sum_{j=0}^{\left[\frac{n}{2}\right]} \frac{\mathcal{H}_{n-2j}^{(m)} (x;\lambda,u)}{(n-2j)!} \frac{(-1)^j y^{2j}}{(2j)!} t^n$$
$$+ it \sum_{n=0}^{\infty} \sum_{j=0}^{\left[\frac{n}{2}\right]} (-1)^j \frac{\mathcal{H}_{n-2j}^{(m)} (x;\lambda,u)}{(n-2j)!} \frac{(-1)^j y^{2j+1}}{(2j+1)!} t^n.$$

After some calculations, we get

$$\sum_{n=0}^{\infty} \mathcal{H}_n^{(m)} (x+iy;\lambda,u) \frac{t^n}{n!}$$
$$= \sum_{n=0}^{\infty} \sum_{j=0}^{\left[\frac{n}{2}\right]} (-1)^j \binom{n}{2j} y^{2j} \mathcal{H}_{n-2j}^{(m)} (x;\lambda,u) \frac{t^n}{n!}$$
$$+ i \sum_{n=0}^{\infty} n \sum_{j=0}^{\left[\frac{n-1}{2}\right]} (-1)^j \binom{n}{2j} y^{2j} \mathcal{H}_{n-1-2j}^{(m)} (x;\lambda,u) \frac{t^n}{n!}.$$

After comparing the coefficients of $\frac{t^n}{n!}$ on both sides of the above equation, we arrive at the following theorem:

Theorem 15. *Let $n, m \in \mathbb{N}_0$. Then, we have*

$$\mathcal{H}_n^{(m)}(x+iy;\lambda,u)\frac{t^n}{n!} = \sum_{j=0}^{\left[\frac{n}{2}\right]}(-1)^j\binom{n}{2j}y^{2j}\mathcal{H}_{n-2j}^{(m)}(x;\lambda,u)$$

$$+in\sum_{j=0}^{\left[\frac{n-1}{2}\right]}(-1)^j\binom{n}{2j}y^{2j}\mathcal{H}_{n-1-2j}^{(m)}(x;\lambda,u).$$

By using (2.37), Kilar and Simsek [58] proved the following theorem:

Theorem 16. *Let $n, m \in \mathbb{N}_0$. Then, we have*

$$\mathcal{H}_n^{(m)}(x+iy;\lambda,u) = \mathcal{H}_n^{(C,m)}(x,y;\lambda,u) + i\mathcal{H}_n^{(S,m)}(x,y;\lambda,u).$$

By using (2.38), we have

$$\sum_{n=0}^{\infty}\mathcal{H}_n^{(m)}(x+iy;\lambda,u)\frac{t^n}{n!}$$

$$= \sum_{n=0}^{\infty}\mathcal{H}_n^{(m)}(\lambda,u)\frac{t^n}{n!}\sum_{n=0}^{\infty}C_n(x,y)\frac{t^n}{n!}$$

$$\times i\sum_{n=0}^{\infty}\mathcal{H}_n^{(m)}(\lambda,u)\frac{t^n}{n!}\sum_{n=0}^{\infty}S_n(x,y)\frac{t^n}{n!}.$$

By using the Cauchy product rule for series, we get

$$\sum_{n=0}^{\infty}\mathcal{H}_n^{(m)}(x+iy;\lambda,u)\frac{t^n}{n!}$$

$$= \sum_{n=0}^{\infty}\sum_{j=0}^{n}\binom{n}{j}\mathcal{H}_j^{(m)}(\lambda,u)C_{n-j}(x,y)\frac{t^n}{n!}$$

$$\times i\sum_{n=0}^{\infty}\sum_{j=0}^{n}\binom{n}{j}\mathcal{H}_j^{(m)}(\lambda,u)S_{n-j}(x,y)\frac{t^n}{n!}.$$

After comparing the coefficients of $\frac{t^n}{n!}$ on both sides of the above equation, we arrive at the following theorem:

Theorem 17. *Let $n, m \in \mathbb{N}_0$. Then, we have*

$$\mathcal{H}_n^{(m)}(x+iy;\lambda,u)=\sum_{j=0}^{n}\binom{n}{j}\mathcal{H}_j^{(m)}(\lambda,u)\left(C_{n-j}(x,y)+iS_{n-j}(x,y)\right).$$

By using (2.39), we get

$$\begin{aligned}\sum_{n=0}^{\infty}&\mathcal{H}_n^{(m)}(x+iy;\lambda,u)\frac{t^{n+m}}{n!}\\ &=\frac{1}{u^m}\sum_{n=0}^{\infty}\mathcal{B}_n^{(m)}\left(\frac{\lambda}{u}\right)\frac{t^n}{n!}\sum_{n=0}^{\infty}C_n(x,y)\frac{t^n}{n!}\\ &\quad+\frac{i}{u^m}\sum_{n=0}^{\infty}\mathcal{B}_n^{(m)}\left(\frac{\lambda}{u}\right)\frac{t^n}{n!}\sum_{n=0}^{\infty}S_n(x,y)\frac{t^n}{n!}.\end{aligned}$$

By applying the Cauchy product rule to the above equation, we obtain

$$\begin{aligned}\sum_{n=0}^{\infty}&\binom{n}{m}m!\mathcal{H}_{n-m}^{(m)}(x+iy;\lambda,u)\frac{t^n}{n!}\\ &=\frac{1}{u^m}\sum_{n=0}^{\infty}\sum_{j=0}^{n}\binom{n}{j}\mathcal{B}_j^{(m)}\left(x;\frac{\lambda}{u}\right)C_{n-j}(x,y)\frac{t^n}{n!}\\ &\quad+\frac{i}{u^m}\sum_{n=0}^{\infty}\sum_{j=0}^{n}\binom{n}{j}\mathcal{B}_j^{(m)}\left(x;\frac{\lambda}{u}\right)S_{n-j}(x,y)\frac{t^n}{n!}.\end{aligned}$$

After comparing the coefficients of $\frac{t^n}{n!}$ on both sides of the above equation, we arrive at the following formula:

$$\begin{aligned}\binom{n}{m}m!\mathcal{H}_{n-m}^{(m)}(x+iy;\lambda,u)=\sum_{j=0}^{n}\binom{n}{j}\mathcal{B}_j^{(m)}\left(\frac{\lambda}{u}\right)\\ \times\frac{C_{n-j}(x,y)+iS_{n-j}(x,y)}{u^m}.\end{aligned}$$

Replacing n by $n+m$ in the above equation, we also get the following theorem:

Theorem 18. *Let $n, m \in \mathbb{N}_0$. Then, we have*

$$\mathcal{H}_n^{(m)}(x+iy;\lambda,u) = \sum_{j=0}^{n+m} \frac{\binom{n+m}{j}}{\binom{n+m}{m} m!} \mathcal{B}_j^{(m)}\left(\frac{\lambda}{u}\right) \times \frac{C_{n+m-j}(x,y) + iS_{n+m-j}(x,y)}{u^m}.$$

By using (2.40), we get

$$\sum_{n=0}^{\infty} \mathcal{H}_n^{(m)}(x+iy;-\lambda,u)\frac{t^n}{n!} = \left(\frac{u-1}{2u}\right)^m \sum_{n=0}^{\infty} \mathcal{E}_n^{(m)}\left(\frac{\lambda}{u}\right)\frac{t^n}{n!} \sum_{n=0}^{\infty} C_n(x,y)\frac{t^n}{n!} + i\left(\frac{u-1}{2u}\right)^m \sum_{n=0}^{\infty} \mathcal{E}_n^{(m)}\left(\frac{\lambda}{u}\right)\frac{t^n}{n!} \sum_{n=0}^{\infty} S_n(x,y)\frac{t^n}{n!}.$$

By applying the Cauchy product rule to the above equation, we obtain

$$\sum_{n=0}^{\infty} \mathcal{H}_n^{(m)}(x+iy;-\lambda,u)\frac{t^n}{n!} = \left(\frac{u-1}{2u}\right)^m \sum_{n=0}^{\infty}\sum_{j=0}^{n} \binom{n}{j} \mathcal{E}_j^{(m)}\left(\frac{\lambda}{u}\right) C_{n-j}(x,y)\frac{t^n}{n!} + i\left(\frac{u-1}{2u}\right)^m \sum_{n=0}^{\infty}\sum_{j=0}^{n} \binom{n}{j} \mathcal{E}_j^{(m)}\left(\frac{\lambda}{u}\right) S_{n-j}(x,y)\frac{t^n}{n!}.$$

After comparing the coefficients of $\frac{t^n}{n!}$ on both sides of the above equation, we arrive at the following theorem:

Theorem 19. *Let $n, m \in \mathbb{N}_0$. Then, we have*

$$\mathcal{H}_{n-m}^{(m)}(x+iy;-\lambda,u) = \left(\frac{u-1}{2u}\right)^m \sum_{j=0}^{n} \binom{n}{j} \mathcal{E}_j^{(m)}\left(x;\frac{\lambda}{u}\right) \times \left(C_{n-j}(x,y) + iS_{n-j}(x,y)\right).$$

Putting $\overrightarrow{v} = \overrightarrow{0}$ in (2.31) and (2.32), Kilar and Simsek [58] showed that

$$\mathfrak{h}_1\left(n, x, y, z; \overrightarrow{0}, r, \frac{a}{u}, 0\right) = 2\left(\frac{u}{1-u}\right)^z \mathcal{H}_n^{(C,z)}(x, y; a, u)$$

and

$$\mathfrak{h}_2\left(n, x, y, z; \overrightarrow{0}, r, \frac{a}{u}, 0\right) = 2\left(\frac{u}{1-u}\right)^z \mathcal{H}_n^{(S,z)}(x, y; a, u).$$

By using (2.36), Kilar and Simsek [58] gave the following generating functions:

$$F_{HC}(t, x, y; \lambda, u) = \frac{1-u}{\lambda e^t - u} e^{xt} \cos(yt) = \sum_{n=0}^{\infty} \mathcal{H}_n^{(C)}(x, y; \lambda, u) \frac{t^n}{n!}$$

and

$$F_{HS}(t, x, y; \lambda, u) = \frac{1-u}{\lambda e^t - u} e^{xt} \sin(yt) = \sum_{n=0}^{\infty} \mathcal{H}_n^{(S)}(x, y; \lambda, u) \frac{t^n}{n!}.$$

Using the above functions, Kilar and Simsek [58] derived the following formula, for $n \in \mathbb{N}_0$:

$$\mathcal{H}_n^{(m)}(x + iy; \lambda, u) = \mathcal{H}_n^{(C,m)}(x, y; \lambda, u) + i\mathcal{H}_n^{(S,m)}(x, y; \lambda, u). \quad (2.41)$$

We [104] defined the generalized Eulerian-type polynomials of order m by means of the following generating function:

$$F_\lambda^{(m)}(t, x; u, a, b, c) = \left(\frac{a^t - u}{\lambda b^t - u}\right)^m c^{xt} = \sum_{n=0}^{\infty} \mathcal{H}_n^{(m)}(x; u; a, b, c; \lambda) \frac{t^n}{n!}. \quad (2.42)$$

We set two parametric kinds of the generalized Eulerian-type polynomials of order m by means of the following generating function:

$$\left(\frac{a^t - u}{\lambda b^t - u}\right)^m c^{xt} \cos(yt \ln c) = \sum_{n=0}^{\infty} \mathcal{H}_n^{(m,C)}(x, y; u; a, b, c; \lambda) \frac{t^n}{n!}$$

and

$$\left(\frac{a^t - u}{\lambda b^t - u}\right)^m c^{xt}\sin(yt\ln c) = \sum_{n=0}^{\infty} \mathcal{H}_n^{(m,S)}(x; u; a, b, c; \lambda)\frac{t^n}{n!}.$$

Prove the following relation:

$$\mathcal{H}_n^{(m)}(x + iy; u; a, b, c; \lambda)$$
$$= \mathcal{H}_n^{(m,C)}(x, y; u; a, b, c; \lambda) + i\mathcal{H}_n^{(m,S)}(x; u; a, b, c; \lambda).$$

Exercise 5. Find explicit formulas for the polynomials $\mathcal{H}_n^{(m,C)}(x, y; u; a, b, c; \lambda)$ and $\mathcal{H}_n^{(m,S)}(x; u; a, b, c; \lambda)$.

Exercise 6. Find relations among the polynomials $C_n(x, y)$, $S_{n-j}(x, y)$, the Apostol-type numbers and polynomials of higher order.

With the aid of the following functional equations:

$$F_{S2}\left(t, k_1; \frac{\lambda}{u}\right) F_{HC}^{(k_1)}(t, x, y; \lambda, u) = k_1!\left(\frac{1-u}{u}\right)^{k_1} F_C(t, x, y),$$

$$F_{S2}\left(t, k_2; \frac{\lambda}{u}\right) F_{HS}^{(k_2)}(t, x, y; \lambda, u) = k_2!\left(\frac{1-u}{u}\right)^{k_2} F_S(t, x, y),$$

$$F_{HC}^{(k_1)}(t, x, y; \lambda, u) = F_{AH}^{(k_1)}(t, x; \lambda, u)\cos(yt)$$

and

$$F_{HS}^{(k_2)}(t, x, y; \lambda, u) = F_{AH}^{(k_2)}(t, x; \lambda, u)\sin(yt),$$

where $k_1, k_2 \in \mathbb{N}_0$, Kilar and Simsek [58] found the following formulas, respectively:

$$C_n(x, y) = k_1!\left(\frac{u}{1-u}\right)^{k_1} \sum_{j=0}^{n} \binom{n}{j} S_2\left(n - j, k_1; \frac{\lambda}{u}\right) \mathcal{H}_j^{(C,k_1)}(x, y; \lambda, u)$$

and

$$S_n(x,y) = k_2! \left(\frac{u}{1-u}\right)^{k_2} \sum_{j=0}^{n} \binom{n}{j} S_2\left(n-j, k_2; \frac{\lambda}{u}\right)$$
$$\times \mathcal{H}_j^{(S,k_2)}(x,y;\lambda,u),$$

$$\mathcal{H}_n^{(C,k_1)}(x,y;\lambda,u) = \sum_{j=0}^{\left[\frac{n}{2}\right]} (-1)^j \binom{n}{2j} \mathcal{H}_{n-2j}^{(k_1)}(x;\lambda,u)\, y^{2j} \tag{2.43}$$

and

$$\mathcal{H}_n^{(S,k_2)}(x,y;\lambda,u) = \sum_{j=0}^{\left[\frac{n-1}{2}\right]} (-1)^j \binom{n}{2j+1} \mathcal{H}_{n-1-2j}^{(k_2)}(x;\lambda,u)\, y^{2j+1}, \tag{2.44}$$

$$\mathcal{H}_n^{(S,k_2)}(x,y;\lambda,u) = \left(\frac{1-u}{\lambda-u}\right)^{k_2} \sum_{j=0}^{\left[\frac{n-1}{2}\right]} (-1)^j \binom{n}{2j+1} y^{2j+1}$$
$$\times H_{n-1-2j}^{(k_2)}\left(x;\frac{u}{\lambda}\right),$$

$$\mathcal{H}_n^{(C,k_1)}(x,y;\lambda,u) = \left(\frac{1-u}{\lambda-u}\right)^{k_1} \sum_{j=0}^{n} \binom{n}{j} H_j^{(k_1)}\left(\frac{u}{\lambda}\right) C_{n-j}(x,y) \tag{2.45}$$

and

$$\mathcal{H}_n^{(S,k_2)}(x,y;\lambda,u) = \left(\frac{1-u}{\lambda-u}\right)^{k_2} \sum_{j=0}^{n} \binom{n}{j} H_j^{(k_2)}\left(\frac{u}{\lambda}\right) S_{n-j}(x,y), \tag{2.46}$$

where $n \in \mathbb{N}_0$.

Exercise 7. Using generating function for $\mathcal{H}_n^{(m)}(x+iy;\lambda,u)$, we have

$$\mathcal{H}_n^{(m)}(x+iy;\lambda,u) = \left(\frac{1-u}{\lambda-u}\right)^{m} \sum_{j=0}^{n} \binom{n}{j} H_j^{(m)}\left(\frac{u}{\lambda}\right) (x+iy)^{n-j}$$

(cf. Ref. [58]). Therefore,

$$\mathcal{H}_n^{(m)}(x+iy;\lambda,u)=\left(\frac{1-u}{\lambda-u}\right)^m\sum_{j=0}^{n}\sum_{v=0}^{n-j}\binom{n-j}{v}\binom{n}{j}\times H_j^{(m)}\left(\frac{u}{\lambda}\right)i^v y^v x^{n-j-v}. \tag{2.47}$$

By using

$$i^{2n}=(-1)^n$$

and

$$i^{2n+1}=(-1)^n i,$$

find the explicit formula for equation (2.47) and the polynomials $\mathcal{H}_n^{(m)}(x+iy;\lambda,u)$.

For $k_1=1,2,3$ and $n=1,2,3$, Kilar and Simsek [58] gave few values of the polynomials $\mathcal{H}_n^{(C,k_1)}(x,y;\lambda,u)$ as follows:

For $k_1=1$, we have

$$\frac{\mathcal{H}_0^{(C,1)}(x,y;\lambda,u)}{u-1}=\frac{1}{u-\lambda},$$

$$\frac{\mathcal{H}_1^{(C,1)}(x,y;\lambda,u)}{u-1}=\frac{\lambda+(u-\lambda)^2x}{(u-\lambda)^2},$$

$$\frac{\mathcal{H}_2^{(C,1)}(x,y;\lambda,u)}{u-1}=\frac{2\lambda^2+\lambda(u-\lambda)^2(1+2x)+(u-\lambda)^3(x^2-y^2)}{(u-\lambda)^3},$$

$$\frac{\mathcal{H}_3^{(C,1)}(x,y;\lambda,u)}{u-1}=\frac{6\lambda^3+6\lambda^2(u-\lambda)(1+x)+\lambda(u-\lambda)^2(1+3x+3x^2-3y^2)+(u-\lambda)^3(x^3-3xy^2)}{(u-\lambda)^4}.$$

For $k_1=2$, we have

$$\mathcal{H}_0^{(C,2)}(x,y;\lambda,u)=\left(\frac{u-1}{u-\lambda}\right)^2,$$

$$\mathcal{H}_1^{(C,2)}(x,y;\lambda,u)=\left(\frac{u-1}{u-\lambda}\right)^2\left(\frac{2\lambda}{u-\lambda}+x\right),$$

$$\mathcal{H}_2^{(C,2)}(x,y;\lambda,u) = \left(\frac{u-1}{u-\lambda}\right)^2 \left(\frac{6\lambda^2}{(u-\lambda)^2} + \frac{2\lambda(1+2x)}{u-\lambda} + x^2 - y^2\right),$$

$$\mathcal{H}_3^{(C,2)}(x,y;\lambda,u) = \left(\frac{u-1}{u-\lambda}\right)^2 \left(\frac{24\lambda^3}{(u-\lambda)^3} + \frac{18\lambda^2(1+x)}{(u-\lambda)^2} + \frac{2\lambda\left(1+3x+3x^2-3y^2\right)}{u-\lambda} + x^3 - 3xy^2\right).$$

For $k_1 = 3$, we have

$$\frac{\mathcal{H}_0^{(C,3)}(x,y;\lambda,u)}{(u-1)^3} = \frac{1}{(u-\lambda)^3},$$

$$\frac{\mathcal{H}_1^{(C,3)}(x,y;\lambda,u)}{(u-1)^3} = \frac{3\lambda + (u-\lambda)x}{(u-\lambda)^4},$$

$$\frac{\mathcal{H}_2^{(C,3)}(x,y;\lambda,u)}{(u-1)^3} = \frac{12\lambda^2 + 3\lambda(u-\lambda)(1+2x) + (u-\lambda)^2\left(x^2-y^2\right)}{(u-\lambda)^5},$$

$$\frac{\mathcal{H}_3^{(C,3)}(x,y;\lambda,u)}{(u-1)^3} = \frac{60\lambda^3 + 36\lambda^2(u-\lambda)(1+x) + 3\lambda(u-\lambda)^2\left(1+3x+3x^2-3y^2\right) + (u-\lambda)^3\left(x^3-3xy^2\right)}{(u-\lambda)^3}.$$

For $k_2 = 1,2,3$ and $n = 1,2,3$, Kilar and Simsek [58] gave few values of the polynomials $\mathcal{H}_n^{(S,k_2)}(x,y;\lambda,u)$ as follows:

When $k_2 = 1$:

$$\frac{\mathcal{H}_0^{(S,1)}(x,y;\lambda,u)}{u-1} = 0,$$

$$\frac{\mathcal{H}_1^{(S,1)}(x,y;\lambda,u)}{u-1} = \frac{y}{u-\lambda},$$

$$\frac{\mathcal{H}_2^{(S,1)}(x,y;\lambda,u)}{u-1} = \frac{2\lambda y + 2(u-\lambda)^2 xy}{(u-\lambda)^2},$$

$$\frac{\mathcal{H}_3^{(S,1)}(x,y;\lambda,u)}{u-1} = \frac{6\lambda^2 y + 3(u-\lambda)^2\lambda(y+2xy) + (u-\lambda)^3\left(3x^2y-y^3\right)}{(u-\lambda)^3}.$$

When $k_2 = 2$:

$$\frac{\mathcal{H}_0^{(S,2)}(x,y;\lambda,u)}{(u-1)^2} = 0,$$

$$\frac{\mathcal{H}_1^{(S,2)}(x,y;\lambda,u)}{(u-1)^2} = \frac{y}{(u-\lambda)^2},$$

$$\frac{\mathcal{H}_2^{(S,2)}(x,y;\lambda,u)}{(u-1)^2} = \frac{4\lambda y + (u-\lambda)\,2xy}{(u-\lambda)^3},$$

$$\frac{\mathcal{H}_3^{(S,2)}(x,y;\lambda,u)}{(u-1)^2} = \frac{18\lambda^2 y + (u-\lambda)\,6\lambda\,(y+2xy) + (u-\lambda)^2\left(3x^2y - y^3\right)}{(u-\lambda)^4}.$$

When $k_2 = 3$:

$$\frac{\mathcal{H}_0^{(S,3)}(x,y;\lambda,u)}{(u-1)^3} = 0,$$

$$\frac{\mathcal{H}_1^{(S,3)}(x,y;\lambda,u)}{(u-1)^3} = \frac{y}{(u-\lambda)^3},$$

$$\frac{\mathcal{H}_2^{(S,3)}(x,y;\lambda,u)}{(u-1)^3} = \frac{6\lambda y + 2\,(u-\lambda)\,xy}{(u-\lambda)^2},$$

$$\frac{\mathcal{H}_3^{(S,3)}(x,y;\lambda,u)}{(u-1)^3} = \frac{36\lambda^2 y + 9\lambda\,(u-\lambda)\,(y+2xy) + (u-\lambda)^2\left(3x^2y - y^3\right)}{(u-\lambda)^5}.$$

With the help of the following functional equations

$$F_{HC}^{(2k_1)}(t,x,y;1,2) = \frac{1}{2^{k_1}} F_a^{(k_1)}(t,x)\cos(yt)$$

and

$$F_{HS}^{(2k_2)}(t,x,y;1,2) = \frac{1}{2^{k_2}} F_a^{(k_2)}(t,x)\sin(yt),$$

Kilar and Simsek [58] also gave the following formulas, respectively, for $n \in \mathbb{N}_0$:

$$H_n^{(C,2k_1)}(x,y;1,2) = \frac{1}{2^{k_1}} \sum_{j=0}^{\left[\frac{n}{2}\right]} (-1)^j \binom{n}{2j} a_{n-2j}^{(k_1)}(x)\, y^{2j} \qquad (2.48)$$

and

$$H_n^{(S,2k_2)}(x,y;1,2)=\frac{1}{2^{k_2}}\sum_{j=0}^{\left[\frac{n-1}{2}\right]}(-1)^j\binom{n}{2j+1}a_{n-1-2j}^{(k_2)}(x)\,y^{2j+1}. \tag{2.49}$$

2.3. *Partial derivative equations of the two parametric kinds of Eulerian-Type polynomials*

Applying partial differential operator to equation (2.33), Kilar and Simsek [58] showed that

$$\begin{aligned}&\frac{\partial^m}{\partial x^m}\left\{\mathcal{H}_n^{(C,k_1)}(x,y;\lambda,u)\right\}\\&=\left(\frac{1-u}{u}\right)^m\sum_{j=0}^{n}\binom{n}{j}\mathcal{B}_j^{(m)}\left(\frac{\lambda}{u}\right)\mathcal{H}_{n-j}^{(C,k_1-m)}(x,y;\lambda,u)\end{aligned} \tag{2.50}$$

and

$$\begin{aligned}&\frac{\partial^m}{\partial x^m}\left\{\mathcal{H}_n^{(C,k_1)}(x,y;\lambda,u)\right\}\\&=\left(\frac{u-1}{2u}\right)^m\sum_{j=0}^{n}\binom{n}{j}\mathcal{G}_j^{(m)}\left(-\frac{\lambda}{u}\right)\\&\quad\times\mathcal{H}_{n-j}^{(C,k_1-m)}(x,y;\lambda,u),\end{aligned} \tag{2.51}$$

$$\begin{aligned}&\frac{\partial^m}{\partial x^m}\left\{\mathcal{H}_n^{(S,k_2)}(x,y;\lambda,u)\right\}\\&=\left(\frac{u-1}{2u}\right)^m\binom{n}{m}m!\sum_{j=0}^{n-m}\binom{n-m}{j}\mathcal{E}_j^{(m)}\left(-\frac{\lambda}{u}\right)\\&\quad\times\mathcal{H}_{n-m-j}^{(S,k_2-m)}(x,y;\lambda,u),\end{aligned} \tag{2.52}$$

where $n,m\in\mathbb{N}$.

2.4. *The twisted generalized q-Frobenius–Euler numbers*

First, we studied on the twisted Bernoulli numbers and polynomials with their interpolation functions involving twisted L-function (cf. Ref. [94]). By using the same motivation of this chapter, we have constructed generating functions for many kinds of twisted-type numbers and polynomials with their twisted interpolation functions. With the aid of this motivation, we also recently defined twisted Dedekind-type sums, twisted Hardy-type sums, twisted special finite sums and others.

In Ref. [96, Lemma 3], we defined the twisted generalized Frobenius–Euler numbers attached to the Dirichlet character with conductor f by means of the following generating function:

Let χ be a primitive Dirichlet character with conductor f Dirichlet character with conductor f and let ξ be r-th root of 1:

$$\sum_{a=1}^{f} \frac{\left(1-u^{rf}\right)\chi(a)\xi^{a}u^{f-a-1}}{\xi^{f}e^{ft}-u^{f}}e^{at} = \sum_{n=0}^{\infty} H_n(u;\chi,\xi,r)\frac{t^n}{n!}. \tag{2.53}$$

Motivation of the generating function in (2.53) is given as follows:

Replacing f with rf and $\chi(a)$ with $\chi(a)\xi^a$ in the following well-known generating function for the generalized Frobenius–Euler numbers attached to the Dirichlet character with conductor f

$$\sum_{a=0}^{f-1} \frac{\left(1-u^{f}\right)\chi(a)u^{f-a-1}}{e^{ft}-u^{f}}e^{at} = \sum_{n=0}^{\infty} H_n(u;\chi)\frac{t^n}{n!}$$

(cf. Refs. [122, 130, 131]) yields

$$\sum_{a=0}^{rf-1} \frac{\left(1-u^{rf}\right)\chi(a)\xi^{a}u^{fr-a-1}}{e^{frt}-u^{rf}}e^{at} = \sum_{n=0}^{\infty} H_n(u;\chi\xi)\frac{t^n}{n!}.$$

Combining the above equation with the following finite sum formula

$$\sum_{a=0}^{rf-1} g(a) = \sum_{a=1}^{f}\sum_{b=0}^{r-1} g(a+bf),$$

we get

$$\begin{aligned}
&\sum_{a=0}^{rf-1} \frac{\left(1-u^{rf}\right)\chi(a)\xi^{a}u^{fr-a-1}}{e^{frt}-u^{rf}}e^{at}\\
&=\sum_{a=1}^{f}\sum_{b=0}^{r-1} \frac{\left(1-u^{rf}\right)\chi(a+bf)\xi^{a+bf}u^{rf-a-fb-1}}{\xi^{f}e^{rft}-u^{rf}}e^{at+bft}\\
&=\sum_{n=0}^{\infty} H_n(u;\chi\xi)\frac{t^n}{n!}.
\end{aligned}$$

Since

$$\chi(a+bf)=\chi(a),$$

we get the following results:

$$\sum_{a=1}^{f} \frac{\left(1-u^{rf}\right)\chi(a)\xi^{a}u^{rf-a-1}}{e^{rft}-u^{rf}}e^{at}\sum_{b=0}^{r-1}\left(\xi^{f}u^{-f}e^{ft}\right)^{b}=\sum_{n=0}^{\infty} H_n(u;\chi\xi)\frac{t^n}{n!}.$$

Combining the above equation with the following well-known geometric progression formula

$$\sum_{b=0}^{r-1} x^{b}=\frac{x^{r}-1}{x-1},$$

we obtain

$$\sum_{a=1}^{f} \frac{\left(1-u^{rf}\right)\chi(a)\xi^{a}u^{rf-a-1}}{e^{rft}-u^{rf}}e^{at}\frac{\xi^{rf}u^{-rf}e^{rft}-1}{\xi^{f}u^{-f}e^{ft}-1}=\sum_{n=0}^{\infty} H_n(u;\chi\xi)\frac{t^n}{n!}.$$

Since

$$\xi^{rf}=1,$$

we obtain the following results:

$$\sum_{a=1}^{f} \frac{\left(1-u^{rf}\right)\chi(a)\xi^{a}u^{rf-a-1}}{e^{rft}-u^{rf}}e^{at}\frac{u^{-rf}e^{rft}-1}{\xi^{f}u^{-f}e^{ft}-1}=\sum_{n=0}^{\infty} H_n(u;\chi\xi)\frac{t^n}{n!}.$$

After some calculations in the above equation, and using

$$H_n(u;\chi\xi):=H_n(u;\chi,\xi,r),$$

we arrive at (2.53).

Using (2.53), we have

$$\sum_{a=1}^{f} \frac{\left(1-u^{rf}\right)\chi(a)\xi^{a-f}u^{f-a-1}}{\xi^f-u^f}\sum_{n=0}^{\infty} H_n\left(\frac{a}{f},\frac{u^f}{\xi^f}\right)\frac{t^n}{n!}$$
$$=\sum_{n=0}^{\infty} H_n(u;\chi,\xi,r)\frac{t^n}{n!}.$$

After comparing the coefficients of $\frac{t^n}{n!}$ on both sides of the above equation, we arrive at the following known formula for the polynomials $H_n(u;\chi,\xi,r)$:

$$H_n(u;\chi,\xi,r)=f^n\sum_{a=1}^{f} \frac{\left(1-u^{rf}\right)\chi(a)\xi^{a-f}u^{f-a-1}}{\xi^f-u^f}H_n\left(\frac{a}{f},\frac{u^f}{\xi^f}\right) \tag{2.54}$$

(cf. Ref. [96, Theorem 1]).

In Ref. [96, (3.2)], we also defined the twisted generalized Frobenius–Euler polynomials attached to the Dirichlet character with conductor f by means of the following generating function:

$$\sum_{a=1}^{f} \frac{\left(1-u^{rf}\right)\chi(a)\xi^{a}u^{f-a-1}}{\xi^f e^{ft}-u^f}e^{(a+x)t}=\sum_{n=0}^{\infty} H_n(x,u;\chi,\xi,r)\frac{t^n}{n!}, \tag{2.55}$$

where χ is a primitive Dirichlet character with conductor f Dirichlet character with conductor f and ξ is r-th root of 1.

By combining (2.55) with (2.53), we get

$$\sum_{n=0}^{\infty} H_n(x,u;\chi,\xi,r)\frac{t^n}{n!}=\sum_{n=0}^{\infty} H_n(u;\chi,\xi,r)\frac{t^n}{n!}\sum_{n=0}^{\infty} x^n\frac{t^n}{n!}.$$

By using the Cauchy product rule, we have

$$\sum_{n=0}^{\infty} H_n(x,u;\chi,\xi,r)\frac{t^n}{n!}=\sum_{n=0}^{\infty}\sum_{j=0}^{n}\binom{n}{j}x^{n-j}H_j(u;\chi,\xi,r)\frac{t^n}{n!}.$$

After comparing the coefficients of $\frac{t^n}{n!}$ on both sides of the above equation, we arrive at the following known formula for the polynomials

$H_n(x, u; \chi, \xi, r)$:

$$H_n(x, u; \chi, \xi, r) = \sum_{j=0}^{n} \binom{n}{j} x^{n-j} H_j(u; \chi, \xi, r) \tag{2.56}$$

(cf. Ref. [96, (3.2)]).

Combining (2.54) with (2.56), we obtain the following computational formula for the polynomials $H_n(x, u; \chi, \xi, r)$:

Theorem 20. *Let χ be a primitive Dirichlet character with conductor f Dirichlet character with conductor f and let ξ be r-th root of 1. Let $n \in \mathbb{N}_0$. Then, we have*

$$H_n(x, u; \chi, \xi, r) = \sum_{j=0}^{n} \binom{n}{j} \sum_{a=1}^{f} \frac{\left(1 - u^{rf}\right) \chi(a) \xi^{a-f} u^{f-a-1} x^{n-j} f^j}{\xi^f - u^f} \times H_j\left(\frac{a}{f}, \frac{u^f}{\xi^f}\right).$$

By constructed generating functions for twisted-type special numbers and polynomials, applying our motivation in Ref. [94] to (1.2) or to the function $F_{u,q}(t, x)$, we [98] construct the following generating function for the twisted generalized q-Frobenius–Euler numbers:

$$F_{\frac{u}{\xi},q,\chi}(t) = \sum_{a=1}^{f} \sum_{b=0}^{r-1} \chi(a) \xi^{a+bf} u^{rf-a-bf} e^{[a+bf]t} F_{u^{rf},q^{rf}}(t[rf]q^{a+bf}),$$

where χ is a primitive Dirichlet character of conductor $f \in \mathbb{N}$ and ξ is an rth root of unity. The above function gives us

$$F_{\frac{u}{\xi},q,\chi}(t) = \left(1 - \frac{1}{u^{rf}}\right) \sum_{a=1}^{f} \sum_{b=0}^{r-1} \chi(a) \xi^{a+bf} u^{rf-a-bf} \sum_{n=0}^{\infty} \frac{e^{[a+bf+rfn]t}}{u^{rfn}}$$

$$= \sum_{n=0}^{\infty} H_{n,\chi,\xi}(u, q) \frac{t^n}{n!}, \tag{2.57}$$

By using the above equation, we arrive at the following result:

Lemma 1 (cf. Ref. [98]). *Let χ be a primitive Dirichlet character of conductor $f \in \mathbb{N}$. Then, we have*

$$F_{u,q,\chi}(t) = \sum_{a=1}^{f} \chi(a) \left(\sum_{n=0}^{\infty} \frac{e^{[a+fn]t}}{u^{f(n-1)+a}} - \sum_{n=0}^{\infty} \frac{e^{[a+fn]t}}{u^{fn+a}} \right). \tag{2.58}$$

The generating function of generalized q-Frobenius–Euler polynomials is given by means of the following functional equation:

$$F_{u,q,\chi}(t,x) = e^{[x]t} F_{u,q,\chi}(q^x t) \tag{2.59}$$

(cf. Ref. [98]).

By combining (1.2) with (2.57), the twisted generalized q-Frobenius–Euler polynomials and numbers are given as follows, respectively:

Let $n \in \mathbb{N}_0$. Let χ be a primitive Dirichlet character of conductor f and ξ be an rth root of unity.

$$H_{n,\chi,\xi}(u,x,q) = [rf]^n \sum_{a=1}^{f} \sum_{b=0}^{r-1} \chi(a) \xi^{a+bf} u^{rf-a-bf} H_n \times \left(u^{rf}, \frac{x+a+bf}{rf}, q^{rf} \right) \tag{2.60}$$

and

$$H_{n,\chi,\xi}(u,q) = [rf]^n \sum_{a=1}^{f} \sum_{b=0}^{r-1} \chi(a) \xi^{a+bf} u^{rf-a-bf} H_n \left(u^{rf}, \frac{a+bf}{rf}, q^{rf} \right)$$

(cf. Ref. [98], see also Ref. [119]).

Exercise 8. Find applications of the polynomials $H_{n,\chi,\xi}(u,x,q)$ and $H_{n,\chi,\xi}(u,x,q)$ of order m in the theory of curves and analytic number theory with the algebraic structures.

2.5. *The exponential Euler spline in terms of the Frobenius–Euler polynomials and numbers*

We now give another important application of the Frobenius–Euler polynomials and numbers on the theory of splines.

The exponential Euler spline was first defined by Schoenberg [90, 91, Lemma 7.1]. The exponential Euler spline is defined in terms of $H_m(x,u)$ and $H_m(0,u)$ in the intervals $[0,1]$ as follows:

$$\frac{H_m(x,u)}{H_m(0,u)}, \tag{2.61}$$

where $u \neq 1$, $u \neq \lambda_v$ $v \in \{1, 2, \ldots, m-1\}$. The extension of

$$\frac{H_m(x,u)}{H_m(0,u)}$$

to a real x may be accomplished by means of the following functional equation:

$$\frac{H_m(x+1,u)}{H_m(0,u)} = u\frac{H_m(x,u)}{H_m(0,u)}$$

(cf. Refs. [90, 91, p. 393]).

By the same method, Schoenberg [90, 91, Lemma 7.1], also defined midpoint Euler–Frobenius polynomials with the aid of B-spline. He also gave characteristic extremum property, Lipschitz condition property, minimizes the norm, Fourier expansion property, much better approximation, etc. of the polynomial $\frac{H_m(x,u)}{H_m(0,u)}$.

Euler's generating function is given by

$$F(z) = \left(\frac{u-1}{u-e^z}\right)\frac{e^{zx}}{z^{n+1}} = \sum_{m=0}^{\infty} \frac{H_m(x,u)}{m!}\frac{z^v}{z^{n+1}}$$

(cf. Ref. [91, Eq. (7.8), p. 398]). The left side $F(z)$ is a meromorphic function having a pole of order $n+1$ at the origin and simple poles at the zeros of $e^z - u$. Setting

$$u^x = \tau e^{i\alpha x},$$

for $x \in \mathbb{R}$. All poles of the function $F(z)$ are at the points

$$u_k = \log\tau + i\alpha + 2k\pi i,$$

where $k \in \mathbb{Z}$ and $u_k \neq 0$, for all k, if $\log\tau = 0$ and $\alpha = 0$, then $u = 1$. It is time to evaluate the following residues of $F(z)$ at the poles u_k:

Residue of $F(z)$ at $z = 0$ is

$$\frac{H_n(x,u)}{n!},$$

and residue of $F(z)$ at u_k is

$$-\frac{u-1}{e^{u_k}}\frac{e^{xu_k}}{u_k^{n+1}} = -\frac{u-1}{u}\frac{e^{xu_k}}{u_k^{n+1}}.$$

Thus, the sum of all the residues of $F(z)$ in the entire finite plane vanishes (cf. Ref. [91, p. 398]).

They [29] also defined the following a new class of modification exponential Euler-type spline of degree n order α:

$$Y(x;\lambda;n,\alpha) = \frac{\mathfrak{u}_n^{(\alpha)}\left(x;-\frac{1}{\lambda}\right)}{\mathfrak{u}_n^{(\alpha)}\left(-\frac{1}{\lambda}\right)}. \tag{2.62}$$

Gun and Simsek [29] also defined

$$s_n(x;\lambda;\alpha) = Y(x;\lambda;n,\alpha) \tag{2.63}$$

if $0 \leq x < 1$. Thus,

$$Y(x;\lambda;n,\alpha) = \frac{H_n^{(\alpha)}(x,\lambda)}{H_n^{(\alpha)}(\lambda)} \tag{2.64}$$

and

$$Y(x;\lambda;n,\alpha) \in C^{n-1}(\mathbb{R})$$

(cf. Ref. [29]).

For $x = 0$, we have

$$Y(0;\lambda;n,\alpha) = 1$$

(cf. Ref. [29]).

Exercise 9. Find Eulerian and Frobenius–Euler polynomials interpolation at the points of a geometric progression.

Exercise 10. Find polynomial spline curves of order n via all polynomials in this chapter.

Exercise 11. Find relations among Eulerian polynomials and Frobenius–Euler polynomials, and asymptotic distributions of arithmetical functions via polynomial approximations.

Exercise 12. How can two parametric kinds of Eulerian-type polynomial splines be constructed.

Exercise 13. Find applications of (2.64) associated with B-spline, the Bezier curves in terms of the Bernstein base function and also (2.61).

3. Interpolation Functions for the Twisted Euler–Frobenius-Type Numbers and Polynomials

Let $s \in \mathbb{C}$, $s = h + ic$ with $h, c \in \mathbb{R}$. Satoh [86] constructed q-l-function. We shall give this function as follows:

For,

$$l_q(u, s) = \sum_{n=0}^{\infty} \frac{1}{u^n [n]^s}, \tag{3.1}$$

where $h > 0$ and $\left|\frac{1}{u}\right| < 1$. Satoh [86] showed that the function $l_q(u, s)$ interpolates q-Euler numbers at negative integers. That is,

$$l_q(u, -k) = \begin{cases} \dfrac{1}{1-u} & \text{if } k = 0, \\[2ex] \dfrac{uH_k(u, q)}{1-u} & \text{if } k \neq 0. \end{cases}$$

For a more detailed information complex and p-adic cases of the function $l_q(u, s)$, see also Koblitz [60], Kim [46], Satoh [86, 87], Tsumura [130, 131] and also Refs. [97–100, 119].

Generalization of the function $l_q(u, s)$ are given as follows:

Definition 3.1 ([86]). For $s \in C$,

$$l_q(u, s, \chi) = \sum_{n=1}^{\infty} \frac{\chi(n)}{u^n [n]^s}. \tag{3.2}$$

Definition 3.2 ([98]). For $s \in C$,

$$l_q\left(\frac{u}{\xi}, s, \chi\right) = \sum_{n=1}^{\infty} \frac{\chi(n)\xi^n}{u^n[n]^s}, \tag{3.3}$$

where ξ is an rth root of unity.

By applying the following Mellin transform to the function $F_{\frac{u}{\xi},q,\chi}(-t)$,

$$\frac{1}{\Gamma(s)}\int_0^{\infty} t^{s-1}F_{\frac{u}{\xi},q,\chi}(-t)dt$$
$$= (u^{rf}-1)\sum_{a=1}^{f}\sum_{b=0}^{r-1}\chi(a)\xi^{a+bf}u^{-a-bf}\sum_{n=0}^{\infty}\frac{1}{u^{rfn}[a+bf+rfn]^s}.$$

Consequently, we have the following definition:

Definition 3.3 (cf. Ref. [119]). Let $s \in \mathbb{C}$ and ξ be an rth root of unity:

$$l_{H,q,\xi}(s,\chi;u) = \sum_{a=1}^{f}\sum_{b=0}^{r-1}\chi(a)\xi^{a+bf}\sum_{n=0}^{\infty}\frac{u^{rf}-1}{u^{rfn+a+bf}[a+bf+rfn]^s}. \tag{3.4}$$

Let $u^{rf} \neq 1$. In Ref. [119], Simsek *et al.* gave the following functional equation:

$$l_q(\frac{u}{\xi}, s, \chi) = \frac{1}{u^{rf}-1} l_{H,q,\xi}(s,\chi;u),$$

where

$$l_{H,q,\xi}(s,\chi;u) = (u^{rf}-1)\sum_{m=1}^{\infty}\frac{\chi(m)\xi^m}{u^m[m]^s}. \tag{3.5}$$

Some special cases of the twisted $l_{H,q,\xi}$:

$$\lim_{q\to 1} l_{H,q,\xi}(s,\chi;u) \to l_{H,\xi}(s,\chi;u) = \left(u^{rf}-1\right)\sum_{m=1}^{\infty}\frac{\chi(m)\xi^m}{u^m m^s}.$$

When f and r are odd integers and $u = -1$, the above function reduces to

$$l_{H,\xi}(s,\chi;-1) = 2\sum_{m=1}^{\infty} \frac{(-1)^{m+1}\chi(m)\xi^m}{m^s}.$$

Putting $\chi \equiv 1$ in the above equation, the above function related to the following Lerch zeta function

$$\Phi(z,s,x) = \sum_{m=0}^{\infty} \frac{z^m}{(m+x)^s},$$

where $x \in \mathbb{C} \setminus \{0,-1,-2,-3,-4,\ldots\}$; $h + ic = s \in \mathbb{C}$ when $|z| < 1$, $h > 1$ when $|z| = 1$ (cf. Ref. [122]).

The function $\Phi(z,s,x)$ satisfies the following integral representation:

$$\Phi(z,s,x) = \frac{1}{\Gamma(s)} \int_0^{\infty} t^{s-1} \frac{e^{-xt}}{1 - ze^{-t}} dt,$$

where $x = w + iy$ with $w, y \in \mathbb{R}$ with $w > 0$, $|z| \leq 1$, $z \neq 1$, $h > 0$; $z = 1$, $h > 1$ (cf. Ref. [122]):

$$l_{H,\xi}(s,1;-1) = \zeta_{\xi}(s) = -2\sum_{m=0}^{\infty} \frac{(-1)^m \xi^m}{(m+1)^s},$$

where $\zeta_{\xi}(s)$ is called twisted Genocchi zeta function. Thus, we have

$$l_{H,\xi}(s,1;-1) = -2\Phi(1,s,-\xi).$$

If $\xi = 1$, then $\zeta_{\xi}(s)$ reduces to the Genocchi zeta function

$$\zeta_E(s) = 2\sum_{m=1}^{\infty} \frac{(-1)^{m+1}}{m^s}$$

(cf. Refs. [35, 36, 43–45, 121–123]).

The function $\zeta_E(s)$ is related to the following Polylogarithm function:

$$\mathrm{Li}_s(z) = \sum_{n=1}^{\infty} \frac{z^n}{n^s},$$

where $|z| < 1$ when $s = 1$ (cf. Refs. [121, 122]),

$$\zeta_E(s) = -2\mathrm{Li}_s(\xi).$$

Interpolation function of the polynomials $H_{k,\xi}(x, u, q)$ is given by the following function $l_{H,q,\xi}(s, \chi; u)$:

$$l_{H,q,\xi}(s, \chi; u) = \frac{1}{[rf]^s} \sum_{a=0}^{f-1} \sum_{b=0}^{r-1} \chi(a) \xi^{a+bf} u^{rf-a-bf} \zeta_{H,q^{rf}} \times \left(s, u^{rf}; \frac{a+bf}{rf}\right),$$

where $s \in \mathbb{C}$ (cf. Ref. [119]).

Substituting $s = -k$ into the function $l_{H,q,\xi}(s, \chi; u)$ and using

$$\zeta_{H,q,\xi}(-k, u; x) = H_{k,\xi}(u, x, q), \tag{3.6}$$

we have

$$l_{H,q,\xi}(-k, \chi; u) = H_{k,\chi,\xi}(u, x, q)$$

(cf. Ref. [119]).

Exercise 14. Find generating functions for twisted Fubini numbers and polynomials (of higher order). Find also their interpolation functions with the aid of the above methods.

3.1. *Dedekind-type sums in terms of the Frobenius–Euler functions*

Under the modular transformation, Dedekind gave the functional equation of the Dedekind eta function. This functional equation involves the following Dedekind sums:

$$s(h,k) = \sum_{a(\mathrm{mod}k)} \left(\left(\frac{a}{k}\right)\right)\left(\left(\frac{ah}{k}\right)\right),$$

where $h, k \in \mathbb{Z}$ with $k > 0$, $(h,k) = 1$, the summation is over a complete residue system (mod$k$) and

$$((x)) = \begin{cases} x - [x] - \frac{1}{2} & \text{if } x \notin \mathbb{Z}, \\ 0 & x \in \mathbb{Z}, \end{cases}$$

$[x]$ is the largest integer $\leq x$ (cf. Refs. [5, 16, 83]).

In Refs. [4, 16], the following generalized Dedekind sums were given:

$$c_s(h,k;p) = \sum_{a(\mathrm{mod}k)} \overline{B}_{p+1-s}\left(\frac{a}{k}\right)\overline{B}_s\left(\frac{ha}{k}\right),$$

where p is a fixed odd integer > 1 and $p+1 \geq s$ with $s \in \mathbb{N}_0$; $(h,k) = 1$, the summation is over a complete residue system (mod k) and $\overline{B}_s(x)$ denotes the Bernoulli function, given by

$$\begin{aligned} \overline{B}_s(x) &= B_s(x - [x]) \\ &= \sum_{j=0}^{s} \binom{s}{j} (x - [x])^{s-j} B_j \end{aligned}$$

(cf. Refs. [4, 5, 83]).

The sums $c_s(h,k;p)$ are derived from the transformation formula for the following function, which is known as the Lambert series, for $|y| < 1$,

$$G_p(y) = \sum_{m=0}^{\infty}\sum_{n=0}^{\infty} \frac{y^{mn}}{n^p},$$

given by Apostol [4].

Carlitz [15, 16] gave the sums $c_s(h,k;p)$ in terms of the Frobenius–Euler numbers (or the Eulerian numbers) as follows:

$$c_{r,s}(h,k;p) = \frac{B_r B_s}{k^{r+s-1}} + \frac{rs}{k^{r+s-1}} \sum_{\eta \neq 1} \frac{H_{r-1}(\eta^h) H_{s-1}(\eta^{-1})}{(1-\eta^h)(1-\eta)},$$

where the summation is over all kth roots of unity distinct from 1. The sum $c_{r,s}(h,k;p)$ is valid for all $r, s \in \mathbb{N}_0$.

We now give a brief construction of Carlitz [16], depending on the generating function method, as follows:

$$\sum_{r=0}^{\infty}\sum_{s=0}^{\infty}k^{r}c_{r,s}(h,k;p)\frac{t^{r}w^{s}}{r!s!}$$
$$=\frac{t}{e^{t}-1}\frac{w}{e^{\frac{w}{k}}-1}+tw\sum_{\eta\neq 1}\frac{1}{(1-\eta^{h})(1-\eta)}$$
$$\times\left(\sum_{r=0}^{\infty}H_{r}\left(\eta^{h}\right)\frac{t^{r}}{r!}\sum_{s=0}^{\infty}k^{-s}H_{s}\left(\eta^{-1}\right)\frac{w^{s}}{s!}\right).$$

After some calculations, one has

$$\sum_{r=0}^{\infty}\sum_{s=0}^{\infty}k^{r}c_{r,s}(h,k;p)\frac{t^{r}w^{s}}{r!s!}$$
$$=\frac{t}{e^{t}-1}\frac{w}{e^{\frac{w}{k}}-1}+wt\sum_{\eta\neq 1}\frac{\eta^{h-1}}{(e^{t}-\eta^{h})\left(e^{\frac{w}{k}}-\eta^{-1}\right)}.$$

Therefore,

$$\sum_{r=0}^{\infty}\sum_{s=0}^{\infty}k^{r}c_{r,s}(h,k;p)\frac{t^{r}w^{s}}{r!s!}=wt\sum_{\eta}\frac{\eta^{h-1}}{(e^{t}-\eta^{h})\left(e^{\frac{w}{k}}-\eta^{-1}\right)},$$

where the summation is now over all kth roots of unity.

Since

$$b_{r,s}(h,k;p)=\sum_{j=0}^{s}(-1)^{s-j}\binom{s}{j}h^{s-j}c_{r+s-j,j}(h,k;p)$$

and

$$c_{r,s}(h,k;p)=\sum_{j=0}^{s}\binom{s}{j}h^{s-j}b_{r+s-j,j}(h,k;p)$$

(cf. Ref. [16]), we also have

$$\sum_{r=0}^{\infty}\sum_{s=0}^{\infty}k^{r}c_{r,s}(h,k;p)\frac{t^{r}w^{s}}{r!s!}=\sum_{r=0}^{\infty}\sum_{s=0}^{\infty}b_{r,s}(h,k;p)\frac{(tk+wh)^{r}}{r!}\frac{w^{s}}{s!}. \tag{3.7}$$

If β runs through the hth roots of unity, then

$$\frac{h}{t^h - 1} = \sum_{\beta} \frac{\beta}{t - \beta} = \sum_{\beta} \frac{1}{t\beta - 1}$$

(cf. Ref. [16]). Using the above elegant formula, we have

$$\frac{h\zeta^h}{e^t - \zeta^h} = \frac{h}{\zeta^{-h}e^t - 1} = \sum_{\beta} \frac{1}{\beta\zeta^{-1}e^{\frac{t}{h}} - 1}$$

which implies

$$\frac{h\zeta^h}{e^t - \zeta^h} = \sum_{\beta} \frac{\zeta}{\beta e^{\frac{t}{h}} - \zeta}.$$

Combining the above formula with (3.7), we arrive at the desired result.

Due to the definition of the Frobenius–Euler numbers and the Bernoulli numbers, we also have the following property of the sums $c_{r,s}(h, k; p)$:

$$c_{r,s}(-k, h; p) = (-1)^s c_{r,s}(k, h; p)$$

(cf. Ref. [16]).

In Ref. [10], Cenkci, Simsek, Can and Kurt defined the following new Dedekind-type sums in terms of the Frobenius–Euler functions:

$$s_{n,u}(h, k) = \frac{1}{k} \sum_{a(\operatorname{mod} k)} au^{-\frac{ha}{k}} \overline{H}_n\left(\frac{ha}{k}, u\right),$$

where $\overline{H}_n(x, u)$ denotes the Frobenius–Euler function, defined by

$$\overline{H}_n(x, u) = H_n(x, u)$$

if $0 < x < 1$ and

$$\overline{H}_n(x + 1, u) = u\overline{H}_n(x, u)$$

(cf. Ref. [18]).

Exercise 15. Find Dedekind-type sums and Hardy-type sums in terms of the twisted Frobenius–Euler functions of higher order with their interpolation functions at negative integers.

References

[1] M. Alkan and Y. Simsek, Generating function for q-Eulerian polynomials and their decomposition and applications, *Fixed Point Theory Appl.*, 72, 2013, 1–14.

[2] M. Alkan and Y. Simsek, The families of L-series associated with decomposition of the generating functions, *Filomat*, 30(7), 2016, 1789–1799.

[3] E. Agyuz, On the convergence properties of generalized Szá sz-Kantorovich type operators involving Frobenious–Euler–Simsek-type polynomials. *Aims Math.*, 9, 2024, 28195–28210.

[4] T. M. Apostol, Generalized Dedekind sums and transformation formulae of certain Lambert series, *Duke Math. J.* 17, 1950, 147–157.

[5] T.M. Apostol, *Modular Functions and Dirichlet Series in Number Theory*, (Springer, Berlin, 1976).

[6] T.M. Apostol, On the Lerch zeta function, *Pac. J. Math.*, 1, 1951, 161–167.

[7] A. Bayad and J. Chikhi, Reduction and duality of the generalized Hurwitz-Lerch zetas, *Fixed Point Theory Appl.*, 82, 2013, 1–14.

[8] A. Bayad and T. Kim, Identities for Apostol-type Frobenius–Euler polynomials resulting from the study of a nonlinear operator, *Russ. J. Math. Phys.*, 23(2), 2016, 164–171.

[9] K. N. Boyadzhiev, Apostol–Bernoulli functions, derivative polynomials and Eulerian polynomials, https://arxiv.org/ftp/arxiv/papers/0710/0710.1124.pdf.

[10] M. Can, M. Cenkci, V. Kurt, Y. Simsek, Twisted Dedekind type sums associated with Barnes' type multiple Frobenius–Euler l-Functions, *Advanc. stud. Contemp. Math.*, 18(2), 2009, 135–160, arXiv:0711.0579v1.

[11] I. N. Cangul, A. S. Cevik, and Y. Simsek, Generalization of q-Apostol-type Eulerian numbers and polynomials, and their interpolation functions, *Adv. Stud. Contemp. Math.*, 25(2), 2015, 211–220.

[12] N. P. Cakic and G. V. Milovanovic, On generalized Stirling numbers and polynomials, *Math. Balk.*, 18, 2004, 241–248.

[13] L. Carlitz, D. P. Roselle, and R. Scoville, Permutations and sequences with repetitions by number of increase, *J. Combin. Th.*, 1, 1966, 350–374.

[14] L. Carlitz, q-Bernoulli numbers and polynomials, *Duke Math. J.*, 15, 1948, 987–1000.

[15] L. Carlitz, Some theorems on generalized Dedekind sums, *Pacific J. Math.*, 3, 1953, 513–522.

[16] L. Carlitz, Dedekind sums and Lambert series, *Proceed. Amer. Math. Soc.*, 5(4), 1954, 580–584.

[17] L. Carlitz, q-Bernoulli and Eulerian numbers, *Trans. Amer. Math. Soc.*, 76, 1954, 332–350.

[18] L. Carlitz, Eulerian numbers and polynomials, *Math. Mag.*, 32(5), 1959, 247–260.

[19] M. Cenkci, M. Can and V. Kurt, p-adic interpolation functions and Kummer-type congruences for q-twisted and q-generalized twisted Euler numbers, *Adv. Stud. Contemp. Math.*, 9, 2004, 203–216.

[20] L. Comtet, *Advanced Combinatorics.* (D. Reidel Publishing Company, Dordrecht, 1974).

[21] G. B. Djordjevic and G. V. Milovanovic, *Special Classes of Polynomials.* (University of Nis, Faculty of Technology, Leskovac, 2014).

[22] R. Dere and Y. Simsek, Normalized polynomials and their multiplication formulas, *Adv. Differ. Equ.*, 31, 2013, 1–10.

[23] A. Erdlyi, W. Magnus, F. Oberhettinger and F. G. Tricomi, *Higher Transcendental Functions, Volume 3.* (McGraw-Hill Book Company, New York, 1955).

[24] L. Euler, *Institutionis Calculi Differentialis, Vol. II.* (Petersburg 1755).

[25] Y. Feng and J. Kozak, On the generalized Euler-Frobenius polynomial, *J. Approx. Theory*, 32(4), 1981, 327–338.

[26] G. Frobenius, *Ober die Bernoullischen Zahlen und die Eulerschen Polynome* (Sitzungsberichte der Preussischen Akad. der Wiss. Phys.-Math. Kl. 809–847. Also in volume 3 of Collected Works, 440–478, 1910).

[27] M. Goubi, An affirmative answer to two questions concerning special case of Simsek numbers and open problems, *Appl. Anal. Discrete Math.*, 14, 2020, 94–105.

[28] M. Goubi, Generating functions for generalization Simsek numbers and their applications, *Appl. Anal. Discrete Math.*, 17(1), 2023, 262–272. https://doi.org/10.2298/AADM200522005G.

[29] D. Gun and Y. Simsek, Modification exponential euler type splines derived from Apostol–Euler numbers and polynomials of complex order, *Appl. Anal. Discrete Math.*, 17, 2023, 197–215.

[30] V. Kac and P. Cheung, *Quantum Calculus*, (Springer-Verlag, New York, 2002).

[31] T. Kim, On a q-analogue of the p-adic log gamma functions and related integrals, *J. Number Theory*, 76, 1999, 320–329.

[32] T. Kim, An invariant p-adic integral associated with Daehee numbers, *Integral Transform. Spec. Funct.*, 13, 2002, 65–69.

[33] T. Kim, On Euler–Barnes multiple zeta functions, *Russian J. Math. Phys.*, 10(3), 2003, 261–267.
[34] T. Kim, q-Euler numbers and polynomials associated with p-adic q-integrals and basic q-zeta function, *Trends Math. (Information Center for Mathematical Sciences)*, 9, 2006, 7–12.
[35] T. Kim, On the analogs of Euler numbers and polynomials associated with p-adic q-integral on Z_p at $q = -1$, *J. Math. Anal. Appl.*, 331, 2007, 779–792.
[36] T. Kim, On the q-extension of Euler and Genocchi numbers, *J. Math. Anal. Appl.*, 326, 2007, 1458–1465.
[37] T. Kim, q-Euler numbers and polynomials associated with p-adic q-integrals, *J. Nonlinear Math. Phys.*, 14, 2007, 15–27.
[38] T. Kim, A note on some formulae for the q-Euler numbers and polynomials, *Proc. Jangjeon Math. Soc.*, 9(2), 2006, 227–232. arXiv:math/0608649.
[39] T. Kim, The modified q-Euler numbers and polynomials, arXiv:math/0702523.
[40] T. Kim, A note on the q-Genocchi numbers and polynomials, *J. Inequal. Appl.*, 2007, Art. ID 71452, 1–8. arXiv:math/0703476.
[41] T. Kim, A note on p-adic q-integral on Z_p associated with q-Euler numbers, *Adv. Stud. Contemp. Math.*, 15, 2007, 133–138. arXiv:math/07064341.
[42] T. Kim, Identities involving Frobenius–Euler polynomials arising from non-linear differential equations, *J. Number Theory*, 132(12), 2012, 2854–2865.
[43] T. Kim, L. C. Jang and H. K. Park, A note on q-Euler and Genocchi numbers, *Proc. Japan Acad.*, 77, 2001, 139–141.
[44] T. Kim and S.-H. Rim, On the twisted q-Euler numbers and polynomials associated with basic q-l-functions, *J. Math. Anal. Appl.*, 336(1), 2007, 738–744.
[45] T. Kim and S.-H. Rim, A new Changhee q-Euler numbers and polynomials associated with p-adic q-integral, *Comput. Math. Appl.*, 54(4), 2007, 484–489.
[46] T. Kim, L. C. Jang, S.-H. Rim and H. K. Pak, On the twisted q-zeta functions and q-Bernoulli polynomials, *Far East J. Appl. Math.*, 13(1), 2003, 13–21.
[47] M. S. Kim, On Euler numbers, polynomials and related p-adic integrals, *J. Number Theory*, 129(9), 2009, 2166–2179.
[48] M. S. Kim and J. W. Son, Analytic properties of the q-Volkenborn integral on the ring of p-adic integers, *Bull. Korean Math. Soc.*, 44(1), 2007, 1–12.
[49] T. Kim, Sums of powers of consecutive q-integers, *Adv. Stud. Contemp. Math.*, 9(1), 2004, 15–18.

[50] T. Kim, A note on exploring the sums of powers of consecutive q-integers, *Adv. Stud. Contemp. Math.*, 11(1), 2005, 137–140.

[51] T. Kim, D. S. Kim, J. Kwon: Analogues of Faulhaber's formula for poly-Bernoulli and type 2 poly-Bernoulli polynomials. *Montes Taurus J. Pure Appl. Math.*, 3(1), 2021, 1–6.

[52] T. Kim, C. S. Ryoo, L. C. Jang, and S. H. Rim, Exploring the sums of powers of consecutive q-integers, Internet. *J. Math. Edu. Sci. Technol.*, 36(8), 2005, 947–956.

[53] T. Kim and C. S. Ryoo, Some identities for Euler and Bernoulli polynomials and their zeros. *Axioms*, 7, 2018, 56.

[54] T. Kim, S.-H. Rim, and Y. Simsek, A note on the alternating sums of powers of consecutive q-integers, *Adv. Stud. Contemp. Math.*, 13(2), 2006, 159–164.

[55] T. Kim, S.-H. Rim, Y. Simsek, and D. Kim, On the analogs of Bernoulli and Euler numbers, related identities and zeta and L-functions, *J. Korean Math. Soc.*, 45(2), 2008, 435–453.

[56] T. Kim, L. Jang, and C. S. Ryoo, Note on q-extensions of Euler numbers and polynomials of higher order, *J. Inequal. Appl.*, 2008, 2008, Article ID: 371295, 1–9.

[57] N. Kilar and Y. Simsek, A new family of Fubini type numbers and polynomials associated with Apostol–Bernoulli numbers and polynomials, *J. Korean Math. Soc.*, 54, 2017, 1605–1621.

[58] N. Kilar and Y. Simsek, Two parametric kinds of Eulerian-type polynomials associated with Euler's formula, *Symmetry* 11, 1–19, (2019).

[59] N. Kilar, A. Bayad and Y. Simsek, Finite sums involving trigonometric functions and special polynomials: Analysis of generating functions and p-Adic integrals, *Appl. Anal. Discrete Math.*, 18, 2024, 452–476. https://doi.org/10.2298/AADM230515010K.

[60] N. Koblitz, A new proof of certain formulas for p-adic L-functions, *Duke Math. J.*, 46(2), 1979, 455–468.

[61] N. Koblitz, On Carlitz's q-Bernoulli numbers, *J. Number Theory*, 14, 1982, 332–339.

[62] I. Kucukoglu, Formulas for q-combinatorial Simsek numbers and polynomials: Analyzing with computational implementations, *Appl. Anal. Discrete Math.*, 19(1), 2025, 185–213. https://doi.org/10.2298/AADM240227005K.

[63] I. Kucukoglu, Generating functions for divisor sums and totative sums arising from combinatorial Simsek numbers, *Filomat*, 38(5), 2024, 1513–1529. https://doi.org/10.2298/FIL2405513K.

[64] I. Kucukoglu and Y. Simsek: Identities and relations on the q-Apostol type Frobenius–Euler numbers and polynomials, *J. Korean Math. Soc.*, 56(1), 2019, 265–284.

[65] I. Kucukoglu and Y. Simsek, Unified presentations of the generating functions for a comprehensive class of numbers and polynomials, *Montes Taurus J. Pure Appl. Math.*, 6(1), 2024, 40–63. Article ID: MTJPAM-D-24-00033.

[66] I. Kucukoglu and Y. Simsek, Some certain families of Catalan-type numbers and polynomials: Analysis of their generating function with their functional equations, *Mathematical Analysis, Differential Equations and Applications*, pp. 485–540. https://doi.org/10.1142/9789811267048_0017 (Springer, 2019).

[67] Q.-M. Luo and H. M. Srivastava, Some relationships between the Apostol–Bernoulli and Apostol–Euler polynomials, *Comput. Math. Appl.*, 10, 2005, 631–642.

[68] Q. M. Luo, Apostol–Euler polynomials of higher order and Gaussian hypergeometric functions, *Taiwanese J. Math.*, 10, 2006, 917–925.

[69] Q. M. Luo and H. M. Srivastava, Some generalizations of the Apostol–Bernoulli and Apostol–Euler polynomials, *J. Math. Anal. Appl.*, 308, 2005, 290–302.

[70] Q. M. Luo and H. M. Srivastava, Some generalizations of the Apostol–Genocchi polynomials and the Stirling numbers of the second kind, *Appl. Math. Comput.*, 217(12), 2011, 5702–5728.

[71] B. Kurt and Y. Simsek, On the generalized Apostol-type Frobenius–Euler polynomials, *Adv. Differ. Equ.*, 1, 2013, 1–9.

[72] M. Masjed-Jamei and W. Koepf, Symbolic computation of some power-trigonometric series. *J. Symb. Comput.*, 80, 2017, 273–284.

[73] N. E. Norlund, *Vorlesungen iiber Differenzenrechnung.* (Springer, Berlin, 1924).

[74] H. Ozden, I. N. Cangul, and Y. Simsek, Multivariate interpolation functions of higher-order q-Euler numbers and their applications, *Abstr. Appl. Anal.*, 2008, 2008, Article ID: 390857, 2008, 1–17.

[75] L. Oussi, On degenerate Simsek and Stirling numbers, *J. Integer Sequences*, 26, 2023, Article 23.5.1.

[76] Rao N, Farid M, Raiz M. On the approximations and symmetric properties of Frobenius–Euler–Simsek polynomials connecting Szasz operators, *Symmetry*, 17(5), 2025, 648. https://doi.org/10.3390/sym17050648.

[77] T. K. Petersen, *Eulerian Numbers.* (Birkhä user Basel, 2015).

[78] J. Riordan, *An Introduction to Combinatorial Analysis.* (John Wiley and Sons, New York, 1958).

[79] S.-H. Rim, T. Kim, A note on q-Euler numbers associated with the basic q-zeta function, *Appl. Math. Lett.*, 20(4), 2007, 366–369.

[80] H. Ozden, Y. Simsek, S.-H. Rim and I. N. Cangul, A note on p-adic q-Euler measure, *Advan. Stud. Contemp. Math.*, 14(2), 2007, 233–239.

[81] H. Peter and P. Jean, Generalizations of Eulerian numbers and their q-analogues, *Nieuw Arch. Wisk.*, 4(9), 1991, 271–298.

[82] P. L. Putzer and M. Hauss, Eulerian numbers with fractional order parameters, *Aequationes Math.*, 46, 1993, 119–142.

[83] H. Rademacher, E. Grosswald, *Dedekind Sums, in: Carus Mathematical Monography*, (Vol. 16. Mathematical Association of America, Washington, DC, 1972).

[84] M. Reimer, *Zur reellen Darstellung periodischer Hermite-Spline-Interpolierender beiaquidistantem Gitter mit Knotenshift.* In J. W. Schmidt, H. Spath (eds.), *Splines in Numerical Analysis*, Akademie-Verlag, Berlin, 1989, pp. 125–134.

[85] S. Roman: *The Umbral Calculus.* (Academic Press, Inc., New York, 1984).

[86] J. Satoh, q-analogue of Riemann's ζ-function and q-Euler Numbers, *J. Number Theory*, 31, 1989, 346–362.

[87] J. Satoh, A construction of q-analogue of Dedekind sums, *Nagoya Math. J.*, 127, 1992, 129–143.

[88] W. Schempp, A contour integral representation of Euler-Frobenius polynomials, *J. Approx. Theory*, 31(3), 1981, 272–278.

[89] W. Schempp, Euler-Frobenius polynomials, *Numer. Methods Approx. Theory*, 7, 1983, 131–138. *Internet. Schriftenreihe Numer. Math.* 67 (1984).

[90] I. J. Schoenberg, Cardinal interpolation and spline functions IV. The exponential Euler splines. In P. L. Butzer, J. P. Kahane, B. Szökefalvi-Nagy (eds.), *Linear Operators and Approximation/Lineare Operatoren und Approximation. International Series of Numerical Mathematics/Internationale Schriftenreihe zur Numerischen Mathematik/Série Internationale D'Analyse Numérique*, Vol. 20. Birkhäuser, Basel, 1972.

[91] I. J. Schoenberg and Carl de Boor (eds.) - I. J. Schoenberg Selected Papers-Birkhäuser Basel (1988).

[92] K. Shiratani, On Euler Numbers, *Mem. Fac. Sci. Kyushu Uni.*, 27, 1975, 1–5.

[93] K. Shiratani and S. Yamamoto, On a p-adic interpolation function for the Euler numbers and its derivatives, *Mem. Fac. Sci. Kyushu Univ.*, 39, 1985, 113–125.

[94] Y. Simsek, Analytic properties of twisted L-functions and twisted Bernoulli numbers. *Adv. Stud. Contemp. Math.*, 11(2), 2005, 205–218.

[95] Y. Simsek, On q-analgue of the twisted L-functions and q-twisted Bernoulli numbers, *J. Korean Math. Soc.*, 40(6), 2003, 963–975.

[96] Y. Simsek, On twisted generalized Euler numbers, *Bull. Korean Math. Soc.*, 41(2), 2004, 299–306.

[97] Y. Simsek, Theorems on twisted L-functions and twisted Bernoulli numbers, *Adv. Stud. Contemp. Math.*, 11(2), 2005, 205–218.

[98] Y. Simsek, q-analogue of the twisted l-series and q-twisted Euler numbers, *J. Number Theory*, 110(2), 2005, 267–278.

[99] Y. Simsek, Twisted (h, q)-Bernoulli numbers and polynomials related to twisted (h, q)-zeta function and L-function, *J. Math. Anal. Appl.*, 324, 2006, 790–804.

[100] Y. Simsek, On p-adic twisted q-L-functions related to generalized twisted Bernoulli numbers, *Russian J. Math. Phys.*, 13, 2006, 327–339.

[101] Y. Simsek, Multiple interpolation functions of higher order (h, q)-Bernoulli numbers, *AIP Conf. Proceed.*, 1048(486), 2008, 1–4.

[102] Y. Simsek, Complete sum of products of (h, q)-extension of Euler polynomials and numbers, *J. Difference Equations Appl.*, 16(11), 2010, 1331–1348.

[103] Y. Simsek, Generating functions for q-Apostol type Frobenius–Euler numbers and polynomials, *Axioms*, 1, 2012, 395–403.

[104] Y. Simsek, Generating functions for generalized Stirling type numbers, Array type polynomials, Eulerian type polynomials and their applications. *Fixed Point Theory Appl.*, 87, 2013, 1–28.

[105] Y. Simsek, Beta-type polynomials and their generating functions, *Appl. Math. Comput.*, 254, 2015, 172–182.

[106] Y Simsek, Identities associated with generalized Stirling type numbers and Eulerian type polynomials, *Math. Comput. Appl.*, 18(3), 2013, 251–263.

[107] Y. Simsek, On generating functions for the special polynomials, *Filomat*, 31(1), 2017, 9–16.

[108] Y. Simsek, New families of special numbers for computing negative order Euler numbers and related numbers and polynomials, *Appl. Anal. Discret. Math.*, 12, 2018, 1–35.

[109] Y. Simsek, Construction of some new families of Apostol-type numbers and polynomials via Dirichlet character and p-adic q-integrals, *Turk. J. Math.*, 42, 2018, 557–577.

[110] Y. Simsek, Explicit formulas for p-adic integrals: Approach to p-adic distribution and some families of special numbers and polynomials, *Montes Taurus J. Pure Appl. Math.*, 1(1), 2019, 1–76.

[111] Y. Simsek, New classes of recurrence relations involving hyperbolic functions, special numbers and polynomials, *Appl. Anal. Discrete Math.*, 15(2), 2021, 426–443.

[112] Simsek, Y. Applications of constructed new families of generating-type functions interpolating new and known classes of polynomials and numbers, *Math. Methods Appl. Sci.*, 44, 2021, 11245–11268.

[113] Simsek, Y. Novel formulas for B-Splines, Bernstein basis functions and special numbers: Approach to derivative and functional equations of generating functions, *Mathematics*, 12(65), 2024, 1–20. https://doi.org/10.3390/math12010065.

[114] Y. Simsek, Formulas and relations for special numbers and polynomials with their generating functions involving characteristic and moment generating functions, *Convex and Variational Analysis with Applications*, (Springer, 2025).

[115] Y. Simsek, Chapter 26: Some families of finite sums associated with interpolation functions for many classes of special numbers and polynomials and hypergeometric series, *Analysis, Geometry, Nonlinear Optimization and Applications*, In Panos M. Pardalos, Themistocles M. Rassias (eds.), World Scientific Publishing Company, New Jersy, London, 2023), https://doi.org/10.1142/9789811261572 0026.

[116] Y. Simsek, analysis of apostol-type numbers and polynomials with their approximations and asymptotic behavior, T. M. Rassias (ed.), *Approximation Theory and Analytic Inequalities*, (Springer Nature, Switzerland AG, 2021) pp. 435–486, https://doi.org/10.1007/978-3-030-60622-0_23.

[117] Y. Simsek, A. Bayad, and V. Lokesha, q-Bernstein polynomials related to q-Frobenius–Euler polynomials, l-functions, and q-Stirling numbers, *Math. Meth. Appl. Sci.*, 35, 2012, 877–884.

[118] Y. Simsek and I. Kucukoglu, Some certain classes of combinatorial numbers and polynomials attached to dirichlet characters: Their construction by p-adic integration and applications to probability distribution functions, In I. N. Parasidis, E. Providas, T. M. Rassias (eds.), *Mathematical Analysis in Interdisciplinary Research.* Springer Optimization and Its Applications, Vol. 179. Springer, Cham. 2021), https://doi.org/10.1007/978-3-030-84721-0_33.

[119] Y. Simsek, V. Kurt and O. Yurekli, On interpolation functions of the twisted generalized Frobenius–Euler numbers, *Adv. Stud. Contemp. Math.*, 15(2), 2007, 187–194.

[120] Y. Simsek, T. Kim, D. W. Park, Y. S. Ro, L. C. Jang and S.-H. Rim, An explicit formula for the multiple Frobenius–Euler numbers and polynomials, *JP J. Algebra Number Theory Appl.*, 4(3), 2004, 519–529.

[121] H. M. Srivastava and J. Choi, *Series Associated with the Zeta and Related Functions*, Kluwer Academic Publishers, Dordrecht, Boston and London, 2001.

[122] H.M. Srivastava, J. Choi, *Zeta and q-Zeta Functions and Associated Series and Integrals.* (Elsevier Science Publishers, Amsterdam, London and New York, 2012).

[123] H. M. Srivastava, T. Kim and Y. Simsek, q-Bernoulli numbers and polynomials associated with multiple q-zeta functions and basic L-series, *Russian J. Math Phys.*, 12(2), 2005, 241–268.

[124] H. M. Srivastava, M. A. Boutiche and Rahmani, Some explicit formulas for the Frobenius–Euler polynomials of higher order, *Appl. Math. Inf. Sci.*, 11, 2017, 621–626.

[125] H. M. Srivastava, M. J. Masjed and M. R. Beyki, A parametric type of the Apostol–Bernoulli, Apostol–Euler and Apostol–Genocchi polynomials, *Appl. Math. Inf. Sci.*, 12, 2018, 907–916.

[126] H. M. Srivastava and C. Kizilates, A parametric kind of the Fubini-type polynomials, *RACSAM*, 113, 2019, 3253–3267. https://doi.org/10.1007/s13398-019-00687-4.

[127] E. Sukruoglu and Y. Simsek, Multiplication formula for the Nörlund sum operator and their applications, *Rev. Real Acad. Cienc. Exactas Fis. Nat. Ser. A-Mat.*, 119, 2025, 68. https://doi.org/10.1007/s13398-025-01730-3.

[128] E. Sukruoglu and Y. Simsek, New approach to construct generating function for Apostol–Bernoulli polynomials via Nörlund sum operator, *Montes Taurus J. Pure Appl. Math.*, 6(2), 2024, 15–23.

[129] E. Sukruoglu and Y. Simsek, Formulas involving moment and interpolation functions for Apostol type polynomials via Nörlund sum and Laplace transform, *J. Math. Anal. Appl.*, 550(1), 2025, 1–18.

[130] H. Tsumura, On a p-adic interpolation of the generalized Euler numbers and its applications, *Tokyo J. Math.*, 10, 1987, 281–293.

[131] H. Tsumura, A note on q-analogues of the Dirichlet series and q-Bernoulli numbers, *J. Number Theory*, 39, 1991, 251–256.

[132] A. Xu, On An open problem of Simsek concerning the computation of a family of special numbers, *Appl. Anal. Discrete Math.*, 13, 2019, 061–072. https://doi.org/10.2298/aadm180804021x.

https://doi.org/10.1142/9789819820498_0020

Chapter 20

Non-Conventional Stacked Three Relator Spaces

Árpád Száz

Institute of Mathematics, University of Debrecen
H–4002 Debrecen, Pf. 400, Hungary
szaz@science.unideb.hu

A subset R of a product set $X \times Y$ is called a relation on X to Y. And, the relations $R^c = (X \times Y) \setminus R$ and $R^{-1} = \{(y, x) : (x, y) \in R\}$ are called the complement and inverse of R, respectively.

For $x \in X$ and $A \subseteq X$, the sets $R(x) = \{y \in Y : (x, y) \in R\}$ and $R[A] = \bigcup_{a \in A} R(a)$ are called the images or neighborhoods of x and A under R, respectively. By using these notations, we may naturally introduce several relations.

For instance, for any $A \subseteq X$ and $B \subseteq Y$, we may naturally write $A \in \mathrm{Cl}_R(B)$ if $R[A] \cap B \neq \emptyset$. Thus, for $x \in X$, we may also naturally write $x \in \mathrm{cl}_R(B)$ if $\{x\} \in \mathrm{Cl}_R(B)$. Moreover, by defining $R^{\triangleright} = \mathrm{cl}_{R^{-1}}$ and $R^{\blacktriangleright} = \mathrm{Cl}_{R^{-1}}$, we can obtain some important operations for relations.

Here, we shall assume that $\mathcal{R}$ is a family of relations on X to Y, $\mathcal{U}$ is a family of relations on $\mathcal{P}(X)$ to Y and $\mathcal{V}$ is a family of relations on $\mathcal{P}(X)$ to $\mathcal{P}(Y)$. In this case, we say that $\mathcal{R}$ is an ordinary relator, $\mathcal{U}$ is a super relator and $\mathcal{V}$ is a hyper relator on X to Y. And, we say that the ordered pair $(X, Y)(\mathcal{R}, \mathcal{U}, \mathcal{V}) = ((X, Y), (\mathcal{R}, \mathcal{U}, \mathcal{V}))$ is a non-conventional three relator space.

Formerly, the $X = Y$ and $\mathrm{card}(\mathcal{U}) = \mathrm{card}(\mathcal{V}) = 1$ particular case of this space has been used to treat the various generalized open sets in a unified way. Now, in addition to the above assumptions, we shall suppose that $\mathcal{D}$ is a stack (ascending system) on Y. And, we shall show that the stacked relator space $(X,Y)(\mathcal{R},\mathcal{U},\mathcal{V})(\mathcal{D}) = ((X,Y),(\mathcal{R},\mathcal{U},\mathcal{V}),\mathcal{D})$ can be used to greatly extend our former results.

Namely, for any $A \subseteq X$ and $B \subseteq Y$, we may also naturally write $A \in \mathrm{Cl}_{(\mathcal{U},\mathcal{D})}(B)$ if $U(A) \cap B \in \mathcal{D}$ for all $U \in \mathcal{U}$. Moreover, we may also naturally define $\mathrm{Cl}_{(\mathcal{R},\mathcal{D})} = \mathrm{Cl}_{(\mathcal{R}^{\triangleright},\mathcal{D})}$, where $\mathcal{R}^{\triangleright} = \{R^{\triangleright} : R \in \mathcal{R}\}$ is already a union-preserving super relator.

And, by using the proper stack $\mathcal{D}_0 = \mathcal{P}(Y) \setminus \{\emptyset\}$, we may also naturally define $\mathrm{Cl}_{\mathcal{U}} = \mathrm{Cl}_{(\mathcal{U},\mathcal{D}_0)}$, and in particular, $\mathrm{Cl}_{\mathcal{R}} = \mathrm{Cl}_{(\mathcal{R},\mathcal{D}_0)}$ too. Thus, $\mathrm{Cl}_{\mathcal{U}}$ is a more general relation than $\mathrm{Cl}_{\mathcal{R}}$. However, in particular, $\mathrm{cl}_{\mathcal{U}} = \mathrm{cl}_{\mathcal{U}^{\triangleleft}}$, where $\mathcal{U}^{\triangleleft} = \{U^{\triangleleft} : U \in \mathcal{U}\}$ is already an ordinary relator. Therefore, much of the extensive theory of ordinary relators cannot be generalized by using super relators instead of the ordinary ones.

1. Some Historical Facts on Topologies

If $\mathcal{T}$ is a family of subsets of a set X such that $\mathcal{T}$ is closed under finite intersections and arbitrary unions, then the family $\mathcal{T}$ is called a *topology* on X, and the ordered pair $X(\mathcal{T}) = (X,\mathcal{T})$ is called a *topological space.*

The members of $\mathcal{T}$ are called the *open subsets* of X. While, the members of $\mathcal{F} = \mathcal{T}^c = \{A^c : A \in \mathcal{T}\}$, where $A^c = X \setminus A$, are called the *closed subsets* of X. Moreover, the members of $\mathcal{T} \cap \mathcal{F}$ are called the *clopen subsets* of X.

Since $\emptyset = \bigcup \emptyset$ and $X = \bigcap \emptyset$, we necessary have $\{\emptyset, X\} \subseteq \mathcal{T} \cap \mathcal{F}$. Therefore, if $\mathcal{T} = \{\emptyset, X\}$, then $\mathcal{T}$ is called *minimal* [130] instead of indiscrete. While, if $\mathcal{T} \cap \mathcal{F} = \{\emptyset, X\}$, then $\mathcal{T}$ is called *connected.* (See Refs. [180] and [173, p. 31].)

For a subset A of $X(\mathcal{T})$, the sets

$$A^{\circ} = \mathrm{int}(A) = \bigcup (\mathcal{T} \cap \mathcal{P}(A)) \quad \text{and} \quad A^{-} = \mathrm{cl}(A) = \mathrm{int}(A^c)^c$$

are called the *interior and closure* of A, respectively.

Thus, $-$ is a *Kuratowski closure operation* on X [76]. That is, $\emptyset^{-} = \emptyset$, and $-$ is *extensive, idempotent and additive* in the sense that, for any $A, B \subseteq X$, we have $A \subseteq A^{-}$, $A^{--} = A^{-}$ and $(A \cup B)^{-} = A^{-} \cup B^{-}$.

In particular, the members of the families

$$\mathcal{E} = \{A \subseteq X : A^{\circ} \neq \emptyset\} \quad \text{and} \quad \mathcal{D} = \{A \subseteq X : A^{-} = X\}$$

are called the *fat and dense subsets* of X, respectively.

In addition, for any $A \subseteq X$, the set $A^{\dagger} = \text{res}(A) = \text{cl}(A) \setminus A$ is called the *residue* of A. And, the members of the family $\mathcal{N} = \{A \subseteq X : A^{-\circ} = \emptyset\}$ are called the *rare (nowhere dense) subsets* of X.

Moreover, a subset of X is called *meager (to be of first category)* if it is a countable union of rare subsets of X. And, the topology $\mathcal{T}$, or the topological space $X(\mathcal{T})$, is called *Baire* if the fat subsets of X are not meager [153, 154].

The open sets as a starting point were first suggested by Tietze [175] and Alexandroff [4], and later standardized by Bourbaki [11] and Kelley [71]. (See Refs. [43] and [173, p. 18].) However, for me, this was the greatest mistake in topology.

Namely, Hausdorff [55], Kuratowski [77], Weil [178], Tukey [176], Efremovič and Švarc [38, 39], Kowalsky [75], Császár [20], Doičinov [35], Herrlich [57] and others [13, 61, 104, 134] offered some more powerful tools other than open sets.

Moreover, by the works of Davis [33], Pervin [117], and Hunsaker and Lindgren [59], all topologies, closures and proximities can be derived from quasi-uniformities. Therefore, they should not be studied separately.

Considering several works on generalized uniformities and their induced structures, the present author in Refs. [139, 140] offered *relators* (families of relations) as the most suitable basic term on which topology and analysis should be based on.

Later, reading the work of Höhle and Kubiak [58] too, he has also observed that relators on $\mathcal{P}(X)$ to Y are somewhat less flexible but more powerful tools than relators on X to Y. (See Refs. [161, 165, 166].)

Now, if $\mathcal{R}$ is a relator on X to Y, $\mathcal{U}$ is a relator on $\mathcal{P}(X)$ to Y and $\mathcal{V}$ is a relator on $\mathcal{P}(X)$ to $\mathcal{P}(Y)$, then we shall say that $\mathcal{R}$ is an *ordinary relator*, $\mathcal{U}$ is a *super relator* and $\mathcal{V}$ is a *hyper relator* on X to Y [125, 169].

Note that a super relation on X to Y is arbitrary subset of $\mathcal{P}(X)\times Y$. While a function of $\mathcal{P}(X)$ to $\mathcal{P}(Y)$ is a particular subset of $\mathcal{P}(X)\times\mathcal{P}(Y)$. Therefore, closures on X will also be rather considered as super relation than functions.

Moreover, if R is an ordinary and U is a super relation on X to Y, then we may naturally define an ordinary relation $U^{\triangleleft}$ and a super relation $R^{\triangleright}$ on X to Y such that $U^{\triangleleft}(x) = U(\{x\})$ and $R^{\triangleright}(A) = R[A]$ for all $x \subseteq X$ and $A \subseteq X$.

Thus, we shall see that the operations $\triangleright$ and $\triangleleft$ form a partial *Galois connection* such that $R^{\triangleright\triangleleft} = R$, but $U^{\circ}(A) = U^{\triangleleft\triangleright}(A) = \bigcup_{x\in A} U(\{x\})$ for all $A \subseteq X$. Therefore, $U^{\circ} = U$ if and only if U is *union-preserving* [161, 169].

2. Some Historical Facts on Generalized Topologies

The notion of a topology has been generalized by a surprisingly great number of mathematicians. For instance, if $\mathcal{T} \subseteq \mathcal{P}(X)$ such that $X \in \mathcal{T}$ and $\mathcal{T}$ is closed under arbitrary unions, then $\mathcal{T}$ is called a *generalized topology* on X.

While, if $\mathcal{T} \subseteq \mathcal{P}(X)$ such that $\emptyset \in \mathcal{T}$ and $X \in \mathcal{T}$, then $\mathcal{T}$ is called a *minimal structure* on X. Generalized topologies and minimal structures were first considered by Lugojan [89] and Maki [90], respectively.

Recently, even the condition $X \in \mathcal{T}$ has also been omitted by Császár [24, 29, 31]. Several authors, publishing in the same Acta, considered his observations to be "the one of the most important developments of general topology".

Each minimal structure or generalized topology $\mathcal{T}$ on X can also be easily derived from the Pervin relator $\mathcal{R}_{\mathcal{T}}$ consisting of all preorder relations $R_A = A^2 \cup (A^c \times X)$ with $A \in \mathcal{T}$. Thus, they should not also be studied separately [157].

If $\mathcal{P}$ and $\mathcal{Q}$ are topologies on X, then by Kelly [72], the ordered pair $X(\mathcal{P},\mathcal{Q}) = (X,(\mathcal{P},\mathcal{Q}))$ is called a *bitopological space.* (See also Refs. [9, 41, 81, 82, 115, 179].) Thus, the space $X(\mathcal{P})$ can be identified with the bispace $X(\mathcal{P},\mathcal{P})$.

Due to Pervin's observation [117], a quasi-uniform space $X(\mathcal{U})$ is already a generalization of a topological space $X(\mathcal{T})$, and a bi-quasi-uniform space $X(\mathcal{U},\mathcal{V})$ is already a generalization of a bitopological space $X(\mathcal{P},\mathcal{Q})$.

More important generalizations of topological spaces are *ideal topological spaces* studied by Kuratowski [77], Vaidyanathaswamy [177], Janković and Hamlett [63] and others [44, 53, 54, 56, 64, 94, 95, 99–101, 131, 182].

If $\mathcal{T}$ is a topology and $\mathcal{J}$ is an *ideal* on X (i.e. a non-void, descending family of subsets of X which is closed under binary unions), then the ordered pair $X(\mathcal{T})(\mathcal{J}) = (X, (\mathcal{T}, \mathcal{J}))$ is called an *ideal topological space.*

In such a space $X(\mathcal{T})(\mathcal{J})$, for any $A \subseteq X$, one defines a "local function"

$$A^* = \{x \in X : \quad \forall \ U \in \mathcal{T}(x) : \quad U \cap A \notin \mathcal{J}\}$$

of A, where $\mathcal{T}(x) = \{U \in \mathcal{T} : \ x \in U\}$.

Moreover, by using this closure-like operation, one also defines

$$\text{cl}^*(A) = A \cup A^*.$$

Thus, cl^* is already a *Kuratowski closure* on X such that $\text{cl}^* = \text{cl}$ if $\mathcal{J} = \{\emptyset\}$.

Bitopological spaces and ideal topological spaces have also been generalized by Császár [28, 30]. He used generalized topologies and *hereditary classes* (descending systems) instead of topologies and ideals, respectively.

However, instead of ideals, it is better to use *filters* and *stacks* (ascending systems) [174]. Namely, if $\mathcal{J}$ is an ideal on X, then its elementwise complement $\mathcal{F} = \mathcal{J}^c = \{A^c : \ A \in \mathcal{J}\}$ is a filter on X such that $\mathcal{J} = \mathcal{F}^c$.

Therefore, for any $V \subseteq X$, we have $V \notin \mathcal{J}$ if and only if $V \notin \mathcal{F}^c$, i.e. $V^c \notin \mathcal{F}$. That is, V is in the generated *grill* $\mathcal{G} = \{A \subseteq X : \ A^c \notin \mathcal{F}\}$, which can also be written in the form $\mathcal{G} = \{A \subseteq X : \quad \forall \ F \in \mathcal{F} : \ A \cap F \neq \emptyset\}$.

In our former paper [168], we have shown that a bi-quasi-uniform space is also a certain generalization of an ideal topological space. However, instead of a bi-quasi-uniform space, it was more convenient to use an arbitrary *birelator space.*

A birelator space is an ordered pair $(X, Y)(\mathcal{R}, \mathcal{S}) = ((X, Y), (\mathcal{R}, \mathcal{S}))$ consisting two sets X, Y and two relators $\mathcal{R}$, $\mathcal{S}$ on X to Y. Here, the relator $\mathcal{S}$ is mainly used to define a stack $\mathcal{D}_{\mathcal{S}} = \{B \subseteq Y : \forall \ S \in \mathcal{S} : \quad X = S^{-1}[B]\}$.

Thus, extending the ideas of Kuratowski [77, p. 62] and Efremovič [38], for any $B \subseteq Y$, we may naturally define

$$\operatorname{Cl}_{(\mathcal{R},\mathcal{S})}(B) = \{A \subseteq X: \quad \forall\ R \in \mathcal{R}: \quad R[A] \cap B \in \mathcal{D}_{\mathcal{S}}\},$$

and in particular, $\operatorname{cl}_{(\mathcal{R},\mathcal{S})}(B) = \{x \in X: \quad \{x\} \in \operatorname{Cl}_{(R,\mathcal{S})}(B)\}$.

Now, for any $B \subseteq Y$, we may also naturally define

$$\operatorname{Int}_{(\mathcal{R},\mathcal{S})}(B) = \operatorname{Cl}_{(\mathcal{R},\mathcal{S})}(B^c)^c = \mathcal{P}(X) \setminus \operatorname{Cl}_{(\mathcal{R},\mathcal{S})}(Y \setminus B)$$

and thus also $\operatorname{int}_{(\mathcal{R},\mathcal{S})}(B) = \operatorname{cl}_{(\mathcal{R},\mathcal{S})}(B^c)^c$.

Thus, we can easily see that

$$\operatorname{Int}_{(\mathcal{R},\mathcal{S})}(B) = \{A \subseteq X: \quad \exists\ \ R \in \mathcal{R}: \quad R[A] \setminus B \notin \mathcal{D}_{\mathcal{S}}\},$$

and thus also $\operatorname{int}_{(\mathcal{R},\mathcal{S})}(B) = \{A \subseteq X: \quad \exists\ R \in \mathcal{R}: \quad R(x) \setminus B \notin \mathcal{D}_{\mathcal{S}}\}$.

Moreover, if in particular $X \neq \emptyset$ and $\mathcal{S}_0 = \{X \times Y\}$, then we can note that $\mathcal{D}_{\mathcal{S}_0} = \mathcal{P}(Y) \setminus \{\emptyset\}$, and thus,

$$\operatorname{Cl}_{\mathcal{R}}(B) = \operatorname{Cl}_{(\mathcal{R},\mathcal{S}_0)}(B) = \{A \subseteq X: \quad \forall\ R \in \mathcal{R}: \quad R[A] \cap B \neq \emptyset\}.$$

In the above definitions, instead of the ordinary relators $\mathcal{R}$ and $\mathcal{S}$, we may easily take a super relator $\mathcal{U}$ on X to Y and an arbitrary stack $\mathcal{D}$ on Y. However, thus we can only get some partial generalizations.

Namely, under the plausible notation $\mathcal{U}^{\triangleleft} = \{U^{\triangleleft}: \quad U \in \mathcal{U}\}$, we already have $\operatorname{cl}_{(\mathcal{U},\mathcal{D})} = \operatorname{cl}_{(\mathcal{U}^{\triangleleft},\mathcal{D})}$. Moreover, each stack $\mathcal{D}$ on Y can also be easily derived from the Pervin preorder relator $\mathcal{R}_{\mathcal{D}}$ [157].

3. Some Historical Facts on Generalized Open Sets

In 1922, a subset A of a closure space $X(-)$ was called *regular open* by Kuratowski [76] if $A = A^{-\circ}$. While, in 1937, a subset A of a topological space $X(\mathcal{T})$ was called regular open by Stone [138] if $A = B^{\circ}$ for some $B \in \mathcal{F}$.

The importance of regular open subsets of $X(\mathcal{T})$ lies mainly in the fact that their family forms a complete *Boolean algebra* [50, p. 66] with respect to the operations defined by $A' = A^{-c}$, $A \wedge B = A \cap B$ and $A \vee B = (A \cup B)''$.

In 1982, a subset A of $X(\mathcal{T})$ was called *preopen* by Mashhour *et al.* [96] if $A \subseteq A^{-\circ}$. However, by Dontchev [36], preopen sets, under different names, were much earlier studied by several mathematicians.

For instance, in 1964, Corson and Michael [19] called a subset A of $X(\mathcal{T})$ *locally dense* if it is a dense subset of some $V \in \mathcal{T}$ in the sense that $A \subseteq V \subseteq A^-$. Moreover, they noted that this property is equivalent to the inclusion $A \subseteq A^{-\circ}$.

This equivalence was later also stated by Jun *et al.* [66]. Moreover, Ganster [45] proved that A is preopen if and only if there exist $V \in \mathcal{T}$ and $B \in \mathcal{D}$ such that $A = V \cap B$. (See also Dontchev [36].)

In 1963, a subset A of $X(\mathcal{T})$ was called *semi-open* by Levine [84] if there exists $V \in \mathcal{T}$ such that $V \subseteq A \subseteq V^-$. First of all, he showed that the set A is semi-open if and only if $A \subseteq A^{\circ -}$.

Moreover, he also proved that if A is a semi-open subset of $X(\mathcal{T})$, then there exist $V \in \mathcal{T}$ and $B \in \mathcal{N}$ such that $A = V \cup B$ and $V \cap B = \emptyset$. In addition, he also noted that the converse statement is false.

Levine's statement closely resembles to a famous stability theorem of Hyers [60] which says that an ε-approximately additive function of one Banach space to another is the sum of an additive function and an ε-small function.

Analogously to the paper of Hyers, Levine's paper has also attracted the interest of a surprisingly great number of mathematicians. For instance, by the Google Scholar, it has been cited by 3640 works.

Moreover, the above statement of Levine was improved by Dlaska *et al.* [34] who observed that a subset A of $X(\mathcal{T})$ is semi-open if and only if there exist $V \in \mathcal{T}$ and $B \subseteq V^\dagger$ such that $A = V \cup B$.

The latter observation was later reformulated, in a more convenient form, by Duszyński and Noiri [37], who noted that a subset A of $X(\mathcal{T})$ is semi-open if and only if there exists $B \subseteq A^{\circ\dagger}$ such that $A = A^\circ \cup B$.

In particular, in 1965 and 1971, Njåstad [108] and Isomichi [62], being not aware of the paper of Levine, studied semi-open sets under the names *β-sets* and *subcondensed sets*, respectively.

Moreover, Njåstad called a subset A of $X(\mathcal{T})$ to be an *α-set* if $A \subseteq A^{\circ - \circ}$. And, he proved that the set A is an α-set if and only if there exist $V \in \mathcal{T}$ and $B \in \mathcal{N}$ such that $A = V \setminus B$.

He also proved that A is an α-set if and only if its intersection with every β-set is a β-set. Thus, the family of all α-sets is a topology. The fact that the semi-open sets form only a generalized topology was already noted by Levine.

A further important property of α-sets was established by Noiri [109] and Reilly and Wamanamurthy [129], in 1984 and 1985, respectively, who proved that a set is α-open if and only if its both preopen and semi-open.

In 1983, the subset A was called *β-open* by Abd El-Monsef *et al.* [1] if $A \subseteq A^{-\circ-}$. Moreover, in 1986, Andrijević [5] used the term *semi-preopen* instead of β-open without knowing of [1].

Actually, Andrijević called a subset A of $X(\mathcal{T})$ to be semi-preopen if there exists a preopen subset V of $X(\mathcal{T})$ such that $V \subseteq A \subseteq V^{-}$. And, he showed that this is equivalent to the inclusion $A \subseteq A^{-\circ-}$.

Moreover, in 1996, a subset A of $X(\mathcal{T})$ was called *b-open* by Andrijević [6] if $A \subseteq A^{\circ-} \cup A^{-\circ}$. And, he proved that A is b-open if and only if there exist a preopen subset B and a semi-open subset C of $X(\mathcal{T})$ such that $A = B \cup C$.

In 1961, a subset A of $X(\mathcal{T})$ was said to have *property Q* by Levine [83] if $A^{\circ-} = A^{-\circ}$. He proved that A has property Q if and only if there exist $V \in \mathcal{T} \cap \mathcal{F}$ and $B \in \mathcal{N}$ such that $A = V \Delta B$. (See also Refs. [14, 17].)

While, in 1991, a subset A of $X(\mathcal{T})$ was called a *δ-set* by Chattopadhyay and Bandyopadhyay [15] if $A^{-\circ} \subseteq A^{\circ-}$. Moreover, in 2001, δ-open sets, under the name *quasi-open sets*, were more systematically studied by Császár [23, 25].

In 1992, Ganster *et al.* [46] proved that A is a δ-set if and only if $A = V \cup N$ for some $V \in \mathcal{T}$ and $B \in \mathcal{N}$. Thus, δ-sets coincide with the *simply open sets* of Biswas [10] and Neubrunnová [107]. (See also [74, 105, 106].)

Actually, such sets were also first studied by Kuratowski [77, p. 69] in a more general framework. By his definition, a subset A of $X(\mathcal{T})$ has to be called *open modulo nowhere dense sets* if there exists $V \in \mathcal{T}$ such that $A \Delta V \in \mathcal{N}$.

In our former papers [126–128], we have shown that the above definitions and several theorems on generalized open sets can be naturally extended to not only generalized topological and closure spaces but also relator spaces.

Now, extending the ideas of Gargouri and Rezgui [49], Premska [119, 120] and our former papers [124, 125], we shall show that the various generalized open sets can even be much better studied in some non-conventional three relator spaces.

For this, we shall assume that $\mathcal{R}$ is an ordinary relator, $\mathcal{U}$ is a super relator and $\mathcal{V}$ is a hyper relator on X. Moreover, we shall assume that $\mathcal{D}$ is a stack on X. Though, instead of $\mathcal{D}$, we could even take an ordinary relator $\mathcal{S}$ on X to Y.

Thus, we shall actually work in a *non-conventional, stacked three relator space* $(X,Y)(\mathcal{R},\mathcal{U},\mathcal{V})(\mathcal{D}) = ((X,Y),(\mathcal{R},\mathcal{U},\mathcal{V}),\mathcal{D})$, instead of a *non-conventional four relator space* $(X,Y)(\mathcal{R},\mathcal{S},\mathcal{U},\mathcal{V}) = ((X,Y),(\mathcal{R},\mathcal{S},\mathcal{U},\mathcal{V}))$.

Our main observation is that

$$\mathcal{T}_{\mathcal{U}} = \{A \subseteq X : \ \ A^c \notin \mathrm{Cl}_{\mathcal{U}}(A)\} = \{A \subseteq X : \ \exists\ U \in \mathcal{U} : \ \ A \subseteq U^{\star}(A)\},$$

where $U^{\star}$ is the dual super relation defined by $U^{\star}(A) = U(A^c)^c$ for all $A \subseteq X$.

Moreover, for any $\mathcal{A} \subseteq X$, we have

$$\begin{aligned}\mathrm{cl}_{\mathcal{V}}(\mathcal{A}) &= \{B \subseteq X : \quad \forall\ V \in \mathcal{V} : \quad V(B) \cap \mathcal{A} \neq \emptyset\} \\ &= \{B \subseteq X : \quad \forall\ V \in \mathcal{V} : \quad \exists\ A \in \mathcal{A} : \quad B \in V(A)\}.\end{aligned}$$

Here, it is an important case when $\mathcal{U} = \{U\}$ and $\mathcal{V} = \{V\}$, with

$$U(A) = \mathrm{cl}_{(\mathcal{R},\mathcal{D})}(\mathrm{int}_{(\mathcal{R},\mathcal{D})}(A))$$

and

$$V(A) = \{B \subseteq X : \ \ A \subseteq B \subseteq \mathrm{cl}_{(\mathcal{R},\mathcal{D})}(A)\}$$

for all $A \subseteq X$.

Namely, in this case, the members of the families $\mathcal{T}_{\mathcal{U}}$ and $\mathrm{cl}_{\mathcal{V}}(\mathcal{T}_{(\mathcal{R},\mathcal{D})})$, with

$$\mathcal{T}_{(\mathcal{R},\mathcal{D})} = \{A \subseteq X : \quad A \subseteq \mathrm{int}_{(\mathcal{R},\mathcal{D})}(A)\},$$

may be naturally considered as some reasonable generalizations of the preopen sets.

4. A Few Basic Facts on Relations

A subset R of a product set $X \times Y$ is called a *relation* on X to Y. In particular, a relation R on X to itself is called a relation on X. And, $R^0 = \Delta_X = \{(x,x) : \ \ x \in X\}$ is called the *identity relation* of X.

If R is a relation on X to Y, then for any $x \in X$ and $A \subseteq X$, the sets $R(x) = \{y \in Y : (x, y) \in R\}$ and $R[A] = \bigcup_{a \in A} R(a)$ are called the *images* or *neighborhoods* of x and A under R, respectively.

If $(x, y) \in R$, then instead of $y \in R(x)$, we may also write xRy. However, instead of $R[A]$, we cannot write $R(A)$. Namely, it may occur that, in addition to $A \subseteq X$, we also have $A \in X$.

Now, the sets $D_R = \{x \in X : R(x) \neq \emptyset\}$ and $R[X]$ are called the *domain* and *range* of R, respectively. If in particular $D_R = X$, then we say that R is a *relation of X to Y*, or that R is a *non-partial relation on X to Y*.

In particular, a relation f on X to Y is called a *function* if for each $x \in D_f$ there exists $y \in Y$ such that $f(x) = \{y\}$. In this case, by identifying singletons with their elements, we may simply write $f(x) = y$ instead of $f(x) = \{y\}$.

Moreover, a function $\star$ of X to itself is called a *unary operation* on X. While, a function $*$ of X^2 to X is called a *binary operation* on X. And, for any $x, y \in X$, we usually write $x^\star$ and $x * y$ instead of $\star(x)$ and $*(x, y) = *((x, y))$.

If R is a relation on X to Y, then a relation S of D_R to Y is called a *selection relation* of R if $S \subseteq R$, i.e. $S(x) \in R(x)$ for all $x \in X$. Thus, by the Axiom of Choice, every relation is the union of its selection functions.

For a relation R on X to Y, we may naturally define two *set-valued functions* φ_R of X to $\mathcal{P}(Y)$ and Φ_R of $\mathcal{P}(X)$ to $\mathcal{P}(Y)$ such that $\varphi_R(x) = R(x)$ for all $x \in X$ and $\Phi_R(A) = R[A]$ for all $A \subseteq X$.

Functions of X to $\mathcal{P}(Y)$ can be naturally identified with relations on X to Y. While functions of $\mathcal{P}(X)$ to $\mathcal{P}(Y)$ are more powerful objects than relations on X to Y. They were briefly called *corelations* on X to Y in Refs. [161, 165, 166].

However, if U is a relation on $\mathcal{P}(X)$ to Y and V is a relation on $\mathcal{P}(X)$ to $\mathcal{P}(Y)$, then it is better to say that U is a *super relation* and V is a *hyper relation on X to Y* [125, 169]. Thus, closures (proximities) [173] are super (hyper) relations.

For a relation R on X to Y, the relations $R^c = (X \times Y) \setminus R$ and $R^{-1} = \{(y, x) : (x, y) \in R\}$ are called the *complement* and the *inverse* of R, respectively. Thus, c is inversion compatible in the sense that $(R^c)^{-1} = (R^{-1})^c$. We can also show that $R^c(x) = R(x)^c = Y \setminus R(x)$ for all $x \in X$, and $R^c[A]^c = \bigcap_{a \in A} R(a)$ for all $A \subseteq X$.

Moreover, $R^{-1}[B] = \{x \in X : R(x) \cap B \neq \emptyset\}$ for all $B \subseteq Y$, and thus, in particular, $D_R = R^{-1}[Y]$.

If R is a relation on X to Y, then $R = \bigcup_{x \in X} (\{x\} \times R(x))$. Therefore, the values $R(x)$, where $x \in X$, uniquely determine R. Thus, a relation R on X to Y can also be naturally defined by specifying $R(x)$ for all $x \in X$.

For instance, if S is a relation on Y to Z, then the *composition relation* $S \circ R$ can be defined such that $(S \circ R)(x) = S[R(x)]$ for all $x \in X$. Thus, it can be easily seen that $(S \circ R)[A] = S[R[A]]$ also holds for all $A \subseteq X$.

While, if S is a relation on Z to W, then the *box product* $R \boxtimes S$ can be defined such that $(R \boxtimes S)(x, z) = R(x) \times S(z)$ for all $x \in X$ and $z \in Z$. Thus, it can be shown that $(R \boxtimes S)[A] = S \circ A \circ R^{-1}$ for all $A \subseteq X \times Z$ [159].

Hence, by taking $A = \{(x, z)\}$, and $A = \Delta_Y$ if $Y = Z$, one can at once see that the box and composition products are actually equivalent tools. However, the box product can be immediately defined for any family of relations.

Now, a relation R on X may be briefly defined to be *reflexive* if $R^0 \subseteq R$ and *transitive* if $R^2 = R \circ R \subseteq R$. Moreover, R may be briefly defined to be *symmetric* if $R^{-1} \subseteq R$, *antisymmetric* if $R \cap R^{-1} \subseteq R^0$ and *total* if $X^2 \subseteq R \cup R^{-1}$.

Thus, a reflexive and transitive (symmetric) relation may be called a *preorder (tolerance) relation.* And, a symmetric (antisymmetric) preorder relation may be called an *equivalence (partial order) relation.*

For any relation R on X, we may also define $R^n = R \circ R^{n-1}$ if $n \in \mathbb{N}$. Moreover, we may also define $R^\infty = \bigcup_{n=0}^{\infty} R^n$. Thus, it can be shown that R^∞ is the smallest preorder relation on X containing R [51].

Now, in contrast to $(R^c)^c = R$ and $(R^{-1})^{-1} = R$, we have $(R^\infty)^\infty = R^\infty$. Moreover, analogously to $(R^c)^{-1} = (R^{-1})^c$, we also have $(R^\infty)^{-1} = (R^{-1})^\infty$. Thus, in particular, R^{-1} is also a preorder on X if R is a preorder on X.

For $A \subseteq X$, the *Pervin relation* $R_A = A^2 \cup (A^c \times X)$ is an important preorder on X [117]. While, for a *pseudo-metric* d on X, the *Weil surrounding* $B_r = \{(x, y) \in X^2 : d(x, y) < r\}$, with $r > 0$, is an important tolerance on X [178].

Note that $S_A = R_A \cap R_A^{-1} = R_A \cap R_{A^c} = A^2 \cup (A^c)^2$ is already an equivalence relation on X. And, more generally if $\mathcal{A}$ is a *cover (partition)* of X, then $S_{\mathcal{A}} = \bigcup_{A \in \mathcal{A}} A^2$ is a tolerance (equivalence) relation on X.

As an important generalization of the Pervin relation R_A, for any $A \subseteq X$ and $B \subseteq Y$, we may also naturally consider the *Hunsaker–Lindgren relation* $R_{(A,B)} = (A \times B) \cup (A^c \times Y)$ [59]. Namely, thus, we evidently have $R_A = R_{(A,A)}$.

The Pervin relations R_A and the Hunsaker–Lindgren relations $R_{(A,B)}$ were actually first used by Davis [33] and Császár [20, p. 42 and 351] in some less explicit and convenient forms, respectively.

5. Some Basic Properties of Super Relations

Notation 1. *In this section, we shall assume that U is a super relation on X to Y.*

Remark 1. Thus, by our former definition, U is actually an ordinary relation on $\mathcal{P}(X)$ to Y, i.e. it is an arbitrary subset of $\mathcal{P}(X) \times Y$.

Moreover, U can be identified with the set-valued function φ_U, defined by $\varphi_U(A) = U(A)$ for all $A \subseteq X$, which is a particular subset of $\mathcal{P}(X) \times \mathcal{P}(Y)$.

Thus, several properties of the super relation U can be easily defined with the help of the set-valued function φ_U. For instance, we may naturally introduce the following:

Definition 1. The super relation U will be called

(1) *increasing* if $U(A) \subseteq U(B)$ for all $A \subseteq B \subseteq X$;
(2) *quasi-increasing* if $U(\{x\}) \subseteq U(A)$ for all $x \in A \subseteq X$;
(3) *union-preserving* if $U(\bigcup \mathcal{A}) = \bigcup_{A \in \mathcal{A}} U(A)$ for all $\mathcal{A} \subseteq \mathcal{P}(X)$.

Thus, we can at once state the following two theorems.

Theorem 1. *The following assertions are equivalent*:

(1) *U is quasi-increasing*;
(2) *$\bigcup_{x \in A} U(\{x\}) \subseteq U(A)$ for all $A \subseteq X$.*

Theorem 2. *The following assertions are equivalent*:

(1) U *is increasing*;
(2) $U(\bigcap \mathcal{A}) \subseteq \bigcap_{A \in \mathcal{A}} U(A)$ *for all* $\mathcal{A} \subseteq \mathcal{P}(X)$;
(3) $\bigcup_{A \in \mathcal{A}} U(A) \subseteq U(\bigcup \mathcal{A})$ *for all* $\mathcal{A} \subseteq \mathcal{P}(X)$.

Moreover, by using Definition 1 and Theorem 2, we can also prove the following:

Theorem 3. *The following assertions are equivalent*:

(1) U *is union-preserving*;
(2) $U(A) = \bigcup_{x \in A} U(\{x\})$ *for all* $A \subseteq X$.

Hint: Note that if $\mathcal{A} \subseteq \mathcal{P}(X)$, then by (2) we have

$$U\left(\bigcup \mathcal{A}\right) = \bigcup_{x \in \bigcup \mathcal{A}} U(\{x\}).$$

Therefore, if $y \in U(\bigcup \mathcal{A})$, then there exists $x \in \bigcup \mathcal{A}$ such that $y \in U(\{x\})$. Thus, in particular, there exists $A_0 \in \mathcal{A}$ such that $x \in A_0$, and so, $\{x\} \subseteq A_0$. Hence, by using the increasingness of U, we can already see that

$$y \in U(\{x\}) \subseteq U(A_0) \subseteq \bigcup_{A \in \mathcal{A}} U(A).$$

Therefore, $U(\bigcup \mathcal{A}) \subseteq \bigcup_{A \in \mathcal{A}} U(A)$ also holds. □

Remark 2. In particular, a super relation U on X to itself may be simply called a super relation on X.

Thus, a super relation U on X may be called *extensive, intensive, involutive* and *idempotent* if $A \subseteq U(A)$, $U(A) \subseteq A$, $U(U(A)) = A$ and $U(U(A)) = U(A)$ for all $A \subseteq X$, respectively.

Moreover, an increasing involutive (idempotent) super relation may be called an *involution (projection) relation.* While, an extensive (intensive) projection relation may be called a *closure (interior) relation.*

6. Relationships between Ordinary and Super Relations

Notation 2. *In this and the following two sections, we shall assume that R and S are ordinary relations, and U and V are super relations on X to Y.*

In Ref. [165], having in mind Galois connections [32, 158], we have introduced the following:

Definition 2. For the ordinary relation R, we define a super relation $R^{\triangleright}$ on X to Y such that

$$R^{\triangleright}(A) = R[A] \quad \text{for all} \quad A \subseteq X.$$

While, for the super relation U, we define an ordinary relation $U^{\triangleleft}$ on X to Y such that

$$U^{\triangleleft}(x) = U(\{x\}) \quad \text{for all} \quad x \in X.$$

The appropriateness of this definition is apparent from the following two theorems whose proofs are included here only for the reader's convenience.

Theorem 4. *$R^{\triangleright} \subseteq U$ implies $R \subseteq U^{\triangleleft}$.*

Proof. If $R^{\triangleright} \subseteq U$, then, in particular, we have

$$R(x) = R[\{x\}] = R^{\triangleright}(\{x\}) \subseteq U(\{x\}) = U^{\triangleleft}(x)$$

for all $x \in X$. Therefore, $R \subseteq U^{\triangleleft}$ also holds. □

Remark 3. From the above proof, we can see that $R(x) = R^{\triangleright\triangleleft}(x)$ for all $x \in X$. Therefore, Theorem 4 can also be proved by using that $R^{\triangleright} \subseteq U$ implies $R^{\triangleright\triangleleft} \subseteq U^{\triangleleft}$.

Theorem 5. *The following assertions are equivalent*:

(1) *U is quasi-increasing*;
(2) *$R \subseteq U^{\triangleleft}$ implies $R^{\triangleright} \subseteq U$ for any relation R on X to Y.*

Hint: If (2) holds, then because of $U^{\triangleleft} \subseteq U^{\triangleleft}$ we have $(U^{\triangleleft})^{\triangleright} \subseteq U$. Therefore, for any $A \subseteq X$, we have $U^{\triangleleft\triangleright}(A) \subseteq U(A)$. Moreover, by using the corresponding definitions, we can see that

$$U^{\lhd\rhd}(A) = (U^{\lhd})^{\rhd}(A) = U^{\lhd}[A] = \bigcup_{x\in A} U^{\lhd}(x) = \bigcup_{x\in A} U(\{x\}).$$

Therefore, $\bigcup_{x\in A} U(\{x\}) \subseteq U(A)$. Thus, by Theorem 1, assertion (1) also holds. □

Now, as an immediate consequence of the above two theorems, we can also state the following:

Corollary 1. *If U is quasi-increasing, then*

$$R^{\rhd} \subseteq U \iff R \subseteq U^{\lhd}.$$

Remark 4. This shows that the operations $\rhd$ and $\lhd$ establish a partial Galois connection between the power sets $\mathcal{P}(X \times Y)$ and $\mathcal{P}(\mathcal{P}(X) \times Y)$.

Therefore, we may also naturally introduce the following:

Definition 3. The super relation

$$U^{\circ} = U^{\lhd\rhd}$$

will be called the *Galois interior* of U.

Thus, by the proof of Theorem 5, we can at once state the following:

Theorem 6. *For any $A \subseteq X$, we have*

$$U^{\circ}(A) = U^{\lhd}[A] = \bigcup_{x\in A} U^{\lhd}(x) = \bigcup_{x\in A} U(\{x\}).$$

Moreover, by letting A to be a singleton, we can also state the following:

Corollary 2. *For any $x \in X$, we have*

$$U^{\circ\lhd}(x) = U^{\circ}(\{x\}) = U^{\lhd}(x) = U(\{x\}).$$

Theorem 6 allows us to easily determine the Galois interior of some very particular super relations.

Example 1. If $U(\{x\}) = \emptyset$ for all $x \in X$, then $U^{\circ}(A) = \emptyset$ for all $A \subseteq X$.

Example 2. If $U(\{x\}) = \{x\}$ for all $x \in X$, then $U^\circ(A) = A$ for all $A \subseteq X$.

Example 3. If $U(A) = A^c$ for all $A \subseteq X$, then for any $A \subseteq X$ we have

$$U^\circ(A) = \begin{cases} \emptyset & \text{if} \quad \operatorname{card}(A) = 0, \\ A^c & \text{if} \quad \operatorname{card}(A) = 1, \\ X & \text{if} \quad \operatorname{card}(A) > 1. \end{cases}$$

Namely, by Theorem 6 and De Morgan's law, we have

$$U^\circ(A) = \bigcup_{x \in A} U(\{x\}) = \bigcup_{x \in A} \{x\}^c = \left(\bigcap_{x \in A} \{x\} \right)^c,$$

whence the required equalities immediately follow.

Remark 5. Note that if U is as in Example 1, then $U^\triangleleft = \emptyset$ and $U^\circ = \emptyset$. While, if U is as in Example 2, then $U^\triangleleft = \Delta_X$ and $U^\circ = \Delta_{\mathcal{P}(X)}$.

However, if U is as in Example 3, then for any $x, y \in X$ we have

$$\begin{aligned} y \in U^\triangleleft(x) &\iff y \in U(\{x\}) \iff y \in \{x\}^c \\ &\iff y \notin \{x\} \iff y \neq x. \end{aligned}$$

Therefore, $U^\triangleleft$ is just the inequality relation on X.

7. Further Theorems on the Operations $\triangleright$, $\triangleleft$ and $\circ$

Several properties of the operations $\triangleright$, $\triangleleft$ and $\circ$ can be immediately derived from the general theory of Galois and Pataki connections [167].

However, because of the simplicity of Definition 2, it is now more convenient to apply some direct proofs to establish the following four theorems.

Theorem 7. *The operations* $\triangleright$, $\triangleleft$ *and* $\circ$ *are increasing.*

Proof. For instance, if $U \subseteq V$, then $U(A) \subseteq V(A)$ for all $A \subseteq X$. Thus, in particular, we also have

$$U^{\triangleleft}(x) = U(\{x\}) \subseteq V(\{x\}) = V^{\triangleleft}(x)$$

for all $x \in X$. Therefore, $U^{\triangleleft} \subseteq V^{\triangleleft}$ also holds. $\square$

Theorem 8. $R^{\triangleright}$ *is a union-preserving super relation on* X *to* Y *such that*

(1) $R^{\triangleright\triangleleft} = R$; (2) $R^{\triangleright\circ} = R^{\triangleright}$.

Proof. By the corresponding definitions, we have

$$R^{\triangleright}(A) = R[A] = \bigcup_{x \in A} R(x) = \bigcup_{x \in A} R[\{x\}] = \bigcup_{x \in A} R^{\triangleright}(\{x\})$$

for all $A \subseteq X$. Thus, by Theorem 3, the super relation $R^{\triangleright}$ is union-preserving.

Moreover, from Remark 3, we can see that $R^{\triangleright\triangleleft}(x) = R(x)$ for all $x \in X$, and thus, $R^{\triangleright\triangleleft} = R$ is also true.

Now, by using Definition 3 and assertion (1), we can also easily see that

$$R^{\triangleright\circ} = (R^{\triangleright})^{\circ} = (R^{\triangleright})^{\triangleleft\triangleright} = (R^{\triangleright\triangleleft})^{\triangleright} = R^{\triangleright}. \qquad \square$$

By using the above two theorems, we can easily establish the following:

Corollary 3. *We have*

$$R \subseteq S \iff R^{\triangleright} \subseteq S^{\triangleright}.$$

Moreover, in addition to Theorem 8, we can also prove the following:

Theorem 9. U° *is a union-preserving super relation on* X *to* Y *such that*

(1) $U^{\circ\triangleleft} = U^{\triangleleft}$; (2) $U^{\circ\circ} = U^{\circ}$.

Proof. From Definition 3, by Theorem 8, it is clear that U° is union-preserving. Moreover, from Corollary 2, we can see that $U^{\circ\triangleleft}(x) = U^{\triangleleft}(x)$ for all $x \in X$, and thus, $U^{\circ\triangleleft} = U^{\triangleleft}$ is also true.

Furthermore, by using Theorem 6, we can easily see that

$$U^{\circ\circ}(A) = (U^{\circ})^{\circ}(A) = \bigcup_{x\in A} U^{\circ}(\{x\}) = \bigcup_{x\in A} U(\{x\}) = U^{\circ}(A)$$

for all $A \subseteq X$, and thus, $U^{\circ\circ} = U^{\circ}$ is also true. □

Now, to characterize union-preserving super relations, we can also prove the following:

Theorem 10. *The following assertions are equivalent:*

(1) $U^{\circ} = U$; (2) *U is union-preserving*;
(3) $U = R^{\triangleright}$ *for some relation R on X to Y.*

Hint: If (2) holds, then by Theorem 6, we can see that

$$U^{\circ}(A) = \bigcup_{x\in A} U(\{x\}) = U(A)$$

for all $A \subseteq X$. Therefore, (1) also holds. □

Now, by using Theorems 7 and 10, we can also easily establish the following:

Corollary 4. *If U and V are union-preserving, then*

$$U \subseteq V \iff U^{\triangleleft} \subseteq V^{\triangleleft}.$$

By using our former results, the following four theorems can also be proved.

Theorem 11. *We have*

(1) $U \subseteq U^{\circ} \iff U(A) \subseteq U^{\triangleleft}[A]$ *for all* $A \subseteq X$;
(2) $U^{\circ} \subseteq U \iff U$ *is quasi-increasing* $\iff U^{\triangleleft}[A] \subseteq U(A)$ *for all* $A \subseteq X$;
(3) $U^{\circ} = U \iff U$ *is union-preserving* $\iff U(A) = U^{\triangleleft}[A]$ *for all* $A \subseteq X$.

Theorem 12. *We have*

(1) $U^{\circ} \subseteq V \implies U^{\circ} \subseteq V^{\circ}$; (2) $U^{\circ} \subseteq V^{\circ} \iff U^{\triangleleft} \subseteq V^{\triangleleft}$;
(3) $U^{\circ} \subseteq V^{\circ} \implies U^{\circ} \subseteq V$ *whenever V is quasi-increasing*;

(4) $U \subseteq V \iff U^\circ \subseteq V^\circ$ *whenever* U *and* V *are union-preserving.*

Theorem 13. *If* $U = R^\triangleright$, *then*

(1) U *is an union-preserving super relation on* X *to* Y *such that* $U^\triangleleft = R$;
(2) U *is the smallest quasi-increasing super relation on* X *to* Y *such that* $R \subseteq U^\triangleleft$;
(3) U *is the largest union-preserving super relation on* X *to* Y *such that* $U^\triangleleft \subseteq R$.

Theorem 14. *If* $R = U^\triangleleft$, *then*

(1) $R^\triangleright \subseteq U$ *if and only if* U *is quasi-increasing;*
(2) $R^\triangleright = U$ *if and only if* U *is union-preserving;*
(3) *if* U *is quasi-increasing, then* R *is the largest relation on* X *to* Y *such that* $R^\triangleright \subseteq U$;
(4) *if* U *is union-preserving, then* R *is the smallest relation on* X *to* Y *such that* $U \subseteq R^\triangleright$.

8. Relationally Defined Inverses of Super Relations

Due to Remark 4, we may also naturally introduce the following:

Definition 4. The super relation

$$U^{\mathbf{-1}} = U^{\triangleleft -1 \triangleright}$$

will be called the *relationally defined inverse* of U.

Remark 6. To feel the necessity of this bold inverse $U^{\mathbf{-1}}$, note that the ordinary inverse U^{-1} of U is not a super relation.

While the ordinary inverse φ_U^{-1} of the associated set-valued function φ_U, which can be identified with U, is usually a hyper relation.

Now, using the corresponding definitions and Theorem 8, we can easily prove the following three theorems.

Theorem 15. *We have*

(1) $R^{\triangleright \mathbf{-1}} = R^{-1 \triangleright}$; (2) $R^{\triangleright \mathbf{-1} \triangleleft} = R^{-1}$.

Proof. By Definition 4 and Theorem 8, we have

$$R^{\triangleright -1} = R^{\triangleright\triangleleft -1\triangleright} = R^{-1\triangleright}, \quad \text{and thus also} \quad R^{\triangleright -1\triangleleft} = R^{-1\triangleright\triangleleft} = R^{-1}.$$

□

Theorem 16. *U^{-1} is a union-preserving super relation on Y to X such that*

(1) $U^{-1\triangleleft} = U^{\triangleleft -1}$; (2) $U^{\circ -1} = U^{-1}$.

Proof. By Definitions 4 and 3 and Theorem 8, we have

$$U^{-1\triangleleft} = U^{\triangleleft -1\triangleright\triangleleft} = U^{\triangleleft -1} \quad \text{and} \quad U^{\circ -1} = U^{\triangleleft\triangleright\triangleleft -1\triangleright} = U^{\triangleleft -1\triangleright} = U^{-1}.$$

□

Remark 7. Note that if $U^{\triangleleft}$ is symmetric, then $U^{-1} = U^{\triangleleft -1\triangleright} = U^{\triangleleft\triangleright} = U^{\circ}$. Thus, if in addition U is union-preserving, then $U^{-1} = U$.

Moreover, if in particular U is as in Example 3, then by Remark 5, the relation $U^{\triangleleft}$ is symmetric. Thus, by the above observation, $U^{-1} = U^{\circ}$.

Theorem 17. *We have*

$$(U^{-1})^{-1} = U^{\circ}.$$

Proof. By the corresponding definitions and Theorem 8, we have

$$(U^{-1})^{-1} = U^{\triangleleft -1\triangleright\triangleleft -1\triangleright} = U^{\triangleleft -1 -1\triangleright} = U^{\triangleleft\triangleright} = U^{\circ}.$$ □

Hence, by using Theorem 10, we can immediately derive the following:

Corollary 5. *The following assertions are equivalent:*

(1) $U = (U^{-1})^{-1}$; (2) *U is union-preserving.*

Moreover, as a counterpart of Theorem 6, we can also prove the following:

Theorem 18. *For any $B \subseteq Y$, we have*

$$U^{-1}(B) = U^{\triangleleft -1}[B]$$
$$= \{x \in X : \ U^{\triangleleft}(x) \cap B \neq \emptyset\} = \{x \in X : \ U(\{x\}) \cap B \neq \emptyset\}.$$

Proof. By the corresponding definitions, we have

$$U^{-1}(B) = (U^{\triangleleft -1\triangleright})(B) = U^{\triangleleft -1}[B].$$

Moreover, it is clear that, for any $x \in X$, we have

$$x \in U^{\lhd-1}[B] \iff U^{\lhd}(x) \cap B \neq \emptyset \iff U(\{x\}) \cap B \neq \emptyset.$$

Therefore, the required equalities are true. □

Remark 8. From the above theorem, by Theorem 16, we can also see that

$$U^{-1}(B) = U^{\lhd-1}[B] = U^{-1\lhd}[B].$$

9. Functionally Defined Compositions of Super Relations

Notation 3. *In this and the following section, we shall assume that R is an ordinary relation and U is super relation on X to Y.*

Moreover, we shall assume that S is an ordinary relation and V is super relation on Y to Z.

Due to the usual identification of the relation U with the function φ_U, we may naturally introduce the following:

Definition 5. The super relation $V \circ U$, defined such that

$$(V \circ U)(A) = V(U(A))$$

for all $A \subseteq X$, will be called the *functionally defined composition* of V and U.

Remark 9. Namely, thus, we can easily check that $\varphi_{V \circ U}(A) = (\varphi_V \circ \varphi_U)(A)$ for all $A \subseteq X$, and thus, $\varphi_{V \circ U} = \varphi_V \circ \varphi_U$ also holds.

The appropriateness of Definition 5 is also quite obvious from the following three theorems and their corollaries.

Theorem 19. *We have*

$$(S \circ R)^{\rhd} = S^{\rhd} \circ R^{\rhd}.$$

Proof. By Definitions 2 and 5, we have

$$(S \circ R)^{\rhd}(A) = (S \circ R)[A] = S[R[A]] = S^{\rhd}(R^{\rhd}(A)) = (S^{\rhd} \circ R^{\rhd})(A)$$

for all $A \subseteq X$. Thus, the required equality is also true. □

From this theorem, by using Definition 3 and Theorem 10, we can derive the following:

Corollary 6. *We have*

(1) $(S \circ U^{\triangleleft})^{\triangleright} = S^{\triangleright} \boldsymbol{\circ} U$ *if* U *is union-preserving;*
(2) $(V^{\triangleleft} \circ R)^{\triangleright} = V \boldsymbol{\circ} R^{\triangleright}$ *if* V *is union-preserving.*

Theorem 20. *If* V *is union-preserving, then*

$$(V \boldsymbol{\circ} U)^{\triangleleft} = V^{\triangleleft} \circ U^{\triangleleft}.$$

Proof. By the corresponding definitions and Theorem 10, we have

$$\begin{aligned}(V \boldsymbol{\circ} U)^{\triangleleft}(x) &= (V \boldsymbol{\circ} U)(\{x\}) = V(U(\{x\})) = V(U^{\triangleleft}(x)) \\ &= V^{\circ}(U^{\triangleleft}(x)) = V^{\triangleleft\triangleright}(U^{\triangleleft}(x)) = V^{\triangleleft}[U^{\triangleleft}(x)] = (V^{\triangleleft} \circ U^{\triangleleft})(x)\end{aligned}$$

for all $x \in X$. Therefore, the required equality is also true. □

Corollary 7. *We have*

(1) $(S^{\triangleright} \boldsymbol{\circ} U)^{\triangleleft} = S \circ U^{\triangleleft}$;
(2) $(V \boldsymbol{\circ} R^{\triangleright})^{\triangleleft} = V^{\triangleleft} \circ R$ *if* V *is union-preserving.*

Theorem 21. *If* V *is union-preserving, then*

$$(V \boldsymbol{\circ} U)^{\boldsymbol{-1}} = U^{\boldsymbol{-1}} \boldsymbol{\circ} V^{\boldsymbol{-1}}.$$

Proof. By Definition 4 and Theorems 20, and 19, we have

$$\begin{aligned}(V \boldsymbol{\circ} U)^{\boldsymbol{-1}} &= (V \boldsymbol{\circ} U)^{\triangleleft-1\triangleright} = (V^{\triangleleft} \circ U^{\triangleleft})^{-1\triangleright} \\ &= (U^{\triangleleft-1} \circ V^{\triangleleft-1})^{\triangleright} = U^{\triangleleft-1\triangleright} \boldsymbol{\circ} V^{\triangleleft-1\triangleright} = U^{\boldsymbol{-1}} \boldsymbol{\circ} V^{\boldsymbol{-1}}.\end{aligned}$$

□

Corollary 8. *We have*

$$(S^{\triangleright} \boldsymbol{\circ} U)^{\boldsymbol{-1}} = U^{\boldsymbol{-1}} \boldsymbol{\circ} S^{-1\triangleright}.$$

Remark 10. By using Definition 5, it can also be easily seen that the functionally defined composition of super relations is associative.

10. Relationally Defined Compositions of Super Relations

Now, analogously to Definition 4, we may also naturally introduce the following:

Definition 6. The super relation

$$V \bullet U = (V^{\lhd} \circ U^{\lhd})^{\rhd}$$

will be called the *relationally defined composition* of V and U.

The appropriateness of this definition is apparent from the following theorems.

Theorem 22. *We have*

(1) $S^{\rhd} \bullet R^{\rhd} = (S \circ R)^{\rhd}$; (2) $(S^{\rhd} \bullet R^{\rhd})^{\lhd} = S \circ R$.

Proof. By Definition 6 and Theorem 8, we have

$$S^{\rhd} \bullet R^{\rhd} = (S^{\rhd\lhd} \circ R^{\rhd\lhd})^{\rhd} = (S \circ R)^{\rhd},$$

and thus also $(S^{\rhd} \bullet R^{\rhd})^{\lhd} = (S \circ R)^{\rhd\lhd} = S \circ R$. □

Theorem 23. $V \bullet U$ *is a union-preserving super relation such that*

(1) $(V \bullet U)^{\lhd} = V^{\lhd} \circ U^{\lhd}$; (2) $V \bullet U = V^{\circ} \mathbin{\text{o}} U^{\circ}$.

Proof. From Definition 6, by Theorem 8, it is clear that $V \bullet U$ is a union-preserving and

$$(V \bullet U)^{\lhd} = (V^{\lhd} \circ U^{\lhd})^{\rhd\lhd} = V^{\lhd} \circ U^{\lhd}.$$

Moreover, by using Theorem 19 and Definition 3, we can see that

$$V \bullet U = (V^{\lhd} \circ U^{\lhd})^{\rhd} = V^{\lhd\rhd} \mathbin{\text{o}} U^{\lhd\rhd} = V^{\circ} \mathbin{\text{o}} U^{\circ}.$$ □

From assertion (2), by using Theorem 10, we can immediately derive the following:

Corollary 9. *If both U and V are union-preserving, then*

$$V \bullet U = V \mathbin{\text{o}} U.$$

Moreover, we can easily see that the relationally defined composition of super relations may greatly differ from the functionally defined one.

Example 4. If $U(A) = A^c$ for all $A \subseteq X$, then by Definition 5, we have

$$(U \circ U)(A) = U(U(A)) = U(A^c) = A^{cc} = A$$

for all $A \subseteq X$.

However, for instance, if $\operatorname{card}(X) > 2$ and $A \subseteq X$ such that $\operatorname{card}(A) = 1$, then by Theorem 23 and Example 3, we already have

$$(U \bullet U)(A) = (U^\circ \circ U^\circ)(A) = U^\circ(U^\circ(A)) = U^\circ(A^c) = X.$$

By using Theorem 23, we can also prove the following two theorems.

Theorem 24. *For any $A \subseteq X$, we have*

$$(V \bullet U)(A) = \bigcup_{x \in A} \bigcup_{y \in U^\triangleleft(x)} V^\triangleleft(y).$$

Proof. By using Theorem 23, we can see that

$$\begin{aligned}(V \bullet U)(A) &= \bigcup_{x \in A} (V \bullet U)(\{x\}) = \bigcup_{x \in A} (V \bullet U)^\triangleleft(x) \\ &= \bigcup_{x \in A} (V^\triangleleft \circ U^\triangleleft)(x) = \bigcup_{x \in A} V^\triangleleft[U^\triangleleft(x)] = \bigcup_{x \in A} \bigcup_{y \in U^\triangleleft(x)} V^\triangleleft(y). \qquad \square\end{aligned}$$

Theorem 25. *We have*

$$(V \bullet U)^{-1} = U^{-1} \circ V^{-1}.$$

Proof. By using Theorems 23, 9, 21 and 16, we can see that

$$(V \bullet U)^{-1} = (V^\circ \circ U^\circ)^{-1} = U^{\circ -1} \circ V^{\circ -1} = U^{-1} \circ V^{-1}. \qquad \square$$

Remark 11. Moreover, by using Theorem 23 and Remark 10, it can be easily seen that the relationally defined composition of super relations is also associative.

11. The Duals of Super and Hyper Relations

Having in mind the relationship between the usual closure and interior operations, we may naturally introduce the following:

Definition 7. For a super relation U on X to Y, we define a *dual super relation* $U^{\star}$ on X to Y such that

$$U^{\star}(A) = U(A^c)^c$$

for all $A \subseteq X$.

Remark 12. Thus, for any ordinary relation R on X to Y, we can note that

$$R^{\star}(A) = R^{\triangleright\star}(A) = R^{\triangleright}(A^c)^c = R[A^c]^c.$$

Moreover, we can easily prove the following four theorems.

Theorem 26. *If U and V are super relations on X to Y, then*

(1) $U = U^{\star\star}$; (2) $U \subseteq V$ *implies* $V^{\star} \subseteq U^{\star}$.

Proof. To prove (2), note that if $U \subseteq V$, then $U(A^c) \subseteq V(A^c)$, and thus,

$$V^{\star}(A) = V(A^c)^c \subseteq U(A^c)^c = U^{\star}(A)$$

for all $A \subseteq X$. Therefore, $V^{\star} \subseteq U^{\star}$ also holds. □

Theorem 27. *If U is a super relation on X to Y, then*

(1) *$U^{\star}$ is increasing if and only if U is increasing;*
(2) *$U^{\star}$ is union-preserving if and only if U is intersection-preserving;*
(3) *$U^{\star}$ is intersection-preserving if and only if U is union-preserving.*

Hint: For instance, if U is union-preserving, then by the corresponding definitions and De Morgan's law, we have

$$U^{\star}\left(\bigcap_{A\in\mathcal{A}} A\right) = U\left(\left(\bigcap_{A\in\mathcal{A}} A\right)^c\right)^c = U\left(\bigcup_{A\in\mathcal{A}} A^c\right)^c$$

$$= \left(\bigcup_{A\in\mathcal{A}} U(A^c)\right)^c = \bigcap_{A\in\mathcal{A}} U(A^c)^c = \bigcap_{A\in\mathcal{A}} U^{\star}(A)$$

for all $\mathcal{A} \subseteq \mathcal{P}(X)$. Therefore, $U^{\star}$ is intersection-preserving.

Thus, the "if part" of assertion (3) is true. Hence, since $U^{\star\star} = U$, it is clear that the "only if part" of assertion (2) is also true. □

Theorem 28. *If U is a super relation on X to Y, then*

(1) *$U^{\star}$ is intensive if and only if U is extensive;*
(2) *$U^{\star}$ is extensive if and only if U is intensive;*
(3) *$U^{\star}$ is involutive if and only if U is involutive;*
(4) *$U^{\star}$ is idempotent if and only if U is idempotent.*

Hint: For instance, if U is idempotent, then by the corresponding definitions

$$U^{\star}(U^{\star}(A)) = U^{\star}(U(A^c)^c) = U(U(A^c))^c = U(A^c)^c = U^{\star}(A)$$

for all $A \subseteq X$. Therefore, $U^{\star}$ is also idempotent.

Thus, the "if part" of assertion (4) is true. Hence, since $U^{\star\star} = U$, it is clear that the "only if part" of assertion (4) is also true. □

Theorem 29. *If U is a super relation on X to Y and V is a super relation on Y to Z, then*

(1) $(V \circ U)^{\star} = V^{\star} \circ U^{\star}$; (2) $(V \bullet U)^{\star} = V^{\circ\star} \circ U^{\circ\star}$.

Proof. By Definitions 7 and 5, we have

$$\begin{aligned}(V \circ U)^{\star}(A) &= (V \circ U)(A^c)^c = V(U(A^c))^c \\ &= V(U^{\star}(A)^c)^c = V^{\star}(U^{\star}(A)) = (V^{\star} \circ U^{\star})(A)\end{aligned}$$

for all $A \subseteq X$. Thus, assertion (1) is true.

Now, by using Theorem 23 and assertion (1), we can also see that

$$(V \bullet U)^{\star} = (V^{\circ} \circ U^{\circ})^{\star} = V^{\circ\star} \circ U^{\circ\star}.$$ □

To determine the super relation $U^{\circ\star}$, we can only prove the following:

Theorem 30. *If R is an ordinary and U is super relation on X to Y, then for any $A \subseteq X$ we have*

(1) $R^{\triangleright\star}(A) = R[A^c]^c = \bigcap_{x \in A^c} R(x)^c$;
(2) $U^{\circ\star}(A) = U^{\triangleleft}[A^c]^c = \bigcap_{x \in A^c} U^{\triangleleft}(x)^c$.

Proof. By the corresponding definitions and De Morgan's law, we have

$$R^{\triangleright\star}(A) = R^{\triangleright}(A^c)^c = R[A^c]^c = \bigcap_{x \in A^c} R(x)^c,$$

and thus also

$$U^{\circ\star}(A) = U^{\triangleleft\triangleright\star}(A) = U^{\triangleleft}[A^c]^c = \bigcap_{x \in A^c} U^{\triangleleft}(x)^c.$$

□

Analogously to Definition 7, we may also naturally introduce the following:

Definition 8. For a hyper relation V on X to Y, we define two *dual hyper relations* $V^{\star}$ and $V^{\bigstar}$ on X to Y such that

$$V^{\star}(A) = V(A^c)^c = \mathcal{P}(Y) \setminus V(A^c)$$

and

$$V^{\bigstar}(A) = [V(A^c)]^c = \{B^c : \quad B \in V(A^c)\}$$

for all $A \subseteq X$.

Remark 13. Thus, some properties of the hyper relations $V^{\star}$ and $V^{\bigstar}$ can also be easily derived from those of the hyper relation V.

Moreover, having in mind the derivations of small closures and interiors from the big ones [140, 145], we may also naturally introduce the following:

Definition 9. For a hyper relation V on X to Y, we define a super relation $V^{\blacktriangleleft}$ on X to Y such that

$$V^{\blacktriangleleft}(A) = \{y \in Y : \quad \{y\} \in V(A)\}$$

for all $A \subseteq X$.

Remark 14. Thus, we may also naturally consider the ordinary relation $V^{\blacktriangleleft\triangleleft}$. In particular, we may define $\operatorname{int}_{\mathcal{R}} = \operatorname{Int}_{\mathcal{R}}^{\blacktriangleleft}$ and $\sigma_{\mathcal{R}} = \operatorname{int}_{\mathcal{R}}^{\triangleleft}$ for any relator $\mathcal{R}$.

In accordance with [165], we shall see that hyper relations can be derived from super, and thus also from ordinary relations and relators in several natural ways.

12. A Few Basic Facts on Relators

A family $\mathcal{R}$ of relations on one set X to another Y is called a *relator on X to Y*, and the ordered pair $(X, Y)(\mathcal{R}) = ((X, Y), \mathcal{R})$ is called a *relator space*. For the origins of this notion, see Refs. [140, 151] and the references in Ref. [140].

If in particular $\mathcal{R}$ is a relator on X to itself, then $\mathcal{R}$ is simply called a *relator on X*. Thus, by identifying singletons with their elements, we may naturally write $X(\mathcal{R})$ instead of $(X, X)(\mathcal{R})$. Namely, $(X, X) = \{\{X\}, \{X, X\}\} = \{\{X\}\}$.

Relator spaces of this simpler type are already substantial generalizations of the various *ordered sets* [32, 163] and *uniform spaces* [42, 71]. However, they are insufficient for several purposes. (See, Refs. [47, 151, 160, 162, 170, 172].)

A relator $\mathcal{R}$ on X to Y, or the relator space $(X, Y)(\mathcal{R})$, is called *simple* if $\mathcal{R} = \{R\}$ for some relation R on X to Y. Simple relator spaces $(X, Y)(R)$ and $X(R)$ were called *formal contexts* and *gosets* in Refs. [47, 163], respectively.

Moreover, a relator $\mathcal{R}$ on X, or the relator space $X(\mathcal{R})$, may, for instance, be naturally called *reflexive* if each member of $\mathcal{R}$ is reflexive on X. Thus, we may also naturally speak of *preorder, tolerance and equivalence relators*.

For instance, for a family $\mathcal{A}$ of subsets of X, the family $\mathcal{R}_{\mathcal{A}} = \{R_A : A \in \mathcal{A}\}$, where $R_A = A^2 \cup (A^c \times X)$, is an important preorder relator on X. Such relators were first used by Pervin [117] and later by Levine [88] and myself [157].

While, for a family $\mathcal{D}$ of *pseudo-metrics* on X, the family $\mathcal{R}_{\mathcal{D}} = \{B_r^d : r > 0, \ d \in \mathcal{D}\}$, where $B_r^d = \{(x, y) : \ d(x, y) < r\}$, is an important tolerance relator on X. Such relators were first considered by Weil [178].

Moreover, if $\mathfrak{S}$ is a family of *covers (partitions)* of X, then the family $\mathcal{R}_{\mathfrak{S}} = \{S_{\mathcal{A}} : \ \mathcal{A} \in \mathfrak{S}\}$, where $S_{\mathcal{A}} = \bigcup_{A \in \mathcal{A}} A^2$, is an important tolerance (equivalence) relator on X. Equivalence relators were first studied by Levine [87].

If $\star$ is a unary operation for relations on X to Y, then for any relator $\mathcal{R}$ on X to Y we may naturally define $\mathcal{R}^\star = \{R^\star : \ R \in \mathcal{R}\}$. However, this plausible notation may cause confusions if $\star$ is a set-theoretic operation.

For instance, for any relator $\mathcal{R}$ on X to Y, we may naturally define the *elementwise complement* $\mathcal{R}^c = \{R^c : \ R \in \mathcal{R}\}$, which may easily be confused with the *global complement* $\mathcal{R}^c = \mathcal{P}(X \times Y) \setminus \mathcal{R}$ of $\mathcal{R}$.

However, for instance, the practical notations $\mathcal{R}^{-1} = \{R^{-1} : \ R \in \mathcal{R}\}$ and $\mathcal{R}^{\infty} = \{R^{\infty} : \ R \in \mathcal{R}\}$, whenever $\mathcal{R}$ is only a relator on X, will certainly not cause confusions in the sequel.

In particular, for a relator $\mathcal{R}$ on X, we may also naturally define $\mathcal{R}^{\partial} = \{S \subseteq X^2 : \ S^{\infty} \in \mathcal{R}\}$. Namely, for any two relators $\mathcal{R}$ and $\mathcal{S}$ on X, we evidently have $\mathcal{R}^{\infty} \subseteq \mathcal{S} \iff \mathcal{R} \subseteq \mathcal{S}^{\partial}$. That is, ∞ and ∂ form a Galois connection.

The operations ∞ and ∂ were introduced by Mala [91, 93] and Pataki [113, 114], respectively. These two former PhD students of mine together with János Kurdics [78, 80], have made substantial developments in the theory of relators.

Moreover, if $*$ is a binary operation for relations, then for any two relators $\mathcal{R}$ and $\mathcal{S}$ we may naturally define $\mathcal{R} * \mathcal{S} = \{R * S : \ R \in \mathcal{R}, S \in \mathcal{S}\}$. However, this notation may again cause confusions if $*$ is a set-theoretic operation.

Therefore, in the former papers, we rather wrote $\mathcal{R} \wedge \mathcal{S} = \{R \cap S : R \in \mathcal{R}, \ S \in \mathcal{S}\}$. Moreover, for instance, we also wrote $\mathcal{R} \triangle \mathcal{R}^{-1} = \{R \cap R^{-1} : \ R \in \mathcal{R}\}$. Thus, $\mathcal{R} \triangle \mathcal{R}^{-1}$ is a symmetric relator such that $\mathcal{R} \triangle \mathcal{R}^{-1} \subseteq \mathcal{R} \wedge \mathcal{R}^{-1}$.

A function $\Box$ of the family of all relators on X to Y is called a *direct (indirect) unary operation for relators* if, for every relator $\mathcal{R}$ on X to Y, the value $\mathcal{R}^{\Box} = \Box(\mathcal{R})$ is a relator on X to Y (on Y to X).

More generally, a function $\mathfrak{F}$ of the family of all relators on X to Y is called a *structure for relators* if, for every relator $\mathcal{R}$ on X to Y, the value $\mathfrak{F}_{\mathcal{R}} = \mathfrak{F}(\mathcal{R})$ is in a power set depending only on X and Y.

Concerning structures and operations for relators, we can freely use some basic terminology on set-to-set functions. However, for closures and projections, we can now also use the terms *refinements and modifications*, respectively.

For instance, c and -1 are *involution operations* for relators. While, ∞ and ∂ are *projection operations* for relators. Moreover, the operation $\Box = c$, ∞ or ∂ is *inversion compatible* in the sense that $(\mathcal{R}^{\Box})^{-1} = (\mathcal{R}^{-1})^{\Box}$.

While, if, for instance, $\operatorname{int}_{\mathcal{R}}(B) = \{x \in X : \exists\, R \in \mathcal{R} : R(x) \subseteq B\}$ for every relator $\mathcal{R}$ on X to Y and $B \subseteq Y$, then the function $\mathfrak{F}$, defined by $\mathfrak{F}(\mathcal{R}) = \operatorname{int}_{\mathcal{R}}$, is a union-preserving structure for relators.

The first basic problem in the theory of relators is that, for any increasing structure $\mathfrak{F}$, we have to find an operation $\square$ for relators such that, for any two relators $\mathcal{R}$ and $\mathcal{S}$ on X to Y, we could have $\mathfrak{F}_{\mathcal{S}} \subseteq \mathfrak{F}_{\mathcal{R}} \iff \mathcal{S} \subseteq \mathcal{R}^{\square}$.

By using *Pataki connections* [113, 167], several closure operations can be derived from union-preserving structures. However, more generally, one can find first the *Galois adjoint* $\mathfrak{G}$ of such a structure $\mathfrak{F}$ and then take $\square_{\mathfrak{F}} = \mathfrak{G} \circ \mathfrak{F}$ [155].

By finding the Galois adjoint of the structure $\mathfrak{F}$, the second basic problem for relators, that which structures can be derived from relators, can also be solved. However, for this, some direct methods can also be well used [145, 157].

Now, for an operation $\square$ for relators, a relator $\mathcal{R}$ on X to Y may be naturally called $\square$-*fine* if $\mathcal{R}^{\square} = \mathcal{R}$. And, for some structure $\mathfrak{F}$ for relators, two relators $\mathcal{R}$ and $\mathcal{S}$ on X to Y may be naturally called $\mathfrak{F}$-*equivalent* if $\mathfrak{F}_{\mathcal{R}} = \mathfrak{F}_{\mathcal{S}}$.

Moreover, for a structure $\mathfrak{F}$ for relators, a relator $\mathcal{R}$ on X to Y may, for instance, be naturally called $\mathfrak{F}$-*simple* if $\mathfrak{F}_{\mathcal{R}} = \mathfrak{F}_{R}$ for some relation R on X to Y. Thus, singleton relators have to be actually called *properly simple*.

13. A Few Basic Facts on Super Relators

Analogously to an *ordinary relator* $\mathcal{R}$ on X to Y, we may also naturally consider a *super relator* $\mathcal{U}$ on X to Y. Actually, instead of the ordinary relators, we have to investigate first the super ones.

Namely, by using Definition 2, for every ordinary relator $\mathcal{R}$ may naturally define a *union-preserving super relator* $\mathcal{R}^{\triangleright} = \{R^{\triangleright} : R \in \mathcal{R}\}$. Moreover, for any structure $\mathfrak{F}$ for for super relators, we may naturally define $\mathfrak{F}_{\mathcal{R}} = \mathfrak{F}_{\mathcal{R}^{\triangleright}}$.

Thus, if $\mathcal{U}$ is an arbitrary super relator on X to Y and $\operatorname{Cl}_{\mathcal{U}}(B) = \{A \subseteq X : \quad \forall\, U \in \mathcal{U} : \quad U(A) \cap B \neq \emptyset\}$ for all $B \subseteq Y$, then it may occur that $\operatorname{Cl}_{\mathcal{U}} \neq \operatorname{Cl}_{\mathcal{R}}$ for every ordinary relator $\mathcal{R}$ on X to Y.

Therefore, the super relators are more powerful tools than the ordinary ones. However, the latter ones are more flexible tools than the former ones. Namely, for every ordinary relator $\mathcal{R}$ on X to Y, we have $\mathrm{Cl}_{\mathcal{R}^{-1}} = \mathrm{Cl}_{\mathcal{R}}^{-1}$.

Fortunately, if $\mathcal{U}$ is a super relator on X to Y, then under the notation $\mathrm{cl}_{\mathcal{U}}(B) = \{x \in X : \ \{x\} \in \mathrm{Cl}_{\mathcal{U}}(B)\}$ with $B \subseteq Y$, we have $\mathrm{cl}_{\mathcal{U}} = \mathrm{cl}_{\mathcal{U}^{\triangleleft}}$, where $\mathcal{U}^{\triangleleft} = \{U^{\triangleleft} : \ U \in \mathcal{U}\}$ is already an ordinary relator on X to Y.

Therefore, most of the theory of relator spaces cannot be generalized by using super relators instead of the ordinary ones. Moreover, there are structures derived from ordinary relators which cannot be easily generalized to the super ones.

For instance, if $\mathcal{R}$ is an ordinary relator on X to Y, then for any $B \subseteq Y$, we may naturally define $\mathrm{Lb}_{\mathcal{R}}(B) = \{A \subseteq X : \ \exists \ R \in \mathcal{R} : A \times B \subseteq R\}$. However, it can be shown that $\mathrm{Lb}_{\mathcal{R}} = \mathrm{Cl}^{c}_{\mathcal{R}^{c}} = \mathrm{Cl}^{\star}_{\mathcal{R}}$ [152].

Moreover, for any function ψ of a relator space $\Gamma(\mathcal{V})$ into Y, we may naturally define

$$\mathrm{Lim}_{\mathcal{R}}(\psi) = \{\varphi \in X^{\Gamma} : \quad \forall \ R \in \mathcal{R} : \ (\varphi, \psi)^{-1}[R] \in \mathcal{E}_{\mathcal{V}}\}$$

and

$$\mathrm{Adh}_{\mathcal{R}}(\psi) = \{\varphi \in X^{\Gamma} : \quad \forall \ R \in \mathcal{R} : \ (\varphi, \psi)^{-1}[R] \in \mathcal{D}_{\mathcal{V}}\},$$

where $(\varphi, \psi)(\gamma) = (\varphi(\gamma), \psi(\gamma))$ for all $\gamma \in \Gamma$, and

$$\mathcal{E}_{\mathcal{V}} = \{A \subseteq \Gamma : \ \mathrm{int}_{\mathcal{V}}(A) \neq \emptyset\} \quad \text{and} \quad \mathcal{D}_{\mathcal{V}} = \{A \subseteq \Gamma : \ \mathrm{cl}_{\mathcal{V}}(A) = \Gamma\}.$$

Thus, the relations $\mathrm{Lim}_{\mathcal{R}}$ and $\mathrm{Adh}_{\mathcal{R}}$, suggested by Efremović and Švarc [39], are usually much stronger tools in the ordinary relator space $(X, Y)(\mathcal{R})$ than the relations $\mathrm{Cl}_{\mathcal{R}}$ and $\mathrm{Int}_{\mathcal{R}}$, suggested by Efremović [38] and Smirnov [136].

Namely, for instance, it can be shown that, for any $A \subseteq X$ and $B \subseteq Y$, we have $A \in \mathrm{Cl}_{\mathcal{R}}(B)$ if and only if there exists a poset $\Gamma(\leq)$ and functions φ and ψ of Γ to A and B, respectively, such that $\varphi \in \mathrm{Lim}_{\mathcal{R}}(\psi)$ [123].

If $\mathcal{U}$ is a super relator on X to Y, then according to our former observation, we may naturally try to define $\mathrm{Lb}_{\mathcal{U}} = \mathrm{Cl}^{\star}_{\mathcal{U}} = \mathrm{Cl}^{c}_{\mathcal{U}^{c}}$ with $\mathcal{U}^{c} = \{U^{c} : \ U \in \mathcal{U}\}$, where $U^{c}(A) = U(A)^{c}$ for all $A \subseteq X$.

Moreover, according to Ref. [123, Theorem 11.3], for any function ψ of a relator space $\Gamma(\mathcal{V})$ into Y, we may naturally define

$\lim_{\mathcal{U}}(\psi) = \bigcap_{D \in \mathcal{D}_{\mathcal{V}}} \mathrm{cl}_{\mathcal{U}}(\psi[D])$. However, a similar definition for the relation $\mathrm{Lim}_{\mathcal{U}}$ cannot be proposed.

In our former definitions of the relations $\mathrm{Lim}_{\mathcal{R}}$ and $\mathrm{Adh}_{\mathcal{R}}$, for an apparently greater generality, the relator space $\Gamma(\mathcal{V})$ can be replaced by a *stack space* $\Gamma(\mathcal{D})$, where $\mathcal{D}$ is a *stack (ascending family)* of subsets of Γ [174].

Moreover, if $\mathcal{U}$ is a super relator on X to Y and $\mathcal{D}$ is a stack on Y, then analogously to Kuratowski [77] and Vaidyanathaswamy [177], for any $B \subseteq Y$, we may also define $\mathrm{Cl}_{(\mathcal{U},\mathcal{D})}(B) = \{A \subseteq X : \quad \forall\, U \in \mathcal{U} : \quad U(A) \cap B \in \mathcal{D}\}$.

However, by Ref. [157, Theorem 3.15], for every stack $\mathcal{D}$ on Y, there exists a preorder relator $\mathcal{S}$ on Y such that $\mathcal{D} = \mathcal{D}_{\mathcal{S}}$. Therefore, stacks can also be actually replaced by preorder relators without any loss of generality.

However, for an easier presentation of our present ideas, we shall rather consider here a *non-conventional, stacked three relator space* $(X, Y)(\mathcal{R}, \mathcal{U}, \mathcal{V})(\mathcal{D})$, where $\mathcal{R}$ is an ordinary, $\mathcal{U}$ is a super and $\mathcal{V}$ is a hyper relator on X to Y.

14. Stacks and Their Duals

Notation 4. *In this and the following section, we shall assume that $\mathcal{D}$ is a stack on Y and*

$$\mathcal{E} = \{E \subseteq Y : \quad E^c \notin \mathcal{D}\}.$$

Remark 15. This means that $\mathcal{D}$ is an *ascending* family of subsets of Y in the sense that $D \in \mathcal{D}$ and $D \subseteq E \subseteq Y$ imply $E \in \mathcal{D}$.

In contrast to $\mathcal{E}$, the elementwise complement $\mathcal{D}^c = \{D^c : \quad D \in \mathcal{D}\}$ of $\mathcal{D}$ is descending. Therefore, $\mathcal{E}$ has to be considered as the dual of $\mathcal{D}$.

For an easy illustration of stacks, we can at once state the following two examples.

Example 5. The family

$$\mathcal{D}_0 = \{D \subseteq Y : \quad D \neq \emptyset\}$$

is the largest stack on Y such that $\emptyset \notin \mathcal{D}_0$. Moreover, its dual family is $\mathcal{E}_0 = \{Y\}$.

Namely, for any $E \subseteq Y$, we have

$$E \in \mathcal{E}_0 \iff E^c \notin \mathcal{D}_0 \iff E^c = \emptyset \iff E = Y.$$

Remark 16. Thus, we have $\mathcal{D}_0 = \mathcal{P}(Y) \setminus \{\emptyset\}$. Moreover, $\mathcal{D}_0$ is the largest proper stack on Y in the sense that $\mathcal{D}_0 \neq \mathcal{P}(Y)$.

Example 6. For any $B \subseteq Y$, the family

$$\mathcal{D}_B = \{D \subseteq Y : \quad B \subseteq D\}$$

is the smallest stack on Y such that $B \in \mathcal{D}_B$. Moreover, its dual family is

$$\mathcal{E}_B = \{E \subseteq Y : \quad B \cap E \neq \emptyset\}.$$

Namely, for any $E \subseteq Y$, we have

$$E \in \mathcal{E}_B \iff E^c \notin \mathcal{D}_B \iff B \not\subseteq E^c \iff B \cap E \neq \emptyset.$$

Remark 17. Thus, we have $\mathcal{D}_Y = \mathcal{E}_0$ and $\mathcal{D}_B = \mathcal{P}^{-1}(B)$. Moreover, we can also note that $\mathcal{D}_\emptyset = \mathcal{P}(X)$ and $\mathcal{E}_\emptyset = \emptyset$.

Now, as an immediate generalization of Example 6, we can also state the following:

Example 7. For any $\mathcal{B} \subseteq \mathcal{P}(Y)$, the family

$$\mathcal{D}_\mathcal{B} = \{D \subseteq Y : \quad \exists\ B \in \mathcal{B} : \quad B \subseteq D\}$$

is the smallest stack on Y such that $\mathcal{B} \subseteq \mathcal{D}_\mathcal{B}$. Moreover, its dual family is

$$\mathcal{E}_\mathcal{B} = \{E \subseteq Y : \quad \forall\ B \in \mathcal{B} : \quad B \cap E \neq \emptyset\}.$$

Remark 18. In this case, we say that $\mathcal{D}_\mathcal{B}$ is generated by $\mathcal{B}$, or that $\mathcal{B}$ is a base of $\mathcal{D}_\mathcal{B}$. Moreover, sometimes we write $\mathcal{B}^* = \mathcal{D}_\mathcal{B}$.

Stacks are common generalizations of *filters* and *grills* [174]. They are also a certain generalization of sequences and nets.

Thus, it is not surprising that stacks allow of a very general theory of convergence and adherence and, thus, that of completeness and compactness properties.

However, the importance of stacks could become completely clear only from the theory of *relator spaces* worked out mainly by the present author and his former PhD students József Mala, János Kurdics and Gergely Pataki.

A relator $\mathcal{R}$ on X to Y is a family of relations on X to Y. Thus, a relator space $(X, Y)(\mathcal{R})$ is a common generalization of not only ordered sets and uniform spaces but also the proximity, closure, topological and convergence ones.

Example 8. If $\mathcal{R}$ is a relator on X to Y, then

$$\mathcal{D}_{\mathcal{R}} = \{D \subseteq Y : \quad \forall\ R \in \mathcal{R} : \quad X = R^{-1}[D]\}$$

is a stack on Y such that

$$\mathcal{D}_{\mathcal{R}} = \{D \subseteq Y : \quad \forall\ x \in X : \quad \forall\ R \in \mathcal{R} : \quad R(x) \cap D \neq \emptyset\}.$$

Therefore, $\mathcal{D}_{\mathcal{R}} = \{D \subseteq Y : \quad X = \mathrm{cl}_{\mathcal{R}}(D)\}$, and thus, the members of the family $\mathcal{D}_{\mathcal{R}}$ are called the *dense subsets* of the relator space $(X, Y)(\mathcal{R})$.

Moreover, the dual family of $\mathcal{D}_{\mathcal{R}}$ is

$$\mathcal{E}_{\mathcal{R}} = \{E \subseteq Y : \ \exists\ x \in X : \quad \exists\ R \in \mathcal{R} : \quad R(x) \subseteq E\}.$$

Therefore, $\mathcal{E}_{\mathcal{R}} = \{E \subseteq Y : \quad \mathrm{int}_{\mathcal{R}}(E) \neq \emptyset\}$, and thus, the members of the family $\mathcal{E}_{\mathcal{R}}$ are called the *fat subsets* of the relator space $(X, Y)(\mathcal{R})$.

The straightforward but not very short proofs of these apparently more complicated statements can be postponed until the reader gets more familiar with the techniques of the theory of relator spaces.

Remark 19. Conversely, by using the Pervin preorder relator

$$\mathcal{R}_{\mathcal{E}} = \{R_E : \quad E \in \mathcal{E}\}, \text{ where } R_E = E^2 \cup (E^c \times Y)$$

and the corresponding results of Ref. [157], we can prove that $\mathcal{D} = \mathcal{D}_{\mathcal{R}_{\mathcal{E}}}$.

Thus, every theorem on stacks, and thus also on filters and grills, can, in principle, be immediately derived from those on relators. However, in this chapter, it seems more convenient to start with stacks instead of relators.

The stack $\mathcal{E}_{\mathcal{R}}$ is frequently a more important tool in the relator space $X(\mathcal{R})$ than the induced *minimal structure* $\tau_{\mathcal{R}}$ and *generalized topology* $\mathcal{T}_{\mathcal{R}}$. In contrast to $\mathcal{D}_{\mathcal{R}}$, the dual family of $\mathcal{T}_{\mathcal{R}}$ must be defined by $\mathcal{F}_{\mathcal{R}} = \mathcal{T}_{\mathcal{R}}^c = \{V^c : \quad V \in \mathcal{T}_R\}$.

15. A Few Basic Theorems on Stacks

Concerning the stack $\mathcal{D}$ and its dual family $\mathcal{E}$, considered in Notation 4, we can easily prove the following two theorems:

Theorem 31. *The family $\mathcal{E}$ is a stack on Y such that*

$$\mathcal{D} = \{D \subseteq Y : \quad D^c \notin \mathcal{E}\}.$$

Proof. For instance, if $E \in \mathcal{E}$, then $E^c \notin \mathcal{D}$. Moreover, if $D \subseteq Y$ such that $E \subseteq D$, then $D^c \subseteq E^c$. Hence, since $E^c \notin \mathcal{D}$ and $\mathcal{D}$ is ascending, we can infer that $D^c \notin \mathcal{D}$. Thus, $D \in \mathcal{E}$ also holds. Therefore, $\mathcal{E}$ is also ascending. □

Theorem 32. *We have*

(1) $\mathcal{D} = \{D \subseteq Y : \quad \forall\ E \in \mathcal{E} : \quad E \cap D \neq \emptyset\}$;
(2) $\mathcal{E} = \{E \subseteq Y : \quad \forall\ D \in \mathcal{D} : \quad E \cap D \neq \emptyset\}$.

Proof. If $E \in \mathcal{E}$ and $D \in \mathcal{D}$, then we necessarily have $E \cap D \neq \emptyset$. Namely, if $E \cap D = \emptyset$, then $D \subseteq E^c$. Hence, since $\mathcal{D}$ is ascending, we can infer that $E^c \in \mathcal{D}$. Thus, by Notation 4, we have $E \notin \mathcal{E}$, which is a contradiction.

While, if, for instance, $E \subseteq Y$ such that $E \cap D \neq \emptyset$ for all $D \in \mathcal{D}$, then we necessarily have $E \in \mathcal{E}$. Namely, if $E \notin \mathcal{E}$, then $E^{cc} \notin \mathcal{E}$. Thus, by Theorem 31, we have $E^c \in \mathcal{D}$ such that $E \cap E^c = \emptyset$. And, this is a contradiction. □

By using the relationships of the families $\mathcal{D}$ and $\mathcal{E}$ and their ascendingness properties, we can also easily prove the following two theorems.

Theorem 33. *The following assertions are equivalent:*

(1) $\emptyset \notin \mathcal{D}$; (2) $\mathcal{E} \neq \emptyset$; (3) $Y \in \mathcal{E}$; (4) $\mathcal{D} \neq \mathcal{P}(Y)$.

Remark 20. Under the notation of Example 8, we can also easily prove that $\emptyset \notin \mathcal{D}_{\mathcal{R}}$ if and only if $\mathcal{R}$ is *non-degenerated* in the sense that both $X \neq \emptyset$ and $\mathcal{R} \neq \emptyset$.

Theorem 34. *The following assertions are equivalent:*

(1) $\emptyset \notin \mathcal{E}$; (2) $\mathcal{D} \neq \emptyset$; (3) $Y \in \mathcal{D}$; (4) $\mathcal{E} \neq \mathcal{P}(Y)$.

Remark 21. Under the notation of Example 8, we can also easily prove that $\emptyset \notin \mathcal{E}_{\mathcal{R}}$ if and only if $\mathcal{R}$ is *non-partial* in the sense $R(x) \neq \emptyset$ for all $x \in X$ and $R \in \mathcal{R}$.

Concerning stacks and their duals, it is also worth proving the following three theorems.

Theorem 35. *If $\mathcal{D}_1$ and $\mathcal{D}_2$ are stacks on Y such that $\mathcal{D}_1 \subseteq \mathcal{D}_2$, then for their duals $\mathcal{E}_1$ and $\mathcal{E}_2$ we have $\mathcal{E}_2 \subseteq \mathcal{E}_1$.*

Theorem 36. *If $(\mathcal{D}_i)_{i \in I}$ is a family of stacks on Y, then $\mathcal{D} = \bigcap_{i \in I} \mathcal{D}_i$ is also a stack on Y such that for the corresponding duals we have $\mathcal{E} = \bigcup_{i \in I} \mathcal{E}_i$.*

Theorem 37. *If $(\mathcal{D}_i)_{i \in I}$ is a family of stacks on Y, then $\mathcal{D} = \bigcup_{i \in I} \mathcal{D}_i$ is also a stack on Y such that for the corresponding duals we have $\mathcal{E} = \bigcap_{i \in I} \mathcal{E}_i$.*

Proof. To prove this equality, note that, for any $E \subseteq Y$, we have

$$E \in \mathcal{E} \iff E^c \notin \mathcal{D} \iff E^c \notin \bigcup_{i \in I} \mathcal{D}_i \iff$$

$$\forall\ i \in I: \ E^c \notin \mathcal{D}_i \iff \forall\ i \in I: \ E \in \mathcal{E}_i \iff E \in \bigcap_{i \in I} \mathcal{E}_i.$$

□

Remark 22. In our former paper [130], a relator $\mathcal{R}$ on X to Y has, for instance, been called

(1) *paratopologically minimal* if $\mathcal{E}_{\mathcal{R}} \subseteq \{Y\}$;
(2) *hyperconnected* if $\mathcal{E}_{\mathcal{R}} \subseteq \mathcal{D}_{\mathcal{R}}$; (3) *irresolvable* if $\mathcal{D}_{\mathcal{R}} \subseteq \mathcal{E}_{\mathcal{R}}$.

Example 9. Thus, if, for instance, $X = \mathbb{R}$ and

$$R_n = \{(x, y) \in X^2 : \quad d(x, y) < n^{-1}\}$$

for all $n \in \mathbb{N}$, then $\mathcal{R} = \{R_n\}_{n \in \mathbb{N}}$ is an important tolerance relator on X, with several useful additional properties, which does not have the above properties.

To see that $\mathcal{R}$ is resolvable, note that $\mathbb{Q} \in \mathcal{D}_{\mathcal{R}}$, but $\mathbb{Q} \notin \mathcal{E}_{\mathcal{R}}$, and thus, $\mathcal{D}_{\mathcal{R}} \not\subseteq \mathcal{E}_{\mathcal{R}}$.

Remark 23. Despite this example, the above curious properties have several interesting characterizations and applications [130].

16. Proximal Closures and Interiors Derived from Super Relators and Stacks

Notation 5. *In this and the following four sections, we shall assume that $\mathcal{U}$ is a super relator on X to Y, and $\mathcal{D}$ is a stack on Y.*

Remark 24. By Remark 19, instead of the assumption that $\mathcal{D}$ is a stack on Y, we could assume here, without loss of generality, that $\mathcal{D} = \mathcal{D}_{\mathcal{S}}$ for some relator $\mathcal{S}$ on X to Y, or even for some preorder relator $\mathcal{S}$ on Y.

The following definition is a generalization of our former similar definitions introduced and investigated in Refs. [125, 140, 168].

Definition 10. We define two hyper relations $\mathrm{Cl}_{(\mathcal{U},\mathcal{D})}$ and $\mathrm{Int}_{(\mathcal{U},\mathcal{D})}$ on Y to X such that

$$\mathrm{Cl}_{(\mathcal{U},\mathcal{D})}(B) = \{A \subseteq X : \ \forall\ U \in \mathcal{U} : \quad U(A) \cap B \in \mathcal{D}\}$$

and

$$\mathrm{Int}_{(\mathcal{U},\mathcal{D})}(B) = \mathrm{Cl}_{(\mathcal{U},\mathcal{D})}(B^c)^c = \mathcal{P}(X) \setminus \mathrm{Cl}_{(\mathcal{U},\mathcal{D})}(Y \setminus B)$$

for all $B \subseteq Y$.

Remark 25. The hyper relations $\mathrm{Cl}_{(\mathcal{U},\mathcal{D})}$ and $\mathrm{Int}_{(\mathcal{U},\mathcal{D})}$ will be called the *proximal closure and interior* generated by the super relator $\mathcal{U}$ and the stack $\mathcal{D}$, respectively.

The origins of these hyper relations go back to the ideal topological spaces of Kuratowski [77] and Vaidyanathaswamy [177] and the proximities and strong inclusions of Efremovič [38] and Smirnov [136].

Concerning the above hyper relations, we can easily prove the following:

Theorem 38. *For any $B \subseteq Y$, we have*

(1) $\operatorname{Cl}_{(\mathcal{U},\mathcal{D})}(B) = \operatorname{Int}_{(\mathcal{U},\mathcal{D})}(B^c)^c$;
(2) $\operatorname{Int}_{(\mathcal{U},\mathcal{D})}(B) = \{A \subseteq X : \quad \exists\ U \in \mathcal{U} : \quad U(A) \setminus B \notin \mathcal{D}\}$.

Proof. By Definition 10, for any $A \subseteq X$, we have

$$A \in \operatorname{Int}_{(\mathcal{U},\mathcal{D})}(B) \iff A \in \operatorname{Cl}_{(\mathcal{U},\mathcal{D})}(B^c)^c \iff A \notin \operatorname{Cl}_{(\mathcal{U},\mathcal{D})}(B^c)$$
$$\iff \exists\ U \in \mathcal{U} : \quad U(A) \cap B^c \notin \mathcal{D}$$
$$\iff \exists\ U \in \mathcal{U} : \quad U(A) \setminus B \notin \mathcal{D}.$$

Therefore, assertion (2) is true.

Assertion (1) is an even more immediate consequence of Definition 10. □

Remark 26. By using the dual $\mathcal{E}$ of $\mathcal{D}$, we can note that

$$U(A) \cap B^c \notin \mathcal{D} \iff (U(A)^c \cup B)^c \notin \mathcal{D} \iff U(A)^c \cup B \in \mathcal{E}.$$

Therefore, assertion (2) can be reformulated in the affirmative form that

$$\operatorname{Int}_{(\mathcal{U},\mathcal{D})}(B) = \{A \subseteq X : \quad \exists\ U \in \mathcal{U} : \quad U(A)^c \cup B \in \mathcal{E}\}.$$

From Definition 10 and Theorem 38, it is clear that the hyper relations $\operatorname{Cl}_{(\mathcal{U},\mathcal{D})}$ and $\operatorname{Int}_{(\mathcal{U},\mathcal{D})}$ are equivalent tools in the *stacked super relator space* $(X, Y)(\mathcal{U})(\mathcal{D})$.

Moreover, we can also easily prove that these hyper relations are increasing.

Theorem 39. *For any $B_1 \subseteq B_2 \subseteq Y$, we have*

(1) $\operatorname{Cl}_{(\mathcal{U},\mathcal{D})}(B_1) \subseteq \operatorname{Cl}_{(\mathcal{U},\mathcal{D}}(B_2)$; (2) $\operatorname{Int}_{(\mathcal{U},\mathcal{D})}(B_1) \subseteq \operatorname{Int}_{(\mathcal{U},\mathcal{D})}(B_2)$.

Proof. By Definition 10, and the fact that $\mathcal{D}$ is ascending, it is clear that

$$A \in \operatorname{Cl}_{(\mathcal{U},\mathcal{D})}(B_1) \implies \forall\ U \in U : \ U(A) \cap B_1 \in \mathcal{D}$$
$$\implies \forall\ U \in U : \ U(A) \cap B_2 \in \mathcal{D} \implies A \in \operatorname{Cl}_{(\mathcal{U},\mathcal{D})}(B_2)$$

and thus also

$$\begin{aligned}B_1 \subseteq B_2 \subseteq Y \implies B_2^c \subseteq B_1^c \subseteq Y \implies \mathrm{Cl}_{(\mathcal{U},\mathcal{D})}(B_2^c) \subseteq \mathrm{Cl}_{(\mathcal{U},\mathcal{D})}(B_1^c)\\ \implies \mathrm{Cl}_{(\mathcal{U},\mathcal{D})}(B_1^c)^c \subseteq \mathrm{Cl}_{(\mathcal{U},\mathcal{D})}(B_2^c)^c\\ \implies \mathrm{Int}_{(\mathcal{U},\mathcal{D})}(B_1) \subseteq \mathrm{Int}_{(\mathcal{U},\mathcal{D})}(B_2).\end{aligned}$$
□

Having in mind the usual closures and interiors in topological spaces, in addition to the above theorem, we must also prove the following:

Theorem 40. *The following assertions are equivalent:*

(1) $\mathcal{U} \neq \emptyset$ *and* $\emptyset \notin \mathcal{D}$;
(2) $\mathrm{Cl}_{(\mathcal{U},\mathcal{D})}(\emptyset) = \emptyset$;
(3) $\mathrm{Cl}_{(\mathcal{U},\mathcal{D})}(\emptyset) \neq \mathcal{P}(X)$;
(4) $\mathrm{Int}_{(\mathcal{U},\mathcal{D})}(Y) \neq \emptyset$;
(5) $\mathrm{Int}_{(\mathcal{U},\mathcal{D})}(Y) = \mathcal{P}(X)$.

Proof. By Definition 10, for any $A \subseteq X$, we have

$$\begin{aligned}A \notin \mathrm{Cl}_{(\mathcal{U},\mathcal{D})}(\emptyset) &\iff \exists\ U \in \mathcal{U}: \quad U(A) \cap \emptyset \notin \mathcal{D}\\ &\iff \exists\ U \in \mathcal{U}: \quad \emptyset \notin \mathcal{D}.\end{aligned}$$

Therefore, assertions (2) and (3) are equivalent to assertion (1).

Moreover, by Definition 10, for any $A \subseteq X$, we also have

$$A \in \mathrm{Int}_{(\mathcal{U},\mathcal{D})}(Y) \iff A \in \mathrm{Cl}_{(\mathcal{U},\mathcal{D})}(Y^c)^c \iff A \notin \mathrm{Cl}_{(\mathcal{U},\mathcal{D})}(\emptyset).$$

Therefore, assertions (4) and (5) are also equivalent to assertions (3) and (2), respectively. □

Now, as an immediate consequence of this theorem, we can also state the following:

Corollary 10. *The following assertions are equivalent:*

(1) $\mathcal{U} = \emptyset$ *or* $\emptyset \in \mathcal{D}$;
(2) $\mathrm{Cl}_{(\mathcal{U},\mathcal{D})}(\emptyset) \neq \emptyset$;
(3) $\mathrm{Cl}_{(\mathcal{U},\mathcal{D})}(\emptyset) = \mathcal{P}(X)$;
(4) $\mathrm{Int}_{(\mathcal{U},\mathcal{D})}(Y) = \emptyset$;
(5) $\mathrm{Int}_{(\mathcal{U},\mathcal{D})}(Y) \neq \mathcal{P}(X)$.

Moreover, in addition to Theorem 40 and its corollary, we can also prove the following:

Theorem 41. *We have*

(1) $\mathrm{Cl}_{(\mathcal{U},\mathcal{D})}(Y) = \{A \subseteq X : \ \forall\, U \in \mathcal{U} : \quad U(A) \in \mathcal{D}\}$;
(2) $\mathrm{Int}_{(\mathcal{U},\mathcal{D})}(\emptyset) = \{A \subseteq X : \ \exists\, U \in \mathcal{U} : \quad U(A) \notin \mathcal{D}\}$.

Proof. By Definition 10, for any $A \subseteq X$, we have

$$A \in \mathrm{Cl}_{(\mathcal{U},\mathcal{D})}(Y) \iff \forall\, U \in \mathcal{U} : \quad U(A) \cap Y \in \mathcal{D}$$
$$\iff \forall\, U \in \mathcal{U} : \quad U(A) \in \mathcal{D}.$$

Therefore, assertion (1) is true.

Assertion (2) can be derived from Theorem 38 quite similarly. □

17. Some Further Theorems on the Relations $\mathrm{Cl}_{(\mathcal{U},\mathcal{D})}$ and $\mathrm{Int}_{(\mathcal{U},\mathcal{D})}$

By using Theorem 39, we can also prove the following three theorems.

Theorem 42. *If $B_1, B_2 \subseteq Y$ such that $B_1 \cap B_2 = \emptyset$, then*

$$\mathrm{Cl}_{(\mathcal{U},\mathcal{D})}(B_1) \cap \mathrm{Int}_{(\mathcal{U},\mathcal{D})}(B_2) = \emptyset.$$

Proof. By using the Theorem 39 and Definition 10, we can see that

$$B_1 \cap B_2 = \emptyset \implies B_1 \subseteq B_2^c \implies \mathrm{Cl}_{(\mathcal{U},\mathcal{D})}(B_1) \subseteq \mathrm{Cl}_{(\mathcal{U},\mathcal{D})}(B_2^c)$$
$$\implies \mathrm{Cl}_{(\mathcal{U},\mathcal{D})}(B_1) \subseteq \mathrm{Int}_{(\mathcal{U},\mathcal{D})}(B_2)^c$$
$$\implies \implies \mathrm{Cl}_{(\mathcal{U},\mathcal{D})}(B_1) \cap \mathrm{Int}_{(\mathcal{U},\mathcal{D})}(B_2) = \emptyset.$$
□

Theorem 43. *If $B = \bigcap_{i \in I} B_i$ for some family $(B_i)_{i \in I}$ of subsets of Y, then*

(1) $\mathrm{Cl}_{(\mathcal{U},\mathcal{D})}(B) \subseteq \bigcap_{i \in I} \mathrm{Cl}_{(\mathcal{U},\mathcal{D})}(B_i)$;
(2) $\mathrm{Int}_{(\mathcal{U},\mathcal{D})}(B) \subseteq \bigcap_{i \in I} \mathrm{Int}_{(\mathcal{U},\mathcal{D})}(B_i)$.

Proof. By the assumption, we have $B \subseteq B_i$, and thus, by Theorem 39, also

$$\mathrm{Cl}_{(\mathcal{U},\mathcal{D})}(B) \subseteq \mathrm{Cl}_{(\mathcal{U},\mathcal{D})}(B_i)$$

for all $i \in I$. Hence, assertion (1) immediately follows. □

Theorem 44. *If* $B = \bigcup_{i \in I} B_i$ *for some family* $(B_i)_{i \in I}$ *of subsets of* Y*, then*

(1) $\bigcup_{i \in I} \mathrm{Cl}_{(\mathcal{U},\mathcal{D})}(B_i) \subseteq \mathrm{Cl}_{(\mathcal{U},\mathcal{D})}(B)$;
(2) $\bigcup_{i \in I} \mathrm{Int}_{(\mathcal{U},\mathcal{D})}(B_i) \subseteq \mathrm{Int}_{(\mathcal{U},\mathcal{D})}(B)$.

Proof. By the assumption, we have $B_i \subseteq B$, and thus, by Theorem 39, also

$$\mathrm{Cl}_{(\mathcal{U},\mathcal{D})}(B_i) \subseteq \mathrm{Cl}_{(\mathcal{U},\mathcal{D})}(B)$$

for all $i \in I$. Hence, assertion (1) immediately follows. □

Remark 27. By using Definition 10, we can note that

$$\begin{aligned} A \in \mathrm{Cl}_{(\mathcal{U},\mathcal{D})}(B) \iff \forall\ U \in \mathcal{U}: \quad & U(A) \cap B \in \mathcal{D} \\ \iff \forall\ U \in \mathcal{U}: \quad & U(A) \cap \bigcup_{i \in I} B_i \in \mathcal{D} \\ \iff \forall\ U \in \mathcal{U}: \quad & \bigcup_{i \in I} (U(A) \cap B_i) \in \mathcal{D}. \end{aligned}$$

Therefore, in assertion (1), the equality cannot be stated even if $\mathrm{card}(\mathcal{U}) = 1$.

Moreover, as an immediate consequence of Definition 10 and Theorem 38, we can also state the following:

Theorem 45. *We have*

(1) $\mathrm{Cl}_{(\mathcal{U},\mathcal{D})} = \bigcap_{U \in \mathcal{U}} \mathrm{Cl}_{(U,\mathcal{D})}$; (2) $\mathrm{Int}_{(\mathcal{U},\mathcal{D})} = \bigcup_{U \in \mathcal{U}} \mathrm{Int}_{(U,\mathcal{D})}$.

Hence, by using some counterparts of Theorem 3, we can infer the following:

Corollary 11. *The mapping*

(1) $\mathcal{U} \mapsto \mathrm{Int}_{(\mathcal{U},\mathcal{D})}$ *is union-preserving;*
(2) $\mathcal{U} \mapsto \mathrm{Cl}_{(\mathcal{U},\mathcal{D})}$ *is intersection-preserving.*

In addition to the above theorems, we can also easily prove the following:

Theorem 46. *If* $\mathcal{D} = \bigcap_{i \in I} \mathcal{D}_i$ *for some family* $(\mathcal{D}_i)_{i \in I}$ *of stacks on* Y, *then*

(1) $\mathrm{Cl}_{(\mathcal{U},\mathcal{D})} = \bigcap_{i \in I} \mathrm{Cl}_{(\mathcal{U},\mathcal{D}_i)}$; (2) $\mathrm{Int}_{(\mathcal{U},\mathcal{D})} = \bigcup_{i \in I} \mathrm{Int}_{(\mathcal{U},\mathcal{D}_i)}$.

Proof. By Definition 10, for any $A \subseteq X$ and $B \subseteq Y$, we have

$$A \in \mathrm{Cl}_{(\mathcal{U},\mathcal{D})}(B) \iff \forall\ U \in \mathcal{U}: \quad U(A) \cap B \in \mathcal{D} \iff$$
$$\forall\ U \in \mathcal{U}: \quad U(A) \cap B \in \bigcap_{i \in I} \mathcal{D}_i$$
$$\iff \forall\ U \in \mathcal{U}: \quad \forall\ i \in I: \ U(A) \cap B \in \mathcal{D}_i$$
$$\iff \forall\ i \in I: \quad \forall\ U \in \mathcal{U}: \ U(A) \cap B \in \mathcal{D}_i$$
$$\iff \forall\ i \in I: \ A \in \mathrm{Cl}_{(\mathcal{U},\mathcal{D}_i)}(B)$$
$$\iff \bigcap_{i \in I} \mathrm{Cl}_{(\mathcal{U},\mathcal{D}_i)}(B) \iff \Big(\bigcap_{i \in I} \mathrm{Cl}_{(\mathcal{U},\mathcal{D}_i)} \Big)(B).$$

Therefore, assertion (1) is true.

Assertion (2) can be proved quite similarly by using Theorems 38 and 36 instead of Definition 10. Moreover, it can also be derived from assertion (1) by using Definition 10 and De Morgan's law. □

Remark 28. If $\mathcal{D} = \bigcup_{i \in I} \mathcal{D}_i$ for some family $(\mathcal{D}_i)_{i \in I}$ of stacks on Y, then using Definition 10, we can note that

$$A \in \mathrm{Cl}_{(\mathcal{U},\mathcal{D})}(B) \iff \forall\ U \in \mathcal{U}: \quad U(A) \cap B \in \mathcal{D}$$
$$\iff \forall\ U \in \mathcal{U}: \quad U(A) \cap B \in \bigcup_{i \in I} \mathcal{D}_i$$
$$\iff \forall\ U \in \mathcal{U}: \quad \exists\ i \in I: \ U(A) \cap B \in \mathcal{D}_i.$$

Therefore, we can only state that $\bigcup_{i\in I} \mathrm{Cl}_{(\mathcal{U},\mathcal{D}_i)} \subseteq \mathrm{Cl}_{(\mathcal{U},\mathcal{D})}$.

However, if $\mathrm{card}(\mathcal{U}) = 1$, then the corresponding equality is also true.

Finally, we note that, by using the hyper relations $\mathrm{Cl}_{(\mathcal{U},\mathcal{D})}$ and $\mathrm{Int}_{(\mathcal{U},\mathcal{D})}$, we may also naturally introduce some further hyper relations.

Definition 11. For any $B \subseteq Y$, we define

$$\mathrm{Bnd}_{(\mathcal{U},\mathcal{D})}(B) = \mathrm{Cl}_{(\mathcal{U},\mathcal{D})}(B) \setminus \mathrm{Int}_{(\mathcal{U},\mathcal{D})}(B).$$

Remark 29. Moreover, if in particular $X = Y$, then for any $A \subseteq X$ we also define

$$\mathrm{Res}_{(\mathcal{U},\mathcal{D})}(A) = \mathrm{Cl}_{(\mathcal{U},\mathcal{D})}(A) \setminus \mathcal{P}(A) \quad \text{and}$$
$$\mathrm{Bor}_{(\mathcal{U},\mathcal{D})}(A) = \mathcal{P}(A) \setminus \mathrm{Int}_{(\mathcal{U},\mathcal{D})}(A).$$

Somewhat differently, the *border, boundary and residue of a set* in neighborhood and closure spaces were already introduced by Hausdorff and Kuratowski [76, pp. 4–5]. (See also Elez and Papaz [40] for a recent treatment.)

18. Some Theorems on the Inverses of the Relations $\mathrm{Cl}_{(\mathcal{U},\mathcal{D})}$ and $\mathrm{Int}_{(\mathcal{U},\mathcal{D})}$

Now, by calling the super relator $\mathcal{U}$ to be *increasing* if each member U of $\mathcal{U}$ is an increasing super relation, we can also prove the following:

Theorem 47. *If $\mathcal{U}$ is increasing, then for any $A_1 \subseteq A_2 \subseteq X$, we have*

(1) $\mathrm{Cl}^{-1}_{(\mathcal{U},\mathcal{D})}(A_1) \subseteq \mathrm{Cl}^{-1}_{(\mathcal{U},\mathcal{D})}(A_2)$; (2) $\mathrm{Int}^{-1}_{(\mathcal{U},\mathcal{D})}(A_2) \subseteq \mathrm{Int}^{-1}_{(\mathcal{U},\mathcal{D})}(A_1)$.

Proof. By the assumption and Definition 10, it is clear that

$$B \in \mathrm{Cl}^{-1}_{(\mathcal{U},\mathcal{D})}(A_1) \implies A_1 \in \mathrm{Cl}_{(\mathcal{U},\mathcal{D})}(B)$$
$$\implies \forall\, U \in \mathcal{U}:\ U(A_1) \cap B \in \mathcal{D} \implies \forall\, U \in \mathcal{U}:\ U(A_2) \cap B \in \mathcal{D}$$
$$\implies A_2 \in \mathrm{Cl}_{(\mathcal{U},\mathcal{D})}(B) \implies B \in \mathrm{Cl}^{-1}_{(\mathcal{U},\mathcal{D})}(A_2).$$

Therefore, assertion (1) is true.

Assertion (2) can be proved quite similarly, by using Theorem 38 instead of Definition 10. Moreover, it can now also be derived from assertion (1) by using Definition 10. □

In addition to this theorem, we can also prove the following:

Theorem 48. *If $U(\emptyset) = \emptyset$ for all $U \in \mathcal{U}$, then the following assertions are equivalent:*

(1) $\mathcal{U} \neq \emptyset$ *and* $\emptyset \notin \mathcal{D}$;

(2) $\mathrm{Cl}^{-1}_{(\mathcal{U},\mathcal{D})}(\emptyset) = \emptyset$;

(3) $\mathrm{Cl}^{-1}_{(\mathcal{U},\mathcal{D})}(\emptyset) \neq \mathcal{P}(Y)$;

(4) $\mathrm{Int}^{-1}_{(\mathcal{U},\mathcal{D})}(\emptyset) \neq \emptyset$:

(5) $\mathrm{Int}^{-1}_{(\mathcal{U},\mathcal{D})}(\emptyset) = \mathcal{P}(Y)$.

Proof. By Definition 10, for any $B \subseteq Y$ we have

$$B \notin \mathrm{Cl}^{-1}_{(\mathcal{U},\mathcal{D})}(\emptyset) \iff \emptyset \notin \mathrm{Cl}_{(\mathcal{U},\mathcal{D})}(B)$$
$$\iff \exists\, U \in \mathcal{U}: \quad U(\emptyset) \cap B \notin \mathcal{D} \iff \exists\, U \in \mathcal{U}: \quad \emptyset \notin \mathcal{D}.$$

Hence, we can see that assertions (2) and (3) are equivalent to assertion (1).

Moreover, by using Definition 10, for any $B \subseteq Y$, we also have

$$B \in \mathrm{Int}^{-1}_{(\mathcal{U},\mathcal{D})}(\emptyset) \iff \emptyset \in \mathrm{Int}_{(\mathcal{U},\mathcal{D})}(B) \iff \emptyset \in \mathrm{Cl}_{(\mathcal{U},\mathcal{D})}(B^c)^c$$
$$\iff \emptyset \notin \mathrm{Cl}_{(\mathcal{U},\mathcal{D})}(B^c) \iff B^c \notin \mathrm{Cl}^{-1}_{(\mathcal{U},\mathcal{D})}(\emptyset).$$

Therefore, assertions (4) and (5) are also equivalent to assertions (3) and (2), respectively. □

Remark 30. Note that the implications (2) $\Longrightarrow$ $\emptyset$ $\Longrightarrow$ $\emptyset \notin \mathrm{Cl}^{-1}_{(\mathcal{U},\mathcal{D})}(\emptyset)$ $\Longrightarrow$ (1), and the equivalences of assertions (2)–(5) do not require do not need any extra condition on $\mathcal{U}$.

Now, as an immediate consequence of Theorem 48, we can also state the following:

Corollary 12. *If $U(\emptyset) = \emptyset$ for all $U \in \mathcal{U}$, then the following assertions are equivalent:*

(1) $\mathcal{U} = \emptyset$ *or* $\emptyset \in \mathcal{D}$;

(2) $\mathrm{Cl}^{-1}_{(\mathcal{U},\mathcal{D})}(\emptyset) \neq \emptyset$;

(3) $\mathrm{Cl}^{-1}_{(\mathcal{U},\mathcal{D})}(\emptyset) = \mathcal{P}(Y)$;

(4) $\mathrm{Int}^{-1}_{(\mathcal{U},\mathcal{D})}(\emptyset) = \emptyset$;

(5) $\mathrm{Int}^{-1}_{(\mathcal{U},\mathcal{D})}(\emptyset) \neq \mathcal{P}(Y)\emptyset$.

Moreover, in addition to Theorem 48 and its corollary, we can also prove the following:

Theorem 49. *If* $U(X) = Y$ *for all* $U \in \mathcal{U}$, *then*

(1) $\operatorname{Int}^{-1}_{(\mathcal{U},\mathcal{D})}(X) = \mathcal{E}$ *if* $\mathcal{U} \neq \emptyset$ *and* $\operatorname{Int}^{-1}_{(\mathcal{U},\mathcal{D})}(X) = \emptyset$ *if* $\mathcal{U} = \emptyset$;

(2) $\operatorname{Cl}^{-1}_{(\mathcal{U},\mathcal{D})}(X) = \mathcal{D}$ *if* $\mathcal{U} \neq \emptyset$ *and* $\operatorname{Cl}^{-1}_{(\mathcal{U},\mathcal{D})}(X) = \mathcal{P}(Y)$ *if* $\mathcal{U} = \emptyset$.

Proof. By the assumption and Definition 10, for any $B \subseteq Y$ we have

$$\begin{aligned} B \in \operatorname{Cl}^{-1}_{(\mathcal{U},\mathcal{D})}(X) \quad &\Longleftrightarrow \quad X \in \operatorname{Cl}_{(\mathcal{U},\mathcal{D})}(B) \\ &\Longleftrightarrow \forall\; U \in \mathcal{U}: \quad U(X) \cap B \in \mathcal{D} \\ &\Longleftrightarrow \forall\; U \in \mathcal{U}: \quad B \in \mathcal{D} \Longleftrightarrow \mathcal{U} = \emptyset \text{or } B \in \mathcal{D}. \end{aligned}$$

Therefore, assertion (2) is true.

Moreover, by Definition 10 and the above observation, we can see that

$$\begin{aligned} B \in \operatorname{Int}^{-1}_{(\mathcal{U},\mathcal{D})}(X) \quad &\Longleftrightarrow \quad X \in \operatorname{Int}_{(\mathcal{U},\mathcal{D})}(B) \Longleftrightarrow X \in \operatorname{Cl}_{(\mathcal{U},\mathcal{D})}(B^c)^c \\ &\Longleftrightarrow X \notin \operatorname{Cl}_{(\mathcal{U},\mathcal{D})}(B^c) \Longleftrightarrow B^c \notin \operatorname{Cl}^{-1}_{(\mathcal{U},\mathcal{D})}(X) \\ &\Longleftrightarrow \mathcal{U} \neq \emptyset \quad \text{and} \quad B^c \notin \mathcal{D} \Longleftrightarrow \mathcal{U} \neq \emptyset \quad \text{and} \quad B \in \mathcal{E}. \end{aligned}$$

Therefore, assertion (1) is also true. □

Now, as some immediate consequences of Theorem 47, we can also state the following two theorems.

Theorem 50. *If* $\mathcal{U}$ *is increasing and* $A = \bigcap_{i \in I} A_i$ *for some family* $(A_i)_{i \in I}$ *of subsets of* X, *then*

(1) $\operatorname{Cl}^{-1}_{(\mathcal{U},\mathcal{D})}(A) \subseteq \bigcap_{i \in I} \operatorname{Cl}^{-1}_{(\mathcal{U},\mathcal{D})}(A_i)$;

(2) $\bigcup_{i \in I} \operatorname{Int}^{-1}_{(\mathcal{U},\mathcal{D})}(A_i) \subseteq \operatorname{Int}^{-1}_{(\mathcal{U},\mathcal{D})}(A)$.

Theorem 51. *If* $\mathcal{U}$ *is increasing and* $A = \bigcup_{i \in I} A_i$ *for some family* $(A_i)_{i \in I}$ *of subsets of* X, *then*

(1) $\bigcup_{i \in I} \operatorname{Cl}^{-1}_{(\mathcal{U},\mathcal{D})}(A_i) \subseteq \operatorname{Cl}^{-1}_{(\mathcal{U},\mathcal{D})}(A)$;

(2) $\operatorname{Int}^{-1}_{(\mathcal{U},\mathcal{D})}(A) \subseteq \bigcap_{i \in I} \operatorname{Int}^{-1}_{(\mathcal{U},\mathcal{D})}(A_i)$.

Moreover, as some immediate consequences of Theorems 45 and 46, we can also state the following two theorems.

Theorem 52. *We have*

(1) $\mathrm{Cl}^{-1}_{(\mathcal{U},\mathcal{D})} = \bigcap_{U \in \mathcal{U}} \mathrm{Cl}^{-1}_{(U,\mathcal{D})}$; (2) $\mathrm{Int}^{-1}_{(\mathcal{U},\mathcal{D})} = \bigcup_{U \in \mathcal{U}} \mathrm{Int}^{-1}_{(U,\mathcal{D})}$.

Corollary 13. *The mapping*

(1) $\mathcal{U} \mapsto \mathrm{Int}^{-1}_{(\mathcal{U},\mathcal{D})}$ *is union-preserving;*
(2) $\mathcal{U} \mapsto \mathrm{Cl}^{-1}_{(\mathcal{U},\mathcal{D})}$ *is intersection-preserving.*

Theorem 53. *If* $\mathcal{D} = \bigcap_{i \in I} \mathcal{D}_i$ *for some family* $(\mathcal{D}_i)_{i \in I}$ *of stacks on* Y, *then*

(1) $\mathrm{Cl}^{-1}_{(\mathcal{U},\mathcal{D})} = \bigcap_{i \in I} \mathrm{Cl}^{-1}_{(\mathcal{U},\mathcal{D}_i)}$; (2) $\mathrm{Int}^{-1}_{(\mathcal{U},\mathcal{D})} = \bigcup_{i \in I} \mathrm{Int}^{-1}_{(\mathcal{U},\mathcal{D}_i)}$.

Remark 31. To feel the importance of the relation $\mathrm{Int}^{-1}_{(\mathcal{U},\mathcal{D})}$, note that, for any $A \subseteq X$,

$$\mathfrak{U}_{(\mathcal{U},\mathcal{D})}(A) = \mathrm{Int}^{-1}_{(\mathcal{U},\mathcal{D})}(A) = \{B \subseteq Y : \ A \in \mathrm{Int}_{(\mathcal{U},\mathcal{D})}(B)\}$$

is just the family of all proximal neighborhoods of the set A in the *stacked super relator space* $(X,Y)(\mathcal{U})(\mathcal{D})$.

19. Topological Closures and Interiors Derived from Super Relators and Stacks

By using Definition 10, we may naturally introduce the following:

Definition 12. We define two super relations $\mathrm{cl}_{(\mathcal{U},\mathcal{D})}$ and $\mathrm{int}_{(\mathcal{U},\mathcal{D})}$ on Y to X such that

$$\mathrm{cl}_{(\mathcal{U},\mathcal{D})}(B) = \{x \in X : \ \{x\} \in \mathrm{Cl}_{(\mathcal{U},\mathcal{D})}(B)\}$$

and

$$\mathrm{int}_{(\mathcal{U},\mathcal{D})}(B) = \{x \in X : \ \{x\} \in \mathrm{Int}_{(\mathcal{U},\mathcal{D})}(B)\}$$

for all $B \subseteq Y$.

Remark 32. The super relations $\mathrm{cl}_{(\mathcal{U},\mathcal{D})}$ and $\mathrm{int}_{(\mathcal{U},\mathcal{D})}$ will be called the *topological closure and interior* generated by the super relator $\mathcal{U}$ and the stack $\mathcal{D}$, respectively.

Thus, by specializing the results of Sections 16 and 17, we can easily establish the following theorems.

However, for some of the forthcoming theorems, it may be more convenient to provide some more convincing direct proofs by using the following:

Theorem 54. *For any $B \subseteq Y$, we have*

(1) $\mathrm{cl}_{(\mathcal{U},\mathcal{D})}(B) = \{x \in X : \ \forall\ U \in \mathcal{U} : \quad U^{\lhd}(x) \cap B \in \mathcal{D}\}$;

(2) $\mathrm{int}_{(\mathcal{U},\mathcal{D})}(B) = \{x \in X : \ \exists\ U \in \mathcal{U} : \quad U^{\lhd}(x) \setminus B \notin \mathcal{D}\}$.

Proof. To prove this, besides using Definition 10 and Theorem 38, recall that in Definition 2, for any super relation U on X to Y, we have introduced an ordinary relation $U^{\lhd}$ on X to Y such that $U^{\lhd}(x) = U(\{x\})$ for all $x \in X$. □

Theorem 55. *For any $B \subseteq Y$, we have*

(1) $\mathrm{cl}_{(\mathcal{U},\mathcal{D})}(B) = \mathrm{int}_{(\mathcal{U},\mathcal{D})}(B^c)^c$; (2) $\mathrm{int}_{(\mathcal{U},\mathcal{D})}(B) = \mathrm{cl}_{(\mathcal{U},\mathcal{D})}(B^c)^c$.

Remark 33. Thus, the super relations $\mathrm{cl}_{(\mathcal{U},\mathcal{D})}$ and $\mathrm{int}_{(\mathcal{U},\mathcal{D})}$ are also equivalent tools in the stacked super relator space $(X,Y)(\mathcal{U})(\mathcal{D})$.

However, these super relations are usually much weaker tools than the hyper relations $\mathrm{Cl}_{(\mathcal{U},\mathcal{D})}$ and $\mathrm{Int}_{(\mathcal{U},\mathcal{D})}$. Namely, in general, we can only prove the following:

Theorem 56. *If $\mathcal{U}$ is quasi-increasing, then for any $A \subseteq X$ and $B \subseteq Y$*

(1) *$A \in \mathrm{Int}_{(\mathcal{U},\mathcal{D})}(B)$ implies $A \subseteq \mathrm{int}_{(\mathcal{U},\mathcal{D})}(B)$;*

(2) *$A \cap \mathrm{cl}_{(\mathcal{U},\mathcal{D})}(B) \neq \emptyset$ implies $A \in \mathrm{Cl}_{(\mathcal{U},\mathcal{D})}(B)$.*

Proof. To prove assertion (1), note that, by Theorems 38 and 54, and the assumption that $U^{\rhd}(x) = U(\{x\}) \subseteq U(A)$ for all $x \in A$,

$$\begin{aligned} A \in \mathrm{Int}_{(\mathcal{U},\mathcal{D})}(B) &\implies \exists\ U \in \mathcal{U} : \quad U(A) \setminus B \notin \mathcal{D} \\ &\implies \exists\ U \in \mathcal{U} : \quad \forall\ x \in A : \ U^{\rhd}(x) \setminus B \notin \mathcal{D} \\ &\implies \forall\ x \in A : \quad \exists\ U \in \mathcal{U} : \ U^{\rhd}(x) \setminus B \notin \mathcal{D} \\ &\implies \forall\ x \in A : \ x \in \mathrm{int}_{(\mathcal{U},\mathcal{D})}(B) \implies A \subseteq \mathrm{int}_{(\mathcal{U},\mathcal{D})}(B). \end{aligned}$$ □

Remark 34. Assertions (1) and (2) can be reformulated in the forms that

(1) $\mathrm{Int}_{(\mathcal{U},\mathcal{D})}(B) \subseteq \mathcal{P}(\mathrm{int}_{(\mathcal{U},\mathcal{D})}(B))$;
(2) $\mathcal{P}^{\star}(\mathrm{cl}_{(\mathcal{U},\mathcal{D})}(B)) \subseteq \mathrm{Cl}_{(\mathcal{U},\mathcal{D})}(B)$,

where $\mathcal{P}^{\star}$ is the dual of the super relation $\mathcal{P}$ taken in the sense of Definition 7.

Theorem 57. *For any* $B_1 \subseteq B_2 \subseteq Y$, *we have*

(1) $\mathrm{cl}_{(\mathcal{U},\mathcal{D})}(B_1) \subseteq \mathrm{cl}_{(\mathcal{U},\mathcal{D}}(B_2)$; (2) $\mathrm{int}_{(\mathcal{U},\mathcal{D})}(B_1) \subseteq \mathrm{int}_{(\mathcal{U},\mathcal{D})}(B_2)$.

Theorem 58. *The following assertions are equivalent:*

(1) $\mathrm{cl}_{(\mathcal{U},\mathcal{D})}(\emptyset) = \emptyset$;
(2) $\mathrm{int}_{(\mathcal{U},\mathcal{D})}(Y) \neq \emptyset$;
(3) $\mathrm{int}_{(\mathcal{U},\mathcal{D})}(Y) = X$;
(4) $X = \emptyset$ *or* ($\mathcal{U} \neq \emptyset$ *and* $\emptyset \notin \mathcal{D}$).

Corollary 14. *The following assertions are equivalent:*

(1) $\mathrm{int}_{(\mathcal{U},\mathcal{D})}(Y) = \emptyset$;
(2) $\mathrm{cl}_{(\mathcal{U},\mathcal{D})}(\emptyset) \neq \emptyset$;
(3) $\mathrm{cl}_{(\mathcal{U},\mathcal{D})}(\emptyset) = X$;
(4) $X \neq \emptyset$ *and* ($\mathcal{U} = \emptyset$ *or* $\emptyset \in \mathcal{D}$).

Theorem 59. *We have*

(1) $\mathrm{cl}_{(\mathcal{U},\mathcal{D})}(Y) = \{x \in X : \ \forall\ U \in \mathcal{U} : \quad U^{\triangleleft}(x) \in \mathcal{D}\}$;
(2) $\mathrm{int}_{(\mathcal{U},\mathcal{D})}(\emptyset) = \{x \in X : \ \exists\ U \in \mathcal{U} : \quad U^{\triangleleft}(x) \notin \mathcal{D}\}$.

Theorem 60. *If* $B_1, B_2 \subseteq Y$ *such that* $B_1 \cap B_2 = \emptyset$, *then*

$$\mathrm{cl}_{(\mathcal{U},\mathcal{D})}(B_1) \cap \mathrm{int}_{(\mathcal{U},\mathcal{D})}(B_2) = \emptyset.$$

Theorem 61. *If* $0B = \bigcap_{i\in I} B_i$ *for some family* $(B_i)_{i\in I}$ *of subsets of* Y, *then*

(1) $\mathrm{cl}_{(\mathcal{U},\mathcal{D})}(B) \subseteq \bigcap_{i\in I} \mathrm{cl}_{(\mathcal{U},\mathcal{D})}(B_i)$;
(2) $\mathrm{int}_{(\mathcal{U},\mathcal{D})}(B) \subseteq \bigcap_{i\in I} \mathrm{int}_{(\mathcal{U},\mathcal{D})}(B_i)$.

Theorem 62. *If* $B = \bigcup_{i\in I} B_i$ *for some family* $(B_i)_{i\in I}$ *of subsets of* Y, *then*

(1) $\bigcup_{i\in I} \mathrm{cl}_{(\mathcal{U},\mathcal{D})}(B_i) \subseteq \mathrm{cl}_{(\mathcal{U},\mathcal{D})}(B)$;
(2) $\bigcup_{i\in I} \mathrm{int}_{(\mathcal{U},\mathcal{D})}(B_i) \subseteq \mathrm{int}_{(\mathcal{U},\mathcal{D})}(B)$.

Theorem 63. *We have*

(1) $\operatorname{cl}_{(\mathcal{U},\mathcal{D})} = \bigcap_{U \in \mathcal{U}} \operatorname{cl}_{(U,\mathcal{D})}$; (2) $\operatorname{int}_{(\mathcal{U},\mathcal{D})} = \bigcup_{U \in \mathcal{U}} \operatorname{int}_{(U,\mathcal{D})}$.

Corollary 15. *The mapping*

(1) $\mathcal{U} \mapsto \operatorname{int}_{(\mathcal{U},\mathcal{D})}$ *is union-preserving;*
(2) $\mathcal{U} \mapsto \operatorname{cl}_{(\mathcal{U},\mathcal{D})}$ *is intersection-preserving.*

Theorem 64. *If* $\mathcal{D} = \bigcap_{i \in I} \mathcal{D}_i$ *for some family* $(\mathcal{D}_i)_{i \in I}$ *of stacks on* Y, *then*

(1) $\operatorname{cl}_{(\mathcal{U},\mathcal{D})} = \bigcap_{i \in I} \operatorname{cl}_{(\mathcal{U},\mathcal{D}_i)}$; (2) $\operatorname{int}_{(\mathcal{U},\mathcal{D})} = \bigcup_{i \in I} \operatorname{int}_{(\mathcal{U},\mathcal{D}_i)}$.

Finally, by specializing Definition 11, we may also naturally introduce the following:

Definition 13. For any $B \subseteq Y$, we define

$$\operatorname{bnd}_{(\mathcal{U},\mathcal{D})}(B) = \{x \in X : \ \{x\} \in \operatorname{Bnd}_{(\mathcal{U},\mathcal{D})}(B)\}.$$

Remark 35. Moreover, if $X = Y$, then for any $A \subseteq X$ we also define

$$\operatorname{res}_{(\mathcal{U},\mathcal{D})}(A) = \{x \in X : \ \{x\} \in \operatorname{Res}_{(\mathcal{U},\mathcal{D})}(A)\}$$

and

$$\operatorname{bor}_{(\mathcal{U},\mathcal{D})}(A) = \{x \in X : \ \{x\} \in \operatorname{Bor}_{(\mathcal{U},\mathcal{D})}(A)\}.$$

Note that the above notions can also be more nicely expressed in terms of the super relations $\operatorname{cl}_{(\mathcal{U},\mathcal{D})}$ and $\operatorname{int}_{(\mathcal{U},\mathcal{D})}$.

However, it is now more important to note that, by using the latter super relations, we may also naturally introduce the following:

Definition 14. For any $y \in Y$, we define

$$\rho_{(\mathcal{U},\mathcal{D})}(y) = \operatorname{cl}_{(\mathcal{U},\mathcal{D})}(\{y\}) \quad \text{and} \quad \sigma_{(\mathcal{U},\mathcal{D})}(y) = \operatorname{int}_{(\mathcal{U},\mathcal{D})}(\{y\}).$$

Remark 36. The ordinary relations $\rho_{(\mathcal{U},\mathcal{D})}$ and $\sigma_{(\mathcal{U},\mathcal{D})}$ will be called the *infinitesimal closure and interior* generated by the super relator $\mathcal{U}$ and the stack $\mathcal{D}$, respectively.

20. Fat and Dense Sets Derived from Super Relators and Stacks

By using Definition 12, we may also naturally introduce the following generalization of our former similar definitions [126, 145, 168].

Definition 15. We define

$$\mathcal{D}_{(\mathcal{U},\mathcal{D})} = \{ D \subseteq Y : \ X = \mathrm{cl}_{(\mathcal{U},\mathcal{D})}(D) \}$$

and

$$\mathcal{E}_{(\mathcal{U},\mathcal{D})} = \{ E \subseteq Y : \ \mathrm{int}_{(\mathcal{U},\mathcal{D})}(E) \neq \emptyset \}.$$

Remark 37. The members of the families $\mathcal{E}_{(\mathcal{U},\mathcal{D})}$ and $\mathcal{D}_{(\mathcal{U},\mathcal{D})}$ will be called the *fat and dense subsets* of the stacked super relator space $(X,Y)(\mathcal{U})(\mathcal{D})$, respectively.

Thus, by using the corresponding results of Section 19, we can easily prove the following theorems.

However, most of the subsequent theorems can be more easily proved with the help of following consequence of Theorem 54.

Theorem 65. *We have*

(1) $\mathcal{D}_{(\mathcal{U},\mathcal{D})} = \{ D \subseteq Y : \ \forall \ x \in X : \quad \forall \ U \in \mathcal{U} : \quad U^{\triangleleft}(x) \cap D \in \mathcal{D} \}$;
(2) $\mathcal{E}_{(\mathcal{U},\mathcal{D})} = \{ E \subseteq Y : \ \exists \ x \in X : \quad \exists \ U \in \mathcal{U} : \quad U^{\triangleleft}(x) \setminus E \notin \mathcal{D} \}$.

Proof. To prove assertion (2), note that, by Definition 15 and Theorem 54, for any $E \subseteq Y$ we have

$$E \in \mathcal{E}_{(\mathcal{U},\mathcal{D})} \iff \mathrm{int}_{(\mathcal{U},\mathcal{D})}(E) \neq \emptyset \iff \exists \ x \in X : \ x \in \mathrm{int}_{(\mathcal{U},\mathcal{D})}(E)$$
$$\iff \exists \ x \in X : \quad \exists \ U \in \mathcal{U} : \quad U^{\triangleleft}(x) \setminus E \notin \mathcal{D}. \qquad \square$$

Theorem 66. *We have*

(1) $\mathcal{D}_{(\mathcal{U},\mathcal{D})} = \{ D \subseteq Y : \ D^c \notin \mathcal{E}_{(\mathcal{U},\mathcal{D})} \}$;
(2) $\mathcal{E}_{(\mathcal{U},\mathcal{D})} = \{ E \subseteq Y : \ E^c \notin \mathcal{D}_{(\mathcal{U},\mathcal{D})} \}$.

Theorem 67. *We have*

(1) $\mathcal{D}_{(\mathcal{U},\mathcal{D})} = \{ D \subseteq Y : \ \forall \ E \in \mathcal{E}_{(\mathcal{U},\mathcal{D})} : \ D \cap E \neq \emptyset \}$;
(2) $\mathcal{E}_{(\mathcal{U},\mathcal{D})} = \{ E \subseteq Y : \ \forall \ D \in \mathcal{D}_{(\mathcal{U},\mathcal{D})} : \ D \cap E \neq \emptyset \}$.

Theorem 68. *The families* $\mathcal{D}_{(\mathcal{U},\mathcal{D})}$ *and* $\mathcal{E}_{(\mathcal{U},\mathcal{D})}$ *are stacks on* Y.

Theorem 69. *The following assertions are equivalent:*

(1) $X \neq \emptyset$, $\mathcal{U} \neq \emptyset$ *and* $\emptyset \notin \mathcal{D}$;
(2) $\emptyset \notin \mathcal{D}_{(\mathcal{U},\mathcal{D})}$; (3) $\mathcal{E}_{(\mathcal{U},\mathcal{D})} \neq \emptyset$; (4) $Y \in \mathcal{E}_{(\mathcal{U},\mathcal{D})}$;
(5) $\mathcal{D}_{(\mathcal{U},\mathcal{D})} \neq \mathcal{P}(Y)$.

Proof. By Definition 15 and Corollary 14, we have

$$(2) \iff X \neq \mathrm{cl}_{(\mathcal{U},\mathcal{D})}(\emptyset) \iff (1). \qquad \square$$

Theorem 70. *The following assertions are equivalent:*

(1) $U^{\triangleleft}(x) \in \mathcal{D}$ *for all* $x \in X$ *and* $U \in \mathcal{U}$;
(2) $\emptyset \notin \mathcal{E}_{(\mathcal{U},\mathcal{D})}$; (3) $\mathcal{D}_{(\mathcal{U},\mathcal{D})} \neq \emptyset$; (4) $Y \in \mathcal{D}_{(\mathcal{U},\mathcal{D})}$;
(5) $\mathcal{E}_{(\mathcal{U},\mathcal{D})} \neq \mathcal{P}(Y)$.

Proof. By Definition 15 and Theorem 59, we have

$$(2) \iff \mathrm{int}_{(\mathcal{U},\mathcal{D})}(\emptyset) = \emptyset \iff \forall\ x \in X:\ \ x \notin \mathrm{int}_{(\mathcal{U},\mathcal{D})}(\emptyset) \iff (1).$$

$\square$

Theorem 71. *We have*

(1) $\mathcal{D}_{(\mathcal{U},\mathcal{D})} = \bigcap_{U \in \mathcal{U}} \mathcal{D}_{(U,\mathcal{D})}$; (2) $\mathcal{E}_{(\mathcal{U},\mathcal{D})} = \bigcup_{U \in \mathcal{U}} \mathcal{E}_{(U,\mathcal{D})}$.

Corollary 16. *The mappings*

(1) $\mathcal{U} \mapsto \mathcal{E}_{(\mathcal{U},\mathcal{D})}$ *is union-preserving;*
(2) $\mathcal{U} \mapsto \mathcal{D}_{(\mathcal{U},\mathcal{D})}$ *is intersection-preserving.*

Remark 38. By using the notation

$$\mathfrak{U}_{(\mathcal{U},\mathcal{D})}(x) = \mathrm{int}_{(\mathcal{U},\mathcal{D})}^{-1}(x) = \{B \subseteq Y:\ \ x \in \mathrm{int}_{(\mathcal{U},\mathcal{D})}(B)\},$$

we can also prove that $\mathcal{E}_{(\mathcal{U},\mathcal{D})} = \bigcup_{x \in X} \mathfrak{U}_{(\mathcal{U},\mathcal{D})}(x)$.

Theorem 72. *If* $\mathcal{D} = \bigcap_{i \in I} \mathcal{D}_i$ *for some family* $(\mathcal{D}_i)_{i \in I}$ *of stacks on* Y, *then*

(1) $\mathcal{D}_{(\mathcal{U},\mathcal{D})} = \bigcap_{i \in I} \mathcal{D}_{(\mathcal{U},\mathcal{D}_i)}$; (2) $\mathcal{E}_{(\mathcal{U},\mathcal{D})} = \bigcup_{i \in I} \mathcal{E}_{(\mathcal{U},\mathcal{D}_i)}$.

Proof. To prove assertion (1), note that, by Definition 15 and Theorem 64, for any $D \subseteq Y$ we have

$$D \in \mathcal{D}_{(\mathcal{U},\mathcal{D})} \iff X = \mathrm{cl}_{(\mathcal{U},\mathcal{D})}(D) \iff X = \bigcap_{i \in I} \mathrm{cl}_{(\mathcal{U},\mathcal{D}_i)}(D)$$

$$\iff \forall\ i \in I: \ X = \mathrm{cl}_{(\mathcal{U},\mathcal{D}_i)}(D) \iff \forall\ i \in I: \ D \in \mathcal{D}_{(\mathcal{U},\mathcal{D}_i)}$$

$$\iff D \in \bigcap_{i \in I} \mathcal{D}_{(\mathcal{U},\mathcal{D}_i)}.$$

□

21. Open and Closed Sets Derived from Super Relators and Stacks

Notation 6. *In this and the following section, we shall assume that $\mathcal{U}$ is a super relator on X, and $\mathcal{D}$ is a stack on X.*

By using Definitions 10 and 12, we may naturally introduce the following generalization of our former similar definitions [126, 145, 168].

Definition 16. We define

$\tau_{(\mathcal{U},\mathcal{D})} = \{A \subseteq X : \ A \in \mathrm{Int}_{(\mathcal{U},\mathcal{D})}(A)\}$;
$\tau\!\!\!\!-_{(\mathcal{U},\mathcal{D})} = \{A \subseteq X : \ A^c \notin \mathrm{Cl}_{(\mathcal{U},\mathcal{D})}(A)\}$;
$\mathcal{T}_{(\mathcal{U},\mathcal{D})} = \{A \subseteq X : \ A \subseteq \mathrm{int}_{(\mathcal{U},\mathcal{D})}(A)\}$;
$\mathcal{F}_{(\mathcal{U},\mathcal{D})} = \{A \subseteq X : \ \mathrm{cl}_{(\mathcal{U},\mathcal{D})}(A) \subseteq A\}$.

Remark 39. The members of the families, $\tau_{(\mathcal{U},\mathcal{D})}$ and $\mathcal{T}_{(\mathcal{U},\mathcal{D})}$, will be called the *proximally open and topologically open subsets* of the stacked super relator space $X(\mathcal{U})(\mathcal{D})$, respectively.

Thus, by using the results of Sections 16 and 19, we can easily establish the following theorems.

Theorem 73. *We have*

(1) $\tau_{(\mathcal{U},\mathcal{D})} = \{A \subseteq X : \quad \exists\ U \in \mathcal{U} : \quad U(A) \setminus A \notin \mathcal{D}\}$;
(2) $\mathcal{T}_{(\mathcal{U},\mathcal{D})} = \{A \subseteq X : \quad \forall\ x \in A : \quad \exists\ U \in \mathcal{U} : \quad U^{\triangleleft}(x) \setminus A \notin \mathcal{D}\}$.

Proof. To prove assertion (2), note that, by Definition 16 and Theorem 54, for any $A \subseteq X$, we have

$$A \in \mathcal{T}_{(\mathcal{U},\mathcal{D})} \iff A \subseteq \operatorname{int}_{(\mathcal{U},\mathcal{D})}(A) \iff \forall\ x \in A: \ \ x \in \operatorname{int}_{(\mathcal{U},\mathcal{D})}(A)$$
$$\iff \forall\ x \in A: \quad \exists\ U \in \mathcal{U}: \quad U^{\triangleleft}(x) \setminus A \notin \mathcal{D}. \qquad \square$$

Theorem 74. *We have*

(1) $\tau_{(\mathcal{U},\mathcal{D})} = \{A^c : \ A \in \tau_{(\mathcal{U},\mathcal{D})}\}$; (2) $\tau_{(\mathcal{U},\mathcal{D})} = \{A^c : \ A \in \digamma_{(\mathcal{U},\mathcal{D})}\}$;
(3) $\mathcal{F}_{(\mathcal{U},\mathcal{D})} = \{A^c : \ A \in \mathcal{T}_{(\mathcal{U},\mathcal{D})}\}$; (4) $\mathcal{T}_{(\mathcal{U},\mathcal{D})} = \{A^c : \ A \in \mathcal{F}_{(\mathcal{U},\mathcal{D})}\}$.

Proof. To prove assertion (1), note that by Definition 16 and Theorem 38, for any $A \subseteq X$, we have

$$A \in \digamma_{(\mathcal{U},\mathcal{D})} \iff A^c \notin \operatorname{Cl}_{(\mathcal{U},\mathcal{D})}(A) \iff A^c \notin \operatorname{Int}_{(\mathcal{U},\mathcal{D})}(A^c)^c$$
$$\iff A^c \in \operatorname{Int}_{(\mathcal{U},\mathcal{D})}(A^c) \iff A^c \in \tau_{(\mathcal{U},\mathcal{D})}. \qquad \square$$

Theorem 75. *If $\mathcal{U}$ is quasi-increasing, then*

(1) $\tau_{(\mathcal{U},\mathcal{D})} \subseteq \mathcal{T}_{(\mathcal{U},\mathcal{D})}$; (2) $\digamma_{(\mathcal{U},\mathcal{D})} \subseteq \mathcal{F}_{(\mathcal{U},\mathcal{D})}$.

Proof. By using Definition 16 and Theorem 56, we can at once see that

$$A \in \tau_{(\mathcal{U},\mathcal{D})} \implies A \in \operatorname{Int}_{(\mathcal{U},\mathcal{D})}(A) \implies A \subseteq \operatorname{int}_{(\mathcal{U},\mathcal{D})}(A)$$
$$\implies A \in \mathcal{T}_{(\mathcal{U},\mathcal{D})}. \qquad \square$$

Theorem 76. *We have*

(1) $\mathcal{T}_{(\mathcal{U},\mathcal{D})} \setminus \{\emptyset\} \subseteq \mathcal{E}_{(\mathcal{U},\mathcal{D})}$; (2) $\mathcal{D}_{(\mathcal{U},\mathcal{D})} \cap \mathcal{F}_{(\mathcal{U},\mathcal{D})} \subseteq \{X\}$.

Proof. For instance, by using Definitions 15 and 16, we can at once see that

$$A \in \mathcal{D}_{(\mathcal{U},\mathcal{D})} \cap \mathcal{F}_{(\mathcal{U},\mathcal{D})} \implies X = \operatorname{cl}_{(\mathcal{U},\mathcal{D})}, \operatorname{cl}_{(\mathcal{U},\mathcal{D})} \subseteq A \implies A = X.$$

$\square$

Remark 40. From assertion (2), by using global complementations, we can easily infer that $\mathcal{F}_{(\mathcal{U},\mathcal{D})} \subseteq (\mathcal{D}_{(\mathcal{U},\mathcal{D})})^c \cup \{X\}$ and $\mathcal{D}_{(\mathcal{U},\mathcal{D})} \subseteq (\mathcal{F}_{(\mathcal{U},\mathcal{D})})^c \cup \{X\}$.

Theorem 77. *For any* $A \subseteq X$,

(1) $\bigcup(\mathcal{T}_{(\mathcal{U},\mathcal{D})} \cap \mathcal{P}(A)) \subseteq \operatorname{int}_{(\mathcal{U},\mathcal{D})}(A)$;

(2) $\mathcal{P}(A) \cap (\mathcal{T}_{(\mathcal{U},\mathcal{D})} \setminus \{\emptyset\}) \neq \emptyset$ *implies* $A \in \mathcal{E}_{(\mathcal{U},\mathcal{D})}$;

(3) $\mathcal{P}[\tau_{(\mathcal{U},\mathcal{D})} \cap \mathcal{P}(A)] \subseteq \operatorname{Int}_{(\mathcal{U},\mathcal{D})}(A)$ *if* $\mathcal{U}$ *is increasing.*

Proof. To prove assertion (3), note that by the corresponding definitions

$$\begin{aligned} B \in \mathcal{P}[\tau_{(\mathcal{U},\mathcal{D})} \cap \mathcal{P}(A)] &\Longrightarrow \exists\; C \in \tau_{(\mathcal{U},\mathcal{D})} \cap \mathcal{P}(A) : \;\; B \in \mathcal{P}(C) \\ &\Longrightarrow \exists\; C \in \mathcal{P}(A) : \;\; B \in \mathcal{P}(C), C \in \tau_{(\mathcal{U},\mathcal{D})} \\ &\Longrightarrow \exists\; C \subseteq X : \;\; B \subseteq C \subseteq A, C \in \operatorname{Int}_{(\mathcal{U},\mathcal{D})}(C). \end{aligned}$$

Hence, by using Theorems 39 and 47, we can infer that

$$B \subseteq C \in \operatorname{Int}_{(\mathcal{U},\mathcal{D})}(C) \subseteq \operatorname{Int}_{(\mathcal{U},\mathcal{D})}(A), \quad \text{and thus also}$$

$$B \in \operatorname{Int}_{(\mathcal{U},\mathcal{D})}(A).$$

□

Theorem 78. *We have*

(1) $\tau_{(\mathcal{U},\mathcal{D})} = \bigcup_{U \in \mathcal{U}} \tau_{(U,\mathcal{D})}$; (2) $\digamma_{(\mathcal{U},\mathcal{D})} = \bigcup_{U \in \mathcal{U}} \digamma_{(U,\mathcal{D})}$.

Corollary 17. *The mappings*

$$\mathcal{U} \mapsto \tau_{(\mathcal{U},\mathcal{D})} \quad \textit{and} \quad \mathcal{U} \mapsto \digamma_{(\mathcal{U},\mathcal{D})}$$

are union-preserving.

Later, we shall see that the same statements do not hold for the topologically open and closed sets. Therefore, we can only prove the following:

Theorem 79. *The mappings*

$$\mathcal{U} \mapsto \mathcal{T}_{(\mathcal{U},\mathcal{D})} \quad \textit{and} \quad \mathcal{U} \mapsto \mathcal{F}_{(\mathcal{U},\mathcal{D})}$$

are increasing.

Corollary 18. *We have*

(1) $\bigcup_{U \in \mathcal{U}} \mathcal{T}_{(U,\mathcal{D})} \subseteq \mathcal{T}_{(\mathcal{U},\mathcal{D})}$; (2) $\bigcup_{U \in \mathcal{U}} \mathcal{F}_{(U,\mathcal{D})} \subseteq \mathcal{F}_{(\mathcal{U},\mathcal{D})}$.

However, by using Theorems 46 and 64, we can now also prove the following:

Theorem 80. *If $\mathcal{D} = \bigcap_{i \in I} \mathcal{D}_i$ for some family $(\mathcal{D}_i)_{i \in I}$ of stacks on Y, then*

(1) $\tau_{(\mathcal{U},\mathcal{D})} = \bigcup_{i \in I} \tau_{(\mathcal{U},\mathcal{D}_i)}$; (2) $\digamma_{(\mathcal{U},\mathcal{D})} = \bigcup_{i \in I} \digamma_{(\mathcal{U},\mathcal{D}_i)}$;

(3) $\bigcup_{i \in I} \mathcal{T}_{(\mathcal{U},\mathcal{D}_i)} \subseteq \mathcal{T}_{(\mathcal{U},\mathcal{D})}$; (4) $\bigcup_{i \in I} \mathcal{F}_{(\mathcal{U},\mathcal{D}_i)} \subseteq \mathcal{F}_{(\mathcal{U},\mathcal{D})}$.

Proof. For instance, by Definition 16 and Theorem 46, for any $A \subseteq X$ we have

$$A \in \tau_{(\mathcal{U},\mathcal{D})} \iff A \in \operatorname{Int}_{(\mathcal{U},\mathcal{D})}(A) \iff A \in \Big(\bigcup_{i \in I} \operatorname{Int}_{(\mathcal{U},\mathcal{D}_i)}\Big)(A)$$

$$\iff A \in \bigcup_{i \in I} \operatorname{Int}_{(\mathcal{U},\mathcal{D}_i)}(A) \iff \exists\ i \in I: \ A \in \operatorname{Int}_{(\mathcal{U},\mathcal{D}_i)}(A)$$

$$\iff \exists\ i \in I: \ A \in \tau_{(\mathcal{U},\mathcal{D}_i)} \iff A \in \bigcup_{i \in I} \tau_{(\mathcal{U},\mathcal{D}_i)}. \qquad \square$$

22. Some Further Theorems on Open and Closed Sets

Theorem 81. *The following assertions are equivalent:*

(1) $\emptyset \in \tau_{(\mathcal{U},\mathcal{D})}$; (2) $X \in \digamma_{(\mathcal{U},\mathcal{D})}$; (3) *there exists $U \in \mathcal{U}$ such that $U(\emptyset) \notin \mathcal{D}$.*

Proof. To prove that (1) $\iff$ (3), note that, by Theorem 73, we have

$$(1) \iff \exists\ U \in \mathcal{U}: \ U(\emptyset) \setminus \emptyset \notin \mathcal{D} \iff (3). \qquad \square$$

Theorem 82. *The following assertions are equivalent:*

(1) $X \in \tau_{(\mathcal{U},\mathcal{D})}$; (2) $\emptyset \in \digamma_{(\mathcal{U},\mathcal{D})}$; (3) $\mathcal{U} \neq \emptyset$ *and* $\emptyset \notin \mathcal{D}$.

Proof. To prove that (1) $\iff$ (3), note that, by Theorem 73, we have

$$(1) \iff \exists\ U \in \mathcal{U}: \ U(X) \setminus X \notin \mathcal{D}$$

$$\iff \exists\ U \in \mathcal{U}: \ \emptyset \notin \mathcal{D} \iff (3). \qquad \square$$

Remark 41. Thus, $\tau_{(\mathcal{U},\mathcal{D})}$ is a *minimal structure* on X if and only if $\emptyset \notin \mathcal{D}$ and there exists $U \in \mathcal{U}$ such that $U(\emptyset) \notin \mathcal{D}$.

Theorem 83. *If $A \subseteq X$ and $B \in \tau_{(\mathcal{U},\mathcal{D})}$ such that $A \cap B = \emptyset$, then*

(1) $B \notin \mathrm{Cl}_{(\mathcal{U},\mathcal{D})}(A)$; (2) $B \in \mathrm{Int}_{(\mathcal{U},\mathcal{D})}(A^c)$.

Proof. By Theorem 42 and Definition 16, we have

$$\mathrm{Cl}_{(\mathcal{U},\mathcal{D})}(A) \cap \mathrm{Int}_{(\mathcal{U},\mathcal{D})}(B) = \emptyset \quad \text{and} \quad B \in \mathrm{Int}_{(\mathcal{U},\mathcal{D})}(B).$$

Therefore, assertion (1) is true. Assertion (2) can be derived from assertion (1) by using Definition 10. □

Theorem 84. *The family*

(1) $\mathcal{T}_{(\mathcal{U},\mathcal{D})}$ *is closed under arbitrary unions;*
(2) $\mathcal{F}_{(\mathcal{U},\mathcal{D})}$ *is closed under arbitrary intersection.*

Proof. If $A_i \in \mathcal{T}_{(\mathcal{U},\mathcal{D})}$ for all $i \in I$ and $A = \bigcup_{i \in I} A_i$, then by using Definition 16 and Theorem 61, we can see that

$$A = \bigcup_{i \in I} A_i \subseteq \bigcup_{i \in I} \mathrm{int}_{(\mathcal{U},\mathcal{D})}(A_i) \subseteq \mathrm{int}_{(\mathcal{U},\mathcal{D})}(A).$$

Thus, by Definition 16, we also have $A \in \mathcal{T}_{(\mathcal{U},\mathcal{D})}$. □

From this theorem, by using that $\bigcup \emptyset = \emptyset$ and $\bigcap \emptyset = X$, we can derive the following:

Corollary 19. *We always have $\emptyset \in \mathcal{T}_{(\mathcal{U},\mathcal{D})}$ and $X \in \mathcal{F}_{(\mathcal{U},\mathcal{D})}$.*

Theorem 85. *The following assertions are equivalent:*

(1) $X \in \mathcal{T}_{(\mathcal{U},\mathcal{D})}$; (2) $\emptyset \in \mathcal{F}_{(\mathcal{U},\mathcal{D})}$;
(3) *either $X = \emptyset$ or ($\mathcal{U} \neq \emptyset$ and $\emptyset \notin \mathcal{D}$).*

Proof. To prove that (1) $\iff$ (3), note that, by Theorem 73, we have

$$\begin{aligned} (1) &\iff \forall\ x \in X: \quad \exists\ U \in \mathcal{U}: \quad U^{\triangleleft}(x) \setminus X \notin \mathcal{D} \\ &\iff \forall\ x \in X: \quad \exists\ U \in \mathcal{U}: \quad \emptyset \notin \mathcal{D} \iff (3). \end{aligned}$$ □

Remark 42. Thus, $\mathcal{T}_{(\mathcal{U},\mathcal{D})}$ is a *generalized topology* on X if and only if either $X = \emptyset$ or ($\mathcal{U} \neq \emptyset$ and $\emptyset \notin \mathcal{D}$).

Theorem 86. *If* $A \subseteq X$ *and* $B \in \mathcal{T}_{(\mathcal{U},\mathcal{D})}$ *such that* $A \cap B = \emptyset$, *then*

(1) $\operatorname{cl}_{(\mathcal{U},\mathcal{D})}(A) \cap B = \emptyset$; (2) $B \subseteq \operatorname{int}_{(\mathcal{U},\mathcal{D})}(A^c)$.

Proof. By Theorem 59 and Definition 16, we have

$$\operatorname{cl}_{(\mathcal{U},\mathcal{D})}(A) \cap \operatorname{int}_{(\mathcal{U},\mathcal{D})}(B) = \emptyset \quad \text{and} \quad B \subseteq \operatorname{int}_{(\mathcal{U},\mathcal{D})}(B).$$

Therefore, assertion (1) is true. Assertion (2) can be derived from assertion (1) by using Theorem 55. □

In addition to Definition 16, we may also naturally introduce the following:

Definition 17. We define

$$\mathcal{N}_{(\mathcal{U},\mathcal{D})} = \{A \subseteq X : \ \operatorname{cl}_{(\mathcal{U},\mathcal{D})}(A) \notin \mathcal{E}_{(\mathcal{U},\mathcal{D})}\}$$

and

$$\mathcal{M}_{(\mathcal{U},\mathcal{D})} = \{A \subseteq X : \ \operatorname{int}_{(\mathcal{U},\mathcal{D})}(A) \in \mathcal{D}_{(\mathcal{U},\mathcal{D})}\}.$$

Remark 43. The members of the family $\mathcal{N}_{(\mathcal{U},\mathcal{D})}$ will be called the *rare or nowhere dense subsets* of the stacked super relator space $X(\mathcal{U})(\mathcal{D})$.

In contrast to $\mathcal{N}_{(\mathcal{U},\mathcal{D})}$, the family $\mathcal{M}_{(\mathcal{U},\mathcal{D})}$ seems to be of no particular importance for us. Its introduction can only be motivated by the following

Theorem 87. *We have*

(1) $\mathcal{M}_{(\mathcal{U},\mathcal{D})} = \{A^c : \ A \in \mathcal{N}_{(\mathcal{U},\mathcal{D})}\}$;
(2) $\mathcal{N}_{(\mathcal{U},\mathcal{D})} = \{A^c : \ A \in \mathcal{M}_{(\mathcal{U},\mathcal{D})}\}$.

Proof. To prove assertion (1), note that by Definition 17 and Theorems 55 and 83, for any $A \subseteq X$, we have

$$A \in \mathcal{M}_{(\mathcal{U},\mathcal{D})} \iff \operatorname{int}_{(\mathcal{U},\mathcal{D})}(A) \in \mathcal{D}_{(\mathcal{U},\mathcal{D})} \iff \operatorname{cl}_{(\mathcal{U},\mathcal{D})}(A^c)^c \in \mathcal{D}_{(\mathcal{U},\mathcal{D})}$$
$$\iff \operatorname{cl}_{(\mathcal{U},\mathcal{D})}(A^c) \notin \mathcal{E}_{(\mathcal{U},\mathcal{D})} \iff A^c \in \mathcal{N}_{(\mathcal{U},\mathcal{D})}.$$

□

Remark 44. Countable unions of rare sets may be briefly called *meager* instead of being of first category.

Moreover, a relator space may be briefly called a *Baire space* if its fat subsets are not meager. (See Refs. [153, 154].)

Analogously to Baire's property, completeness, compactness and Lebesgue properties can also be nicely treated in relator spaces. (See Refs. [142, 147, 150].)

23. Proximal Closures and Interiors Derived from Super Relators

Notation 7. *In this and the following two sections, we shall assume that $\mathcal{U}$ is a super relator on X to Y.*

From Definition 10, by using the stack $\mathcal{D}_0 = \mathcal{P}(Y) \setminus \{\emptyset\}$, we may naturally derive the following:

Definition 18. We define

$$\mathrm{Cl}_{\mathcal{U}} = \mathrm{Cl}_{(\mathcal{U},\mathcal{D}_0)} \quad \text{and} \quad \mathrm{Int}_{\mathcal{U}} = \mathrm{Int}_{(\mathcal{U},\mathcal{D}_0)} \,.$$

Remark 45. The hyper relations $\mathrm{Cl}_{\mathcal{U}}$ and $\mathrm{Int}_{\mathcal{U}}$ will be called the *proximal closure and interior* generated by the super relator $\mathcal{U}$, respectively.

Thus, from the results of Sections 16, and 17, we can easily derive the following theorems.

Theorem 88. *For any $B \subseteq Y$, we have*

(1) $\mathrm{Int}_{\mathcal{U}}(B) = \{A \subseteq X : \ \exists\ U \in \mathcal{U} : \quad U(A) \subseteq B\}$;

(2) $\mathrm{Cl}_{\mathcal{U}}(B) = \{A \subseteq X : \ \forall\ U \in \mathcal{U} : \quad U(A) \cap B \neq \emptyset\}$.

Proof. By Definition 18 and Theorem 38, for any $A \subseteq X$ we have

$$\begin{aligned} A \in \mathrm{Int}_{\mathcal{U}}(B) &\iff A \in \mathrm{Int}_{(\mathcal{U},\mathcal{D}_0)}(B) \\ &\iff \exists\ U \in \mathcal{U} : \quad U(A) \setminus B \notin \mathcal{D}_0 \\ &\iff \exists\ U \in \mathcal{U} : \quad U(A) \setminus B = \emptyset \\ &\iff \exists\ U \in \mathcal{U} : \quad U(A) \subseteq B. \end{aligned}$$

Thus, assertion (1) is true. The proof of assertion (2) is even more simple. □

Theorem 89. *For any $B \subseteq Y$, we have*

(1) $\mathrm{Cl}_{\mathcal{U}}(B) = \mathrm{Int}_{\mathcal{U}}(B^c)^c$; (2) $\mathrm{Int}_{\mathcal{U}}(B) = \mathrm{Cl}_{\mathcal{U}}(B^c)^c$.

Theorem 90. *For any $B_1 \subseteq B_2 \subseteq Y$, we have*

(1) $\mathrm{Cl}_{\mathcal{U}}(B_1) \subseteq \mathrm{Cl}_{\mathcal{U}}(B_2)$; (2) $\mathrm{Int}_{\mathcal{U}}(B_1) \subseteq \mathrm{Int}_{\mathcal{U}}(B_2)$.

Theorem 91. *The following assertions are equivalent:*

(1) $\mathrm{Cl}_{\mathcal{U}}(\emptyset) = \emptyset$; (2) $\mathcal{U} \neq \emptyset$; (3) $\mathrm{Int}_{\mathcal{U}}(Y) \neq \emptyset$;
(4) $\mathrm{Int}_{\mathcal{U}}(Y) = \mathcal{P}(X)$.

Corollary 20. *The following assertions are equivalent:*

(1) $\mathrm{Int}_{\mathcal{U}}(Y) = \emptyset$; (2) $\mathcal{U} = \emptyset$; (3) $\mathrm{Cl}_{\mathcal{U}}(\emptyset) \neq \emptyset$;
(4) $\mathrm{Cl}_{\mathcal{U}}(\emptyset) = \mathcal{P}(X)$.

Theorem 92. *We have*

(1) $\mathrm{Cl}_{\mathcal{U}}(Y) = \{A \subseteq X : \ \forall\ U \in \mathcal{U} : \quad U(A) \neq \emptyset\}$;

(2) $\mathrm{Int}_{\mathcal{U}}(\emptyset) = \{A \subseteq X : \ \exists\ U \in \mathcal{U} : \quad U(A) = \emptyset\}$.

Theorem 93. *If $B_1, B_2 \subseteq Y$ such that $B_1 \cap B_2 = \emptyset$, then*

$$\mathrm{Cl}_{\mathcal{U}}(B_1) \cap \mathrm{Int}_{\mathcal{U}}(B_2) = \emptyset.$$

Theorem 94. *If $B = \bigcap_{i \in I} B_i$ for some family $(B_i)_{i \in I}$ of subsets of Y, then*

(1) $\mathrm{Cl}_{\mathcal{U}}(B) \subseteq \bigcap_{i \in I} \mathrm{Cl}_{\mathcal{U}}(B_i)$; (2) $\mathrm{Int}_{\mathcal{U}}(B) \subseteq \bigcap_{i \in I} \mathrm{Int}_{\mathcal{U}}(B_i)$.

Theorem 95. *If $B = \bigcup_{i \in I} B_i$ for some family $(B_i)_{i \in I}$ of subsets of Y, then*

(1) $\bigcup_{i \in I} \mathrm{Cl}_{\mathcal{U}}(B_i) \subseteq \mathrm{Cl}_{\mathcal{U}}(B)$; (2) $\bigcup_{i \in I} \mathrm{Int}_{\mathcal{U}}(B_i) \subseteq \mathrm{Int}_{\mathcal{U}}(B)$.

Theorem 96. *If $\mathrm{card}(\mathcal{U}) = 1$, then*

(1) $\mathrm{Cl}_{\mathcal{U}}$ *is union-preserving*; (2) $\mathrm{Int}_{\mathcal{U}}$ *is intersection-preserving.*

Proof. If U is a super relation on X to Y and $B = \bigcup_{i \in I} B_i$ for some family $(B_i)_{i \in I}$ of subsets of Y, then by using Theorem 88, we can see that

$$A \in \mathrm{Cl}_U(B) \iff U(A) \cap B \neq \emptyset \iff U(A) \cap \bigcup_{i \in I} B_i \neq \emptyset$$

$$\iff \bigcup_{i \in I} (U(A) \cap B_i) \neq \emptyset \iff \exists\, i \in I: \; U(A) \cap B_i \neq \emptyset$$

$$\iff \exists\, i \in I: \; A \in \mathrm{Cl}_U(B_i) \iff A \in \bigcup_{i \in I} \mathrm{Cl}_U(B_i).$$

Therefore, now $\mathrm{Cl}_U(B) = \bigcup_{i \in I} \mathrm{Cl}_U(B_i)$ also holds. □

Theorem 97. *We have*

(1) $\mathrm{Cl}_{\mathcal{U}} = \bigcap_{U \in \mathcal{U}} \mathrm{Cl}_{\mathcal{U}}$; (2) $\mathrm{Int}_{\mathcal{U}} = \bigcup_{U \in \mathcal{U}} \mathrm{Int}_{\mathcal{U}}$.

Corollary 21. *The mapping*

(1) $\mathcal{U} \mapsto \mathrm{Int}_{\mathcal{U}}$ *is union-preserving;*
(2) $\mathcal{U} \mapsto \mathrm{Cl}_{\mathcal{U}}$ *is intersection-preserving.*

Theorem 98. *If $\mathcal{U}$ is increasing, then for any $A_1 \subseteq A_2 \subseteq X$, we have*

(1) $\mathrm{Cl}_{\mathcal{U}}^{-1}(A_1) \subseteq \mathrm{Cl}_{\mathcal{U}}^{-1}(A_2)$; (2) $\mathrm{Int}_{\mathcal{U}}^{-1}(A_2) \subseteq \mathrm{Int}_{\mathcal{U}}^{-1}(A_1)$.

Theorem 99. *If $U(\emptyset) = \emptyset$ for all $U \in \mathcal{U}$, then the following assertions are equivalent:*

(1) $\mathrm{Cl}_{\mathcal{U}}^{-1}(\emptyset) = \emptyset$; (2) $\mathcal{U} \neq \emptyset$; (3) $\mathrm{Int}_{\mathcal{U}}^{-1}(\emptyset) \neq \emptyset$:
(4) $\mathrm{Int}_{\mathcal{U}}^{-1}(\emptyset) = \mathcal{P}(Y)$.

Remark 46. Note that the implications

$$(1) \iff (4) \implies (3) \quad \text{and} \quad (1) \implies \emptyset \notin \mathrm{Cl}_{\mathcal{U}}^{-1}(\emptyset) \implies (2)$$

do not need any extra condition on $\mathcal{U}$.

Corollary 22. *If $U(\emptyset) = \emptyset$ for all $U \in \mathcal{U}$, then the following assertions are equivalent:*

(1) $\mathrm{Int}_{\mathcal{U}}^{-1}(\emptyset) = \emptyset$; (2) $\mathcal{U} = \emptyset$; (3) $\mathrm{Cl}_{\mathcal{U}}^{-1}(\emptyset) \neq \emptyset$;
(4) $\mathrm{Cl}_{\mathcal{U}}^{-1}(\emptyset) = \mathcal{P}(Y)$.

Theorem 100. *If* $U(X) = Y$ *for all* $U \in \mathcal{U}$, *then*

(1) $\operatorname{Int}_{\mathcal{U}}^{-1}(X) = \mathcal{E}$ *if* $\mathcal{U} \neq \emptyset$ *and* $\operatorname{Int}_{\mathcal{U}}^{-1}(X) = \emptyset$ *if* $\mathcal{U} = \emptyset$;
(2) $\operatorname{Cl}_{\mathcal{U}}^{-1}(X) = \mathcal{D}$ *if* $\mathcal{U} \neq \emptyset$ *and* $\operatorname{Cl}_{\mathcal{U}}^{-1}(X) = \mathcal{P}(Y)$ *if* $\mathcal{U} = \emptyset$.

Theorem 101. *If* $A = \bigcap_{i \in I} A_i$ *for some family* $(A_i)_{i \in I}$ *of subsets of* X, *then*

(1) $\operatorname{Cl}_{\mathcal{U}}^{-1}(A) \subseteq \bigcap_{i \in I} \operatorname{Cl}_{\mathcal{U}}^{-1}(A_i)$; (2) $\bigcup_{i \in I} \operatorname{Int}_{\mathcal{U}}^{-1}(A_i) \subseteq \operatorname{Int}_{\mathcal{U}}^{-1}(A)$.

Theorem 102. *If* $A = \bigcup_{i \in I} A_i$ *for some family* $(A_i)_{i \in I}$ *of subsets of* X, *then*

(1) $\bigcup_{i \in I} \operatorname{Cl}_{\mathcal{U}}^{-1}(A_i) \subseteq \operatorname{Cl}_{\mathcal{U}}^{-1}(A)$; (2) $\operatorname{Int}_{\mathcal{U}}^{-1}(A) \subseteq \bigcap_{i \in I} \operatorname{Int}_{\mathcal{U}}^{-1}(A_i)$.

Theorem 103. *We have*

(1) $\operatorname{Cl}_{\mathcal{U}}^{-1} = \bigcap_{U \in \mathcal{U}} \operatorname{Cl}_{U}^{-1}$; (2) $\operatorname{Int}_{\mathcal{U}}^{-1} = \bigcup_{U \in \mathcal{U}} \operatorname{Int}_{U}^{-1}$.

Corollary 23. *The mapping*

(1) $\mathcal{U} \mapsto \operatorname{Int}_{\mathcal{U}}^{-1}$ *is union-preserving;*
(2) $\mathcal{U} \mapsto \operatorname{Cl}_{\mathcal{U}}^{-1}$ *is intersection-preserving.*

24. Topological Closures and Interiors Derived from Super Relators

From Definition 12, by using the stack $\mathcal{D}_0$, we may naturally derive the following:

Definition 19. We define

$$\operatorname{cl}_{\mathcal{U}} = \operatorname{cl}_{(\mathcal{U}, \mathcal{D}_0)} \quad \text{and} \quad \operatorname{int}_{\mathcal{U}} = \operatorname{int}_{(\mathcal{U}, \mathcal{D}_0)}.$$

Remark 47. The super relations $\operatorname{cl}_{\mathcal{U}}$ and $\operatorname{int}_{\mathcal{U}}$ will be called the *topological closure and interior* generated by the super relator $\mathcal{U}$, respectively.

Thus, by Definitions 12 and 18, we also have the following:

Theorem 104. *For any* $B \subseteq Y$, *we have*

(1) $\operatorname{cl}_{\mathcal{U}}(B) = \{x \in X : \{x\} \in \operatorname{Cl}_{\mathcal{U}}(B)\}$;
(2) $\operatorname{int}_{\mathcal{U}}(B) = \{x \in X : \{x\} \in \operatorname{Int}_{\mathcal{U}}(B)\}$.

Proof. By Definitions 19, 12 and 18, for any $x \in X$, we have

$$x \in \mathrm{cl}_{\mathcal{U}}(B) \iff x \in \mathrm{cl}_{(\mathcal{U},\mathcal{D}_0)}(B) \iff \{x\} \in \mathrm{Cl}_{(\mathcal{U},\mathcal{D}_0)}(B) \iff \{x\} \in \mathrm{Cl}_{\mathcal{U}}(B).$$

□

Moreover, by using Theorems 54 and 55, we can easily establish the following two theorems.

Theorem 105. *For any $B \subseteq Y$, we have*

(1) $\mathrm{int}_{\mathcal{U}}(B) = \{x \in X : \ \exists\ U \in \mathcal{U} : \ U^{\lhd}(x) \subseteq B\}$;
(2) $\mathrm{cl}_{\mathcal{U}}(B) = \{x \in X : \ \forall\ U \in \mathcal{U} : \ U^{\lhd}(x) \cap B \neq \emptyset\}$.

Theorem 106. *For any $B \subseteq Y$, we have*

(1) $\mathrm{cl}_{\mathcal{U}}(B) = \mathrm{int}_{\mathcal{U}}(B^c)^c$; (2) $\mathrm{int}_{\mathcal{U}}(B) = \mathrm{cl}_{\mathcal{U}}(B^c)^c$.

Furthermore, by using Theorems 105 and 106, we can prove the following:

Theorem 107. *For any $B \subseteq Y$, we have*

(1) $\mathrm{cl}_{\mathcal{U}}(B) = \bigcap_{U \in \mathcal{U}} U^{-1}(B)$; (2) $\mathrm{int}_{\mathcal{U}}(B) = \bigcup_{U \in \mathcal{U}} U^{-1\star}(B)$.

Proof. By Theorems 105 and 18, for any $x \in X$ we have

$$x \in \mathrm{cl}_{\mathcal{U}}(B) \iff \forall\ U \in \mathcal{U} : \ \ U^{\lhd}(x) \cap B \neq \emptyset \iff \forall\ U \in \mathcal{U} : \ \ x \in U^{-1}(B) \iff x \in \bigcap_{U \in \mathcal{U}} U^{-1}(B).$$

Therefore, assertion (1) is true.

From assertion (1), by using Theorem 106 and Definition 7, we can see that

$$\mathrm{int}_{\mathcal{U}}(B) = \mathrm{cl}_{\mathcal{U}}(B^c)^c = \left(\bigcap_{U \in \mathcal{U}} U^{-1}(B^c) \right)^c = \bigcup_{U \in \mathcal{U}} U^{-1}(B^c)^c = \bigcup_{U \in \mathcal{U}} U^{-1\star}(B),$$

and thus assertion (2) is also true. □

Now, by using the results of Section 19 and Theorem 96, we can also easily establish the following theorems.

Theorem 108. *If $\mathcal{U}$ is quasi-increasing, then for any $A \subseteq X$ and $B \subseteq Y$*

(1) $A \in \text{Int}_{\mathcal{U}}(B)$ *implies* $A \subseteq \text{int}_{\mathcal{U}}(B)$;
(2) $A \cap \text{cl}_{\mathcal{U}}(B) \neq \emptyset$ *implies* $A \in \text{Cl}_{\mathcal{U}}(B)$.

Remark 48. Assertions (1) and (2), can again be reformulated in the concise forms that

(1) $\text{Int}_{(\mathcal{U},\mathcal{D})}(B) \subseteq \mathcal{P}(\text{int}_{(\mathcal{U},\mathcal{D})}(B))$;
(2) $\mathcal{P}^{\star}(\text{cl}_{(\mathcal{U},\mathcal{D})}(B)) \subseteq \text{Cl}_{(\mathcal{U},\mathcal{D})}(B)$.

Theorem 109. *For any $B_1 \subseteq B_2 \subseteq Y$, we have*

(1) $\text{cl}_{\mathcal{U}}(B_1) \subseteq \text{cl}_{\mathcal{U}}(B_2)$; (2) $\text{int}_{\mathcal{U}}(B_1) \subseteq \text{int}_{\mathcal{U}}(B_2)$.

Theorem 110. *The following assertions are equivalent:*

(1) $\text{cl}_{\mathcal{U}}(\emptyset) = \emptyset$; (2) $\mathcal{U} \neq \emptyset$; (3) $\text{int}_{\mathcal{U}}(Y) \neq \emptyset$; (4) $\text{int}_{\mathcal{U}}(Y) = X$.

Corollary 24. *The following assertions are equivalent:*

(1) $\text{int}_{\mathcal{U}}(Y) = \emptyset$; (2) $\mathcal{U} = \emptyset$; (3) $\text{cl}_{\mathcal{U}}(\emptyset) \neq \emptyset$; (4) $\text{cl}_{\mathcal{U}}(\emptyset) = X$.

Theorem 111. *We have*

(1) $\text{cl}_{\mathcal{U}}(Y) = \{x \in X : \ \forall\ U \in \mathcal{U} : \quad U^{\triangleleft}(x) \neq \emptyset\}$;
(2) $\text{int}_{\mathcal{U}}(\emptyset) = \{x \in X : \ \exists\ U \in \mathcal{U} : \quad U^{\triangleleft}(x) = \emptyset\}$.

Theorem 112. *If $B_1, B_2 \subseteq Y$ such that $B_1 \cap B_2 = \emptyset$, then*

$$\text{cl}_{\mathcal{U}}(B_1) \cap \text{int}_{\mathcal{U}}(B_2) = \emptyset.$$

Theorem 113. *If $B = \bigcap_{i \in I} B_i$ for some family $(B_i)_{i \in I}$ of subsets of Y, then*

(1) $\text{cl}_{\mathcal{U}}(B) \subseteq \bigcap_{i \in I} \text{cl}_{\mathcal{U}}(B_i)$; (2) $\text{int}_{\mathcal{U}}(B) \subseteq \bigcap_{i \in I} \text{int}_{\mathcal{U}}(B_i)$.

Theorem 114. *If $B = \bigcup_{i \in I} B_i$ for some family $(B_i)_{i \in I}$ of subsets of Y, then*

(1) $\bigcup_{i \in I} \text{cl}_{\mathcal{U}}(B_i) \subseteq \text{cl}_{\mathcal{U}}(B)$; (2) $\bigcup_{i \in I} \text{int}_{\mathcal{U}}(B_i) \subseteq \text{int}_{\mathcal{U}}(B)$.

Theorem 115. *If* $\operatorname{card}(\mathcal{U}) = 1$, *then*

(1) $\operatorname{cl}_{\mathcal{U}}$ *is union-preserving*; (2) $\operatorname{int}_{\mathcal{U}}$ *is intersection-preserving.*

Theorem 116. *We have*

(1) $\operatorname{cl}_{\mathcal{U}} = \bigcap_{U \in \mathcal{U}} \operatorname{cl}_U$; (2) $\operatorname{int}_{\mathcal{U}} = \bigcup_{U \in \mathcal{U}} \operatorname{int}_U$.

Corollary 25. *The mapping*

(1) $\mathcal{U} \mapsto \operatorname{int}_{\mathcal{U}}$ *is union-preserving*;
(2) $\mathcal{U} \mapsto \operatorname{cl}_{\mathcal{U}}$ *is intersection-preserving.*

Finally, we note that, by specializing Definitions 13 and 14, we may also naturally introduce the following two definitions.

Definition 20. We define

$$\operatorname{bnd}_{\mathcal{U}} = \operatorname{bnd}_{(\mathcal{U}, \mathcal{D}_0)}.$$

Remark 49. Moreover, if in particular $X = Y$, then we may also define

$$\operatorname{res}_{\mathcal{U}} = \operatorname{res}_{(\mathcal{U}, \mathcal{D}_0)} \quad \text{and} \quad \operatorname{bor}_{\mathcal{U}} = \operatorname{bor}_{(\mathcal{U}, \mathcal{D}_0)}.$$

Definition 21. We define

$$\rho_{\mathcal{U}} = \rho_{(\mathcal{U}, \mathcal{D}_0)} \quad \text{and} \quad \sigma_{\mathcal{U}} = \sigma_{(\mathcal{U}, \mathcal{D}_0)}.$$

Thus, by taking $B = \{y\}$ in Theorem 107, we can easily establish the following:

Theorem 117. *We have*

$$\rho_{\mathcal{U}} = \bigcap \mathcal{U}^{\triangleleft -1} = \left(\bigcap \mathcal{U}^{\triangleleft} \right)^{-1}.$$

25. Fat and Dense Sets Derived from Super Relators

From Definition 15, by using the stack $\mathcal{D}_0$, we may naturally derive the following:

Definition 22. We define

$$\mathcal{D}_{\mathcal{U}} = \mathcal{D}_{(\mathcal{U}, \mathcal{D}_0)} \quad \text{and} \quad \mathcal{E}_{\mathcal{U}} = \mathcal{E}_{(\mathcal{U}, \mathcal{D}_0)}.$$

Remark 50. The members of the families $\mathcal{E}_{\mathcal{U}}$ and $\mathcal{D}_{\mathcal{U}}$ will be called the *fat and dense subsets* of the super relator space $(X, Y)(\mathcal{U})$, respectively.

Thus, by the corresponding definitions, we can at once state the following:

Theorem 118. *We have*

(1) $\mathcal{D}_{\mathcal{U}} = \{D \subseteq Y : \ X = \mathrm{cl}_{\mathcal{U}}(D)\}$;
(2) $\mathcal{E}_{\mathcal{U}} = \{E \subseteq Y : \ \mathrm{int}_{\mathcal{U}}(E)\} \neq \emptyset\}$.

Proof. By Definitions 22, 15 and 19, for any $D \subseteq Y$ we have

$$D \in \mathcal{D}_{\mathcal{U}} \iff D \in \mathcal{D}_{(\mathcal{U}, \mathcal{D}_0)} \iff X = \mathrm{cl}_{(\mathcal{U}, \mathcal{D}_0)}(D)$$
$$\iff X = \mathrm{cl}_{\mathcal{U}}(D).$$
□

Moreover, by specializing the results of Section 20, we can easily establish the following theorems.

Theorem 119. *We have*

(1) $\mathcal{E}_{\mathcal{U}} = \{E \subseteq Y : \quad \exists\ x \in X : \quad \exists\ U \in \mathcal{U} : \quad U^{\triangleleft}(x) \subseteq E\}$;
(2) $\mathcal{D}_{\mathcal{U}} = \{D \subseteq Y : \quad \forall\ x \in X : \quad \forall\ U \in \mathcal{U} : \quad U^{\triangleleft}(x) \cap D \neq \emptyset\}$.

Theorem 120. *We have*

(1) $\mathcal{D}_{\mathcal{U}} = \{D \subseteq Y : \quad \forall\ U \in \mathcal{U} : \quad X = U^{\triangleleft-1}[D]\}$;
(2) $\mathcal{E}_{\mathcal{U}} = \{E \subseteq Y : \quad \exists\ U \in \mathcal{U} : \quad X \neq U^{\triangleleft-1}[E^c]\}$.

Proof. Assertion (1) can be derived from not only Theorem 119 but also Theorem 107.

Namely, by Definitions 4 and 2, we have $U^{-1}(D) = U^{\triangleleft-1\triangleright}(D) = U^{\triangleleft-1}[D]$ for all $D \subseteq Y$ and $U \in \mathcal{U}$. □

Theorem 121. *We have*

(1) $\mathcal{D}_{\mathcal{U}} = \{D \subseteq Y : \ D^c \notin \mathcal{E}_{\mathcal{U}}\}$; (2) $\mathcal{E}_{\mathcal{U}} = \{E \subseteq Y : \ E^c \notin \mathcal{D}_{\mathcal{U}}\}$.

Theorem 122. *We have*

(1) $\mathcal{D}_{\mathcal{U}} = \{D \subseteq Y : \ \forall\ E \in \mathcal{E}_{\mathcal{U}} : \ D \cap E \neq \emptyset\}$;
(2) $\mathcal{E}_{\mathcal{U}} = \{E \subseteq Y : \ \forall\ D \in \mathcal{D}_{\mathcal{U}} : \ D \cap E \neq \emptyset\}$.

Theorem 123. *The families $\mathcal{D}_{\mathcal{U}}$ and $\mathcal{E}_{\mathcal{U}}$ are stacks on Y.*

Theorem 124. *The following assertions are equivalent:*

(1) $X \neq \emptyset$ *and* $\mathcal{U} \neq \emptyset$;
(2) $\emptyset \notin \mathcal{D}_{\mathcal{U}}$; (3) $\mathcal{E}_{\mathcal{U}} \neq \emptyset$; (4) $Y \in \mathcal{E}_{\mathcal{U}}$; (5) $\mathcal{D}_{\mathcal{U}} \neq \mathcal{P}(Y)$.

Remark 51. Property (1) can be briefly expressed by saying that the super relator $\mathcal{U}$ is non-degenerated.

Theorem 125. *The following assertions are equivalent:*

(1) $U^{\triangleleft}(x) \neq \emptyset$ *for all* $x \in X$ *and* $U \in \mathcal{U}$;
(2) $\emptyset \notin \mathcal{E}_{\mathcal{U}}$; (3) $\mathcal{D}_{\mathcal{U}} \neq \emptyset$; (4) $Y \in \mathcal{D}_{\mathcal{U}}$; (5) $\mathcal{E}_{\mathcal{U}} \neq \mathcal{P}(Y)$.

Remark 52. Property (1) can be briefly expressed by saying that ordinary relator $\mathcal{U}^{\triangleleft}$ is non-partial.

Theorem 126. *We have*

(1) $\mathcal{D}_{\mathcal{U}} = \bigcap_{U \in \mathcal{U}} \mathcal{D}_U$; (2) $\mathcal{E}_{\mathcal{U}} = \bigcup_{U \in \mathcal{U}} \mathcal{E}_U$.

Corollary 26. *The mapping*

(1) $\mathcal{U} \mapsto \mathcal{E}_{\mathcal{U}}$ *is union-preserving;*
(2) $\mathcal{U} \mapsto \mathcal{D}_{\mathcal{U}}$ *is intersection-preserving.*

Remark 53. By using Remark 38 and the notation $\mathfrak{U}_{\mathcal{U}} = \mathfrak{U}_{(\mathcal{U}, \mathcal{D}_0)}$, we can also state that $\mathcal{E}_{\mathcal{U}} = \bigcup_{x \in X} \mathfrak{U}_{\mathcal{U}}(x)$.

26. Open and Closed sets Derived from Super Relators

Notation 8. *In this section, we shall assume that* $\mathcal{U}$ *is a super relator on* X.

From Definition 16, by using the stack $\mathcal{D}_0$, we may naturally derive the following:

Definition 23. We define

$$\tau_{\mathcal{U}} = \tau_{(\mathcal{U}, \mathcal{D}_0)}; \quad \tau\!\!\!\!-_{\mathcal{U}} = \tau\!\!\!\!-_{(\mathcal{U}, \mathcal{D}_0)} \quad \text{and} \quad \mathcal{T}_{\mathcal{U}} = \mathcal{T}_{(\mathcal{U}, \mathcal{D}_0)}; \quad \mathcal{F}_{\mathcal{U}} = \mathcal{F}_{(\mathcal{U}, \mathcal{D}_0)}.$$

Remark 54. The members of the families, $\tau_{\mathcal{U}}$ and $\mathcal{T}_{\mathcal{U}}$, will be called the *proximally open and topologically open subsets* of the super relator space $X(\mathcal{U})$, respectively.

Thus, by the corresponding definitions, we can at once state the following:

Theorem 127. *We have*

(1) $\tau_{\mathcal{U}} = \{A \subseteq X : \ A \in \operatorname{Int}_{\mathcal{U}}(A)\}$;
(2) $\tau_{\mathcal{U}} = \{A \subseteq X : \ A^c \notin \operatorname{Cl}_{\mathcal{U}}(A)\}$;
(3) $\mathcal{T}_{\mathcal{U}} = \{A \subseteq X : \ A \subseteq \operatorname{int}_{\mathcal{U}}(A)\}$;
(4) $\mathcal{F}_{\mathcal{U}} = \{A \subseteq X : \operatorname{cl}_{\mathcal{U}}(A) \subseteq A\}$.

Proof. For instance, by Definitions 23, 16 and 18, for any $A \subseteq X$, we have

$$A \in \tau_{\mathcal{U}} \iff A \in \tau_{(\mathcal{U},\mathcal{D}_0)} \iff A \in \operatorname{Int}_{(\mathcal{U},\mathcal{D}_0)}(A) \iff A \in \operatorname{Int}_{\mathcal{U}}(A).$$

□

Moreover, by specializing the results of Sections 21 and 22, we can easily establish the following theorems.

Theorem 128. *We have*

(1) $\tau_{\mathcal{U}} = \{A \subseteq X : \ \exists\ U \in \mathcal{U} : \ U(A) \subseteq A\}$;
(2) $\mathcal{T}_{\mathcal{U}} = \{A \subseteq X : \ \forall\ x \in A : \ \exists\ U \in \mathcal{U} : \ U^{\triangleleft}(x) \subseteq A\}$.

Theorem 129. *We have*

(1) $\digamma_{\mathcal{U}} = \{A^c : A \in \tau_{\mathcal{U}}\}$; (2) $\tau_{\mathcal{U}} = \{A^c : A \in \digamma_{\mathcal{U}}\}$;
(3) $\mathcal{F}_{\mathcal{U}} = \{A^c : \ A \in \mathcal{T}_{\mathcal{U}}\}$; (4) $\mathcal{T}_{\mathcal{U}} = \{A^c : \ A \in \mathcal{F}_{\mathcal{U}}\}$.

Theorem 130. *If $\mathcal{U}$ is quasi-increasing, then*

(1) $\tau_{\mathcal{U}} \subseteq \mathcal{T}_{\mathcal{U}}$; (2) $\digamma_{\mathcal{U}} \subseteq \mathcal{F}_{\mathcal{U}}$.

Remark 55. If card $(\mathcal{U}) = 1$ and $\mathcal{U}$ is union-preserving, then the corresponding equalities can also be stated.

Theorem 131. *We have*

(1) $\mathcal{T}_{\mathcal{U}} \setminus \{\emptyset\} \subseteq \mathcal{E}_{\mathcal{U}}$; (2) $\mathcal{D}_{\mathcal{U}} \cap \mathcal{F}_{\mathcal{U}} \subseteq \{X\}$.

Remark 56. Hence, by using global complementations, we can easily infer that $\mathcal{F}_{\mathcal{U}} \subseteq (\mathcal{D}_{\mathcal{U}})^c \cup \{X\}$ and $\mathcal{D}_{\mathcal{U}} \subseteq (\mathcal{F}_{\mathcal{U}})^c \cup \{X\}$.

Theorem 132. *For any* $A \subseteq X$,

(1) $\bigcup(\mathcal{T}_{\mathcal{U}} \cap \mathcal{P}(A)) \subseteq \operatorname{int}_{\mathcal{U}}(A)$;
(2) $\mathcal{P}(A) \cap (\mathcal{T}_{\mathcal{U}} \setminus \{\emptyset\}) \neq \emptyset$ *implies* $A \in \mathcal{E}_{\mathcal{U}}$;
(3) $\mathcal{P}[\tau_{\mathcal{U}} \cap \mathcal{P}(A)] \subseteq \operatorname{Int}_{\mathcal{U}}(A)$ *if* $\mathcal{U}$ *is increasing.*

Remark 57. Conditions in order that the converse implication and the corresponding equalities could hold will be given later.

Theorem 133. *The following assertions are equivalent:*

(1) $\emptyset \in \tau_{\mathcal{U}}$; (2) $X \in \digamma_{\mathcal{U}}$;
(3) *there exists* $U \in \mathcal{U}$ *such that* $U(\emptyset) = \emptyset$.

Theorem 134. *The following assertions are equivalent:*

(1) $X \in \tau_{\mathcal{U}}$; (2) $\emptyset \in \digamma_{\mathcal{U}}$; (3) $\mathcal{U} \neq \emptyset$.

Remark 58. Thus, $\tau_{\mathcal{U}}$ is a *minimal structure* on X if and only if there exists $U \in \mathcal{U}$ such that $U(\emptyset) = \emptyset$.

Theorem 135. *If* $A \subseteq X$ *and* $B \in \tau_{\mathcal{U}}$ *such that* $A \cap B = \emptyset$, *then*

(1) $B \notin \operatorname{Cl}_{\mathcal{U}}(A)$; (2) $B \in \operatorname{Int}_{\mathcal{U}}(A^c)$.

Theorem 136. *The family*

(1) $\mathcal{T}_{\mathcal{U}}$ *is closed under arbitrary unions;*
(2) $\mathcal{F}_{\mathcal{U}}$ *is closed under arbitrary intersection.*

Corollary 27. *We always have* $\emptyset \in \mathcal{T}_{\mathcal{U}}$ *and* $X \in \mathcal{F}_{\mathcal{U}}$.

Theorem 137. *The following assertions are equivalent:*

(1) $X \in \mathcal{T}_{\mathcal{U}}$; (2) $\emptyset \in \mathcal{F}_{\mathcal{U}}$; (3) *either* $X = \emptyset$ *or* $\mathcal{U} \neq \emptyset$.

Remark 59. Thus, $\mathcal{T}_{\mathcal{U}}$ is a *generalized topology* on X if and only if either $X = \emptyset$ or $\mathcal{U} \neq \emptyset$.

Theorem 138. *If* $A \subseteq X$ *and* $B \in \mathcal{T}_{\mathcal{U}}$ *such that* $A \cap B = \emptyset$, *then*

(1) $\operatorname{cl}_{\mathcal{U}}(A) \cap B = \emptyset$; (2) $B \subseteq \operatorname{int}_{\mathcal{U}}(A^c)$.

In addition to Definition 23, we may also naturally introduce the following:

Definition 24. We define

$$\mathcal{N}_{\mathcal{U}} = \mathcal{N}_{(\mathcal{U},\mathcal{D}_0)} \quad \text{and} \quad \mathcal{M}_{\mathcal{U}} = \mathcal{M}_{(\mathcal{U},\mathcal{D}_0)}.$$

Remark 60. The members of the family $\mathcal{N}_{\mathcal{U}}$ will be called the *rare or nowhere dense subsets* of the super relator space $X(\mathcal{U})$.

Thus, by Definitions 17, 19 and 22, we can at once state the following:

Theorem 139. *We have*

(1) $\mathcal{N}_{\mathcal{U}} = \{A \subseteq X : \; \mathrm{cl}_{\mathcal{U}}(A) \notin \mathcal{E}_{\mathcal{U}}\}$;
(2) $\mathcal{M}_{\mathcal{U}} = \{A \subseteq X : \; \mathrm{int}_{\mathcal{U}}(A) \in \mathcal{D}_{\mathcal{U}}\}$.

Moreover, by Theorem 87, we can also at once state the following:

Theorem 140. *We have*

(1) $\mathcal{M}_{\mathcal{U}} = \{A^c : \; A \in \mathcal{N}_{\mathcal{U}}\}$; (2) $\mathcal{N}_{\mathcal{U}} = \{A^c : \; A \in \mathcal{M}_{\mathcal{U}}\}$.

27. Proximal Closures and Interiors Derived from Ordinary Relators and Stacks

Notation 9. *In this and the following two sections, we shall assume that $\mathcal{R}$ is an ordinary relator on X to Y, and $\mathcal{D}$ is a stack on Y.*

From Definition 10, by using the union-preserving super relator $\mathcal{R}^{\triangleright} = \{R^{\triangleright} : \; R \in \mathcal{R}\}$, we may naturally derive the following:

Definition 25. We define

$$\mathrm{Cl}_{(\mathcal{R},\mathcal{D})} = \mathrm{Cl}_{(\mathcal{R}^{\triangleright},\mathcal{D})} \quad \text{and} \quad \mathrm{Int}_{(\mathcal{R},\mathcal{D})} = \mathrm{Int}_{(\mathcal{R}^{\triangleright},\mathcal{D})}\,.$$

Remark 61. The hyper relations $\mathrm{Cl}_{(\mathcal{R},\mathcal{D})}$ and $\mathrm{Int}_{(\mathcal{R},\mathcal{D})}$ will be called the *proximal closure and interior* generated by the ordinary relator $\mathcal{R}$ and the stack $\mathcal{D}$, respectively.

Thus, from the results of Sections 16, 17 and 18, we can immediately derive the following theorems.

Theorem 141. *For any $B \subseteq Y$, we have*

(1) $\operatorname{Cl}_{(\mathcal{R},\mathcal{D})}(B) = \{A \subseteq X : \quad \forall\ R \in \mathcal{R} : \ R[A] \cap B \in \mathcal{D}\}$;
(2) $\operatorname{Int}_{(\mathcal{R},\mathcal{D})}(B) = \{A \subseteq X : \quad \exists\ R \in \mathcal{R} : \ R[A] \setminus B \notin \mathcal{D}\}$.

Remark 62. Due to Remark 26, we can also state that

$$\operatorname{Int}_{(\mathcal{R},\mathcal{D})}(B) = \{A \subseteq X : \quad \exists\ R \in \mathcal{R} : \ R[A]^c \cup B \in \mathcal{E}\}.$$

Theorem 142. *We have*

(1) $\operatorname{Cl}_{(\mathcal{R},\mathcal{D})}(B) = \operatorname{Int}_{(\mathcal{U},\mathcal{D})}(B^c)^c$; (2) $\operatorname{Int}_{(\mathcal{R},\mathcal{D})}(B) = \operatorname{Cl}_{(\mathcal{R},\mathcal{D})}(B^c)^c$.

Theorem 143. *For any $B_1 \subseteq B_2 \subseteq Y$, we have*

(1) $\operatorname{Cl}_{(\mathcal{R},\mathcal{D})}(B_1) \subseteq \operatorname{Cl}_{(\mathcal{R},\mathcal{D}}(B_2)$; (2) $\operatorname{Int}_{(\mathcal{R},\mathcal{D})}(B_1) \subseteq \operatorname{Int}_{(\mathcal{R},\mathcal{D})}(B_2)$.

Theorem 144. *The following assertions are equivalent:*

(1) $\operatorname{Cl}_{(\mathcal{R},\mathcal{D})}(\emptyset) = \emptyset$; (2) $\mathcal{R} \neq \emptyset$ *and* $\emptyset \notin \mathcal{D}$;
(3) $\operatorname{Int}_{(\mathcal{R},\mathcal{D})}(Y) \neq \emptyset$; (4) $\operatorname{Int}_{(\mathcal{R},\mathcal{D})}(Y) = \mathcal{P}(X)$.

Corollary 28. *The following assertions are equivalent:*

(1) $\operatorname{Int}_{(\mathcal{R},\mathcal{D})}(Y) = \emptyset$; (2) $\mathcal{R} = \emptyset$ *or* $\emptyset \in \mathcal{D}$;
(3) $\operatorname{Cl}_{(\mathcal{R},\mathcal{D})}(\emptyset) \neq \emptyset$; (4) $\operatorname{Cl}_{(\mathcal{R},\mathcal{D})}(\emptyset) = \mathcal{P}(X)$.

Theorem 145. *We have*

(1) $\operatorname{Int}_{(\mathcal{R},\mathcal{D})}(\emptyset) = \{A \subseteq X : \ \exists\ R \in \mathcal{R} : \ R[A] \notin \mathcal{D}\}$;
(2) $\operatorname{Cl}_{(\mathcal{R},\mathcal{D})}(Y) = \{A \subseteq X : \ \forall\ R \in \mathcal{R} : \ R[A] \in \mathcal{D}\}$.

Theorem 146. *If $B_1, B_2 \subseteq Y$ such that $B_1 \cap B_2 = \emptyset$, then*

$$\operatorname{Cl}_{(\mathcal{R},\mathcal{D})}(B_1) \cap \operatorname{Int}_{(\mathcal{R},\mathcal{D})}(B_2) = \emptyset.$$

Theorem 147. *If $B = \bigcap_{i \in I} B_i$ for some family $(B_i)_{i \in I}$ of subsets of Y, then*

(1) $\operatorname{Cl}_{(\mathcal{R},\mathcal{D})}(B) \subseteq \bigcap_{i \in I} \operatorname{Cl}_{(\mathcal{R},\mathcal{D})}(B_i)$;
(2) $\operatorname{Int}_{(\mathcal{R},\mathcal{D})}(B) \subseteq \bigcap_{i \in I} \operatorname{Int}_{(\mathcal{R},\mathcal{D})}(B_i)$.

Theorem 148. *If* $B = \bigcup_{i \in I} B_i$ *for some family* $(B_i)_{i \in I}$ *of subsets of* Y*, then*

(1) $\bigcup_{i \in I} \mathrm{Cl}_{(\mathcal{R},\mathcal{D})}(B_i) \subseteq \mathrm{Cl}_{(\mathcal{R},\mathcal{D})}(B)$;
(2) $\bigcup_{i \in I} \mathrm{Int}_{(\mathcal{R},\mathcal{D})}(B_i) \subseteq \mathrm{Int}_{(\mathcal{R},\mathcal{D})}(B)$.

Theorem 149. *We have*

(1) $\mathrm{Cl}_{(\mathcal{R},\mathcal{D})} = \bigcap_{R \in \mathcal{R}} \mathrm{Cl}_{(R,\mathcal{D})}$; (2) $\mathrm{Int}_{(\mathcal{R},\mathcal{D})} = \bigcup_{R \in \mathcal{R}} \mathrm{Int}_{(R,\mathcal{D})}$.

Corollary 29. *The mapping*

(1) $\mathcal{R} \mapsto \mathrm{Int}_{(\mathcal{R},\mathcal{D})}$ *is union-preserving;*
(2) $\mathcal{R} \mapsto \mathrm{Cl}_{(\mathcal{R},\mathcal{D})}$ *is intersection-preserving.*

Theorem 150. *If* $\mathcal{D} = \bigcap_{i \in I} \mathcal{D}_i$ *for some family* $(\mathcal{D}_i)_{i \in I}$ *of stacks on* Y*, then*

(1) $\mathrm{Cl}_{(\mathcal{R},\mathcal{D})} = \bigcap_{i \in I} \mathrm{Cl}_{(\mathcal{R},\mathcal{D}_i)}$; (2) $\mathrm{Int}_{(\mathcal{R},\mathcal{D})} = \bigcup_{i \in I} \mathrm{Int}_{(\mathcal{R},\mathcal{D}_i)}$.

Theorem 151. *For any* $A_1 \subseteq A_2 \subseteq X$*, we have*

(1) $\mathrm{Cl}^{-1}_{(\mathcal{R},\mathcal{D})}(A_1) \subseteq \mathrm{Cl}^{-1}_{(\mathcal{R},\mathcal{D})}(A_2)$; (2) $\mathrm{Int}^{-1}_{(\mathcal{R},\mathcal{D})}(A_2) \subseteq \mathrm{Int}^{-1}_{(\mathcal{R},\mathcal{D})}(A_1)$.

Theorem 152. *The following assertions are equivalent:*

(1) $\mathrm{Cl}^{-1}_{(\mathcal{R},\mathcal{D})}(\emptyset) = \emptyset$; (2) $\mathcal{R} \neq \emptyset$ *and* $\emptyset \notin \mathcal{D}$;
(3) $\mathrm{Int}^{-1}_{(\mathcal{R},\mathcal{D})}(\emptyset) \neq \emptyset$; (4) $\mathrm{Int}^{-1}_{(\mathcal{R},\mathcal{D})}(\emptyset) = \mathcal{P}(Y)$.

Corollary 30. *The following assertions are equivalent:*

(1) $\mathrm{Int}^{-1}_{(\mathcal{R},\mathcal{D})}(\emptyset) = \emptyset$; (2) $\mathcal{R} = \emptyset$ *or* $\emptyset \in \mathcal{D}$;
(3) $\mathrm{Cl}^{-1}_{(\mathcal{R},\mathcal{D})}(\emptyset) \neq \emptyset$; (4) $\mathrm{Cl}^{-1}_{(\mathcal{R},\mathcal{D})}(\emptyset) = \mathcal{P}(Y)$.

Theorem 153. *If* $R[X] = Y$ *for all* $R \in \mathcal{R}$*, then*

(1) $\mathrm{Int}^{-1}_{(\mathcal{R},\mathcal{D})}(X) = \mathcal{E}$ *if* $\mathcal{R} \neq \emptyset$ *and* $\mathrm{Int}^{-1}_{(\mathcal{R},\mathcal{D})}(X) = \emptyset$ *if* $\mathcal{R} = \emptyset$;

(2) $\mathrm{Cl}^{-1}_{(\mathcal{R},\mathcal{D})}(X) = \mathcal{D}$ *if* $\mathcal{R} \neq \emptyset$ *and* $\mathrm{Cl}^{-1}_{(\mathcal{R},\mathcal{D})}(X) = \mathcal{P}(Y)$ *if* $\mathcal{R} = \emptyset$.

Theorem 154. *If* $A=\bigcap_{i\in I}A_i$ *for some family* $(A_i)_{i\in I}$ *of subsets of* X, *then*

(1) $\mathrm{Cl}^{-1}_{(\mathcal{R},\mathcal{D})}(A)\subseteq\bigcap_{i\in I}\mathrm{Cl}^{-1}_{(\mathcal{R},\mathcal{D})}(A_i)$;

(2) $\bigcup_{i\in I}\mathrm{Int}^{-1}_{(\mathcal{R},\mathcal{D})}(A_i)\subseteq\mathrm{Int}^{-1}_{(\mathcal{R},\mathcal{D})}(A)$.

Theorem 155. *If* $A=\bigcup_{i\in I}A_i$ *for some family* $(A_i)_{i\in I}$ *of subsets of* X, *then*

(1) $\bigcup_{i\in I}\mathrm{Cl}^{-1}_{(\mathcal{R},\mathcal{D})}(A_i)\subseteq\mathrm{Cl}^{-1}_{(\mathcal{R},\mathcal{D})}(A)$;

(2) $\mathrm{Int}^{-1}_{(\mathcal{R},\mathcal{D})}(A)\subseteq\bigcap_{i\in I}\mathrm{Int}^{-1}_{(\mathcal{R},\mathcal{D})}(A_i)$.

Theorem 156. *We have*

(1) $\mathrm{Cl}^{-1}_{(\mathcal{R},\mathcal{D})}=\bigcap_{R\in\mathcal{R}}\mathrm{Cl}^{-1}_{(R,\mathcal{D})}$; (2) $\mathrm{Int}^{-1}_{(\mathcal{R},\mathcal{D})}=\bigcup_{R\in\mathcal{R}}\mathrm{Int}^{-1}_{(R,\mathcal{D})}$.

Corollary 31. *The mapping*

(1) $\mathcal{R}\mapsto\mathrm{Int}^{-1}_{(\mathcal{R},\mathcal{D})}$ *is union-preserving*;

(2) $\mathcal{R}\mapsto\mathrm{Cl}^{-1}_{(\mathcal{R},\mathcal{D})}$ *is intersection-preserving.*

Theorem 157. *If* $\mathcal{D}=\bigcap_{i\in I}\mathcal{D}_i$ *for some family* $(\mathcal{D}_i)_{i\in I}$ *of stacks on* Y, *then*

(1) $\mathrm{Cl}^{-1}_{(\mathcal{R},\mathcal{D})}=\bigcap_{i\in I}\mathrm{Cl}^{-1}_{(\mathcal{R},\mathcal{D}_i)}$; (2) $\mathrm{Int}^{-1}_{(\mathcal{R},\mathcal{D})}=\bigcup_{i\in I}\mathrm{Int}^{-1}_{(\mathcal{R},\mathcal{D}_i)}$.

28. Topological Closures and Interiors Derived from Ordinary Relators and Stacks

From Definition 12, by using the super relator $\mathcal{R}^{\triangleright}$, we may naturally derive the following:

Definition 26. We define

$$\mathrm{cl}_{(\mathcal{R},\mathcal{D})}=\mathrm{cl}_{(\mathcal{R}^{\triangleright},\mathcal{D})}\quad\text{and}\quad\mathrm{int}_{(\mathcal{R},\mathcal{D})}(B)=\mathrm{int}_{(\mathcal{R}^{\triangleright},\mathcal{D})}\,.$$

Remark 63. The super relations $\mathrm{cl}_{(\mathcal{R},\mathcal{D})}$ and $\mathrm{int}_{(\mathcal{R},\mathcal{D})}$ will be called the *topological closure and interior* generated by the ordinary relator $\mathcal{U}$ and the stack $\mathcal{D}$, respectively.

Thus, by the corresponding definitions, we can at once state the following:

Theorem 158. *For any $B \subseteq Y$, we have*

(1) $\text{cl}_{(\mathcal{R},\mathcal{D})}(B) = \{x \in X : \ \{x\} \in \ \text{Cl}_{(\mathcal{R},\mathcal{D})}(B)\}$;
(2) $\text{int}_{(\mathcal{R},\mathcal{D})}(B) = \{x \in X : \ \{x\} \in \text{Int}_{(\mathcal{R},\mathcal{D})}(B)\}$.

Moreover, by specializing the results of Section 19, we can easily establish the following theorems.

Theorem 159. *For any $B \subseteq Y$, we have*

(1) $\text{cl}_{(\mathcal{R},\mathcal{D})}(B) = \{x \in X : \ \forall \ R \in \mathcal{R} : \ R(x) \cap B \in \mathcal{D}\}$;
(2) $\text{int}_{(\mathcal{R},\mathcal{D})}(B) = \{x \in X : \ \exists \ R \in \mathcal{R} : \ R(x) \setminus B \notin \mathcal{D}\}$.

Theorem 160. *For any $B \subseteq Y$, we have*

(1) $\text{cl}_{(\mathcal{R},\mathcal{D})}(B) = \text{int}_{(\mathcal{R},\mathcal{D})}(B^c)^c$; (2) $\text{int}_{(\mathcal{R},\mathcal{D})}(B) = \text{cl}_{(\mathcal{R},\mathcal{D})}(B^c)^c$.

Theorem 161. *For any $A \subseteq X$ and $B \subseteq Y$,*

(1) $A \in \text{Int}_{(\mathcal{R},\mathcal{D})}(B)$ *implies* $A \subseteq \text{int}_{(\mathcal{R},\mathcal{D})}(B)$;
(2) $A \cap \text{cl}_{(\mathcal{R},\mathcal{D})}(B) \neq \emptyset$ *implies* $A \in \text{Cl}_{(\mathcal{R},\mathcal{D})}(B)$.

Remark 64. Assertions (1) and (2), can be reformulated in the forms that

(1) $\text{Int}_{(\mathcal{R},\mathcal{D})}(B) \subseteq \mathcal{P}(\text{int}_{(\mathcal{R},\mathcal{D})}(B))$;
(2) $\mathcal{P}^\star(\text{cl}_{(\mathcal{R},\mathcal{D})}(B)) \subseteq \text{Cl}_{(\mathcal{U},\mathcal{D})}(B)$.

Theorem 162. *For any $B_1 \subseteq B_2 \subseteq Y$, we have*

(1) $\text{cl}_{(\mathcal{R},\mathcal{D})}(B_1) \subseteq \text{cl}_{(\mathcal{R},\mathcal{D}}(B_2)$; (2) $\text{int}_{(\mathcal{R},\mathcal{D})}(B_1) \subseteq \text{int}_{(\mathcal{R},\mathcal{D})}(B_2)$.

Theorem 163. *The following assertions are equivalent:*

(1) $\text{cl}_{(\mathcal{R},\mathcal{D})}(\emptyset) = \emptyset$; (2) $\mathcal{R} \neq \emptyset$ *and* $\emptyset \notin \mathcal{D}$;
(3) $\text{int}_{(\mathcal{R},\mathcal{D})}(Y) \neq \emptyset$; (4) $\text{int}_{(\mathcal{R},\mathcal{D})}(Y) = X$.

Corollary 32. *The following assertions are equivalent:*

(1) $\text{int}_{(\mathcal{R},\mathcal{D})}(Y) = \emptyset$; (2) $\mathcal{R} = \emptyset$ *or* $\emptyset \in \mathcal{D}$;
(3) $\text{cl}_{(\mathcal{R},\mathcal{D})}(\emptyset) \neq \emptyset$; (4) $\text{cl}_{(\mathcal{R},\mathcal{D})}(\emptyset) = X$.

Theorem 164. *We have*

(1) $\operatorname{cl}_{(\mathcal{R},\mathcal{D})}(Y)=\{x\in X:\ \forall\ R\in\mathcal{R}:\ \ R(x)\in\mathcal{D}\}$;
(2) $\operatorname{int}_{(\mathcal{R},\mathcal{D})}(\emptyset)=\{x\in X:\ \exists\ R\in\mathcal{R}:\ \ R(x)\notin\mathcal{D}\}$.

Theorem 165. *If $B_1, B_2 \subseteq Y$ such that $B_1 \cap B_2 = \emptyset$, then*

$$\operatorname{cl}_{(\mathcal{R},\mathcal{D})}(B_1)\cap\operatorname{int}_{(\mathcal{R},\mathcal{D})}(B_2)=\emptyset.$$

Theorem 166. *If $B=\bigcap_{i\in I}B_i$ for some family $(B_i)_{i\in I}$ of subsets of Y, then*

(1) $\operatorname{cl}_{(\mathcal{R},\mathcal{D})}(B)\subseteq\bigcap_{i\in I}\operatorname{cl}_{(\mathcal{R},\mathcal{D})}(B_i)$;
(2) $\operatorname{int}_{(\mathcal{R},\mathcal{D})}(B)\subseteq\bigcap_{i\in I}\operatorname{int}_{(\mathcal{R},\mathcal{D})}(B_i)$.

Theorem 167. *If $B=\bigcup_{i\in I}B_i$ for some family $(B_i)_{i\in I}$ of subsets of Y, then*

(1) $\bigcup_{i\in I}\operatorname{cl}_{(\mathcal{R},\mathcal{D})}(B_i)\subseteq\operatorname{cl}_{(\mathcal{R},\mathcal{D})}(B)$;
(2) $\bigcup_{i\in I}\operatorname{int}_{(\mathcal{R},\mathcal{D})}(B_i)\subseteq\operatorname{int}_{(\mathcal{R},\mathcal{D})}(B)$.

Theorem 168. *We have*

(1) $\operatorname{cl}_{(\mathcal{R},\mathcal{D})}=\bigcap_{R\in\mathcal{R}}\operatorname{cl}_{(R,\mathcal{D})}$; (2) $\operatorname{int}_{(\mathcal{R},\mathcal{D})}=\bigcup_{R\in\mathcal{R}}\operatorname{int}_{(R,\mathcal{D})}$.

Corollary 33. *The mapping*

(1) $\mathcal{R}\mapsto\operatorname{int}_{(\mathcal{R},\mathcal{D})}$ *is union-preserving;*
(2) $\mathcal{R}\mapsto\operatorname{cl}_{(\mathcal{R},\mathcal{D})}$ *is intersection-preserving.*

Theorem 169. *If $\mathcal{D}=\bigcap_{i\in I}\mathcal{D}_i$ for some family $(\mathcal{D}_i)_{i\in I}$ of stacks on Y, then*

(1) $\operatorname{cl}_{(\mathcal{R},\mathcal{D})}=\bigcap_{i\in I}\operatorname{cl}_{(\mathcal{R},\mathcal{D}_i)}$; (2) $\operatorname{int}_{(\mathcal{R},\mathcal{D})}=\bigcup_{i\in I}\operatorname{int}_{(\mathcal{R},\mathcal{D}_i)}$.

Now, by specializing Definitions 13 and 14, we may also naturally introduce the following two definitions.

Definition 27. We define

$$\operatorname{bnd}_{(\mathcal{R},\mathcal{D})}=\operatorname{bnd}_{(\mathcal{R}^{\triangleright},\mathcal{D})}\,.$$

Remark 65. Moreover, if in particular $X = Y$, then we may also define

$$\operatorname{res}_{(\mathcal{R},\mathcal{D})} = \operatorname{res}_{(\mathcal{R}^{\triangleright},\mathcal{D})} \quad \text{and} \quad \operatorname{bor}_{(\mathcal{U},\mathcal{D})} = \operatorname{bor}_{(\mathcal{R}^{\triangleright},\mathcal{D})}\,.$$

Definition 28. We define

$$\rho_{(\mathcal{R},\mathcal{D})} = \rho_{(\mathcal{R}^{\triangleright},\mathcal{D})} \quad \text{and} \quad \sigma_{(\mathcal{R},\mathcal{D})} = \sigma_{(\mathcal{R}^{\triangleright},\mathcal{D})}.$$

Remark 66. The ordinary relations $\rho_{(\mathcal{R},\mathcal{D})}$ and $\sigma_{(\mathcal{R},\mathcal{D})}$ will be called the *infinitesimal closure and interior* generated by the ordinary relator $\mathcal{R}$ and the stack $\mathcal{D}$, respectively.

29. Fat and Dense Sets Derived from Ordinary Relators and Stacks

From Definition 15, by using the super relator $\mathcal{R}^{\triangleright}$ we may naturally derive the following:

Definition 29. We define

$$\mathcal{D}_{(\mathcal{R},\mathcal{D})} = \mathcal{D}_{(\mathcal{R}^{\triangleright},\mathcal{D})} \quad \text{and} \quad \mathcal{E}_{(\mathcal{R},\mathcal{D})} = \mathcal{E}_{(\mathcal{R}^{\triangleright},\mathcal{D})}.$$

Remark 67. The members of the families $\mathcal{E}_{(\mathcal{R},\mathcal{D})}$ and $\mathcal{D}_{(\mathcal{R},\mathcal{D})}$ will be called the *fat and dense subsets* of the stacked ordinary relator space $(X, Y)(\mathcal{R})(\mathcal{D})$, respectively.

Thus, by the corresponding definitions, we can at once state the following:

Theorem 170. *We have*

(1) $\mathcal{D}_{(\mathcal{R},\mathcal{D})} = \{D \subseteq Y : \ X = \operatorname{cl}_{(\mathcal{R},\mathcal{D})}(D)\}$;
(2) $\mathcal{E}_{(\mathcal{R},\mathcal{D})} = \{E \subseteq Y : \ \operatorname{int}_{(\mathcal{R},\mathcal{D})}(E) \neq \emptyset\}$.

Moreover, by using the results of Section 20, we can easily establish the following theorems.

Theorem 171. *We have*

(1) $\mathcal{D}_{(\mathcal{R},\mathcal{D})} = \{D \subseteq Y : \ \forall\ x \in X : \quad \forall\ R \in \mathcal{R} : \ R(x) \cap D \in \mathcal{D}\}$;
(2) $\mathcal{E}_{(\mathcal{R},\mathcal{D})} = \{E \subseteq Y : \ \exists\ x \in X : \quad \exists\ R \in \mathcal{R} : \ R(x) \setminus E \notin \mathcal{D}\}$.

Theorem 172. *We have*

(1) $\mathcal{D}_{(\mathcal{R},\mathcal{D})} = \{ D \subseteq Y : \ D^c \notin \mathcal{E}_{(\mathcal{R},\mathcal{D})} \}$;
(2) $\mathcal{E}_{(\mathcal{R},\mathcal{D})} = \{ E \subseteq Y : \ E^c \notin \mathcal{D}_{(\mathcal{R},\mathcal{D})} \}$.

Theorem 173. *We have*

(1) $\mathcal{D}_{(\mathcal{R},\mathcal{D})} = \{ D \subseteq Y : \ \forall \ E \in \mathcal{E}_{(\mathcal{R},\mathcal{D})} : \ D \cap E \neq \emptyset \}$;
(2) $\mathcal{E}_{(\mathcal{R},\mathcal{D})} = \{ E \subseteq Y : \ \forall \ D \in \mathcal{D}_{(\mathcal{R},\mathcal{D})} : \ D \cap E \neq \emptyset \}$.

Theorem 174. *The families $\mathcal{D}_{(\mathcal{R},\mathcal{D})}$ and $\mathcal{E}_{(\mathcal{R},\mathcal{D})}$ are stacks on Y.*

Theorem 175. *The following assertions are equivalent:*

(1) $X \neq \emptyset$, $\mathcal{R} \neq \emptyset$ *and* $\emptyset \notin \mathcal{D}$;
(2) $\emptyset \notin \mathcal{D}_{(\mathcal{R},\mathcal{D})}$; (3) $\mathcal{E}_{(\mathcal{R},\mathcal{D})} \neq \emptyset$; (4) $Y \in \mathcal{E}_{(\mathcal{R},\mathcal{D})}$;
(5) $\mathcal{D}_{(\mathcal{R},\mathcal{D})} \neq \mathcal{P}(Y)$.

Theorem 176. *The following assertions are equivalent:*

(1) $R(x) \in \mathcal{D}$ *for all* $x \in X$ *and* $R \in \mathcal{R}$;
(2) $\emptyset \notin \mathcal{E}_{(\mathcal{R},\mathcal{D})}$; (3) $\mathcal{D}_{(\mathcal{R},\mathcal{D})} \neq \emptyset$; (4) $Y \in \mathcal{D}_{(\mathcal{R},\mathcal{D})}$;
(5) $\mathcal{E}_{(\mathcal{R},\mathcal{D})} \neq \mathcal{P}(Y)$.

Theorem 177. *We have*

(1) $\mathcal{D}_{(\mathcal{R},\mathcal{D})} = \bigcap_{R \in \mathcal{R}} \mathcal{D}_{(R,\mathcal{D})}$; (2) $\mathcal{E}_{(\mathcal{R},\mathcal{D})} = \bigcup_{R \in \mathcal{R}} \mathcal{E}_{(R,\mathcal{D})}$.

Corollary 34. *The mapping*

(1) $\mathcal{R} \mapsto \mathcal{E}_{(\mathcal{R},\mathcal{D})}$ *is union-preserving;*
(2) $\mathcal{R} \mapsto \mathcal{D}_{(\mathcal{R},\mathcal{D})}$ *is intersection-preserving.*

Remark 68. By using Remark 38 and the the notation $\mathfrak{U}_{(\mathcal{R},\mathcal{D})} = \mathfrak{U}_{(\mathcal{R}^{\triangleright},\mathcal{D})}$, we can also state that $\mathcal{E}_{(\mathcal{R},\mathcal{D})} = \bigcup_{x \in X} \mathfrak{U}_{(\mathcal{R},\mathcal{D})}(x)$.

Theorem 178. *If $\mathcal{D} = \bigcap_{i \in I} \mathcal{D}_i$ for some family $(\mathcal{D}_i)_{i \in I}$ of stacks on Y, then*

(1) $\mathcal{D}_{(\mathcal{R},\mathcal{D})} = \bigcap_{i \in I} \mathcal{D}_{(\mathcal{R},\mathcal{D}_i)}$; (2) $\mathcal{E}_{(\mathcal{R},\mathcal{D})} = \bigcup_{i \in I} \mathcal{E}_{(\mathcal{R},\mathcal{D}_i)}$.

30. Open and Closed Sets Derived from Ordinary Relators and Stacks

Notation 10. *In this section, we shall assume that $\mathcal{R}$ is an ordinary relator on X, and $\mathcal{D}$ is a stack on X.*

From Definition 16, by using the super relator $\mathcal{R}^{\triangleright}$, we may naturally derive the following:

Definition 30. We define

$$\tau_{(\mathcal{R},\mathcal{D})}=\tau_{(\mathcal{R}^{\triangleright},\mathcal{D})};\quad \digamma_{(\mathcal{R},\mathcal{D})}=\digamma_{(\mathcal{R}^{\triangleright},\mathcal{D})}$$

and

$$\mathcal{T}_{(\mathcal{R},\mathcal{D})}=\mathcal{T}_{(\mathcal{R}^{\triangleright},\mathcal{D})};\quad \mathcal{F}_{(\mathcal{R},\mathcal{D})}=\mathcal{F}_{(\mathcal{R}^{\triangleright},\mathcal{D})}.$$

Remark 69. The members of the families, $\tau_{(\mathcal{R},\mathcal{D})}$ and $\mathcal{T}_{(\mathcal{R},\mathcal{D})}$, will be called the *proximally open and topologically open subsets* of the stacked ordinary relator space $X(\mathcal{R})(\mathcal{D})$, respectively.

Thus, by the corresponding definitions, we can at once state the following:

Theorem 179. *We have*

(1) $\tau_{(\mathcal{R},\mathcal{D})}=\{A\subseteq X:\ A\in \operatorname{Int}_{(\mathcal{R},\mathcal{D})}(A)\}$;
(2) $\digamma_{(\mathcal{R},\mathcal{D})}=\{A\subseteq X:\ A^c\notin \operatorname{Cl}_{(\mathcal{R},\mathcal{D})}(A)\}$;
(3) $\mathcal{T}_{(\mathcal{R},\mathcal{D})}=\{A\subseteq X:\ A\subseteq \operatorname{int}_{(\mathcal{R},\mathcal{D})}(A)\}$;
(4) $\mathcal{F}_{(\mathcal{R},\mathcal{D})}=\{A\subseteq X:\ \operatorname{cl}_{(\mathcal{R},\mathcal{D})}(A)\subseteq A\}$.

Moreover, by using the results of Sections 21 and 22, we can easily establish the following theorems.

Theorem 180. *We have*

(1) $\tau_{(\mathcal{R},\mathcal{D})}=\{A\subseteq X:\quad \exists\ R\in\mathcal{R}:\ R[A]\setminus A\notin\mathcal{D}\}$;
(2) $\mathcal{T}_{(\mathcal{R},\mathcal{D})}=\{A\subseteq X:\quad \forall\ x\in A:\quad \exists\ R\in\mathcal{R}:\ R(x)\setminus A\notin\mathcal{D}\}$.

Theorem 181. *We have*

(1) $\digamma_{(\mathcal{R},\mathcal{D})}=\{A^c:\ A\in\tau_{(\mathcal{R},\mathcal{D})}\}$; (2) $\tau_{(\mathcal{R},\mathcal{D})}=\{A^c:\ A\in\digamma_{(\mathcal{R},\mathcal{D})}\}$;
(3) $\mathcal{F}_{(\mathcal{R},\mathcal{D})}=\{A^c:\ A\in\mathcal{T}_{(\mathcal{R},\mathcal{D})}\}$; (4) $\mathcal{T}_{(\mathcal{R},\mathcal{D})}=\{A^c:\ A\in\mathcal{F}_{(\mathcal{R},\mathcal{D})}\}$.

Theorem 182. *We have*

(1) $\mathscr{T}_{(\mathcal{R},\mathcal{D})} \subseteq \mathcal{T}_{(\mathcal{R},\mathcal{D})}$; (2) $\mathscr{F}_{(\mathcal{R},\mathcal{D})} \subseteq \mathcal{F}_{(\mathcal{R},\mathcal{D})}$.

Remark 70. If $\operatorname{card}(\mathcal{R}) = 1$, then the corresponding equalities can also be stated.

Theorem 183. *We have*

(1) $\mathcal{T}_{(\mathcal{R},\mathcal{D})} \setminus \{\emptyset\} \subseteq \mathcal{E}_{(\mathcal{R},\mathcal{D})}$; (2) $\mathcal{D}_{(\mathcal{R},\mathcal{D})} \cap \mathcal{F}_{(\mathcal{R},\mathcal{D})} \subseteq \{X\}$.

Remark 71. From assertion (2), by using global complementations, we can easily infer that $\mathcal{F}_{(\mathcal{R},\mathcal{D})} \subseteq (\mathcal{D}_{(\mathcal{R},\mathcal{D})})^c \cup \{X\}$ and $\mathcal{D}_{(\mathcal{R},\mathcal{D})} \subseteq (\mathcal{F}_{(\mathcal{R},\mathcal{D})})^c \cup \{X\}$.

Theorem 184. *For any* $A \subseteq X$,

(1) $\bigcup(\mathcal{T}_{(\mathcal{R},\mathcal{D})} \cap \mathcal{P}(A)) \subseteq \operatorname{int}_{(\mathcal{R},\mathcal{D})}(A)$;
(2) $\mathcal{P}[\mathscr{T}_{(\mathcal{R},\mathcal{D})} \cap \mathcal{P}(A)] \subseteq \operatorname{Int}_{(\mathcal{R},\mathcal{D})}(A)$;
(3) $\mathcal{P}(A) \cap (\mathcal{T}_{(\mathcal{R},\mathcal{D})} \setminus \{\emptyset\}) \neq \emptyset$ *implies* $A \in \mathcal{E}_{(\mathcal{R},\mathcal{D})}$.

Theorem 185. *The following assertions are equivalent:*

(1) $\emptyset \in \mathscr{T}_{(\mathcal{R},\mathcal{D})}$; (2) $X \in \mathscr{F}_{(\mathcal{R},\mathcal{D})}$;
(3) $X \in \mathscr{T}_{(\mathcal{R},\mathcal{D})}$; (4) $\emptyset \in \mathscr{F}_{(\mathcal{R},\mathcal{D})}$; (5) $\mathcal{R} \neq \emptyset$ *and* $\emptyset \notin \mathcal{D}$.

Remark 72. Thus, $\mathscr{T}_{(\mathcal{U},\mathcal{D})}$ is a *minimal structure* on X if and only if $\mathcal{R} \neq \emptyset$ and $\emptyset \notin \mathcal{D}$.

Theorem 186. *If* $A \subseteq X$ *and* $B \in \mathscr{T}_{(\mathcal{R},\mathcal{D})}$ *such that* $A \cap B = \emptyset$, *then*

(1) $B \notin \operatorname{Cl}_{(\mathcal{R},\mathcal{D})}(A)$; (2) $B \in \operatorname{Int}_{(\mathcal{R},\mathcal{D})}(A^c)$.

Theorem 187. *The family*

(1) $\mathcal{T}_{(\mathcal{R},\mathcal{D})}$ *is closed under arbitrary unions;*
(2) $\mathcal{F}_{(\mathcal{R},\mathcal{D})}$ *is closed under arbitrary intersection.*

Corollary 35. *We always have* $\emptyset \in \mathcal{T}_{(\mathcal{R},\mathcal{D})}$ *and* $X \in \mathcal{F}_{(\mathcal{R},\mathcal{D})}$.

Theorem 188. *The following assertions are equivalent:*

(1) $X \in \mathcal{T}_{(\mathcal{R},\mathcal{D})}$; (2) $\emptyset \in \mathcal{F}_{(\mathcal{R},\mathcal{D})}$;
(3) *either* $X = \emptyset$ *or* ($\mathcal{R} \neq \emptyset$ *and* $\emptyset \notin \mathcal{D}$).

Remark 73. Thus, $\mathcal{T}_{(\mathcal{R},\mathcal{D})}$ is a *generalized topology* on X if and only if either $X=\emptyset$ or ($\mathcal{R}\neq\emptyset$ and $\emptyset\notin\mathcal{D}$).

Theorem 189. *If $A\subseteq X$ and $B\in\mathcal{T}_{(\mathcal{R},\mathcal{D})}$ such that $A\cap B=\emptyset$, then*

(1) $\operatorname{cl}_{(\mathcal{R},\mathcal{D})}(A)\cap B=\emptyset$; (2) $B\subseteq\operatorname{int}_{(\mathcal{R},\mathcal{D})}(A^c)$.

Now, by specializing Definition 17, we may also have the following:

Definition 31. We define

$$\mathcal{N}_{(\mathcal{R},\mathcal{D})}=\mathcal{N}_{(\mathcal{R}^{\triangleright},\mathcal{D})}\quad\text{and}\quad\mathcal{M}_{(\mathcal{R},\mathcal{D})}=\mathcal{M}_{(\mathcal{R}^{\triangleright},\mathcal{D})}.$$

Remark 74. The members of the family $\mathcal{N}_{(\mathcal{R},\mathcal{D})}$ will be called the *rare or nowhere dense subsets* of the stacked ordinary relator space $X(\mathcal{R})(\mathcal{D})$.

Thus, by Definitions 17, 26 and 29, we can at once state the following:

Theorem 190. *We have*

(1) $\mathcal{N}_{(\mathcal{R},\mathcal{D})}=\{A\subseteq X:\ \operatorname{cl}_{(\mathcal{R},\mathcal{D})}(A)\notin\mathcal{E}_{(\mathcal{R},\mathcal{D})}\}$;
(2) $\mathcal{M}_{(\mathcal{R},\mathcal{D})}=\{A\subseteq X:\ \operatorname{int}_{(\mathcal{R},\mathcal{D})}(A)\in\mathcal{D}_{(\mathcal{R},\mathcal{D})}\}$.

Moreover, by Theorem 87, we can also at once state the following:

Theorem 191. *We have*

(1) $\mathcal{M}_{(\mathcal{R},\mathcal{D})}=\{A^c:\ A\in\mathcal{N}_{(\mathcal{R},\mathcal{D})}\}$;
(2) $\mathcal{N}_{(\mathcal{R},\mathcal{D})}=\{A^c:\ A\in\mathcal{M}_{(\mathcal{R},\mathcal{D})}\}$.

31. Proximal Closures and Interiors Derived from Ordinary Relators

Notation 11. *In this and the following two sections, we shall assume that $\mathcal{R}$ is an ordinary relator on X to Y.*

From Definition 18, by using the super relator $\mathcal{R}^{\triangleright}$, we may naturally derive the following:

Definition 32. We define

$$\operatorname{Cl}_{\mathcal{R}}=\operatorname{Cl}_{\mathcal{R}^{\triangleright}}\quad\text{and}\quad\operatorname{Int}_{\mathcal{R}}=\operatorname{Int}_{\mathcal{R}^{\triangleright}}.$$

Remark 75. The hyper relations $\mathrm{Cl}_{\mathcal{R}}$ and $\mathrm{Int}_{\mathcal{R}}$ will be called the *proximal closure and interior* generated by the ordinary relator $\mathcal{R}$, respectively.

Thus, by the corresponding definitions, we can at once state the following:

Theorem 192. *We have*

(1) $\mathrm{Cl}_{\mathcal{R}} = \mathrm{Cl}_{(\mathcal{R},\mathcal{D}_0)}$; (2) $\mathrm{Int}_{\mathcal{R}} = \mathrm{Int}_{(\mathcal{R},\mathcal{D}_0)}$.

Proof. For instance, by Definitions 32, 18 and 25, we can see that

$$\mathrm{Cl}_{\mathcal{R}} = \mathrm{Cl}_{\mathcal{R}^{\triangleright}} = \mathrm{Cl}_{(\mathcal{R}^{\triangleright},\mathcal{D}_0)} = \mathrm{Cl}_{(\mathcal{R},\mathcal{D}_0)}\,. \qquad \square$$

Therefore, the following theorems can be derived from not only the results of Section 23 but also those of Section 27.

Theorem 193. *For any $B \subseteq Y$, we have*

(1) $\mathrm{Int}_{\mathcal{R}}(B) = \{A \subseteq X : \ \exists\ R \in \mathcal{R} : \ R[A] \subseteq B\}$;
(2) $\mathrm{Cl}_{\mathcal{R}}(B) = \{A \subseteq X : \ \forall\ R \in \mathcal{R} : \ R[A] \cap B \neq \emptyset\}$.

Theorem 194. *For any $B \subseteq Y$, we have*

(1) $\mathrm{Cl}_{\mathcal{R}}(B) = \mathrm{Int}_{\mathcal{R}}(B^c)^c$; (2) $\mathrm{Int}_{\mathcal{R}}(B) = \mathrm{Cl}_{\mathcal{R}}(B^c)^c$.

Theorem 195. *For any $B_1 \subseteq B_2 \subseteq Y$, we have*

(1) $\mathrm{Cl}_{\mathcal{R}}(B_1) \subseteq \mathrm{Cl}_{\mathcal{R}}(B_2)$; (2) $\mathrm{Int}_{\mathcal{R}}(B_1) \subseteq \mathrm{Int}_{\mathcal{R}}(B_2)$.

Theorem 196. *The following assertions are equivalent:*

(1) $\mathrm{Cl}_{\mathcal{R}}(\emptyset) = \emptyset$; (2) $\mathcal{R} \neq \emptyset$; (3) $\mathrm{Int}_{\mathcal{R}}(Y) \neq \emptyset$;
(4) $\mathrm{Int}_{\mathcal{R}}(Y) = \mathcal{P}(X)$.

Corollary 36. *The following assertions are equivalent:*

(1) $\mathrm{Int}_{\mathcal{R}}(Y) = \emptyset$; (2) $\mathcal{R} = \emptyset$; (3) $\mathrm{Cl}_{\mathcal{R}}(\emptyset) \neq \emptyset$;
(4) $\mathrm{Cl}_{\mathcal{R}}(\emptyset) = \mathcal{P}(X)$.

Theorem 197. *We have*

(1) $\mathrm{Cl}_{\mathcal{R}}(Y) = \{A \subseteq X : \ \forall\ R \in \mathcal{R} : \ R[A] \neq \emptyset\}$;
(2) $\mathrm{Int}_{\mathcal{R}}(\emptyset) = \{A \subseteq X : \ \exists\ R \in \mathcal{R} : \ R[A] = \emptyset\}$.

Theorem 198. *If* $B_1, B_2 \subseteq Y$ *such that* $B_1 \cap B_2 = \emptyset$, *then*

$$\mathrm{Cl}_{\mathcal{R}}(B_1) \cap \mathrm{Int}_{\mathcal{R}}(B_2) = \emptyset.$$

Theorem 199. *If* $B = \bigcap_{i \in I} B_i$ *for some family* $(B_i)_{i \in I}$ *of subsets of* Y, *then*

(1) $\mathrm{Cl}_{\mathcal{R}}(B) \subseteq \bigcap_{i \in I} \mathrm{Cl}_{\mathcal{R}}(B_i)$; (2) $\mathrm{Int}_{\mathcal{R}}(B) \subseteq \bigcap_{i \in I} \mathrm{Int}_{\mathcal{R}}(B_i)$.

Theorem 200. *If* $B = \bigcup_{i \in I} B_i$ *for some family* $(B_i)_{i \in I}$ *of subsets of* Y, *then*

(1) $\bigcup_{i \in I} \mathrm{Cl}_{\mathcal{R}}(B_i) \subseteq \mathrm{Cl}_{\mathcal{R}}(B)$; (2) $\bigcup_{i \in I} \mathrm{Int}_{\mathcal{R}}(B_i) \subseteq \mathrm{Int}_{\mathcal{R}}(B)$.

Theorem 201. *If* $\mathrm{card}(\mathcal{R}) = 1$, *then*

(1) $\mathrm{Cl}_{\mathcal{R}}$ *is union-preserving;* (2) $\mathrm{Int}_{\mathcal{R}}$ *is intersection-preserving.*

Theorem 202. *We have*

(1) $\mathrm{Cl}_{\mathcal{R}} = \bigcap_{R \in \mathcal{R}} \mathrm{Cl}_R$; (2) $\mathrm{Int}_{\mathcal{R}} = \bigcup_{R \in \mathcal{R}} \mathrm{Int}_R$.

Corollary 37. *The mapping*

(1) $\mathcal{R} \mapsto \mathrm{Int}_{\mathcal{R}}$ *is union-preserving;* (2) $\mathcal{R} \mapsto \mathrm{Cl}_{\mathcal{R}}$ *is intersection-preserving.*

Theorem 203. *We have*

(1) $\mathrm{Cl}_{\mathcal{R}^{-1}} = \mathrm{Cl}_{\mathcal{R}}^{-1}$; (2) $\mathrm{Int}_{\mathcal{R}^{-1}} = \mathcal{C}_Y \circ \mathrm{Int}_{\mathcal{R}}^{-1} \circ \mathcal{C}_X$.

Proof. To prove (1), not that, for any $A \subseteq X$ and $B \subseteq Y$, we have

$$B \in \mathrm{Cl}_{\mathcal{R}^{-1}}(A) \iff \forall\ R \in \mathcal{R}: \ R^{-1}[B] \cap A \neq \emptyset \iff$$

$$\forall\ R \in \mathcal{R}: \ B \cap R[A] \neq \emptyset \iff A \in \mathrm{Cl}_{\mathcal{R}}(B) \iff B \in \mathrm{Cl}_{\mathcal{R}}^{-1}(A).$$

□

Remark 76. Hence, we can see that, despite their equivalence, closures are sometimes more convenient tools than interiors.

Moreover, this theorem, together with some further theorems, will show that ordinary relators are less general but more flexible tools than super relators.

Recall that, for a super relation U on X to Y, instead of its ordinary inverse U^{-1}, we had to consider its relationally generated inverse $\boldsymbol{U^{-1}} = U^{\triangleleft -1 \triangleright}$.

32. Topological Closures and Interiors Derived from Ordinary Relators

From Definition 19, by using the super relator $\mathcal{R}^{\triangleright}$, we may naturally derive the following:

Definition 33. We define

$$\mathrm{cl}_{\mathcal{R}} = \mathrm{cl}_{\mathcal{R}^{\triangleright}} \quad \text{and} \quad \mathrm{int}_{\mathcal{R}} = \mathrm{int}_{\mathcal{R}^{\triangleright}}\,.$$

Remark 77. The super relations $\mathrm{cl}_{\mathcal{R}}$ and $\mathrm{int}_{\mathcal{R}}$ will be called the *topological closure and interior* generated by the ordinary relator $\mathcal{R}$, respectively.

Thus, by Definitions 19 and 26, we can at once state the following:

Theorem 204. *We have*

(1) $\mathrm{cl}_{\mathcal{R}} = \mathrm{cl}_{(\mathcal{R},\mathcal{D}_0)}$; (2) $\mathrm{int}_{\mathcal{R}} = \mathrm{int}_{(\mathcal{R},\mathcal{D}_0)}$.

Therefore, the following theorems can be derived from not only the results of Section 24 but also those of Section 28.

Theorem 205. *For any $B \subseteq Y$, we have*

(1) $\mathrm{int}_{\mathcal{R}}(B) = \{x \in X : \ \exists \ \ R \in \mathcal{R} : \ R(x) \subseteq B\}$;
(2) $\mathrm{cl}_{\mathcal{R}}(B) = \{x \in X : \ \forall \ R \in \mathcal{R} : \ R(x) \cap B \neq \emptyset\}$.

Theorem 206. *For any $B \subseteq Y$, we have*

(1) $\mathrm{cl}_{\mathcal{R}}(B) = \mathrm{int}_{\mathcal{R}}(B^c)^c$; (2) $\mathrm{int}_{\mathcal{R}}(B) = \mathrm{cl}_{\mathcal{R}}(B^c)^c$.

Theorem 207. *For any $B \subseteq Y$, we have*

(1) $\mathrm{cl}_{\mathcal{R}}(B) = \bigcap_{R\in\mathcal{R}} R^{-1}[B]$; (2) $\mathrm{int}_{\mathcal{R}}(B) = \bigcup_{R\in\mathcal{R}} R^{-1}[B^c]^c$.

Proof. To prove assertion (1), note that, by Definition 33 and Theorems 107 and 15, we have

$$\mathrm{cl}_{\mathcal{R}}(B) = \mathrm{cl}_{\mathcal{R}^{\triangleright}}(B) = \bigcap_{R\in\mathcal{R}} R^{\triangleright-1}(B) = \bigcap_{R\in\mathcal{R}} R^{-1\triangleright}(B) = \bigcap_{R\in\mathcal{R}} R^{-1}[B].$$

□

Theorem 208. *For any* $A \subseteq X$ *and* $B \subseteq Y$,

(1) $A \in \operatorname{Int}_{\mathcal{R}}(B)$ *implies* $A \subseteq \operatorname{int}_{\mathcal{R}}(B)$;
(2) $A \cap \operatorname{cl}_{\mathcal{R}}(B) \neq \emptyset$ *implies* $A \in \operatorname{Cl}_{\mathcal{R}}(B)$.

Remark 78. Assertions (1) and (2) can be reformulated in the forms that

(1) $\operatorname{Int}_{\mathcal{R}}(B) \subseteq \mathcal{P}(\operatorname{int}_{\mathcal{R}}(B))$; (2) $\mathcal{P}^{\star}(\operatorname{cl}_{\mathcal{R}}(B)) \subseteq \operatorname{Cl}_{\mathcal{R}}(B)$.

Theorem 209. *For any* $B_1 \subseteq B_2 \subseteq Y$, *we have*

(1) $\operatorname{cl}_{\mathcal{R}}(B_1) \subseteq \operatorname{cl}_{\mathcal{R}}(B_2)$; (2) $\operatorname{int}_{\mathcal{R}}(B_1) \subseteq \operatorname{int}_{\mathcal{R}}(B_2)$.

Theorem 210. *The following assertions are equivalent:*

(1) $\operatorname{cl}_{\mathcal{R}}(\emptyset) = \emptyset$; (2) $\mathcal{R} \neq \emptyset$; (3) $\operatorname{int}_{\mathcal{R}}(Y) \neq \emptyset$:
(4) $\operatorname{int}_{\mathcal{R}}(Y) = X$.

Corollary 38. *The following assertions are equivalent:*

(1) $\operatorname{int}_{\mathcal{R}}(Y) = \emptyset$; (2) $\mathcal{R} = \emptyset$; (3) $\operatorname{cl}_{\mathcal{R}}(\emptyset) \neq \emptyset$;
(4) $\operatorname{cl}_{\mathcal{R}}(\emptyset) = X$.

Theorem 211. *We have*

(1) $\operatorname{cl}_{\mathcal{R}}(Y) = \{x \in X : \ \forall\ R \in \mathcal{R} : \ \ R(x) \neq \emptyset\}$;
(2) $\operatorname{int}_{\mathcal{R}}(\emptyset) = \{x \in X : \exists\ R \in \mathcal{R} : \ \ R(x) = \emptyset\}$.

Theorem 212. *If* $B_1, B_2 \subseteq Y$ *such that* $B_1 \cap B_2 = \emptyset$, *then*

$$\operatorname{cl}_{\mathcal{R}}(B_1) \cap \operatorname{int}_{\mathcal{R}}(B_2) = \emptyset.$$

Theorem 213. *If* $B = \bigcap_{i \in I} B_i$ *for some family* $(B_i)_{i \in I}$ *of subsets of* Y, *then*

(1) $\operatorname{cl}_{\mathcal{R}}(B) \subseteq \bigcap_{i \in I} \operatorname{cl}_{\mathcal{R}}(B_i)$; (2) $\operatorname{int}_{\mathcal{R}}(B) \subseteq \bigcap_{i \in I} \operatorname{int}_{\mathcal{R}}(B_i)$.

Theorem 214. *If* $B = \bigcup_{i \in I} B_i$ *for some family* $(B_i)_{i \in I}$ *of subsets of* Y, *then*

(1) $\bigcup_{i \in I} \operatorname{cl}_{\mathcal{R}}(B_i) \subseteq \operatorname{cl}_{\mathcal{R}}(B)$; (2) $\bigcup_{i \in I} \operatorname{int}_{\mathcal{R}}(B_i) \subseteq \operatorname{int}_{\mathcal{R}}(B)$.

Theorem 215. *If* $\operatorname{card}(\mathcal{R}) = 1$, *then*

(1) $\operatorname{cl}_{\mathcal{R}}$ *is union-preserving;*
(2) $\operatorname{int}_{\mathcal{R}}$ *is intersection-preserving.*

Theorem 216. *We have*

(1) $\operatorname{cl}_{\mathcal{R}} = \bigcap_{R \in \mathcal{R}} \operatorname{cl}_R$; (2) $\operatorname{int}_{\mathcal{R}} = \bigcup_{R \in \mathcal{R}} \operatorname{int}_R$.

Corollary 39. *The mapping*

(1) $\mathcal{R} \mapsto \operatorname{int}_{\mathcal{R}}$ *is union-preserving;*
(2) $\mathcal{U} \mapsto \operatorname{cl}_{\mathcal{R}}$ *is intersection-preserving.*

Finally, we note that, by specializing Definitions 20 and 21 we may also naturally introduce the following two definitions.

Definition 34. We define

$$\operatorname{bnd}_{\mathcal{R}} = \operatorname{bnd}_{\mathcal{R}^{\triangleright}}.$$

Remark 79. Moreover, if in particular $X = Y$, then we may also define

$$\operatorname{res}_{\mathcal{R}} = \operatorname{res}_{\mathcal{R}^{\triangleright}} \quad \text{and} \quad \operatorname{bor}_{\mathcal{R}} = \operatorname{bor}_{\mathcal{R}^{\triangleright}}.$$

Definition 35. We define

$$\rho_{\mathcal{R}} = \rho_{\mathcal{R}^{\triangleright}} \quad \text{and} \quad \sigma_{\mathcal{R}} = \sigma_{\mathcal{R}^{\triangleright}}.$$

Thus, by using Theorem 117, we can also prove the following:

Theorem 217. *We have*

$$\rho_{\mathcal{R}} = \bigcap \mathcal{R}^{-1} = \Big(\bigcap \mathcal{R}\Big)^{-1}.$$

Proof. By using Definition 35, Theorems 117 and 8, we can see that

$$\rho_{\mathcal{R}} = \rho_{\mathcal{R}^{\triangleright}} = \bigcup \mathcal{R}^{\triangleright \triangleleft -1} = \bigcup \mathcal{R}^{-1}. \qquad \square$$

33. Fat and Dense Sets Derived from Ordinary Relators

From Definition 22, by using the super relator $\mathcal{R}^{\triangleright}$, we may naturally derive the following:

Definition 36. We define

$$\mathcal{D}_{\mathcal{R}} = \mathcal{D}_{\mathcal{R}^{\triangleright}} \quad \text{and} \quad \mathcal{E}_{\mathcal{R}} = \mathcal{E}_{\mathcal{R}^{\triangleright}}.$$

Remark 80. The members of the families $\mathcal{E}_{\mathcal{R}}$ and $\mathcal{D}_{\mathcal{R}}$ will be called the *fat and dense subsets* of the ordinary relator space $(X, Y)(\mathcal{R})$, respectively.

Thus, by the corresponding definitions, we can at once state the following two theorems.

Theorem 218. *We have*

(1) $\mathcal{D}_{\mathcal{R}} = \mathcal{D}_{(\mathcal{R}, \mathcal{D}_0)}$; (2) $\mathcal{E}_{\mathcal{R}} = \mathcal{E}_{(\mathcal{R}, \mathcal{D}_0)}$.

Theorem 219. *We have*

(1) $\mathcal{D}_{\mathcal{R}} = \{ D \subseteq Y : \ X = \mathrm{cl}_{\mathcal{R}}(D) \}$;
(2) $\mathcal{E}_{\mathcal{R}} = \{ E \subseteq Y : \ \mathrm{int}_{\mathcal{R}}(E) \neq \emptyset \}$.

Moreover, by using the results of Section 25, we can easily establish the following theorems.

Theorem 220. *We have*

(1) $\mathcal{E}_{\mathcal{R}} = \{ E \subseteq Y : \quad \exists \ x \in X : \quad \exists \ R \in \mathcal{R} : \ R(x) \subseteq E \}$;
(2) $\mathcal{D}_{\mathcal{R}} = \{ D \subseteq Y : \quad \forall \ x \in X : \quad \forall \ R \in \mathcal{R} : \ R(x) \cap D \neq \emptyset \}$.

Theorem 221. *We have*

(1) $\mathcal{D}_{\mathcal{R}} = \{ D \subseteq Y : \quad \forall \ R \in \mathcal{R} : \ X = R^{-1}[D] \}$;
(2) $\mathcal{E}_{\mathcal{U}} = \{ E \subseteq Y : \quad \exists \ R \in \mathcal{R} : \ X \neq R^{-1}[E^c] \}$.

Theorem 222. *We have*

(1) $\mathcal{D}_{\mathcal{R}} = \{ D \subseteq Y : \ D^c \notin \mathcal{E}_{\mathcal{R}} \}$;
(2) $\mathcal{E}_{\mathcal{R}} = \{ E \subseteq Y : \ E^c \notin \mathcal{D}_{\mathcal{R}} \}$.

Theorem 223. *We have*

(1) $\mathcal{D}_{\mathcal{R}} = \{ D \subseteq Y : \ \forall \ E \in \mathcal{E}_{\mathcal{R}} : D \cap E \neq \emptyset \}$;
(2) $\mathcal{E}_{\mathcal{R}} = \{ E \subseteq Y : \ \forall \ D \in \mathcal{D}_{\mathcal{R}} : \ D \cap E \neq \emptyset \}$.

Theorem 224. *The families $\mathcal{D}_{\mathcal{R}}$ and $\mathcal{E}_{\mathcal{R}}$ are stacks on Y.*

Theorem 225. *The following assertions are equivalent:*

(1) $X \neq \emptyset$ *and* $\mathcal{R} \neq \emptyset$;
(2) $\emptyset \notin \mathcal{D}_{\mathcal{R}}$; (3) $\mathcal{E}_{\mathcal{R}} \neq \emptyset$; (4) $Y \in \mathcal{E}_{\mathcal{R}}$; (5) $\mathcal{D}_{\mathcal{R}} \neq \mathcal{P}(Y)$.

Remark 81. Property (1) can be briefly expressed by saying that the ordinary relator $\mathcal{R}$ is non-degenerated.

Theorem 226. *The following assertions are equivalent:*

(1) $R(x) \neq \emptyset$ *for all* $x \in X$ *and* $R \in \mathcal{R}$;
(2) $\emptyset \notin \mathcal{E}_{\mathcal{R}}$; (3) $\mathcal{D}_{\mathcal{U}} \neq \emptyset$; (4) $Y \in \mathcal{D}_{\mathcal{R}}$; (5) $\mathcal{E}_{\mathcal{R}} \neq \mathcal{P}(Y)$.

Remark 82. Property (1) can be briefly expressed by saying that the ordinary $\mathcal{R}$ is non-partial.

Theorem 227. *We have*

(1) $\mathcal{D}_{\mathcal{R}} = \bigcap_{R \in \mathcal{R}} \mathcal{D}_R$; (2) $\mathcal{E}_{\mathcal{R}} = \bigcup_{R \in \mathcal{R}} \mathcal{E}_R$.

Corollary 40. *The mapping*

(1) $\mathcal{R} \mapsto \mathcal{E}_{\mathcal{R}}$ *is union-preserving;*
(2) $\mathcal{R} \mapsto \mathcal{D}_{\mathcal{R}}$ *is intersection-preserving.*

Remark 83. By using Remark 53 and the notation $\mathfrak{U}_{\mathcal{R}} = \mathfrak{U}_{\mathcal{R}^{\triangleright}}$, we can also state that $\mathcal{E}_{\mathcal{R}} = \bigcup_{x \in X} \mathfrak{U}_{\mathcal{R}}(x)$.

Example 10. If $\leq$ is a relation on X, then for any $A \subseteq X$ we have

(1) $A \in \mathcal{E}_{\leq}$ if and only if there exists $x \in X$ such that $y \in A$ for all $y \geq x$;
(2) $A \in \mathcal{D}_{\leq}$ if and only if for each $x \in X$ there exists $y \in A$ such that $y \geq x$.

Therefore, $\mathcal{E}_{\leq}$ and $\mathcal{D}_{\leq}$ are just the families of all *residual* and *cofinal* subsets of the goset (generalized ordered set) $X(\leq)$, respectively.

34. Open and Closed sets Derived from Ordinary Relators

Notation 12. *In this and the following section, we shall assume that* $\mathcal{R}$ *is an ordinary relator on* X.

From Definition 23, by using the super relator $\mathcal{R}^{\triangleright}$, we may naturally derive the following:

Definition 37. We define

$$\tau_{\mathcal{R}} = \tau_{\mathcal{R}^{\triangleright}};\ \tau\!\!\!-_{\mathcal{R}} = \tau\!\!\!-_{\mathcal{R}^{\triangleright}};\quad \text{and}\quad \mathcal{T}_{\mathcal{R}} = \mathcal{T}_{\mathcal{R}^{\triangleright}};\ \mathcal{F}_{\mathcal{R}} = \mathcal{F}_{\mathcal{R}^{\triangleright}}.$$

Remark 84. The members of the families, $\tau_{\mathcal{R}}$ and $\mathcal{T}_{\mathcal{R}}$, will be called the *proximally open and topologically open subsets* of the ordinary relator space $X(\mathcal{R})$, respectively.

The families $\tau_{\mathcal{R}}$ and $\tau\!\!\!-_{\mathcal{R}}$ were first explicitly used by the present author in Refs. [143, 145]. While the practical notation $\tau\!\!\!-_{\mathcal{R}}$ has been suggested by J. Kurdics who first noted that connectedness is a particular case of well-chainedness [78, 80, 114].

Thus, by Definitions 23 and 30, we can at once state the following:

Theorem 228. *We have*

(1) $\tau_{\mathcal{R}} = \tau_{(\mathcal{R},\mathcal{D}_0)}$; (2) $\tau\!\!\!-_{\mathcal{R}} = \tau\!\!\!-_{(\mathcal{R},\mathcal{D}_0)}$;
(3) $\mathcal{T}_{\mathcal{R}} = \mathcal{T}_{(\mathcal{R},\mathcal{D}_0)}$; (4) $\mathcal{F}_{\mathcal{R}} = \mathcal{F}_{(\mathcal{R},\mathcal{D}_0)}$.

Moreover, by using the results of Section 26, we can easily establish the following theorems.

Theorem 229. *We have*

(1) $\tau_{\mathcal{R}} = \{A \subseteq X :\ A \in \operatorname{Int}_{\mathcal{R}}(A)\}$;
(2) $\tau\!\!\!-_{\mathcal{R}} = \{A \subseteq X :\ A^c \notin \operatorname{Cl}_{\mathcal{R}}(A)\}$;
(3) $\mathcal{T}_{\mathcal{R}} = \{A \subseteq X :\ A \subseteq \operatorname{int}_{\mathcal{R}}(A)\}$;
(4) $\mathcal{F}_{\mathcal{R}} = \{A \subseteq X : \operatorname{cl}_{\mathcal{R}}(A) \subseteq A\}$.

Theorem 230. *We have*

(1) $\tau_{\mathcal{R}} = \{A \subseteq X :\ \ \exists\ R \in \mathcal{R} :\ R[A] \subseteq A\}$;
(2) $\mathcal{T}_{\mathcal{R}} = \{A \subseteq X :\ \ \forall\ x \in A :\ \exists\ R \in \mathcal{R} :\ R(x) \subseteq A\}$.

Theorem 231. *We have*

(1) $\digamma_{\mathcal{R}} = \tau_{\mathcal{R}^{-1}}$; (2) $\tau_{\mathcal{R}} = \digamma_{\mathcal{R}^{-1}}$.

Proof. By Theorems 229, 203 and 194, for any $A \subseteq X$, we have

$$\begin{aligned} A \in \digamma_{\mathcal{R}} &\iff A^c \notin \mathrm{Cl}_{\mathcal{R}}(A) \iff A \notin \mathrm{Cl}_{\mathcal{R}}^{-1}(A^c) \\ &\iff A \notin \mathrm{Cl}_{\mathcal{R}^{-1}}(A^c) \iff A \in \mathrm{Cl}_{\mathcal{R}^{-1}}(A^c)^c \\ &\iff A \in \mathrm{Int}_{\mathcal{R}^{-1}}(A) \iff A \in \tau_{\mathcal{R}^{-1}}. \end{aligned}$$

□

Theorem 232. *We have*

(1) $\digamma_{\mathcal{R}} = \{A^c : A \in \tau_{\mathcal{R}}\}$; (2) $\tau_{\mathcal{R}} = \{A^c : A \in \digamma_{\mathcal{R}}\}$;
(3) $\mathcal{F}_{\mathcal{R}} = \{A^c : A \in \mathcal{T}_{\mathcal{R}}\}$; (4) $\mathcal{T}_{\mathcal{R}} = \{A^c : A \in \mathcal{F}_{\mathcal{R}}\}$.

Theorem 233. *We have*

(1) $\tau_{\mathcal{R}} \subseteq \mathcal{T}_{\mathcal{R}}$; (2) $\digamma_{\mathcal{R}} \subseteq \mathcal{F}_{\mathcal{R}}$.

Remark 85. If $\mathrm{card}(\mathcal{R}) = 1$, then the corresponding equalities can also be stated.

Theorem 234. *We have*

(1) $\mathcal{T}_{\mathcal{R}} \setminus \{\emptyset\} \subseteq \mathcal{E}_{\mathcal{R}}$; (2) $\mathcal{D}_{\mathcal{R}} \cap \mathcal{F}_{\mathcal{R}} \subseteq \{X\}$.

Remark 86. Hence, by using global complementations, we can again infer that $\mathcal{F}_{\mathcal{R}} \subseteq (\mathcal{D}_{\mathcal{R}})^c \cup \{X\}$ and $\mathcal{D}_{\mathcal{R}} \subseteq (\mathcal{F}_{\mathcal{R}})^c \cup \{X\}$.

Theorem 235. *For any* $A \subseteq X$,

(1) $\bigcup(\mathcal{T}_{\mathcal{R}} \cap \mathcal{P}(A)) \subseteq \mathrm{int}_{\mathcal{R}}(A)$;
(2) $\mathcal{P}[\tau_{\mathcal{R}} \cap \mathcal{P}(A)] \subseteq \mathrm{Int}_{\mathcal{R}}(A)$;
(3) $\mathcal{P}(A) \cap (\mathcal{T}_{\mathcal{T}} \setminus \{\emptyset\}) \neq \emptyset$ *implies* $A \in \mathcal{E}_{\mathcal{R}}$.

Remark 87. The fat sets are frequently more convenient tools than the open ones. For instance, if $\leq$ is a relation on X, then $\mathcal{T}_{\leq}$ and $\mathcal{E}_{\leq}$ are the families of all *ascending and residual subsets* of the goset $X(\leq)$, respectively.

This fact, stressed first by the present author at the Seventh Prague Topological Symposium in 1991, can also be well seen from the following:

Example 11. If in particular $X = \mathbb{R}$ and

$$R(x) = \{x - 1\} \cup [x, +\infty[$$

for all $x \in X$, then R is a reflexive relation on X such that $\mathcal{T}_R = \{\emptyset, X\}$, but $\mathcal{E}_R$ is quite a large family.

35. Some Further Theorems on Open and Closed Sets

Theorem 236. *The following assertions are equivalent:*

(1) $\emptyset \in \tau_{\mathcal{R}}$; (2) $X \in \tau\!\!\!-_{\mathcal{R}}$; (5) $\mathcal{U} \neq \emptyset$.
(3) $X \in \tau_{\mathcal{U}}$; (4) $\emptyset \in \tau\!\!\!-_{\mathcal{U}}$;

Remark 88. Thus, $\tau_{\mathcal{R}}$ is a *minimal structure* on X if and only if $\mathcal{R} \neq \emptyset$.

Theorem 237. *If $A \subseteq X$ and $B \in \tau_{\mathcal{R}}$ such that $A \cap B = \emptyset$, then*

(1) $B \notin \mathrm{Cl}_{\mathcal{R}}(A)$; (2) $B \in \mathrm{Int}_{\mathcal{R}}(A^c)$.

Theorem 238. *The family*

(1) *$\mathcal{T}_{\mathcal{R}}$ is closed under arbitrary unions;*
(2) *$\mathcal{F}_{\mathcal{R}}$ is closed under arbitrary intersection.*

Corollary 41. *We always have $\emptyset \in \mathcal{T}_{\mathcal{R}}$ and $X \in \mathcal{F}_{\mathcal{R}}$.*

Theorem 239. *The following assertions are equivalent:*

(1) $X \in \mathcal{T}_{\mathcal{R}}$; (2) $\emptyset \in \mathcal{F}_{\mathcal{R}}$; (3) *either $X = \emptyset$ or $\mathcal{R} \neq \emptyset$.*

Remark 89. Thus, $\mathcal{T}_{\mathcal{R}}$ is a *generalized topology* on X if and only if either $X = \emptyset$ or $\mathcal{R} \neq \emptyset$.

Theorem 240. *If* $A \subseteq X$ *and* $B \in \mathcal{T}_{\mathcal{R}}$ *such that* $A \cap B = \emptyset$, *then*

(1) $\operatorname{cl}_{\mathcal{R}}(A) \cap B = \emptyset$; (2) $B \subseteq \operatorname{int}_{\mathcal{R}}(A^c)$.

Theorem 241. *We have*

(1) ${\scriptstyle\mathcal{T}}_{\mathcal{R}} = \bigcup_{R \in \mathcal{R}} {\scriptstyle\mathcal{T}}_R$; (2) ${\scriptstyle\mathcal{F}}_{\mathcal{R}} = \bigcup_{R \in \mathcal{R}} {\scriptstyle\mathcal{F}}_R$.

Corollary 42. *The mappings*

$$\mathcal{R} \mapsto {\scriptstyle\mathcal{T}}_{\mathcal{R}} \quad \textit{and} \quad \mathcal{R} \mapsto {\scriptstyle\mathcal{F}}_{\mathcal{R}}$$

are union-preserving.

The following example shows that the same statements, do not, in general, hold for the topologically open and closed sets.

Example 12. If $\operatorname{card}(X) > 2$ and $x_1, x_2 \in X$ such that $x_1 \neq x_2$, and

$$R_i = \{x_i\}^2 \cup (\{x_i\}^c)^2$$

for all $i = 1, 2$, then $\mathcal{R} = \{R_1, R_2\}$ is an equivalence relator on X such that

$$\{x_1, x_2\} \in \mathcal{T}_{\mathcal{R}} \setminus (\mathcal{T}_{R_1} \cup \mathcal{T}_{R_2}) \quad \text{and thus,} \quad \mathcal{T}_{\mathcal{R}} \not\subseteq \mathcal{T}_{R_1} \cup \mathcal{T}_{R_2}.$$

Therefore, we can only prove the following:

Theorem 242. *The mappings*

$$\mathcal{R} \mapsto \mathcal{T}_{\mathcal{R}} \quad \textit{and} \quad \mathcal{R} \mapsto \mathcal{F}_{\mathcal{R}}$$

are increasing.

Corollary 43. *We have*

(1) $\bigcup_{R \in \mathcal{R}} \mathcal{T}_R \subseteq \mathcal{T}_{\mathcal{R}}$; (2) $\bigcup_{R \in \mathcal{R}} \mathcal{F}_R \subseteq \mathcal{F}_{\mathcal{R}}$.

Remark 90. However, if $\mathcal{R} \neq \emptyset$, then by using the topological refinement

$$\mathcal{R}^{\wedge} = \{S \subseteq X^2 : \quad \forall\ x \in X : \ x \in \operatorname{int}_{\mathcal{R}}(S(x))\}$$

of $\mathcal{R}$, it can be shown that $\operatorname{Int}_{\mathcal{R}^{\wedge}}(A) = \mathcal{P}(\operatorname{int}_{\mathcal{R}}(A))$ for all $A \subseteq X$. Therefore,

$$\mathcal{T}_{\mathcal{R}} = \tau_{\mathcal{R}^{\wedge}} = \bigcup_{R \in \mathcal{R}^{\wedge}} \tau_R = \bigcup_{R \in \mathcal{R}^{\wedge}} \mathcal{T}_R .$$

Finally, we note that, by specializing Definition 24, we may also introduce the following:

Definition 38. We define

$$\mathcal{N}_{\mathcal{R}} = \mathcal{N}_{\mathcal{R}^{\triangleright}} \quad \text{and} \quad \mathcal{M}_{\mathcal{R}} = \mathcal{M}_{\mathcal{R}^{\triangleright}} .$$

Remark 91. The members of the family $\mathcal{N}_{\mathcal{R}}$ will be called the *rare or nowhere dense subsets* of the ordinary relator space $X(\mathcal{R})$.

Thus, by Definitions 24 and 31, we can state the following:

Theorem 243. *We have*

(1) $\mathcal{N}_{\mathcal{R}} = \mathcal{N}_{(\mathcal{R}, \mathcal{D}_0)}$; (2) $\mathcal{M}_{\mathcal{R}} = \mathcal{M}_{(\mathcal{R}, \mathcal{D}_0)}$.

Moreover, by Theorems 139 and 140, we can state the following two theorems.

Theorem 244. *We have*

(1) $\mathcal{N}_{\mathcal{R}} = \{ A \subseteq X : \ \operatorname{cl}_{\mathcal{R}}(A) \notin \mathcal{E}_{\mathcal{R}} \}$;
(2) $\mathcal{M}_{\mathcal{R}} = \{ A \subseteq X : \ \operatorname{int}_{\mathcal{R}}(A) \in \mathcal{D}_{\mathcal{R}} \}$.

Theorem 245. *We have*

(1) $\mathcal{M}_{\mathcal{R}} = \{ A^c : \ A \in \mathcal{N}_{\mathcal{R}} \}$;
(2) $\mathcal{N}_{\mathcal{R}} = \{ A^c : \ A \in \mathcal{M}_{\mathcal{R}} \}$.

36. Relationships Between Closures and Interiors Derived from Ordinary and Super Relators

Notation 13. *In this section, we shall assume that $\mathcal{U}$ is a super relator on X to Y, and $\mathcal{D}$ is a stack on Y.*

Theorem 246. *We have*

(1) $\operatorname{Cl}_{(\mathcal{U}^{\triangleleft},\mathcal{D})} = \operatorname{Cl}_{(\mathcal{U}^{\circ},\mathcal{D})}$; (2) $\operatorname{Int}_{(\mathcal{U}^{\triangleleft},\mathcal{D})} = \operatorname{Int}_{(\mathcal{U}^{\circ},\mathcal{D})}$.

Proof. By Definitions 25 and 3, we have

$$\operatorname{Cl}_{(\mathcal{U}^{\triangleleft},\mathcal{D})} = \operatorname{Cl}_{(\mathcal{U}^{\triangleleft\triangleright},\mathcal{D})} = \operatorname{Cl}_{(\mathcal{U}^{\circ},\mathcal{D})}\,.$$

Thus, assertion (1) is true.

Assertion (2) can be proved quite similarly. Moreover, it can also be derived from assertion (1) by using Theorem 142 and Definition 10. □

Theorem 247. *We have*

(1) $\operatorname{cl}_{(\mathcal{U}^{\triangleleft},\mathcal{D})} = \operatorname{cl}_{(\mathcal{U}^{\circ},\mathcal{D})} = \operatorname{cl}_{(\mathcal{U},\mathcal{D})}$;
(2) $\operatorname{int}_{(\mathcal{U}^{\triangleleft},\mathcal{D})} = \operatorname{int}_{(\mathcal{U}^{\circ},\mathcal{D})} = \operatorname{int}_{(\mathcal{U},\mathcal{D})}$.

Proof. By Definitions 26 and 3, we have

$$\operatorname{cl}_{(\mathcal{U}^{\triangleleft},\mathcal{D})} = \operatorname{cl}_{(\mathcal{U}^{\triangleleft\triangleright},\mathcal{D})} = \operatorname{cl}_{(\mathcal{U}^{\circ},\mathcal{D})}\,.$$

Thus, the first part of assertion (1) is true.

Moreover, by Theorems 54 and 10, for any $x \in X$ and $B \subseteq Y$, we have

$$x \in \operatorname{cl}_{(\mathcal{U}^{\circ},\mathcal{D})}(B) \iff \forall\ U \in \mathcal{U}:\quad U^{\circ\triangleleft}(x) \cap B \in \mathcal{D}$$
$$\iff \forall\ U \in \mathcal{U}:\quad U^{\triangleleft}(x) \cap B \in \mathcal{D} \iff x \in \operatorname{cl}_{(\mathcal{U},\mathcal{D})}(B).$$

Therefore, $\operatorname{cl}_{(\mathcal{U}^{\circ},\mathcal{D})}(B) = \operatorname{cl}_{(\mathcal{U},\mathcal{D})}(B)$ for all $B \subseteq Y$, and thus the second part of assertion (1) is also true.

Assertion (2) can be proved quite similarly. Moreover, it can also be derived from assertion (1) by using Theorems 160 and 55. □

Remark 92. The first part of assertion (1) can also be derived from assertion (1) of Theorem 246.

Namely, by Theorems 158 and 246 and Definition 12, for any $x \in X$ and $B \subseteq Y$, we have

$$x \in \operatorname{cl}_{(\mathcal{U}^{\triangleleft},\mathcal{D})}(B) \iff \{x\} \in \operatorname{Cl}_{(\mathcal{U}^{\triangleleft},\mathcal{D})}(B)$$
$$\iff \{x\} \in \operatorname{Cl}_{(\mathcal{U}^{\circ},\mathcal{D})}(B) \iff x \in \operatorname{cl}_{(\mathcal{U}^{\circ},\mathcal{D})}(B).$$

Theorem 248. *We have*

(1) $\mathcal{D}_{(\mathcal{U}^{\triangleleft},\mathcal{D})} = \mathcal{D}_{(\mathcal{U}^{\circ},\mathcal{D})} = \mathcal{D}_{(\mathcal{U},\mathcal{D})}$;
(2) $\mathcal{E}_{(\mathcal{U}^{\triangleleft},\mathcal{D})} = \mathcal{E}_{(\mathcal{U}^{\circ},\mathcal{D})} = \mathcal{E}_{(\mathcal{U},\mathcal{D})}$.

Proof. By Definitions 25 and 3, we have

$$\mathcal{D}_{(\mathcal{U}^{\triangleleft},\mathcal{D})} = \mathcal{D}_{(\mathcal{U}^{\triangleleft\triangleright},\mathcal{D})} = \mathcal{D}_{(\mathcal{U}^{\circ},\mathcal{D})}.$$

Thus, the first part of assertion (1) is true.

Moreover, by Definition 15 and Theorem 247, for any $B \subseteq Y$, we have

$$B \in \mathcal{D}_{(\mathcal{U}^{\circ},\mathcal{D})} \iff X = \mathrm{cl}_{(\mathcal{U}^{\circ},\mathcal{D})}(B)$$
$$\iff X = \mathrm{cl}_{(\mathcal{U},\mathcal{D})}(B) \iff B \in \mathcal{D}_{(\mathcal{U},\mathcal{D})}.$$

Thus, the second part of assertion (1) is also true.

Assertion (2) can be proved quite similarly. Moreover, it can also be derived from assertion (1) by using Theorems 172 and 83. □

From the above three theorems, by using the stack $\mathcal{D}_0$ and Theorems 192, 204 and 218 and Definition 18, 19 and 22, we can immediately derive the following:

Corollary 44. *We have*

(1) $\mathrm{Cl}_{\mathcal{U}^{\triangleleft}} = \mathrm{Cl}_{\mathcal{U}^{\circ}}$;
(2) $\mathrm{Int}_{\mathcal{U}^{\triangleleft}} = \mathrm{Int}_{\mathcal{U}^{\circ}}$;
(3) $\mathrm{cl}_{\mathcal{U}^{\triangleleft}} = \mathrm{cl}_{\mathcal{U}^{\circ}} = \mathrm{cl}_{\mathcal{U}}$;
(4) $\mathrm{int}_{\mathcal{U}^{\triangleleft}} = \mathrm{int}_{\mathcal{U}^{\circ}} = \mathrm{int}_{\mathcal{U}}$;
(5) $\mathcal{D}_{\mathcal{U}^{\triangleleft}} = \mathcal{D}_{\mathcal{U}^{\circ}} = \mathcal{D}_{\mathcal{U}}$;
(6) $\mathcal{E}_{\mathcal{U}^{\triangleleft}} = \mathcal{E}_{\mathcal{U}^{\circ}} = \mathcal{E}_{\mathcal{U}}$.

37. Relationships Between Open and Closed Sets Derived from Ordinary and Super Relators

Notation 14. *In this section, we shall assume that $\mathcal{U}$ is a super relator on X, an $\mathcal{D}$ is a stack on X.*

Theorem 249. *We have*

(1) $\mathcal{F}_{(\mathcal{U}^{\triangleleft},\mathcal{D})} = \mathcal{F}_{(\mathcal{U}^{\circ},\mathcal{D})}$; (2) $\mathcal{T}_{(\mathcal{U}^{\triangleleft},\mathcal{D})} = \mathcal{T}_{(\mathcal{U}^{\circ},\mathcal{D})}$.

Proof. By Definitions 30 and 3, we have

$$\mathscr{F}_{(\mathcal{U}^{\triangleleft},\mathcal{D})} = \mathscr{F}_{(\mathcal{U}^{\triangleleft\triangleright},\mathcal{D})} = \mathscr{F}_{(\mathcal{U}^{\circ},\mathcal{D})}.$$

Thus, assertion (1) is true.

Assertion (2) can be proved quite similarly. Moreover, it can also be derived from assertion (1) by using Theorems 181 and 74. □

Remark 93. Assertion (1) can also be naturally derived from assertion (1) of Theorem 246.

Namely, by Theorems 179 and 246 and Definition 16, for any $A \subseteq X$, we have

$$A \in \mathscr{F}_{(\mathcal{U}^{\triangleleft},\mathcal{D})} \iff A^c \notin \mathrm{Cl}_{(\mathcal{U}^{\triangleleft},\mathcal{D})}(A) \iff A^c \notin \mathrm{Cl}_{(\mathcal{U}^{\circ},\mathcal{D})}(A)$$
$$\iff A \in \mathscr{F}_{(\mathcal{U}^{\circ},\mathcal{D})}.$$

Theorem 250. *We have*

(1) $\mathcal{F}_{(\mathcal{U}^{\triangleleft},\mathcal{D})} = \mathcal{F}_{(\mathcal{U}^{\circ},\mathcal{D})} = \mathcal{F}_{(\mathcal{U},\mathcal{D})}$;
(2) $\mathcal{T}_{(\mathcal{U}^{\triangleleft},\mathcal{D})} = \mathcal{T}_{(\mathcal{U}^{\circ},\mathcal{D})} = \mathcal{T}_{(\mathcal{U},\mathcal{D})}$.

Proof. By Definitions 30 and 3, we have

$$\mathcal{F}_{(\mathcal{U}^{\triangleleft},\mathcal{D})} = \mathcal{F}_{(\mathcal{U}^{\triangleleft\triangleright},\mathcal{D})} = \mathcal{F}_{(\mathcal{U}^{\circ},\mathcal{D})}.$$

Thus, the first part of assertion (1) is true.

Moreover, by Definition 16 and Theorem 237, for any $A \subseteq X$, we have

$$A \in \mathcal{F}_{(\mathcal{U}^{\circ},\mathcal{D})} \iff \mathrm{cl}_{(\mathcal{U}^{\circ},\mathcal{D})}(A) \subseteq A \iff \mathrm{cl}_{(\mathcal{U},\mathcal{D})}(A) \subseteq A$$
$$\iff A \in \mathcal{F}_{(\mathcal{U},\mathcal{D})}.$$

Thus, the second part of assertion (1) is also true.

Assertion (2) can be proved quite similarly. Moreover, it can also be derived from assertion (1) by using Theorems 181 and 74. □

Theorem 251. *We have*

(1) $\mathcal{N}_{(\mathcal{U}^{\triangleleft},\mathcal{D})} = \mathcal{N}_{(\mathcal{U}^{\circ},\mathcal{D})} = \mathcal{N}_{(\mathcal{U},\mathcal{D})}$;
(2) $\mathcal{M}_{(\mathcal{U}^{\triangleleft},\mathcal{D})} = \mathcal{M}_{(\mathcal{U}^{\circ},\mathcal{D})} = \mathcal{M}_{(\mathcal{U},\mathcal{D})}$.

Proof. By Definitions 31 and 3, we have

$$\mathcal{N}_{(\mathcal{U}^{\triangleleft},\mathcal{D})} = \mathcal{N}_{(\mathcal{U}^{\triangleleft\triangleright},\mathcal{D})} = \mathcal{N}_{(\mathcal{U}^{\circ},\mathcal{D})}.$$

Thus, the first part of assertion (1) is true.

Moreover, by Definition 17 and Theorems 237 and 248, for any $A \subseteq X$, we have

$$A \in \mathcal{N}_{(\mathcal{U}^{\circ},\mathcal{D})} \iff \mathrm{cl}_{(\mathcal{U}^{\circ},\mathcal{D})}(A) \notin \mathcal{E}_{(\mathcal{U}^{\circ},\mathcal{D})}$$
$$\iff \mathrm{cl}_{(\mathcal{U},\mathcal{D})}(A) \notin \mathcal{E}_{(\mathcal{U},\mathcal{D})} \iff A \in \mathcal{N}_{(\mathcal{U},\mathcal{D})}.$$

Thus, the second part of assertion (1) is also true.

Assertion (2) can be proved quite similarly. Moreover, it can also be derived from assertion (1) by using Theorems 191 and 87. □

From the above three theorems, by using the stack $\mathcal{D}_0$, Theorems 228 and 243 and Definition 23 and 24, we can immediately derive the following:

Corollary 45. *We have*

(1) $\mathcal{F}_{\mathcal{U}^{\triangleleft}} = \mathcal{F}_{\mathcal{U}^{\circ}}$; (2) $\mathcal{T}_{\mathcal{U}^{\triangleleft}} = \mathcal{T}_{\mathcal{U}^{\circ}}$;
(3) $\mathcal{F}_{\mathcal{U}^{\triangleleft}} = \mathcal{F}_{\mathcal{U}^{\circ}} = \mathcal{F}_{\mathcal{U}}$; (4) $\mathcal{T}_{\mathcal{U}^{\triangleleft}} = \mathcal{T}_{\mathcal{U}^{\circ}} = \mathcal{T}_{\mathcal{U}}$;
(5) $\mathcal{N}_{\mathcal{U}^{\triangleleft}} = \mathcal{N}_{\mathcal{U}^{\circ}} = \mathcal{N}_{\mathcal{U}}$; (6) $\mathcal{M}_{\mathcal{U}^{\triangleleft}} = \mathcal{M}_{\mathcal{U}^{\circ}} = \mathcal{M}_{\mathcal{U}}$.

Remark 94. To feel the importance of our former results, note that in contrast to the super relator $\mathcal{U}$, the super relator $\mathcal{U}^{\circ}$ is already union-preserving.

Moreover, $\mathcal{U}^{\triangleleft}$ is an ordinary relator. Therefore, by using super relators instead of the ordinary ones, the structures cl, $\mathcal{D}$, $\mathcal{F}$ and $\mathcal{N}$ cannot be generalized.

The fact that the structures Cl and $\mathcal{F}$ can be generalized by using super relators instead of the ordinary ones can be demonstrated by some examples.

38. An Illustrating Example

Notation 15. *In this section, we shall assume that* $X = \{1, 2\}$ *and* U *is a super relation on* X *such that, for any* $A \subseteq X$, *we have*

$$U(A) = A \quad \text{if} \quad A \neq X \quad \text{and} \quad U(A) = \{1\} \quad \text{if} \quad A = X.$$

Remark 95. Hence, we can see that $U(\{2\}) = \{2\} \not\subseteq \{1\} = U(X)$. Thus, the super relation U is not even quasi-increasing.

Moreover, to illustrate our former definitions and results, we can easily prove the following very particular theorems.

Theorem 252. *We have*

(1) $U^{\triangleleft}(x) = \{x\}$ *for all* $x \in X$;
(2) $U^{\circ}(A) = U^{-1}(A) = A$ *for all* $A \subseteq X$.

Proof. By using Definition 2 and Theorems 6 and 18, we can see that

$$U^{\triangleleft}(x) = U(\{x\}) = \{x\}$$

for all $x \in X$, and

$$U^{\circ}(A) = U^{\triangleleft}[A] = A \quad \text{and} \quad U^{-1}(A) = U^{\triangleleft -1}[A] = U^{\triangleleft}[A] = A$$

for all $A \subseteq X$. □

Remark 96. Note that the ordinary inverse U^{-1} of the relation U is a relation on X to $\mathcal{P}(X)$ such that

$$U^{-1}(1) = \{\{1\}, X\} \quad \text{and} \quad U^{-1}(2) = \{\{2\}\}.$$

Theorem 253. *We have*

(1) $\mathrm{Int}_U(\emptyset) = \{\emptyset\}$, $\mathrm{Int}_U(X) = \mathcal{P}(X)$,
$\mathrm{Int}_U(\{1\}) = \{\emptyset, \{1\}, X\}$, $\mathrm{Int}_U(\{2\}) = \{\emptyset, \{2\}\}$.

(2) $\mathrm{int}_U(A) = A$ *for all* $A \subseteq X$; (3) $\tau_U = \mathcal{T}_{\mathcal{R}} = \mathcal{P}(X)$;

(4) $\mathcal{E}_U = \mathcal{P}(X) \setminus \{\emptyset\}$; (5) $\mathcal{N}_U = \{\emptyset\}$.

Proof. To check this, note that by Theorems 88, 104, 118 and 127, for any $x \in X$ and $A, B \subseteq X$, we have

(a) $B \in \mathrm{Int}_U(A) \iff U(B) \subseteq A$;
(b) $x \in \mathrm{int}_U(A) \iff \{x\} \in \mathrm{Int}_U(A)$;
(c) $A \in \mathcal{E}_U \iff \mathrm{int}_U(A) \neq \emptyset$;
(d) $A \in \tau_U \iff A \in \mathrm{Int}_U(A)$;
(e) $A \in \mathcal{T}_U \iff A \subseteq \mathrm{int}_U(A)$. □

Remark 97. From assertion (1), we can see that, now in addition to Theorem 96, the hyper relation Int_U is also union-preserving.

Moreover, from assertion (1) of Theorem 253, we can also easily infer the following:

Corollary 46. *We have*

(1) $\operatorname{Int}_U^{-1}(\emptyset) = \mathcal{P}(X)$, (2) $\operatorname{Int}_U^{-1}(X) = \{\{1\}, X\}$;
(3) $\operatorname{Int}_U^{-1}(\{1\}) = \{\{1\}, X\}$, (4) $\operatorname{Int}_U^{-1}(\{2\}) = \{\{2\}, X\}$.

Remark 98. Hence, we can see that $\operatorname{Int}_U^{-1}(\{2\}) \not\subseteq \operatorname{Int}_U^{-1}(X)$. Thus, the hyper relation $\operatorname{Int}_U^{-1}$ is not even quasi-increasing.

However, it is now more important to note that $\operatorname{Int}_U^{-1}(X) \not\subseteq \operatorname{Int}_U^{-1}(\{2\})$ also holds. Thus, the hyper relation $\operatorname{Int}_U^{-1}$ is not decreasing.

Hence, by Theorem 47, we can see that $\operatorname{Int}_U \neq \operatorname{Int}_{\mathcal{V}}$ for every increasing super relator $\mathcal{V}$ on X. However, by using Theorem 253, we can actually prove the following:

Theorem 254. *For any quasi-increasing super relator $\mathcal{V}$ on X, we have*

$$\operatorname{Int}_U(\{1\}) \neq \operatorname{Int}_{\mathcal{V}}(\{1\}).$$

Proof. Assume, on the contrary, that there exists a quasi-increasing super relator $\mathcal{V}$ on X such that

$$\operatorname{Int}_U(\{1\}) = \operatorname{Int}_{\mathcal{V}}(\{1\}),$$

and thus, $\operatorname{Int}_U(\{1\}) \subseteq \operatorname{Int}_{\mathcal{V}}(\{1\})$. Hence, since $X \in \operatorname{Int}_U(\{1\})$, we can infer that $X \in \operatorname{Int}_{\mathcal{V}}(\{1\})$. Thus, by Theorem 88, there exists $V \in \mathcal{V}$ such that $V(X) \subseteq \{1\}$. Hence, since V is quasi-increasing, we can infer that

$$V(\{2\}) \subseteq V(X) \subseteq \{1\},$$

and thus, $\{2\} \in \operatorname{Int}_{\mathcal{V}}(\{1\})$. Now, since $\{2\} \notin \operatorname{Int}_{\mathcal{V}}(\{1\})$, we can also state that

$$\operatorname{Int}_{\mathcal{V}}(\{1\}) \not\subseteq \operatorname{Int}_U(\{1\}),$$

and thus, $\operatorname{Int}_U(\{1\}) \neq \operatorname{Int}_{\mathcal{V}}(\{1\})$. This contradiction proves the assertion of the theorem. □

Now, as an immediate consequence of this theorem, we can also state the following:

Corollary 47. *For any ordinary relator $\mathcal{R}$ on X, we have*

$$\mathrm{Int}_U(\{1\}) \neq \mathrm{Int}_{\mathcal{R}}(\{1\}).$$

Proof. If $\mathcal{R}$ is an ordinary relator on X, then by Theorem 8 the super relator $\mathcal{R}^{\triangleright}$ is union-preserving, and thus, in particular, it is quasi-increasing. Moreover, by Definition 32, we have $\mathrm{Int}_{\mathcal{R}} = \mathrm{Int}_{\mathcal{R}^{\triangleright}}$. Thus, Theorem 254 can be applied to obtain the require assertion. □

Remark 99. Thus, in particular, we have $\mathrm{Int}_U \neq \mathrm{Int}_{\mathcal{R}}$ for every ordinary relator $\mathcal{R}$ on X. This shows that super relators can generate more general proximal interiors than the ordinary ones.

Note that, to prove a quite similar statement for the families of the generated proximally open sets, by Remarks 55 and 85, an even more simple super relation can be applied.

For instance, if $Y = \{1\}$ and V is a super relation on Y, such that $V(\emptyset) = X$, then we can already note that $\tau_V = \{X\}$, and thus, $\tau_V \neq \tau_{\mathcal{S}}$ for any ordinary relator $\mathcal{S}$ on Y.

Finally, we note that, by using Theorem 252, we can also prove the following:

Theorem 255. *If $\diamond \in \{\triangleleft, \circ, -1\}$, then*

(1) $\mathrm{Int}_{U^\diamond}(A) = \mathcal{P}(A)$ *for all* $A \subseteq X$;
(2) $\mathrm{int}_{U^\diamond}(A) = A$ *for all* $A \subseteq X$;
(3) $\tau_{U^\diamond} = \mathcal{T}_{U^\diamond} = \mathcal{P}(X)$;
(4) $\mathcal{E}_{U^\diamond} = \mathcal{P}(X) \setminus \{\emptyset\}$;
(5) $\mathcal{N}_{\mathcal{U}^\diamond} = \{\emptyset\}$.

39. Reflexive and Quasi-Topological Relators

Notation 16. *In this and the following section, we shall assume that $\mathcal{R}$ is an ordinary relator on X.*

Definition 39. The relator $\mathcal{R}$ is called *reflexive* if each member R of $\mathcal{R}$ is a reflexive relation on X.

Remark 100. Thus, the following assertions are equivalent:

(1) $\mathcal{R}$ is reflexive;
(2) $x \in R(x)$ for all $x \in X$ and $R \in \mathcal{R}$;
(3) $A \subseteq R[A]$ for all $A \subseteq X$ and $R \in \mathcal{R}$.

The importance of reflexive relators is also apparent from the following two theorems.

Theorem 256. *The following assertions are equivalent:*

(1) $\mathcal{R}$ *is reflexive*; (2) $\rho_{\mathcal{R}}$ *is reflexive*;
(3) $A \subseteq \mathrm{cl}_{\mathcal{R}}(A)$ *for all* $A \subseteq X$; (4) $\mathrm{int}_{\mathcal{R}}(A) \subseteq A$ *for all* $A \subseteq X$.

Remark 101. Therefore, if $\mathcal{R}$ is reflexive, then

(1) $\mathcal{F}_{\mathcal{R}} = \{A \subseteq X : \ A = \mathrm{cl}_{\mathcal{R}}(A)\}$;
(2) $\mathcal{T}_{\mathcal{R}} = \{A \subseteq X : \ A = \mathrm{int}_{\mathcal{R}}(A)\}$.

Theorem 257. *The following assertions are equivalent:*

(1) $\mathcal{R}$ *is reflexive*;
(2) $A \in \mathrm{Int}_{\mathcal{R}}(B)$ *implies* $A \subseteq B$ *for all* $A, B \subseteq X$;
(3) $A \cap B \neq \emptyset$ *implies* $A \in \mathrm{Cl}_{\mathcal{R}}(B)$ *for all* $A, B \subseteq X$.

Remark 102. In addition to the above two theorems, it is also worth mentioning that if $\mathcal{R}$ is a reflexive ordinary relator on X, then

(1) $\mathrm{Int}_{\mathcal{R}}$ is transitive;
(2) $B \in \mathrm{Cl}_{\mathcal{R}}(A)$ implies $\mathcal{P}(X) = \mathrm{Cl}_{\mathcal{R}}(A)^c \cup \mathrm{Cl}_{\mathcal{R}}^{-1}(B)$;
(3) $\mathrm{int}_{\mathcal{R}}(\mathrm{bor}_{\mathcal{R}}(A)) = \emptyset$ and $\mathrm{int}_{\mathcal{R}}(\mathrm{res}_{\mathcal{R}}(A)) = \emptyset$ for all $A \subseteq X$.

Thus, for instance, for any $A \subseteq X$ we have $\mathrm{res}_{\mathcal{R}}(A) \in \mathcal{T}_{\mathcal{R}}$ if and only if $A \in \mathcal{F}_{\mathcal{R}}$.

In contrast to the reflexivity property of $\mathcal{R}$, we may naturally introduce several useful symmetry and transitivity properties of $\mathcal{R}$ [141, 143, 144].

However, it is now more important to note that, in addition to Definition 39, we may also naturally introduce the following:

Definition 40. The relator $\mathcal{R}$ is called

(1) *quasi-topological* if $x \in \mathrm{int}_{\mathcal{R}}(\mathrm{int}_{\mathcal{R}}(R(x)))$ for all $x \in X$ and $R \in \mathcal{R}$;

(2) *topological* if for any $x \in X$ and $R \in \mathcal{R}$ there exists $V \in \mathcal{T}_{\mathcal{R}}$ such that $x \in V \subseteq R(x)$.

Remark 103. Note that, for any $x \in X$ and $R \in \mathcal{R}$, we have $R(x) \subseteq R(x)$, and thus $x \in \operatorname{int}_{\mathcal{R}}(R(x))$. But, despite this, $\mathcal{R}$ need not be quasi-topological.

However, if $\mathcal{R}$ is *topologically transitive* in the sense that for each $x \in X$ and $R \in \mathcal{R}$ there exists $S, T \in \mathcal{R}$ such that

$$(T \circ S)(x) \subseteq R(x), \quad \text{and thus } T[S(x)] \subseteq R(x),$$

then $S(x) \in \operatorname{Int}_{\mathcal{R}}(R(x))$, and thus,

$$S(x) \subseteq \operatorname{int}_{\mathcal{R}}(R(x)), \text{ and thus, } x \in \operatorname{int}_{\mathcal{R}}(S(x)) \subseteq \operatorname{int}_{\mathcal{R}}(\operatorname{int}_{\mathcal{R}}(R(x))).$$

Therefore, $\mathcal{R}$ is quasi-topological.

The importance of quasi-topological relators is also apparent from the following theorems.

Theorem 258. *The following assertions are equivalent:*

(1) $\mathcal{R}$ *is quasi-topological;*
(2) $\operatorname{int}_{\mathcal{R}}(R(x)) \in \mathcal{T}_{\mathcal{R}}$ *for all* $x \in X$ *and* $R \in \mathcal{R}$;
(3) $\operatorname{cl}_{\mathcal{R}}(A) \in \mathcal{F}_{\mathcal{R}}$ *for all* $A \subseteq X$;
(4) $\operatorname{int}_{\mathcal{R}}(A) \in \mathcal{T}_{\mathcal{R}}$ *for all* $A \subseteq X$.

Remark 104. Hence, for instance, we can see that the ordinary relator $\mathcal{R}$ is quasi-topological if and only if the super relation $\operatorname{cl}_{\mathcal{R}}$ is *upper semi-idempotent* in the sense that $\operatorname{cl}_{\mathcal{R}}(\operatorname{cl}_{\mathcal{R}}(A)) \subseteq \operatorname{cl}_{\mathcal{R}}(A)$ for all $A \subseteq X$.

Theorem 259. *The following assertions are equivalent:*

(1) $\mathcal{R}$ *is topological;* (2) $\mathcal{R}$ *is reflexive and quasi-topological.*

Remark 105. By Theorem 258, the relator $\mathcal{R}$ may be called *weakly (strongly) quasi-topological* if $\rho_{\mathcal{R}}(x) \in \mathcal{F}_{\mathcal{R}}$ $(R(x) \in \mathcal{T}_{\mathcal{R}})$ for all $x \in X$ and $R \in \mathcal{R}$.

Moreover, by Theorem 259, the relator $\mathcal{R}$ may be called *weakly (strongly) topological* if it is reflexive and weakly (strongly) quasi-topological.

40. Some Further Theorems on Topological Relators

The following theorem, together with Theorem 229, shows that in a topological relator space $X(\mathcal{R})$, the family $\mathcal{T}_{\mathcal{R}}$ and the relation $\text{int}_{\mathcal{R}}$ are equivalent tools.

Theorem 260. *The following assertions are equivalent:*

(1) $\mathcal{R}$ *is topological;*
(2) $\text{int}_{\mathcal{R}}(A) = \bigcup(\mathcal{T}_{\mathcal{R}} \cap \mathcal{P}(A))$ *for all* $A \subseteq X$;
(3) $\text{cl}_{\mathcal{R}}(A) = \bigcap(\mathcal{F}_{\mathcal{R}} \cap \mathcal{P}^{-1}(A))$ *for all* $A \subseteq X$.

Hence, by using Theorems 235, 222 and 232, we can derive the following:

Corollary 48. *If* $\mathcal{R}$ *is topological, then for any* $A \subset X$, *we have*

(1) $A \in \mathcal{E}_{\mathcal{R}}$ *if and only if there exists* $V \in \mathcal{T}_{\mathcal{R}} \setminus \{\emptyset\}$ *such that* $V \subseteq A$;
(2) $A \in \mathcal{D}_{\mathcal{R}}$ *if and only if for all* $W \in \mathcal{F}_{\mathcal{R}} \setminus \{X\}$ *we have* $A \setminus W \neq \emptyset$.

Remark 106. By Remark 18, assertion (1) can be briefly expressed in the form that $\mathcal{E}_{\mathcal{R}} = (\mathcal{T}_{\mathcal{R}} \setminus \{\emptyset\})^*$. That is, $\mathcal{T}_{\mathcal{R}} \setminus \{\emptyset\}$ is base of $\mathcal{E}_{\mathcal{R}}$.

However, it is now more important to note that we can also prove the following:

Theorem 261. *The following assertions are equivalent:*

(1) $\mathcal{R}$ *is topological;*
(2) $\mathcal{R}$ *is topologically equivalent to a preorder relator.*

Proof. To prove the implication (1) $\Longrightarrow$ (2), note that if (1) holds, then by Definition 40, for any $x \in X$ and $R \in \mathcal{R}$, there exists $V \in \mathcal{T}_{\mathcal{R}}$ such that $x \in V \subseteq R(x)$. Thus, by using the Pervin preorder relator

$$\mathcal{S} = \mathcal{R}_{\mathcal{T}_{\mathcal{R}}} = \{R_V : \ V \in \mathcal{T}_{\mathcal{R}}\}, \text{ where } R_V = V^2 \cup (V^c \times X),$$

we can show that $\text{int}_{\mathcal{R}}(A) = \text{int}_{\mathcal{S}}(A)$ for all $A \subseteq X$, and thus also $\text{int}_{\mathcal{R}} = \text{int}_{\mathcal{S}}$. $\square$

Remark 107. The above theorem can also be proved by using the *topological refinement relator*

$$\mathcal{R}^{\wedge} = \{S \subseteq X^2 : \quad \forall\ x \in X : \ x \in \text{int}_{\mathcal{R}}(S(x))\}$$

and its preorder modification $\mathcal{R}^{\wedge\infty} = \{S^{\infty} : S \in \mathcal{R}^{\wedge}\}$, considered already in Refs. [91, 93, 113, 146].

Remark 108. The importance of these relators lies mainly in the facts that if $\mathcal{R} \neq \emptyset$, then

(1) $\mathcal{S} = \mathcal{R}^{\wedge}$ is the largest relator on X such that $\operatorname{int}_{\mathcal{S}} = \operatorname{int}_{\mathcal{R}}$;
(2) $\mathcal{S} = \mathcal{R}^{\wedge\infty}$ is the largest preorder relator on X such that $\mathcal{T}_{\mathcal{S}} = \mathcal{T}_{\mathcal{R}}$.

Remark 109. If $\operatorname{card}(X) > 2$, then by using the equivalence relator $\mathcal{R} = \{X^2\}$ Mala [91, Example 5.3] proved that there does not exist a largest relator $\mathcal{S}$ on X such that $\mathcal{T}_{\mathcal{R}} = \mathcal{T}_{\mathcal{S}}$.

Moreover, Pataki [113, Example 7.2] proved that $\mathcal{T}_{\mathcal{R}^{\wedge\partial}} \not\subseteq \mathcal{T}_{\mathcal{R}}$ and $\wedge\partial$ is not idempotent. (Actually, it can be proved that $\mathcal{R}^{\wedge\partial\wedge} \not\subseteq \mathcal{R}^{\wedge\partial}$ also holds [155, Example 10.11].)

In addition to Theorem 258, it is also worth proving the following:

Theorem 262. *The following assertions are equivalent:*

(1) $\mathcal{R}$ *is quasi-topological;*
(2) $\mathcal{R} \subseteq (\mathcal{R}^{\wedge} \circ \mathcal{R})^{\wedge}$; (3) $\mathcal{R}^{\wedge} \subseteq (\mathcal{R}^{\wedge} \circ \mathcal{R}^{\wedge})^{\wedge}$.

Remark 110. The equivalence of assertions (1) and (3) can be briefly expressed by stating that $\mathcal{R}$ is quasi-topological if and only if its topological refinement $\mathcal{R}^{\wedge}$ is topologically transitive.

In particular, we can easily prove the following:

Theorem 263. *For any $R \in \mathcal{R}$, the following assertions are equivalent:*

(1) R *is quasi-topological;* (2) R *is transitive.*

Hence, it is clear that, even more specially, we can also state the following:

Corollary 49. *A relation $R \in \mathcal{R}$ is topological if and only if it is a preorder relation.*

Remark 111. Analogously to Definition 40, the relator $\mathcal{R}$ may be called

(1) *quasi-proximal* if $A \in \operatorname{Int}_{\mathcal{R}}[\tau_{\mathcal{R}} \cap \operatorname{Int}_{\mathcal{R}}(R[A])]$ for all $A \subseteq X$ and $R \in \mathcal{R}$;
(2) *proximal* if for any $A \subseteq X$ and $R \in \mathcal{R}$ there exists $V \in \tau_{\mathcal{R}}$ such that $A \subseteq V \subseteq R[A]$.

Thus, in addition to the counterparts of our former theorems, we can prove that $\mathcal{R}$ is topological if and only if its topological refinement $\mathcal{R}^{\wedge}$ is proximal.

Therefore, several theorems on topological relators can be derived from those on the proximal ones by using that $\tau_{\mathcal{R}^{\wedge}} = \mathcal{T}_{\mathcal{R}}$ whenever $\mathcal{R} \neq \emptyset$.

41. A Few Basic Facts on Filtered Relators

Notation 17. *In this section, we shall assume that $\mathcal{R}$ is an ordinary relator on X to Y.*

The following definition was first investigated in Refs. [141, 143].

Definition 41. The relator $\mathcal{R}$ is called

(1) *proximally filtered* if for any $A \subseteq X$ and $R, S \in \mathcal{R}$ there exists $T \in \mathcal{R}$ such that $T[A] \subseteq R[A] \cap S[A]$;
(2) *topologically filtered* if for any $x \in X$ and $R, S \in \mathcal{R}$ there exists $T \in \mathcal{R}$ such that $T(x) \subseteq R(x) \cap S(x)$.

Remark 112. By using the binary operation $\wedge$ and the closure operation $\wedge$ on relators, the latter property can be expressed in some concise forms.

For instance, we can see that $\mathcal{R}$ is topologically filtered if and only if any one of the properties $\mathcal{R} \wedge \mathcal{R} \subseteq \mathcal{R}^{\wedge}$, $(\mathcal{R} \wedge \mathcal{R})^{\wedge} = \mathcal{R}^{\wedge}$ and $\mathcal{R}^{\wedge} \wedge \mathcal{R}^{\wedge} = \mathcal{R}^{\wedge}$ holds.

However, in general, we only have $(R \cap S)[A] \subseteq R[A] \cap S[A]$. Therefore, the corresponding proximal filteredness properties are, unfortunately, not equivalent.

Despite this, we can easily prove the following theorem which shows the appropriateness of the above proximal filteredness property.

Theorem 264. *The following assertions are equivalent:*

(1) *$\mathcal{R}$ is proximally filtered;*
(2) $\mathrm{Cl}_{\mathcal{R}}(A \cup B) = \mathrm{Cl}_{\mathcal{R}}(A) \cup \mathrm{Cl}_{\mathcal{R}}(B)$ *for all* $A, B \subseteq Y$;
(3) $\mathrm{Int}_{\mathcal{R}}(A \cap B) = \mathrm{Int}_{\mathcal{R}}(A) \cap \mathrm{Int}_{\mathcal{R}}(B)$ *for all* $A, B \subseteq Y$.

Proof. To prove that (3) implies (1), note that if $A \subseteq X$ and $R, S \in \mathcal{R}$, then by Theorem 193 we have $A \in \mathrm{Int}_{\mathcal{R}}(R[A])$ and $A \in \mathrm{Int}_{\mathcal{R}}(S[A])$. Therefore, if (3) holds, then we also have $A \in \mathrm{Int}_{\mathcal{R}}(R[A] \cap S[A])$. Thus, by Theorem 193, there exists $T \in \mathcal{R}$ such that $T[A] \subseteq R[A] \cap S[A]$. □

Analogously to this theorem, we can also prove the following theorem, which can also be derived from the above theorem.

Theorem 265. *The following assertions are equivalent:*

(1) *$\mathcal{R}$ is topologically filtered;*
(2) $\mathrm{cl}_{\mathcal{R}}(A \cup B) = \mathrm{cl}_{\mathcal{R}}(A) \cup \mathrm{cl}_{\mathcal{R}}(B)$ *for all* $A, B \subseteq Y$;
(3) $\mathrm{int}_{\mathcal{R}}(A \cap B) = \mathrm{int}_{\mathcal{R}}(A) \cap \mathrm{int}_{\mathcal{R}}(B)$ *for all* $A, B \subseteq Y$.

The importance of topologically filtered relators is also apparent from the next.

Theorem 266. *If $\mathcal{R}$ is topologically filtered, then for any $A, B \subseteq X$ we have*

(1) $\mathrm{cl}_{\mathcal{R}}(A) \cap \mathrm{int}_{\mathcal{R}}(B) = \mathrm{cl}_{\mathcal{R}}(A \cap B) \cap \mathrm{int}_{\mathcal{R}}(B)$;
(2) $\mathrm{int}_{\mathcal{R}}(A \cup B) \cup \mathrm{cl}_{\mathcal{R}}(B) = \mathrm{int}_{\mathcal{R}}(A) \cup \mathrm{cl}_{\mathcal{R}}(B)$.

Proof. Assume that $x \in A^- \cap B^\circ$ and $R \in \mathcal{R}$. Then, since $x \in B^\circ$, there exists $S \in \mathcal{R}$ such that $S(x) \subseteq B$. Moreover, since $\mathcal{R}$ is topologically filtered, there exists $T \in \mathcal{R}$ such that $T(x) \subseteq R(x) \cap S(x)$. Furthermore, since $x \in A^-$, there exists $y \in A$ such that $y \in T(x)$. Hence, we can already infer that

$$y \in A \cap T(x) \subseteq A \cap S(x) \subseteq A \cap B \quad \text{and} \quad y \in T(x) \subseteq R(x).$$

Therefore, $R(x) \cap (A \cap B) \neq \emptyset$ for all $R \in \mathcal{R}$, and thus, $x \in (A \cap B)^-$ also holds.

Hence, we can see that $A^- \cap B^\circ \subseteq (A \cap B)^-$ and thus also

$$A^- \cap B^\circ = A^- \cap B^\circ \cap B^\circ \subseteq (A \cap B)^- \cap B^\circ.$$

Moreover, by using Theorem 209, we can see that

$$(A \cap B)^- \cap B^\circ \subseteq A^- \cap B^\circ.$$

Therefore, $A^- \cap B^\circ = (A \cap B)^- \cap B^\circ$, and thus assertion (1) is true.

Now, by applying assertion (1) to the sets A^c and B^c and using De Morgan's laws and Theorem 206, we can easily see that assertion (2) is also true. □

Now, as a trivial, but important consequence of this theorem, we can also state the following:

Corollary 50. *If $\mathcal{R}$ is topologically filtered, then for any $A, B \subseteq X$ we have*

(1) $\mathrm{cl}_{\mathcal{R}}(A) \cap \mathrm{int}_{\mathcal{R}}(B) \subseteq \mathrm{cl}_{\mathcal{R}}(A \cap B)$;

(2) $\mathrm{int}_{\mathcal{R}}(A \cup B) \subseteq \mathrm{int}_{\mathcal{R}}(A) \cup \mathrm{cl}_{\mathcal{R}}(B)$.

42. Some More Particular Theorems on Topologically Filtered Relators

Notation 18. *In this and the following section, we shall already assume that $\mathcal{R}$ is an ordinary relator on X.*

Now, in addition to Theorem 265, we can also prove the following:

Theorem 267. *If $\mathcal{R}$ is topologically filtered, $A, B \subseteq X$ and there exists $V \in \mathcal{T}_{\mathcal{R}} \cap \mathcal{F}_{\mathcal{R}}$ such that $A \subseteq V$ and $B \subseteq V^c$, then*

$$\mathrm{int}_{\mathcal{R}}(A \cup B) = \mathrm{int}_{\mathcal{R}}(A) \cup \mathrm{int}_{\mathcal{R}}(B).$$

Proof. By Theorem 214, we have $A^\circ \cup B^\circ \subseteq (A \cup B)^\circ$. Therefore, we need actually prove the converse inclusion.

For this, note that if $x \in (A \cup B)^\circ$, then by Theorem 205 there exists $R \in \mathcal{R}$ such that $R(x) \subseteq A \cup B$.

Now, if $x \in V$, then by Theorem 230 there exists $S \in \mathcal{R}$ such that $S(x) \subseteq V$. Moreover, by Definition 41, there exists $T \in \mathcal{R}$ such that $T(x) \subseteq R(x) \cap S(x)$. Hence, we can already infer that

$$\begin{aligned} T(x) &\subseteq R(x) \cap S(x) \subseteq (A \cup B) \cap V \\ &= (A \cap V) \cup (B \cap V) = A \cup \emptyset = A. \end{aligned}$$

Therefore, if $x \in V$, then $x \in A^\circ$.

A quite similar argument shows that if $x \in V^c$, then $x \in B^\circ$. Therefore, in both cases, we have $x \in A^\circ \cup B^\circ$. Thus, the inclusion $(A \cup B)^\circ \subseteq A^\circ \cup B^\circ$ is also true. □

Remark 113. This theorem is a straightforward generalization of Ref. [83, Lemma 7] of Levine.

More difficult conditions for the dual equality $(A \cap B)^- = A^- \cap B^-$ to hold were given by Gottschalk [52] and Jung and Nam [67, 68].

Concerning the latter problem, we could only prove the following:

Theorem 268. *If* $\mathcal{R}$ *is non-void and reflexive, and*

(1) $\operatorname{cl}_{\mathcal{R}}(\{x\} \cap B) = \operatorname{cl}_{\mathcal{R}}(\{x\}) \cap \operatorname{cl}_{\mathcal{R}}(B)$ *for all* $x \in X$ *and* $A, B \subseteq X$,

then

(2) $\mathcal{T}_{\mathcal{R}} = \mathcal{P}(X)$; (3) $\mathcal{R}^{\wedge\infty} = \mathcal{P}(X^2)^{\infty}$.

Proof. If assertion (2) does not hold, then there exists $A \subseteq X$ such that $A \notin \mathcal{T}_{\mathcal{R}}$. Hence, by using Theorem 232, we can infer that $B = A^c \notin \mathcal{F}_{\mathcal{R}}$. Therefore, by Theorem 229, we have $B^- \not\subseteq B$. Thus, there exists $x \in B^-$ such that $x \notin B$. Hence, by using Theorems 256 and 210 and the assumption (1), we can arrive at the contradiction that

$$x \in \{x\}^- \cap B^- = (\{x\} \cap B)^- = \emptyset^- = \emptyset.$$

Therefore, assertion (2) is true.

Now, to prove assertion (3), it remains only to note that, by Theorem 230 and [130, Theorem 15.5], and the definitions of the operations $\wedge$ and ∞, we have

$$(2) \iff \mathcal{T}_{\mathcal{R}} = \mathcal{T}_{\Delta_X} \iff \mathcal{R}^{\wedge\infty} = \{\Delta_X\}^{\wedge\infty} \iff (3). \qquad \square$$

From Theorem 266, by using Theorem 229, we can immediately derive the following:

Theorem 269. *If* $\mathcal{R}$ *is topologically filtered, then*

(1) $\operatorname{cl}_{\mathcal{R}}(A) \cap B = \operatorname{cl}_{\mathcal{R}}(A \cap B) \cap B$ *for all* $A \subseteq X$ *and* $B \in \mathcal{T}_{\mathcal{R}}$;

(2) $\operatorname{int}_{\mathcal{R}}(A \cup B) \cup B = \operatorname{int}_{\mathcal{R}}(A) \cup B$ *for all* $A \subseteq X$ *and* $B \in \mathcal{F}_{\mathcal{R}}$.

Proof. To derive assertion (1) from that of Theorem 266, note that if $B \in \mathcal{T}_{\mathcal{R}}$, then by Theorem 229 we have $B \subseteq B^{\circ}$, and thus also $B^{\circ} \cap B = B$. $\square$

Now, as a trivial but important consequence of this theorem we can also state the following:

Corollary 51. *If $\mathcal{R}$ is topologically filtered, then*

(1) $\mathrm{cl}_{\mathcal{R}}(A) \cap B \subseteq \mathrm{cl}_{\mathcal{R}}(A \cap B)$ *for all* $A \subseteq X$ *and* $B \in \mathcal{T}_{\mathcal{R}}$;
(2) $\mathrm{int}_{\mathcal{R}}(A \cup B) \subseteq \mathrm{int}_{\mathcal{R}}(A) \cup B$ *for all* $A \subseteq X$ *and* $B \in \mathcal{F}_{\mathcal{R}}$.

Remark 114. The importance of the inclusion $A^- \cap B \subseteq (A \cap B)^-$, with B being open, was first observed by Kuratowski [77, p. 45].

Later, Császár [21–23, 25, 27, 29] and Sivagami [135] assumed it as an axiom for an increasing set-to-set function γ.

Corollary 51 is again more important than Theorem 269. Namely, by using it and our former theorems on topological relators, we can already prove the following:

Theorem 270. *If $\mathcal{R}$ is topological and topologically filtered, then*

(1) $\mathrm{cl}_{\mathcal{R}}(A \cap B) = \mathrm{cl}_{\mathcal{R}}(\mathrm{cl}_{\mathcal{R}}(A) \cap B)$ *for all* $A \subseteq X$ *and* $B \in \mathcal{T}_{\mathcal{R}}$;
(2) $\mathrm{int}_{\mathcal{R}}(A \cup B) = \mathrm{int}_{\mathcal{R}}(\mathrm{int}_{\mathcal{R}}(A) \cup B)$ *for all* $A \subseteq X$ *and* $B \in \mathcal{F}_{\mathcal{R}}$.

Proof. To prove (1), note that if $A \subseteq X$ and $B \in \mathcal{T}_{\mathcal{R}}$, then by Corollary 51 we have $A^- \cap B \subseteq (A \cap B)^-$. Hence, by using Theorems 209, 258 and 229, we can infer that

$$(A^- \cap B)^- \subseteq (A \cap B)^{--} \subseteq (A \cap B)^-.$$

On the other hand, by Theorems 259 and 256, we have $A \subseteq A^-$, and thus also $A \cap B \subseteq A^- \cap B$. Hence, by Theorem 209, we can infer that $(A \cap B)^- \subseteq (A^- \cap B)^-$. Therefore, the corresponding equality is also true. □

From this theorem, by using Theorem 219, we can immediately derive the following:

Corollary 52. *If $\mathcal{R}$ is topological and topologically filtered, then*

(1) $\mathrm{cl}_{\mathcal{R}}(A \cap B) = \mathrm{cl}_{\mathcal{R}}(B)$ *for all* $A \in \mathcal{D}_{\mathcal{R}}$ *and* $B \in \mathcal{T}_{\mathcal{R}}$;
(2) $\mathrm{int}_{\mathcal{R}}(A \cup B) = \mathrm{int}_{\mathcal{R}}(A)$ *for all* $A \in \mathcal{E}_{\mathcal{R}}^c$ *and* $B \in \mathcal{F}_{\mathcal{R}}$.

Now, by modifying an argument of Levine [85], we can also prove the following:

Theorem 271. *If $\mathcal{R}$ is non-void and topological, and $A \subseteq X$, then*

(1) $\operatorname{cl}_{\mathcal{R}}(A \cap B) = \operatorname{cl}_{\mathcal{R}}(B)$ *for all* $B \in \mathcal{T}_{\mathcal{R}}$ *implies that* $A \in \mathcal{D}_{\mathcal{R}}$;
(2) $\operatorname{int}_{\mathcal{R}}(A \cup B) = \operatorname{int}_{\mathcal{R}}(A)$ *for all* $B \in \mathcal{F}_{\mathcal{R}}$ *implies that* $A^c \in \mathcal{E}_{\mathcal{R}}$.

Proof. For instance, if $A \notin \mathcal{D}_{\mathcal{R}}$, then by Theorem 219 there exists $x \in X$ such that $x \notin A^-$. Thus, by Theorem 205, there exists $R \in \mathcal{R}$ such that $A \cap R(x) = \emptyset$. Moreover, by Definition 40, there exists $B \in \mathcal{T}_{\mathcal{R}}$ such that $x \in B \subseteq R(x)$. Thus, we also have $A \cap B = \emptyset$.

Hence, by using the assumption of (1) and Theorem 210, we can infer that $B^- = (A \cap B)^- = \emptyset^- = \emptyset$. On the other hand, from $x \in B$, by using Theorem 256, we can infer that $x \in \{x\}^- \subseteq B^-$, and thus $B^- \neq \emptyset$. This contradiction proves (1). □

Remark 115. If $\mathcal{R}$ is only non-void and reflexive, and $A \subseteq X$ such that $\operatorname{cl}_{\mathcal{R}}(A \cap R(x)) = \operatorname{cl}_{\mathcal{R}}(R(x))$ for all $x \in X$ and $R \in \mathcal{R}$, then we can even more easily prove that $A \in \mathcal{D}_{\mathcal{R}}$.

43. A Few Basic Facts on Quasi-Filtered Relators

Since $R \subseteq R^\infty$ for every $R \in \mathcal{R}$, in addition to Definition 41, we may also naturally introduce the following:

Definition 42. The relator $\mathcal{R}$ is called

(1) *quasi-proximally filtered* if for any $A \subseteq X$ and $R, S \in \mathcal{R}$ there exists $T \in \mathcal{R}$ such that $T[A] \subseteq R^\infty[A] \cap S^\infty[A]$;
(2) *quasi-topologically filtered* if for any $x \in X$ and $R, S \in \mathcal{R}^\wedge$ there exists $T \in \mathcal{R}$ such that $T(x) \subseteq R^\infty(x) \cap S^\infty(x)$.

Remark 116. Analogously to Remark 112, property (2) can also be reformulated in some concise forms.

For instance, we can see that $\mathcal{R}$ is quasi-topologically filtered if and only if $\mathcal{R}^{\wedge\infty} \wedge \mathcal{R}^{\wedge\infty} \subseteq \mathcal{R}^\wedge$, $(\mathcal{R}^{\wedge\infty} \wedge \mathcal{R}^{\wedge\infty})^{\wedge\infty} = \mathcal{R}^{\wedge\infty}$ or $\mathcal{R}^{\wedge\infty} \wedge \mathcal{R}^{\wedge\infty} = \mathcal{R}^{\wedge\infty}$.

However, it is now more important to note that, by using some former results on relators, we can also prove the following two theorems which show the appropriateness of the above quasi-filteredness properties.

Theorem 272. *The following assertions are equivalent:*

(1) $\mathcal{R}$ *is a quasi-proximally filtered;*
(2) $\digamma_{\mathcal{R}}$ *is closed under binary unions;*
(3) $\tau_{\mathcal{R}}$ *is closed under binary intersections.*

Theorem 273. *The following assertions are equivalent:*

(1) $\mathcal{R}$ *is a quasi-topologically filtered;*
(2) $\mathcal{F}_{\mathcal{R}}$ *is closed under binary unions;*
(3) $\mathcal{T}_{\mathcal{R}}$ *is closed under binary intersections.*

Remark 117. In this respect, it is also worth mentioning that if $\mathcal{R}$ is a relator on X to Y, then the family $\mathcal{E}_{\mathcal{R}}$ is closed under binary intersections if and only if $\mathcal{R}$ is *quasi-directed* in the sense that for any $x, y \in X$ and $R, S \in \mathcal{R}$ we have $R(x) \cap S(y) \in \mathcal{E}_{\mathcal{R}}$.

Now, by using Theorem 272, we can also easily prove the following:

Theorem 274. *If* $\mathcal{R}$ *is proximal and quasi-proximally filtered, then* $\mathcal{R}$ *is proximally filtered.*

Proof. Suppose that $A \subseteq X$ and $R, S \in \mathcal{R}$. Then, by Remark 111, there exist $U, V \in \tau_{\mathcal{R}}$ such that $A \subseteq U \subseteq R[A]$ and $A \subseteq V \subseteq S[A]$. Moreover, by Theorem 273, we can state that $U \cap V \in \tau_{\mathcal{R}}$. Therefore, by Theorem 230, there exists $T \in \mathcal{R}$ such that $T[U \cap V] \subseteq U \cap V$. Hence, we can already see that

$$T[A] \subseteq T[U \cap V] \subseteq U \cap V \subseteq R[A] \cap S[A].$$

Thus, $\mathcal{R}$ is proximally filtered. □

Moreover, by using Theorem 273, we can quite similarly prove the following:

Theorem 275. *If* $\mathcal{R}$ *is topological and quasi-topologically filtered, then* $\mathcal{R}$ *is topologically filtered.*

The following example shows that even a quasi-proximally filtered relator need not be topologically filtered.

Example 13. If $X = \{1, 2, 3\}$ and R_i is relation on X, for each $i = 1, 2$, such that

$$R_i(1) = \{1, i+1\} \quad \text{and} \quad R_i(2) = R_i(3) = \{2, 3\},$$

then $\mathcal{R} = \{R_1, R_2\}$ is reflexive and quasi-proximally filtered but not topologically filtered and quasi-topological.

By using Theorem 230, we can see that $\tau_{\mathcal{R}} = \mathcal{T}_{\mathcal{R}} = \{\emptyset, \{2, 3\}, X\}$. Thus, by Theorems 272 and 273, the relator $\mathcal{R}$ is both quasi-proximally and quasi-topologically filtered.

Moreover, we can note that $R_i(1) \not\subseteq R_1(1) \cap R_2(1)$ for each $i = 1, 2$. Therefore, $\mathcal{R}$ is not topologically filtered, and thus by Theorems 275 it is also not quasi-topological.

Remark 118. In this respect, it is also worth noting that if

$$R(1) = X \quad \text{and} \quad R(2) = R(3) = \{2, 3\},$$

then R is already a preorder relation on X such that $\tau_R = \mathcal{T}_R = \{\emptyset, \{2, 3\}, X\}$.

The importance of Corollary 51 is also apparent from the following:

Theorem 276. *If $\mathcal{R}$ is quasi-topological and topologically filtered, then the family $\mathcal{N}_{\mathcal{R}}$ is closed under binary unions.*

Proof. Suppose that $A, B \in \mathcal{N}_{\mathcal{R}}$. Then, by Theorem 258, we have $B^- \in \mathcal{F}_{\mathcal{R}}$. Hence, by using Theorem 265, Corollary 51 and Theorems 243 and 219, we can see that

$$(A \cup B)^{-\circ} = (A^- \cup B^-)^\circ \subseteq A^{-\circ} \cup B^- = \emptyset \cup B^- = B^-.$$

Moreover, by Theorem 258, we have $(A \cup B)^{-\circ} \in \mathcal{T}_{\mathcal{R}}$. Hence, by using Theorems 229 and 209, and also our former observations, we can see that

$$(A \cup B)^{-\circ} \subseteq (A \cup B)^{-\circ\circ} \subseteq B^{-\circ} = \emptyset.$$

Therefore, $(A \cup B)^{-\circ} = \emptyset$, and thus $A \cup B \in \mathcal{N}_{\mathcal{R}}$ also holds. □

Now, by using this theorem, we can also easily establish the following:

Corollary 53. *If $\mathcal{R}$ is non-void, non-partial, quasi-topological and topologically filtered, then $\mathcal{N}_{\mathcal{R}}$ is an ideal on X.*

Proof. By Theorems 243, 219 and 209, it is clear that the family $\mathcal{N}_{\mathcal{R}}$ is always descending. Moreover, by Theorems 210 and 211, we can also note that $\emptyset^{-\circ} = \emptyset^{\circ} = \emptyset$. Therefore, $\emptyset \in \mathcal{N}_{\mathcal{R}}$, and thus $\mathcal{N}_{\mathcal{R}} \neq \emptyset$. Furthermore, from Theorem 276, we know that $\mathcal{N}_{\mathcal{R}}$ is closed under binary unions. □

Remark 119. It can be shown that the following assertions are equivalent:

(1) X is a *topologically regular open subset* of the relator space $X(\mathcal{R})$ in the sense that $\operatorname{int}_{\mathcal{R}}(\operatorname{cl}_{\mathcal{R}}(X)) = X$;
(2) $\mathcal{R}$ is *locally non-partial* in the sense that for each $x \in X$ there exists $R \in \mathcal{R}$ such that for any $y \in R(x)$ and $S \in \mathcal{R}$ we have $S(y) \neq \emptyset$.

Therefore, if X is non-void and $\mathcal{R}$ is locally non-partial, then $X \notin \mathcal{N}_{\mathcal{R}}$, and thus $\mathcal{N}_{\mathcal{R}} \neq \mathcal{P}(X)$.

While, if $\mathcal{R}$ is quasi-topological and $A \in \mathcal{N}_{\mathcal{R}}$, then by using Theorems 258 and 209, we can also see that $A^{--\circ} \subseteq A^{-\circ} = \emptyset$. Therefore, $A^{--\circ} = \emptyset$, and thus $A^{-} \in \mathcal{N}_{\mathcal{R}}$ also holds.

The importance of quasi-topologically filtered relators is also apparent from the following:

Theorem 277. *If $\mathcal{R}$ is topological and quasi-topologically filtered, then for any $A \in \mathcal{T}_{\mathcal{R}}$ we have $\operatorname{res}_{\mathcal{R}}(A) \in \mathcal{F}_{\mathcal{R}} \setminus \mathcal{E}_{\mathcal{R}}$.*

Proof. Namely, by Theorem 232, we have $A^{c} \in \mathcal{F}_{\mathcal{R}}$. Moreover, by Theorem 258, we have $A^{-} \in \mathcal{F}_{\mathcal{R}}$. Hence, by using Theorem 273, we can see that

$$A^{\dagger} = A^{-} \setminus A = A^{-} \cap A^{c} \in \mathcal{F}_{\mathcal{R}}.$$

Moreover, by using Theorems 275, 265, 206, 259 and 256, we can also see that

$$A^{\dagger\circ} = (A^{-} \cap A^{c})^{\circ} = A^{-\circ} \cap A^{c\circ} = A^{-\circ} \cap A^{-c} \subseteq A^{-} \cap A^{-c} = \emptyset,$$

and thus $A^{\dagger\circ} = \emptyset$. Therefore, $A^{\dagger} \notin \mathcal{E}_{\mathcal{R}}$, and thus $A^{\dagger} \in \mathcal{F}_{\mathcal{R}} \setminus \mathcal{E}_{\mathcal{R}}$. □

From this theorem, it is clear that in particular we also have the following:

Corollary 54. *If $\mathcal{R}$ is topological and quasi-topologically filtered, then for any $A \in \mathcal{T}_{\mathcal{R}}$ we have $\operatorname{res}_{\mathcal{R}}(A) \in \mathcal{N}_{\mathcal{R}}$.*

Remark 120. If $\mathcal{R}$ is topological and $A \in \mathcal{T}_{\mathcal{R}}$, then by Theorems 229, 259 and 256, we have $A = A^{\circ}$. Therefore, $A^{\ddagger} = \text{bnd}_{\mathcal{R}}(A) = A^{-} \setminus A^{\circ} = A^{-} \setminus A = A^{\dagger}$.

44. Proximally Closed Sets in Super Relator Spaces

Notation 19. *In this and the following two sections, we shall assume that $\mathcal{R}$ is an ordinary relator, $\mathcal{U}$ is a super relator and $\mathcal{D}$ is a stack on X.*

The importance of the dual super relator $\mathcal{U}^{\star} = \{U^{\star} : \ U \in \mathcal{U}\}$ is also apparent from the following:

Theorem 278. *We have*

(1) $\mathcal{F}_{\mathcal{U}} = \{A \subseteq X : \quad \exists\ U \in \mathcal{U} : \quad A \subseteq U^{\star}(A)\}$;
(2) $\mathcal{F}_{\mathcal{U}^{\star}} = \{A \subseteq X : \quad \exists\ U \in \mathcal{U} : \quad A \subseteq U(A)\}$.

Proof. By Theorems 129 and 128 and Definition 7, we have

$$A \in \mathcal{F}_{\mathcal{U}} \iff A^{c} \in \mathcal{T}_{\mathcal{U}} \iff \exists\ U \in \mathcal{U} : \quad U(A^{c}) \subseteq A^{c}$$
$$\iff \exists\ U \in \mathcal{U} : \ A \subseteq U(A^{c})^{c} \iff \exists\ U \in \mathcal{U} : \ A \subseteq U^{\star}(A).$$

Therefore, assertion (1) is true.

Now, assertion (2) can be easily derived from assertion (1), by writing $\mathcal{U}^{\star}$ in place of $\mathcal{U}$. □

From this theorem, by taking $\mathcal{U} = \{U\}$, we can at once derive the following:

Corollary 55. *For any $U \in \mathcal{U}$, we have*

(1) $\mathcal{F}_{U} = \{A \subseteq X : \ A \subseteq U^{\star}(A)\}$;
(2) $\mathcal{F}_{U^{\star}} = \{A \subseteq X : \ A \subseteq U(A)\}$.

Moreover, from Theorem 278, we can also easily derive the following:

Theorem 279. *We have*

(1) $\mathcal{F}_{\mathcal{R}} = \{A \subseteq X : \quad \exists\ R \in \mathcal{R} : A \subseteq R^{\star}[A]\}$;
(2) $\mathcal{F}_{\mathcal{R}^{\star}} = \{A \subseteq X : \quad \exists\ R \in \mathcal{R} : A \subseteq R[A]\}$.

Proof. To prove assertion (1), note that, by Definition 37, Theorem 278 and Remark 12, for any $A \subseteq X$ we have

$$A \in \digamma_{\mathcal{R}} \iff A \in \digamma_{\mathcal{R}^{\triangleright}} \iff \exists\ R \in \mathcal{R}: \ A \subseteq R^{\triangleright\star}(A)$$
$$\iff \exists\ R \in \mathcal{R}: \ A \subseteq R^{\star}(A).$$ □

Thus, in particular, we can also state the following:

Corollary 56. *For any $R \in \mathcal{R}$, we have*

(1) $\digamma_R = \{A \subseteq X: \ A \subseteq R^{\star}[A]\}$;
(2) $\digamma_{R^{\star}} = \{A \subseteq X: \ A \subseteq R[A]\}$.

Remark 121. If $U \in \mathcal{U}$ is an union-preserving, then by Remark 55 we have $\digamma_U = \mathcal{F}_U$.

While, if $R \in \mathcal{R}$, then by Theorem 231 and Remark 85 we have $\digamma_R = \tau_{R^{-1}} = \mathcal{T}_{R^{-1}}$.

However, the importance of Theorem 278 lies mainly in the fact that it allows of a general unified treatment of certain generalized open sets.

Example 14. If $U \in \mathcal{U}$ such that

$$U(A) = \mathrm{cl}_{(\mathcal{R},\mathcal{D})}(\mathrm{int}_{(\mathcal{R},\mathcal{D})}(A))$$

for all $A \subseteq X$, then

(1) $\digamma_U = \{A \subseteq X: \ A \subseteq \mathrm{int}_{(\mathcal{R},\mathcal{D})}(\mathrm{cl}_{(\mathcal{R}}(A))\}$;
(2) $\digamma_{U^{\star}} = \{A \subseteq X: \ A \subseteq \mathrm{cl}_{(\mathcal{R},\mathcal{D})}(\mathrm{int}_{(\mathcal{R},\mathcal{D})}(A))\}$.

Namely, by using the notations

$$A^{-} = \mathrm{cl}_{(\mathcal{R},\mathcal{D})}(A) \quad \text{and} \quad A^{\circ} = \mathrm{int}_{(\mathcal{R},\mathcal{D})}(A),$$

and Definition 7 and Theorem 160, we can see that

$$U^{\star}(A) = U(A^c)^c = A^{c\circ-c} = A^{-c-c} = A^{-\circ} = \mathrm{int}_{(\mathcal{R},\mathcal{D})}(\mathrm{cl}_{(\mathcal{R},\mathcal{D})}(A))$$

for all $A \subseteq X$. Thus, by Corollary 55, assertion (1) is also true.

Remark 122. Thus, if U is as in the above example and $\mathcal{D} = \mathcal{D}_0$, then $\digamma_U$ and $\digamma_{U^{\star}}$ are just the families $\mathcal{T}_{\mathcal{R}}^{p}$ and $\mathcal{T}_{\mathcal{R}}^{s}$ of all *topologically preopen and semi-open subsets* of the relator space $X(\mathcal{R})$ considered in Ref. [127].

Remark 123. By assuming that

$$U(A) = \mathrm{cl}_{(\mathcal{R},\mathcal{D})}(\mathrm{int}_{(\mathcal{R},\mathcal{D})}(\mathrm{cl}_{(\mathcal{R},\mathcal{D})}(A)))$$

for all $A \subseteq X$, we can quite similarly get some generalizations the families $\mathcal{T}_{\mathcal{R}}^{\alpha}$ and $\mathcal{T}_{\mathcal{R}}^{\beta}$ of all *topologically α-open and β-open subsets* of the relator space $X(\mathcal{R})$ considered also in Ref. [126].

Remark 124. Moreover, for instance, we may also naturally take

$$U(A) = \mathrm{cl}_{(\mathcal{R},\mathcal{D})}(\mathrm{int}_{(\mathcal{R},\mathcal{D})}(A)) \cup \mathrm{int}_{(\mathcal{R},\mathcal{D})}(\mathrm{cl}_{(\mathcal{R},\mathcal{D})}(A))$$

for all $A \subseteq X$, to obtain some generalizations of the families $\mathcal{T}_{\mathcal{R}}^{a}$ and $\mathcal{T}_{\mathcal{R}}^{b}$ of all *topologically α-open and β-open subsets* of the relator space $X(\mathcal{R})$ considered also in Ref. [126].

Remark 125. However, if for instance

$$U(A) = \mathrm{cl}_{(\mathcal{R},\mathcal{D})}[\mathrm{Int}_{(\mathcal{R},\mathcal{D})}(A)]$$

for all $A \subseteq X$, then the super relation $U^{\star}$ and thus the family $\mathcal{F}_U$ cannot be similarly described.

45. Some Set-Theoretic Properties of the Families $\mathcal{F}_{\mathcal{U}}$ and $\mathcal{F}_{\mathcal{U}^{\star}}$

To show that, in some important particular cases, the inclusions $\mathcal{T}_{(\mathcal{R},\mathcal{D})} \subseteq \mathcal{F}_{\mathcal{U}}$ and $\mathcal{T}_{(\mathcal{R},\mathcal{D})} \subseteq \mathcal{F}_{\mathcal{U}^{\star}}$ are true, in addition to Theorem 278 and its corollary, we must also prove the following two closely related theorems and their corollaries.

Theorem 280. *If for each $A \in \mathcal{T}_{(\mathcal{R},\mathcal{D})}$ there exists $U \in \mathcal{U}$ such that*

$$\mathrm{int}_{(\mathcal{R},\mathcal{D})}(A) \subseteq U(A),$$

then $\mathcal{T}_{(\mathcal{R},\mathcal{D})} \subseteq \mathcal{F}_{\mathcal{U}^{\star}}$.

Proof. If $A \in \mathcal{T}_{(\mathcal{R},\mathcal{D})}$, then by Theorem 179, we have

$$A \subseteq \mathrm{int}_{(\mathcal{R},\mathcal{D})}(A).$$

Moreover, by the assumption of the theorem, there exists $U \in \mathcal{U}$ such that

$$\mathrm{int}_{(\mathcal{R},\mathcal{D})}(A) \subseteq U(A).$$

Therefore, we also have $A \subseteq U(A)$. Hence, by Theorem 278, we can see that $A \in \tau_{U^\star}$. Thus, the required inclusion is also true. $\square$

Corollary 57. *If* $U \in \mathcal{U}$ *such that* $\operatorname{int}_{(\mathcal{R},\mathcal{D})} \subseteq U$, *then* $\mathcal{T}_{(\mathcal{R},\mathcal{D})} \subseteq \tau_{U^\star}$.

Theorem 281. *If for each* $A \in \mathcal{F}_{(\mathcal{R},\mathcal{D})}$ *there exists* $U \in \mathcal{U}$ *such that*

$$U(A) \subseteq \operatorname{cl}_{(\mathcal{R},\mathcal{D})}(A),$$

then $\mathcal{T}_{(\mathcal{R},\mathcal{D})} \subseteq \tau_{\mathcal{U}}$.

Proof. If $A \in \mathcal{T}_{(\mathcal{R},\mathcal{D})}$, then by Theorem 181 we have $A^c \in \mathcal{F}_{(\mathcal{R},\mathcal{D})}$. Thus, by the assumption of the theorem, there exists $U \in \mathcal{U}$ such that

$$U(A^c) \subseteq \operatorname{cl}_{(\mathcal{R},\mathcal{D})}(A^c).$$

Hence, by using Theorem 160 and Definition 7, we can infer that

$$\operatorname{int}_{(\mathcal{R},\mathcal{D})}(A) = \operatorname{cl}_{(\mathcal{R},\mathcal{D})}(A^c)^c \subseteq U(A^c)^c = U^\star(A).$$

Thus, Theorem 280 can be applied to the super relator $\mathcal{U}^\star$ to obtain the required inclusion. $\square$

Corollary 58. *If* $U \in \mathcal{U}$ *such that* $U \subseteq \operatorname{cl}_{(\mathcal{R},\mathcal{D})}$, *then* $\mathcal{T}_{(\mathcal{R},\mathcal{D})} \subseteq \tau_U$.

Remark 126. Due to Theorems 281 and 280, the members of the families $\tau_{\mathcal{U}^\star}$ and $\tau_{\mathcal{U}}$ may be naturally considered as some generalized topologically open sets.

In particular, the $\mathcal{D} = \mathcal{D}_0$ particular cases of Corollaries 58 and 57 allow us to easily establish the following:

Example 15. If $U \in \mathcal{U}$ such that

$$U(A) = \operatorname{cl}_{\mathcal{R}}(\operatorname{int}_{\mathcal{R}}(A))$$

for all $A \subseteq X$, and $\mathcal{R}$ is reflexive, then

(1) $\mathcal{T}_{\mathcal{R}} \subseteq \tau_U$; (2) $\mathcal{T}_{\mathcal{R}} \subseteq \tau_{U^\star}$.

Namely, by Theorem 256, we have

$$\operatorname{int}_{\mathcal{R}}(A) \subseteq A \quad \text{and} \quad A \subseteq \operatorname{cl}_{\mathcal{R}}(A)$$

for all $A \subseteq X$. Hence, by using Theorem 209, we can infer that

$$\operatorname{int}_{\mathcal{R}}(A) \subseteq \operatorname{cl}_{\mathcal{R}}(\operatorname{int}_{\mathcal{R}}(A)) = U(A) \quad \text{and}$$
$$U(A) = \operatorname{cl}_{\mathcal{R}}(\operatorname{int}_{\mathcal{R}}(A)) \subseteq \operatorname{cl}_{\mathcal{R}}(A)$$

for all $A \subseteq X$. Therefore, by Corollaries 58 and 57, inclusions (1) and (2) hold.

Remark 127. Quite similar arguments can also be applied whenever U is as in Remark 123 or 124 with $\mathcal{D} = \mathcal{D}_0$.

By using Theorem 278, or Theorems 128, 134 and 133, we can easily establish the following two theorems.

Theorem 282. *The following assertions are equivalent:*

(1) $\emptyset \in \mathcal{T}_{\mathcal{U}}$; (2) $\emptyset \in \mathcal{T}_{\mathcal{U}^\star}$; (5) $\mathcal{T}_{\mathcal{U}^\star} \neq \emptyset$.
(3) $\mathcal{U} \neq \emptyset$; (4) $\mathcal{T}_{\mathcal{U}} \neq \emptyset$;

Theorem 283. *The following assertions are true:*

(1) $X \in \mathcal{T}_{\mathcal{U}}$ *if and only if there exists* $U \in \mathcal{U}$ *such that* $U(\emptyset) = \emptyset$;
(2) $X \in \mathcal{T}_{\mathcal{U}^\star}$ *if and only if there exists* $U \in \mathcal{U}$ *such that* $U(X) = X$.

Proof. By Theorem 278, we have

$$X \in \mathcal{T}_{\mathcal{U}} \iff \exists\ U \in \mathcal{U}: \quad X \subseteq U^\star(X).$$

Hence, since

$$X \subseteq U^\star(X) \iff U^\star(X) = X \iff U(X^c)^c = \emptyset^c \iff U(\emptyset) = \emptyset,$$

we can see that assertion (1) is true. □

Now, as an immediate consequence of the above two theorems, we can also state the following:

Corollary 59. *The following assertions are true:*

(1) $\mathcal{T}_{\mathcal{U}}$ *is a minimal structure on* X *if and only if there exists* $U \in \mathcal{U}$ *such that* $U(\emptyset) = \emptyset$;

(2) *$\mathcal{T}_{\mathcal{U}^\star}$ is a minimal structure on X if and only if there exists $U \in \mathcal{U}$ such that $U(X) = X$.*

Moreover, in addition to Example 14, we can also state the following:

Example 16. If U is as in Example 14 and

(1) $\mathcal{R} \neq \emptyset$ and $\emptyset \notin \mathcal{D}$; (2) $R(x) \in \mathcal{D}$ for all $x \in X$ and $R \in \mathcal{R}$;

then $\mathcal{T}_U$ and $\mathcal{T}_{U^\star}$ are minimal structures on X.

Namely, now by Theorems 164 and 163, we have

$$U(\emptyset) = \mathrm{cl}_{(\mathcal{R},\mathcal{D})}(\mathrm{int}_{(\mathcal{R},\mathcal{D})}(\emptyset)) = \mathrm{cl}_{(\mathcal{R},\mathcal{D})}(\emptyset) = \emptyset$$

and

$$U(X) = \mathrm{cl}_{(\mathcal{R},\mathcal{D})}(\mathrm{int}_{(\mathcal{R},\mathcal{D})}(X)) = \mathrm{cl}_{(\mathcal{R},\mathcal{D})}(X) = X.$$

Thus, Corollary 59 can be applied to get the required assertion.

Remark 128. Thus, if in particular U is as in Example 15 and $\mathcal{R}$ is non-void and non-partial, then $\mathcal{T}_{\mathcal{U}}$ and $\mathcal{T}_{\mathcal{U}^\star}$ are minimal structures on X.

Namely, because of $\mathcal{D}_0 = \mathcal{P}(X) \setminus \{\emptyset\}$, we have $\emptyset \notin \mathcal{D}_0$. Moreover, for any $x \in X$ and $R \in \mathcal{R}$, we have $R(x) \notin \mathcal{D}_0$ if and only if $R(x) = \emptyset$.

46. Some Further Set-Theoretic Properties of the Families $\mathcal{T}_U$ and $\mathcal{T}_{U^\star}$

In addition to Theorems 282 and 283, we can also easily prove the following:

Theorem 284. *If $U \in \mathcal{U}$ such that U is increasing, then the families $\mathcal{T}_U$ and $\mathcal{T}_{U^\star}$ are closed under arbitrary unions.*

Proof. If $\mathcal{A} \subseteq \mathcal{T}_{U^\star}$, then by Corollary 55, for each $A \in \mathcal{A}$, we have $A \subseteq U(A)$. Hence, by using the increasingness of U, we can infer that

$$\bigcup \mathcal{A} = \bigcup_{A \in \mathcal{A}} A \subseteq \bigcup_{A \in \mathcal{A}} U(A) \subseteq \bigcup_{A \in \mathcal{A}} U\left(\bigcup \mathcal{A}\right) \subseteq U\left(\bigcup \mathcal{A}\right).$$

Therefore, by Corollary 55, we have $\bigcup \mathcal{A} \in \mathcal{T}_{U^\star}$. Thus, the family $\mathcal{T}_{U^\star}$ is closed under arbitrary unions.

Now, to prove the same assertion for the family $\mathcal{F}_U$, it is enough to note only that if U is increasing, then by Theorem 27 its dual $U^\star$ is also increasing. □

Now, by Theorem 283, we can also state the following:

Corollary 60. *If $U \in \mathcal{U}$ such that U is increasing, then*

(1) *$\mathcal{F}_U$ is a generalized topology on X if and only if $U(\emptyset) = \emptyset$;*
(2) *$\mathcal{F}_{U^\star}$ is a generalized topology on X if and only if $U(X) = X$.*

Thus, as an improvement of Example 16, we can also state the following:

Example 17. If U is as in Example 16, then the families $\mathcal{F}_U$ and $\mathcal{F}_{U^\star}$ are actually generalized topologies on X.

Namely, by Theorem 209, the super relations $\mathrm{cl}_{\mathcal{R}}$ and $\mathrm{int}_{\mathcal{R}}$ are increasing, and thus the super relation $U = \mathrm{cl}_{\mathcal{R}} \circ \mathrm{int}_{\mathcal{R}}$ is also increasing.

Analogously to Theorem 284, we can also prove the following:

Theorem 285. *If $U \in \mathcal{U}$ such that U is a union-preserving (resp. intersection-preserving), then the family $\mathcal{F}_U$ (resp. $\mathcal{F}_{U^\star}$) is closed under arbitrary unions and intersections.*

However, the latter theorem cannot be applied to the super relation U considered in Examples 15 and 16. Therefore, in addition to Theorem 284, it is more important to prove the following:

Theorem 286. *If $A \in \mathcal{F}_{\mathcal{U}^\star}$ and $B \in \mathcal{T}_{(\mathcal{R},\mathcal{D})}$ such that*

$$U(A) \cap \mathrm{int}_{(\mathcal{R},\mathcal{D})}(B) \subseteq U(A \cap B),$$

for all $U \in \mathcal{U}$, then $A \cap B \in \mathcal{F}_{\mathcal{U}^\star}$.

Proof. If $A \in \mathcal{F}_{\mathcal{U}^\star}$, then by Theorem 278 there exists $U \in \mathcal{U}$ such that $A \subseteq U(A)$. Moreover, if $B \in \mathcal{T}_{(\mathcal{R},\mathcal{D})}$, then by Theorem 179 we have $B \subseteq \mathrm{int}_{(\mathcal{R},\mathcal{D})}(B)$. Hence, by using the assumption of the theorem, we can infer that

$$A \cap B \subseteq U(A) \cap \mathrm{int}_{(\mathcal{R},\mathcal{D})}(B) \subseteq U(A \cap B).$$

Therefore, by Theorem 278, we have $A \cap B \in \mathcal{F}_{\mathcal{U}^\star}$. □

Now, analogously to this theorem, or applying this theorem to the super relator $\mathcal{U}^\star$, we can also easily establish the following:

Theorem 287. *If $A \in \tau_{\mathcal{U}}$ and $B \in \mathcal{T}_{(\mathcal{R},\mathcal{D})}$ such that*

$$U^\star(A) \cap \operatorname{int}_{(\mathcal{R},\mathcal{D})}(B) \subseteq U^\star(A \cap B),$$

for all $U \in \mathcal{U}$, then $A \cap B \in \tau_{\mathcal{U}}$.

Remark 129. Moreover, as another reformulation of Theorem 286, we can also state that if $A \in \tau_{\mathcal{U}}$ and $B \in \mathcal{F}_{(\mathcal{R},\mathcal{D})}$ such that

$$U(A \cup B) \subseteq U(A) \cup \operatorname{cl}_{(\mathcal{R},\mathcal{D})}(B)$$

for all $U \in \mathcal{U}$, then $A \cup B \in \tau_{\mathcal{U}}$.

Namely, if $A, B \subseteq X$, then by using Definition 7 and Theorem 160, we can see that

$$\begin{aligned} U^\star(A) \cap \operatorname{int}_{(\mathcal{R},\mathcal{D})}(B) \subseteq U^\star(A \cap B) &\iff U^\star(A) \cap B^\circ \subseteq U^\star(A \cap B) \\ \iff U(A^c)^c \cap B^{c-c} \subseteq U((A \cap B)^c)^c &\iff (U(A^c) \cup B^{c-})^c \\ \subseteq U(A^c \cup B^c)^c \iff U(A^c \cup B^c) &\subseteq U(A^c) \cup B^{c-} \\ \iff U(A^c \cup B^c) \subseteq U(A^c) \cup \operatorname{cl}_{(\mathcal{R},\mathcal{D})}(B^c). \end{aligned}$$

Moreover, by using Theorem 181, we can also see that

$$A \in \tau_{\mathcal{U}}, \quad B \in \mathcal{T}_{(\mathcal{R},\mathcal{D})} \iff A^c \in \tau_{\mathcal{U}}, \quad B^c \in \mathcal{F}_{(\mathcal{R},\mathcal{D})}$$

and

$$A \cap B \in \tau_{\mathcal{U}} \iff (A \cap B)^c \in \tau_{\mathcal{U}} \iff A^c \cup B^c \in \tau_{\mathcal{U}}.$$

Therefore, Theorem 287 can be applied to obtain the required assertion.

Theorems 286 and 287 and Remark 129 can already be applied to the super relation U consider in Example 15.

Example 18. If U is as in Example 15 and $\mathcal{R}$ is topologically filtered, then

(1) $A \in \tau_{\mathcal{U}}$ and $B \in \mathcal{T}_{\mathcal{R}}$ imply $A \cap B \in \tau_{\mathcal{U}}$;
(2) $A \in \tau_{\mathcal{U}^\star}$ and $B \in \mathcal{T}_{\mathcal{R}}$ imply $A \cap B \in \tau_{\mathcal{U}^\star}$.

Namely, if $A, B \subseteq X$ and $\mathcal{R}$ is topologically filtered, then by using the notations

$$A^- = \text{cl}_{\mathcal{R}}(A) \quad \text{and} \quad A^\circ = \text{int}_{\mathcal{R}}(A),$$

and Theorem 265 and Corollary 50, we can see that

$$(A \cap B)^\circ = A^\circ \cap B^\circ \quad \text{and} \quad A^- \cap B^\circ \subseteq (A \cap B)^-.$$

Therefore, for any $A \subseteq X$ and $B \in \mathcal{T}_{\mathcal{R}}$, we have

$$\begin{aligned} U(A) \cap \text{int}_{\mathcal{R}}(B) &= A^{\circ-} \cap B^\circ \subseteq (A^\circ \cap B)^- \\ &\subseteq (A^\circ \cap B^\circ)^- = (A \cap B)^{\circ-} = U(A \cap B) \end{aligned}$$

and

$$\begin{aligned} U^\star(A) \cap \text{int}_{\mathcal{R}}(B) &= A^{-\circ} \cap B^\circ = (A^- \cap B)^\circ \\ &\subseteq (A^- \cap B^\circ)^\circ \subseteq (A \cap B)^{-\circ} = U^\star(A \cap B). \end{aligned}$$

Thus, Theorems 286 and 287 can be applied to obtain assertions (1) and (2).

Remark 130. If U is as in Example 15 and $\mathcal{R}$ is topologically filtered, then by Remark 129, we can also state that $A \in \tau_{\mathcal{U}}$ and $B \in \mathcal{F}_{\mathcal{R}}$ imply $A \cup B \in \tau_{\mathcal{U}}$.

Namely, by Theorem 265 and Corollary 50, we can also see that

$$(A \cup B)^- = A^- \cup B^- \quad \text{and} \quad (A \cup B)^\circ \subseteq A^\circ \cup B^-.$$

Therefore, for any $A \subseteq X$ and $B \in \mathcal{T}_{\mathcal{R}}$, we have

$$\begin{aligned} U(A \cup B) &= (A \cup B)^{\circ-} \subseteq (A^\circ \cup B^-)^- \\ &\subseteq (A^\circ \cup B)^- = A^{\circ-} \cup B^- = U(A) \cup \text{cl}_{\mathcal{R}}(B). \end{aligned}$$

Thus, Remark 129 can be applied to obtain the required assertion.

47. Topological Closures of Families of Sets

Notation 20. *In this and the following two sections, we shall assume that $\mathcal{R}$ is an ordinary relator, $\mathcal{V}$ is a hyper relator and $\mathcal{D}$ is a stack on X.*

Definition 43. For any $\mathcal{A} \subseteq \mathcal{P}(X)$, we define

$$\mathcal{A}^k = \mathcal{A}^{k_\mathcal{V}} = \mathrm{cl}_\mathcal{V}(\mathcal{A}) \quad \text{and} \quad \mathcal{A}^\ell = \mathcal{A}^{\ell_\mathcal{V}} = \mathrm{cl}_{\mathcal{V}^{-1}}(\mathcal{A}).$$

Thus, by Theorem 207, we can at once state the following:

Theorem 288. *For any $\mathcal{A} \subseteq \mathcal{P}(X)$, we have*

(1) $\mathcal{A}^\ell = \bigcap_{V \in \mathcal{V}} V[\mathcal{A}]$; (2) $\mathcal{A}^k = \bigcap_{V \in \mathcal{V}} V^{-1}[\mathcal{A}]$.

From equality (1), we can immediately derive the following:

Theorem 289. *For any $\mathcal{A} \subseteq \mathcal{P}(X)$ and $B \subseteq X$, the following assertions are equivalent:*

(1) $B \in \mathcal{A}^\ell$;
(2) *for each $V \in \mathcal{V}$ we have $B \in V[\mathcal{A}]$;*
(3) *for each $V \in \mathcal{V}$ there exists $A \in \mathcal{A}$ such that $B \in V(A)$.*

While, from the equality $\mathcal{A}^k = \mathrm{cl}_\mathcal{V}(\mathcal{A})$, we can immediately derive the following:

Theorem 290. *For any $\mathcal{A} \subseteq \mathcal{P}(X)$ and $B \subseteq X$, the following assertions are equivalent:*

(1) $B \in \mathcal{A}^k$;
(2) *for each $V \in \mathcal{V}$ we have $V(B) \cap \mathcal{A} \neq \emptyset$;*
(3) *for each $V \in \mathcal{V}$ there exists $A \in \mathcal{A}$ such that $A \in V(B)$.*

The importance of Definition 43 is apparent from the following consequence of Theorems 289 and 290.

Example 19. If $V \in \mathcal{V}$ such that

$$V(A) = \{B \subseteq X : \ A \subseteq B \subseteq \mathrm{cl}_{(\mathcal{R},\mathcal{D})}(A)\}$$

for all $A \subseteq X$, then for any $\mathcal{A} \subseteq \mathcal{P}(X)$ and $B \subseteq X$ we have

(1) $B \in \mathcal{A}^{\ell_V}$ if and only if there exists $A \in \mathcal{A}$ such that $A \subseteq B \subseteq \mathrm{cl}_{(\mathcal{R},\mathcal{D})}(A)$;
(2) $B \in \mathcal{A}^{k_V}$ if and only if there exists $A \in \mathcal{A}$ such that $B \subseteq A \subseteq \mathrm{cl}_{(\mathcal{R},\mathcal{D})}(B)$.

Remark 131. By Theorem 158, the inclusion $A \subseteq \text{cl}_{(\mathcal{R},\mathcal{D})}(B)$ means only that each $x \in A$ and $R \in \mathcal{R}$ we have $R(x) \cap B \in \mathcal{D}$.

The latter condition, in the $\mathcal{D} = \mathcal{D}_0 = \mathcal{P}(X) \setminus \{\emptyset\}$ particular case, means only that $R(x) \cap B \neq \emptyset$. That is, there exists $y \in B$ such that $y \in R(x)$.

Remark 132. Moreover, if $A \subseteq \text{cl}_{(\mathcal{R},\mathcal{D})}(B)$ and $A \neq \emptyset$, then we can note that $A \cap \text{cl}_{(\mathcal{R},\mathcal{D})}(B) = A \neq \emptyset$.

Hence, by using Theorem 161, we can infer that $A \in \text{Cl}_{(\mathcal{R},\mathcal{D})}(B)$. That is, A is, in a certain sense, near to B.

Remark 133. However, it is now more important to note that if V is as in Example 19 with $\mathcal{D} = \mathcal{D}_0$, then $\mathcal{T}_{\mathcal{R}}^{\ell_V}$ and $\mathcal{T}_{\mathcal{R}}^{k_V}$ are just the families $\mathcal{T}_{\mathcal{R}}^{q}$ and $\mathcal{T}_{\mathcal{R}}^{ps}$ of all *topologically quasi-open and pseudo-open subsets* of the relator space $X(\mathcal{R})$ considered in Ref. [126].

Remark 134. In Refs. [124, 126], concerning the families $\mathcal{T}_U$ and $\mathcal{T}_{\mathcal{R}}^{k_V}$ mentioned in Remarks 123 and 133, we have proved that

(1) $\mathcal{T}_{\mathcal{R}}^{k_V} \subseteq \mathcal{T}_U$ is always true;
(2) $\mathcal{T}_{\mathcal{R}} \subseteq \mathcal{T}_{\mathcal{R}}^{k_V} \subseteq \mathcal{T}_U^{k_V}$ if $\mathcal{R}$ is reflexive;
(3) $\mathcal{T}_U^{k_V} \subseteq \mathcal{T}_U$ if $\mathcal{R}$ is quasi-topological;
(4) $\mathcal{T}_{\mathcal{R}}^{k_V} = \mathcal{T}_U^{k_V} = \mathcal{T}_U$ if $\mathcal{R}$ is topological.

Remark 135. In Ref. [124], we have also proved that if $\mathcal{R}$ is topological, then $\mathcal{A} = \mathcal{T}_U$ is actually the smallest subset of $\mathcal{P}(X)$ such that $\mathcal{T}_{\mathcal{R}} \subseteq \mathcal{A}$ and $\mathcal{A}^{k_V} \subseteq \mathcal{A}$.

Moreover, if $\mathcal{R}$ is topological and topologically filtered, then for any $B \subseteq X$ we have $B \in \mathcal{T}_U$ if and only if there exist $A \in \mathcal{T}_{\mathcal{R}}$ and $D \in \mathcal{D}_{\mathcal{R}}$ such that $B = A \cap D$.

Remark 136. In this respect, it is curious that if $\mathcal{R}$ is topological and topologically filtered, then for any $B \in \mathcal{T}_{U^\star}$ there exist $A \in \mathcal{T}_{\mathcal{R}}$ and $N \in \mathcal{N}_{\mathcal{R}}$ such that $B = A \cup N$ and $A \cap N = \emptyset$.

However, the converse statement need not be true. Moreover, the genuine characterizations of $\mathcal{T}_{U^\star}$, established in Ref. [127], do not require the relator $\mathcal{R}$ to be topologically filtered.

48. Some Basic Properties of the Operations k and ℓ

By using Theorems 289 and 290, or rather some general theorems on the topological closures derived from ordinary relators, we can easily establish several useful properties of the operations $k = k_{\mathcal{V}}$ and $\ell = \ell_{\mathcal{V}}$ introduced in Definition 43.

For instance, we shall only mention here the following three basic theorems.

Theorem 291. *We have*

(1) $\emptyset^k = \emptyset$ *if and only if* $\mathcal{V} \neq \emptyset$;
(2) $\mathcal{P}(X)^k = \mathcal{P}(X)$ *if and only if* $\mathcal{V}$ *is non-partial.*

Theorem 292. *The following assertions hold:*

(1) k *is always increasing;*
(2) k *is additive if and only if* $\mathcal{V}$ *is topologically filtered.*

Theorem 293. *The following assertions hold:*

(1) k *is extensive if and only if* $\mathcal{V}$ *is reflexive;*
(2) k *is upper quasi-idempotent if and only if* $\mathcal{V}$ *is quasi-topological.*

Remark 137. In addition to above statements, we can also prove that the operation $k_{\mathcal{V}}$ is union-preserving if and only if the hyper relator $\mathcal{V}$ is topologically simple in the sense that $\mathrm{cl}_{\mathcal{V}} = \mathrm{cl}_V$ for some hyper relation V on X.

Instead of proving the above statements, we shall rather consider here the following two illustrative examples in greater details.

Example 20. If $V \in \mathcal{V}$ such that

$$V(A) = \{B \subseteq X : \ A \subseteq B \subseteq \mathrm{cl}_{\mathcal{R}}(A)\}$$

for all $A \subseteq X$, then the following assertions are equivalent:

(1) V is reflexive; (2) V is non-partial; (3) $\mathcal{R}$ is reflexive.

To prove the implication (2) $\Longrightarrow$ (3), note that if (2) holds then for any $A \subseteq X$ we have $V(A) \neq \emptyset$. Therefore, there exists $B \subseteq X$ such that $B \in V(A)$. Hence, by the definition of V, we can infer that $A \subseteq B \subseteq \mathrm{cl}_{\mathcal{R}}(A)$. Therefore, for any $A \subseteq X$, we have $A \subseteq \mathrm{cl}_{\mathcal{R}}(A)$. Thus, by Theorem 256, assertion (3) also holds.

Example 21. If the relator $\mathcal{R}$ is quasi-topological, then the hyper relation V considered in Example 20 is transitive, and thus in particular it is also quasi-topological.

Namely, if $B \in V(A)$ and $C \in V(B)$, then by the definition of V, we have

$$A \subseteq B \subseteq \operatorname{cl}_{\mathcal{R}}(A) \quad \text{and} \quad B \subseteq C \subseteq \operatorname{cl}_{\mathcal{R}}(B).$$

Thus, in particular $A \subseteq C$. Moreover, by Theorems 209 and 258, we also have

$$C \subseteq \operatorname{cl}_{\mathcal{R}}(B) \subseteq \operatorname{cl}_{\mathcal{R}}(\operatorname{cl}_{\mathcal{R}}(A)) \subseteq \operatorname{cl}_{\mathcal{R}}(A).$$

Hence, by the definition of V, we can already see that $C \in V(A)$. Therefore,

$$B \in V(A) \quad \text{and} \quad C \in V(B) \quad \text{imply} \quad C \in V(A),$$

and thus V is transitive.

From Theorem 263, we can see that V is quasi-topological if and only if it is transitive. However, if V is quasi-topological, then it is certainly not true that $\mathcal{R}$ is also quasi-topological.

Remark 138. If in particular $\mathcal{R}$ is topological, then by Theorem 259 $\mathcal{R}$ is both reflexive and quasi-topological. Therefore, by Examples 20 and 21, the super relation V considered in Example 20 is both reflexive and quasi-topological. Thus, by Theorem 293 and Remark 137, we can state that k_V is a union-preserving closure operation on X.

To prove the same property of the relation ℓ_V, we have to note that if V is a preorder, then V^{-1} is also a preorder. Thus, in particular V^{-1} is also reflexive and quasi-topological.

49. Some Set-Theoretic Properties of the Families $\mathcal{A}^{\ell}$ and $\mathcal{A}^{k}$

Definition 44. The hyper relator $\mathcal{V}$ will be called

(1) *empty-set-stable* if $\emptyset \in V(\emptyset)$ for all $V \in \mathcal{V}$;
(2) *ground-set-stable* if $X \in V(X)$ for all $V \in \mathcal{V}$.

Thus, by using Theorems 289 and 290, we can easily prove the two theorems.

Theorem 294. *If $\mathcal{V}$ is empty-set-stable, then for any $\mathcal{A} \subseteq \mathcal{P}(X)$*

(1) $\emptyset \in \mathcal{A}$ *implies* $\emptyset \in \mathcal{A}^{\ell}$; (2) $\emptyset \in \mathcal{A}$ *implies* $\emptyset \in \mathcal{A}^{k}$.

Proof. To prove (1), note that by Theorem 289 we have $\emptyset \in \mathcal{A}^{\ell}$ if and only if for each $V \in \mathcal{V}$ there exists $A \in \mathcal{A}$ such that $\emptyset \in V(A)$. Therefore, if $\emptyset \in \mathcal{A}$ and $\emptyset \in V(\emptyset)$ for all $V \in \mathcal{V}$, then $\emptyset \in \mathcal{A}^{\ell}$. □

Theorem 295. *If $\mathcal{V}$ is ground-set-stable, then for any $\mathcal{A} \subseteq \mathcal{P}(X)$*

(1) $X \in \mathcal{A}$ *implies* $X \in \mathcal{A}^{\ell}$; (2) $X \in \mathcal{A}$ *implies* $X \in \mathcal{A}^{k}$.

Proof. To prove (2), note that by Theorem 290 we have $X \in \mathcal{A}^{k}$ if and only if for each $V \in \mathcal{V}$ there exists $A \in \mathcal{A}$ such that $A \in V(X)$. Therefore, if $X \in \mathcal{A}$ and $X \in V(X)$ for all $V \in \mathcal{V}$, then $X \in \mathcal{A}^{k}$. □

Now, as an immediate consequence of the above two theorems, we can also state the following:

Corollary 61. *If $\mathcal{V}$ is both empty-set-stable and ground-set-stable and $\mathcal{A}$ is a minimal structure on X, then $\mathcal{A}^{\ell}$ and $\mathcal{A}^{k}$ are also minimal structures on X.*

The appropriateness of Definition 44 is also apparent from the following:

Example 22. If V is as in Example 20, then

(1) V is empty-set-stable;
(2) V is ground-set-stable if and only if $\mathcal{R}$ is non-partial.

To prove (2), note that by the corresponding definitions and Theorem 226 we have

$$\begin{aligned} X \in V(X) &\iff X \subseteq X \subseteq \mathrm{cl}_{\mathcal{R}}(X) \\ &\iff X = \mathrm{cl}_{\mathcal{R}}(X) \iff X \in \mathcal{D}_{\mathcal{R}} \iff \mathcal{R} \text{ is non-partial.} \end{aligned}$$

Remark 139. Therefore, if V is as in Example 20, then

(1) $\emptyset \in \mathcal{T}_{\mathcal{R}}^{\ell_V}$ and $\emptyset \in \mathcal{T}_{\mathcal{R}}^{k_V}$ are always true;
(2) $X \in \mathcal{T}_{\mathcal{R}}^{\ell_V}$ and $X \in \mathcal{T}_{\mathcal{R}}^{k_V}$ if $\mathcal{R}$ is non-partial and either $X = \emptyset$ or $\mathcal{R} \neq \emptyset$.

To prove (2), note that if $\mathcal{R}$ is non-partial, then by Example 22 the relation V is ground-set-stable. While, if either $X = \emptyset$ or $\mathcal{R} \neq \emptyset$, then by Theorem 239 we have $X \in \mathcal{T}_{\mathcal{R}}$. Thus, Theorem 295 can be applied.

In addition to Definition 44, it is also worth introducing the following:

Definition 45. The hyper relator $\mathcal{V}$ will be called *union-compatible* if $V \in \mathcal{V}$ and $B_i \in V(A_i)$ for all $i \in I$ imply that

$$\bigcup_{i \in I} B_i \in V\left(\bigcup_{i \in I} A_i\right).$$

Remark 140. The intersection-compatibility of $\mathcal{V}$ is to be defined quite similarly.

Thus, by letting $I = \emptyset$, we can see that if $\mathcal{V}$ is union-compatible (intersection-compatible), then $\mathcal{V}$ is in particular empty-set-stable (ground-set-stable).

Theorem 296. *If $\mathcal{V}$ is union-compatible, then for any $\mathcal{A} \subseteq \mathcal{P}(X)$*

(1) *$\mathcal{A}^{\ell}$ is closed under unions if $\mathcal{A}$ is closed under unions;*
(2) *$\mathcal{A}^{k}$ is closed under unions if $\mathcal{A}$ is closed under unions.*

Proof. To prove (1), note that if $B_i \in \mathcal{A}^{\ell}$ for all $i \in I$, and $V \in \mathcal{V}$, then by Theorem 289, for each $i \in I$, there exists $A_i \in \mathcal{A}$ such that $B_i \in V(A_i)$. Hence, if $\mathcal{A}$ is closed under unions, we can infer that $\bigcup_{i \in I} A_i \in \mathcal{A}$. Moreover, since $\mathcal{V}$ is union-compatible, we can also state that $\bigcup_{i \in I} B_i \in V\left(\bigcup_{i \in I} A_i\right)$. Therefore, by Theorem 289, we also have $\bigcup_{i \in I} B_i \in \mathcal{A}^{\ell}$. □

Now, as an immediate consequence of Theorems 295 and 296, we can also state the following:

Corollary 62. *If $\mathcal{V}$ is both union-compatible and ground-set-stable and $\mathcal{A}$ is a generalized topology on X, then $\mathcal{A}^{\ell}$ and $\mathcal{A}^{k}$ are also generalized topologies on X.*

Example 23. If V is as in Example 20, then V is union-compatible.

Namely, if $B_i \in V(A_i)$ for all $i \in I$, then $A_i \subseteq B_i \subseteq \mathrm{cl}_{\mathcal{R}}(A_i)$ for all $i \in I$. Hence, by using the increasingness of $\mathrm{cl}_{\mathcal{R}}$, we can infer that

$$\bigcup_{i\in I} A_i \subseteq \bigcup_{i\in I} B_i \subseteq \bigcup_{i\in I} \mathrm{cl}_{\mathcal{R}}(B_i) \subseteq \mathrm{cl}_{\mathcal{R}}\left(\bigcup_{i\in I} A_i\right).$$

Thus, $\bigcup_{i\in I} B_i \in V\left(\bigcup_{i\in I} A_i\right)$ also holds.

Remark 141. Therefore, if V is as in Example 20, then the families $\mathcal{T}_{\mathcal{R}}^{\ell_V}$ and $\mathcal{T}_{\mathcal{R}}^{k_V}$ are closed under unions. Thus, if in particular $\mathcal{R}$ is non-void and non-partial, then the above families are generalized topologies.

Namely, by Example 23, the relation V is union-compatible. Moreover, by Theorem 238, the family $\mathcal{T}_{\mathcal{R}}$ is closed under unions. Therefore, by Theorems 296 the families $\mathcal{T}_{\mathcal{R}}^{\ell_V}$ and $\mathcal{T}_{\mathcal{R}}^{k_V}$ are also closed under unions. Moreover, if $\mathcal{R}$ is non-void and non-partial, then by Remark 139 we also have $X \in \mathcal{T}_{\mathcal{R}}^{\ell_V}$ and $X \in \mathcal{T}_{\mathcal{R}}^{k_V}$.

Now, in addition to Example 23, we can also easily establish the following:

Example 24. If V is as in Example 20, then V need not even be finitely intersection-compatible.

Namely, if for instance X and $\mathcal{R}$ are as in Example 9, then by taking

$$A_1 = \mathbb{Q},\ A_2 = \mathbb{Q}^c \quad \text{and} \quad B_1 = B_2 = X,$$

we can see that

$$A_i \subseteq B_i \subseteq \mathrm{cl}_{\mathcal{R}}(A_i), \text{ and thus } B_i \in V(A_i)$$

for $i = 1, 2$. However,

$$B_1 \cap B_2 \not\subseteq \mathrm{cl}_{\mathcal{R}}(A_1 \cap A_2), \text{ and thus } B_1 \cap B_2 \notin V(A_1 \cap A_2).$$

Remark 142. Therefore, it is not surprising that if X and $\mathcal{R}$ are as in Example 9 and V is as in Example 20, then the families $\mathcal{T}_{\mathcal{R}}^{\ell_V}$ and $\mathcal{T}_{\mathcal{R}}^{k_V}$ are not closed even under binary intersections.

Namely, if for instance

$$A = [0,1],\ B = [1,2] \quad \text{and} \quad C = \mathbb{Q},\ D = \{1\} \cup \mathbb{Q}^c,$$

then we can easily see that

$$A, B \in \mathcal{T}_{\mathcal{R}}^{\ell_V},\ C, D \in \mathcal{T}_{\mathcal{R}}^{k_V}, \quad \text{but} \quad A \cap B \notin \mathcal{T}_{\mathcal{R}}^{\ell_V},\ C \cap D \notin \mathcal{T}_{\mathcal{R}}^{k_V}.$$

References

[1] M. E. Abd El-Monsef, S. N. El-Deeb and R. A. Mahmoud, β–open sets and β–continuous mappings, *Bull. Fac. Sci. Assiut Univ.*, 12, 1983, 77–90.

[2] S. Acharjee, I. L. Reilly and D. J. Sarma, On compactness via bI–open sets in ideal bitopological spaces, *Bull. Transilvania Univ. Brasov*, 12, 2019, 1–8.

[3] S. Acharjee, M. Th. Rassias and Á. Száz, Upper and Lower Semicontinuous Relations in Relator Spaces, In: P. M. Pardalos and Th. M. Rassias (Eds.), *Analysis, Geometry, Nonlinear Optimization and Applications* 9 (World Scientific, New Jersey, 2023).

[4] P. Alexandroff, Zur Begründung der n-dimensionalen mengentheorischen Topologie, *Math. Ann.*, 94, 1925, 296–308.

[5] D. Andrijević, Semi-preopen sets, *Mat. Vesnik*, 38, 1986, 24–32.

[6] D. Andrijević, On b–open sets, *Mat. Vesnik*, 48, 1996, 59–64.

[7] F. G. Arenas, Alexandroff spaces, *Acta Math. Univ. Commen.*, 68, 1999, 17–25.

[8] H. Arianpoor, Preorder relators and generalized topologies, *J. Lin. Top. Algebra*, 5, 2016, 271–277.

[9] Z. Balogh, Relative compactness and recent common generalizations of metric and locally compact spaces, *Fund. Math.*, 100, 1978, 165–177.

[10] N. Biswas, On some mappings in topological spaces, *Bull. Cal. Math. Soc.*, 61, 1969, 127–135.

[11] N. Bourbaki, *General Topology, Chapters 1–4*, (Springer-Verlag, Berlin, 1989).

[12] P. Cameron, J. G. Hocking and S. A. Naimpally, Nearness – A better approach to continuity and limits, *Amer. Math. Montly*, 81, 1974, 739–745.

[13] E. Čech, *Topological Spaces*, (Academia, Prague, 1966).

[14] T. A. Chapman, A further note on closure and interior operators, *Amer. Math. Monthly*, 69, 1962, 524–529.

[15] Ch. Chattopadhyay and Ch. Bandyopadhyay, On structure of δ-sets, *Bull. Calcutta Math. Soc.*, 83, 1991, 281–290.

[16] M. Chicourrat and C. D. Horváth, Pretopologies, preuniformities and probabilistic metric spaces, *Acta Math. Hungar.*, 110, 2006, 91–116.

[17] H. Choda and K. Matoba, On a theorem of Levine, *Proc. Japan Acad.*, 37, 1961, 462–463.

[18] K. Ciesielski and J. Jasinski, Topologies making a given ideal nowhere dense or meager, *Topology Appl.*, 63, 1995, 277–298.

[19] H. H. Corson and E. Michael, Metrizability of certain countable unions, *Illinois J. Math.*, 8, 1964, 351–360.
[20] Á. Császár, *Foundations of General Topology*, (Pergamon Press, London, 1963).
[21] Á. Császár, Generalized open sets, *Acta Math. Hungar.*, 75, 1997, 65–87.
[22] Á. Császár, On the γ–interior and γ–closure of a set, *Acta Math. Hungar.*, 80, 1998, 89–93.
[23] Á. Császár, γ–quasi-open sets, *Studia Sci. Math. Hungar.*, 38, 2001, 171–176.
[24] Á. Császár, Generalized topology, generalized continuity, *Acta Math. Hungar.*, 96, 2002, 351–357.
[25] Á. Császár, Remarks on γ-quasi-open sets, *Studia Sci. Math. Hungar.*, 39, 2002, 137–141.
[26] Á. Császár, Generalized open sets in generalized topologies, *Acta Math. Hungar.*, 106, 2005, 53–66.
[27] Á. Császár, Further remarks on the formula for the γ–interior, *Acta Math. Hungar.*, 113, 2006, 325–332.
[28] Á. Császár, Modification of generalized topologies via hereditary classes, *Acta Math. Hungar.*, 115, 2007, 29–36.
[29] Á. Császár, Remarks on quasi-topologies, *Acta Math. Hungar.*, 119, 2008, 197–200.
[30] Á. Császár, Mixed constructions for generalized topologies, *Acta Math. Hungar.*, 122, 2009, 153–159.
[31] Á. Császár, Weak structures, *Acta Math. Hungar.*, 131, 2011, 193–195.
[32] B. A. Davey and H. A. Priestley, *Introduction to Lattices and Order*, (Cambridge University Press, Cambridge, 2002).
[33] A. S. Davis, Indexed systems of neighbordoods for general topological spaces, *Amer. Math. Monthly*, 68, 1961, 886–893.
[34] K. Dlaska, N. Ergun and M. Ganster, On the topology generated by semi-regular sets. *Indian J. Pure Appl. Math.*, 25, 1994, 1163–1170.
[35] D. Doičinov, A unified theory of topological spaces, proximity spaces and uniform spaces (Russian), *Dokl. Acad. Nauk SSSR*, 156, 1964, 21–24.
[36] J. Dontchev, *Survey on preopen sets, Meetings on Topological Spaces, Theory and Applications*, Yatsushiro College of Technology, Kumamoto, Japan, p. 18 (1998).
[37] Z. Duszyński and T. Noiri, Semi-open, semi-closed sets and semi-continuity of functions, *Math. Pannon.*, 23, 2012, 195–200.
[38] V. A. Efremovič, The geometry of proximity (Russian), *Mat. Sb.*, 31, 1952, 189–200.

[39] V. A. Efremović and A. S. Švarc, A new definition of uniform spaces, Metrization of proximity spaces (Russian), *Dokl. Acad. Nauk. SSSR*, 89, 1953, 393–396.
[40] N. Elez and O. Papaz, The new operators in topological spaces, *Math. Moravica*, 17, 2013, 63–68.
[41] P. Flether, H. B. Hoyle and C. W. Patty, The comparison of topologies, *Duke Math. J.*, 36, 1969, 325–331.
[42] P. Fletcher and W. F. Lindgren, *Quasi-Uniform Spaces*, (Marcel Dekker, New York, 1982).
[43] G. B. Folland, A tale of topology, *Amer. Math. Monthly*, 117, 2010, 663–672.
[44] G. Freud, Ein Beitrag zu dem Satze von Cantor und Bendixson, *Acta Math. Hungar.*, 9, 1958, 333–336.
[45] M. Ganster, Preopen sets and resolvable spaces, *Kyungpook J.*, 27, 1987, 135–143.
[46] M. Ganster, I. L. Reilly and M. K. Vamanamurthy, Remarks on locally closed sets, *Math. Pannon.*, 3, 1992, 107–113.
[47] B. Ganter and R. Wille, *Formal Concept Analysis*, (Springer-Verlag, Berlin, 1999).
[48] B. Garai and C. Bandyopadhyay, On parirwise hyperconnected spaces, *Soochow J. Math.*, 27, 2001, 391–399.
[49] R. Gargouri and A. Rezgui, A unification of weakening of open and closed subsets in a topological spaces, *Bull. Malays. Math. Sci. Soc.*, 40, 2017, 1219–1230.
[50] S. Givant and P. Halmos, *Introduction to Boolean Algebras*, (Springer-Verlag, Berlin, 2009).
[51] T. Glavosits, Generated preorders and equivalences, *Acta Acad. Paed. Agrienses, Sect. Math.*, 29, 2002, 95–103.
[52] W. H. Gottschalk, Intersection and closure, *Proc. Amer. Math. Soc.*, 4, 1953, 470–473.
[53] T. R. Hamlett and D. Janković, Ideals in topological spaces and the set operator ψ, *Bollettino U.M.I.*, (7)4-B, 1990, 863–874.
[54] H. Hashimoto, On the $*$topology and its applications, *Fund. Math.*, 91, 1976, 5–10.
[55] F. Hausdorff, *Grundzüge der Mengenlehre*, (De Gruyter, Berlin, 1914).
[56] E. Hayashi, Topologies defined by local properties, *Math. Ann.*, 156, 1964, 205–215.
[57] H. Herrlich, Topological structures, *Math. Centre Tracts*, 52, 1974, 59–122.
[58] U. Höhle and T. Kubiak, On regularity of sup-preserving maps: Generalizing Zareckiĭ's theorem, *Semigroup Forum*, 83, 2011, 313–319.

[59] W. Hunsaker and W. Lindgren, Construction of quasi-uniformities. *Math. Ann.*, 188, 1970, 39–42.
[60] D. H. Hyers, On the stability of the linear functional equation, *Proc. Nat. Acad. Sci. U.S.A*, 27, 1941, 222–224.
[61] J. R. Isbell, *Uniform Spaces*, (Amer. Math. Soc., Providence, 1964).
[62] Y. Isomichi, New concept in the theory of topological spaces–Supercondensed set, subcondensed set and condensed set, *Pacific J. Math.*, 38, 1971, 657–668.
[63] D. Janković and T. R. Hamlett, New topologies from old via ideals, *Amer. Math. Monthly*, 97, 1990, 295–310.
[64] S. Jafari and N. Rajesh, Generalized closed sets with respect to an ideal, *European J. Pure Appl. Math.*, 4, 2011, 147–151.
[65] S. N. Jothi and P. Thangavelu, Topology between two sets, *J. Math. Sci. Comp. Apll.*, 1, 2011, 95–107.
[66] Y. B. Jun, S. W. Jeong, H. j. Lee and J. W. Lee, Applications of pre-open sets, *Appl. Gen. Top.*, 9, 2008, 213–228.
[67] S.-M. Jung, Interiors and closures of sets and applications, *Int. J. Pure Math.*, 3, 2016, 41–45.
[68] S.-M. Jung and D. Nam, Some properties of interior and closure in general topology, *Mathematics*, 7, 2019, 624.
[69] A. Kandil, M. M. Yakout and A. Zakaria, Generalized rough sets via ideals, *Fuzzy Math. Inf.*, 5, 2013, 525–532.
[70] A. Kandil, O. A. El-Tantawy, S. A. El-Sheikh and A. Zakaria, On $\mathcal{J}$-proximity spaces, *Appl. Math. Inf. Sci. Lett.*, 4, 2016, 79–84.
[71] J. L. Kelley, *General Topology*, (Van Nostrand Reinhold Company, New York, 1955).
[72] J. C. Kelly, Bitopological spaces, *Proc. London Math. Soc.*, 13, 1963, 71–89.
[73] H. Kenyon, Two theorems about relations, *Trans. Amer. Math. Soc.*, 107, 1963, 1–9.
[74] V. L. Kljushin, Al bayati and J. H. Hussein, On simply-open sets (Russian), *Vestnik UDC*, 3, 2011, 34–38.
[75] H.-J. Kowalsky, *Topologische Räumen*, (Birkhäuser, Basel, 1960).
[76] K. Kuratowski, Sur l'opération $\overline{A}$ de l'analysis situs, *Fund. Math.*, 3, 1922, 182–199. (An English translation: On the operation $\overline{A}$ in analysis situs, prepared by M. Bowron in 2010, is available on the Internet.)
[77] K. Kuratowski, *Topology I.*, (Academic Press, New York, 1966).
[78] J. Kurdics, A note on connection properties, *Acta Math. Acad. Paedagog. Nyházi.*, 12, 1990, 57–59.
[79] J. Kurdics and Á. Száz, Well-chained relator spaces, *Kyungpook Math. J.*, 32, 1992, 263–271.

[80] J. Kurdics and Á. Száz, Well-chainedness characterizations of connected relators, *Math. Pannon.*, 4, 1993, 37–45.
[81] E. P. Lane, Bitopological spaces and quasi-uniform spaces, *Proc. London. Math. Soc.*, 17, 1967, 241–256.
[82] E. P. Lane, Quasi-proximities and bitopological spaces, *Potugal. Math.*, 28, 1969, 151–159.
[83] N. Levine, On the commutivity of the closure and interior operators in topological spaces, *Amer. Math. Monthly*, 68, 1961, 474–477.
[84] N. Levine, Semi-open sets and semi-continuity in topological spaces, *Amer. Math. Monthly*, 70, 1963, 36–41.
[85] N. Levine, Some remarks on the closure operator in topological spaces, *Amer. Math. Monthly*, 70, 1963, 553.
[86] N. Levine, Dense topologies, *Amer. Math. Monthly*, 75, 1968, 847–852.
[87] N. Levine, On uniformities generated by equivalence relations, *Rend. Circ. Mat. Palermo*, 18, 1969, 62–70.
[88] N. Levine, On Pervin's quasi uniformity, *Math. J. Okayama Univ.*, 14, 1970, 97–102.
[89] S. Lugojan, Topologii generalizate, *Stud. Cerc. Mat.*, 34, 1982, 348–360.
[90] H. Maki, On generalizing semi-open and preopen sets, *Report for Meeting on Topological Spaces Theory and its Applications*, Yatsushiro College of Technology, pp. 13–18 (1996).
[91] J. Mala, Relators generating the same generalized topology, *Acta Math. Hungar.*, 60, 1992, 291–297.
[92] J. Mala and Á. Száz, Properly topologically conjugated relators, *Pure Math. Appl. Ser. B*, 3, 1992, 119–136.
[93] J. Mala and Á. Száz, Modifications of relators, *Acta Math. Hungar.*, 77, 1997, 69–81.
[94] R. Manoharan and P. Thangevalu, Some new sets and topologies in ideal topological spaces, *Chin. J. Math.*, 2013, 6.
[95] N. F. G. Martin, Generalized condensation point, *Duke Math. J.*, 28, 1961, 507–514.
[96] A. S. Mashhour, M. E. Abd El-Monsef and S. N. El-Deeb, On precontinuous and weak precontinuous mappings, *Proc. Math. Phys. Soc. Egypt*, 53, 1982, 47–53.
[97] P. M. Mathew, On hyperconnected spaces, *Indian J. Pure Appl. Math.*, 19, 1988, 1180–1184.
[98] Ž. M. Mitrinović and I. F. Berković, The induced relations on a power set, *Filomat (Niš)*, 9, 1995, 857–865.
[99] S. Modak, Some new topologies on ideal topological spaces, *Proc. Natl. Acad. Sci. India*, 82, 2012, 233–243.

[100] S. Modak, Topology of grill filter space and continuity, *Bol. Soc. Paran. Mat.*, 31, 2013, 219–230.

[101] S. Modak and C. Bandyopadhyay, A note on ψ-operator, *Bull. Malays. Math. Sci. Soc.*, 30, 2007, 43–48.

[102] S. Modak and A. A. Al-Omari, Generalized closed sets in binary ideal topological spaces, *J. Chungcheong Math. Soc.*, 31, 2018, 183–191.

[103] S. A. Naimpally and B. D. Warrack, *Proximity Spaces*, (Cambridge University Press, Cambridge, 1970).

[104] H. Nakano and K. Nakano, Connector theory, *Pacific J. Math.*, 56, 1975, 195–213.

[105] A. A. Nasef and R. Mareay, More on simply open sets and its applications, *South Asian J. Math.*, 5, 2015, 100–108.

[106] A. A. Nasef and R. Mareay, Ideals and some applications of simply open sets, *J. Adv. Math.*, 13, 2017, 7264–7271.

[107] A. Neubrunnová, On transfinite sequences of certain types of functions, *Acta Fac. Rer. Natur. Univ. Com. Math.*, 30, 1975, 121–126.

[108] O. Njåstad, On some classes of nearly open sets, *Pacific J. Math.*, 15, 1965, 195–213.

[109] T. Noiri, Hyperconnectedness and preopen sets, *Rev. Roum. Math. Pures Appl.*, 29, 1984, 329–334.

[110] T. Noiri, Properties of hyperconnected spaces, *Acta Math. Hungar.*, 66, 1995, 147–154.

[111] S. O'Neill, *Two Topologies are Better Than One*, Department of Computer Science, University of Warwick, p. 27 (1995).

[112] G. Pataki, Supplementary notes to the theory of simple relators, *Radovi Mat.*, 9, 1999, 101–118.

[113] G. Pataki, On the extensions, refinements and modifications of relators, *Math. Balk.*, 15, 2001, 155–186.

[114] G. Pataki and Á. Száz, A unified treatment of well-chainedness and connectedness properties, *Acta Math. Acad. Paedagog. Nyházi. (N.S.)*, 19, 2003, 101–165.

[115] C. W. Patty, Bitopological spaces, *Duke Math. J.*, 34, 1967, 387–391.

[116] W. J. Pervin, Uniformizations of neighborhood axioms, *Math. Ann.*, 147, 1962, 313–315.

[117] W. J. Pervin, Quasi-uniformization of topological spaces, *Math. Ann.*, 147, 1962, 316–317.

[118] W. J. Pervin, Quasi-proximities for topological spaces, *Math. Ann.*, 150, 1963, 325–326.

[119] E. Przemska, The lattices of families of regular sets in topological spaces, *Math. Slovaca*, 70, 2020, 477–488.

[120] E. Przemska, Regular sets in topological spaces, A Short Communication prepared for "The 34th International Summer Conference on

Real Functions Theory", Mathematical Institute, Slovak Academy of Sciences, Bratislava and Košice, p. 5 (2020).

[121] E. Przemska, A unified treatment of generalized closed sets in topological spaces, *Tatra Mt. Math. Publ.*, 85, 2023, 121–154.

[122] E. Przemska, On the unified concept of generalizations of Λ–sets, *Symmetry*, 16, 2024, 527.

[123] M. Th. Rassias and Á. Száz, Basic tools and continuity-like properties in relator spaces, *Contribut. Math.*, 3, 2021, 77–106.

[124] Th. M. Rassias and Á. Száz, A General Framework for Studying Certain Generalized Topologically Open Sets in Relator Spaces. In: Th. M. Rassias (Ed.), *Nonlinear Analysis, Differential Equations, and Applications*, Springer Optim. Appl. 173, (Springer, Cham, 2021), pp. 415–491.

[125] Th. M. Rassias and Á. Száz, Ordinary, Super and Hyper Relators Can Be Used To Treat the Various Generalized Open Sets in a Unified Way, In: N. J. Daras and Th. M. Rassias (Eds.), *Approximation and Computation in Science and Engineering*, Springer Optim. Appl. 180, (Springer, Cham, 2022), pp. 709–782.

[126] Th. M. Rassias, M. M. Salih and Á. Száz, Characterizations of generalized topologically open sets in relator spaces, *Montes Taurus J. Pure Appl. Math.*, 3, 2021, 39–94.

[127] Th. M. Rassias, M. M. Salih and Á. Száz, Set-Theoretic Properties of Generalized Topologically Open Sets in Relator Spaces, In: I. N. Parasidis, E. Providas and Th. M. Rassias (Eds.), *Mathematical Analysis in Interdisciplinary Research*, Springer Optim. Appl. 179, (Springer, Cham, 2021), pp. 661–730.

[128] Th. M. Rassias, M. M. Salih and Á. Száz, Characterizations and Set Theoretic Properties of Some Generalized Open and Fat Sets in Relator Spaces, In: P. M. Pardalos and Th. M. Rasias (Eds.), *Geometry and Non-Convex Optimization*, Springer Optim. Appl. 223, (Springer Cham, 2025), pp. 613–709.

[129] I. L. Reilly and M. K. Vamanamurthy, On α-continuity in topological spaces, *Acta Math. Hungar.*, 45, 1985, 27–32.

[130] M. M. Salih and Á. Száz, Generalizations of some ordinary and extreme connectedness properties of topological spaces to relator spaces, *Electron. Res. Arch.*, 28, 2020, 471–548.

[131] P. Samuels, A topology formed from a given topology and ideal, *J. London Math. Soc.*, 10, 1975, 409–416.

[132] D. J. Sarma and S. Acharjee, Weakly semi-I-open sets in bitopological spaces, *Trans. Natl. Acad. Sci. Azerb. Ser. Phys.-Tech. Math. Sci. Mathematics*, 41, 2021, 71–77.

[133] S. Selim, T. Noiri and S. Modak, Ideals and the associated filters on topological spaces, *Eurasian Bull. Math. EBM*, 2, 2019, 80–85.

[134] W. Sierpinski, *General Topology*, Mathematical Expositions 7, (University of Toronto Press, Toronto, 1956).

[135] P. Sivagami, Remarks on γ–interior, *Acta Math. Hungar.*, 119, 2008, 81–94.

[136] Yu. M. Smirnov, On proximity spaces (Russian), *Math. Sb.*, 31, 1952, 543–574.

[137] T. Speer, A short study of Alexandroff spaces, arXiv:0708.2136v1 [math.GN] 16 Aug 2007.

[138] M. H. Stone, Application of the theory of Boolean rings to general topology, *Trans. Amer. Math. Soc.*, 41, 1937, 374–481.

[139] Á. Száz, Coherences instead of convergences. *Proceedings of the Conference on Convergence and Generalized Functions (Katowice, 1983)*, Institute of Mathematics, Polish Academy of Sciences (Warsaw, 1984), pp. 141–148.

[140] Á. Száz, Basic tools and mild continuities in relator spaces, *Acta Math. Hungar.*, 50, 1987, 177–201.

[141] Á. Száz, Directed, topological and transitive relators, *Publ. Math. Debrecen*, 35, 1988, 179–196.

[142] Á. Száz, Lebesgue relators, *Monatsh. Math.*, 110, 1990, 315–319.

[143] Á. Száz, Relators, Nets and Integrals. *Unfinished doctoral thesis*, Debrecen, p. 126 (1991).

[144] Á. Száz, Inverse and symmetric relators, *Acta Math. Hungar.*, 60, 1992, 157–176.

[145] Á. Száz, Structures derivable from relators, *Singularité*, 3, 1992, 14–30.

[146] Á. Száz, Refinements of relators, *Tech. Rep., Inst. Math., Univ. Debrecen*, 76, 1993, 19.

[147] Á. Száz, Cauchy nets and completeness in relator spaces, *Colloq. Math. Soc. János Bolyai*, 55, 1993, 479–489.

[148] Á. Száz, Neighbourhood relators, *Bolyai Soc. Math. Stud.*, 4, 1995, 449–465.

[149] Á. Száz Topological characterizations of relational properties, *Grazer Math. Ber.*, 327, 1996, 37–52.

[150] Á. Száz, Uniformly, proximally and topologically compact relators, *Math. Pannon.*, 8, 1997, 103–116.

[151] Á. Száz, Somewhat continuity in a unified framework for continuities of relations, *Tatra Mt. Math. Publ.*, 24, 2002, 41–56.

[152] Á. Száz, Upper and lower bounds in relator spaces, *Serdica Math. J.*, 29, 2003, 239–270.

[153] Á. Száz, An extension of Baire's category theorem to relator spaces, *Math. Morav.*, 7, 2003, 73–811.
[154] Á. Száz, Rare and meager sets in relator spaces, *Tatra Mt. Math. Publ.*, 28, 2004, 75–95.
[155] Á. Száz, Galois-type connections on power sets and their applications to relators, *Tech. Rep., Inst. Math., Univ. Debrecen*, 2005/2, p. 38.
[156] Á. Száz, Supremum properties of Galois–type connections, *Comment. Math. Univ. Carolin.*, 47, 2006, 569–583.
[157] Á. Száz, Minimal structures, generalized topologies, and ascending systems should not be studied without generalized uniformities, *Filomat (Niš)*, 21, 2007, 87–97.
[158] Á. Száz, Galois type connections and closure operations on preordered sets, *Acta Math. Univ. Comenian. (N.S.)*, 78, 2009, 1–21.
[159] Á. Száz, Inclusions for compositions and box products of relations, *J. Int. Math. Virt. Inst.*, 3, 2013, 97–125.
[160] Á. Száz, Lower semicontinuity properties of relations in relator spaces, *Adv. Stud. Contemp. Math. (Kyungshang)*, 23, 2013, 107–158.
[161] Á. Száz, A particular Galois connection between relations and set functions, *Acta Univ. Sapientiae, Math.*, 6, 2014, 73–91.
[162] Á. Száz, Generalizations of Galois and Pataki connections to relator spaces, *J. Int. Math. Virtual Inst.*, 4, 2014, 43–75.
[163] Á. Száz, Basic tools, increasing functions, and closure operations in generalized ordered sets, In P. M. Pardalos and Th. M. Rassias (Eds.), *Contributions in Mathematics and Engineering: In Honor of Constantion Caratheodory*, (Springer, Switzerland, 2016), pp. 551–616.
[164] Á. Száz, The closure-interior Galois connection and its applications to relational equations and inclusions, *J. Int. Math. Virt. Inst.*, 8, 2018, 181–224.
[165] Á. Száz, Correlations are more powerful tools than relations, In Th. M. Rassias (ed.), *Applications of Nonlinear Analysis*, Springer Optim. Appl. 134, (Springer, Cham, 2018), pp. 711–779.
[166] Á. Száz, Relationships between inclusions for relations and inequalities for correlations, *Math. Pannon.*, 26, 2018, 15–31.
[167] Á. Száz, Galois and Pataki connections on generalized ordered sets, *Earthline J. Math. Sci.*, 2, 2019, 283–323.
[168] Á. Száz, Birelator spaces are natural generalizations of not only bitopological spaces, but also ideal topological spaces, In Th. M. Rassias, P. M. Pardalos (eds.), *Mathematical Analysis and Applications*, Springer Optim. Appl. 154, (Springer, Cham, 2019), pp. 543–586.
[169] Á. Száz, Super and hyper products of super relations, *Tatra Mt. Math. Publ.*, 78, 2021, 1–34.

[170] Á. Száz, Contra continuity properties of relations in relator spaces. In S. Acharjee (ed.), *Advances in Topology and Their Interdisciplinary Applications*, (Springer Nature, Cham, 2023), pp. 19–74.
[171] Á. Száz, Four general continuity properties, for pairs of functions, relations and relators, whose particular cases could be investigated by hundreds of mathematicians, *Int. J. Math. Stat. Oper. Res.*, 3, 2023, 135–154.
[172] Á. Száz and A. Zakaria, Mild Continuity Properties of Relations and Relators in Relator Spaces, In P. M. Pardalos, Th. M. Rassias (eds.), *Essays in Mathematics and its Applications*: In Honor of Vladimir Arnold, (Springer, Switzerland, 2016), pp. 439–511.
[173] W. J. Thron, *Topological Structures*, (Holt, Rinehart and Winston, New York, 1966).
[174] W. J. Thron, Proximity structures and grills, *Math. Ann.*, 206, 1973, 35–62.
[175] H. Tietze, Beiträge zur allgemeinen Topologie I. Axiome für verschiedene Fassungen des Umgebungsbegriffs, *Math. Ann.*, 88, 1923, 290–312.
[176] J. W. Tukey, *Convergence and Uniformity in Topology*, Ann. Math. Studies 2, (Princeton University Press, Princeton, 1940).
[177] R. Vaidyanathaswamy, The localization theory in set-topology, *Proc. Indian Acad. Sci.*, 20, 1945, 51–61.
[178] A. Weil, Sur les espaces á structure uniforme et sur la topologie générale. In: *Actualités Scientifiques et Industrielles, 551* (Herman and Cie, Paris, 1937).
[179] J. D. Weston, On the comparison of topologies, *J. London Math. Soc.*, 32, 1957, 342–354.
[180] R. L. Wilder, Evolution of the topological concept of "connected", *Amer. Math. Monthly*, 85, 1978, 720–726.
[181] I. Zuzčák, Generalized topological spaces, *Math. Slovaca*, 33, 1983, 249–256.
[182] I. Zvina, On i-topological spaces: Generalization of the concept of a topological space via ideals, *Appl. Gen. Top.*, 7, 2006, 51–66.

https://doi.org/10.1142/9789819820498_0021

Chapter 21

PPF-Dependent Fixed Points for Multivalued Geometric Contractions

Mihai Turinici
A. Myller Mathematical Seminar,
A. I. Cuza University, 700506 Iaşi, Romania
mturi@uaic.ro

A geometric version is given for the PPF-dependent fixed point result involving multivalued Wardowski-type contractions due to Hammad and De la Sen [Mathematics, 2019, 7, 52].

Keywords: Metric space, Picard operator, fixed and coincidence point, PPF dependence, geometric contraction, constant functions class.

1. Introduction

Throughout this exposition, the axiomatic system in use is

> (ZF − AC + DC) = the Zermelo–Fraenkel system (ZF), with (AC) (Axiom of Choice) substituted by (DC) (Dependent Choice principle).

Some technical aspects concerning the construction of (ZF) may be found in Cohen [11, Chapter 2] and the references therein. The remaining ones involving (DC) and a lot of basic facts to be considered in this setting were treated along the 2020 book by Turinici [56, Section 11].

Let X be a non-empty set. Call the subset Y of X, *almost singleton* (in short: *asingleton*), provided $[y_1, y_2 \in Y$ implies $y_1 = y_2]$; and *singleton* if, in addition, Y is non-empty; note that in this case $Y = \{y\}$, for some $y \in X$. Take a metric $d : X \times X \to R_+ := [0, \infty[$ over X; as well as a selfmap $T \in \mathcal{F}(X)$. [Here, for each couple A, B of non-empty sets, $\mathcal{F}(A, B)$ denotes the class of all functions from A to B; when $A = B$, we write $\mathcal{F}(A)$ in place of $\mathcal{F}(A, A)$]. Denote $\text{Fix}(T) = \{x \in X; x = Tx\}$; each point of this set is referred to as *fixed* under T. The determination of such elements is to be performed under the directions below, comparable with the ones in Rus [45, Chapter 2, Section 2.2] and Turinici [51]:

(pic-0) We say that T is *fix-asingleton*, if $\text{Fix}(T)$ is an asingleton; and *fix-singleton*, if $\text{Fix}(T)$ is a singleton.
(pic-1) We say that T is a *Picard operator* (modulo d) if, for each $x \in X$, the iterative sequence $(T^n x; n \geq 0)$ is d-Cauchy.
(pic-2) We say that T is a *strong Picard operator* (modulo d) if, for each $x \in X$, the iterative sequence $(T^n x; n \geq 0)$ is d-convergent with $\lim_n (T^n x) \in \text{Fix}(T)$.

A basic result in this area is the 1922 one due to Banach [4]. Call T, $(d; k)$-*contractive* (where $k \geq 0$), if

$$d(Tx, Ty) \leq kd(x, y), \quad \text{for all} \quad x, y \in X.$$

Theorem 1. *Assume that T is $(d; k)$-contractive, for some $k \in [0, 1[$. In addition, let X be d-complete. Then,*

(11-a) *T is fix-singleton:* $\text{Fix}(T) = \{z\}$, *for some $z \in X$;*
(11-b) *T is a strong Picard operator (modulo d): precisely, $T^n x \xrightarrow{d} z$ as $n \to \infty$, for each $x \in X$.*

This result (referred to as *Banach fixed point principle* (in short: (B-fpp)) found some basic applications to the operator equations theory. As a consequence, many extensions of it were proposed. From the perspective of this exposition, the following ones are of interest:

(I) *Structural contractive extensions*: The original Banach contractive property is taken in a generalized way, as

$$F(d(Tx,Ty),d(x,y),d(x,Tx),d(y,Ty),d(x,Ty),d(Tx,y)) \leq 0,$$
for all $x, y \in X$, $x\nabla y$,

where $F : R^6_+ \to R$ is a function and $\nabla \subseteq X \times X$ is a *relation* over X. Note that, in the case of

$\nabla = X \times X$ (the *trivial relation* over X),

some consistent lists of these may be found in the survey papers by Reich [41] and Rhoades [42]; further aspects were developed in Leader [32], Piticari [39, Chapter II] and Turinici [49]. On the other hand, the case of

∇ is *reflexive* and *transitive* (hence, a *quasi-order*)

was considered in the 1986 paper by Turinici [50]. Two decades later, these results have been re-discovered — in a (partially) ordered context — by Ran and Reurings [40]; see also Nieto and Rodriguez–Lopez [36]. Finally, the case

∇ is *amorphous* (i.e. it has no regularity properties at all)

was discussed via graph techniques by Jachymski [22] and (in a general context) by Samet and Turinici [46]; see also Bojor [6]. Some other aspects involving additional convergence structures may be found in Kasahara [26].

(II) *Non-self extensions*: T is no longer a selfmap. In 1977, Bernfeld *et al.* [5] introduced the concept of *PPF (past-present-future)-dependent fixed point* for *non-self* mappings (whose domain is distinct from their range). Furthermore, the quoted authors established some PPF-dependent fixed point theorems for certain contraction mappings of this type. The obtained results — completed with the methodological directions in Dhage [12, 13] — were ultimately viewed as useful tools in the study of existence and uniqueness questions for solutions of nonlinear functional differential/integral equations which may depend upon the past history, present data and future evolution. As a consequence, this theory attracted a lot of contributors in the area; see, for instance, Chuasuk and Kaewcharoen [9], Drici *et al.* [14], Hussain *et al.* [21], Kaewcharoen [24], Kutbi and Sintunavarat [31] or the references therein. However, according to a

2014 paper by Turinici [52], in all recent statements of this type — based on *algebraical closeness* of the associated Razumikhin class — we may arrange for these PPF-dependent fixed points belonging to the *constant class* of ambient functional space. It is our aim of the present exposition to show that this reduction process includes as well a geometric version of the multivalued PPF-dependent fixed points result in Hammad and De la Sen [19]. Further aspects will be discussed elsewhere.

2. Plane Relations

In the following, some preliminary facts about plane relations will be given.

(A) Let $\Omega \subseteq R_+^0 \times R_+^0$ be a relation over R_+^0; as a rule, we write $(t, s) \in \Omega$ as $t\Omega s$. The following basic (global) properties upon this object are considered:

(u-diag) Ω is *upper diagonal*: $t\Omega s$ implies $t < s$
(1-decr) Ω is *first variable decreasing*:
$t_1, t_2, s \in R_+^0$, $t_1 \geq t_2$ and $t_1\Omega s$ imply $t_2\Omega s$
(2-incr) Ω is *second variable increasing*:
$t, s_1, s_2 \in R_+^0$, $s_1 \leq s_2$ and $t\Omega s_1$ imply $t\Omega s_2$.

The class of all upper diagonal relations will be denoted as $\text{udiag}(R_+^0)$. Note that all subsequent constructions are being handled in this setting. In particular, the following sequential properties for upper diagonal relations Ω are considered:

(matk-1) Ω in *Matkowski admissible*:
$(t_n; n \geq 0)$ in R_+^0 and $(t_{n+1}\Omega t_n, \forall n)$ imply $\lim_n t_n = 0$;
(matk-2) Ω in *strongly Matkowski admissible*:
$(t_n; n \geq 0)$ in R_+^0 and $(t_{n+1}\Omega t_n, \forall n)$ imply $\sum_n t_n < \infty$.

The following local condition involving our relations is — in particular — useful for discussing the former of these properties:

(g-mk) (Ω has the *geometric Meir–Keeler property*)
$\forall \varepsilon > 0, \exists \delta > 0$: $t\Omega s$, $\varepsilon < s < \varepsilon + \delta \Longrightarrow t \leq \varepsilon$.

(This condition is related to the developments in Meir and Keeler [34].) Precisely, the following auxiliary fact in Turinici [56, Section 15] is available.

Proposition 1. *Under these conditions, one has in (ZF − AC + DC) the following:*

(21-1) (for each $\Omega \in \mathrm{udiag}(R^0_+)$*):*
Ω is geometric Meir–Keeler implies Ω is Matkowski admissible
(21-2) (for each first variable decreasing $\Omega \in \mathrm{udiag}(R^0_+)$*):*
Ω is Matkowski admissible implies Ω is geometric Meir–Keeler.

Hence, summing up,

(21-3) (for each first variable decreasing $\Omega \in \mathrm{udiag}(R^0_+)$*):*
Ω is geometric Meir–Keeler iff Ω is Matkowski admissible.

Given the upper diagonal relation Ω over R^0_+, let us introduce the convention:

(a-mk) Ω has the *asymptotic Meir–Keeler property*:
there are no strictly descending sequences (t_n) and (s_n) in R^0_+ and no elements ε in R^0_+, with
$((t_n, s_n) \in \Omega, \forall n)$ and $(t_n \to \varepsilon+, s_n \to \varepsilon+)$.

Here, for each sequence (r_n) in R and each element $r \in R$, we denoted

$r_n \to r+$ (resp., $r_n \to r-$), when $r_n \to r$ and
$[r_n > r$ (resp., $r_n < r$), $\forall n]$.

The following equivalence result is valid. (See the above reference for details.)

Proposition 2. *The following generic relationship is valid (for an arbitrary relation* $\Omega \in \mathrm{udiag}(R^0_+)$*), in the reduced system (ZF − AC + DC):*

asymptotic Meir–Keeler property is equivalent with
geometric Meir–Keeler property.

(B) A first basic example of (upper diagonal) Matkowski admissible and strongly Matkowski admissible relations is constructed as follows.

Denote by $\mathcal{F}(re)(R^0_+, R)$ the family of all functions $\chi \in \mathcal{F}(R^0_+, R)$ with

χ is *regressive*: $\chi(t) < t$, for all $t \in R^0_+$.

For each $\chi \in \mathcal{F}(re)(R^0_+, R)$, define the (associated) *Matkowski* relation $\Omega = \Omega[\chi]$ in $R^0_+ \times R^0_+$ as

$(t, s \in R^0_+)$: $(t, s) \in \Omega$ iff $t \leq \chi(s)$.

Note that, by the regressive condition, Ω is upper diagonal. This convention is related to the developments in Matkowski [33]; see also Boyd and Wong [7].

Further properties of this associated relation are available under some local and sequential conditions upon χ. These may be written as follows:

(MK-a) χ is *Meir–Keeler admissible*:
for each $\varepsilon > 0$, there exists $\delta > 0$, such that
$\varepsilon < s < \varepsilon + \delta \Longrightarrow \chi(s) \leq \varepsilon$;
(M-a) χ is *Matkowski admissible*:
for each $(t_n; n \geq 0)$ in R^0_+ with $(t_{n+1} \leq \chi(t_n), \forall n)$ we have $\lim_n t_n = 0$;
(str-M-a) χ is *strongly Matkowski admissible*:
for each $(t_n; n \geq 0)$ in R^0_+ with $(t_{n+1} \leq \chi(t_n), \forall n)$ we have $\sum_n t_n < \infty$.

Some important examples of such functions may be given along the following lines. For any $\varphi \in \mathcal{F}(re)(R^0_+, R)$ and any $s \in R^0_+$, put

$\Lambda^+\varphi(s) = \inf_{0<\varepsilon<s} \Phi(s+)(\varepsilon)$;
where $\Phi(s+)(\varepsilon) = \sup \varphi(]s, s+\varepsilon[)$.
$\Lambda^\pm\varphi(s) = \inf_{0<\varepsilon<s} \Phi(s\pm)(\varepsilon)$;
where $\Phi(s\pm)(\varepsilon) = \sup \varphi(]s-\varepsilon, s+\varepsilon[)$.

From the regressive property of φ, these limit quantities fulfill

$$-\infty \leq \Lambda^+\varphi(s) \leq \Lambda^\pm\varphi(s) \leq s, \ \ \forall s \in R^0_+;$$

but the case of extremal values being attained cannot be avoided.

Call $\varphi \in \mathcal{F}(re)(R^0_+, R)$, *Boyd–Wong admissible* [7] provided

(bw-adm) $\Lambda^+\varphi(s) < s$, for all $s > 0$.

In particular, $\varphi \in \mathcal{F}(re)(R^0_+, R)$ is Boyd–Wong admissible when it is *upper semicontinuous at the right* on R^0_+:

$$\Lambda^+\varphi(s) \le \varphi(s), \text{ for each } s \in R^0_+.$$

This, e.g., is fulfilled if φ is *continuous at the right* on R^0_+; for, in such a case,

$$\Lambda^+\varphi(s) = \varphi(s), \text{ for each } s \in R^0_+.$$

On the other hand, $\varphi \in \mathcal{F}(re)(R^0_+, R)$ is Boyd–Wong admissible when

φ is *strongly Boyd–Wong admissible*: $\Lambda^{\pm}\varphi(s) < s$, $\forall s \in R^0_+$.

Some partial relationships between these conditions may be expressed as follows. (See, for instance, Turinici [56, Section 13].)

Proposition 3. *For each $\chi \in \mathcal{F}(re)(R^0_+, R)$,*

(23-1) Matkowski admissible is equivalent with Meir–Keeler admissible
(23-2) Boyd–Wong admissible implies Matkowski admissible (or, equivalently: Meir–Keeler admissible).

The following connections between the properties of the function $\chi \in \mathcal{F}(re)(R^0_+, R)$ and the ones of attached relation $\Omega[\chi]$ are available. (Since the verification is immediate, we do not give details.)

Theorem 2. *Under the precise conventions, the following conclusions (involving our upper diagonal relation $\Omega := \Omega[\chi]$) are valid:*

(21-a) Ω is first variable decreasing;
(21-b) Ω is second variable increasing, when χ is increasing;
(21-c) Ω has the Meir–Keeler property whenever χ is Meir–Keeler admissible (or, equivalently: Matkowski admissible);
(21-d) Ω is Matkowski admissible when χ is Matkowski admissible (or, equivalently: Meir–Keeler admissible);
(21-e) Ω is strongly Matkowski admissible whenever χ is strongly Matkowski admissible.

We complete these facts with a criterion for the strong Matkowski admissible property. Precisely, let $\chi \in \mathcal{F}(R^0_+,]0, 1[)$ be a function, and $I \otimes \chi$ stand for the "product" object

$(I \otimes \chi)(t) = t\chi(t)$, $t \in R^0_+$; clearly, $I \otimes \chi \in \mathcal{F}(re)(R^0_+)$.

We are interested in getting such a property for this last function.

Given $h \in \mathcal{F}(R^0_+)$, $\gamma \in]0, \infty]$, let us say that h is γ-*locally normal* provided

(locn-1) $h(.)$ is decreasing on $]0, \gamma[$;
(locn-2) $H(t) := \int_0^t h(\xi)d\xi < \infty$, for each $t \in]0, \gamma[$.

Note that, by the former condition (locn-1),

$\int_0^t h(\xi)d\xi := \lim_{s\to 0+} \int_s^t g(\xi)d\xi$ exists in $R_+ \cup \{\infty\}$, for each $t \in]0, \gamma[$,

so, the latter condition (locn-2) is meaningful; moreover,

$H(.)$ is strictly increasing on $]0, \gamma[$ $(t_1 < t_2 < \gamma \Longrightarrow H(t_1) < H(t_2))$.

Given $g \in \mathcal{F}(R^0_+)$, $\gamma \in]0, \infty]$, let us say that g is γ-*locally subnormal*, provided

$g(t) \leq h(t)$, $t \in]0, \gamma[$, where $h \in \mathcal{F}(R^0_+)$ is γ-locally normal.

Then, given $\chi \in \mathcal{F}(R^0_+,]0, 1[)$, let us associate it the function

$(g \in \mathcal{F}(R^0_+))$: $g(t) = 1/(1 - \chi(t))$, $t > 0$; in short: $g = 1/(1 - \chi)$.

Proposition 4. *Let the function $\chi \in \mathcal{F}(R^0_+,]0, 1[)$ be such that*

(I) χ is unitary Boyd-Wong admissible $(\Lambda^+\chi(s) < 1, \forall s > 0)$
(II) the associated function $g = 1/(1 - \chi)$ in $\mathcal{F}(R^0_+)$ is γ-locally subnormal, for some $\gamma \in]0, \infty]$.

Then, $I \otimes \chi$ is strongly Matkowski admissible (see above).

Proof. By the imposed condition,

$g(t) \leq h(t)$, $t \in]0, \gamma[$, for some γ-locally normal function $h \in \mathcal{F}(R^0_+)$.

Moreover, by the very definitions above,

$H(.)$ is continuous on $]0, \gamma[$ and $H(0+) := \lim_{t\to 0+} H(t) = 0$.

Let the sequence $(t_n; n \geq 0)$ in R^0_+ be such that

(iter) $(\forall n)$: $t_{n+1} \leq t_n\chi(t_n)$; or, equivalently, $t_{n+1}/t_n \leq \chi(t_n)$.

Clearly, $(t_n; n \geq 0)$ is strictly descending; whence, $\tau := \lim_n t_n$ exists in R_+. Suppose by contradiction that $\tau > 0$. Passing to superior limit in (iter) gives (by a direct evaluation)

$1 = \limsup_n (t_{n+1}/t_n) \leq \limsup_n \chi(t_n) \leq \Lambda^+\chi(\tau) < 1$; contradiction.

Hence, the working hypothesis above is not acceptable so that

$\tau = 0$; i.e.: $t_n \to 0$ as $n \to \infty$.

Note that, as a consequence of this,

there exists $k := k(\gamma) \in N$, such that $(t_n < \gamma, \forall n \geq k)$.

Having these precise, let $i \geq k$ be arbitrary fixed. By the above choice,

$t_i - t_i\chi(t_i) \leq t_i - t_{i+1}$; whence, $t_i \leq (t_i - t_{i+1})/(1 - \chi(t_i))$.

Combining with $g \leq h$ and the decreasing property of $h(.)$ yields

$t_i \leq (t_i - t_{i+1})g(t_i) \leq (t_i - t_{i+1})h(t_i) \leq H(t_i) - H(t_{i+1})$.

Taking the relation between extremal members gives (by the preceding fact)

$\sum\{t_n; n \geq k\} \leq H(t_k) - H(0+) < \infty$;

whence the series $\sum_n t_n$ converges. The proof is complete. □

A particular case of this result is the following. Let us say that $\chi \in \mathcal{F}(R^0_+,]0,1[)$ is *Mizoguchi–Takahashi admissible* (in short: *MT-admissible*), provided

(MT-ad) $\Lambda^+\chi(s) < 1, \forall s \geq 0$.

Proposition 5. *Let the function* $\chi \in \mathcal{F}(R^0_+,]0,1[)$ *be MT-admissible. Then,* $I \otimes \chi$ *is strongly Matkowski admissible (see above).*

Proof. Let $g = 1/(1 - \chi)$ be the associated with χ function; clearly, $g \in \mathcal{F}(R^0_+)$. By MT-admissible property at $s = 0$, there exists $\gamma > 0$ and $\lambda \in]0,1[$, such that

(for each $t \in]0,\gamma[$): $\chi(t) \leq \lambda$; whence, $g(t) \leq \mu := 1/(1 - \lambda)$.

The function $h \in \mathcal{F}(R^0_+)$ defined as

$$h(t) = \mu,\ 0 < t < \gamma;\ h(t) = g(t),\ t \geq \gamma$$

is clearly γ-locally normal. This, along with $(g(t) \leq h(t),\ t \in R^0_+)$, tells us that $g(.)$ is γ-locally subnormal so that the preceding statement is applicable to the underlying context. The proof is thereby complete. □

Note, finally, that these results are not the most general in the area; but, they may be useful in practice. Further aspects are to be found in Timofte [48].

(B) Another basic example of Matkowski admissible and strongly Matkowski admissible (upper diagonal) relations is constructed along the following lines. Let (ψ, φ) be a pair of functions over $\mathcal{F}(R^0_+, R)$, fulfilling

(R-couple) (ψ, φ) is a ***Rhoades couple***:
ψ is increasing and φ is *strictly positive* $(\varphi(R^0_+) \subseteq R^0_+)$.

Define the (associated) *Rhoades* relation $\Omega = \Omega[\psi, \varphi]$ in $R^0_+ \times R^0_+$, as

$$(t, s \in R^0_+)\text{: } (t, s) \in \Omega \text{ iff } \psi(t) \leq \psi(s) - \varphi(s).$$

Note that, by the Rhoades couple condition, Ω is upper diagonal. In fact, let $t, s \in R^0_+$ be such that

$$(t, s) \in \Omega;\text{ i.e.: } \psi(t) \leq \psi(s) - \varphi(s).$$

By the strict positivity of φ, one gets $\psi(t) < \psi(s)$, and this, along with the increasing property of ψ shows that $t < s$; whence, the assertion. The introduced convention is related to the developments in Rhoades [43]; see also Dutta and Choudhury [16].

Further properties of this associated relation are available under some extra sequential conditions upon (ψ, φ). The former of these writes

(rs-pos) φ is *right sequentially positive*:
for each strictly descending sequence $(t_n; n \geq 0)$ in R^0_+
and each $\varepsilon > 0$ with $t_n \to \varepsilon+$, we must have $\limsup_n(\varphi(t_n)) > 0$.

The next one is a pair of asymptotic regular properties involving our couple. Namely, let us say that (ψ, φ) is *finitely asymptotic regular*, provided

(f-a-reg-1) $\psi(0+0)(:= \lim_{t\to 0+}\psi(t)) > -\infty$;
(f-a-reg-2) φ is *asymptotic expansive*:
for each strictly descending sequence $(t_n; n \geq 0)$ in R_+^0 with $\sum_n \varphi(t_n) < \infty$, we have $\sum_n t_n < \infty$.

Also, let us say that (ψ, φ) is *infinitely asymptotic regular*, provided

(i-a-reg-1) $\psi(0+0) = -\infty$ and φ is increasing;
(i-a-reg-2) $\exists \mu > 1$, such that (ψ, φ) is *asymptotic μ-dominating*:
for each strictly descending sequence $(t_n; n \geq 0)$ in R_+^0 with $\lim_n t_n = 0$ and $\sum_n [\varphi(t_n)/(1+\psi(t_0)-\psi(t_{n+1}))]^\mu < \infty$, we have $\sum_n t_n < \infty$.

Theorem 3. *Let (ψ, φ) be a Rhoades couple and $\Omega := \Omega[\psi, \varphi]$ stand for its associated relation. Then,*

(22-a) Ω is first variable decreasing;
(22-b) Ω is second variable increasing if $\psi - \varphi$ is increasing;
(22-c) Ω is Meir–Keeler (hence: Matkowski admissible) if φ is asymptotic positive;
(22-d) Ω is strongly Matkowski admissible if (ψ, φ) is finitely-asymptotic regular, or infinitely-asymptotic regular.

Proof. (i), (ii): Evident.

(iii): Suppose that Ω does not have the asymptotic Meir–Keeler property: There exist strictly descending sequences (t_n) and (s_n) in R_+^0 and elements ε in R_+^0, with

$$(t_n \Omega s_n, \forall n) \text{ and } (t_n \to \varepsilon+,\ s_n \to \varepsilon+).$$

The former of these means

$$(\forall n)\colon \psi(t_n) \leq \psi(s_n) - \varphi(s_n), \text{ or, equivalently,}$$
$$(0 <)\varphi(s_n) \leq \psi(s_n) - \psi(t_n).$$

Passing to the superior limit as $n \to \infty$, and noting that

$$\lim_n \psi(s_n) = \lim_n \psi(t_n) = \psi(\varepsilon + 0),$$

one gets $\limsup_n \varphi(t_n) = 0$, in contradiction with the right sequentially positive property of φ. Consequently, the (negative) working assumption cannot hold, proving that Ω has the asymptotic Meir–Keeler property.

(iv): Let $(t_n; n \geq 0)$ be a sequence in R_+^0, with

$t_{n+1} \Omega t_n$ (i.e. $\psi(t_{n+1}) \leq \psi(t_n) - \varphi(t_n)$), for all n;
note that (by the above) (t_n) is strictly descending.

There are two alternatives to be discussed.

Alter 1. Suppose that (ψ, φ) is finitely asymptotic regular. By the posed hypothesis about (t_n), one gets

$$\varphi(t_n) \leq \psi(t_n) - \psi(t_{n+1}), \text{ for all } n.$$

On the other hand,

$$\psi(0+0) > -\infty; \text{ whence, } \sum_n (\psi(t_n) - \psi(t_{n+1})) < \infty.$$

Combining these, we have

$$\sum_n \varphi(t_n) < \infty; \text{ whence, } \sum_n t_n < \infty,$$

if we remember that φ is asymptotic expansive.

Alter 2. Suppose that (ψ, φ) is infinitely asymptotic regular. Remember that, by the working hypothesis about (t_n), one derives that (t_n) is strictly descending; in addition, the underlying hypothesis writes

$$\varphi(t_n) \leq \psi(t_n) - \psi(t_{n+1}), \text{ for all } n.$$

Adding the first $(n+1)$ relations we derive (via $\varphi =$ increasing)

(add-in) $(n+1)\varphi(t_n) \leq \psi(t_0) - \psi(t_{n+1}) \leq 1 + \psi(t_0) - \psi(t_{n+1})$,
$\forall n$.

On the other hand, as $(t_n; n \geq 0)$ is strictly descending, $\tau := \lim_n t_n$ exists in R_+. We claim that

$$\tau = 0; \text{ i.e. } t_n \to 0 \text{ as } n \to \infty.$$

In fact, assume by contradiction that

$$\tau > 0; \text{ whence } \varphi(\tau) > 0 \text{ (as } \varphi \text{ is strictly positive).}$$

From the above inequality, one gets (via $\varphi =$ increasing)

$$(n+1)\varphi(\tau) \leq \psi(t_0) - \psi(\tau), \text{ for all } n.$$

Passing to limit as $n \to \infty$, we derive

$$\infty \leq \psi(t_0) - \psi(\tau) < \infty; \text{ contradiction;}$$

hence, the claim.

Now, by an appropriate multiplication of the (additively generated) inequality (add-in), one gets

$$[\varphi(t_n)/(1 + \psi(t_0) - \psi(t_{n+1}))]^\mu \leq 1/(n+1)^\mu, \text{ for all } n \geq 0.$$

As $\sum_n 1/(n+1)^\mu < \infty$, it follows that

$$\sum_n [\varphi(t_n)/(1 + \psi(t_0) - \psi(t_{n+1}))]^\mu < \infty; \text{ whence, } \sum_n t_n < \infty,$$

if we remember that (ψ, φ) is asymptotic μ-dominating; hence the claim. □

Note that further extensions of these facts are available, via simulation techniques such as in Roldán *et al.* [44]; we will discuss these facts elsewhere.

3. Multivalued Geometric Contractions

In the following, a lot of fixed point results for a class of multivalued geometric contractions will be given, as well as some useful examples. The setting of our investigations is (ZF − AC + DC) [the reduced Zermelo–Fraenkel system].

(A) Let (X, d) be a metric space. Denote

$$cb(X) = \text{the class of all (nonempty) closed bounded parts of } X.$$

Further, let the point to set distance (in short: *ps-distance*) be introduced as

$$d(x, Y) = \inf\{d(x, y); y \in Y\}, \; x \in X, \; Y \in \exp(X).$$

Note that, this object is *non-expansive*, in the sense

$(\forall Y \in \exp(X))$: the map $x \mapsto d(x, Y)$ is *nonexpansive*
$(|d(x_1, Y) - d(x_2, Y)| \leq d(x_1, x_2), \; \forall x_1, x_2 \in X)$.

Let $\Delta : cb(X) \times cb(X) \to R_+$ be the mapping

$$\Delta(Y, Z) = \sup\{d(y, Z); y \in Y\},\ Y, Z \in cb(X).$$

It is not hard to see (cf. Kuratowski [29, Chapter II, Section 21]) that

(hp-1) (Δ is *reflexive*): $\Delta(Y, Y) = 0$, $\forall Y \in cb(X)$;
(hp-2) (Δ is *triangular*):
$\Delta(Y, W) \le \Delta(Y, Z) + \Delta(Z, W)$, $Y, Z, W \in cb(X)$;
(hp-3) (Δ is *almost sufficient*): $\Delta(Y, Z) = 0$ iff $Y \subseteq Z$.

The mapping $(Y, Z) \mapsto \Delta(Y, Z)$ enjoys two properties of a metric; we call it the *Hausdorff–Pompeiu almost metric* of $cb(X)$. Moreover, its associated map

$$\tilde{\Delta}(Y, Z) = \max\{\Delta(Y, Z), \Delta(Z, Y)\},\ Y, Z \in cb(X)$$

has all properties of a metric over $cb(X)$; it will be referred to as the *Hausdorff–Pompeiu metric* of the space $cb(X)$.

The following properties of the introduced objects are almost evident; see, for instance, Nadler [35].

Proposition 6. *Let $Y, Z \in cb(X)$ and $\varepsilon > 0$ satisfy $\Delta(Y, Z) < \varepsilon$. Then,*

for every $u \in Y$ there is $v \in Z$ with $d(u, v) < \varepsilon$.

Proposition 7. *Let $Y, Z \in cb(X)$, $\tau \ge 0$, be such that*

for every $u \in Y$ there is a $v \in Z$ with $d(u, v) \le \tau$.

Then, necessarily, $\Delta(Y, Z) \le \tau$.

By a *multivalued map* over X, we mean any map $T : X \to cb(X)$. As usual, we identify T with its *graph* in $X \times X$ (that acts as a *relation* over X), defined as

$(x, y \in X)$: xTy iff $y \in Tx$.

Let us say that $z \in X$ is a fixed point of T, iff

$z \in Tz$ (i.e.: zTz), or, equivalently, $d(z, Tz) = 0$ (as Tz is d-closed);

the set of all such points will be denoted as $\mathrm{Fix}(T)$.

It is our aim in the sequel to determine elements of Fix(T) by means of orbital iterative sequences and contractive type techniques. Some preliminaries are in order. Let $x_0 \in X$ be arbitrary fixed. As the relation T is proper, it follows — by the Dependent Choice principle — that there may be constructed an orbital iterative sequence $(x_n; n \geq 0)$, so as

$$(x_n, x_{n+1}) \in T \text{ (i.e. } x_{n+1} \in Tx_n\text{), for all } n \geq 0.$$

This will be referred to as a x_0-*orbital sequence*; when x_0 is generic, then (x_n) will be termed an *orbital sequence* (in short: *o-sequence*). As a rule, all sufficient (regularity) conditions to be used involve such sequences; precisely,

(reg-1) Call X, *(o,d)-complete*, provided

$$(z_n) = \text{orbital } d\text{-Cauchy sequence implies } (z_n) \text{ is } d\text{-convergent.}$$

(reg-2) Call T, *(o,d)-closed*, provided

$$((z_n) = \text{orbital sequence, } z \in X, \text{ and } z_n \xrightarrow{d} z) \text{ imply } (z, z) \in T.$$

A basic particular case when such a property holds is to be described as follows. Let us introduce the system of applications [for $x, y \in X$]

(ap-1) $A(x, y) = \min\{\Delta(Tx, Ty), d(y, Ty)\}$;
(ap-2) $B_1(x, y) = (1/2)[d(x, Tx) + d(y, Ty)]$;
$B_2(x, y) = (1/2)[d(x, Ty) + d(y, Tx)]$;
$B(x, y) = \max\{d(x, y), B_1(x, y), B_2(x, y)\}$.

Given $T : X \to cb(X)$, let us introduce the concept

(AB-nex) T is (A, B)-*non-expansive*:
$A(x, y) \leq B(x, y)$, $\forall x, y \in X$, $x \neq y$.

Note that, the following property is a sufficient condition for this:

(De-nex) T is (Δ, d)-*non-expansive*:
$\Delta(Tx, Ty) \leq d(x, y)$, $\forall x, y \in X$, $x \neq y$.

Proposition 8. *Suppose that the multivalued map $T \in \mathcal{F}(X, cb(X))$ is (A, B)-non-expansive. Then, necessarily, T is (o,d)-closed.*

Proof. Let (x_n) be an orbital sequence in X, and z be an element of X, with $x_n \xrightarrow{d} z$; we have to conclude that $(z,z) \in T$. Two alternatives occur.

Alter 1. Suppose that there exists an index k, such that

$$(\forall n)\colon n > k \text{ implies } x_n = z.$$

Then, evidently, $(z,z) \in T$, where $z = x_{k+1}$, and, from this, we are done.

Alter 2. Suppose that the opposite situation is holding:

For each k, there exists $n > k$ such that $x_n \neq z$.

In this case, there must be a strictly ascending sequence of ranks $(i(n); n \geq 0)$, with

the subsequence $(y_n := x_{i(n)}; n \geq 0)$ of $(x_n; n \geq 0)$ fulfills $(y_n \xrightarrow{d} z \text{ as } n \to \infty)$ and $(y_n \neq z, \text{ for all } n)$.

The (AB)-non-expansive condition is applicable to (y_n, z), for all n, and gives

$$\text{(AB-con)} \quad A(y_n, z) \leq B(y_n, z), \ \forall n.$$

But, according to definition (and the orbital property),

$$\begin{aligned}
&A(y_n, z) = \min\{\Delta(Ty_n, Tz), d(z, Tz)\} \\
&= \min\{\Delta(Tx_{i(n)}, Tz), d(z, Tz)\} \geq \min\{d(x_{i(n)+1}, Tz), d(z, Tz)\}, \\
&B_1(y_n, z) = (1/2)[d(y_n, Ty_n) + d(z, Tz)] \\
&= (1/2)[d(x_{i(n)}, Tx_{i(n)}) + d(z, Tz)] \leq \\
&(1/2)[d(x_{i(n)}, x_{i(n)+1}) + d(z, Tz)]; \\
&B_2(y_n, z) = (1/2)[d(y_n, Tz) + d(z, Ty_n)] \\
&= (1/2)[d(x_{i(n)}, Tz) + d(z, Tx_{i(n)})] \leq \\
&(1/2)[d(x_{i(n)}, Tz) + d(z, x_{i(n)+1})], \text{ for all } n.
\end{aligned}$$

This, along with the relation (AB-con), gives (for the same ranks)

$$\begin{aligned}
&\min\{d(x_{i(n)+1}, Tz)), d(z, Tz)\} \leq \max\{d(x_{i(n)}, z), \\
&(1/2)[d(x_{i(n)}, x_{i(n)+1}) + d(z, Tz)], (1/2)[d(x_{i(n)}, Tz) \\
&+ d(z, x_{i(n)+1})]\}.
\end{aligned}$$

Passing to limit as $n \to \infty$, yields (by the non-expansiveness of ps-distance)

$$d(z, Tz) \leq (1/2)d(z, Tz); \text{ whence, } d(z, Tz) = 0,$$

and the assertion follows. □

As a completion of these, we describe the metrical-type contractive conditions to be considered. Let Ω be an upper diagonal relation over R^0_+. We say that the multivalued map $T : X \to cb(X)$ is *iteratively Ω-contractive*, provided

(contr) for each $(x, y) \in T$ with $d(x, y) > 0$, there exists $z \in Ty$ with $(d(y, z), d(x, y)) \in \Omega$, whenever $d(y, z) > 0$.

The following fixed point result involving these data — valid, as precise, in the reduced system (ZF − AC + DC) — is useful for our purposes.

Theorem 4. *Suppose that the multivalued map $T : X \to cb(X)$ and the upper diagonal relation Ω over R^0_+ are such that*

(i) T is iteratively Ω-contractive and Ω is strongly Matkowski admissible.

Further, assume that

(ii) X is (o,d)-complete and T is (o,d)-closed.

Then, T has at least a fixed point in X.

Proof. Suppose by contradiction that this assertion is not true:

(non-fp) $(\forall x \in X)$: $x \notin Tx$; that is, $(x, y) \in T$ implies $d(x, y) > 0$.

Let $\mathcal{R}$ denote the relation over (the graph of) T, introduced as

$(x_1, y_1)\mathcal{R}(x_2, y_2)$ iff $y_1 = x_2$ (hence, $d(x_1, x_2), d(y_1, y_2) > 0$), and $(d(y_1, y_2), d(x_1, x_2)) \in \Omega$.

We claim that

$\mathcal{R}$ is proper over T:
for each $(x_1, y_1) \in T$ there exists $(x_2, y_2) \in T$ with (x_1, y_1) $\mathcal{R}(x_2, y_2)$.

In fact, let $(x_1, y_1) \in T$ be arbitrary fixed; hence (by the above), $d(x_1, y_1) > 0$. By the contractive condition, there exists, for the pair $(x_1, y_1) \in T$ (hence, $y_1 \in Tx_1$), some point $y_2 \in Ty_1$, with

$$(d(y_1, y_2), d(x_1, y_1)) \in \Omega \text{ (possible, in view of } d(y_1, y_2) > 0).$$

But then, evidently,

$$(x_1, y_1)\mathcal{R}(x_2, y_2), \text{ where } x_2 = y_1,$$

and the assertion follows. By the Dependent Choice principle we derive that, given $(x_0, y_0) \in T$, there exists a sequence $((x_n, y_n); n \geq 0)$ in T, with

$$(\forall n)\colon (x_n, y_n)\mathcal{R}(x_{n+1}, y_{n+1}); \text{ that is,}$$
$$y_n = x_{n+1} \text{ and } d(y_n, y_{n+1})\Omega d(x_n, x_{n+1}).$$

This means that there exists a sequence $(x_n; n \geq 0)$ in $X \times X$, with

$$\text{(iter)} \quad (\forall n)\colon (x_n, x_{n+1}) \in T \text{ and } \rho_{n+1}\Omega\rho_n,$$

where $(\rho_n = d(x_n, x_{n+1}); n \geq 0)$. By the choice of Ω, we have that

$$\text{(d-C)} \quad \sum_n d(x_n, x_{n+1}) = \sum_n \rho_n < \infty; \text{ whence, } (x_n) \text{ is } d\text{-Cauchy.}$$

As X is (o,d)-complete, it results that

$$\text{(conv)} \quad x_n \xrightarrow{d} z \text{ as } n \to \infty, \text{ for some } z \in X.$$

This, along with (x_n) being orbital, yields (in view of T is (o,d)-closed)

$$zTz; \text{ that is, } z \in \text{Fix}(T); \text{ a contradiction.}$$

Hence, (non-fp) is rejected, and the proof is complete. □

Note that an extended setting of these developments is possible, under the lines developed by Hitzler [20, Chapter 1, Section 1.2]. On the other hand, this result may be stated in the context of relational metric spaces, and then, our main result may be viewed as a partial multivalued extension of a related statement in Turinici [55]. We shall discuss all these facts elsewhere.

(B) Technically speaking, the obtained result is a conditional one; because no information about the (o,d)-closed property for T is provided. For this reason, it would be useful for us having concrete realizations of this property via contractive conditions, including the auxiliary (o,d)-criterion above. Two basic particular cases are entering in our discussion.

Case 1. The first case is essentially connected with the Matkowski-type constructions of upper diagonal relations we already sketched in a previous place.

Let $\Upsilon \in \mathcal{F}(R^0_+,]0,1[)$ be a function, and $I \otimes \Upsilon$ stand for the product function

$$(I \otimes \Upsilon)(t) = t\Upsilon(t),\ t \in R^0_+;\ \text{clearly, } I \otimes \Upsilon \in \mathcal{F}(re)(R^0_+).$$

Then, let us introduce the parametric family of functions in $\mathcal{F}(R^0_+,]0,1[)$

$$(\Upsilon_\beta(t) = \beta\Upsilon(t) + 1 - \beta; t \in R^0_+),\ \beta \in]0,1];\ \text{clearly, } \Upsilon_1 = \Upsilon.$$

By this convention, the following property holds

$(\forall \beta \in]0,1[)$: $(\Upsilon(t) < \Upsilon_\beta(t) < 1,\ t \in R^0_+)$, and $I \otimes \Upsilon_\beta \in \mathcal{F}(re)(R^0_+)$.

Let us say that Υ is *unitary strongly Matkowski admissible*, provided

$I \otimes \Upsilon$ is strongly Matkowski admissible: for each sequence $(t_n; n \geq 0)$ in R^0_+ with $(t_{n+1} \leq (I \otimes \Upsilon)(t_n),\ \forall n)$, we have $\sum_n t_n < \infty$.

Starting from this, we may introduce the following generic property:

Υ is *unitary upper strongly Matkowski admissible*:
Υ_β is unitary strongly Matkowski admissible, for some $\beta \in]0,1[$.

By the relations above, we have the following inclusion:

Υ is unitary upper strongly Matkowski admissible implies
Υ is unitary strongly Matkowski admissible.

The reciprocal inclusion is not in general true.

Let again $T : X \to cb(X)$ be a multivalued map over X. Remember that we associated it the system of applications [for $x, y \in X$]:

(ap-1) $A(x,y) = \min\{\Delta(Tx,Ty), d(y,Ty)\}$;
(ap-2) $B_1(x,y) = (1/2)[d(x,Tx) + d(y,Ty)]$;
$B_2(x,y) = (1/2)[d(x,Ty) + d(y,Tx)]$;
$B(x,y) = \max\{d(x,y), B_1(x,y), B_2(x,y)\}$.

Given the function $\Upsilon \in \mathcal{F}(R^0_+,]0,1[)$, let us introduce the concept

T is $(A, B; \Upsilon)$-*contractive*:
$A(x,y) \leq \Upsilon(d(x,y))B(x,y)$, for all $x, y \in X$, $x \neq y$.

The following fixed point theorem involving these data is available.

Theorem 5. *Suppose that $T : X \to cb(X)$ is $(A, B; \Upsilon)$-contractive, where (A, B) is as before and $\Upsilon \in \mathcal{F}(R^0_+,]0,1[)$ is unitary upper strongly Matkowski admissible. In addition, let X be (o,d)-complete. Then, T has a fixed point in X.*

Proof. There are two parts to be checked.

Part 1. By the very definition of our contractive condition,

T is (A, B)-non-expansive: $A(x,y) \leq B(x,y)$, $\forall x, y \in X$, $x \neq y$.

This, by a previous result, tells us that T is (o,d)-closed.

Part 2. Let $\beta \in]0,1[$ be the number given by the unitary upper strongly Matkowski admissible property for Υ. Note that, under the choice of Υ_β, the relation Ω over R^0_+, given as

$(t, s \in R^0_+)$: $t\Omega s$ iff $t \leq (I \otimes \Upsilon_\beta)(s)$,

is upper diagonal and strongly Matkowski admissible. On the other hand, from the contractive condition, the construction of Υ_β, and $(d(x,y) > 0 \Longrightarrow B(x,y) > 0)$,

T is *strongly* $(A, B; \Upsilon_\beta)$-*contractive*:
$A(x,y) < \Upsilon_\beta(d(x,y))B(x,y)$, for all $x, y \in X$, $x \neq y$.

Suppose (by contradiction) that T does not admit fixed points

(non-fp) $x \notin Tx$ (whence: $d(x, Tx) > 0$), for all $x \in X$.

We show that this cannot be in agreement with the conclusions of Theorem 4.

To begin with, we claim that

(Cla-1) T is iteratively Ω-contractive, that is, (under (non-fp)):
$\forall (x,y) \in T$, $\exists z \in Ty$, with $d(y,z) \leq (I \otimes \Upsilon_\beta)(d(x,y))$.

In fact, let $(x,y) \in T$ be given; hence,

$d(x,y) \geq d(x, Tx) > 0$ (as $y \in Tx$).

Further, by the definition of Hausdorff–Pompeiu almost metric

$$\Delta(Tx,Ty) \geq d(y,Ty);\ \text{whence, } A(x,y) = d(y,Ty).$$

This, by the strong Υ_β-contractive property above, gives

$$d(y,Ty) < \Upsilon_\beta(d(x,y))B(x,y).$$

Consequently, by the infimum definition, there must be some $z \in Ty$ with

$$\text{(Cla-2)}\quad d(y,z) < \Upsilon_\beta(d(x,y))B(x,y).$$

But, according to definition,

$$\begin{aligned}
&B_1(x,y) = (1/2)[d(x,Tx) + d(y,Ty)]\\
&\leq (1/2)[d(x,y) + d(y,z)] \leq \max\{d(x,y), d(y,z)\},\\
&B_2(x,y) = (1/2)[d(x,Ty) + d(y,Tx)] = (1/2)d(x,Ty)\\
&\leq (1/2)d(x,z) \leq (1/2)[d(x,y) + d(y,z)] \leq \max\{d(x,y), d(y,z)\},\\
&B(x,y) = \max\{d(x,y), B_1(x,y), B_2(x,y)\}\\
&\leq \max\{d(x,y), d(y,z)\}.
\end{aligned}$$

Replacing these into strong contractive condition (Cla-2), we derive

$$\text{(Cla-3)}\quad d(y,z) < \Upsilon_\beta(d(x,y)) \max\{d(x,y), d(y,z)\}.$$

The alternative $d(x,y) \leq d(y,z)$ cannot be accepted; for, in this case,

$$d(y,z) < \Upsilon_\beta(d(x,y))d(y,z);\ \ \text{impossible, via } \Upsilon_\beta(d(x,y)) < 1.$$

So, the alternative $d(x,y) > d(y,z)$ must be accepted; whence, in view of (Cla-3),

$$d(y,z) < \Upsilon_\beta(d(x,y))d(x,y) = (I \otimes \Upsilon_\beta)(d(x,y)),$$

and (Cla-1) is proved. In this case, by the same way as the one proposed in Theorem 4, there exists a sequence (x_n) in X, with

$$\text{(iter-a)}\quad (\forall n)\text{: } (x_n, x_{n+1}) \in T \text{ and } \rho_{n+1} \leq (I \otimes \Upsilon_\beta)(\rho_n),$$

where $(\rho_n = d(x_n, x_{n+1}); n \geq 0)$. For, let $\mathcal{Q}$ denote the relation over (the graph of) T, introduced as

$$\begin{aligned}
&(x_1, y_1)\mathcal{Q}(x_2, y_2) \text{ iff } y_1 = x_2 \text{ (hence, } d(x_1,x_2), d(y_1,y_2) > 0\text{)},\\
&\text{and } d(y_1,y_2) \leq (I \otimes \Upsilon_\beta)(d(x_1,x_2)).
\end{aligned}$$

We claim that

$\mathcal{Q}$ is proper over T:
for each $(x_1, y_1) \in T$ there exists $(x_2, y_2) \in T$ with $(x_1, y_1)\mathcal{Q}(x_2, y_2)$.

In fact, let $(x_1, y_1) \in T$ be arbitrary fixed; hence (by the above), $d(x_1, y_1) > 0$. Taking (Cla-1) into account, there exists, for $(x_1, y_1) \in T$ (hence, $y_1 \in Tx_1$), some point $y_2 \in Ty_1$, with $d(y_1, y_2) \leq (I \otimes \Upsilon_\beta)(d(x_1, y_1))$. But then,

$(x_1, y_1)\mathcal{Q}(x_2, y_2)$, where $x_2 = y_1$,

and the assertion follows. By the Dependent Choice principle it then follows that, given $(x_0, y_0) \in T$, there exists a sequence $((x_n, y_n); n \geq 0)$ in T, with

$(\forall n)$: $(x_n, y_n)\mathcal{Q}(x_{n+1}, y_{n+1})$; that is,
$y_n = x_{n+1}$ and $d(y_n, y_{n+1}) \leq (I \otimes \Upsilon_\beta)(d(x_n, x_{n+1}))$,

and, from this, we are done. As Υ_β is strongly Matkowski admissible, we have that

(d-C) $\sum_n d(x_n, x_{n+1}) = \sum_n \rho_n < \infty$; whence, (x_n) is d-Cauchy.

Moreover, as X is (o,d)-complete, it results that

(conv) $x_n \xrightarrow{d} z$ as $n \to \infty$, for some $z \in X$.

Finally, by the preceding part, T is (o,d)-closed. Putting these together, it results that Theorem 4 is applicable here; wherefrom, Fix(T) is non-empty. This, however, cannot be in agreement with the (non-fp) hypothesis. Hence, this last assertion cannot be accepted; and the proof is complete. □

A useful variant of the obtained result may be stated as follows. Let us say that the function $\Upsilon \in \mathcal{F}(R^0_+,]0,1[)$ is *Mizoguchi–Takahashi admissible* (in short: *MT-admissible*), when

$\Lambda^+\Upsilon(s) < 1$, for all $s \geq 0$.

Note that, under such a context, one has, for each $\beta \in]0,1[$,

$\Lambda^+\Upsilon_\beta(s) = \beta\Lambda^+\Upsilon(s) + 1 - \beta < 1$, $s \geq 0$; whence (see above)
$I \otimes \Upsilon_\beta$ is strongly Matkowski admissible,

and this, by a previous auxiliary fact, yields

Υ is MT-admissible implies Υ is unitary
upper strongly Matkowski admissible.

In other words, Theorem 5 applies to such functions and yields the following:

Theorem 6. *Suppose that $T : X \to cb(X)$ is $(A, B; \Upsilon)$-contractive, where (A, B) is as before and $\Upsilon \in \mathcal{F}(R^0_+,]0, 1[)$ is MT-admissible. In addition, let X be (o,d)-complete. Then, T has at least a fixed point in X.*

In particular, when Δ is substituted by $\widetilde{\Delta}$ (the Hausdorff–Pompeiu metric), the corresponding version of this result includes directly the related statement in Du [15]. Note that the constant function in $\mathcal{F}(R^0_+,]0, 1[)$

$$\Upsilon = \Gamma_\alpha, \text{ where } (\Gamma_\alpha(t) = \alpha,\ t \in R^0_+), \text{ for some } \alpha \in]0, 1[$$

is MT-admissible; hence, unitary upper strongly Matkowski admissible. Replacing this in Theorem 6 yields the well-known fixed point result in Nadler [35].

Case 2. The second case has a strong connection with the Rhoades-type constructions of upper-diagonal relations described in a previous place.

Let (F, G) be a functional couple over $\mathcal{F}(R^0_+, R)$, with

(nor) (F, G) is *normal*: F is increasing and
G is strictly positive ($G(t) > 0$, $\forall t \in R^0_+$).

The following extra conditions will be considered here:

(ex-1) (F is *right continuous*) $F(t + 0) = F(t)$, for each $t > 0$;
(ex-2) (G is *right sequentially positive*)
$\limsup_n G(t_n) > 0$, for each strictly descending sequence (t_n) in R^0_+ and each $\varepsilon > 0$ with $t_n \to \varepsilon+$;
(ex-3) ((F, G) is *admissible*)
each strictly descending sequence (t_n) in R^0_+ with $t_n \to 0+$ and $(G(t_n) \le F(t_n) - F(t_{n+1}),\ \forall n)$ fulfills $\sum_n t_n < \infty$.

The last condition above deserves a special attention. Let $F : R^0_+ \to R$ be a function with

(W-1) F is increasing and *vertical* ($F(0+0) = -\infty$);
(W-2) F is *regular*: $\lim_{t\to 0+} t^k F(t) = 0$, for some $k \in]0,1[$;

we then say that F is a *Wardowski function*.

Proposition 9. *Let $F : R^0_+ \to R$ be a Wardowski function and $C > 0$ be a constant. Then, (F, C) is admissible.*

Proof. Let the number $k \in]0,1[$ be given by the Wardowski property of F. Then, let the strictly descending sequence (t_n) in R^0_+ be such that $t_n \to 0+$ and $(C \le F(t_n) - F(t_{n+1}),\ \forall n)$; we have to establish that $\sum_n t_n < \infty$. By the posed condition,

$$(\forall n)\text{: } Cn \le F(t_0) - F(t_n)\text{; whence } Cnt_n^k \le (F(t_0) - F(t_n))t_n^k.$$

In addition, the regularity condition tells us that there exist $\beta > 0$ and some index $n(\beta)$, such that

$$(\forall n \ge n(\beta))\text{: } Cnt_n^k \le \beta\text{; that is: } t_n \le (\beta/Cn)^{1/k}.$$

This, along with the convergence of the series $\sum_{n\ge 1} n^{-1/k}$, ends the argument. □

Under these preliminaries, we may now pass to the announced statement. Let $T : X \to cb(X)$ be a multivalued map over X. Further, let the couple of applications (A, B) be the previous one [for $x, y \in X$]

(ap-1) $A(x, y) = \min\{\Delta(Tx, Ty), d(y, Ty)\}$;
(ap-2) $B_1(x, y) = (1/2)[d(x, Tx) + d(y, Ty)]$;
$B_2(x, y) = (1/2)[d(x, Ty) + d(y, Tx)]$;
$B(x, y) = \max\{d(x, y), B_1(x, y), B_2(x, y)\}$.

Given the normal couple (F, G) over $\mathcal{F}(R^0_+, R)$, let us say that T is $(A, B; F, G)$*-contractive*, provided

(T-con) $F(A(x, y)) \le F(B(x, y)) - G(d(x, y))$,
whenever $A(x, y) > 0$, $d(x, y) > 0$.

Theorem 7. *Let the multivalued map T over X be $(A, B; F, G)$-contractive, where (A, B) is as before, and (F, G) is a normal couple over $\mathcal{F}(R^0_+, R)$. In addition, suppose that*

F is right continuous and $(F, \lambda G)$ is admissible, for some $\lambda \in]0,1[$.

Then, the following conclusions hold

(34-a) T is (o,d)-closed:
(z_n) = orbital sequence in X, $z \in X$, and $z_n \xrightarrow{d} z$, imply $(z,z) \in T$;
(34-b) If, in addition, X is (o,d)-complete, then T has a fixed point in X.

Proof. There are two parts to be verified.

Part 1. Let $x, y \in X$ be such that $x \neq y$; hence, $d(x,y) > 0$. For the moment, $B(x,y) > 0$ (by the very definition of B). If $A(x,y) = 0$, then, evidently, $A(x,y) < B(x,y)$. Suppose now that $A(x,y) > 0$. From the contractive condition, we have

$F(A(x,y)) \leq F(B(x,y)) - G(d(x,y)) < F(B(x,y))$;
hence (as F = increasing), $A(x,y) < B(x,y)$.

Summing up, one derives that T is (A,B)-non-expansive. Combining with a previous auxiliary fact, we have that T is (o,d)-closed, and the first conclusion follows.

Part 2. Let, in addition, X be (o,d)-complete; we intend to show that T has a fixed point in X. Suppose not

(non-fp) $x \notin Tx$ (whence: $d(x,Tx) > 0$), for all $x \in X$.

Let $\lambda \in]0,1[$ be the number given by the admissible condition we imposed. Note that, by the very choice of $(F, \lambda G)$, the associated relation over R^0_+

$(t, s \in R^0_+)$: $t\Omega s$ iff $F(t) \leq F(s) - \lambda G(s)$

is upper diagonal and strongly Matkowski admissible. Further, it is clear that, for each $(x,y) \in T$, we have

$\Delta(Tx, Ty) \geq d(y, Ty)$; hence, $A(x,y) = d(y,Ty)$;

as well as (by the contractive condition)

(contr) $F(d(y,Ty)) \leq F(B(x,y)) - G(d(x,y))$; hence,
$F(d(y,Ty)) < F(B(x,y)) - \lambda G(d(x,y))$.

Part 3. We claim that the following assertion is valid

(Kla) for each $(x, y) \in T$ there exists $z \in Ty$ with $F(d(y, z)) \leq F(d(x, y)) - \lambda G(d(x, y))$.

Indeed, as $d(y, Ty) = \inf\{d(y, z); z \in Ty\}$, and F is right continuous, it follows — by the second half of (contr) — that there must be some $z \in Ty$, such that

(eva-1) $F(d(y, z)) < F(B(x, y)) - \lambda G(d(x, y))$.

But, according to the definition,

$$\begin{aligned}
&B_1(x, y) = (1/2)[d(x, Tx) + d(y, Ty)] \\
&\leq (1/2)[d(x, y) + d(y, z)] \leq \max\{d(x, y), d(y, z)\}, \\
&B_2(x, y) = (1/2)[d(x, Ty) + d(y, Tx)] = (1/2)d(x, Ty) \\
&\leq (1/2)d(x, z) \leq (1/2)[d(x, y) + d(y, z)] \leq \max\{d(x, y), d(y, z)\}, \\
&B(x, y) = \max\{d(x, y), B_1(x, y), B_2(x, y)\} \\
&\leq \max\{d(x, y), d(y, z)\}.
\end{aligned}$$

Replacing these into (eva-1) gives (under F=increasing)

(eva-2) $F(d(y, z)) < F(\max\{d(x, y), d(y, z)\}) - \lambda G(d(x, y))$.

The alternative $d(x, y) \leq d(y, z)$ cannot be accepted; for, in this case,

$F(d(y, z)) < F(d(y, z)) - \lambda G(d(x, y)))$; impossible, via $\lambda G(d(x, y)) > 0$.

So, the alternative $d(x, y) > d(y, z)$ must be accepted; wherefrom, in view of (eva-2), our initial assertion (Kla) follows. In this case, by the same way as the one proposed in Theorem 4, there exists a sequence (x_n) in X, with

(iter-a) $(\forall n)$: $(x_n, x_{n+1}) \in T$ and $F(\rho_{n+1}) \leq F(\rho_n) - \lambda G(\rho_n)$,

where $(\rho_n = d(x_n, x_{n+1}); n \geq 0)$. For, let $\mathcal{Q}$ denote the relation over (the graph of) T, introduced as

$(x_1, y_1)\mathcal{Q}(x_2, y_2)$ iff $y_1 = x_2$ (hence, $d(x_1, x_2), d(y_1, y_2) > 0$) and $F(d(y_1, y_2)) \leq F(d(x_1, x_2)) - \lambda G(d(x_1, x_2))$.

We claim that

$\mathcal{Q}$ is proper over T:
for each $(x_1, y_1) \in T$ there exists $(x_2, y_2) \in T$ with $(x_1, y_1)\mathcal{Q}(x_2, y_2)$.

In fact, let $(x_1, y_1) \in T$ be arbitrary fixed; hence, $d(x_1, y_1) > 0$. Taking (Kla) into account, there exists, $(x_1, y_1) \in T$ (hence, $y_1 \in Tx_1$), some point $y_2 \in Ty_1$, with $F(d(y_1, y_2)) \leq F(d(x_1, y_1)) - \lambda G(d(x_1, y_1))$. But then, evidently,

$(x_1, y_1)\mathcal{Q}(x_2, y_2)$, where $x_2 = y_1$,

and the assertion follows. By the Dependent Choice principle one derives that, given $(x_0, y_0) \in T$, there must be a sequence $((x_n, y_n); n \geq 0)$ in T, with

$(x_n, y_n)\mathcal{Q}(x_{n+1}, y_{n+1})$, for all n,

and from this, we are done. As $(F, \lambda G)$ is admissible, we have that

(d-C) $\sum_n d(x_n, x_{n+1}) = \sum_n \rho_n < \infty$; whence, (x_n) is d-Cauchy.

Combining with [X is (o,d)-complete], it results that

(conv) $x_n \overset{d}{\longrightarrow} z$ as $n \to \infty$, for some $z \in X$.

This, along with the previous part, yields $(z, z) \in T$; that is: $z \in Tz$; and the second part follows as well. □

In particular, when

$T \in \mathcal{F}(X, cb(X))$ is $(\Delta, d; F, G)$-contractive and $\lambda = 1/2$

the obtained result extends in a direct way the related one in Amini-Harandi [3]. Another particular case may be obtained under the preceding observations concerning the admissible property. Precisely, we have the following:

Theorem 8. *Let the multivalued map $T : X \to cb(X)$ be $(\Delta, d; F, C)$-contractive, where $F : R^0_+ \to R$ is Wardowski, right continuous and $C > 0$ is a constant. Then, T has a fixed point in X.*

Proof. By a previous result,

$(F, \lambda C)$ is admissible, for each $\lambda \in]0, 1[$.

This shows that the preceding result applies; and our conclusion follows. □

The obtained results extend in a direct way the ones in Acar *et al.* [1]; see also Sgroi and Vetro [47]. Note that an extended metrical setting of these developments is possible, under the lines developed by Hitzler [20, Chapter 1, Section 1.2] and Turinici [54]. On the other hand, this result may be stated in the context of relational metric spaces; and then, its corresponding version includes the statement in Javahernia *et al.* [23]. Finally, it is worth noting that the compact version of our main results is, technically, identical with the univalued version of the same, developed by Krasnoselskii and Stetsenko [28], re-discovered by Rhoades [43] and refined by Dutta and Choudhury [16], Wardowski [58, 59], Vujaković *et al.* [57] and Turinici [53]; see also the survey paper by Karapinar *et al.* [25]. We shall discuss all these facts elsewhere.

4. Razumikhin Classes

Let $(E, ||.||)$ be a Banach space and $d(.,.)$ be the induced by norm metric on E:

$$d(x,y) = ||x-y||,\ x,y \in E;$$

hence, (E,d) is a complete metric space. Further, let $I = [a,b]$ be a closed bounded real interval and $E_0 := C(I,E)$ stand for the class of all continuous functions $\varphi : I \to E$, endowed with the supremum norm

$$||\varphi||_0 = \sup\{||\varphi(t)||; t \in I\},\ \varphi \in E_0.$$

Then, let $D(.,.)$ stand for the induced by norm metric on E_0

$$D(\varphi,\xi) = ||\varphi - \xi||_0,\ \varphi,\xi \in E_0;$$

clearly, (E_0, D) is a complete metric space.

For each $u \in E$, let Ku denote the associated object

(K-def) $((Ku)(t) = u;\ t \in I)$ (the *constant function* of E_0 generated by u)

and $\mathcal{K}$ stand for the collection of all these:

$\mathcal{K} := K(E) = \{Ku; u \in E\}$ (referred to as: the *constant class* of E_0).

The following properties of this class are immediate, so, we do not give details.

Proposition 10. *Under the above conventions,*

(41-1) $K(u + v) = Ku + Kv$, $K(\lambda u) = \lambda Ku$, $\forall u, v \in E$, $\forall \lambda \in R$; *hence,* $K(.)$ *is a linear operator and* $\mathcal{K}$ *is a linear subspace of* E_0;
(41-2) K *is an isometry* ($d(u, v) = D(Ku, Kv)$, $\forall u, v \in E$); *hence (combining with the above)* $u \mapsto Ku$ *is an algebraic and topological isomorphism between* $(E, ||.||)$ *and* $(\mathcal{K}, ||.||_0)$.

Let in the following $\mathcal{S} : E_0 \to E$ be an operator; we call it *normal*, provided

(nor-1) $\mathcal{S}$ is linear; that is, additive ($\mathcal{S}(\varphi + \xi) = \mathcal{S}(\varphi) + \mathcal{S}(\xi)$, $\forall \varphi, \xi \in E_0$) and homogeneous ($\mathcal{S}(\lambda\varphi) = \lambda\mathcal{S}(\varphi)$, $\forall \lambda \in R$, $\forall \varphi \in E_0$);
(nor-2) $(\mathcal{S}, K)$ is invertible: $\mathcal{S}(Ku) = u$, $u \in E$.

The *Razumikhin class* in E_0 attached to this operator is defined as

(r-cl) $\mathrm{Raz}(\mathcal{S}) = \{\varphi \in E_0; ||\varphi||_0 = ||\mathcal{S}\varphi||\}$.

Note that this class is non-empty because it includes the constant class $\mathcal{K}$. For, given $u \in E$, we have

$||Ku||_0 = ||u||$, $||\mathcal{S}Ku|| = ||u||$; whence, $Ku \in \mathrm{Raz}(\mathcal{S})$.

The following property of this class is useful for our purposes. (See also Turinici [52] for a number of related facts.)

Proposition 11. *Under these conventions:*
(42-1) *The Razumikhin class* $\mathrm{Raz}(\mathcal{S})$ *is homogeneous, in the sense*

$$\mathrm{Raz}(\mathcal{S}) = \lambda\mathrm{Raz}(\mathcal{S}),\ \forall \lambda \in R \setminus \{0\};$$

(42-2) *In particular, the Razumikhin class* $\mathrm{Raz}(\mathcal{S})$ *is symmetric:*

$$\mathrm{Raz}(\mathcal{S}) = -\mathrm{Raz}(\mathcal{S})\ (\varphi \in \mathrm{Raz}(\mathcal{S}) \iff -\varphi \in \mathrm{Raz}(\mathcal{S})).$$

Proof. (i) Let $\lambda \in R \setminus \{0\}$ be arbitrary fixed. Given $\mu \in \{\lambda, 1/\lambda\}$ and $\varphi \in \text{Raz}(\mathcal{S})$, denote $\xi = \mu\varphi$. By definition, $||\xi||_0 = |\mu| \cdot ||\varphi||_0$, $||\mathcal{S}(\xi)|| = |\mu| \cdot ||\mathcal{S}(\varphi)||$. This, along with the choice of φ, gives $||\xi||_0 = ||\mathcal{S}(\xi)||$; i.e., $\xi \in \text{Raz}(\mathcal{S})$, so that $\lambda\text{Raz}(\mathcal{S}) \subseteq \text{Raz}(\mathcal{S})$, $(1/\lambda)\text{Raz}(\mathcal{S}) \subseteq \text{Raz}(\mathcal{S})$. On the other hand, for each $\psi \in \text{Raz}(\mathcal{S})$, we have $(1/\lambda)\psi \in \text{Raz}(\mathcal{S})$ so that $\psi = \lambda(1/\lambda)\psi \in \lambda\text{Raz}(\mathcal{S})$; whence, $\text{Raz}(\mathcal{S}) \subseteq \lambda\text{Raz}(\mathcal{S})$. Putting these together, the claim follows.

(ii) Evident, by the above. □

In the following, a basic condition will be introduced at the level of this class:

(a-clo) $\text{Raz}(\mathcal{S})$ is *algebraically closed*: $\varphi, \xi \in \text{Raz}(\mathcal{S}) \Longrightarrow \varphi - \xi \in \text{Raz}(\mathcal{S})$.

The following characterization of this concept is almost immediate by the above auxiliary facts, so we do not give details.

Proposition 12. *The following conditions are equivalent:*

(43-1) $\text{Raz}(\mathcal{S})$ *is algebraically closed;*
(43-2) $\text{Raz}(\mathcal{S})$ *is additive:* $\varphi, \xi \in \text{Raz}(\mathcal{S}) \Longrightarrow \varphi + \xi \in \text{Raz}(\mathcal{S})$;
(43-3) $\text{Raz}(\mathcal{S})$ *is a linear subspace of* E_0.

Remember that, by the above,

$\mathcal{K} \subseteq \text{Raz}(\mathcal{S})$ (each constant function belongs to $\text{Raz}(\mathcal{S})$).

So, it is natural to ask whether the converse inclusion holds. A positive answer to this is true under the introduced regularity assumption. Precisely, we have the following:

Proposition 13. *Suppose that the Razumikhin class* $\text{Raz}(\mathcal{S})$ *is algebraically closed; or, equivalently, a linear subspace. Then, necessarily,* $\text{Raz}(\mathcal{S}) = \mathcal{K}$.

Proof. Suppose that $\text{Raz}(\mathcal{S})$ includes strictly $\mathcal{K}$ and let φ be some function with

$\varphi \in \text{Raz}(\mathcal{S}) \setminus \mathcal{K}$; i.e., $\varphi \neq Ku$, for all $u \in E$.

The function $\xi := Ku$ where $u := \mathcal{S}\varphi$ is an element of the class $\mathcal{K}$, hence, of the Razumikhin class $\text{Raz}(\mathcal{S})$ as well. In addition,

$\mathcal{S}\xi = \mathcal{S}Ku = u = \mathcal{S}\varphi$ (by this very definition).

As Raz($\mathcal{S}$) is algebraically closed, $\delta := \varphi - \xi$ is in Raz($\mathcal{S}$), with, in addition, $\mathcal{S}\delta = \mathcal{S}\varphi - \mathcal{S}\xi = 0$. This, by definition, gives

$||\delta||_0 = ||\mathcal{S}\delta|| = 0$; or, equivalently: $\varphi = Ku \in \mathcal{K}$;

in contradiction with the choice of φ. Hence, Raz($\mathcal{S}$) = $\mathcal{K}$, as claimed. □

As a consequence of this, the following properties are valid:
(Conseq-1) The following inclusion is holding:

> Raz($\mathcal{S}$) is algebraically closed implies Raz($\mathcal{S}$)($= \mathcal{K}$) is topologically closed, in the sense: Raz($\mathcal{S}$) is a closed (hence complete) subspace of E_0.

This tells us that, in the fixed point theory over Raz($\mathcal{S}$), the topological closeness hypothesis upon it — as in Dhage [12, 13] — is superfluous and may be removed.

(Conseq-2) All reasonings upon the Razumikhin class Raz($\mathcal{S}$) are trivially reducible to the ones involving the constant class $\mathcal{K}$.

Remark 1. A different proof of these facts is available, by means of

> whenever Raz($\mathcal{S}$) is algebraically closed, the map $\varphi \mapsto \mathcal{S}\varphi$ is an isometry between Raz($\mathcal{S}$) and $\mathcal{K}$.

In the following, some basic examples of normal operators $\mathcal{S}$ are given.

Example 1. Let $c \in I$ be an arbitrary fixed point. The operator

$\mathcal{S}_c : E_0 \to E$: $\mathcal{S}_c(\varphi) = \varphi(c)$, $\varphi \in E_0$,

is linear and $(\mathcal{S}_c, K)$ is invertible; hence, $\mathcal{S}_c$ is normal. Moreover, the Razumikhin class of functions in E_0 attached to $\mathcal{S}_c$ has the representation

$\text{Raz}(\mathcal{S}_c) = \{\varphi \in E_0; ||\varphi||_0 = ||\varphi(c)||\}$.

In fact, these are the only examples of normal operator and Razumikhin class to be used until now.

Example 2. Let $C = (c_n; n \geq 0)$ be a sequence in I and $\Lambda = (\lambda_n; n \geq 0)$ be a sequence in R_+ with $\sum_n \lambda_n = 1$. The operator

$$\mathcal{S}_{(C,\Lambda)} : E_0 \to E\text{: } \mathcal{S}_{(C,\Lambda)}(\varphi) = \sum_n \lambda_n \varphi(c_n),\ \varphi \in E_0$$

is well defined because φ is bounded on I. In addition, $\mathcal{S}_{(C,\Lambda)}$ is linear and $\mathcal{S}_{(C,\Lambda)}, K)$ is invertible; hence, $\mathcal{S}_{(C,\Lambda)}$ is normal. Moreover, the Razumikhin class of functions in E_0 attached to $\mathcal{S}_{(C,\Lambda)}$ has the representation

$$\mathrm{Raz}(\mathcal{S}_{(C,\Lambda)}) = \{\varphi \in E_0; ||\varphi||_0 = ||\mathcal{S}_{(C,\Lambda)}(\varphi)||\}.$$

This class of normal operators and Razumikhin classes was never used until now in practice. In particular, when the sequence $(c_n; n \geq 0)$ is constant and $(\lambda_0 = 1; \lambda_n = 0,\ \forall n \geq 1)$, this operator reduces to the preceding one.

Example 3. Let $\int$ stands for the Bochner integral over E. The operator

$$\mathcal{S}_{\int} : E_0 \to E\text{: } \mathcal{S}_{\int}(\varphi) = (1/(b-a)) \int_a^b \varphi(t)dt,\ \varphi \in E_0$$

is well defined because φ is continuous and bounded on I. In addition, $\mathcal{S}_{\int}$ is linear and $(\mathcal{S}_{\int}, K)$ is invertible; hence, $\mathcal{S}_{\int}$ is normal. Moreover, the Razumikhin class of functions in E_0 attached to $\mathcal{S}_{\int}$ is defined as

$$\mathrm{Raz}(\mathcal{S}_{\int}) = \{\varphi \in E_0; ||\varphi||_0 = ||\mathcal{S}_{\int}(\varphi)||\}.$$

As before, this class of normal operators and Razumikhin classes was never used until now in practice.

5. PPF-Dependent Fixed Points

Let $(E, ||.||)$ be a Banach space; and $d(.,.)$ be the (complete) induced by norm metric on E. Further, taking the closed bounded real interval $I = [a, b]$, let $E_0 := C(I, E)$ stand for the Banach space of all continuous functions in $\mathcal{F}(I, E)$, endowed with the (complete) supremum norm $||.||_0$, and the (complete) induced by norm metric $D(.,.)$. Remember that, for each $u \in E$, Ku denotes the attached constant function $(Ku(t) = u; t \in I)$, and $\mathcal{K} = K(E)$ stands for the family of all these, referred to as the *constant class* of E_0.

Let $\mathcal{S} : E_0 \to E$ be a *normal* operator, in the sense

(nor-1) $\mathcal{S}$ is linear (additive, homogeneous);
(nor-2) $(\mathcal{S}, K)$ is *invertible*: $\mathcal{S}(Ku) = u$, $u \in E$.

Then, let the map $\mathcal{T} : E_0 \to cb(E)$ be given; referred to as a *multivalued map* over (E_0, E). We say that $\varphi \in E_0$ is a *coincidence point* of $\mathcal{S}$ and $\mathcal{T}$, when $\mathcal{S}\varphi \in \mathcal{T}\varphi$. To determine these points, the following construction will be in effect. Let $T : E \to cb(E)$ be the (associated to $\mathcal{T}$) multivalued map of E introduced as

(T-def) $Tx = \mathcal{T}(Kx)$, $x \in E$.

Proposition 14. *Under these conventions, the following are valid:*

(51-1) *If $x \in E$ is a fixed point of T, then $\xi := Kx \in \mathcal{K}$ is a coincidence point of $\mathcal{S}$ and $\mathcal{T}$;*
(51-2) *Conversely, if $\zeta := Kz \in \mathcal{K}$ is a coincidence point of $\mathcal{S}$ and $\mathcal{T}$, then $z \in E$ is a fixed point of T.*

Proof. (i) If $x \in E$ is a fixed point of T, we have

$\mathcal{S}(Kx) = x \in Tx = \mathcal{T}(Kx)$, or, equivalently,
$\mathcal{S}\xi \in \mathcal{T}\xi$, where $\xi := Kx \in \mathcal{K}$,

which tells us that $\xi \in \mathcal{K}$ is a coincidence point of $\mathcal{S}$ and $\mathcal{T}$.

(ii) Suppose that $\zeta := Kz \in \mathcal{K}$ is a coincidence point of $\mathcal{S}$ and $\mathcal{T}$; i.e., $\mathcal{S}(\zeta) \in \mathcal{T}(\zeta)$. This yields (by the properties of $\mathcal{S}$) $z = \mathcal{S}(Kz) \in \mathcal{T}(Kz) = Tz$ so that $z \in E$ is a fixed point of T. □

Remark 2. The following completions of these facts are to be noted.

Given $\varphi_0 \in E_0$, call the sequence $(\varphi_n; n \geq 0)$ in E_0, $(\varphi_0, \mathcal{S}, \mathcal{T})$-*orbital* provided

$$\mathcal{T}\varphi_n \ni \mathcal{S}\varphi_{n+1}, \text{ for all } n \geq 0.$$

(com-1): Given the $(\varphi_0, \mathcal{S}, \mathcal{T})$-orbital sequence $(\varphi_n = Kx_n; n \geq 0)$ in $\mathcal{K}$, we have (cf. our previous constructions)

$$Tx_n = \mathcal{T}\varphi_n \ni \mathcal{S}\varphi_{n+1} = \mathcal{S}(Kx_{n+1}) = x_{n+1}, \forall n \geq 0,$$

so that the supporting sequence $(x_n; n \geq 0)$ in E is a x_0-orbital one, according to the formula $(Tx_n \ni x_{n+1}; n \geq 0)$.

(com-2): Conversely, given the x_0-orbital sequence $(x_n; n \geq 0)$ in E, the functional sequence $(\varphi_n := Kx_n; n \geq 0)$ in $\mathcal{K}$ is $(\varphi_0, \mathcal{S}, \mathcal{T})$-orbital, via

$$\mathcal{T}\varphi_n = \mathcal{T}(Kx_n) = Tx_n \ni x_{n+1} = \mathcal{S}(Kx_{n+1}) = \mathcal{S}\varphi_{n+1}, \forall n.$$

Under these preliminaries, we may now establish our main results in this exposition. Let $\mathcal{S} : E_0 \to E$ and $\mathcal{T} : E_0 \to cb(E)$ be as before.

(I) Let us construct a system of applications as: for $\varphi, \xi \in E_0$,

$$\begin{aligned}
A^*(\varphi,\xi) &= \min\{\Delta(\mathcal{T}\varphi,\mathcal{T}\xi), d(\mathcal{S}\xi,\mathcal{T}\xi)\};\\
B_1^*(\varphi,\xi) &= (1/2)[d(\mathcal{S}\varphi,\mathcal{T}\varphi) + d(\mathcal{S}\xi,\mathcal{T}\xi)];\\
B_2^*(\varphi,\xi) &= (1/2)[d(\mathcal{S}\varphi,\mathcal{T}\xi) + d(\mathcal{S}\xi,\mathcal{T}\varphi)];\\
B^*(\varphi,\xi) &= \max\{D(\varphi,\xi), B_1^*(\varphi,\xi), B_2^*(\varphi,\xi)\}.
\end{aligned}$$

Then, given the function $\Upsilon \in \mathcal{F}(R^0_+,]0,1[)$, let us introduce the concept

$(\mathcal{T},\mathcal{S})$ is $(A^*, B^*; \Upsilon)$-*contractive*:
$A^*(\varphi,\xi) \le \Upsilon(D(\varphi,\xi))B^*(\varphi,\xi)$, for all $\varphi, \xi \in E_0$, $D(\varphi,\xi) > 0$.

In particular, when $[\varphi = Kx,\ \xi = Ky$, for $x, y \in E]$, the introduced functions become (in view of our previous notations)

$$\begin{aligned}
A^*(Kx,Ky) &= \min\{\Delta(Tx,Ty), d(y,Ty)\} = A(x,y);\\
B_1^*(Kx,Ky) &= (1/2)[d(x,Tx) + d(y,Ty)] = B_1(x,y);\\
B_2^*(Kx,Ky) &= (1/2)[d(x,Ty) + d(y,Tx)] = B_2(x,y);\\
B^*(Kx,Ky) &= \max\{d(x,y), B_1(x,y), B_2(x,y)\} = B(x,y);
\end{aligned}$$

and the contractive condition becomes

T is $(A, B; \Upsilon)$-*contractive*:
$A(x,y) \le \Upsilon(d(x,y))B(x,y)$, for all $x, y \in E$, $d(x,y) > 0$.

The following fixed point theorem involving these data is available.

Theorem 9. *Let the normal map $\mathcal{S} : E_0 \to E$, the multivalued map $\mathcal{T}$ over (E_0, E), and the function $\Upsilon \in \mathcal{F}(R^0_+,]0,1[)$ be such that $(\mathcal{T},\mathcal{S})$ is $(A^*, B^*; \Upsilon)$-contractive and Υ is unitary upper strongly Matkowski admissible. Then, $(\mathcal{S},\mathcal{T})$ has a coincidence point in $\mathcal{K}$.*

Proof. Let T be associated with $\mathcal{T}$ mapping; clearly, T is multivalued over E. By the imposed conditions, Theorem 5 is applicable to these data, and then, the conclusion follows. □

Now, as precise, the unitary upper strongly Matkowski admissible property for $\Upsilon \in \mathcal{F}(R^0_+,]0,1[)$ is assured upon an MT-admissible property of the same. As a consequence, the following particular variant of this result is valid.

Theorem 10. *Let the normal map $\mathcal{S} : E_0 \to E$, the multivalued map $\mathcal{T}$ over (E_0, E), and the function $\Upsilon \in \mathcal{F}(R^0_+,]0,1[)$ be such that $(\mathcal{T}, \mathcal{S})$ is $(A^*, B^*; \Upsilon)$-contractive and Υ is MT-admissible. Then, $(\mathcal{S}, \mathcal{T})$ has a coincidence point in $\mathcal{K}$.*

(II) Let the normal couple (F, G) over $\mathcal{F}(R^0_+, R)$ be given. Two contractive concepts appear.

(II-1) Let us say that $(\mathcal{T}, \mathcal{S})$ is $(\Delta, d; F, G)$-*contractive*, provided

(TS-211) $F(\Delta(\mathcal{T}\varphi, \mathcal{T}\xi)) \leq F(d(\mathcal{S}\varphi, \mathcal{S}\xi)) - G(d(\mathcal{S}\varphi, \mathcal{S}\xi))$,
whenever $\Delta(\mathcal{T}\varphi, \mathcal{T}\xi) > 0$, $d(\mathcal{S}\varphi, \mathcal{S}\xi) > 0$.

In particular, when $[\varphi = Kx,\ \xi = Ky$, for $x, y \in E]$, this contractive condition becomes (via $(\mathcal{S}, K) =$ invertible)

(T-212) $F(\Delta(Tx, Ty)) \leq F(d(x, y)) - G(d(x, y))$,
whenever $\Delta(Tx, Ty) > 0$, $d(x, y) > 0$,

so, the attached to $\mathcal{T}$ multivalued map $T : E \to cb(E)$ is $(\Delta, d; F, G)$-contractive.

(II-2) Let us say that $\mathcal{T}$ is $(\Delta, D; F, G)$-*contractive*, provided

(T-221) $F(\Delta(\mathcal{T}\varphi, \mathcal{T}\xi)) \leq F(D(\varphi, \xi)) - G(D(\varphi, \xi))$,
whenever $\Delta(\mathcal{T}\varphi, \mathcal{T}\xi) > 0$, $D(\varphi, \xi) > 0$.

In particular, when $[\varphi = Kx,\ \xi = Ky$, for $x, y \in E]$, this contractive condition becomes (in view of $K =$ isometry)

(T-222) $F(\Delta(Tx, Ty)) \leq F(d(x, y)) - G(d(x, y))$,
whenever $\Delta(Tx, Ty) > 0$, $d(x, y) > 0$,

so the attached to $\mathcal{T}$ multivalued map $T : E \to cb(E)$ is (again) endowed with the $(\Delta, d; F, G)$-contractive property.

Theorem 11. *Let the normal map $\mathcal{S} : E_0 \to E$, the multivalued map $\mathcal{T}$ over (E_0, E) and the normal couple (F, G) over $\mathcal{F}(R^0_+, R)$ be such that one of the following alternatives is holding:*

(i) *$(\mathcal{T}, \mathcal{S})$ is $(\Delta, d; F, G)$-contractive;*
(ii) *$\mathcal{T}$ is $(\Delta, D; F, G)$-contractive.*

In addition, suppose that

(iii) *F is right continuous and $(F, \lambda G)$ is admissible, for some $\lambda \in]0, 1[$.*

Then, $(\mathcal{S}, \mathcal{T})$ has a coincidence point in $\mathcal{K}$.

Proof. Let T be associated with $\mathcal{T}$ mapping; as precise, T is multivalued over E. By the imposed conditions, Theorem 7 is applicable to these data, and then, the conclusion follows. □

A basic particular case of this result is obtainable under the preceding observations concerning the admissible property. Precisely, we have the following:

Theorem 12. *Let the normal map $\mathcal{S} : E_0 \to E$, the multivalued map $\mathcal{T}$ over (E_0, E), the Wardowski right continuous $F \in \mathcal{F}(R^0_+, R)$ and the constant $C > 0$ be such that one of the following alternatives is holding:*

(i) *$(\mathcal{T}, \mathcal{S})$ is $(\Delta, d; F, C)$-contractive;*
(ii) *$\mathcal{T}$ is $(\Delta, D; F, C)$-contractive.*

Then, $(\mathcal{S}, \mathcal{T})$ has a coincidence point in $\mathcal{K}$.

Proof. Let T be associated to $\mathcal{T}$ mapping; clearly, T is multivalued over E. By the imposed conditions, Theorem 8 is applicable here, and the conclusion follows. □

In particular, when $\mathcal{T}$ is univalent and $\mathcal{S} = \mathcal{S}_c$ (see above), these main results are comparable with some contributions due to Agarwal *et al.* [2], Cho *et al.* [8], Ćirić *et al.* [10], Farajzadeh and Kaewcharoen [17], Farajzadeh *et al.* [18], Khan and Jhade [27], Kutbi *et al.* [30], Paknazar *et al.* [37], Parvaneh *et al.* [38], among many others. Further aspects will be delineated elsewhere.

References

[1] O. Acar, G. Durmaz, and G. Minak, Generalized multi-valued F-contractions on complete metric spaces, *Bull. Iran. Math. Soc.*, 40, 2014, 1469–1478.

[2] R. P. Agarwal, P. Kumam and W. Sintunavarat, PPF dependent fixed point theorems for an α_c-admissible non-self mapping in the Razumikhin class, *Fixed Point Th. Appl.*, 2013, 2013, 280.

[3] A. Amini-Harandi, Fixed and coupled fixed points of a new type of set-valued contractive mappings in complete metric spaces, *Fixed Point Th. Appl.*, 2012, 2012, 215.

[4] S. Banach, Sur les opérations dans les ensembles abstraits et leur application aux équations intégrales, *Fund. Math.*, 3, 1922, 133–181.

[5] S. R. Bernfeld, V. Lakshmikantham, and Y. M. Reddy, Fixed point theorems of operators with PPF dependence in Banach spaces, *Applicable Anal.*, 6, 1977, 271–280.

[6] F. Bojor, Fixed points of Kannan mappings in metric spaces endowed with a graph, *An. Şt. Univ. Constanţa (Mat.),* 20, 2012, 31–40.

[7] D. W. Boyd and J. S. W. Wong, On nonlinear contractions, *Proc. Amer. Math. Soc.*, 20, 1969, 458–464.

[8] Y. J. Cho, S. M. Kang, and P. Salimi, Some PPF dependent fixed point theorems for generalized $\alpha - F$-contractions in Banach spaces and applications, *Mathematics,* 6, 2018, 267.

[9] P. Chuasuk and A. Kaewcharoen, Fixed point results for generalized rational type contractions with PPF dependence, *Int. J. Pure Appl. Math.*, 96, 2014, 197–212.

[10] L. B. Ćirić, S. M. Alsulami, P. Salimi and P. Vetro, PPF dependent fixed point results for triangular α_c-admissible mappings, *The Sci. World J.*, 2014, 2014, Article ID 673647.

[11] P. J. Cohen, *Set Theory and the Continuum Hypothesis*, (Benjamin, New York, 1966).

[12] B. C. Dhage, Fixed point theorems with PPF dependence and functional differential equations, *Fixed Point Th.*, 13, 2012, 439–452.

[13] B. Dhage, On some common fixed point theorems with PPF dependence in Banach spaces, *J. Nonlin. Sci. Appl.*, 5, 2012, 220–232.

[14] Z. Drici, F. A. McRae, and J. Vasundara Devi, Fixed point theorem in partially ordered metric spaces for operators with PPF dependence, *Nonlin. Anal.*, 67, 2007, 641–647.

[15] W.-S. Du, Simultaneous generalizations of fixed point theorems of Mizoguchi-Takahashi type, Nadler type, Banach type, Kannan type, and Chatterjea type, *Nonlin. Anal. Diff. Eqs.*, 5, 2017, 171–180.

[16] P. N. Dutta and B. S. Choudhury, A generalisation of contraction principle in metric spaces, *Fixed Point Th. Appl.*, 2008, 2008, Article ID 406368.

[17] A. Farajzadeh and A. Kaewcharoen, On fixed point theorems for mappings with PPF dependence, *J. Ineq. Appl.*, 2014, 2014, 372.

[18] A. Farajzadeh, A. Kaewcharoen, and S. Plubtieng, PPF dependent fixed point theorems for multi-valued mappings in Banach spaces, *Bull. Iranian Math. Soc.*, 42, 2016, 1583–1596.
[19] H. A. Hammad and M. De la Sen, PPF dependent fixed point results for new multi-valued generalized F-contraction in the Razumikhin class with an application, *Mathematics,* 7, 2019, 52.
[20] P. Hitzler, *Generalized metrics and topology in logic programming semantics, PhD Thesis*, (Natl. Univ. Ireland, Univ. College Cork, 2001).
[21] N. Hussain, S. Khaleghizadeh, P. Salimi, and F. Akbar, New fixed point results with PPF dependence in Banach spaces endowed with a graph, *Abstr. Appl. Anal.*, 2013, 2013, Article ID 827205.
[22] J. Jachymski, The contraction principle for mappings on a metric space with a graph, *Proc. Amer. Math. Soc.*, 136, 2008, 1359–1373.
[23] M. Javahernia, A. Razani, and F. Khojasteh, Fixed point of multi-valued contractions via manageable functions and Liu's generalization, *Cogent Math.*, 3, 2016, 1276818.
[24] A. Kaewcharoen, PPF dependent common fixed point theorems for mappings in Banach spaces, *J. Inequal. Appl.*, 2013, 2013, 287.
[25] E. Karapinar, A. Fulga, and R. P. Agarwal, A survey: F-contractions with related fixed point results, *J. Fixed Point Th. Appl.*, 2020, 20–69.
[26] S. Kasahara, On some generalizations of the Banach contraction theorem, *Publ. Res. Inst. Math. Sci. Kyoto Univ.*, 12, 1976, 427–437.
[27] M. S. Khan and P. K. Jhade, On a fixed point theorem with PPF dependence in the Razumikhin class, *Gazi Univ. J. Science,* 28, 2015, 211–219.
[28] M. A. Krasnoselskii and V. Ya. Stetsenko, Towards a theory of equations with concave operators (Russian), *Sib. Math. Zh.*, 10, 1969, 565–572.
[29] K. Kuratowski, *Topology,* (Academic Press, New York, 1966).
[30] M. A. Kutbi, N. Hussain, and S. Khaleghizadeh, New PPF-dependent fixed point theorems for Suzuki type GF-contractions, *J. Function Spaces,* 2015, 2015, Article ID 136306.
[31] M. A. Kutbi and W. Sintunavarat, On sufficient conditions for the existence of past-present-future dependent fixed point in the Razumikhin class and application, *Abstr. Appl. Anal.*, 2014, 2014, Article ID 342687.
[32] S. Leader, Fixed points for general contractions in metric spaces, Math. *Japonica,* 24, 1979, 17–24.
[33] J. Matkowski, Integrable solutions of functional equations, *Dissert. Math.*, Vol. 127. Polish Sci. Publ., Warsaw, 1975.
[34] A. Meir and E. Keeler, A theorem on contraction mappings, *J. Math. Anal. Appl.*, 28, 1969, 326–329.

[35] S. B. Nadler Jr., Multi-valued contraction mappings, *Pacific J. Math.*, 30, 1969, 475–488.
[36] J. J. Nieto and R. Rodriguez-Lopez, Contractive mapping theorems in partially ordered sets and applications to ordinary differential equations, *Order*, 22, 2005, 223–239.
[37] M. Paknazar, M. A. Kutbi, M. Demma, and P. Salimi, PPF dependent fixed point in modified Razumikhin class with applications, *J. Nonlin. Sci. Appl.*, 9, 2016, 270–286.
[38] V. Parvaneh, H. Hosseinzadeh, N. Hussain, and L. B. Ćirić, PPF-dependent fixed point results for hybrid rational and Suzuki-Edelstein type contractions in Banach spaces, *Filomat*, 30, 2016, 1339–1351.
[39] M. Piticari, *Successive approximations method and fixed point principle* (Romanian), *Diss. Thesis*, (A. I. Cuza Univ., Iaşi, 1982).
[40] A. C. M. Ran and M. C. Reurings, A fixed point theorem in partially ordered sets and some applications to matrix equations, *Proc. Amer. Math. Soc.*, 132, 2004, 1435–1443.
[41] S. Reich, Fixed points of contractive functions, *Boll. Un. Mat. Ital.*, 5, 1972, 26–42.
[42] B. E. Rhoades, A comparison of various definitions of contractive mappings, *Trans. Amer. Math. Soc.*, 226, 1977, 257–290.
[43] B. E. Rhoades, Some theorems on weakly contractive maps, *Nonlin. Anal.*, 47, 2001, 2683–2693.
[44] A.-F. Roldán, E. Karapinar, C. Roldán, and J. Martinez-Moreno, Coincidence point theorems on metric spaces via simulation method, *Fixed Point Th. Appl.*, 2015, 2015, 98.
[45] I. A. Rus, *Generalized Contractions and Applications*, (Cluj University Press, Cluj-Napoca, 2001).
[46] B. Samet and M. Turinici, Fixed point theorems on a metric space endowed with an arbitrary binary relation and applications, *Commun. Math. Anal.*, 13, 2012, 82–97.
[47] M. Sgroi and C. Vetro, Multi-valued F-contractions and the solution of certain functional and integral equations, *Filomat*, 27, 2013, 1259–1268.
[48] V. Timofte, New tests for positive iteration series, *Real Analysis Exchange*, 30, 2004/2005, 799–812.
[49] M. Turinici, Fixed points of implicit contraction mappings, *An. Şt. Univ. "A. I. Cuza" Iaşi (S I-a, Mat)*, 22, 1976, 177–180.
[50] M. Turinici, Fixed points for monotone iteratively local contractions, *Dem. Math.*, 19, 1986, 171–180.
[51] M. Turinici, Ran-Reurings theorems in ordered metric spaces, *J. Indian Math. Soc.*, 78, 2011, 207–214.

[52] M. Turinici, PPF dependent fixed points in A-closed Razumikhin classes, *Romai J.*, 10(No. 1), 2014, 205–220.

[53] M. Turinici, Implicit contractive maps in ordered metric spaces, Topics in Mathematical Analysis and Applications. T. M. Rassias and L. Tóth (eds.), (Springer Intl. Publ., Switzerland, 2014), pp. 715–746.

[54] M. Turinici, *Contraction maps in pseudometric structures, Essays in Mathematics and its Applications.* T. M. Rassias and P. M. Pardalos (eds.), (Springer Intl. Publ., Switzerland, 2016), pp. 513–562.

[55] M. Turinici, *Meir-Keeler sequential contractions and applications, Differential and Integral Inequalities.* D. Andrica and T. M. Rassias (eds.), (Springer Nature, Switzerland, 2019), pp. 761–829.

[56] M. Turinici, Reports in Metrical Fixed Point Theory, (Pim Editorial House, Iaşi, 2020).

[57] J. Vujaković, S. Mitrović, M. Pavlović and S. Radenović, On recent results concerning F-contraction in generalized metric spaces, *Mathematics*, 8, 2020, 767.

[58] D. Wardowski, Fixed points of a new type of contractive mappings in complete metric spaces, *Fixed Point Th. Appl.*, 2012, 2012, 94.

[59] D. Wardowski, Solving existence problems via F-contractions, *Proc. Amer. Math. Soc.*, 146, 2018, 1585–1598.

https://doi.org/10.1142/9789819820498_0022

Chapter 22

Multiple Coercivity over Quasi-Ordered Quasi-Metric Spaces

Mihai Turinici
A. Myller Mathematical Seminar, A. I. Cuza University, 700506 Iaşi, Romania
mturi@uaic.ro

A quasi-ordered quasi-metric multiple coercivity result is established for non-smooth functionals fulfilling Palais–Smale conditions. The core of this approach is an asymptotic statement involving such objects, obtainable from a quasi-ordered quasi-metric variational principle comparable with the related one in Turinici [Romai J., 2011, 7(2), 169–179].

Keywords: Quasi-ordered quasi-metric space, local maximality principle, multiple coercive functional, slope, Palais–Smale condition, directional derivative.

1. Introduction

Let $(X, \|.\|)$ be a (real) Banach space. Given a (proper) functional $F \in \mathcal{F}(X, R \cup \{\infty\})$, we say that it is *coercive*, provided

(coer) $F(u) \to \infty$, whenever $u \to \infty$ (in the sense: $\|u\| \to \infty$).

[Here, for each couple A, B of nonempty sets, $\mathcal{F}(A, B)$ denotes the class of all functions from A to B; when $A = B$, we write $\mathcal{F}(A)$ in place of $\mathcal{F}(A, A)$.] Sufficient conditions for such a property are obtainable in a *differential* setting, by means of the celebrated 1964 Palais–Smale technique [36]. A typical result in this direction is due to Caklovic *et al.* [11]; it states that, whenever

(G-lsc) F is Gâteaux differentiable and lower semicontinuous (lsc),

the relation (coer) is deductible under a Palais–Smale requirement such as

(PS) each sequence (v_n) in X with $(F(v_n))$=bounded and $F'(v_n) \to 0$ (in X^*) has a convergent (in X) subsequence.

(Here, $(X^*, \|.\|)$ is the *topological dual* of X.) Note that (G-lsc) holds when $F \in C^1(X)$; hence, their statement includes Brezis–Nirenberg's [8]. An extension of it was obtained by Goeleven [20], under (G-lsc) substituted by the representation

(G-lsc-conv) $F = F_1 + F_2$,
where F_1 is Gâteaux differentiable lsc and F_2 is (proper) convex lsc,

and the Palais–Smale condition (PS) being adapted to this decomposition. Further enlargements of (G-lsc-conv) were given by Motreanu and Motreanu [31], under the general decomposition

(loc-L-conv) $F = F_1 + F_2$, where F_1 is locally Lipschitz (hence continuous), and F_2 is (proper) convex lsc,

and a Palais–Smale condition such as in Motreanu and Panagiotopoulos [34, Chapter 3]. Finally, some other statements may be found in Bae *et al.* [3] and Xu [60]; see also Turinici [47–49, 53].

The basic tools of all these are, essentially, Ekeland Variational Principle [19] (in short: (EVP)) and Zhong Variational Principle [61] (in short: (ZVP)), and as such, F must be lsc (on X). So, we may ask whether this is removable; an appropriate answer is available in a quasi-order quasi-metrical context. As far as we know, the first result in this direction is the 2002 one due to Motreanu and Turinici [35], based on the monotone Ekeland Variational Principle in Turinici [45]; some related developments of this theory are to be found in Motreanu *et al.* [33] and Turinici [51, 54]. It is our aim in the following to show that further refinements of this result to such structures are possible. The basic tool of these investigations is an asymptotic-type statement involving such functionals, obtained via "local" quasi-order quasi-metric versions of Ekeland Variational Principle comparable with the one in Turinici [55]. And the specific tool is a quasi-order quasi-metric version of the *slope* concept introduced by DeGiorgi *et al.* [17]. Note, finally, that multivalued and functional extensions of these results are available — under the described methods — by following the lines in Turinici [52, 57]; further aspects will be delineated elsewhere.

2. Dependent Choice Principles

Throughout this exposition, the axiomatic system in use is Zermelo–Fraenkel's (abbreviated: (ZF)), as described by Cohen [13, Chapter 2]. The notations and basic facts to be considered in this system are more or less standard. Some important ones are discussed in the following.

(A) Let X be a non-empty set. By a *relation* over X, we mean any (non-empty) part $\mathcal{R}$ of $X \times X$; then, $(X, \mathcal{R})$ will be referred to as a *relational structure*. For simplicity, we sometimes write $(x, y) \in \mathcal{R}$ as $x\mathcal{R}y$. Note that $\mathcal{R}$ may be regarded as a mapping between X and $\exp[X]$ (=the class of all subsets in X). In fact, denote

$$X(x, \mathcal{R}) = \{y \in X; x\mathcal{R}y\} \text{ (the } \textit{section} \text{ of } \mathcal{R} \text{ through } x), \ x \in X;$$

then, the desired mapping representation is $[\mathcal{R}(x) = X(x, \mathcal{R}), x \in X]$. Call $\mathcal{R}$, *proper* when $\mathcal{R}(x) \neq \emptyset$, for all $x \in X$. Note that, in such a case, $\mathcal{R}$ appears as a mapping between X and $\exp(X)$ (= the class of all nonempty parts in X); this will be also referred to as $(X, \mathcal{R})$ is a *proper* relational structure. A basic example of such an object is $\mathcal{I} = \{(x, x); x \in X\}$ (the *identical relation* over X).

Given the relations $\mathcal{R}$, $\mathcal{S}$ over X, define their *product* $\mathcal{R} \circ \mathcal{S}$ as

$$(x, z) \in \mathcal{R} \circ \mathcal{S}, \text{ if there exists } y \in X \text{ with } (x, y) \in \mathcal{R}, \ (y, z) \in \mathcal{S}.$$

Also, for each relation $\mathcal{R}$ in X, denote

$$\mathcal{R}^{-1} = \{(x, y) \in X \times X; (y, x) \in \mathcal{R}\} \text{ (the } \textit{inverse} \text{ of } \mathcal{R}).$$

Finally, given the relations $\mathcal{R}$ and $\mathcal{S}$ on X, let us say that $\mathcal{R}$ is *coarser* than $\mathcal{S}$ (or, equivalently: $\mathcal{S}$ is *finer* than $\mathcal{R}$), provided

$\mathcal{R} \subseteq \mathcal{S}$; which means: $x\mathcal{R}y$ implies $x\mathcal{S}y$.

Given a relation $\mathcal{R}$ on X, the following properties are to be discussed here:

(P1) $\mathcal{R}$ is *reflexive*: $\mathcal{I} \subseteq \mathcal{R}$;
(P2) $\mathcal{R}$ is *irreflexive*: $\mathcal{R} \cap \mathcal{I} = \emptyset$;
(P3) $\mathcal{R}$ is *transitive*: $\mathcal{R} \circ \mathcal{R} \subseteq \mathcal{R}$;
(P4) $\mathcal{R}$ is *symmetric*: $\mathcal{R}^{-1} = \mathcal{R}$;
(P5) $\mathcal{R}$ is *antisymmetric*: $\mathcal{R}^{-1} \cap \mathcal{R} \subseteq \mathcal{I}$.

This yields the classes of relations to be used; the following ones are important for our developments:

(C1) $\mathcal{R}$ is a *quasi-order* (reflexive and transitive);
(C2) $\mathcal{R}$ is a *strict order* (irreflexive and transitive);
(C3) $\mathcal{R}$ is an *equivalence* (reflexive, transitive, symmetric);
(C4) $\mathcal{R}$ is a *(partial) order* (reflexive, transitive, antisymmetric);
(C5) $\mathcal{R}$ is *trivial* (which means: $\mathcal{R} = X \times X$).

A basic example of relational structure is to be constructed as follows. Let $N = \{0, 1, \cdots\}$ be the set of *natural* numbers, endowed with the usual addition and (partial) order; note that

$(N, \leq)$ is well ordered: any (nonempty) subset of N has a first element.

Further, denote for $p, q \in N$, $p \leq q$,

$$N[p, q] = \{n \in N; p \leq n \leq q\},\ N]p, q[= \{n \in N; p < n < q\},$$
$$N[p, q[= \{n \in N; p \leq n < q\},\ N]p, q] = \{n \in N; p < n \leq q\},$$

as well as, for $r \in N$:

$$N[r, \infty[= \{n \in N; r \leq n\},\ N]r, \infty[= \{n \in N; r < n\}.$$

For each $r \geq 1$, $N[0, r[= N(r, >)$ is referred to as the *initial interval* (in N) induced by r. Any set P with $P \sim N$ (in the sense: there exists a bijection from P to N) will be referred to as *effectively denumerable.* In addition, given some natural number $n \geq 1$, any (non-empty) set Q with $Q \sim N(n, >)$ will be said to be n-*finite*; when n is generic here, we say that Q is *finite*. Finally, the (non-empty) set Y is called (at most) *denumerable*, iff it is either effectively denumerable or finite.

Let X be a non-empty set. By a *sequence* in X, we mean any mapping $x : N \to X$, where $N = \{0, 1, \ldots\}$ is the set of *natural* numbers. For simplicity reasons, it will be useful to denote it as $(x(n); n \geq 0)$, or $(x_n; n \geq 0)$; moreover, when no confusion can arise, we further simplify this notation as $(x(n))$ or (x_n), respectively. Also, any sequence $(y_n := x_{i(n)}; n \geq 0)$ with

$(i(n); n \geq 0)$ is *strictly ascending* (hence, $i(n) \to \infty$ as $n \to \infty$)

will be referred to as a *subsequence* of $(x_n; n \geq 0)$. Note that, under such a convention, the relation "subsequence of" is transitive; i.e.

(z_n)=subsequence of (y_n) and (y_n)=subsequence of (x_n)
imply (z_n)=subsequence of (x_n).

(B) Remember that, an outstanding part of (ZF) is the *Axiom of Choice* (abbreviated: (AC)); which, in a convenient manner, may be written as follows:

(AC) For each couple (J, X) of non-empty sets and each function $F : J \to \exp(X)$, there exists a *selective function* $f : J \to X$ (characterized as: $f(\nu) \in F(\nu)$, for each $\nu \in J$).

Sometimes, when the index set J is denumerable, the existence of such a selective function may be determined by using a weaker form of (AC), called *Dependent Choice* principle (in short: (DC)). Some preliminaries are needed. For each natural number $k \geq 1$, call the map $F : N(k, >) \to X$, a *k-sequence*; if $k \geq 1$ is generic, we talk about a *finite* sequence. The following result, referred to as the *Finite Dependent Choice property* (in short: (DC-fin)), is available in the *strongly reduced* Zermelo–Fraenkel system (ZF-AC). Given $a \in X$, let us say that the k-sequence $F : N(k, >) \to X$ (where $k \geq 2$) is *$(a, \mathcal{R})$-iterative*, provided

$F(0) = a$ and $F(i)\mathcal{R}F(i + 1)$, for all $i \in N(k - 1, >)$.

Proposition 1. *Let the relational structure $(X, \mathcal{R})$ be proper. Then, for each $k \geq 2$, the following property holds:*

(P(k)) for each $a \in X$, there exists an $(a, \mathcal{R})$-iterative k-sequence.

Proof. Clearly, $(P(2))$ is true; just take $b \in \mathcal{R}(a)$ and define $F : N(2,>) \to X$ as $F(0) = a$, $F(1) = b$. Assume that $(P(k))$ is true, for some $k \geq 2$; we claim that $(P(k+1))$ is true as well. In fact, let $F : N(k,>) \to X$ be an $(a,\mathcal{R})$-iterative k-sequence, assured by hypothesis. As $\mathcal{R}$ is proper, $\mathcal{R}(F(k-1))$ is non-empty; let u be some element of it. The map $G : N(k+1,>) \to X$ introduced as

$$G(i) = F(i), i \in N(k,>);\ G(k) = u$$

is an $(a,\mathcal{R})$-iterative $(k+1)$-sequence, and then, we are done. □

Now, it is natural to see what happens when k "tends to infinity". At a first glance, the following *Dependent Choice* principle (in short: (DC)) is obtainable in (ZF-AC) from this "limit" process. Given $a \in X$, let us say that the sequence $(x_n; n \geq 0)$ in X is $(a;\mathcal{R})$-*iterative*, provided

$$x_0 = a \text{ and } [x_n \mathcal{R} x_{n+1} \text{ (meaning: } x_{n+1} \in \mathcal{R}(x_n)),\ \forall n].$$

Proposition 2. *Let the relational structure $(X,\mathcal{R})$ be proper. Then, for each $a \in X$, there is at least an $(a,\mathcal{R})$-iterative sequence in X.*

Formally, the "argument" involved here consists in the possibility of constructing an (infinite) sequence via finite sequences. This, ultimately, cannot be done under the precise context, so the limit process in question does not work in our strongly reduced system (ZF-AC); whence, (DC) is not obtainable from its axioms. On the other hand, this principle — proposed, independently, by Bernays [5] and Tarski [38] — is deductible from (AC) but not conversely; cf. Wolk [59]. Denote, for simplicity,

(ZF − AC + DC) (the *reduced Zermelo-Fraenkel system*)
=the strongly reduced Zermelo-Fraenkel system (ZF-AC),
completed with the Dependent Choice principle (DC).

According the developments in Moskhovakis [30, Chapter 8] and Schechter [37, Chapter 6], the *reduced* system (ZF − AC + DC) is large enough so as to cover the "usual" mathematics; see also Moore [29, Appendix 2, Table 4]. So, determining various equivalents of (DC) may be not without profit.

(C) Let $(\mathcal{R}_n; n \geq 0)$ be a sequence of relations on X. Given $a \in X$, let us say that the sequence $(x_n; n \geq 0)$ in X is $(a; (\mathcal{R}_n; n \geq 0))$-*iterative*, provided

$$x_0 = a \text{ and } [x_n \mathcal{R}_n x_{n+1} \text{ (meaning: } x_{n+1} \in \mathcal{R}_n(x_n)), \forall n].$$

The following *Diagonal Dependent Choice* principle (in short: (DDC)) is available.

Proposition 3. *Let $(\mathcal{R}_n; n \geq 0)$ be a sequence of proper relations on X. Then, for each $a \in X$, there exists an $(a; (\mathcal{R}_n; n \geq 0))$-iterative sequence in X.*

Clearly, (DDC) includes (DC) to which it reduces when $(\mathcal{R}_n; n \geq 0)$ is constant. The reciprocal of this is also true. In fact, letting the premises of (DDC) hold, put $P = N \times X$, and let $\mathcal{S}$ be the relation over P introduced as

$$\mathcal{S}(i, x) = \{i + 1\} \times \mathcal{R}_i(x), \ \ (i, x) \in P.$$

It will suffice applying (DC) to $(P, \mathcal{S})$ and $b := (0, a) \in P$ to get the conclusion in the statement; we do not give details.

Summing up, (DDC) is provable in (ZF − AC + DC). This is valid as well for its variant, referred to as the *Selected Dependent Choice* principle (in short: (SDC)).

Proposition 4. *Let the map $F : N \to \exp(X)$ and the relation $\mathcal{R}$ over X fulfill*

$$(\forall n \in N)\colon \mathcal{R}(x) \cap F(n+1) \neq \emptyset, \ \ \forall x \in F(n).$$

Then, for each $a \in F(0)$, there exists a sequence $(x(n); n \geq 0)$ in X, with

$$x(0) = a, \ x(n) \in F(n), \ x(n+1) \in \mathcal{R}(x(n)), \ \forall n.$$

As before, (SDC) $\Longrightarrow$ (DC) ($\Longleftrightarrow$ (DDC)); just take $(F(n) = X; n \geq 0)$. But, the reciprocal is also true, in the sense (DDC) $\Longrightarrow$ (SDC). This follows from the following:

Proof. (Proposition 4) Let the premises of (SDC) be true. Define a sequence of relations $(\mathcal{R}_n; n \geq 0)$ over X as follows: for each $n \geq 0$,

$$\mathcal{R}_n(x) = \mathcal{R}(x) \cap F(n+1), \ \ \text{if } x \in F(n);$$
$$\mathcal{R}_n(x) = \{x\}, \text{ otherwise } (x \in X \setminus F(n)).$$

Clearly, $\mathcal{R}_n$ is proper for all $n \geq 0$. So, by (DDC), we have that, for the starting $a \in F(0)$, there exists an $(a, (R_n; n \geq 0))$-iterative sequence $(x(n); n \geq 0)$ in X. Combining with the very definition above gives the conclusion. □

In particular, when $\mathcal{R} = X \times X$, the regularity condition imposed in (SDC) holds. The corresponding variant of the underlying statement is just (AC(N)) (=the *Denumerable Axiom of Choice*). Precisely, we have the following:

Proposition 5. *Let $F : N \to \exp(X)$ be a function. Then, for each $a \in F(0)$ there exists a function $f : N \to X$ with $f(0) = a$ and $(f(n) \in F(n), \forall n)$.*

As a consequence of these, (DC) $\Longrightarrow$ (AC(N)) in (ZF-AC). A direct verification is obtainable by an application of (DC) to the relational structure $(Q, \mathcal{T})$, where

$$Q = N \times X,\ \mathcal{T}(n, x) = \{n + 1\} \times F(n + 1),\ \ n \in N,\ x \in X;$$

we do not give details. The reciprocal inclusion is not true; see, for instance, Moskhovakis [30, Chapter 8, Section 8.25].

3. Pseudometric Structures

In the following, some preliminary facts about convergent and Cauchy sequences in pseudometric spaces are being discussed.

(A) Let X be a non-empty set. By a *pseudometric* over X, we mean any map $d : X \times X \to R_+$. For an easy reference, we list the conditions to be used:

(ref) d is *reflexive*: $d(x, x) = 0$, for each $x \in X$;
(suf) d is *sufficient*: $x, y \in X$, $d(x, y) = 0$ imply $x = y$;
(F-tri) d is *Frink triangular*:
$\forall \varepsilon > 0, \exists \delta > 0$, such that: $d(x, y), d(y, z) < \delta$ imply $d(x, z) < \varepsilon$;
(B-tri) d is *Bakhtin triangular*:
there exists $\beta \geq 1$, such that $d(x, z) \leq \beta(d(x, y) + d(y, z))$, $\forall x, y, z \in X$;
(tri) d is *triangular*: $d(x, z) \leq d(x, y) + d(y, z)$, $\forall x, y, z \in X$;
(sym) d is *symmetric*: $d(x, y) = d(y, x)$, for all $x, y \in X$.

For the moment, we only assume that the reflexive property holds. Then, d is called a *r-pseudometric* on X, and (X, d) is referred to as a *r-pseudometric space.*

Define a d-convergence and a d-Cauchy structure on X as follows. Let us say that the sequence (x_n) in X, *d-converges* to $x \in X$ (and write: $x_n \xrightarrow{d} x$), when

$d(x_n, x) \to 0$ as $n \to \infty$; that is,
$\forall \varepsilon > 0, \exists p = p(\varepsilon), \forall n$: $(p \leq n \Longrightarrow d(x_n, x) < \varepsilon)$; or, equivalently,
$\forall \varepsilon > 0, \exists p = p(\varepsilon), \forall n$: $(p \leq n \Longrightarrow d(x_n, x) \leq \varepsilon)$.

Then, x is called a *d-limit* of (x_n); the set of all these will be denoted as $d - \lim_n(x_n)$ [or, $\lim_n(x_n)$, when d is understood]; if such elements exist, we say that (x_n) is *d-convergent.* Clearly, the introduced convergence $(\xrightarrow{d})$ has the following properties:

(conve-1) ($(\xrightarrow{d})$ is *hereditary*)
if $x_n \xrightarrow{d} x$, then $y_n \xrightarrow{d} x$, for each subsequence (y_n) of (x_n);
(conve-2) ($(\xrightarrow{d})$ is *reflexive*)
$\forall u \in X$, the constant sequence $(x_n = u; n \geq 0)$ fulfills $x_n \xrightarrow{d} u$;

so, it fulfills the general requirements in Kasahara [26].

Let us say that the sequence $(x_n; \geq 0)$ in X is *d-Cauchy*, when

$d(x_m, x_n) \to 0$ as $m, n \to \infty$, $m \leq n$; that is,
$\forall \varepsilon > 0, \exists q = q(\varepsilon), \forall (m, n)$: $(q \leq m \leq n \Longrightarrow d(x_m, x_n) < \varepsilon)$;
or, equivalently,
$\forall \varepsilon > 0, \exists q = q(\varepsilon), \forall (m, n)$: $(q < m < n \Longrightarrow d(x_m, x_n) \leq \varepsilon)$.

The class of all these will be indicated as $Cauchy(d)$; some basic properties of it are described in the following:

(Cauchy-1) ($Cauchy(d)$ is *hereditary*)
(x_n) is d-Cauchy implies (y_n) is d-Cauchy,
for each subsequence (y_n) of (x_n)
(Cauchy-2) ($Cauchy(d)$ is *reflexive*)
$\forall u \in X$, the constant sequence $(x_n = u; n \geq 0)$ is d-Cauchy,

so this concept fulfills the general requirements in Turinici [50].

Finally, call the couple $((\xrightarrow{d}), Cauchy(d))$, a *conv-Cauchy structure* induced by d. The following optional conditions about this structure (that, ultimately, appear as conditions upon d) are to be considered:

(conv-Cauchy-1) d is *regular*:
each d-convergent sequence in X is d-Cauchy;
(conv-Cauchy-2) d is *complete*:
each d-Cauchy sequence in X is d-convergent.

Note that, by the ambient r-pseudometric setting, neither of these properties is attainable, in general.

(B) Let (X, d) be a r-pseudometric space. In the following, an *adherence* concept over $\exp[X]$ will be introduced, as well as its associated *closure operator*.

Given $Y \in \exp[X]$ and $w \in X$, let us define the couple of properties:

(adhe-1) w is *d-sequentially adherent* to Y when
$w \in d - \lim_n(y_n)$, for some sequence (y_n) of Y;
(adhe-2) w is *d-non-sequentially adherent* to Y when
for each $\varepsilon > 0$, $Y(\varepsilon, w) := \{y \in Y; d(y, w) < \varepsilon\}$ is non-empty.

Proposition 6. *We have, in (ZF* $-$ *AC* $+$ *DC), for* Y *and* w *as before*

d-sequentially adherent is equivalent with d-non-sequentially adherent.

Proof. The left to right inclusion is clear, by the very definition of convergence. Concerning the right to left inclusion, let $(\varepsilon_n; n \geq 0)$ be a strictly descending sequence in R_+^0 with $\varepsilon_n \to 0$ as $n \to \infty$, and put $(G(n) = Y(\varepsilon_n, w); n \geq 0)$; this is a mapping from N to $\exp(X)$. By the Denumerable Axiom of Choice (AC(N)) (deductible in (ZF $-$ AC $+$ DC)), there exists a selective function $g : N \to X$ with $(g(n) \in G(n), \forall n)$. It is now clear that $(y_n = g(n); n \geq 0)$ is a sequence in Y with $w \in d - \lim_n(y_n)$, and we are done. □

By definition, the class of all *d-adherent* points $w \in X$ (in the sense of (adhe-1) or (adhe-2)) will be denoted as $\text{cl}(Y)$ and referred to as the *closure* (or, *adherence*) of Y. The selfmap $Y \mapsto \text{cl}(Y)$ of $\exp[X]$ is a *semi-closure* over X, in the following sense:

(cl-1) $\text{cl}(\emptyset) = \emptyset$, $\text{cl}(X) = X$;
(cl-2) $Y \subseteq Z$ implies $\text{cl}(Y) \subseteq \text{cl}(Z)$;

(cl-3) $\mathrm{cl}(U \cup V) = \mathrm{cl}(U) \cup \mathrm{cl}(V)$, $U, V \in \exp[X]$;
(cl-4) $Y \subseteq \mathrm{cl}(Y)$, for each $Y \in \exp[X]$.

Unfortunately, $Y \mapsto \mathrm{cl}(Y)$ is not involutive; i.e.

(cl-inv) $\mathrm{cl}(\mathrm{cl}(Y)) = \mathrm{cl}(Y)$, for each $Y \in \exp[X]$

is not in general true so that $Y \mapsto \mathrm{cl}(Y)$ is not a closure over X as in Kuratowski [27, Chapter I, Section 4]. Further, let us say that $Y \in \exp[X]$ is *d-closed*, provided

the d-limit of each sequence in Y is included in Y.

It is not hard to see that

(cl-char) (for each $Y \in \exp[X]$): Y is d-closed iff $Y = \mathrm{cl}(Y)$.

In fact, the right to left inclusion is clear, by definition. For the left to right inclusion, let $w \in \mathrm{cl}(Y)$ be arbitrary fixed. By definition,

$w \in d - \lim_n(y_n)$, for some sequence (y_n) of Y;

and this, by the imposed condition, yields $w \in Y$. By the arbitrariness of our underlying point, one gets

$Y \supseteq \mathrm{cl}(Y)$; hence, $Y = \mathrm{cl}(Y)$,

proving our assertion.

Returning to the involutive question above, we may ask of under which extra conditions upon our r-pseudometric is this property valid. The most general one is

(F-tri) d is *Frink triangular*:
$\forall \varepsilon > 0$, $\exists \delta > 0$, such that $d(x, y), d(y, z) < \delta$, imply $d(x, z) < \varepsilon$.

We then say that d is a *Frink r-pseudometric* on X, and (X, d) is referred to as a *Frink r-pseudometric space.*

Proposition 7. *Let (X, d) be a Frink r-pseudometric space. Then, the following properties are valid:*

(32-1) $\mathrm{cl}(Y)$ *is d-closed, for each* $Y \in \exp[X]$;
(32-2) the selfmap $Y \mapsto \mathrm{cl}(Y)$ *of* $\exp[X]$ *is involutive:*
$\mathrm{cl}(\mathrm{cl}(Y)) = \mathrm{cl}(Y)$, *for each* $Y \in \exp[X]$;

(32-3) the selfmap $Y \mapsto \mathrm{cl}(Y)$ *of* $\exp[X]$ *is a closure operator over* X*, as in Kuratowski* [27, *Chapter* I, *Section* 4].

Proof. (i): Let $w \in \mathrm{cl}(\mathrm{cl}(Y))$ be arbitrary fixed. Given $\varepsilon > 0$, let $\delta > 0$ be the number assured by the Frink triangular property. By the non-sequential definition of closure, there exists $v \in \mathrm{cl}(Y)$ such that $d(v, w) < \delta$. Further, again by the underlying definition, there exists $u \in Y$ with $d(u, v) < \delta$. This, via Frink triangular inequality, gives $d(u, w) < \varepsilon$. Combining with the arbitrariness of $\varepsilon > 0$, we derive $w \in \mathrm{cl}(Y)$, proving that $\mathrm{cl}(Y)$ is d-closed.

(ii): Evident, by the preceding characterization of d-closeness.

(iii): Evident, by the preceding stage and the properties of closure operator. □

(C) Let X be a nonempty set. By a *quasi-metric* over X we shall mean any mapping $e : X \times X \to R_+$ with the properties

(qm-1) e is *reflexive sufficient*: $e(x, x) = 0$ iff $x = y$;
(qm-2) e is *triangular*: $e(x, z) \leq e(x, y) + e(y, z)$, $\forall x, y, z \in X$;

in this case, the couple (X, e) will be referred to as a *quasi-metric space*. Clearly, any quasi-metric is a Bakhtin one; hence, a Frink one.

Technically speaking, a quasi-metric has all properties of a metric, excepting symmetry; hence, in particular, any metric is a quasi-metric. The converse is not in general true, as shown by the following:

Example 1. Put $X = \{1, 2, 3\}$, and define the mapping $e : X \times X \to R_+$ as

$$e(1,1) = e(2,2) = e(3,3) = 0;$$
$$e(1,2) = e(1,3) = 1,\ e(2,3) = 2;\ e(2,1) = e(3,1) = e(3,2) = 1.$$

Clearly, $e(.,.)$ is a quasi-metric on X. On the other hand, $e(.,.)$ is not a metric on X because $e(2,3) = 2$, $e(3,2) = 1$.

Finally, note that, given a quasi-metric e on X, its *conjugate map*

$$(e^* : X \times X \to R_+):\ e^*(x, y) = e(y, x),\ x, y \in X$$

is a quasi-metric too. On the other hand, the associated (with e and e^*) mapping $d : X \times X \to R_+$, given as

$d(x, y) = \max\{e(x, y), e^*(x, y)\}$, $x, y \in X$ (in short: $d = \max(e, e^*)$),

is a symmetric quasi-metric; hence, a (standard) metric on X.

The concept of quasi-metric seems to have a pretty long tradition in metrical spaces theory. For example, in his 2001 PhD Thesis, Hitzler [23, Chapter 1, Section 1.2] introduced this notion as a useful tool for the topological study of logic semantics. Later, in a 2004 paper, Turinici [46] used the same concept [referred to as: *reflexive triangular sufficient pseudometric*] with the aim of establishing a Caristi–Kirk fixed point theorem over such structures. Further aspects may be found in Cobzaş [12].

Suppose that we introduced a quasi-metric $e(.,.)$ (on X). A natural convergence and Cauchy structure over X were already defined. It is our aim in the following to discuss a lot of specific properties appearing in this context.

(I) Let us say that e is *separated*, when

(qm-sep) $(\xrightarrow{e})$ is *separated*:
$e - \lim_n(x_n)$ is an asingleton, for each sequence $(x_n; n \geq 0)$ in X.

(Here, $Y \in \exp[X]$ is called *asingleton*, iff either $Y = \emptyset$ or $Y =$ singleton.) In this case, note that — given the e-convergent sequence (x_n) — we must have

$\lim_n(x_n) = \{z\}$ (written as: $\lim_n(x_n) = z$), for some $z \in X$.

Remark 1. The separated property of a quasi-metric $e(.,.)$ includes its sufficiency. In fact, let $u, v \in X$ be such that $e(u, v) = 0$. The constant sequence $(x_n = u; n \geq 0)$ fulfills $x_n \xrightarrow{e} u$, $x_n \xrightarrow{e} v$ so that (by the separated property), $u = v$. On the other hand, when, in addition,

$e(.,.)$ is symmetric (hence, $e(.,.)$ is a metric on X),

the separated property holds. In fact, let (x_n) be a sequence in X with $x_n \xrightarrow{e} u$, $x_n \xrightarrow{e} v$. As e is triangular and symmetric,

$e(u, v) \leq e(x_n, u) + e(x_n, v)$, $\forall n$,

so by a limit process, $e(u, v) = 0$, and this, via $e =$ sufficient, yields $u = v$.

(II) Let us say that e is *first variable continuous*, provided

(fv-cont) for each $y \in X$, the map $x \mapsto e(x,y)$
is (sequentially) continuous: $x_n \xrightarrow{e} x$ implies $e(x_n, y) \to e(x,y)$.

For example, if e = metric, this condition is fulfilled. In fact, a stronger property holds; as results from the following:

Proposition 8. *Suppose that* $e(.,.)$ *is a metric on* X. *Then, we have the following:*

(33-1) The mapping $(x,y) \mapsto e(x,y)$ *is e-Lipschitz, in the sense*

$|e(x,y) - e(u,v)| \le e(x,u) + e(y,v)$, $\forall (x,y),(u,v) \in X \times X$.

(33-2) As a consequence, this map is e-continuous; i.e.

$x_n \xrightarrow{e} x$, $y_n \xrightarrow{e} y$ *imply* $e(x_n, y_n) \to e(x,y)$.

The proof is immediate, by the properties of $e(.,.)$, so we do not give details.

Remark 2. Returning to the quasi-metric setting, let us consider the following (apparently weaker) form of (fv-cont):

(fv-lsc) e is *first variable lsc*: for each $y \in X$, the map $x \mapsto e(x,y)$ is (sequentially) e-lsc: $x_n \xrightarrow{e} x$ implies $\liminf_n e(x_n, y) \ge e(x,y)$.

Note that, in many concrete circumstances to be considered, this variant works; so, it is legitimate to ask of what is the reason of introducing its strong version. The answer to this is very simple: by the conditions imposed upon $e(.,.)$, the weaker form in question is equivalent with its initial (strong) counterpart. In fact, assume that e is first variable lsc, and fix some $y \in X$. Given the sequence (x_n) in X and the point $x \in X$ with $x_n \xrightarrow{e} x$, we have (by the triangular inequality)

$e(x_n, y) \le e(x_n, x) + e(x,y)$, for all n;
wherefrom (as $e(x_n, x) \to 0$ as $n \to \infty$): $\limsup_n e(x_n, y) \le e(x,y)$.

Combining with the working assumption about e gives

$e(x,y) \le \liminf_n e(x_n, y) \le \limsup_n e(x_n, y) \le e(x,y)$;
or, equivalently: $\lim_n e(x_n, y) = e(x,y)$;

which tells us that $e(.,.)$ is first variable continuous. Hence, passing from (fv-cont) to (fv-lsc) does not bring any generality relative to the precise framework.

Finally, we show by an example — patterned after Alsulami *et al.* [1] — that such extra conditions imposed upon our quasi-metrics are effective.

Example 2. Let $X = R_+$, and $(g(x,y) = |x-y|;\ x,y \in X)$ stand for the usual metric over it. Further, let $\beta \in]0,1[$ be fixed in the sequel. Define a mapping $e : X \times X \to R_+$ as

$$(e(x,y) = \beta g(x,y), \text{ if } x < y),\ \ (e(x,y) = g(x,y), \text{ if } x \geq y).$$

Clearly, $e(.,.)$ is a quasi-metric over X. However, it is not a metric because, for example,

$$e(1,2) = \beta < 1 = e(2,1).$$

Nevertheless, its convergence structure is equivalent with the one induced by the usual metric, in view of

$$\beta g(x,y) \leq e(x,y) \leq g(x,y),\ \forall x,y \in X.$$

In fact, as a direct consequence of this,

$$(\forall(x_n), \forall x)\text{: } x_n \xrightarrow{e} x \text{ iff } x_n \xrightarrow{g} x;$$

wherefrom, e is separated. Moreover, for each $y \in X$, the partial map (in $\mathcal{F}(X,R_+)$) $\varphi_y := e(.,y)$ defined as

$$(\varphi_y(x) = \beta g(x,y), \text{ if } x < y),\ \ (\varphi_y(x) = g(x,y), \text{ if } x \geq y)$$

is continuous with respect to the usual convergence structure of X; hence (by the above), with respect to the convergence structure attached to e; so that $e(.,.)$ is first variable (sequentially) continuous.

(D) In the following, a relative type construction is proposed for obtaining a large class of quasi-metrics. Let the locally Riemann integrable function $b : R_+ \to R_+^0$ be given; note that, in this case,

the *primitive* $(B(t) = \int_0^t b(s)ds;\ t \in R_+)$ exists as an element in $\mathcal{F}(R_+)$.

Suppose in the following that

(norm) (b, B) is *normal*: b is decreasing and $B(\infty) = \infty$.

In particular, we have the (integral) representation

(int-rep) $\int_p^q b(\xi)d\xi = (q-p)\int_0^1 b(p+\tau(q-p))d\tau$, when $0 \leq p < q < \infty$.

Some basic facts involving this couple are being collected in the following:

Theorem 1. *The following are valid:*

(31-a) B is a continuous order isomorphism of R_+; hence, so is B^{-1};
(31-b) $b(s) \leq (B(s) - B(t))/(s-t) \leq b(t), \forall t, s \in R_+, t < s$;
(31-c) B is almost concave:
$t \mapsto [B(t+s) - B(t)]$ is decreasing on R_+, $\forall s \in R_+$;
(31-d) B is concave: $B(t + \lambda(s-t)) \geq B(t) + \lambda(B(s) - B(t))$, for all $t, s \in R_+$ with $t < s$ and all $\lambda \in [0, 1]$;
(31-e) B is sub-additive (hence B^{-1} is super-additive).

The proof is immediate, by the (norm) above, so we do not give details. Note that the properties in (31-c) and (31-d) are equivalent to each other, under (31-a). This follows at once from the (non-differential) mean value theorem in Bantaş and Turinici [4].

Now, let X be some non-empty set and $d : X \times X \to R_+$ be a quasi-metric over it. Further, let $\Gamma : X \to R_+$ be chosen as

(a-nexp) Γ is *almost d-nonexpansive*: $\Gamma(x) - \Gamma(y) + d(x, y) \geq 0$, $\forall x, y \in X$.

Define a pseudometric $e := [d; B, \Gamma]$ over X as

(e-expli) $e(x, y) = B(\Gamma(x) + d(x, y)) - B(\Gamma(x))$, $x, y \in X$.

This may be viewed as an "explicit" formula; the "implicit" version of it is

(e-impli) $d(x, y) = B^{-1}(B(\Gamma(x)) + e(x, y)) - \Gamma(x)$, $x, y \in X$.

We will establish some properties of this map, useful in the sequel.

First, the "metrical" nature of $(x, y) \mapsto e(x, y)$ is of interest.

Theorem 2. *The mapping* $(x, y) \mapsto e(x, y)$ *is a quasi-metric over* X.

Proof. The reflexivity and sufficiency are clear, by Theorem 1 (first part), so it remains to establish the triangular property. Let $x, y, z \in X$ be arbitrary fixed. The triangular property of $d(.,.)$ yields [via Theorem 1 (first part)]

$$e(x, z) \leq B(\Gamma(x) + d(x, y) + d(y, z)) - B(\Gamma(x) + d(x, y)) + e(x, y).$$

On the other hand, the almost d-non-expansiveness of Γ gives

$\Gamma(x) + d(x, y) \geq \Gamma(y)$, so (by Theorem 1 (third part))
$B(\Gamma(x) + d(x, y) + d(y, z)) - B(\Gamma(x) + d(x, y)) \leq e(y, z)$.

Combining with the previous relation yields our desired conclusion. □

By definition, e will be called the *Zhong metric* attached to d and the couple (B, Γ). The following properties of (d, e) are immediate (via Theorem 1):

Proposition 9. *Under the prescribed conventions,*

(34-1) $b(\Gamma(x) + d(x, y))d(x, y) \leq e(x, y) \leq b(\Gamma(x))d(x, y)$, $\forall x, y \in X$;
(34-2) $e(x, y) \leq B(d(x, y))$, $\forall x, y \in X$; *hence* $x_n \xrightarrow{d} y$ *implies* $x_n \xrightarrow{e} y$.

A basic property of $e(.,.)$ to be checked is d-compatibility. Given the couple (d, e) of quasi-metrics over X, we say that e is *d-compatible* provided

(comp-1) each e-Cauchy sequence is d-Cauchy too;
(comp-2) $y \mapsto e(x, y)$ is d-lsc, for each $x \in X$.

Theorem 3. *The Zhong metric* $e(.,.)$ *is d-compatible (see above).*

Proof. (i): We first check (comp-2), which may be written as

$$(e(x, y_n) \leq \lambda, \forall n) \text{ and } y_n \xrightarrow{d} y \text{ imply } e(x, y) \leq \lambda.$$

So, let x, (y_n), λ and y be as in the premise of this relation. By Proposition 9, we have $y_n \xrightarrow{e} y$ as $n \to \infty$. Moreover (as e is triangular),

$$e(x, y) \leq e(x, y_n) + e(y_n, y) \leq \lambda + e(y_n, y), \text{ for all } n.$$

It will suffice passing to limit as $n \to \infty$ to get the desired conclusion.

(ii): Further, we claim that (comp-1) holds too, in the following stronger sense:

$$\text{(for each sequence) } d\text{-Cauchy} \Longleftrightarrow e\text{-Cauchy}.$$

The left to right implication is clear, via Proposition 9. For the right to left one, assume that (x_n) is an e-Cauchy sequence in X. As e=triangular,

$$e(x_i, x_j) \leq \mu, \text{ for all } (i, j) \text{ with } i \leq j, \text{ and some } \mu \geq 0.$$

This, along with the implicit definition of e, yields

$$d(x_0, x_i) = B^{-1}(B(\Gamma(x_0)) + e(x_0, x_i)) - \Gamma(x_0) \leq$$
$$B^{-1}(B(\Gamma(x_0)) + \mu) - \Gamma(x_0), \ \forall i \geq 0;$$

wherefrom (by the choice of Γ),

$$\Gamma(x_i) \leq \Gamma(x_0) + d(x_0, x_i) \leq B^{-1}(B(\Gamma(x_0)) + \mu)$$
$$\text{(hence } B(\Gamma(x_i)) \leq B(\Gamma(x_0)) + \mu), \text{ for all } i \geq 0.$$

Putting these facts together yields (again via implicit definition of e)

$$\Gamma(x_i) + d(x_i, x_j) = B^{-1}(B(\Gamma(x_i)) + e(x_i, x_j)) \leq$$
$$\nu := B^{-1}(B(\Gamma(x_0)) + 2\mu), \text{ for all } (i, j) \text{ with } i \leq j.$$

And this, via Proposition 9, gives (for the same pairs (i, j))

$$e(x_i, x_j) \geq b(\Gamma(x_i) + d(x_i, x_j))d(x_i, x_j) \geq b(\nu)d(x_i, x_j).$$

But then, the d-Cauchy property of (x_n) is clear, and the proof is complete. □

Note finally that this relative construction $d \mapsto [d; B, \Gamma]$ of quasi-metrics may be viewed as an iterative process of generating such objects and yields a lot of interesting problems. But, for the moment, this will suffice.

4. Local Maximality Principles

Let X be a non-empty set. By a *pseudometric* over X, we mean any map $d : X \times X \to R_+$. For an easy reference, we list the conditions to be used in the sequel:

(ref) d is *reflexive*: $d(x,x) = 0$, for each $x \in X$;
(suf) d is *sufficient*: $x, y \in X$, $d(x,y) = 0$ imply $x = y$;
(F-tri) d is *Frink triangular*:
$\forall \varepsilon > 0, \exists \delta > 0$, such that: $d(x,y), d(y,z) < \delta$ imply $d(x,z) < \varepsilon$;
(B-tri) d is *Bakhtin triangular*:
there exists $\beta \geq 1$, such that $d(x,z) \leq \beta(d(x,y) + d(y,z))$, $\forall x, y, z \in X$;
(tri) d is *triangular*: $d(x,z) \leq d(x,y) + d(y,z)$, $\forall x, y, z \in X$;
(sym) d is *symmetric*: $d(x,y) = d(y,x)$, for all $x, y \in X$.

(A) For the moment, we only assume that the reflexive property holds. Then, d will be called a *r-pseudometric* on X, and (X,d) will be referred to as a *r-pseudometric space.* Further, let $(\preceq)$ be a *quasi-order* (i.e. reflexive and transitive relation) over X; the triple $(X, d, \preceq)$ will be referred to as a *quasi-ordered r-pseudometric space.*

Remember that the sequence (x_n) in X is called *d-Cauchy* when

$d(x_m, x_n) \to 0$ as $m, n \to \infty$, $m \leq n$; that is,
$\forall \varepsilon > 0, \exists n(\varepsilon)$, such that $n(\varepsilon) \leq p \leq q \Longrightarrow d(x_p, x_q) \leq \varepsilon$;
or, equivalently (by the reflexive property)
$\forall \varepsilon > 0, \exists n(\varepsilon)$, such that $n(\varepsilon) < p < q \Longrightarrow d(x_p, x_q) \leq \varepsilon$.

Further, let us say that (x_n) is *d-asymptotic,* if

$d(x_n, x_{n+1}) \to 0$ as $n \to \infty$; which means
$\forall \varepsilon > 0, \exists n(\varepsilon)$, such that $n(\varepsilon) \leq p \Longrightarrow d(x_p, x_{p+1}) \leq \varepsilon$.

Then, for each $M \in \exp(X)$, let us consider the global conditions:

(C-reg) $(M, d, \preceq)$ is *Cauchy regular*:
each $(\preceq)$-ascending sequence in M is d-Cauchy;
(A-reg) $(M, d, \preceq)$ is *asymptotic regular*:
each $(\preceq)$-ascending sequence in M is d-asymptotic.

As each d-Cauchy sequence is d-asymptotic too, it follows that (C-reg) $\Longrightarrow$ (A-reg). The reverse implication also holds, in the following sense:

Proposition 10. *We have in (ZF-AC), for each $M \in \exp(X)$,*

$$\text{(A-reg)} \Longrightarrow \text{(C-reg); whence, (A-reg)} \Longleftrightarrow \text{(C-reg).}$$

Proof. Suppose that (A-reg) holds, but for some $(\preceq)$-ascending sequence (x_n), the d-Cauchy property fails; i.e. (for some $\varepsilon > 0$)

$$C(n) = \{(p, q) \in N \times N; n < p < q, d(x_p, x_q) > \varepsilon\} \neq \emptyset, \forall n.$$

Denote, for simplicity,

$$p(n) = \min \mathrm{Dom}(C(n)),\ q(n) = \min(C(n)(p(n)),\ n \in N;$$

clearly, no choice techniques are used in this construction. Fix some rank $i(0)$. By this assumption, there exist $i(1) = p(i(0))$, $i(2) = q(i(0))$ with $i(0) < i(1) < i(2)$, $d(x_{i(1)}, x_{i(2)}) > \varepsilon$. Further, given the rank $i(2)$, there exist $i(3) = p(i(2))$, $i(4) = q(i(2))$ with $i(2) < i(3) < i(4)$, $d(x_{i(3)}, x_{i(4)}) > \varepsilon$, and so on. By induction, we get a $(\preceq)$-ascending subsequence $(y_n = x_{i(n)}; n \geq 0)$ of (x_n) with $d(y_{2n+1}, y_{2n+2}) > \varepsilon$, for all n. This contradicts (A-reg); hence, the claim. □

Having these precise, define for each $M \in \exp(X)$,

(reg) $(M, d, \preceq)$ is *regular*, when it fulfills
one of the equivalent properties (C-reg) or (A-reg).

Technically speaking, this regularity concept is pretty stringent; because, under a rs-pseudometric condition upon d, it induces a strong property of our quasi-order over the ambient set. Precisely, the following statement holds:

Proposition 11. *Let the quasi-ordered rs-pseudometric space $(X, d, \preceq)$ and the (non-empty) subset M of X be such that $(M, d, \preceq)$ is regular. Then,*

$(\preceq)$ is antisymmetric (hence, a (partial) order) on M.

Proof. (cf. Hamel [22, Chapter 4, Section 4.1]) Let $u, v \in M$ be such that $u \preceq v$ and $v \preceq u$. The sequence $(y_{2n} = u,\ y_{2n+1} = v;$

$n \geq 0$) is ascending; hence, d-asymptotic by hypothesis. This yields $d(u, v) = 0$; whence (as d is rs-pseudometric), $u = v$. □

Returning to the r-pseudometric setting, a basic extension of the introduced regular concept may be obtained under the following lines. For each $\varepsilon > 0$, let $\mathcal{R}(\varepsilon)$ stand for the relation over M:

$(x, y \in M)$: $x\mathcal{R}(\varepsilon)y$ iff $x \preceq y$ and
$(\forall u, v \in M)$: $y \preceq u \preceq v$ implies $d(u, v) \leq \varepsilon$.

Proposition 12. *We have in (ZF – AC + DC), for each* $M \in \exp(X)$

(43-1) *If* $(M, d, \preceq)$ *is regular, then* $(M, d, \preceq)$ *is weakly regular:*
($\forall\varepsilon > 0$*):* $\mathcal{R}(\varepsilon)$ *is proper (*$\mathcal{R}(\varepsilon)(x) \neq \emptyset$, $\forall x \in M$*).*

(43-2) *If* $(M, d, \preceq)$ *is weakly regular, then,* $(M, d, \preceq)$ *is weakly sequentially regular: for each* $u \in M$*, and each strictly descending sequence* $(\varepsilon_n; n \geq 0)$ *in* R^0_+ *with* $\varepsilon_n \to 0$*, there exists a* $(\preceq)$*-ascending sequence* $(u_n; n \geq 0)$ *in* M*, with*
$u \preceq u_0$*, and [(*$\forall n \geq 1$*), (*$\forall y, z \in M$*)]:* $u_n \preceq y \preceq z \Longrightarrow d(y, z) \leq \varepsilon_n$.

Proof. There are two steps to be passed.

Step 1. Assume this would be false; that is (for some $\varepsilon > 0$ and $x_0 \in M$)

for each $y \in M(x_0, \preceq)$ there exist $u, v \in M$ with $y \preceq u \preceq v$, $d(u, v) > \varepsilon$.

This, by definition, yields (for the same (ε, x_0))

$$\forall y \in M(x_0, \preceq),\ \exists(u, v) \in (\preceq)\text{: } y \preceq u,\ d(u, v) > \varepsilon,$$

where $(\preceq) := \{(a, b) \in M \times M; a \preceq b\}$. Put $Q := \{(a, b) \in (\preceq); x_0 \preceq a\}$, and fix $(y_0, y_1) \in Q$ (for example, $y_0 = y_1 = x_0$). Define a relation $\mathcal{S} = \mathcal{S}(\varepsilon)$ on Q as

$$(a_1, b_1)\mathcal{S}(a_2, b_2) \text{ if and only if } b_1 \preceq a_2,\ d(a_2, b_2) > \varepsilon.$$

From the imposed condition, $Q((a, b), \mathcal{S}) \neq \emptyset$, $\forall(a, b) \in Q$. So, by the Dependent Choice principle (DC), it follows that for the starting point $w_0 = (y_0, y_1)$ in Q, there exists a sequence

$(w_n := (y_{2n}, y_{2n+1}); n \geq 0)$ in Q, with

$w_n \mathcal{S} w_{n+1}$, for all n; hence, by definition,
$y_{2n+1} \preceq y_{2n+2}$, $d(y_{2n+2}, y_{2n+3}) > \varepsilon$, for all n.

As a consequence, $(y_n; n \geq 0)$ is $(\preceq)$-ascending and not d-asymptotic, in contradiction with the regularity of $(M, d, \preceq)$; hence, the claim.

Step 2. Suppose that $(M, d, \preceq)$ is weakly regular, and let $(\varepsilon_n; n \geq 0)$ be a strictly descending sequence in R^0_+ with $\varepsilon_n \to 0$; clearly, the relational sequence $(\mathcal{R}(\varepsilon_n); n \geq 0)$ is proper. Fix some $u \in X$, as well as an element $u_0 \in X(u, \preceq)$. By the Diagonal Dependent Choice principle (DDC), we have that there exists a sequence $(u_n; n \geq 0)$ in X, with

$(u_n \mathcal{R}(\varepsilon_{n+1}) u_{n+1}, \forall n)$; hence, (u_n) is $(\preceq)$-ascending and
$((\forall n), (\forall y, z \in X))$: $u_{n+1} \leq y \leq z \Longrightarrow d(y, z) \leq \varepsilon_{n+1}$,

and, from this, we are done. □

(B) Let again $(X, d, \preceq)$ be a quasi-ordered r-pseudometric space. In the following, various convergence structures are defined, as well as their attached closure operators. Then, the relationships between these are discussed.

Define a d-convergence structure on X under the above discussed way. Precisely, let us say that the sequence (x_n) in X, *d-converges* to $x \in X$ (and write: $x_n \xrightarrow{d} x$) iff $d(x_n, x) \to 0$ as $n \to \infty$; that is

$\forall \varepsilon > 0$, $\exists p = p(\varepsilon)$, $\forall n$: $(p \leq n \Longrightarrow d(x_n, x) < \varepsilon)$; or, equivalently,
$\forall \varepsilon > 0$, $\exists p = p(\varepsilon)$, $\forall n$: $(p \leq n \Longrightarrow d(x_n, x) \leq \varepsilon)$.

Then, x is called a *d-limit* of (x_n); the set of these will be denoted as $d - \lim_n(x_n)$ [or, $\lim_n(x_n)$, when d is understood]; if it is non-empty, we say that (x_n) is *d-convergent.*

Given $Y \in \exp[X]$, let us say that $w \in X$ is a *$(d, \preceq)$-adherence* point of it, when

$w \in d - \lim_n(x_n)$, for some $(\preceq)$-ascending sequence (x_n) of Y;

the class of all these will be denoted as $\mathrm{cl}_{(\preceq)}(Y)$. It is not hard to see that $Y \mapsto \mathrm{cl}_{(\preceq)}(Y)$ is a *semi-closure* over X, in the sense

(docl-1) $\mathrm{cl}_{(\preceq)}(\emptyset) = \emptyset$, $\mathrm{cl}_{(\preceq)}(X) = X$;
(docl-2) $Y \subseteq Z$ implies $\mathrm{cl}_{(\preceq)}(Y) \subseteq \mathrm{cl}_{(\preceq)}(Z)$;

(docl-3) $\text{cl}_{(\preceq)}(U \cup V) = \text{cl}_{(\preceq)}(U) \cup \text{cl}_{(\preceq)}(V)$, $U, V \in \exp[X]$;
(docl-4) $Y \subseteq \text{cl}_{(\preceq)}(Y)$, for each $Y \in \exp[X]$.

In fact, (docl-1), (docl-2) and (docl-4) are clear. Concerning (docl-3), the right to left inclusion is clear, via (docl-2). For the left to right inclusion, let U, V be non-empty subsets of X, and $w \in \text{cl}_{(\preceq)}(U \cup V)$ be arbitrary fixed; hence,

$w \in d-\lim_n(x_n)$, for some $(\preceq)$-ascending sequence (x_n) of $U \cup V$.

By definition, there exists a subsequence $(y_n := x_{i(n)}; n \geq 0)$ of $(x_n; n \geq 0)$, with

(y_n) is a $(\preceq)$-ascending sequence in U or V (and $y_n \xrightarrow{d} w$).

This yields

$w \in \text{cl}_{(\preceq)}(U)$ or $w \in \text{cl}_{(\preceq)}(V)$; hence, $w \in \text{cl}_{(\preceq)}(U) \cup \text{cl}_{(\preceq)}(V)$,

proving our assertion. Unfortunately, $Y \mapsto \text{cl}_{(\preceq)}(Y)$ is not involutive; i.e.

(docl-inv) $\text{cl}_{(\preceq)}(\text{cl}_{(\preceq)}(Y)) = \text{cl}_{(\preceq)}(Y)$, for each $Y \in \exp[X]$

is not in general true so that $Y \mapsto \text{cl}_{(\preceq)}(Y)$ is not a closure over X as in Kuratowski [**?**, Chapter I, Section 4].

For each subset $Y \in \exp[X]$, let us introduce the concept

Y is $(d, \preceq)$-*closed*:
the d-limit of each $(\preceq)$-ascending sequence in Y is included in Y.

It is not hard to see that

(ocl-char) (for each $Y \in \exp[X]$): Y is $(d, \preceq)$-closed iff $Y = \text{cl}_{(\preceq)}(Y)$.

In fact, the right to left inclusion is clear, by definition. For the left to right inclusion, let $w \in \text{cl}_{(\preceq)}(Y)$ be arbitrary fixed. By definition,

$w \in d - \lim_n(x_n)$, for some $(\preceq)$-ascending sequence (x_n) of Y,

and this, by the imposed condition, yields $w \in Y$. By the arbitrariness of our underlying point, one gets

$Y \supseteq \text{cl}_{(\preceq)}(Y)$; hence, $Y = \text{cl}_{(\preceq)}(Y)$, proving our assertion.

Denote, for simplicity

$\mathcal{K}[d,\preceq]$=the class of all $(d,\preceq)$-closed subsets in X.

Some basic properties of this class are listed below:

(cotop-1) $\emptyset, X \in \mathcal{K}[d,\preceq]$;
(cotop-2) the intersection of any subset in $\mathcal{K}[d,\preceq]$ belongs to $\mathcal{K}[d,\preceq]$;
(cotop-3) the union of any finite subset of $\mathcal{K}[d,\preceq]$ is in $\mathcal{K}[d,\preceq]$.

The family $\mathcal{K}[d,\preceq]$ is therefore a *cotopology* on X. So, it may generate a closure operator $\mathrm{kl}_{(\preceq)}$: $\exp[X] \to \exp[X]$, as: for each $M \in \exp[X]$,

$\mathrm{kl}_{(\preceq)}(M)$= the intersection of all $Y \in \mathcal{K}[d,\preceq]$ with $M \subseteq Y$.

The basic properties of this closure operator are concentrated in the following:

Proposition 13. *Under these conventions, we have*

(44-1) *the selfmap* $Y \mapsto \mathrm{kl}_{(\preceq)}(Y)$ *is a closure operator, as in Kuratowski* [27, *Chapter I, Section* 4], *with the supplementary properties*

(p1) $\mathrm{kl}_{(\preceq)}(M) \in \mathcal{K}[d,\preceq]$, *for each* $M \in \exp[X]$,
(p2) $\mathrm{kl}_{(\preceq)}(M) = M$, *for each* $M \in \mathcal{K}[d,\preceq]$

(44-2) *in addition, the following inclusion is valid:*
(p3) $\mathrm{cl}_{(\preceq)}(M) \subseteq \mathrm{kl}_{(\preceq)}(M)$, *for each* $M \in \exp[X]$.

Proof. (i): Clearly, (p1) and (p2) hold by definition. Further, we claim that

(kl-1) $\mathrm{kl}_{(\preceq)}(\emptyset) = \emptyset$, $\mathrm{kl}_{(\preceq)}(X) = X$;
(kl-2) $Y \subseteq Z$ implies $\mathrm{kl}_{(\preceq)}(Y) \subseteq \mathrm{kl}_{(\preceq)}(Z)$;
(kl-3) $\mathrm{kl}_{(\preceq)}(U \cup V) = \mathrm{kl}_{(\preceq)}(U) \cup \mathrm{kl}_{(\preceq)}(V)$, $U, V \in \exp[X]$;
(kl-4) $Y \subseteq \mathrm{kl}_{(\preceq)}(Y)$, $\forall Y \in \exp[X]$;
(kl-5) $\mathrm{kl}_{(\preceq)}(\mathrm{kl}_{(\preceq)}(Y)) = \mathrm{kl}_{(\preceq)}(Y)$, for each $Y \in \exp[X]$.

In fact, (kl-1), (kl-2) and (kl-4) are clear, by definition, and (kl-5) is immediate via (p1)+(p2). Concerning (kl-3), the right to left inclusion is clear, via (kl-2). For the reverse inclusion, let $U, V \in \exp[X]$

be arbitrary fixed. By (kl-4),

$$U \subseteq \mathrm{kl}_{(\preceq)}(U),\ V \subseteq \mathrm{kl}_{(\preceq)}(V);\ \text{hence, } U \cup V \subseteq \mathrm{kl}_{(\preceq)}(U) \cup \mathrm{kl}_{(\preceq)}(V).$$

This, along with $\mathrm{kl}_{(\preceq)}(U) \cup \mathrm{kl}_{(\preceq)}(V)$ being an element of $\mathcal{K}[d, \preceq]$, gives

$$\mathrm{kl}_{(\preceq)}(U \cup V) \subseteq \mathrm{kl}_{(\preceq)}(U) \cup \mathrm{kl}_{(\preceq)}(V);\ \text{hence, the conclusion.}$$

(ii): Given $M \in \exp[X]$, we have that $M \subseteq Y := \mathrm{kl}_{(\preceq)}(M) \in \mathcal{K}[d, \preceq]$. Let $w \in \mathrm{cl}_{(\preceq)}(M)$ be arbitrary fixed. By definition, there exists a $(\preceq)$-ascending sequence (x_n) in $M \subseteq Y$ with $x_n \xrightarrow{d} w$. This, via $Y \in \mathcal{K}[d, \preceq]$, gives $w \in Y$, and our claim follows. The proof is complete. □

Remark 3. The inclusion between these operators may be strict. This follows from the semi-closure operator $Y \mapsto \mathrm{cl}_{(\preceq)}(Y)$ being not involutive, in general.

From a theoretical perspective, the closure operator $Y \mapsto \mathrm{kl}_{(\preceq)}(Y)$ has an essential role in our $(\preceq)$-maximal results. However, under the described general setting, this operator cannot give us a consistent variational help in practice; because, for a fixed $Y \in \exp[X]$, the points of $\mathrm{kl}_{(\preceq)}(Y)$ are not limit attainable from the points of Y. So, it is natural asking of under which extra conditions upon d, this is somehow possible. As our previous developments suggest, d must be a r-pseudometric subjected to the scale of conditions

(F-B) d is Frink (resp., Bakhtin) triangular;

we then say that d is a *Frink (resp., Bakhtin) r-pseudometric* on X, and (X, d) is a *Frink (resp, Bakhtin) r-pseudometric space.*

Proposition 14. *Suppose that $(X, d, \preceq)$ is a quasi-ordered Frink (resp., Bakhtin) r-pseudometric space. Then, the double inclusion is valid:*

(ocl-kl-cl) $\mathrm{cl}_{(\preceq)}(M) \subseteq \mathrm{kl}_{(\preceq)}(M) \subseteq \mathrm{cl}(M)$, *for each* $M \in \exp[X]$.

Proof. The former inclusion is clear, by the above, so it remains to establish the latter one. Let $M \in \exp[X]$ be arbitrary fixed. By a previous auxiliary fact,

$\mathrm{cl}(M)$ is d-closed; hence, $(d, \preceq)$-closed, so $M \subseteq \mathrm{cl}(M) \in \mathcal{K}[d, \preceq]$.

This, under the definition of our closure operator, yields the needed conclusion. □

Returning to the general r-pseudometric setting, let us introduce a *quasi-order convergence* ($\xrightarrow{\preceq}$) over X as follows: Given the sequence (x_n) in X and the point $x \in X$, put

$$x_n \xrightarrow{\preceq} x \text{ if } \exists m(x) \in N, \text{ such that } n \geq m(x) \text{ implies } x_n \preceq x,$$

also referred to as: x is a $(\preceq)$*-limit* of (x_n). The class of all these will be denoted as $(\preceq) - \lim_n(x_n)$; when it is non-empty, we say that (x_n) is $(\preceq)$*-convergent.*

Remark 4. The following fact is useful in the sequel. Given the sequence (x_n) in X and the point $u \in X$, let us say that (x_n) is $(\preceq)$-*bounded* by u (or: $u \in X$ is an $(\preceq)$*-upper bound* of (x_n)), provided

$$(x_n \preceq x, \forall n); \text{ in short: } (x_n; n \geq 0) \preceq x.$$

The set of all such x will be denoted as $\text{ubd}(x_n)$ (the *upper bound* of (x_n)); if it is non-empty, we will say that (x_n) is $(\preceq)$*-bounded.* Under these conventions, we have, for each $(\preceq)$-ascending sequence (x_n) in X and each $x \in X$,

(prop-1) $x_n \xrightarrow{\preceq} x$ iff (x_n) is $(\preceq)$-bounded by x (that is, $x \in \text{ubd}(x_n)$);
(prop-2) $(\preceq) - \lim_n(x_n)$ is identical with $\text{ubd}(x_n)$.

The proof is immediate, by the involved facts, so we do not give details.

Further, as a combination of d-convergence and $(\preceq)$-convergence, define the *product* convergence structure ($\xrightarrow{d\preceq}$) over X, according to

$$x_n \xrightarrow{d\preceq} x \text{ when } x_n \xrightarrow{d} x \text{ and } x_n \xrightarrow{\preceq} x.$$

In this case, x is called a $(d \preceq)$*-limit* of (x_n); the set of all these will be denoted as $(d \preceq) - \lim_n(x_n)$; if such elements exist, we say that (x_n) is $(d \preceq)$*-convergent.*

Finally, let us say a few words about the completeness concepts to be considered here at the level of quasi-ordered r-pseudometric spaces $(M, d, \preceq)$, where $M \in \exp(X)$ is arbitrary fixed. The most useful for us are the following:

(ful-com) $(M, d, \preceq)$ is *fully complete*:
each $(\preceq)$-ascending d-Cauchy sequence (x_n) in M is $(d\ \preceq)$-convergent
(there exists $x \in X$ such that $x_n \xrightarrow{d} x$ and $x_n \xrightarrow{\preceq} x$);
(com) $(M, d, \preceq)$ is complete:
each $(\preceq)$-ascending d-Cauchy sequence (x_n) in M is d-convergent
(there exists $x \in X$ such that $x_n \xrightarrow{d} x$).

To clarify the relationships between these, we need one more condition:

(c-dom) $(M, d, \preceq)$ is *conv-dominated*: the d-limit of each $(\preceq)$-ascending sequence in M is a $(\preceq)$-limit (hence, a $(\preceq)$-upper bound) of it.

Then, evidently,

(for each structure $(M, d, \preceq)$ as before):
(complete and conv-dominated) implies fully complete.

A sufficient condition for conv-dominated property writes (cf. Turinici [56])

$X(x, \preceq) := \{y \in X; x \preceq y\}$ is $(d, \preceq)$-closed, for each $x \in X$;
referred to as: $(\preceq)$ is d-selfclosed on X.

(C) Under these preliminaries, we are now in position to state a lot of maximality statements with a practical meaning. Some preliminaries are needed.

Given $M \in \exp(X)$, call $z \in X$,

(max-1) $(d, \preceq)$-*maximal* over M, if $(u, v \in M,\ z \preceq u \preceq v) \Longrightarrow d(u, v) = 0$;
(max-2) $(\preceq)$-*maximal* over M, if $(u \in M,\ z \preceq u) \Longrightarrow z = u$.

Two basic facts about these concepts will be useful for us.

(I) The former of these involves the hereditary property.

Proposition 15. *The introduced maximal concepts are hereditary, in the sense, for each couple* (M, P) *over* $\exp[X]$ *with* $M \supseteq P$ *and each* $z \in X$,

(he-1) z *is* $(d, \preceq)$*-maximal over* M *implies* z *is* $(d, \preceq)$*-maximal over* P;

(he-2) z *is* $(\preceq)$*-maximal over* M *implies* z *is* $(\preceq)$*-maximal over* P.

The verification of these is immediate, by definition, so we do not give the details.

(II) The latter of these tells us that, under pretty general conditions upon d, the underlying properties are ultimately equivalent. Precisely, suppose that

(r-s) d is *reflexive sufficient*: $d(x, y) = 0$ iff $x = y$;

we then say that d is a *rs-pseudometric* over X.

Proposition 16. *Let the quasi-ordered rs-pseudometric space* $(X, d, \preceq)$ *be given. Then, the following equivalence is true:*

for each $Y \in \exp(X)$ *and each* $z \in Y$:
z *is* $(d, \preceq)$*-maximal over* Y *iff* z *is* $(\preceq)$*-maximal over* Y.

Proof. The left to right inclusion follows via

$z \preceq w \in Y \implies z \preceq z \preceq w$ (and $z, w \in Y$) $\implies$
$d(z, w) = 0 \implies z = w$ (as d is rs-pseudometric).

On the other hand, the right to left inclusion follows as well, by the very definition of $(\preceq)$-maximal element, combined with

$(u, v \in Y,\ z \preceq u \preceq v) \implies (z \preceq u \in Y)$ and $(z \preceq v \in Y) \implies$
$(z = u$ and $z = v) \implies u = v \implies (d(u, v) = 0)$.

This ends the argument. □

Returning to the r-pseudometric setting, the first result in this series is a $(d, \preceq)$-maximal statement (referred to as: Kang–Park r-pseudometric maximal principle; in short: (KP-rp-mp)).

Theorem 4. *Let the quasi-ordered r-pseudometric space* $(X, d, \preceq)$ *and the subset* $M \in \exp(X)$ *be such that*

(41-i) $(M, d, \preceq)$ *is regular: each* $(\preceq)$*-ascending sequence in* M *is d-Cauchy;*
(41-ii) $(M, d, \preceq)$ *is fully complete: each* $(\preceq)$*-ascending d-Cauchy sequence in* M *is* $(d \preceq)$*-convergent (in* X*).*

Then, the following conclusion holds in $(ZF - AC + DC)$:

$\forall u \in M$, $\exists v \in \mathrm{cl}_{(\preceq)}(M)$, with $u \preceq v$ and v is $(d, \preceq)$-maximal over M.

Proof. By regularity (and an auxiliary fact), $(M, d, \preceq)$ is weakly sequentially regular. Hence, given $u \in M$, there exists a $(\preceq)$-ascending sequence (u_n) in M with

(41-a) $u \preceq u_0$, and $[(\forall n \geq 1), (\forall y, z \in M)]$: $u_n \preceq y \preceq z \Longrightarrow d(y, z) \leq 2^{-n}$.

In particular, this tells us that (u_n) is d-Cauchy (in M). Combining with the full completeness hypothesis, there exists $v \in X$ such that

(41-b) $u_n \xrightarrow{d\preceq} v$; so that: $u_n \xrightarrow{d} v$ and $u_n \xrightarrow{\preceq} v$.

For the moment, (41-b) (the first half) yields $v \in \mathrm{cl}_{(\preceq)}(M)$. Then, by (41-a) (the first half) and (41-b) (the second half) we get $u \preceq v$. Finally, let $y, z \in M$ be such that $v \preceq y \preceq z$. By (41-a) (the second half)

$(\forall n \geq 1)$: $u_n \preceq y \preceq z$; whence, $d(y, z) \leq 2^{-n}$,

and, from this, $d(y, z) = 0$. Combining with the arbitrariness of our underlying couple, one derives that v is $(d, \preceq)$-maximal over M; as claimed. □

Formally, this result is comparable with the 1990 one in Kang and Park [25]. However, its basic lines were already set up in the papers by Turinici [41, 44]. Note that both these results extend the 1976 Brezis–Browder ordering principle [7]. Further aspects may be found in Altman [2].

The second result in this series is a $(\preceq)$-maximal statement (referred to as: Zorn–Bourbaki rs-pseudometric maximal principle (in short: (ZB-rsp-mp)).

Theorem 5. *Let the quasi-ordered rs-pseudometric space $(X, d, \preceq)$ and the subset $Z \in \exp(X)$ be such that*

(42-i) *Z is $(d, \preceq)$-closed (that is: $Z = \mathrm{cl}_{(\preceq)}(Z)$);*
(42-ii) *$(Z, d, \preceq)$ is regular: each $(\preceq)$-ascending sequence in Z is d-Cauchy;*
(42-iii) *$(Z, d, \preceq)$ is fully complete: each $(\preceq)$-ascending d-Cauchy sequence in Z is $(d \preceq)$-convergent (in X).*

Further, take the subset $M \subseteq Z$. Then, for each $u \in M$, there exists some other point $v \in Z$, with

(42-a) $u \preceq v$ *and* v *is* $(\preceq)$*-maximal over* Z;
(42-b) $u \preceq v$ *and* v *is* $(\preceq)$*-maximal over* M.

Proof. By the admitted hypotheses (42-ii) and (42-iii) upon $(Z, d, \preceq)$, the Kang–Park r-pseudometric maximal principle (KP-rp-mp) is applicable to these data. As a consequence of this, we have that, given the subset $M \subseteq Z$, the following conclusion holds in (ZF − AC + DC):

$\forall u \in M \subseteq Z$, $\exists v \in \text{cl}_{(\preceq)}(Z)$, with $u \preceq v$ and v is $(d, \preceq)$-maximal over Z.

Taking (42-i) into account, we have (by the written conclusion)

$v \in Z$ and v is $(d, \preceq)$-maximal over Z; hence v is $(\preceq)$-maximal over Z
(if we remember that d is rs-pseudometric).

But then, via $M \subseteq Z$ (and a preceding fact), $v \in Z$ is $(\preceq)$-maximal in M, and the conclusion follows. □

The obtained Zorn–Bourbaki rs-pseudometric maximal principle (ZB-rsp-mp) has, essentially, a theoretical importance only. However, from a practical viewpoint, this result is not very useful — in this form — because the points of Z cannot be attained via points of M; so we may ask of whether this is possible in some way. It is our aim in the following to show that a positive answer is available, by taking our rs-pseudometric d and the underlying subset Z according to

(F-tri) d is Frink triangular (hence, a Frink rs-pseudometric);
(kl-M) $Z = \text{kl}_{(\preceq)}(M)$, where $\text{kl}_{(\preceq)}(.)$ is the introduced closure operator.

Precisely, the following $(\preceq)$-maximal statement (referred to as: Zorn–Bourbaki Frink rs-pseudometric maximal principle (in short: (ZB-Frsp-mp)) is available.

Theorem 6. *Let the quasi-ordered Frink rs-pseudometric space* $(X, d, \preceq)$ *and the subset* $Y \in \exp(X)$ *be such that*

(43-i) Y *is* $(d, \preceq)$*-closed (that is:* $Y = \text{cl}_{(\preceq)}(Y)$*);*

(43-ii) $(Y, d, \preceq)$ *is regular: each* $(\preceq)$*-ascending sequence in* Y *is* d*-Cauchy;*
(43-iii) $(Y, d, \preceq)$ *is fully complete: each* $(\preceq)$*-ascending* d*-Cauchy sequence in* Y *is* $(d \preceq)$*-convergent (in* X*).*

Further, take the subset $M \subseteq Y$ *and denote* $Z := \mathrm{kl}_{(\preceq)}(M)$*. Then,*

(43-a) $M \subseteq Z$*,* $Z \subseteq Y$ *and* $Z \subseteq \mathrm{cl}(M)$*; hence,* $M \subseteq Z \subseteq Y \cap \mathrm{cl}(M)$*;*

(43-b) *for each* $u \in M$*, there exists some other point* $v \in Z$*, with*

(43-b1) $u \preceq v$ *and* $v \in Y \cap \mathrm{cl}(M)$*;*
(43-b2) v *is* $(\preceq)$*-maximal over* Z*;*
(43-b3) v *is* $(\preceq)$*-maximal over* M*.*

Proof. (i): Clearly, $M \subseteq Z$, by definition. Further,

$Z \subseteq Y$ because $M \subseteq Y$ and Y is $(d, \preceq)$-closed.

On the other hand, by an auxiliary fact above,

$Z \subseteq \mathrm{cl}(M)$ (because $M \subseteq \mathrm{cl}(M)$ and $\mathrm{cl}(M)$ is $(d, \preceq)$-closed),

so by simply combining these, the first part is proved.

(ii): As $Z := \mathrm{kl}_{(\preceq)}(M)$, we have (see above) $Z \in \mathcal{K}[d, \preceq]$ so that

(43-i-Z) Z is $(d, \preceq)$-closed (that is: $Z = \mathrm{cl}_{(\preceq)}(Z)$).

On the other hand, by a hereditary argument we have (under $Z \subseteq Y$)

(43-ii-Z) $(Z, d, \preceq)$ is regular: each $(\preceq)$-ascending sequence in Z is d-Cauchy;
(43-iii-Z) $(Z, d, \preceq)$ is fully complete: each $(\preceq)$-ascending d-Cauchy sequence in Z is $(d \preceq)$-convergent (in X).

And then, under the Zorn–Bourbaki rs-pseudometric maximal principle (ZB-rsp-mp), our conclusions follows. □

(D) Some basic particular cases of these corresponds to the Frink triangular property imposed to our rs-pseudometric d being assured by some related conditions involving such objects.

(I) Suppose in the following that our ambient rs-pseudometric d fulfills:

(B-tri) d is *Bakhtin triangular* (hence, a Bakhtin rs-pseudometric).

As a direct consequence of Zorn–Bourbaki Frink rs-pseudometric maximal principle (ZB-Frsp-mp), one gets the following statement, referred to as Zorn–Bourbaki Bakhtin rs-pseudometric maximal principle (in short: (ZB-Brsp-mp)).

Theorem 7. *Let the quasi-ordered Bakhtin rs-pseudometric space* $(X, d, \preceq)$ *and the subset* $Y \in \exp(X)$ *be such that*

(44-i) Y *is* $(d, \preceq)$*-closed (that is:* $Y = \mathrm{cl}_{(\preceq)}(Y)$*);*
(44-ii) $(Y, d, \preceq)$ *is regular: each* $(\preceq)$*-ascending sequence in* Y *is* d*-Cauchy;*
(44-iii) $(Y, d, \preceq)$ *is fully complete: each* $(\preceq)$*-ascending* d*-Cauchy sequence in* Y *is* $(d \preceq)$*-convergent (in* X*).*

Further, take the subset $M \subseteq Y$ *and denote* $Z := \mathrm{kl}_{(\preceq)}(M)$*. Then,*
(44-a) $M \subseteq Z$*,* $Z \subseteq Y$ *and* $Z \subseteq \mathrm{cl}(M)$*; hence,* $M \subseteq Z \subseteq Y \cap \mathrm{cl}(M)$*;*
(44-b) *for each* $u \in M$*, there exists some other point* $v \in Z$*, with*

(44-b1) $u \preceq v$ *and* $v \in Y \cap \mathrm{cl}(M)$*;*
(44-b2) v *is* $(\preceq)$*-maximal over* Z*;*
(44-b3) v *is* $(\preceq)$*-maximal over* M*.*

(II) A basic particular case of this corresponds to the rs-pseudometric d fulfilling

(tri) d is *triangular*: $d(x, z) \leq d(x, y) + d(y, z)$, $\forall x, y, z \in X$,

when (see above) d is a *quasi-metric* on X. As a direct consequence of Zorn–Bourbaki Bakhtin rs-pseudometric maximal principle (ZB-Brsp-mp), one gets the following statement, referred to as Zorn–Bourbaki quasi-metric maximal principle (in short: (ZB-qm-mp)).

Theorem 8. *Let the quasi-ordered quasi-metric space* $(X, d, \preceq)$ *and the subset* $Y \in \exp(X)$ *be such that*

(45-i) Y *is* $(d, \preceq)$*-closed (that is:* $Y = \mathrm{cl}_{(\preceq)}(Y)$*);*
(45-ii) $(Y, d, \preceq)$ *is regular: each* $(\preceq)$*-ascending sequence in* Y *is* d*-Cauchy;*
(45-iii) $(Y, d, \preceq)$ *is fully complete: each* $(\preceq)$*-ascending* d*-Cauchy sequence in* Y *is* $(d \preceq)$*-convergent (in* X*).*

Further, take the subset $M \subseteq Y$ *and denote* $Z := \mathrm{kl}_{(\preceq)}(M)$*. Then,*

(45-a) $M \subseteq Z$, $Z \subseteq Y$ *and* $Z \subseteq \mathrm{cl}(M)$*; hence,* $M \subseteq Z \subseteq Y \cap \mathrm{cl}(M)$;
(45-b) *for each* $u \in M$*, there exists some other point* $v \in Z$*, with*

(45-b1) $u \preceq v$ *and* $v \in Y \cap \mathrm{cl}(M)$;
(45-b2) v *is* $(\preceq)$*-maximal over* Z;
(45-b3) v *is* $(\preceq)$*-maximal over* M.

In particular, when, in addition

(sym) d is *symmetric* (hence, a *metric* on X),

this last results may be viewed as a metrical variant of the Zorn-Bourbaki maximal principle [6]. An early version of it was formulated in Turinici [40]; note that it includes the one due to Dancs *et al.* [16]. In particular, when $M = X$, the underlying result is just the ordering principle in Kang and Park [25]. Moreover, by the characterization of regularity condition, this statement includes as well the related one in Turinici [56]. Finally, all these are ultimately equivalent with the Brezis–Browder's ordering principle [7]; we do not give details. Further aspects may be found in the papers by Turinici [39, 43].

(E) A basic application of these facts is to *local* monotone variational principles. Let X be a nonempty set; and $(\leq)$ be a quasi-order on it. Then, let $d(.,.)$ be a pseudometric over X; endowed with

(r-s) d is *reflexive sufficient*: $d(x, y) = 0$ iff $x = y$;
(tri) d is *triangular*: $d(x, z) \leq d(x, y) + d(y, z)$, $\forall x, y, z \in X$;

as precise, it will be referred to as a *quasi-metric* over X. All quasi-order notions to be used refer to the *dual* $(\geq)$, but these may be also formulated in terms of our initial relation. Precisely, take the structure $(X, d, \geq)$ according to

(com) $(X, d, \geq)$ is *complete*: each $(\geq)$-ascending d-Cauchy sequence (x_n) in X is d-convergent; also referred to as: d is $(\geq)$-*complete;*
(conv-dom) $(X, d, \geq)$ is *conv-dominated*:
the limit of each $(\geq)$-ascending sequence in X is a $(\geq)$-upper bound of it;
also referred to as: $(\geq)$ is *d-selfclosed.*

In addition, let the function $\varphi : X \to R \cup \{\infty\}$ be such that

(inf-prop) φ is *proper* ($\mathrm{Dom}(\varphi) \neq \emptyset$);
(ge-lsc) φ is $(d, \geq)$-*lsc* on X;
($[\varphi \leq t] := \{x \in X; \varphi(x) \leq t\}$ is $(d, \geq)$-closed, $\forall t \in R$).

A basic particular case of this last condition upon φ is the one described as follows:

Proposition 17. *Suppose that $(X, d, \geq)$ is conv-dominated and φ is $(\leq)$-increasing ($x \leq y$ implies $\varphi(x) \leq \varphi(y)$). Then, necessarily, φ is $(d, \geq)$-lsc on X.*

Proof. Let $t \in R$, $x \in X$, and the sequence (x_n) in $[\varphi \leq t]$ be such that

(x_n) is $(\geq)$-ascending and $x_n \xrightarrow{d} x$ as $n \to \infty$.

By the conv-dominated assumption,

$(x_n \geq x, \forall n)$; whence, $(\varphi(x_n) \geq \varphi(x), \forall n)$ [as φ is $(\leq)$-increasing].

But then, $\varphi(x) \leq t$, and the conclusion follows. □

Returning to our general setting, let the subset $G \in \exp(X)$ be taken as

(phi-adm) (G, φ) is *admissible*: $G \cap \mathrm{Dom}(\varphi) \neq \emptyset$ and $\inf \varphi(\mathrm{cl}(G)) > -\infty$.

The following (local type) variational statement [referred to as: quasi-order Ekeland Variational Principle on quasi-metric spaces; in short: (EVP-qo-qms)] is a basic tool for our future developments.

Theorem 9. *Under the precise general assumptions, take some $u \in G \cap \mathrm{Dom}(\varphi)$. There exists then some point $v \in X$, with the properties*

(46-a) *$v \in \mathrm{cl}(G)$ and $v \in \mathrm{Dom}(\varphi)$; hence $v \in \mathrm{cl}(G) \cap \mathrm{Dom}(\varphi)$;*
(46-b) *$u \geq v, d(u, v) \leq \varphi(u) - \varphi(v)$ (hence $\varphi(u) \geq \varphi(v)$);*
(46-c) *$d(v, x) > \varphi(v) - \varphi(x)$, for each $x \in G(v, \geq) \setminus \{v\}$.*

Proof. Let $(\sqsubseteq)$ stand for the quasi-order on X

$x \sqsubseteq y$ iff $x \leq y$, $\varphi(x) \leq \varphi(y)$.

As usual, we denote by $(\sqsupseteq)$ the dual of $(\sqsubseteq)$:

$$x \sqsupseteq y \text{ iff } y \sqsubseteq x; \text{ that is: } x \geq y, \ \varphi(x) \geq \varphi(y).$$

Then, let $(\preceq)$ stand for the relation (over X):

$$x \preceq y \text{ iff } x \geq y, \ d(x,y) + \varphi(y) \leq \varphi(x).$$

Clearly, $(\preceq)$ is a quasi-order on X, with, in addition,

$(\preceq)$ is coarser than $(\sqsupseteq)$: $x \preceq y$ implies $x \sqsupseteq y$.

Moreover, $(\preceq)$ is antisymmetric — hence, a (partial) order — on $\mathrm{Dom}(\varphi)$, as it can be directly seen. Denote

$$Y = \mathrm{cl}(G)(u, \sqsupseteq); \text{ hence, } u \in Y \subseteq \mathrm{cl}(G) \cap \mathrm{Dom}(\varphi);$$

note that, by the above, $(\preceq)$ is a (partial) order on Y.

We claim that conditions of Zorn–Bourbaki quasi-metric maximal principle (ZB-qm-mp) are fulfilled over $(X, d, \preceq)$, with respect to the subset Y.

Step 1. Let (x_n) be a $(\preceq)$-ascending sequence in Y; that is, (via $Y \subseteq \mathrm{Dom}(\varphi)$)

(asc-1) $i \leq j$ implies $x_i \preceq x_j$; hence, $x_i \geq x_j$, $d(x_i, x_j) \leq \varphi(x_i) - \varphi(x_j)$.

(1-1) The sequence $(\varphi(x_n))$ is descending and (via $Y \subseteq \mathrm{cl}(G)$) bounded from below; hence a Cauchy one. This, along with (asc-1) shows that (x_n) is a $(\preceq)$-ascending d-Cauchy sequence; wherefrom, $(Y, d, \preceq)$ is regular.

(1-2) Under the same premise, (x_n) is a $(\geq)$-ascending sequence in Y. This, along with $(X, d, \geq)$ being complete, gives us an element $x \in X$ with $x_n \xrightarrow{d} x$ as $n \to \infty$; which proves that $(Y, d, \preceq)$ is complete (in X).

Step 2. Let again (x_n) be a $(\preceq)$-ascending sequence in Y; that is, condition (asc-1) holds, as well as

(asc-2) $(\forall n)$: $u \sqsupseteq x_n$ (that is, $u \geq x_n$, $\varphi(u) \geq \varphi(x_n)$).

As precise, there exists an element $x \in X$ with $x_n \xrightarrow{d} x$ as $n \to \infty$.

(2-1) By the remarks above, (x_n) is, in particular, a $(\geq)$-ascending sequence in Y with $x_n \xrightarrow{d} x$ as $n \to \infty$. Note that, in view of $Y \subseteq \text{cl}(G)$, we must have $x \in \text{cl}(G)$. On the other hand, let the index $i \geq 0$ be arbitrary fixed. Clearly,

(rela-1) $(\forall m \geq i)$: $x_i \geq x_m$, $d(x_i, x_m) \leq \varphi(x_i) - \varphi(x_m)$ (whence, $\varphi(x_i) \geq \varphi(x_m)$).

As $(X, d, \geq)$ is conv-dominated and φ is $(d, \geq)$-lsc, one derives (via (asc-1))

(rela-2) $x_i \geq x$ and $\varphi(x_i) \geq \varphi(x)$; that is, $x_i \sqsupseteq x$.

This, in the particular case of $i = 0$, yields

$u \sqsupseteq x$; that is, $x \in Y$,

proving that Y is $(d, \preceq)$-closed and $(Y, d, \preceq)$ is complete.

(2-2) Returning to the general case (relative to $i \geq 0$) a combination of (rela-1) and (rela-2) yields

$(\forall m \geq i)$: $d(x_i, x_m) \leq \varphi(x_i) - \varphi(x)$.

Taking the triangular inequality into account, we get

$(\forall m \geq i)$: $d(x_i, x) \leq d(x_i, x_m) + d(x_m, x) \leq \varphi(x_i) - \varphi(x) + d(x_m, x)$.

Passing to limit as $m \to \infty$, one derives (again via (rela-2))

$(\forall i)$: $x_i \geq x$ and $d(x_i, x) \leq \varphi(x_i) - \varphi(x)$; that is, $x_i \preceq x$,

proving that $(Y, d, \preceq)$ is conv-dominated; hence, fully complete.

Summing up, all conditions in Zorn–Bourbaki quasi-metric maximal principle (ZB-qm-mp) are indeed fulfilled over $(X, d, \preceq)$, with respect to the subset $Y = \text{cl}(G)(u, \sqsupseteq)$. In this case, put

$M = G \cap Y$, $Z = \text{kl}_{(\preceq)}(M)$; hence, $u \in M$ (by the choice of Y).

By the first conclusion of the underlying maximal principle,

(con-1) $M \subseteq Z$, $Z \subseteq Y$ and $Z \subseteq \text{cl}(M)$; hence, $M \subseteq Z \subseteq Y \cap \text{cl}(M)$.

Moreover, by the second conclusion of the underlying maximal principle it follows that, for the starting $u \in M$, there exists some other point $v \in Z$, with

(con-2a) $u \preceq v$ and $v \in Y \cap \mathrm{cl}(M)$;
(con-2b) v is $(\preceq)$-maximal over Z;
(con-2c) v is $(\preceq)$-maximal over M.

We now show that, from these, one derives all needed conclusions.

(C-1) The first conclusion in the statement is clear, by (con-2a) (the second part), and the immediate relations

$$Y \subseteq \mathrm{Dom}(\varphi) \text{ and } \mathrm{cl}(M) \subseteq \mathrm{cl}(G).$$

(C-2) The second conclusion in the statement follows via (con-2a) (the first part).

(C-3) To establish the third conclusion in the statement, we proceed by contradiction. Precisely, assume that the underlying conclusion would be false:

there exists $x \in G(v, \geq) \setminus \{v\}$ with $d(v, x) \leq \varphi(v) - \varphi(x)$.

By the very definition of $(\preceq)$, the obtained point $x \in G$ has the properties

$v \preceq x$ and $v \neq x$ (hence, $d(v, x) > 0$, as d is a quasi-metric).

In addition, as $(\preceq)$ is coarser that $(\sqsupseteq)$, the same point $x \in G$ fulfills

$v \sqsupseteq x$; whence (via $u \sqsupseteq v$), $u \sqsupseteq x$; that is, $x \in Y$; so, $x \in G \cap Y = M$.

But then, in view of $v \neq x$, the $(\preceq)$-maximality of v in M would be contradicted. Hence, the working hypothesis cannot be accepted; and our third conclusion in the statement is retainable as well. The proof is thereby complete. □

As already precise, the $(d, \geq)$-lsc condition upon φ is obtainable (under the conv-dominated assumption) under

φ is $(\leq)$-increasing ($x \leq y \Longrightarrow \varphi(x) \leq \varphi(y)$).

In this case, Theorem 9 may be viewed as a (local and) quasi-metric version of the monotone variational principle in Turinici [45]. Further aspects may be found in Hyers *et al.* [24, Chapter 5].

(F) A basic particular case corresponds to the choice

(triv-qo) $(\leq) = (\geq) = X \times X$ (=the *trivial quasi-order* on X).

Let X be a non-empty set, and $d(.,.)$ be a quasi-metric over X. Suppose that the general condition holds

(com) (X, d) is complete: each d-Cauchy sequence (x_n) in X is d-convergent; also referred to as d is *complete.*

In addition, let the function $\varphi : X \to R \cup \{\infty\}$ be such that

(prop) φ is proper ($\mathrm{Dom}(\varphi) \neq \emptyset$);
(lsc) φ is d-lsc on X ($[\varphi \leq t] := \{x \in X; \varphi(x) \leq t\}$ is d-closed, $\forall t \in R$).

Finally, let the (non-empty) subset G of X be taken according to

(phi-adm) G is φ*-admissible*: $G \cap \mathrm{Dom}(\varphi) \neq \emptyset$ and inf $\varphi(\mathrm{cl}(G)) > -\infty$.

The following (local-type) variational statement [referred to as: Ekeland Variational Principle on quasi-metric spaces; in short: (EVP-qms)] is available.

Theorem 10. *Under the precise general assumptions, take some* $u \in G \cap \mathrm{Dom}(\varphi)$. *There exists then some point* $v \in X$, *with the properties*

(47-a) $v \in \mathrm{cl}(G)$ *and* $v \in \mathrm{Dom}(\varphi)$; *hence* $v \in \mathrm{cl}(G) \cap \mathrm{Dom}(\varphi)$;
(47-b) $d(u, v) \leq \varphi(u) - \varphi(v)$ *(hence* $\varphi(u) \geq \varphi(v)$*);*
(47-c) $d(v, x) > \varphi(v) - \varphi(x)$, *for each* $x \in G \setminus \{v\}$.

Proof. We show that the preceding statement is applicable here (under the trivial choice of $(\leq)$). In fact, under such a context,

(47-d) $(X, d, X \times X)$ is complete and conv-dominated;
(47-e) φ is proper and $(d, X \times X)$-lsc;

hence the claim. Adding the conclusions of that statement, we are done. □

In particular, letting d = metric and $G = X$ in (EVP-qms), one gets the following variational statement [referred to as Ekeland Variational Principle on metric spaces; in short: (EVP-m)].

Theorem 11. *Let (X, d) be a complete metric space, and $\varphi : X \to R \cup \{\infty\}$ be proper, d-lsc, bounded from below. Then, for each $u \in \mathrm{Dom}(\varphi)$, there exists some point $v \in \mathrm{Dom}(F)$, with the properties*

(48-a) $d(u, v) \leq \varphi(u) - \varphi(v)$ *(hence $\varphi(u) \geq \varphi(v)$);*
(48-b) $d(v, x) > \varphi(v) - \varphi(x)$*, for each $x \in X \setminus \{v\}$.*

This (metric) principle, due to Ekeland [19], found some basic applications to control and optimization, generalized differential calculus, critical point theory and global analysis; we refer to the quoted paper for a survey of these. So, it cannot be surprising that, soon after its formulation, many extensions of (EVP-m) were proposed. For example, the *dimensional* way of extension refers to the ambient space (R) of $\varphi(X)$ being substituted by a (topological or not) vector space. An account of the results in this area is to be found in Goepfert *et al.* [21, Chapter 3]. The metrical extension of the same consists in conditions imposed upon our metric being relaxed. Some of these extensions were already stated; for the remaining ones, we refer to Hyers *et al.* [24, Chapter 5]; see also Turinici [45].

By the developments above, we therefore have the following implications:

(DC) $\Longrightarrow$ (KP-rp-mp) $\Longrightarrow$ (ZB-rsp-mp) $\Longrightarrow$ (ZB-Frsp-mp);
(ZB-Frsp-mp) $\Longrightarrow$ (ZB-Brsp-mp) $\Longrightarrow$ (ZB-qm-mp);
(ZB-qm-mp) $\Longrightarrow$ (EVP-qo-qms) $\Longrightarrow$ (EVP-qms) $\Longrightarrow$ (EVP-m).

So, we may ask whether these may be reversed. Clearly, the natural setting for solving this problem is the strongly reduced system (ZF-AC).

Let X be a non-empty set; and $(\leq)$ be a (partial) order on it. We say that $(\leq)$ has the *inf-lattice* property, provided $\inf\{x, y\}$ exists, for all $x, y \in X$. Remember that $z \in X$ is a $(\leq)$-*maximal* element if $X(z, \leq) = \{z\}$; the class of all these points will be denoted as $\max(X, \leq)$. Call $(\leq)$, a *Zorn order* when

$\max(X, \leq)$ is non-empty and *cofinal* in X
(for each $u \in X$ there exists a $(\leq)$-maximal $v \in X$ with $u \leq v$).

Further aspects are to be described in a metric setting. Let $d(.,.)$ be a metric over X; and $\varphi : X \to R_+$ be some function. Then, the natural choice for $(\le)$ above is

$$x \le_{(d,\varphi)} y \text{ iff } d(x,y) \le \varphi(x) - \varphi(y);$$

referred to as the *Brøndsted order* [9] attached to (d,φ). Denote

$$X(x,\rho) = \{u \in X; d(x,u) < \rho\},\ x \in X,\ \rho > 0$$

[the *open sphere* with center x and radius ρ]. Call (X,d) *discrete* when

for each $x \in X$ there exists $\rho = \rho(x) > 0$ such that $X(x,\rho) = \{x\}$.

Note that, under such an assumption, any function $\psi : X \to R$ is continuous over X. However, this is not extendable to the *d-Lipschitz property*

$$|\psi(x) - \psi(y)| \le Ld(x,y),\ x,y \in X, \text{ for some } L > 0;$$

hence, all the more, to the *d-non-expansive property* $(L = 1)$.

Now, the statement below is a particular case of (EVP-m):

Theorem 12. *Let the metric space (X,d) and the function $\varphi : X \to R_+$ satisfy*

(49-i) *(X,d) is discrete bounded and complete;*
(49-ii) *$(\le_{(d,\varphi)})$ has the inf-lattice property;*
(49-iii) *φ is d-nonexpansive and $\varphi(X)$ is countable.*

Then, $(\le_{(d,\varphi)})$ is a Zorn order.

We shall refer to it as: the discrete Lipschitz countable version of (EVP-m) (in short: (EVP-m-dLc)). Clearly, (EVP-m) $\Longrightarrow$ (EVP-m-dLc). The remarkable fact is that this last principle yields (DC); and completes the circle between all these.

Proposition 18. *We have the inclusion (EVP-m-dLc) $\Longrightarrow$ (DC) [in the strongly reduced system (ZF-AC)]. So (by the above),*

- **(49-1)** *the maximal/variational principles (KP-rp-mp), (ZB-rsp-mp), (ZB-Frsp-mp), (ZB-Brsp-mp), (ZB-qm-mp), (EVP-qo-qms), (EVP-qms), and (EVP-m) are all equivalent with (DC); hence, mutually equivalent;*
- **(49-2)** *each intermediary maximal/variational statement (VP), fulfilling (DC) $\Longrightarrow$ (VP) $\Longrightarrow$ (EVP-m) is equivalent with both (DC) and (EVP-m).*

For a complete proof, see Turinici [56]. In particular, when the inf-lattice and nonexpansive properties are ignored in (EVP-m-dLc), the last result above reduces to the one in Brunner [10]. Note that, in the same particular setting, a different proof of the underlying inclusion was provided in Dodu and Morillon [18]. Further aspects may be found in Schechter [37, Chapter 19, Section 19.51].

5. Main Results

Let $(X, d, \leq)$ be quasi-ordered quasi-metric space. The basic conditions to be imposed involve the dual structure $(X, d, \geq)$; and have the form

(bcond-1) $(X, d, \geq)$ is *complete*: each $(\geq)$-ascending d-Cauchy sequence in X is d-convergent; also referred to as d is $(\geq)$-*complete*;
(bcond-2) $(X, d, \geq)$ is *conv-dominated*:
the limit of each $(\geq)$-ascending sequence in X is a $(\geq)$-upper bound of it;
also referred to as $(\geq)$ is d-*selfclosed.*

In addition to these, let us accept the essential specific local condition

(bcond-3) $(X, d, \geq)$ is *full*:
$(\forall u \in X, \forall \delta > 0)$: $X[u, \delta; \geq] := X[u, \delta] \cap X(u, \geq)$
has a non-empty intersection with $X \setminus \{u\}$,
where $X[u, \delta] = \{x \in X; d(u, x) \leq \delta\}$, $X(u, \geq) = \{x \in X; u \geq x\}$.

(A) Passing to the remaining specific conditions, the first group involves a sequence of maps $(\Gamma_i : X \to R; i \geq 0)$ to be introduced here. Precisely, assume that

(ga-1) $(\forall i)$: $\sup[\Gamma_i(X)] = \infty$; hence
$[\Gamma_i \geq \sigma] := \{x \in X; \Gamma_i(x) \geq \sigma\}$ is nonempty, $\forall \sigma \in R$;
(ga-2) $(\forall i)$: $(\exists(\lambda_i, \mu_i), 0 < \lambda_i < 1 < \mu_i)$, such that
$(\forall \rho \geq \mu_i)$: $x \in [\Gamma_i \geq \rho]$ and $d(x, y) \leq \lambda_i$ imply $y \in [\Gamma_i \geq \rho - \mu_i]$;
(ga-3) $\nu := \liminf_i(\lambda_i/2\mu_i) > 0$ (hence, $0 < \nu \leq 1/2 < 1$).

Remark 5. The following consequences of these facts are useful for us

(501) $(\forall i, \forall \tau \geq \mu_i)$: $u \in [\Gamma_i \geq \tau]$ implies $X[u, \lambda_i] \subseteq [\Gamma_i \geq \tau - \mu_i]$;
(501-a) $(\forall i, \forall \tau \geq \mu_i)$: $\mathrm{cl}[\Gamma_i \geq \tau] \subseteq [\Gamma_i \geq \tau - \mu_i]$.

In fact, the former is clear, by (ga-2). Concerning the latter, note that, given $v \in \mathrm{cl}[\Gamma_i \geq \tau]$, there exists $u \in [\Gamma_i \geq \tau]$ with $d(u, v) \leq \lambda_i$. Combining with (ga-2), yields $v \in [\Gamma_i \geq \tau - \mu_i]$; hence the assertion.

The second group of conditions refers to some functional $F : X \to R \cup \{\infty\}$. Namely, suppose that

(F-1) F is proper ($\mathrm{Dom}(F) \neq \emptyset$);
(F-2) F is $(d, \geq)$-lsc on X
($[F \leq t] := \{x \in X; F(x) \leq t\}$ is $(d, \geq)$-closed, $\forall t \in R$).

The last group consists in a combined requirement involving $(F; (\Gamma_i))$. To state it, we need a convention. Let $\Gamma : X \to R$ be a mapping. Denote

$$J(F; \Gamma)(r) = \inf F([\Gamma \geq r]) = \inf\{F(y); \Gamma(y) \geq r\},\ r \in R.$$

By this convention,

$$J(F; \Gamma)(.) \text{ is increasing from } R \text{ to } R \cup \{-\infty, \infty\}.$$

This, in the particular case of the sequence (Γ_i) above, tells us that

$$(\forall i)\text{: } J(F; \Gamma_i)(.) \text{ is increasing from } R \text{ to } R \cup \{-\infty, \infty\}.$$

We are now in position to formulate our last condition involving $(F; (\Gamma_i))$; namely,

(F-Ga) $(F; (\Gamma_i))$ is asymptotic uniformly bounded below:
there exist $\alpha \geq 0$, $m \in N$, $F_* \in R$, such that $(J(F; \Gamma_i)(\alpha) \geq F_*$, $\forall i \geq m)$.

Note that, without any loss, one may assume that $m = 0$; that is,

(F-Ga-1) $(F; (\Gamma_i))$ is uniformly bounded below:
there exist $\alpha \geq 0$, $F_* \in R$, such that $(J(F; \Gamma_i)(\alpha) \geq F_*$, $\forall i \geq 0)$;

just pass to the (truncated) subsequence $(\widetilde{\Gamma}_i := \Gamma_{m+i}; i \geq 0)$ for getting this. Moreover, again without loss, one may assume that $\alpha = 0$ in this condition:

(F-Ga-2) $(F; (\Gamma_i))$ is global uniformly bounded below:
there exists $F_* \in R$, such that $(J(F; \Gamma_i)(0) \geq F_*, \forall i \geq 0)$;

just pass to the translated sequence $(\Delta_i(.) = \Gamma_i(.) - \alpha; i \geq 0)$ to derive this variant.

Remark 6. A basic consequence of this fact writes

(502) $(\forall i, \forall \rho \geq \mu_i)$: $\inf F(\mathrm{cl}[\Gamma_i \geq \rho]) \geq F_*$.

In fact, let $i \in N$ and $\rho \geq \mu_i$ be arbitrary fixed. Then (by a previous observation),

$\mathrm{cl}[\Gamma_i \geq \rho] \subseteq [\Gamma_i \geq \rho - \mu_i]$; whence, $\inf F(\mathrm{cl}[\Gamma_i \geq \rho]) \geq \inf F([\Gamma_i \geq \rho - \mu_i]) = J(F; \Gamma_i)(\rho - \mu_i) \geq J(F; \Gamma_i)(0) \geq F_*$;

wherefrom, all is clear.

(B) With these preliminaries, we may now pass to the formulation of our problem. The (extended) functional sequence

$J(F; \Gamma_i)(\sigma) := \inf F([\Gamma_i \geq \sigma]),\ i \in N,\ \sigma \in R_+$

is well defined over $\mathcal{F}(R_+, R \cup \{\infty\})$, via (ga-1) and

$J(F; \Gamma_i)(\sigma) \geq J(F; \Gamma_i)(0) \geq F_* > -\infty,\ \forall i \in N,\ \forall \sigma \geq 0$.

Moreover, by this very convention,

$(\forall i)$: $J(F; \Gamma_i)(.)$ is increasing on R_+:
$0 \leq \sigma_1 \leq \sigma_2$ implies $J(F; \Gamma_i)(\sigma_1) \leq J(F; \Gamma_i)(\sigma_2)$.

Further, the (extended) space asymptotic function

$J(F; (\Gamma_i))(\sigma) := \liminf_i J(F; \Gamma_i)(\sigma),\ \sigma \geq 0$

is well defined in $\mathcal{F}(R_+, R \cup \{\infty\})$, in view of

(503) $J(F; (\Gamma_i))(\sigma) \geq F_* > -\infty,\ \forall \sigma \geq 0$.

In addition, by the properties of lim inf operator,

$J(F;(\Gamma_i))(.)$ is increasing on R_+:
$0 \le \sigma_1 \le \sigma_2$ implies $J(F;(\Gamma_i))(\sigma_1) \le J(F;(\Gamma_i))(\sigma_2)$.

Finally, as a direct consequence of this, the asymptotic limit

$$J(F;(\Gamma_i))(\infty) := \lim_{\sigma\to\infty} J(F;(\Gamma_i))(\sigma) = \sup_{\sigma\ge 0} J(F;(\Gamma_i))(\sigma)$$

exists as an element of $R \cup \{\infty\}$, in view of

$$\text{(503-a)} \quad F_* \le J(F;(\Gamma_i))(\sigma) \le J(F;(\Gamma_i))(\infty) \le \infty, \ \forall \sigma \ge 0.$$

When $J_\infty := J(F;(\Gamma_i))(\infty) = \infty$, the functional F will be referred to as (Γ_i)-*coercive*. It is our aim in the following to get sufficient conditions in order that such a property be attained. These, as a rule, require a *differential* setting.

Given the proper function $F : X \to R \cup \{\infty\}$, denote for each $u \in \mathrm{Dom}(F)$,

$$\text{(slo-1)} \quad \nabla_{(\ge)} F(u) = \limsup \left\{ \frac{F(u) - F(x)}{d(u,x)};\ d(u,x) \to 0+, u \ge x \right\};$$

that is,

$$\nabla_{(\ge)} F(u) = \inf_{\delta>0} \sup \left\{ \frac{F(u) - F(x)}{d(u,x)}; x \in X[u,\delta;\ge] \setminus \{u\} \right\},$$

as well as

$$\text{(slo-2)} \quad |\nabla_{(\ge)}| F(u) = \max \{0, \nabla_{(\ge)} F(u)\}.$$

Clearly, the right member of (slo-1) is meaningful, by (bcond-3); wherefrom, (slo-1) and (slo-2) are well defined. The objects in question are a quasi-metric version of the ones introduced (in a trivial quasi-order and Banach space context) by DeGiorgi *et al.* [17]; we call these, the *relative/absolute* $(d,\ge)$-*slope* of F at u. The usefulness of these particular tools for the critical point theory (over Banach spaces) was underlined by Corvellec *et al.* [15]. Here, we shall establish that the introduced slope concepts are appropriate for our quasi-order quasi-metrical setting as well.

The following local type statement is a basic step toward the desired answer. For each $\beta > 0$, $i \in N$, define the concepts

(admi) (β, i) is *admissible*: the class $A(\beta, i)$ of all $r \in R_+$ with $r \geq 1/\beta$ and
$J_\infty - \beta^2 < J(F; \Gamma_i)(r) \leq J(F; \Gamma_i)(4r) < J_\infty + \beta^2$ is non-empty;
(lim-admi) (β, i) is *limit admissible*: the class $B(\beta, i)$ of all couples $(r, v) \in R_+ \times \mathrm{Dom}(F)$ with $r \in A(\beta, i)$, $\Gamma_i(v) \geq r$, $|F(v) - J_\infty| < \beta^2$, and $|\nabla_{(\geq)}|F(v) \leq \beta$, is non-empty;

clearly, both these are effective only when $-\infty < J_\infty < \infty$. Further, let us say that the subset M of N is *cofinal* in N, provided

(cofi) for each $n \in N$ there exists $m \in M$ with $n < m$.

Theorem 13. *Suppose that*

(fin-Joo) $J_\infty := J(F; (\Gamma_i))(\infty) < \infty$ *(hence,* $-\infty < F_* \leq J_\infty < \infty$*).*

Then, the following is true, in (ZF − AC + DC):

(51-a) *for each* $\beta \in]0, \nu[$*, the subset of all* $p \in N$ *with* $(\beta, p) =$ *admissible is cofinal in* N*;*
(51-b) *for each* $\beta \in]0, \nu[$*, the subset of all* $p \in N$ *with* $(\beta, p) =$ *limit admissible is cofinal in* N*.*

Proof. Let $\beta \in]0, \nu[$ be fixed in the sequel; hence,

$0 < \beta < \liminf_i (\lambda_i / 2\mu_i)$.

From the definition of lim inf operator, there exists some rank $m(\beta)$, such that

$(\forall i \geq m(\beta))$: $0 < \beta < \lambda_i / 2\mu_i$; whence (cf. (ga-2)) $(1/\beta) > \mu_i > (\lambda_i/2) > \beta$.

Part 1. For the moment, we have by definition

$(\forall i \geq m(\beta))$: $\sup_{\sigma \geq 0} J(F; (\Gamma_i))(\sigma) = J(F; (\Gamma_i))(\infty) = J_\infty > J_\infty - \beta^2$.

Combining with the increasing property of $r \mapsto J(F; (\Gamma_i))(r)$, one gets

(504) there exists $r \geq 1/\beta$, such that $J(F; (\Gamma_i))(r) > J_\infty - \beta^2$; meaning: $\liminf_i J(F; \Gamma_i)(r) = J(F; (\Gamma_i))(r) > J_\infty - \beta^2$.

This, by the definition of lim inf operator, tells us that there must be some index $k(\beta) \geq m(\beta)$, such that

(504-a) $(\forall i \geq k(\beta))$: $J(F;\Gamma_i)(r) > J_\infty - \beta^2$; hence, $J(F;\Gamma_i)(4r) \geq J(F;\Gamma_i)(r) > J_\infty - \beta^2$.

On the other hand, by definition (and the choice of J_∞)

$$\liminf_i J(F;\Gamma_i)(4r) = J(F;(\Gamma_i))(4r) \leq J_\infty < J_\infty + \beta^2.$$

So, again combining with the definition of lim inf operator, we derive that

(504-b) $\forall p \geq k(\beta)$, $\exists q \in N$, with $p < q$ and $J(F;\Gamma_q)(4r) < J_\infty + \beta^2$.

Putting these last relations together, we derive that

(504-c) $\forall p \geq k(\beta)$, $\exists q \in N$, with $p < q$ and $J_\infty - \beta^2 < J(F;\Gamma_q)(r) \leq J(F;\Gamma_q)(4r) < J_\infty + \beta^2$; which, along with $r \geq 1/\beta$, tells us that $r \in A(\beta, q)$ (whence, $A(\beta, q) \neq \emptyset$); proving that (β, q) is admissible,

and, from this, our first conclusion in the statement follows.

Part 2. It will be sufficient verifying that

(AB-gen) $(\forall \beta \in]0, \nu[, \forall p \geq k(\beta))$:
(β, p) is admissible implies (β, p) is limit admissible

to establish our second conclusion. Precisely, we claim that

(AB-r) $(\forall \beta \in]0, \nu[, \forall p \geq k(\beta))$:
$r \in A(\beta, p)$ implies $(r, v) \in B(\beta, p)$, for some $v \in \mathrm{Dom}(F)$.

To verify the written inclusion, fix some $r \in A(\beta, p)$; that is,

(admi) $r \geq 1/\beta$ (hence, $r \geq 1/\beta > \mu_p$, in view of $p \geq k(\beta)$), and $J_\infty - \beta^2 < J(F;\Gamma_p)(r) \leq J(F;\Gamma_p)(4r) < J_\infty + \beta^2$.

Further, define the data

$$G_p := [\Gamma_p \geq 2r], \ \varphi := (1/\beta)F.$$

Taking (F-Ga-2) into account gives (cf. a previous fact)

$$\inf F(\mathrm{cl}(G_p)) = \inf F(\mathrm{cl}[\Gamma_p \geq 2r]) \geq \inf F_p([\Gamma \geq 2r - \mu_p]) = J(F;\Gamma_p)(2r - \mu_p) \geq J(F;\Gamma_p)(r) \geq F_* > -\infty.$$

On the other hand, by (admi) and the infimum definition,

there exists $u_p \in [\Gamma_p \geq 4r]$, such that $F(u_p) < J_\infty + \beta^2$;
hence, $u_p \in [\Gamma_p \geq 4r] \cap \mathrm{Dom}(F) \subseteq G_p \cap \mathrm{Dom}(F)$.

Combining these with conditions (bcond-1)+(bcond-2) and (F-1)+(F-2), it results that the quasi-order Ekeland Variational Principle on quasi-metric spaces (EVP-qo-qms) applies to the data

(505) $(X, d, \geq)$, $G_p := [\Gamma_p \geq 2r]$, and $\varphi := (1/\beta)F$.

So, for the starting point $u_p \in G_p \cap \mathrm{Dom}(F)$, there must be some $v_p \in X$, with

(505-a) $v_p \in \mathrm{cl}(G_p)$ and $v_p \in \mathrm{Dom}(F)$; hence, $v_p \in \mathrm{cl}(G_p) \cap \mathrm{Dom}(F)$;
(505-b) $u_p \geq v_p$, $\beta d(u_p, v_p) \leq F(u_p) - F(v_p)$ (hence $F(u_p) \geq F(v_p)$);
(505-c) $\beta d(v_p, x) > F(v_p) - F(x)$, for all $x \in G_p(v_p, \geq) \setminus \{v_p\}$.

By (505-a), we have [via $(r \geq 1/\beta > \mu_p)$, (admi), and an auxiliary fact]

(505-d) $v_p \in [\Gamma \geq 2r - \mu_p] \subseteq [\Gamma \geq r]$ so that
$\Gamma_p(v_p) \geq r$ and $F(v_p) \geq J(F; \Gamma_p)(r) > J_\infty - \beta^2$.

This, along with (505-b), gives (combining with (admi) and the definition of u_p)

(505-e) $J_\infty - \beta^2 < F(v_p) \leq F(u_p) < J_\infty + \beta^2$,

proving that v_p is an element in $\mathrm{Dom}(F)$ fulfilling the first and second parts of (lim-adm). Moreover, again via (505-b), one gets (via $p \geq k(\beta)$)

(505-f) $d(u_p, v_p) \leq (1/\beta)2\beta^2 = 2\beta < \lambda_p$, so [by (ga-2)]
$v_p \in [\Gamma_p \geq 4r - \mu_p] \subseteq [\Gamma \geq 3r] \subseteq G_p$,

which improves the evaluation (505-d) of v_p. Finally, by (ga-2) (and (bcond-3)),

$(\forall \delta \in]0, \lambda_p])$: $X[v_p, \delta] \subseteq X[v_p, \lambda_p] \subseteq [\Gamma_p \geq 3r - \mu_p] \subseteq [\Gamma_p \geq 2r] = G_p$;
whence, $X[v_p, \delta; \geq] \subseteq G_p(v_p, \geq)$.

This, along with (505-c), gives

$$\beta \geq \sup\left\{\frac{F(v_p)-F(x)}{d(v_p,x)}; x \in X[v_p,\delta;\geq]\setminus\{v_p\}\right\}, \ \forall \delta \in]0,\lambda];$$

whence

$$\beta \geq \ \nabla_{(\geq)}F(v_p) = \inf_{\delta>0}\sup\left\{\frac{F(v_p)-F(x)}{d(v_p,x)}; x \in X[v_p,\delta;\geq]\setminus\{v_p\}\right\},$$

and the third part of (lim-admi) holds as well. The proof is complete. □

The obtained result may now be used to get an asymptotic type statement that will be directly used further.

Theorem 14. *Suppose that*

(fin-Joo) $J_\infty := J(F;(\Gamma_i))(\infty) < \infty$ *(hence,* $-\infty < F_* \leq J_\infty < \infty$*).*

There exists then, in ($ZF - AC + DC$*), a strictly ascending sequence* $(i(n); n \geq 0)$ *in* N*, and a sequence* $(v_n; n \geq 0)$ *in* X*, with*

(52-a) *(*$v_n \in \mathrm{Dom}(F)$*,* $\forall n$*), and* $\Gamma_{i(n)}(v_n) \to \infty$ *as* $n \to \infty$*;*
(52-b) $F(v_n) \to J_\infty$ *and* $|\nabla_{(\geq)}|F(v_n) \to 0$ *as* $n \to \infty$*.*

Proof. Remember that, given $(\beta, i) \in R_+^0 \times N$, we introduced the concept

(lim-admi) (β, i) is *limit admissible*: the subset $B(\beta, i)$ of all $(r, v) \in R_+ \times \mathrm{Dom}(F)$ with $r \geq 1/\beta$, $\Gamma_i(v) \geq r$, $|F(v) - J_\infty| < \beta^2$, $|\nabla_{(\geq)}|F(v) \leq \beta$ is nonempty.

Denote further

$\mathcal{L}$=the class of all limit admissible couples $(\beta, i) \in R_+^0 \times N$.

By the preceding result we have that, under the finiteness condition above,

(ladm-cofi) for each $\beta \in]0, \nu[$, the subset (of N) $\mathcal{L}^{-1}(\beta) := \{q \in N; (\beta, q) \in \mathcal{L}\}$ is cofinal in N;

or, equivalently, by definition,

(ladm-cofi-1) for each $\beta \in]0, \nu[$ and each $p \in N$, there exists $q \in N(p, <)$ with $q \in \mathcal{L}^{-1}(\beta)$.

Let (β_n) be a strictly descending to zero sequence in $]0, \nu[$; for example, $(\beta_n = 2^{-n-1}\nu; n \geq 0)$. We construct a strictly ascending sequence of ranks $(i(n); n \geq 0)$ according to the iterative procedure

$$i(0) = \min \mathcal{L}^{-1}(\beta_0);\ i(n+1) = \min[N(i(n), <) \cap \mathcal{L}^{-1}(\beta_{n+1})], \quad n \geq 0.$$

This, by definition, means

$((\beta_n, i(n)); n \geq 0)$ is a sequence in $\mathcal{L}$, with
$(i(n); n \geq 0)$= strictly ascending in N.

Note that, for the moment, this construction is carried out upon (ZF-AC).

By definition, $(B(\beta_n, i(n)); n \geq 0)$ is a sequence over $\exp(R_+ \times X)$. This, in combination with the Denumerable Axiom of Choice (AC(N)) (deductible in (ZF − AC + DC)), tells us that there exists a sequence (r_n, v_n) in $R_+ \times X$, with

$$(r_n, v_n) \in B(\beta_n, i(n)), \text{ for all } n;$$

that is, the following limit admissible properties hold

(LA-1) $(\forall n)$: $r_n \geq 1/\beta_n$, $v_n \in \mathrm{Dom}(F)$, and $\Gamma_{i(n)}(v_n) \geq r_n$;
(LA-2) $(\forall n)$: $|F(v_n) - J_\infty| < \beta_n^2$, and $|\nabla_{(\geq)}|F(v_n) \leq \beta_n$.

As a direct consequence of these, one gets (via $\lim_n \beta_n = 0$, $\lim_n r_n = \infty$)

$$\lim_n \Gamma_{i(n)}(v_n) = \infty,\ \lim_n F(v_n) = J_\infty,\ \lim_n |\nabla_{(\geq)}|F(v_n) = 0.$$

But then, the strictly ascending sequence $(i(n); n \geq 0)$ in N and the sequence $(v_n; n \geq 0)$ in X fulfill all needed conclusions. The proof is thereby complete. □

We are now in position to give the promised answer to our coercivity question. Let us consider the "hybrid" condition upon $(F; (\Gamma_i))$

(PS;ge) for each strictly ascending sequence $(i(n); n \geq 0)$ in N and each sequence $(v_n; n \geq 0)$ in X fulfilling ($v_n \in \mathrm{Dom}(F)$, for each $n \geq 0$) as well as

$(F(v_n); n \geq 0)$= convergent, and $|\nabla_{(\geq)}|F(v_n) \to 0$ as $n \to \infty$, we have that $(\Gamma_{i(n)}(v_n); n \geq 0)$ has a bounded subsequence.

This will be referred to as a Palais–Smale condition (modulo $(\geq)$) upon $(F; (\Gamma_i))$.

Theorem 15. *Suppose that (in addition to our general and specific conditions)*

$(F; (\Gamma_i))$ *satisfies a Palais–Smale condition (modulo* $(\geq)$*).*

Then, F *is* (Γ_i)*-coercive (see above).*

Proof. If, by absurd, F is not (Γ_i)-coercive, the relation (fin-Joo) must be true. By Theorem 14, we have promised a strictly ascending sequence $(i(n); n \geq 0)$ in N and a sequence $(v_n; n \geq 0)$ in X, with the properties (52-a) + (52-b). Note that, as a consequence of the latter,

$(F(v_n))$ is convergent and $|\nabla_{(\geq)}|F(v_n) \to 0$ as $n \to \infty$,

and this, via former conclusion (first half) and (PS;ge), tells us that

$(\Gamma_{i(n)}(v_n); n \geq 0)$ must have a bounded subsequence.

On the other hand, as a consequence of the former conclusion (second half),

each subsequence of $(\Gamma_{i(n)}(v_n); n \geq 0)$ is not bounded.

The contradiction at which we arrived shows that our working assumption is not acceptable, and then, our coerciveness property follows. □

In particular, when $d(.,.)$ is a metric on X and $(\leq)$ is the trivial quasi-order on X, Theorem 15 may be viewed as a complement of the result in Motreanu and Motreanu [32]. On the other hand, when $(\leq)$ is a (non-trivial) quasi-order on X and (Γ_i) is a constant sequence, Theorem 15 includes the statement in Motreanu *et al.* [33], refined in Turinici [53, 58]. Further aspects will be discussed elsewhere.

6. Differential Aspects

The obtained results are quasi-metric in nature; so, it would be useful having linear variants of them, expressed in a quasi-norm differential setting.

(A) Let X be a (real) vector space. We say that the function $\Delta : X \to R_+$ is a *quasi-norm* on X, when

(qn-1) Δ is reflexive-sufficient: $\Delta(x) = 0$ iff $x = 0$;
(qn-2) Δ is *subadditive*: $\Delta(x + y) \leq \Delta(x) + \Delta(y)$, for all $x, y \in X$;
(qn-3) Δ is *directional continuous*: $\Delta(\tau x) \to 0$ as $\tau \to 0+$, $\forall x \in X$.

Remark 7. Concerning the last condition, let us consider the family of directional functions (in $\mathcal{F}(R_+)$)

$$\Delta_x(\tau) = \Delta(\tau x),\ \tau \in R_+,\ x \in X.$$

By the properties of Δ, we have

$(\forall x \in X)$: $\Delta_x(0) = 0$ and $\Delta_x(.)$ is sub-additive
$(\Delta_x(\tau_1 + \tau_2) \leq \Delta_x(\tau_1) + \Delta_x(\tau_2),\ \forall \tau_1, \tau_2 \in R_+)$.

Then, the directional continuous property means

$$(\forall x \in X)\text{: } \Delta_x(\tau) \to 0 = \Delta_x(0) \text{ as } \tau \to 0+.$$

In this perspective, it would be natural to ask whether the sub-additive property suffices for the underlying limit condition being retainable. Some related aspects may be found in Matkowski and Swiatkowski [28].

Clearly, each norm $x \mapsto ||x||$ on X is a quasi-norm. The reciprocal of this is not in general true. An appropriate example certifying our claim is in fact the linear version of the previous one due to Alsulami *et al.* [1]. For completeness reasons, we will rephrase it, under our linear setting.

Example 3. Take $X = R$; and let $x \mapsto |x|$ stand for the usual modulus function; it acts as a norm over X. Further, take some $\beta \in]0, 1[$, and define a mapping $\Delta : R \to R_+$ as

$$(\Delta(t) = \beta|t|, \text{ if } t < 0),\ \ (\Delta(t) = |t|, \text{ if } t \geq 0).$$

Clearly, $\Delta(.)$ is reflexive-sufficient and subadditive. Moreover, Δ is directional continuous, in view of

$$\Delta(\tau x) = \beta\tau|x| \to 0 \text{ as } \tau \to 0+, \text{ when } x < 0;$$
$$\Delta(\tau x) = \tau|x| \to 0 \text{ as } \tau \to 0+, \text{ when } x \geq 0.$$

Hence, summing up, $\Delta(.)$ is a quasi-norm over X. Unfortunately, $\Delta(.)$ is not a norm; because, for instance, $\Delta(-1) = \beta < 1 = \Delta(1)$.

Given the quasi-norm $\Delta(.)$ over X, the mapping (in $\mathcal{F}(X \times X, R_+)$)

$$d(x,y) = \Delta(x-y),\ x,y \in X$$

defines a quasi-metric on X, which, in addition, fulfills

d is invariant to translations: $d(x,y) = d(x+a, y+a)$, $x,y,a \in X$.

Then, let K be some part of X with

K is a *convex cone* of X ($\xi K + \eta K \subseteq K$, for all $\xi, \eta \geq 0$), with the *non-degenerate* property ($K \neq \{0\}$).

Denote by $(\leq)$ and $(\geq)$ the associated with K quasi-orders:

$$x \leq y \text{ iff } y - x \in K;\ x \geq y \text{ iff } x - y \in K.$$

We are now ready to formulate the basic conditions needed in the sequel. As a rule, the quasi-order $(\geq)$ will be essentially used here.

(I) The first condition to be considered reads

(X;ge;com) $(X, d, \geq)$ is complete:
each $(\geq)$-ascending d-Cauchy sequence (x_n) in X is d-convergent.

Note that, by the invariance to translation property, this is equivalent with

(-K;ge;com) $(-K, d, \geq)$ is complete:
each $(\geq)$-ascending d-Cauchy sequence (y_n) in $(-K)$ is d-convergent;

just pass to the translated sequence $(y_n = x_n - x_0; n \geq 0)$ to establish this.

(II) The second condition to be considered reads

(X;conv-dom) $(X, d, \geq)$ is conv-dominated: the d-limit of each $(\geq)$-ascending sequence in X is a $(\geq)$-upper bound of it.

Again by the invariance to translation property, this condition is equivalent with

(-K;conv-dom) $(-K, d, \geq)$ is conv-dominated: the d-limit of each $(\geq)$-ascending sequence in $(-K)$ is a $(\geq)$-upper bound of it.

In fact, a stronger conclusion holds in our context.

Proposition 19. *The following are equivalent:*

(61-1) $(X, d, \geq)$ *is conv-dominated (see above);*
(61-2) $(-K)$ *is* $(d, \geq)$*-closed:*
the d-limit of each $(\geq)$*-ascending sequence in* $(-K)$ *is in* $(-K)$.

Proof. (i): Suppose that $(X, d, \geq)$ is conv-dominated. Given the $(\geq)$-ascending sequence (x_n) in $-K$ and the point $x \in X$ be such that $x_n \xrightarrow{d} x$ as $n \to \infty$, we have (by the imposed condition) that

$(0 \geq) x_n \geq x, \forall n$; whence, $0 \geq x$; meaning: $x \in -K$,

proving that $(-K)$ is $(d, \geq)$-closed.

(ii): Conversely, suppose that $(-K)$ is $(d, \geq)$-closed. Given the $(\geq)$-ascending sequence (x_n) in X and the point $x \in X$ with $x_n \xrightarrow{d} x$, we have that

(for each $i \geq 0$): the translated sequence $(y_n^i := x_{n+i} - x_i; n \geq 0)$ is in $(-K)$ and converges to $y_i := x - x_i$ as $n \to \infty$.

This, by the imposed condition, yields

(for each $i \geq 0$): $y_i \in -K$; that is, $x_i \geq x$,

proving that $(X, d, \geq)$ is conv-dominated. □

Having these precise, denote for simplicity

$K_0 = K \cap X_0$, where $X_0 = X \setminus \{0\}$;
clearly, $K_0 \neq \emptyset$ (as K is non-degenerate).

We claim that the following property holds:

(full) $(X, d, \geq)$ is *full*:
$(\forall u \in X, \forall \delta > 0)$: $X[u, \delta; \geq] := X[u, \delta] \cap X(u, \geq)$
has a non-empty intersection with $X \setminus \{u\}$,

where $X[u, \delta] = \{x \in X; d(u, x) \leq \delta\}$, $X(u, \geq) = \{x \in X; u \geq x\}$.

In fact, given $h \in K_0$, we have (by the directional continuous property)

(p-1) $u - \tau h \in X[u, \delta] \cap (X \setminus \{u\})$, $\forall \tau \in]0, \eta]$, where $\eta > 0$ is small;
(p-2) $u - \tau h \in X(u, \geq) \cap (X \setminus \{u\})$, $\forall \tau > 0$,

so, putting these together,

(p-3) $u - \tau h \in X[u, \delta] \cap X(u, \geq) \cap (X \setminus \{u\})$, $\forall \tau \in]0, \eta]$,

proving the assertion.

Under the introduced setting, take some sequence of maps $(\Gamma_i : X \to R; i \geq 0)$ with the properties

(ga-1) $(\forall i)$: $\sup[\Gamma_i(X)] = \infty$; hence
$[\Gamma_i \geq \sigma] := \{x \in X; \Gamma_i(x) \geq \sigma\}$ is nonempty, $\forall \sigma \in R$;
(ga-2) $(\forall i)$: $(\exists(\lambda_i, \mu_i), 0 < \lambda_i < 1 < \mu_i)$, such that
$(\forall \rho \geq \mu_i)$: $x \in [\Gamma_i \geq \rho]$ and $d(x, y) \leq \lambda_i$ imply $y \in [\Gamma_i \geq \rho - \mu_i]$;
(ga-3) $\nu := \liminf_i(\lambda_i/2\mu_i) > 0$ (hence, $0 < \nu \leq 1/2 < 1$).

Further, take some function F in $\mathcal{F}(X, R \cup \{\infty\})$, with

(F-1) F is proper ($\mathrm{Dom}(F) \neq \emptyset$);
(F-2) F is $(d, \geq)$-lsc on X;
($[F \leq t] := \{x \in X; F(x) \leq t\}$ is $(d, \geq)$-closed, $\forall t \in R$).

Finally, assume that the combined property holds

(F-Ga) $(F; (\Gamma_i))$ is global uniformly bounded below:
there exists $F_* \in R$ such that $(J(F; \Gamma_i)(0) \geq F_*, \forall i)$.

Here, for each map $\Gamma \in \mathcal{F}(X, R)$, we denoted

$$J(F; \Gamma)(r) = \inf F[\Gamma \geq r] = \inf\{F(y); \Gamma(y) \geq r\},\ r \in R.$$

The problem to be solved is the one in our preceding section. Note that, by the imposed conditions, Theorem 13 and Theorem 14 hold for our data; as well as their direct consequence, Theorem 15.

(B) In the light of these, the addressed question amounts to expressing this quasi-normed version of Theorem 15 in a standard differential setting. Given the proper function $F : X \to R \cup \{\infty\}$, denote for each $u \in \mathrm{Dom}(F)$

$$\Theta F(u)(h) = \limsup_{\tau \to 0+} \frac{F(u) - F(u - \tau h)}{\Delta(\tau h)},\ h \in K_0.$$

This object [referred to as the *h-directional derivative* of F at u] always exists, as an element of $R \cup \{-\infty, \infty\}$. As a consequence, for each $u \in \mathrm{Dom}(F)$, the quantities

(n-slo-1) $\Lambda_{(-K)}F(u) = \sup\{\Theta F(u)(h); h \in K_0\}$;
(n-slo-2) $|\Lambda_{(-K)}|F(u) = \max\{0, \Lambda_{(-K)}F(u)\}$

are meaningful; after a previous convention, we call these the *quasi-normed relative/absolute $(-K)$-slope* of F at u.

Concerning the relationships between this concept and the related one involving the associate quasi-order $(\geq)$, the following auxiliary fact will be useful.

Proposition 20. *We have, for the arbitrary fixed $u \in \mathrm{Dom}(F)$,*

$$\Lambda_{(-K)}F(u) \leq \nabla_{(\geq)}F(u); \quad \textit{hence,} \quad |\Lambda_{(-K)}|F(u) \leq |\nabla_{(\geq)}|F(u).$$

Proof. Let $h \in K_0$ be arbitrary fixed. Given $\delta > 0$, there exists — by a previous observation — some $\eta \in]0, \delta[$, with

$$0 < d(u, u - \tau h) = \Delta(\tau h) < \delta,\ \forall \tau \in]0, \eta[.$$

This, by definition, yields

$$\Theta F(u)(h) \leq \sup\left\{\frac{F(u) - F(u - \tau h)}{\Delta(\tau h)}; 0 < \tau < \eta\right\} \leq$$
$$\sup\left\{\frac{F(u) - F(x)}{d(u, x)}; x \in X[u, \delta; \geq] \setminus \{u\}\right\}.$$

Passing to supremum upon $h \in K_0$ gives

$$\Lambda_{(-K)}F(u) \leq \sup\left\{\frac{F(u) - F(x)}{d(u, x)}; x \in X[u, \delta; \geq] \setminus \{u\}\right\}, \text{ for all } \delta > 0;$$

wherefrom, passing to infimum upon $\delta > 0$, we derive the desired fact. □

Having these precise, let us consider the differential type condition

(PS;-K) for each strictly ascending sequence $(i(n); n \geq 0)$ in N and each sequence $(v_n; n \geq 0)$ in X fulfilling ($v_n \in \mathrm{Dom}(F)$, for each $n \geq 0$) as well as
$(F(v_n); n \geq 0)=$ convergent, and $|\Lambda_{(-K)}|F(v_n) \to 0$ as $n \to \infty$,
we have that $(\Gamma_{i(n)}(v_n); n \geq 0)$ has a bounded subsequence.

This will be referred to as a Palais–Smale differential condition (modulo $(-K)$) upon $(F; (\Gamma_i))$.

Theorem 16. *Suppose that (in addition) $(F; (\Gamma_i))$ satisfies a Palais–Smale differential condition (modulo $(-K)$). Then, F is (Γ_i)-coercive.*

Proof. By the auxiliary statement above, we have

Palais–Smale differential condition (modulo $(-K)$) upon $(F; (\Gamma_i))$ implies
Palais–Smale condition (modulo $(\geq)$) upon $(F; (\Gamma_i))$.

In fact, let the strictly ascending sequence $(i(n); n \geq 0)$ and the sequence $(v_n; n \geq 0)$ be as in the premise of Palais–Smale condition (modulo $(\geq)$) upon $(F; (\Gamma_i))$. From the auxiliary statement in question,

$$|\nabla_{(\geq)}|F(v_n) \to 0 \text{ as } n \to \infty \text{ implies } |\Lambda_{(-K)}|F(v_n) \to 0 \text{ as } n \to \infty$$

so that by the differential Palais–Smale condition (modulo $(-K)$) upon $(F; (\Gamma_i))$,

$$(\Gamma_{i(n)}(v_n); n \geq 0) \text{ has a bounded subsequence,}$$

and the assertion follows. Summing up, conditions of Theorem 15 are fulfilled by these data, and this concludes the argument. □

In particular, when the quasi-norm $\Delta(.)$ is a norm on X, and $K = X$ (hence, its attached quasi-order is the trivial one) Theorem 16 may be viewed as a complement of the result in Motreanu and Motreanu [32]. On the other hand, when (Γ_i) is a constant sequence, Theorem 16 includes the related statement in Motreanu and Turinici [35], obtained by means of the monotone variational principle in Turinici [45]. An extension of these facts to quasi-ordered gauge spaces is possible by following the lines in Bae *et al.* [3], completed with the ones in Turinici [42]. Some concrete applications of these facts to critical point theory may be found in Corvellec [14] and the references therein.

References

[1] H. H. Alsulami, E. Karapinar, F. Khojasteh and A.-F. Roldán, A proposal to the study of contractions in quasi-metric spaces, *Discrete Dyn. Nat. Soc.*, 2014, 2014, Article ID 269286.

[2] M. Altman, A generalization of the Brezis-Browder principle on ordered sets, *Nonlin. Anal.*, 6, 1982, 157–165.

[3] J.-S. Bae, S.-H. Cho and J.-J. Kim, An Ekeland type variational principle on gauge spaces with applications to fixed point theory, drop theory and coercivity, *Bull. Korean Math. Soc.*, 48, 2011, 1023–1032.

[4] G. Bantaş and M. Turinici, Mean value theorems via division methods, *An. Şt. Univ. A. I. Cuza Iaşi (Sect. I-a: Mat.)*, 40, 1994, 135–150.

[5] P. Bernays, A system of axiomatic set theory: Part III. Infinity and enumerability analysis, *J. Symbolic Logic*, 7, 1942, 65–89.

[6] N. Bourbaki, Sur le théorème de Zorn, *Archiv Math.*, 2, 1949/1950, 434–437.

[7] H. Brezis and F. E. Browder, A general principle on ordered sets in nonlinear functional analysis, *Advances Math.*, 21, 1976, 355–364.

[8] H. Brezis and L. Nirenberg, Remarks on finding critical points, *Commun. Pure Appl. Math.*, 44, 1991, 939–963.

[9] A. Brøndsted, Fixed points and partial orders, *Proc. Amer. Math. Soc.*, 60, 1976, 365–366.

[10] N. Brunner, Topologische Maximalprinzipien, *Zeitschr. Math. Logik Grundl. Math.*, 33, 1987, 135–139.

[11] L. Caklovic, S. Li and M. Willem, A note on Palais-Smale condition and coercivity, *Diff. Int. Equations*, 3, 1990, 799–800.

[12] S. Cobzaş, Fixed points and completeness in metric and in generalized metric spaces, Arxiv, 1508-05173-v5, 7 Oct 2019.

[13] P. J. Cohen, Set Theory and the Continuum Hypothesis, (Benjamin, New York, 1966).

[14] J. N. Corvellec, A note on coercivity of lower semicontinuous functions and nonsmooth critical point theory, *Serdica Math. J.*, 22, 1996, 57–68.

[15] J. N. Corvellec, M. DeGiovanni and M. Marzocchi, Deformation properties for continuous functionals and critical point theory, *Topol. Meth. Nonlin. Analysis*, 1, 1993, 151–171.

[16] S. Dancs, M. Hegedus and P. Medvegyev, A general ordering and fixed-point principle in complete metric space, *Acta Sci. Math. (Szeged)*, 46, 1983, 381–388.

[17] E. DeGiorgi, A. Marino and M. Tosques, Problemi di evoluzione in spazi metrici e curve di massima pendenza, *Rend. Accad. Naz. Lincei (Serie 8)*, 68, 1980, 180–187.

[18] J. Dodu and M. Morillon, The Hahn-Banach property and the Axiom of Choice, *Math. Logic Quarterly*, 45, 1999, 299–314.

[19] I. Ekeland, Nonconvex minimization problems, *Bull. Amer. Math. Soc. (New Series)*, 1, 1979, 443–474.

[20] D. Goeleven, A note on Palais-Smale condition in the sense of Szulkin, *Diff. Int. Equations*, 6, 1993, 1041–1043.

[21] A. Goepfert, H. Riahi, C. Tammer and C. Zălinescu, Variational Methods in Partially Ordered Spaces, *Canad. Math. Soc. Books Math. Vol. 17*, (Springer, New York, 2003).

[22] A. Hamel, Variational Principles on Metric and Uniform Spaces, (Habilitation Thesis), (Martin-Luther University, Halle-Wittenberg, 2005).

[23] P. Hitzler, Generalized metrics and topology in logic programming semantics, PhD Thesis, Natl. Univ. Ireland, Univ. College Cork, 2001.

[24] D. H. Hyers, G. Isac and T. M. Rassias, *Topics in Nonlinear Analysis and Applications*, (World Sci. Publ., Singapore, 1997).

[25] B. G. Kang and S. Park, On generalized ordering principles in nonlinear analysis, *Nonlin. Anal.*, 14, 1990, 159–165.

[26] S. Kasahara, On some generalizations of the Banach contraction theorem, *Publ. Res. Inst. Math. Sci. Kyoto Univ.*, 12, 1976, 427–437.

[27] K. Kuratowski, *Topology*, (Academic Press, New York, 1966).

[28] J. Matkowski and T. Swiatkowski, On subadditive functions, *Proc. Amer. Math. Soc.*, 119, 1993, 187–197.

[29] G. H. Moore, *Zermelo's Axiom of Choice: Its Origin, Development and Influence*, (Springer, New York, 1982).

[30] Y. Moskhovakis, *Notes on Set Theory*, (Springer, New York, 2006).

[31] D. Motreanu and V. V. Motreanu, Coerciveness property for a class of nonsmooth functionals, Zeitschr. *Anal. Anwendungen (J. Analysis Appl.)*, 19, 2000, 1087–1093.

[32] D. Motreanu and V. V. Motreanu, Coercivity properties for sequences of lower semicontinuous functions on metric spaces, *Abstr. Appl. Anal.*, 2013, 2013, Article ID 268650.

[33] D. Motreanu, V. V. Motreanu and M. Turinici, Coerciveness property for conical nonsmooth functionals, *J. Optim. Th. Appl.*, 145, 2010, 148–163.

[34] D. Motreanu and P. D. Panagiotopoulos, *Minimax Theorems and Qualitative Properties of the Solutions of Hemivariational Inequalities*, (Kluwer Acad. Publ., Dordrecht, 1999).
[35] V. V. Motreanu and M. Turinici, Coercivity properties for monotone functionals, *Note Mat.*, 21, 2002, 83–91.
[36] R. S. Palais and S. Smale, A generalized Morse theory, *Bull. Amer. Math. Soc.*, 70, 1964, 165–171.
[37] E. Schechter, *Handbook of Analysis and its Foundation*, (Academic Press, New York, 1997).
[38] A. Tarski, Axiomatic and algebraic aspects of two theorems on sums of cardinals, *Fund. Math.*, 35, 1948, 79–104.
[39] M. Turinici, Maximal elements in a class of order complete metric spaces, *Math. Japonica*, 25, 1980, 511–517.
[40] M. Turinici, Maximality principles and mean value theorems, *An. Acad. Brasil. Cienc.*, 53, 1981, 653–655.
[41] M. Turinici, A generalization of Brezis-Browder's ordering principle, *An. Şt. Univ. A. I. Cuza Iaşi (Sect. I-a: Mat)*, 28 1982, 11–16.
[42] M. Turinici, Mapping theorems via contractor directions in metrizable locally convex spaces, *Bull. Acad. Pol. Sci. (Ser. Sci. Math.)*, 30, 1982, 161–166.
[43] M. Turinici, A generalization of Altman's ordering principle, *Proc. Amer. Math. Soc.*, 90, 1984, 128–132.
[44] M. Turinici, Pseudometric extensions of the Brezis-Browder ordering principle, *Math. Nachrichten*, 130, 1987, 91–103.
[45] M. Turinici, A monotone version of the variational Ekeland's principle, *An. Şt. Univ. A. I. Cuza Iaşi (Sect. I-a: Mat.)*, 36, 1990, 329–352.
[46] M. Turinici, Pseudometric versions of the Caristi-Kirk fixed point theorem, *Fixed Point Th.*, 5, 2004, 147–161.
[47] M. Turinici, Function variational principles and coercivity, *J. Math. Anal. Appl.*, 304, 2005, 236–248.
[48] M. Turinici, Gauge variational principles and normed coercivity, *An. Şt. Univ. A. I. Cuza Iaşi (Sect. I-a: Mat.)*, 52, 2006, 251–275.
[49] M. Turinici, Maximality principles: Theory and practice, *Sc. Annals UASVM Iaşi*, 49, 2006, 323–360.
[50] M. Turinici, Function pseudometric VP and applications, *Bul. Inst. Polit. Iaşi (Sect.: Mat., Mec. Teor., Fiz.)*, 53(57), 2007, 393–411.
[51] M. Turinici, Functional monotone VP and metrical coercivity, *Libertas Math.*, 27, 2007, 1–16.
[52] M. Turinici, Multivalued variational principles and normed coercivity, *Libertas Math.*, 30, 2010, 1–17.
[53] M. Turinici, Function variational principles and coercivity over normed spaces, *Optimization*, 59, 2010, 199–222.

[54] M. Turinici, Coercivity properties for order nonsmooth functionals, *An. Şt. Univ. Ovidius Constanţa (Mat.)*, 19, 2011, 297–312.
[55] M. Turinici, Normed coercivity for monotone functionals, *Romai J.*, 7(2), 2011, 169–179.
[56] M. Turinici, Sequential maximality principles, *Mathematics Without Boundaries.* T. M. Rassias and P. M. Pardalos (eds.), pp. 515–548, (Springer, New York, 2014).
[57] M. Turinici, Function variational principles and normed minimizers, *Computational Mathematics and Variational Analysis.* N. J. Daras and T. M. Rassias (eds.), pp. 513–536, (Springer Nature, Switzerland, 2020).
[58] M. Turinici, Metrical coercivity for monotone functionals, *Analysis, Geometry, Nonlinear Optimization, and Applications.* T. M. Rassias and P. M. Pardalos (eds.), Ch. 28, (World Sci. Publishing, Singapore, 2022).
[59] E. S. Wolk, On the principle of dependent choices and some forms of Zorn's lemma, *Canad. Math. Bull.*, 26, 1983, 365–367.
[60] H.-K. Xu, On the Palais-Smale condition for nondifferentiable functionals, *Taiwanese J. Math.*, 4, 2000, 627–634.
[61] C. K. Zhong, A generalization of Ekeland's variational principle and application to the study of the relation between the weak P.S. condition and coercivity, *Nonlin. Anal.*, 29, 1997, 1421–1431.

https://doi.org/10.1142/9789819820498_0023

Chapter 23

A Multidimensional Hardy–Hilbert's Inequality Involving One Partial Sum

Bicheng Yang[*,¶] **and Michael Th. Rassias**[†,‖]

[*]*School of Mathematics, Guangdong University of Education, Guangzhou, Guangdong 510303, P. R. China*
[†]*Department of Mathematics and Engineering Sciences, Hellenic Military Academy, 16673 Vari Attikis, Greece*
[¶]*bcyang@gdei.edu.cn, bcyang818@163.com*
[‖]*mthrassias@yahoo.com*

By means of the weight functions and the idea of introduced parameters, using the techniques of real analysis, a multidimensional Hardy–Hilbert's inequality involving one partial sum is obtained. The equivalent statements of the best value related to parameters are considered. Some corollaries are obtained.

Keywords: Weight function, parameter, Beta function, partial sum, multidimensional Hardy–Hilbert's inequality.

1. Introduction

Suppose that $p > 1, \frac{1}{p} + \frac{1}{q} = 1, a_m, b_n \geq 0, 0 < \sum_{m=1}^{\infty} a_m^p < \infty$ and $0 < \sum_{n=1}^{\infty} b_n^q < \infty$. We have the following Hardy–Hilbert's inequality with the best value $\frac{\pi}{\sin(\pi/p)}$ (cf. Ref. [1], Theorem 315):

$$\sum_{m=1}^{\infty}\sum_{n=1}^{\infty}\frac{a_m b_n}{m+n} < \frac{\pi}{\sin(\frac{\pi}{p})}\left(\sum_{m=1}^{\infty} a_m^p\right)^{\frac{1}{p}}\left(\sum_{n=1}^{\infty} b_n^q\right)^{\frac{1}{q}}. \tag{1.1}$$

If $f(x), g(y) \geq 0$, $0 < \int_0^{\infty} f^p(x)dx < \infty$ and $0 < \int_0^{\infty} g^q(y)dy < \infty$, then we have the integral analogue of (1.1) with the same best value named in Hardy–Hilbert's integral inequality as follows (cf. Ref. [1], Theorem 316):

$$\int_0^{\infty}\int_0^{\infty}\frac{f(x)g(y)}{x+y}dxdy < \frac{\pi}{\sin(\frac{\pi}{p})}\left(\int_0^{\infty} f^p(x)dx\right)^{\frac{1}{p}}\left(\int_0^{\infty} g^q(y)dy\right)^{\frac{1}{q}}. \tag{1.2}$$

In 2006, by means of the Euler–Maclaurin summation formula and the techniques of real analysis, Krnić *et al.* [2] gave the following extension of (1.1):

$$\sum_{m=1}^{\infty}\sum_{n=1}^{\infty}\frac{a_m b_n}{(m+n)^{\lambda}} < B(\lambda_1,\lambda_2)\left[\sum_{m=1}^{\infty} m^{p(1-\lambda_1)-1}a_m^p\right]^{\frac{1}{p}}\left[\sum_{n=1}^{\infty} n^{q(1-\lambda_2)-1}b_n^q\right]^{\frac{1}{q}}, \tag{1.3}$$

where $\lambda_1, \lambda_2 \in (0, 2], \lambda_1 + \lambda_2 = \lambda \in (0, 4]$, the constant factor $B(\lambda_1, \lambda_2)$ is the best value, and

$$B(u,v) = \int_0^{\infty}\frac{t^{u-1}}{(1+t)^{u+v}}dt, u, v > 0 \tag{1.4}$$

is the Beta function. In 2019, by applying (1.3) and Abel's partial summation formula, Adiyasuren *et al.* [3] obtained an extended application of (1.3) involving two partial sums. In 2020, Mo *et al.* [4] gave an extension of (1.2) involving two upper limit functions. Inequalities (1.1)–(1.2) with their extensions played an important role in analysis and its applications (cf. Refs. [5–15]). In 2016–2017, Hong *et al.* [16, 17] considered a few equivalent conditions of the extension of (1.1) (resp. (1.2)) with several parameters related to the best value. Some other research results were provided by [18–20].

In 2023, Hong *et al.* [21] gave a more accurate multidimensional half-discrete Hilbert-type inequality involving one derivative function of m-order and [22] gave an extended inequality with the same kernel involving one multiple upper limit function. Some dependent results were published by Refs. [23–30].

In this chapter, following the way of Refs. [21, 22], by using the weight functions, the idea of introduced parameters and the techniques of real analysis, a multidimensional discrete Hardy–Hilbert's inequality with the kernel as $\frac{1}{(u(m)+||k||_\beta^\alpha)^\lambda}$ $(\alpha, \lambda > 0)$ involving one partial sum is obtained. The equivalent statements of the best value related to parameters are considered, and some corollaries are provided.

2. Transfer Formula and Some Lemmas

In what follows, we assume that

(H1). $p > 1$ $(q > 1), \frac{1}{p} + \frac{1}{q} = 1, \alpha, \beta \in \mathbf{R}_+ := (0, \infty)$, $m_0, n \in \mathbf{N} := \{1, 2, \ldots\}$, $\lambda > 0, \lambda_1, \lambda_2 \in (0, \lambda), \lambda_2 \leq \frac{n}{\alpha}$, $u(x), u'(x) > 0$, $(u(x))^{\lambda_1} u'(x)$ is decreasing in $x \in (m_0 - 1, \infty)$ with $u(\infty) = \infty$, $\widehat{\lambda}_1 = \frac{\lambda - \lambda_2}{p} + \frac{\lambda_1}{q}$, $\widehat{\lambda}_2 = \frac{\lambda - \lambda_1}{q} + \frac{\lambda_2}{p}$, $a_k \geq 0, A_m^{(0)} := a_m$, $A_m^{(1)} := \sum_{k=m_0}^m a_k$ $(k, m \in \mathbf{N}_{m_0} := \{m_0, m_0 + 1, \ldots\})$, satisfying $A_m^{(1)} = o(e^{tu(m)})$ $(t > 0; m \to \infty)$, $b_k \geq 0$ $(k = (k_1, \ldots, k_n) \in \mathbf{N}^n)$, $||k||_\beta := (\sum_{j=1}^n k_j^\beta)^{\frac{1}{\beta}}$,

$$0 < \sum_{m=m_0}^\infty \frac{(u'(m))^{1-p+pi}}{(u(m))^{p(i-1+\widehat{\lambda}_1)+1}} (A_m^{(i)})^p < \infty \quad (i \in \{0, 1\}), \text{ and}$$

$$0 < \sum_k ||k||_\beta^{q(n-\alpha\widehat{\lambda}_2)-n} b_k^q < \infty,$$

where $\sum_k h(k) := \sum_{k_n=1}^\infty \cdots \sum_{k_1=1}^\infty h(k_1, \ldots, k_n)$.

Remark 1. (i) In view of the assumption H1, for $i \in \{0,1\}$, $u'(x) > 0$, $(u(x))^{\lambda_1+i-1}u'(x)$ $(= \frac{1}{(u(x))^{1-i}}[(u(x))^{\lambda_1}u'(x)])$ is also decreasing in $x \in (m_0-1,\infty)$.

(ii) For $\gamma \in (0,1), m_0 = 1, u_1(x) = x^\gamma, x \in (0,\infty)$, $\lambda_1 \in \left(0, \frac{1}{\gamma}-1\right]$, we have $u_1(x) > 0, u_1'(x) = \gamma x^{\gamma-1} > 0, u_1(\infty) = \lim_{x\to\infty} x^\gamma = \infty$, and $(u_1(x))^{\lambda_1}u_1'(x) = \gamma x^{(\lambda_1+1)\gamma-1}$ is obviously decreasing in $x \in (0,\infty)$.

(iii) For $\gamma \in (0,1), m_0 = 2, u_2(x) = \ln^\gamma x, x \in (1,\infty)$, $\lambda_1 \in \left(0, \frac{1}{\gamma}-1\right]$, we have $u_2(x) > 0, u_2'(x) = \frac{\gamma}{x}\ln^{\gamma-1} x > 0, u_2(\infty) = \lim_{x\to\infty} \ln^\gamma x = \infty$, and $(u_2(x))^{\lambda_1}u_2'(x) = \frac{\gamma}{x}\ln^{(\lambda_1+1)\gamma-1} x$ is still decreasing in $x \in (1,\infty)$.

If $M > 0, \psi(u)$ $(u > 0)$ is a non-negative measurable function, then we have the following transfer formula (cf. Ref. [5], (9.1.5)):

$$\int\cdots\int_{\{y\in\mathbf{R}_+^n;0<\sum_{j=1}^n\left(\frac{y_j}{M}\right)^\beta\le 1\}} \psi\left(\sum_{j=1}^n\left(\frac{y_j}{M}\right)^\beta\right) dy_1\cdots dy_n$$

$$= \frac{M^n\Gamma^n(\frac{1}{\beta})}{\beta^n\Gamma(\frac{n}{\beta})}\int_0^1 \psi(u)u^{\frac{n}{\beta}-1}du. \tag{2.1}$$

(i) For $||y||_\beta = M[\sum_{j=1}^n(\frac{y_j}{M})^\beta]^{\frac{1}{\beta}}, \psi(u) = \varphi(Mu^{\frac{1}{\beta}})$, by (2.1), we have

$$\int_{\mathbf{R}_+^n} \varphi(||y||_\beta)dy = \lim_{M\to\infty}\int\cdots\int_{\{y\in\mathbf{R}_+^n;0<\sum_{j=1}^n\left(\frac{y_j}{M}\right)^\beta\le 1\}} \varphi$$

$$\times\left(M\left[\sum_{j=1}^n\left(\frac{y_j}{M}\right)^\beta\right]^{\frac{1}{\beta}}\right)dy_1\cdots dy_n$$

$$= \lim_{M\to\infty}\frac{M^n\Gamma^n(\frac{1}{\beta})}{\beta^n\Gamma(\frac{n}{\beta})}\int_0^1 \varphi(Mu^{\frac{1}{\beta}})u^{\frac{n}{\beta}-1}du$$

$$= \frac{\Gamma^n(\frac{1}{\beta})}{\beta^{n-1}\Gamma(\frac{n}{\beta})}\int_0^\infty \varphi(v)v^{n-1}dv\,(v = Mu^{\frac{1}{\beta}}). \tag{2.2}$$

(ii) For $\varphi(||y||_\beta) = \varphi(Mu^{\frac{1}{\beta}}) = 0$, for $u = \sum_{i=1}^n \left(\frac{y_i}{M}\right)^\beta < \left(\frac{b}{M}\right)^\beta$ $(b > 0)$, i.e. $||y||_\beta = Mu^{\frac{1}{\beta}} < b$, by (2.2), it follows that

$$\begin{aligned}
&\int_{\{y\in\mathbf{R}_+^n;||y||_\beta\geq b\}} \varphi(||y||_\beta)dy \\
&= \lim_{M\to\infty} \frac{M^n\Gamma^n(\frac{1}{\beta})}{\beta^n\Gamma(\frac{n}{\beta})} \int_{(\frac{b}{M})^\beta}^1 \varphi(Mu^{\frac{1}{\beta}})u^{\frac{n}{\beta}-1}du \\
&= \frac{\Gamma^n(\frac{1}{\beta})}{\beta^{n-1}\Gamma(\frac{n}{\beta})} \int_b^\infty \varphi(v)v^{n-1}dv. \qquad (2.3)
\end{aligned}$$

Lemma 1. *For $c \in \mathbf{R}_+$, we have*

$$\frac{\left[1 - \frac{\alpha c(n-1)}{\alpha c+1}\right]\Gamma^n(\frac{1}{\beta})}{\alpha c\beta^{n-1}\Gamma(\frac{n}{\beta})} < \sum_k ||k||_\beta^{-\alpha c-n} < \frac{2^{\alpha c}\Gamma^n(\frac{1}{\beta})}{\alpha c\beta^{n-1}\Gamma(\frac{n}{\beta})}. \qquad (2.4)$$

Proof. By Hermite–Hadamard's inequality (cf. Ref. [31]) and (2.3) (for $b = \frac{1}{2}, \varphi(v) = v^{-\alpha c-n}$), it follows that

$$\begin{aligned}
\sum_k ||k||_\beta^{-\alpha c-n} &< \int_{\{y\in\mathbf{R}_+^n;y_j\geq\frac{1}{2}\}} ||y||_\beta^{-\alpha c-n}dy \\
&\leq \int_{\{y\in\mathbf{R}_+^n;||y||_\beta\geq\frac{1}{2}\}} ||y||_\beta^{-\alpha c-n}dy \\
&= \frac{\Gamma^n\left(\frac{1}{\beta}\right)}{\beta^{n-1}\Gamma\left(\frac{n}{\beta}\right)} \int_{\frac{1}{2}}^\infty v^{-\alpha c-n}v^{n-1}dv \\
&= \frac{\Gamma^n\left(\frac{1}{\beta}\right)}{\beta^{n-1}\Gamma\left(\frac{n}{\beta}\right)} \int_{\frac{1}{2}}^\infty v^{-\alpha c-1}dv = \frac{2^{\alpha c}\Gamma^n\left(\frac{1}{\beta}\right)}{\alpha c\beta^{n-1}\Gamma\left(\frac{n}{\beta}\right)}.
\end{aligned}$$

By the decreasingness property of series and (2.2) (for $\varphi(v) = (v+1)^{-\alpha c-n}$), we have

$$\sum_k ||k||_\beta^{-\alpha c-n} > \int_{\{y\in\mathbf{R}_+^n; y_j\geq 1\}} ||y||_\beta^{-\alpha c-n} dy = \int_{\mathbf{R}_+^n} ||y+1||_\beta^{-\alpha c-n} dy$$

$$= \frac{\Gamma^n\left(\frac{1}{\beta}\right)}{\beta^{n-1}\Gamma\left(\frac{n}{\beta}\right)} \int_0^\infty (v+1)^{-\alpha c-n} v^{n-1} dv$$

$$= \frac{\Gamma^n\left(\frac{1}{\beta}\right)}{\beta^{n-1}\Gamma\left(\frac{n}{\beta}\right)} \left[\int_0^\infty (v+1)^{-\alpha c-1} dv - \int_0^\infty \frac{(v+1)^{n-1} - v^{n-1}}{(v+1)^{\alpha c+n}} dv\right]$$

$$= \frac{\Gamma^n\left(\frac{1}{\beta}\right)}{\alpha c\beta^{n-1}\Gamma\left(\frac{n}{\beta}\right)}(1 - A(n)),$$

where $A(n) := \alpha c \int_0^\infty \frac{(v+1)^{n-1}-v^{n-1}}{(v+1)^{\alpha c+n}} dv \geq 0$, with $A(1) = 0$.

For $n \geq 2$, by the differentiation mid-value theorem, there exists a $\theta \in (0, 1)$, such that

$$A(n) = \alpha c(n-1) \int_0^\infty \frac{(v+\theta)^{n-2}}{(v+1)^{\alpha c+n}} dv$$

$$\leq \alpha c(n-1) \int_0^\infty \frac{(v+1)^{n-2}}{(v+1)^{\alpha c+n}} dv$$

$$= \alpha c(n-1) \int_0^\infty \frac{dv}{(v+1)^{\alpha c+2}} = \frac{\alpha c(n-1)}{\alpha c+1},$$

satisfying $A(1) = 0$. Hence, we have (2.4).

This proves the lemma. □

Lemma 2. *With regard to the assumption H1, if $s \in (0,\infty), s_1 \in (0, s), (u(x))^{s_1-1}u'(x)$ is decreasing in (m_0-1,∞), $s_2 \in (0, \frac{n}{\alpha}] \cap (0, s)$, then we define the following weight functions:*

$$\omega_s(s_1, k) := ||k||_\beta^{\alpha(s-s_1)} \sum_{m=m_0}^{\infty} \frac{(u(m))^{s_1-1}u'(m)}{(u(m) + ||k||_\beta^\alpha)^s} \quad (k \in \mathbf{N}^n), \qquad (2.5)$$

$$\varpi_s(s_2,m) := (u(m))^{s-s_2}\sum_k \frac{||k||_\beta^{\alpha s_2-n}}{(u(m)+||k||_\beta^\alpha)^s}\ (m\in\mathbf{N}_{m_0}). \quad (2.6)$$

The following inequalities are value:

$$\omega_s(s_1,k) < B(s_1,s-s_1)\ \ (k\in\mathbf{N}^n), \quad (2.7)$$

$$\varpi_s(s_2,m) < \frac{\Gamma^n\left(\frac{1}{\beta}\right)}{\alpha\beta^{n-1}\Gamma\left(\frac{n}{\beta}\right)}B(s_2,s-s_2)\ \ (m\in\mathbf{N}_{m_0}). \quad (2.8)$$

Proof. In view of the decreasingness property of series, setting $v = \frac{u(x)}{||k||_\beta^\alpha}$, we find

$$\omega_s(s_1,k) < ||k||_\beta^{\alpha(s-s_1)}\int_{m_0-1}^\infty \frac{(u(x))^{s_1-1}u'(x)}{(u(x)+||k||_\beta^\alpha)^s}dx$$

$$\leq \int_0^\infty \frac{v^{s_1-1}}{(v+1)^s}dv = B(s_1,s-s_1),$$

and then we have (2.7).

For $\alpha s_2 \leq n$, still by the decreasingness property of series and (2.2), for $\varphi(v) = \frac{v^{\alpha s_2-n}}{(u(m)+v^\alpha)^s}$, we have

$$\varpi_s(s_2,m) < (u(m))^{s-s_2}\int_{\mathbf{R}_+^n} \frac{||y||_\beta^{\alpha s_2-n}}{(u(m)+||y||_\beta^\alpha)^s}dy$$

$$= \frac{\Gamma^n\left(\frac{1}{\beta}\right)}{\beta^{n-1}\Gamma\left(\frac{n}{\beta}\right)}(u(m))^{s-s_2}\int_0^\infty \frac{v^{\alpha s_2-n}}{(u(m)+v^\alpha)^s}v^{n-1}dv$$

$$= \frac{\Gamma^n\left(\frac{1}{\beta}\right)}{\beta^{n-1}\Gamma\left(\frac{n}{\beta}\right)}(u(m))^{s-s_2}\int_0^\infty \frac{v^{\alpha s_2-1}}{(u(m)+v^\alpha)^s}dv$$

$$= \frac{\Gamma^n\left(\frac{1}{\beta}\right)}{\alpha\beta^{n-1}\Gamma\left(\frac{n}{\beta}\right)}\int_0^\infty \frac{t^{s_2-1}}{(1+t)^s}dt\ (t = v^\alpha/u(m))$$

$$= \frac{\Gamma^n\left(\frac{1}{\beta}\right)}{\alpha\beta^{n-1}\Gamma\left(\frac{n}{\beta}\right)}B(s_2,s-s_2),$$

and then we have (2.8).

This proves the lemma. □

Lemma 3. *With regard to the assumption H1, for $t > 0$, $i \in \{0, 1\}$, we have the following inequality:*

$$\sum_{m=m_0}^{\infty} e^{-tu(m)} a_m \leq t^i \sum_{m=m_0}^{\infty} e^{-tu(m)} (u'(m))^i A_m^{(i)}. \tag{2.9}$$

Proof. For $i = 0$, since $a_m = A_m^{(0)}$, (2.9) keeps the form of an equality; for $i = 1$, since $A_m^{(1)} e^{-tu(m)} = o(1)$ $(t > 0; m \to \infty)$, by Abel's partial summation formula, we find

$$\begin{aligned}\sum_{m=m_0}^{\infty} e^{-tu(m)} a_m &= \lim_{m\to\infty} A_m^{(1)} e^{-tu(m)} + \sum_{m=m_0}^{\infty} A_m^{(1)} (e^{-tu(m)} - e^{-tu(m+1)}) \\ &= \sum_{m=m_0}^{\infty} A_m^{(1)} (e^{-tu(m)} - e^{-tu(m+1)}).\end{aligned}$$

We set function $f(x) := e^{-tu(x)}, x \in (m_0 - 1, \infty)$. Then, we find $f'(x) := -te^{-tu(x)} u'(x) = -th(x)$, where

$$h(x) := e^{-tu(x)} u'(x) = e^{-tu(x)} (u^{\lambda_1}(x) u'(x)) u^{-\lambda_1}(x)$$

is decreasing in $(m_0 - 1, \infty)$. By the differentiation mid-value theorem, there exists a $\theta_m \in (0, 1)$, such that

$$\begin{aligned}\sum_{m=m_0}^{\infty} e^{-tu(m)} a_m &= - \sum_{m=m_0}^{\infty} A_m^{(1)} [f(m+1) - f(m)] \\ &= - \sum_{m=m_0}^{\infty} A_m^{(1)} f'(m + \theta_m) = t \sum_{m=m_0}^{\infty} A_m^{(1)} h(m + \theta_m) \\ &\leq t \sum_{m=m_0}^{\infty} h(m) A_m^{(1)} = t \sum_{m=m_0}^{\infty} e^{-tu(m)} u'(m) A_m^{(1)}.\end{aligned}$$

Then we have (2.9).

This proves the lemma. □

Lemma 4. *With regard to the assumption H1, for $i \in \{0,1\}$, we have the following inequality:*

$$I_{\lambda+i} := \sum_k \sum_{m=m_0}^{\infty} \frac{(u'(m))^i}{(u(m)+||k||_\beta^\alpha)^{\lambda+i}} A_m^{(i)} b_k$$

$$< \left(\frac{\Gamma^n\left(\frac{1}{\beta}\right) B(\lambda_2, \lambda+i-\lambda_2)}{\alpha\beta^{n-1}\Gamma\left(\frac{n}{\beta}\right)} \right)^{\frac{1}{p}} B^{\frac{1}{q}}(\lambda_1+i, \lambda-\lambda_1)$$

$$\times \left[\sum_{m=m_0}^{\infty} \frac{(u'(m))^{1-p+pi}(A_m^{(i)})^p}{(u(m))^{p(i-1+\widehat{\lambda}_1)+1}} \right]^{\frac{1}{p}} \left[\sum_k ||k||_\beta^{q(n-\alpha\widehat{\lambda}_2)-n} b_k^q \right]^{\frac{1}{q}}. \tag{2.10}$$

Proof. By Hōlder's inequality (cf. Ref. [31]), we have

$$I_{\lambda+i} = \sum_k \sum_{m=m_0}^{\infty} \frac{1}{(u(m)+||k||_\beta^\alpha)^{\lambda+i}} \left[\frac{(u(m))^{(1-\lambda_1-i)/q} A_m^{(i)}}{||k||_\beta^{(n-\alpha\lambda_2)/p}(u'(m))^{-i+(1/q)}} \right]$$

$$\times \left[\frac{||k||_\beta^{(n-\alpha\lambda_2)/p} b_k}{(u(m))^{(1-\lambda_1-i)/q}(u'(m))^{-1/q}} \right]$$

$$\leq \left\{ \sum_{m=m_0}^{\infty} \sum_k \frac{1}{(u(m)+||k||_\beta^\alpha)^{\lambda+i}} \frac{(u(m)^{(1-\lambda_1-i)(p-1)}(A_m^{(i)})^p}{||k||_\beta^{n-\alpha\lambda_2}(u'(m))^{-pi+p-1}} \right\}^{\frac{1}{p}}$$

$$\times \left\{ \sum_k \sum_{m=m_0}^{\infty} \frac{1}{(u(m)+||k||_\beta^\alpha)^{\lambda+i}} \frac{||k||_\beta^{(n-\alpha\lambda_2)(q-1)} b_k^q}{(u(m))^{1-\lambda_1-i}(u'(m))^{-1}} \right\}^{\frac{1}{q}}$$

$$= \left\{ \sum_{m=m_0}^{\infty} \left[(u(m))^{\lambda+i-\lambda_2} \sum_k \frac{||k||_\beta^{\alpha\lambda_2-n}}{(u(m)+||k||_\beta^\alpha)^{\lambda+i}} \right] \right.$$

$$\left. \times \frac{(u'(m))^{1-p+pi}(A_m^{(i)})^p}{(u(m))^{p(i-1+\widehat{\lambda}_1)+1}} \right\}^{\frac{1}{p}}$$

$$\times \left\{ \sum_k \left[||k||_\beta^{\alpha(\lambda-\lambda_1)} \sum_{m=m_0}^{\infty} \frac{(u(m)^{\lambda_1+i-1}u'(m)}{(u(m)+||k||_\beta^\alpha)^{\lambda+i}} \right] \right.$$

$$\left. \times ||k||_\beta^{q(n-\alpha\widehat{\lambda}_2)-n} b_k^q \right\}^{\frac{1}{q}}$$

$$= \left[\sum_{m=m_0}^{\infty} \varpi_{\lambda+i}(\lambda_2, m) \frac{(u'(m))^{1-p+pi}}{(u(m))^{p(i-1+\widehat{\lambda}_1)+1}} (A_m^{(i)})^p \right]^{\frac{1}{p}}$$

$$\times \left[\sum_k \omega_{\lambda+i}(\lambda_1+i, y) ||k||_\beta^{q(n-\alpha\widehat{\lambda}_2)-n} b_k^q \right]^{\frac{1}{q}}.$$

Then by (2.7), (2.8) and Remark 1(i), for $s = \lambda + i > 0$, $s_1 = \lambda_1 + i \in (i, \lambda + i) \subset (0, \lambda + i)$, $s_2 = \lambda_2 \in (0, \frac{n}{\alpha}] \cap (0, \lambda + i)$, and $(u(x))^{s_1-1} u'(x) = (u(x))^{(\lambda_1+i)-1}u'(x)$ being decreasing in $(m_0 - 1, \infty)$, in view of H1, we have (2.10).

This proves the lemma. □

3. Main Results

Theorem 1. *With regard to the assumption H1, for $i \in \{0, 1\}$, we have the following multidimensional Hardy–Hilbert's inequality involving one partial sum:*

$$I := \sum_k \sum_{m=m_0}^{\infty} \frac{a_m b_k}{(u(m)+||k||_\beta^\alpha)^\lambda}$$

$$< \lambda^i \left(\frac{\Gamma^n\left(\frac{1}{\beta}\right) B(\lambda_2, \lambda+i-\lambda_2)}{\alpha\beta^{n-1}\Gamma\left(\frac{n}{\beta}\right)} \right)^{\frac{1}{p}} B^{\frac{1}{q}}(\lambda_1+i, \lambda-\lambda_1)$$

$$\times \left[\sum_{m=m_0}^{\infty} \frac{(u'(m))^{1-p+pi}(A_m^{(i)})^p}{(u(m))^{p(i-1+\widehat{\lambda}_1)+1}} \right]^{\frac{1}{p}} \left[\sum_k ||k||_\beta^{q(n-\alpha\widehat{\lambda}_2)-n} b_k^q \right]^{\frac{1}{q}}. \tag{3.1}$$

In particular, for $\lambda = \lambda_1 + \lambda_2$, *we have*

$$0 < \sum_{m=m_0}^{\infty} \frac{(u'(m))^{1-p+pi}}{(u(m))^{p(i-1+\lambda_1)+1}} (A_m^{(i)})^p < \infty,$$

$$0 < \sum_k ||k||_\beta^{q(n-\alpha\lambda_2)-n} b_k^q < \infty,$$

and the following inequality:

$$\sum_k \sum_{m=m_0}^{\infty} \frac{a_m b_k}{(u(m) + ||k||_\beta^\alpha)^\lambda} < \lambda^i \left(\frac{\Gamma^n\left(\frac{1}{\beta}\right)}{\alpha\beta^{n-1}\Gamma\left(\frac{n}{\beta}\right)} \right)^{\frac{1}{p}} B(\lambda_1 + i, \lambda_2)$$
$$\times \left[\sum_{m=m_0}^{\infty} \frac{(u'(m))^{1-p+pi}(A_m^{(i)})^p}{(u(m))^{p(i-1+\lambda_1)+1}} \right]^{\frac{1}{p}} \left[\sum_k ||k||_\beta^{q(n-\alpha\lambda_2)-n} b_k^q \right]^{\frac{1}{q}}. \tag{3.2}$$

Proof. By the following expression of the Gamma function:

$$\frac{1}{(u(m) + ||k||_\beta^\alpha)^\lambda} = \frac{1}{\Gamma(\lambda)} \int_0^\infty t^{\lambda-1} e^{-(u(m)+||k||_\beta^\alpha)t} dt,$$

equation (2.9) and Lebesgue term by term theorem (cf. Ref. [32]), we have

$$\begin{aligned}
I &= \frac{1}{\Gamma(\lambda)} \sum_k \sum_{m=m_0}^{\infty} a_m b_k \left[\int_0^\infty t^{\lambda-1} e^{-(u(m)+||k||_\beta^\alpha)t} dt \right] \\
&= \frac{1}{\Gamma(\lambda)} \int_0^\infty t^{\lambda-1} \left(\sum_{m=m_0}^{\infty} e^{-xu(m)} a_m \right) \sum_k e^{-||k||_\beta^\alpha t} b_k dt \\
&\leq \frac{1}{\Gamma(\lambda)} \int_0^\infty t^{\lambda-1} \left[t^i \sum_{m=m_0}^{\infty} e^{-xu(m)} (u'(m))^i A_m^{(i)} \right] \sum_k e^{-||k||_\beta^\alpha t} b_k dt \\
&= \frac{1}{\Gamma(\lambda)} \sum_k \sum_{m=m_0}^{\infty} (u'(m))^i A_m^{(i)} b_k \left[\int_0^\infty t^{(\lambda+i)-1} e^{-(u(m)+||k||_\beta^\alpha)t} dt \right] \\
&= \frac{\Gamma(\lambda+i)}{\Gamma(\lambda)} \sum_k \sum_{m=m_0}^{\infty} \frac{(u'(m))^i A_m^{(i)} b_k}{(u(m) + ||k||_\beta^\alpha)^{\lambda+i}} = \lambda^i I_{\lambda+i}.
\end{aligned}$$

Then by (2.10), we have (3.1). For $\lambda = \lambda_1 + \lambda_2$ in (3.1), we have (3.2).

This proves the theorem. □

Theorem 2. *With regard to the assumption H1, if $i \in \{0,1\}$, $\lambda_1 + \lambda_2 = \lambda$, then the constant factor*

$$\lambda^i \left(\frac{\Gamma^n\left(\frac{1}{\beta}\right) B(\lambda_2, \lambda + i - \lambda_2)}{\alpha\beta^{n-1}\Gamma\left(\frac{n}{\beta}\right)} \right)^{\frac{1}{p}} B^{\frac{1}{q}}(\lambda_1 + i, \lambda - \lambda_1)$$

in (3.1) *is the best value.*

Proof. We need to prove that the constant factor in (3.2) is the best value for $i \in \{0,1\}$. For any $0 < \varepsilon < p\lambda_1$, we set

$$\widetilde{A}_m^{(0)} := \widetilde{a}_m = (u(m))^{\lambda_1 - \frac{\varepsilon}{p} - 1} u'(m),$$

$$\widetilde{A}_m^{(1)} := \sum_{k=m_0}^{m} \widetilde{a}_k = \sum_{k=m_0}^{m} (u(k))^{(\lambda_1 - \frac{\varepsilon}{p}) - 1} u'(k), m \in \mathbf{N}_{m_0}, \quad \text{and}$$

$$\widetilde{b}_k := ||k||_\beta^{\alpha(\lambda_2 - \frac{\varepsilon}{q}) - n}, \quad k \in \mathbf{N}^n.$$

Since both $(u(x))^{(\lambda_1 - \frac{\varepsilon}{p}) - 1} u'(x) (= (u(x))^{-\frac{\varepsilon}{p} - 1}[(u(x))^{\lambda_1} u'(x)])$ and $(u(x))^{-\varepsilon-1} u'(x)$ $(= (u(x))^{-\lambda_1 - \varepsilon - 1}[(u(x))^{\lambda_1} u'(x)])$ are strictly decreasing in $(m_0 - 1, \infty)$, we find

$$0 < \widetilde{A}_m^{(1)} < \int_{m_0 - 1}^{m} (u(x))^{(\lambda_1 - \frac{\varepsilon}{p}) - 1} u'(x) dx \le \frac{(u(m))^{\lambda_1 - \frac{\varepsilon}{p}}}{\lambda_1 - \frac{\varepsilon}{p}},$$

satisfying $\widetilde{A}_m^{(1)} = o(e^{tu(m)})$ $(t > 0; m \to \infty)$. For $i \in \{0,1\}$, it follows that

$$\widetilde{A}_m^{(i)} \le \frac{1}{(\lambda_1 - \frac{\varepsilon}{p})^i} (u(m))^{\lambda_1 - \frac{\varepsilon}{p} + i - 1} (u'(m))^{1-i} \ (m \in \mathbf{N}_{m_0}) \ \text{ and}$$

$$\sum_{m=m_0}^{\infty} \frac{(u'(m))^{1-p+pi} (\widetilde{A}_m^{(i)})^p}{(u(m))^{p(i-1+\lambda_1)+1}} \le \frac{1}{(\lambda_1 - \frac{\varepsilon}{p})^{ip}} \sum_{m=m_0}^{\infty} \frac{u'(m)}{(u(m))^{\varepsilon+1}}$$

$$< \frac{1}{(\lambda_1 - \frac{\varepsilon}{p})^{ip}} \left[\frac{u'(m_0)}{(u(m_0))^{\varepsilon+1}} + \int_{m_0}^{\infty} \frac{u'(x)}{(u(x))^{\varepsilon+1}} dx \right]$$

$$= \frac{1}{(\lambda_1 - \frac{\varepsilon}{p})^{ip}} \left[d + \frac{1}{\varepsilon (u(m_0))^{\varepsilon}} \right] \left(d := \frac{u'(m_0)}{(u(m_0))^{\varepsilon+1}} \right). \tag{3.3}$$

If there exists a positive constant M, with

$$M \le \lambda^i \left(\frac{\Gamma^n\left(\frac{1}{\beta}\right)}{\alpha \beta^{n-1} \Gamma\left(\frac{n}{\beta}\right)} \right)^{\frac{1}{p}} B(\lambda_1 + i, \lambda_2),$$

such that (3.2) is valid when we replace the constant factor by M, then in particular, by (3.3) and (2.4) (for $c = \varepsilon$), we have

$$\widetilde{I} := \sum_k \sum_{m=m_0}^{\infty} \frac{\widetilde{a}_m \widetilde{b}_k}{(u(m) + ||k||_\beta^\alpha)^\lambda}$$

$$< M \left[\sum_{m=m_0}^{\infty} \frac{(u'(m))^{1-p+pi} (\widetilde{A}_m^{(i)})^p}{(u(m))^{p(i-1+\lambda_1)+1}} \right]^{\frac{1}{p}} \left[\sum_k ||k||_\beta^{q(n-\alpha\lambda_2)-n} \widetilde{b}_k^q \right]^{\frac{1}{q}}$$

$$= M \left[\sum_{m=m_0}^{\infty} \frac{(u'(m))^{1-p+pi} (\widetilde{A}_m^{(i)})^p}{(u(m))^{p(i-1+\lambda_1)+1}} \right]^{\frac{1}{p}} \left(\sum_k ||k||_\beta^{-\alpha\varepsilon-n} \right)^{\frac{1}{q}}$$

$$< \frac{M}{\varepsilon \left(\lambda_1 - \frac{\varepsilon}{p}\right)^i} \left[\varepsilon d + \frac{1}{(u(m_0))^{\varepsilon}} \right]^{\frac{1}{p}} \left(\frac{2^{\alpha\varepsilon} \Gamma^n\left(\frac{1}{\beta}\right)}{\alpha \beta^{n-1} \Gamma\left(\frac{n}{\beta}\right)} \right)^{\frac{1}{q}}.$$

By the decreasing property of series and (2.4), we have

$$\widetilde{I} = \sum_k ||k||_\beta^{\alpha(\lambda_2 - \frac{\varepsilon}{q})-n} \sum_{m=m_0}^{\infty} \frac{(u(m))^{\lambda_1 - \frac{\varepsilon}{p} - 1} u'(m)}{(u(m) + ||k||_\beta^\alpha)^\lambda}$$

$$> \sum_k ||k||_\beta^{\alpha(\lambda_2 - \frac{\varepsilon}{q})-n} \int_{m_0}^{\infty} \frac{(u(x))^{\lambda_1 - \frac{\varepsilon}{p} - 1} u'(x)}{(u(x) + ||k||_\beta^\alpha)^\lambda} dx$$

$$= \sum_k ||k||_\beta^{-\alpha\varepsilon-n} \left[\int_0^{\infty} \frac{v^{\lambda_1 - \frac{\varepsilon}{p} - 1}}{(1+v)^\lambda} dv - \int_0^{u(m_0)/||k||_\beta^\alpha} \frac{v^{\lambda_1 - \frac{\varepsilon}{p} - 1}}{(1+v)^\lambda} dv \right]$$

$$> \sum_k ||k||_\beta^{-\alpha\varepsilon - n}\left[B\left(\lambda_1 - \frac{\varepsilon}{p}, \lambda_2 + \frac{\varepsilon}{p}\right) - \int_0^{u(m_0)/||k||_\beta^\alpha} v^{\lambda_1 - \frac{\varepsilon}{p} - 1} dv\right]$$

$$= B\left(\lambda_1 - \frac{\varepsilon}{p}, \lambda_2 + \frac{\varepsilon}{p}\right)\sum_k ||k||_\beta^{-\alpha\varepsilon - n} - \frac{(u(m_0))^{\lambda_1 - \frac{\varepsilon}{p}}}{\lambda_1 - \frac{\varepsilon}{p}}$$

$$\times \sum_k ||k||_\beta^{-\alpha\left(\lambda_1 + \frac{\varepsilon}{q}\right) - n}$$

$$> B(\lambda_1 - \frac{\varepsilon}{p}, \lambda_2 + \frac{\varepsilon}{p})\frac{\left[1 - \frac{\alpha\varepsilon(n-1)}{\alpha\varepsilon + 1}\right]\Gamma^n\left(\frac{1}{\beta}\right)}{\alpha\varepsilon\beta^{n-1}\Gamma\left(\frac{n}{\beta}\right)} - O(1).$$

Based on the above results, we have the following inequality:

$$B(\lambda_1 - \frac{\varepsilon}{p}, \lambda_2 + \frac{\varepsilon}{p})\frac{[1 - \frac{\alpha\varepsilon(n-1)}{\alpha\varepsilon+1}]\Gamma^n(\frac{1}{\beta})}{\alpha\beta^{n-1}\Gamma(\frac{n}{\beta})} - \varepsilon O(1)$$

$$< \varepsilon\widetilde{I} < \frac{M}{(\lambda_1 - \frac{\varepsilon}{p})^i}\left[\varepsilon d + \frac{1}{(u(m_0))^\varepsilon}\right]^{\frac{1}{p}}\left(\frac{2^{\alpha\varepsilon}\Gamma^n(\frac{1}{\beta})}{\alpha\beta^{n-1}\Gamma(\frac{n}{\beta})}\right)^{\frac{1}{q}}.$$

For $\varepsilon \to 0^+$, in view of the continuity of the Beta function, we have

$$\frac{B(\lambda_1, \lambda_2)\Gamma^n(\frac{1}{\beta})}{\alpha\beta^{n-1}\Gamma(\frac{n}{\beta})} \le \frac{M}{\lambda_1^i}\left(\frac{\Gamma^n(\frac{1}{\beta})}{\alpha\beta^{n-1}\Gamma(\frac{n}{\beta})}\right)^{\frac{1}{q}}.$$

Since $i \in (0,1\}$, $\lambda^i B(\lambda_1 + i, \lambda_2) = \lambda_1^i B(\lambda_1, \lambda_2)$, it follows that

$$\lambda^i\left(\frac{\Gamma^n(\frac{1}{\beta})}{\alpha\beta^{n-1}\Gamma(\frac{n}{\beta})}\right)^{\frac{1}{p}} B(\lambda_1 + i, \lambda_2)$$

$$= \lambda_1^i\left(\frac{\Gamma^n(\frac{1}{\beta})}{\alpha\beta^{n-1}\Gamma(\frac{n}{\beta})}\right)^{\frac{1}{p}} B(\lambda_1, \lambda_2) \le M.$$

Therefore,

$$M = \lambda^i\left(\frac{\Gamma^n(\frac{1}{\beta})}{\alpha\beta^{n-1}\Gamma(\frac{n}{\beta})}\right)^{\frac{1}{p}} B(\lambda_1 + i, \lambda_2)$$

is the best value of (3.2) (namely, for $\lambda_1 + \lambda_2 = \lambda$ in (3.1)).

This proves the theorem. □

Theorem 3. *With regard to the assumption H1, if the constant factor in* (3.1) *is the best possible, then for* $\lambda - \lambda_1 - \lambda_2 \le 0$, *we have* $\lambda_1 + \lambda_2 = \lambda$.

Proof. For $\widehat{\lambda}_1 = \frac{\lambda-\lambda_2}{p} + \frac{\lambda_1}{q} = \frac{\lambda-\lambda_1-\lambda_2}{p} + \lambda_1$, $\widehat{\lambda}_2 = \frac{\lambda-\lambda_1}{q} + \frac{\lambda_2}{p} = \frac{\lambda-\lambda_1-\lambda_2}{q} + \lambda_2$, $0 < \lambda_1, \lambda_2 < \lambda$, we find $\widehat{\lambda}_1 + \widehat{\lambda}_2 = \lambda$ with $0 < \widehat{\lambda}_1, \widehat{\lambda}_2 < \lambda$. For $\lambda - \lambda_1 - \lambda_2 \le 0$, we observe that

$$(u(x))^{\widehat{\lambda}_1} u'(x) = (u(x))^{\frac{\lambda-\lambda_1-\lambda_2}{p}} [(u(x))^{\lambda_1} u'(x)]$$

is decreasing in $(m_0 - 1, \infty)$, such that $\widehat{\lambda}_2 \le \lambda_2 \le \frac{n}{\alpha}$.

By Hölder's inequality (cf. Ref. [31]), we obtain

$$\begin{aligned}
&B(\widehat{\lambda}_1 + i, \widehat{\lambda}_2) \\
&= \int_0^\infty \frac{u^{\widehat{\lambda}_1+i-1} du}{(1+u)^{\lambda+i}} = \int_0^\infty \frac{(u^{\frac{\lambda-\lambda_2+i-1}{p}})(u^{\frac{\lambda_1+i-1}{q}})}{(1+u)^{\lambda+i}} du \\
&\le \left[\int_0^\infty \frac{u^{\lambda-\lambda_2+i-1}}{(1+u)^{\lambda+i}} du\right]^{\frac{1}{p}} \left[\int_0^\infty \frac{u^{\lambda_1+i-1}}{(1+u)^{\lambda+i}} du\right]^{\frac{1}{q}} \\
&= \left[\int_0^\infty \frac{v^{\lambda_2-1}}{(1+v)^{\lambda+i}} dv\right]^{\frac{1}{p}} \left[\int_0^\infty \frac{u^{\lambda_1+i-1}}{(1+u)^{\lambda+i}} du\right]^{\frac{1}{q}} \\
&= B^{\frac{1}{p}}(\lambda_2, \lambda + i - \lambda_2) B^{\frac{1}{q}}(\lambda_1 + i, \lambda - \lambda_1). \qquad (3.4)
\end{aligned}$$

Since the constant factor

$$\lambda^i \left(\frac{\Gamma^n(\frac{1}{\beta}) B(\lambda_2, \lambda + i - \lambda_2)}{\alpha \beta^{n-1} \Gamma(\frac{n}{\beta})} \right)^{\frac{1}{p}} B^{\frac{1}{q}}(\lambda_1 + i, \lambda - \lambda_1)$$

in (3.1) is the best value, comparing with the constant factors in (3.1) and (3.2) (for $\lambda_1 = \widehat{\lambda}_1, \lambda_2 = \widehat{\lambda}_2$), we have

$$\begin{aligned}
&\lambda^i \left(\frac{\Gamma^n(\frac{1}{\beta}) B(\lambda_2, \lambda + i - \lambda_2)}{\alpha \beta^{n-1} \Gamma(\frac{n}{\beta})} \right)^{\frac{1}{p}} B^{\frac{1}{q}}(\lambda_1 + i, \lambda - \lambda_1) \\
&\le \lambda^i \left(\frac{\Gamma^n(\frac{1}{\beta})}{\alpha \beta^{n-1} \Gamma(\frac{n}{\beta})} \right)^{\frac{1}{p}} B(\widehat{\lambda}_1 + i, \widehat{\lambda}_2),
\end{aligned}$$

namely,

$$B(\widehat{\lambda}_1 + i, \widehat{\lambda}_2) \geq B^{\frac{1}{p}}(\lambda_2, \lambda + i - \lambda_2) B^{\frac{1}{q}}(\lambda_1 + i, \lambda - \lambda_1).$$

Hence, (3.4) keeps the form of an equality. The necessary and sufficient condition for taking an equal sign is that there exist constants A and B, such that they are not both zero, and (cf. Ref. [31]) $Au^{\lambda-\lambda_2+i-1} = Bu^{\lambda_1+i-1}$ a.e. in $\mathbf{R}_+$. Assuming that $A \neq 0$, we have $u^{\lambda-\lambda_2-\lambda_1} = \frac{B}{A}$ a.e. in $\mathbf{R}_+$. It follows that $\lambda - \lambda_2 - \lambda_1 = 0$, and then $\lambda_1 + \lambda_2 = \lambda$.

This proves the theorem. □

Remark 2. For $\gamma \in (0, 1)$, in view of Remark 1(ii), (iii), both $u_1(x) = x$, $(x \in (0, \infty); m_0 = 1)$ and $u_2(x) = \ln^\gamma x$ $(x \in (1, \infty); m_0 = 2)$ satisfy for using Theorems 1–3.

Corollary 1. *For $i = 0$ in (3.1), we have the following inequality:*

$$\begin{aligned} I &:= \sum_k \sum_{m=m_0}^{\infty} \frac{a_m b_k}{(u(m) + \|k\|_\beta^\alpha)^\lambda} \\ &< \left(\frac{\Gamma^n(\frac{1}{\beta}) B(\lambda_2, \lambda - \lambda_2)}{\alpha \beta^{n-1} \Gamma(\frac{n}{\beta})} \right)^{\frac{1}{p}} B^{\frac{1}{q}}(\lambda_1, \lambda - \lambda_1) \\ &\quad \times \left[\sum_{m=m_0}^{\infty} \frac{(u'(m))^{1-p} a_m^p}{(u(m))^{p(\widehat{\lambda}_1 - 1)+1}} \right]^{\frac{1}{p}} \left[\sum_k \|k\|_\beta^{q(n-\alpha\widehat{\lambda}_2)-n} b_k^q \right]^{\frac{1}{q}}. \end{aligned} \tag{3.5}$$

In particular, for $\lambda_1 + \lambda_2 = \lambda$, we have

$$\begin{aligned} \sum_k \sum_{m=m_0}^{\infty} \frac{a_m b_k}{(u(m) + \|k\|_\beta^\alpha)^\lambda} &< \left(\frac{\Gamma^n\left(\frac{1}{\beta}\right)}{\alpha \beta^{n-1} \Gamma\left(\frac{n}{\beta}\right)} \right)^{\frac{1}{p}} B(\lambda_1, \lambda_2) \\ &\quad \times \left[\sum_{m=m_0}^{\infty} \frac{(u'(m))^{1-p} a_m^p}{(u(m))^{p(\lambda_1 - 1)+1}} \right]^{\frac{1}{p}} \left[\sum_k \|k\|_\beta^{q(n-\alpha\lambda_2)-n} b_k^q \right]^{\frac{1}{q}}. \end{aligned} \tag{3.6}$$

Corollary 2. *If* $\lambda_1 + \lambda_2 = \lambda$, *then the constant factor*

$$\left(\frac{\Gamma^n(\frac{1}{\beta})B(\lambda_2, \lambda - \lambda_2)}{\alpha\beta^{n-1}\Gamma(\frac{n}{\beta})}\right)^{\frac{1}{p}} B^{\frac{1}{q}}(\lambda_1, \lambda - \lambda_1)$$

in (3.5) *is the best value. On the other hand, if the same constant factor in* (3.5) *is the best value, then for* $\lambda - \lambda_1 - \lambda_2 \leq 0$, *we have* $\lambda_1 + \lambda_2 = \lambda$.

Remark 3. Inequality (3.1) (resp. (3.2)) is an extended application of 3.5) (resp. (3.6)).

4. Conclusion

In this chapter, following the way of Refs. [21, 22], by means of the weight functions, the idea of introduced parameters, the techniques of real analysis and Abel's partial summation formula, a multidimensional discrete Hardy–Hilbert's inequality with the kernel as $\frac{1}{(u(m)+||k||_{\beta}^{\alpha})^{\lambda}}$ $(\alpha, \lambda > 0)$ involving one partial sum is obtained in Theorem 1. The equivalent statements of the best value related to several parameters in the new inequality are given in Theorems 2 and 3. Some particular results are provided in Corollaries 1 and 2.

Acknowledgment

This work is supported by the National Science Foundation of China (No. 61772140).

References

[1] G. H. Hardy, J. E. Littlewood and G. Polya, *Inequalities*, Cambridge University Press, Cambridge, 1934.

[2] M. Krnić and J. Pečarić, Extension of Hilbert's inequality. *J. Math. Anal. Appl.*, 324(1), 2006, 150–160.

[3] V. Adiyasuren, T. Batbold and E. A. Azar, A new discrete Hilbert-type inequality involving partial sums. *J. Inequal. App.*, 2019, 2019, 127.

[4] H. M. Mo and B. C. Yang, On a new Hilbert-type integral involving the upper limit functions. *J. Inequal. App.*, 2020, 2020, 5.

[5] B. C. Yang, *The Norm of Operator and Hilbert-type Inequality*, Science Press, Beijing, 2009.

[6] M. Krnić and J. Pečarić, General Hilbert's and Hardy's inequalities. *Math. Inequal. Appl.*, 8(1), 2005, 29–51.

[7] I. Perić and P. Vuković, Multiple Hilbert's type inequalities with a homogeneous kernel, *Banach J. Math. Anal.*, 5(2), 2011, 33–43.

[8] Q. L. Huang, A new extension of Hardy-Hilbert-type inequality. *J. Inequal. Appl.*, 2015, 2015, 397.

[9] B. He, A multiple Hilbert-type discrete inequality with a new kernel and best possible constant factor. *J. Math. Anal. Appl.*, 431, 2015, 902–990.

[10] J. S. Xu, Hardy-Hilbert's inequalities with two parameters. *Adv. Math.*, 36(2), 2007, 63–76.

[11] Z. T. Xie, Z. Zeng and Y. F. Sun, A new Hilbert-type inequality with the homogeneous kernel of degree -2. *Adv. Appl. Math. Sci.*, 12(7), 2013, 391–401.

[12] Z. Zeng, K. Raja Rama Gandhi and Z. T. Xie, A new Hilbert-type inequality with the homogeneous kernel of degree -2 and with the integral. *Bull. Math. Anal. Appl.*, 3(1), 2014, 11–20.

[13] D. M. Xin, A Hilbert-type integral inequality with the homogeneous kernel of zero degree. *Math. Theory Appl.*, 30(2), 2010, 70–74.

[14] L. E. Azar, The connection between Hilbert and Hardy inequalities. *J. Inequal. Appl.*, 2013, 2013, 452

[15] V. Adiyasuren, T. Batbold and M. Krnić, Hilbert–type inequalities involving differential operators, the best constants and applications. *Math. Inequal. Appl.*, 18(2015), 2015, 111–124.

[16] Y. Hong and Y. M. Wen, A necessary and sufficient condition of that Hilbert type series inequality with homogeneous kernel has the best constant factor. *Ann. Math.*, 37A(3), 2016, 329–336.

[17] Y. Hong, On the structure character of Hilbert's type integral inequality with homogeneous kernel and applications. *J. Jilin Univ. Sci. Ed.*, 55(2), 2017, 189–194.

[18] Y. Hong, Q. L. Huang, B. C. Yang and J. Q. Liao, The necessary and sufficient conditions for the existence of a kind of Hilbert-type multiple integral inequality with the non-homogeneous kernel and its applications. *J. Inequal. Appl.*, 2017, 2017, 316.

[19] Q. Chen, B. He, Y. Hong and Z. Li, Equivalent parameter conditions for the validity of half-discrete Hilbert-type multiple integral inequality with generalized homogeneous kernel. *J. Funct. Spaces*, 2020, 2020, 7414861.

[20] B. He, Y. Hong and Z. Li, Conditions for the validity of a class of optimal Hilbert type multiple integral inequalities with non-homogeneous. *J. Inequal. Appl.*, 2021, 2021, 64.
[21] Y. Hong, Y. R. Zhong and B. C. Yang, On a more accurate half-discrete multidimensional Hilbert-type inequality involving one derivative function of m-order. *J. Inequal. Appl.*, 2023, 2023, 74.
[22] Y. Hong, Y. R. Zhong and B. C. Yang, A more accurate half-discrete multidimensional Hilbert-type inequality involving one multiple upper limit function. *Axioms*, 2023, 12, 211.
[23] M. Th. Rassias and B. C. Yang. A multidimensional half-discrete Hilbert-type inequality and the Riemann zeta function. *Appl. Math. Comput.*, 225, 2013, 263–277.
[24] M. Th. Rassias and B. C. Yang. On a multidimensional half C discrete Hilbert C type inequality related to the hyperbolic cotangent function. *Appl. Math. Comput.*, 242, 2014, 800–813.
[25] M. Th. Rassias and B. C. Yang. On a multidimensional Hilbert-type integral inequality associated to the gamma function. *Appl. Math. Comput.*, 249(2014), 408–418.
[26] B. C. Yang and J. Q. Liao, *Parameterized Multidimensional Hilbert-type Inequalities*, Scientific Rescarch Publishing, 2020, USA.
[27] Yiyun Li, Yanru Zhong and B. C. Yang, Equivalent statements of two multidimensional Hilbert-type integral inequalities involving one with parameters, *Axioms*, 2023, 12, 956.
[28] J. Q. Liao and B. C. Yang. A new reverse half-discrete Hilbert-type inequality with one partial sum involving one derivative function of higher order. *Open Math.*, 2023, 21: 20230139.
[29] A. Z. Wang and B. C. Yang. A new half-discrete Hilbert-type inequality involving one multiple upper limit function and one partial sums. *Ann. Math.*, 45A(1), 2024, 25–38.
[30] Z. H. Zeng and B. C. Yang. A Hardy-Hilbert-type integral inequality involving the derivative functions of n-order. *J. South China Norm. Univ. Nat. Sci. Ed.*, 56(4), 2024, 123–128.
[31] J. C. Kuang, *Applied Inequalities*, Shandong Science and Technology Press, Jinan, 2004.
[32] J. C. Kuang. *Real and Functional Analysis (Continuation)*, Vol. 2. Higher Education Press, Beijing, 2015.

https://doi.org/10.1142/9789819820498_0024

Chapter 24

Equivalent Properties of the Reverse Hardy-Type Integral Inequalities in the Whole Plane

Michael Th. Rassias[*,¶], **Bicheng Yang**[‡,‖] and **Andrei Raigorodskii**[§,**]

[*]*Department of Mathematics and Engineering Sciences, Hellenic Military Academy, 16673 Vari Attikis, Greece*
[‡]*Department of Mathematics, Guangdong University of Education, Guangzhou, Guangdong 510303, P. R. China*
[§]*Moscow Institute of Physics and Technology 141700 Dolgoprudny Institutskiy per, d. 9, Russia*
[¶]*mthrassias@yahoo.com*
[‖]*bcyang@gdei.edu.cn,bcyang818@163.com*
[**]*raigorodsky@yandex-team.ru*

By using the way of real analysis and the weight functions, a few equivalent conditions of two kinds of the reverse Hardy-type integral inequalities with the general non-homogeneous kernel in the whole plane are obtained. The constant factors are proved to be the best possible. As applications, a few equivalent conditions of two kinds of the reverse Hardy-type integral inequalities with the general homogeneous kernel in the whole plane are deduced. We also consider some particular cases and a few examples.

Keywords: Hardy-type integral inequality, weight function, parameter, equivalent form, reverse.

1. Introduction

If $f(x), g(y) \geq 0$, $0 < \int_0^\infty f^2(x)dx < \infty$ and $0 < \int_0^\infty g^2(y)dy < \infty$, then we have the following Hilbert's integral inequality with the best possible constant factor π (see Ref. [1]):

$$\int_0^\infty \int_0^\infty \frac{f(x)g(y)}{x+y} dxdy < \pi \left(\int_0^\infty f^2(x)dx \int_0^\infty g^2(y)dy \right)^{\frac{1}{2}}. \quad (1.1)$$

Recently, by using the weight functions, a lot of generalizations of (1.1) were given by two books (see Refs. [2, 3]). Some Hilbert-type inequalities with the homogenous kernels and non-homogenous kernels were provided by Refs. [4–9]. In 2017, Hong [10] also gave an equivalent condition between a Hilbert-type inequalities with the homogenous kernel and some parameters. Some other kinds of Hilbert-type inequalities were obtained by Refs. [11–16]. Most of them are built in the quarter plane of the first quadrant.

Using the way of real analysis, in 2007, Yang [20] gave a Hilbert-type integral inequality in the whole plane as follows:

$$\int_{-\infty}^\infty \int_{-\infty}^\infty \frac{f(x)g(y)}{(1+e^{x+y})^\lambda} dxdy$$
$$< B\left(\frac{\lambda}{2}, \frac{\lambda}{2}\right) \left(\int_{-\infty}^\infty e^{-\lambda x} f^2(x)dx \int_{-\infty}^\infty e^{-\lambda y} g^2(y)dy \right)^{\frac{1}{2}}, \quad (1.2)$$

where the constant factor $B(\frac{\lambda}{2}, \frac{\lambda}{2})$ is the best possible ($\lambda > 0$, $B(u, v)$ is the beta function) (see Ref. [21]). He *et al.* [22–35] proved some new Hilbert-type integral inequalities in the whole plane with the best possible constant factors.

In this chapter, by using the way of real analysis and the weight functions, a few equivalent conditions of two kinds of the reverse Hardy-type integral inequalities with the general non-homogeneous kernel in the whole plane are obtained. The constant factors are proved to be the best possible. As applications, a few equivalent conditions of two kinds of the reverse Hardy-type integral inequalities with the general homogeneous kernel in the whole plane are deduced. We also consider some particular cases and a few examples.

2. Two Lemmas

In the following, we suppose that $0 < p < 1, q < 0, \frac{1}{p} + \frac{1}{q} = 1, \sigma_1, \sigma \in \mathbf{R} = (-\infty, \infty)$, $M_1, M_2 > 0$, $h(u)$ is a non-negative measurable function in $\mathbf{R}$, and

$$K^{(1)}(\sigma) := \int_0^1 (h(-u) + h(u))u^{\sigma-1} du\,, \tag{2.1}$$

$$K^{(2)}(\sigma) := \int_1^\infty (h(-u) + h(u))u^{\sigma-1} du\,. \tag{2.2}$$

Lemma 1. *If $K^{(1)}(\sigma) < \infty$, for any non-negative measurable functions $f(x)$ and $g(y)$ in $\mathbf{R}$, the following inequality*

$$\int_{-\infty}^{\infty} g(y) \left(\int_{-\frac{1}{|y|}}^{\frac{1}{|y|}} h(xy) f(x) dx \right) dy$$
$$\geq M_1 \left[\int_{-\infty}^{\infty} |x|^{p(1-\sigma)-1} f^p(x) dx \right]^{\frac{1}{p}} \left[\int_{-\infty}^{\infty} |y|^{q(1-\sigma_1)-1} g^q(y) dy \right]^{\frac{1}{q}} \tag{2.3}$$

holds true, then we have $\sigma_1 = \sigma$ and $K^{(1)}(\sigma) \geq M_1$.

Proof. If $\sigma_1 < \sigma$, then for $n \in \mathbf{N}$, we set the following two functions:

$$f_n(x) := \begin{cases} |x|^{\sigma + \frac{1}{pn} - 1}, 0 < |x| \leq 1 \\ 0, |x| > 1 \end{cases},$$

$$g_n(y) := \begin{cases} 0, 0 < |y| < 1 \\ |y|^{\sigma_1 - \frac{1}{qn} - 1}, y \geq 1 \end{cases},$$

and find

$$J_1 := \left[\int_{-\infty}^{\infty} |x|^{p(1-\sigma)-1} f_n^p(x)dx\right]^{\frac{1}{p}} \left[\int_{-\infty}^{\infty} |y|^{q(1-\sigma_1)-1} g_n^q(y)dy\right]^{\frac{1}{q}}$$

$$= \left[\int_{-1}^{1} |x|^{p(1-\sigma)-1}|x|^{p(\sigma+\frac{1}{pn}-1)}dx\right]^{\frac{1}{p}}$$

$$\times \left[\int_{\{y;|y|\geq 1\}} |y|^{q(1-\sigma_1)-1}|y|^{q(\sigma_1-\frac{1}{qn}-1)}dy\right]^{\frac{1}{q}}$$

$$= \left(2\int_{0}^{1} x^{\frac{1}{n}-1}dx\right)^{\frac{1}{p}} \left(2\int_{1}^{\infty} y^{-\frac{1}{n}-1}dy\right)^{\frac{1}{q}} = 2n. \tag{2.4}$$

We obtain

$$I_1 := \int_{-\infty}^{\infty} g_n(y)\left(\int_{-\frac{1}{|y|}}^{\frac{1}{|y|}} h(xy)f_n(x)dx\right)dy$$

$$= \int_{-\infty}^{-1}\left(\int_{\frac{1}{y}}^{\frac{-1}{y}} h(xy)|x|^{\sigma+\frac{1}{pn}-1}dx\right)(-y)^{\sigma_1-\frac{1}{qn}-1}dy$$

$$+\int_{1}^{\infty}\left(\int_{\frac{-1}{y}}^{\frac{1}{y}} h(xy)|x|^{\sigma+\frac{1}{pn}-1}dx\right)y^{\sigma_1-\frac{1}{qn}-1}dy$$

$$= \int_{1}^{\infty}\left[\int_{\frac{-1}{y}}^{\frac{1}{y}}(h(-xy)+h(xy))|x|^{\sigma+\frac{1}{pn}-1}dx\right]y^{\sigma_1-\frac{1}{qn}-1}dy$$

$$= \int_{1}^{\infty}\left[\int_{-1}^{1}(h(-u)+h(u))|u|^{(\sigma+\frac{1}{pn})-1}du\right]y^{(\sigma_1-\sigma)-\frac{1}{n}-1}dy, \tag{2.5}$$

and then by (2.3), we have

$$2K^{(1)}\left(\sigma+\frac{1}{pn}\right)\frac{1}{\sigma-\sigma_1+\frac{1}{n}}$$

$$= \left[\int_{-1}^{1}(h(-u)+h(u))|u|^{(\sigma+\frac{1}{pn})-1}du\right]\int_{1}^{\infty} y^{\sigma_1-\sigma-\frac{1}{n}-1}dy$$

$$= I_1 \geq M_1 J_1 = 2M_1 n. \tag{2.6}$$

Since $\left\{(h(-u)+h(u))u^{\sigma+\frac{1}{pn}-1}\right\}_{n=1}^{\infty}$ $(u \in (0,1))$ is a non-negative and increasing sequence, by Levi theorem (cf. Ref. [45]), it follows that

$$K^{(1)}\left(\sigma+\frac{1}{pn}\right) = \int_0^1 (h(-u)+h(u))u^{\sigma+\frac{1}{pn}-1}du$$
$$\to \int_0^1 (h(-u)+h(u))u^{\sigma-1}du = K^{(1)}(\sigma)(n\to\infty),$$

and then by (2.6), we have $\infty > \frac{2K^{(1)}(\sigma)}{\sigma-\sigma_1} \geq \infty$, which is a contradiction.

If $\sigma_1 > \sigma$, then for $n \geq \frac{1}{(\sigma_1-\sigma)|q|}$ $(n \in \mathbf{N})$, we set the following two functions:

$$\widetilde{f}_n(x) := \begin{cases} 0, 0<|x|<1 \\ |x|^{\sigma-\frac{1}{pn}-1}, |x|\geq 1 \end{cases},$$

$$\widetilde{g}_n(y) := \begin{cases} |y|^{\sigma_1+\frac{1}{qn}-1}, 0<|y|\leq 1 \\ 0, |y|>1 \end{cases},$$

and find

$$\widetilde{J}_1 := \left[\int_{-\infty}^{\infty} |x|^{p(1-\sigma)-1}\widetilde{f}_n^p(x)dx\right]^{\frac{1}{p}} \left[\int_{-\infty}^{\infty} y^{q(1-\sigma_1)-1}\widetilde{g}_n^q(y)dy\right]^{\frac{1}{q}}$$
$$= \left[2\int_1^{\infty} x^{p(1-\sigma)-1}x^{p(\sigma-\frac{1}{pn}-1)}dx\right]^{\frac{1}{p}}$$
$$\times \left[2\int_0^1 y^{q(1-\sigma_1)-1}y^{q(\sigma_1+\frac{1}{qn}-1)}dy\right]^{\frac{1}{q}}$$
$$= \left(2\int_1^{\infty} x^{-\frac{1}{n}-1}dx\right)^{\frac{1}{p}} \left(2\int_0^1 y^{\frac{1}{n}-1}dy\right)^{\frac{1}{q}} = 2n.$$

We obtain

$$\widetilde{I}_1 := \int_{-\infty}^{\infty} \widetilde{f}_n(x)\left(\int_{-\frac{1}{|x|}}^{\frac{1}{|x|}} h(xy)\widetilde{g}_n(y)dy\right)dx$$
$$= \int_{-\infty}^{-1}\left(\int_{\frac{1}{x}}^{\frac{-1}{x}} h(xy)|y|^{\sigma_1+\frac{1}{qn}-1}dy\right)(-x)^{\sigma-\frac{1}{pn}-1}dx$$

$$+\int_1^\infty\left(\int_{-\frac{1}{x}}^{\frac{1}{x}} h(xy)|y|^{\sigma_1+\frac{1}{qn}-1}dy\right)x^{\sigma-\frac{1}{pn}-1}dx$$

$$=\int_1^\infty\left[\int_{-\frac{1}{x}}^{\frac{1}{x}}(h(-xy)+h(xy))|y|^{\sigma_1+\frac{1}{qn}-1}dy\right]x^{\sigma-\frac{1}{pn}-1}dx$$

$$=\int_1^\infty\left[\int_{-1}^{1}(h(-u)+h(u))|u|^{\sigma_1+\frac{1}{qn}-1}du\right]x^{(\sigma-\sigma_1)-\frac{1}{n}-1}dx, \tag{2.7}$$

and then by Fubini theorem (cf. Ref. [45]) and (2.3), it follows that

$$2K^{(1)}\left(\sigma_1+\frac{1}{qn}\right)\frac{1}{\sigma_1-\sigma+\frac{1}{n}}$$

$$=\left[\int_{-1}^{1}(h(-u)+h(u))|u|^{\sigma_1+\frac{1}{qn}-1}du\right]\int_1^\infty x^{\sigma-\sigma_1-\frac{1}{n}-1}dx$$

$$=\widetilde{I}_1=\int_{-\infty}^{\infty}\widetilde{g}_n(y)\left(\int_{\frac{-1}{|y|}}^{\frac{1}{|y|}}h(xy)\widetilde{f}_n(x)dx\right)dy\geq M_1\widetilde{J}_1=2M_1n. \tag{2.8}$$

Since $\sigma_1+\frac{1}{qn}\geq\sigma$, we have

$$0\leq K^{(1)}\left(\sigma_1+\frac{1}{qn}\right)=\int_0^1(h(-u)+h(u))u^{\sigma_1+\frac{1}{qn}-1}du$$

$$\leq\int_0^1(h(-u)+h(u))u^{\sigma-1}du=K^{(1)}(\sigma)<\infty,$$

and then by (2.8), for $n\to\infty$, it follows that $\infty>\frac{2K^{(1)}(\sigma)}{\sigma_1-\sigma}\geq\infty$, which is a contradiction.

Hence, we conclude that $\sigma_1=\sigma$.

For $\sigma_1=\sigma$, we reduce (2.6) as follows:

$$K^{(1)}\left(\sigma+\frac{1}{pn}\right)=\int_0^1(h(-u)+h(u))u^{\sigma+\frac{1}{pn}-1}du\geq M_1.$$

Still by Levi theorem, for $n\to\infty$, we have $K^{(1)}(\sigma)\geq M_1$.

The lemma is proved. □

Lemma 2. *If $K^{(2)}(\sigma) < \infty$, for any non-negative measurable functions $f(x)$ and $g(y)$ in* **R**, *the following inequality*

$$\int_{-\infty}^{\infty} g(y) \left(\int_{\{x;|x|\geq \frac{1}{|y|}\}} h(xy) f(x) dx \right) dy$$
$$\geq M_2 \left[\int_{-\infty}^{\infty} |x|^{p(1-\sigma)-1} f^p(x) dx \right]^{\frac{1}{p}} \left[\int_{-\infty}^{\infty} |y|^{q(1-\sigma_1)-1} g^q(y) dy \right]^{\frac{1}{q}} \tag{2.9}$$

holds true, then we have $\sigma_1 = \sigma$ and $K^{(2)}(\sigma) \geq M_2$.

Proof. If $\sigma_1 > \sigma$, then for $n \in \mathbf{N}$, we set two functions $\widetilde{f}_n(x)$ and $\widetilde{g}_n(y)$ as in Lemma 1 and find

$$\widetilde{J}_1 = \left[\int_{-\infty}^{\infty} |x|^{p(1-\sigma)-1} \widetilde{f}_n^p(x) dx \right]^{\frac{1}{p}} \left[\int_{-\infty}^{\infty} |y|^{q(1-\sigma_1)-1} \widetilde{g}_n^q(y) dy \right]^{\frac{1}{q}} = 2n.$$

We obtain

$$\widetilde{I}_2 := \int_{-\infty}^{\infty} \widetilde{g}_n(y) \left(\int_{\{x;|x|\geq \frac{1}{|y|}\}} h(xy) \widetilde{f}_n(x) dx \right) dy$$
$$= \int_{-1}^{0} \left(\int_{\{x;|x|\geq \frac{-1}{y}\}} h(xy) |x|^{\sigma - \frac{1}{pn} - 1} dx \right) (-y)^{\sigma_1 + \frac{1}{qn} - 1} dy$$
$$+ \int_{0}^{1} \left(\int_{\{x;|x|\geq \frac{1}{y}\}} h(xy) |x|^{\sigma - \frac{1}{pn} - 1} dx \right) y^{\sigma_1 + \frac{1}{qn} - 1} dy$$
$$= \int_{0}^{1} \left[\int_{\{x;|x|\geq \frac{1}{y}\}} (h(-xy) + h(xy)) |x|^{\sigma - \frac{1}{pn} - 1} dx \right] y^{\sigma_1 + \frac{1}{qn} - 1} dy$$
$$= \int_{0}^{1} \left[\int_{\{u;|u|\geq 1\}} (h(-u) + h(u)) |u|^{\sigma - \frac{1}{pn} - 1} du \right] y^{(\sigma_1 - \sigma) + \frac{1}{n} - 1} dy,$$

and then by (2.9), it follows that

$$2K^{(2)} \left(\sigma - \frac{1}{pn} \right) \frac{1}{\sigma_1 - \sigma + \frac{1}{n}} = \widetilde{I}_2 \geq M_2 \widetilde{J}_1 = 2M_2 n. \tag{2.10}$$

Since $\{(h(-u)+h(u))u^{\sigma-\frac{1}{pn}-1}\}_{n=1}^{\infty}$ $(u \in (1,\infty))$ is a non-negative and increasing function sequence, by Levi theorem (cf. Ref. [45]), it follows that

$$K^{(2)}\left(\sigma - \frac{1}{pn}\right) = \int_1^{\infty} (h(-u) + h(u))u^{\sigma-\frac{1}{pn}-1}du$$
$$\to \int_1^{\infty} (h(-u) + h(u))u^{\sigma-1}du = K^{(2)}(\sigma)\ (n \to \infty),$$

and then by (2.10), we have $\infty > \frac{2K^{(2)}(\sigma)}{\sigma_1-\sigma} \geq \infty$, which is a contradiction.

If $\sigma_1 < \sigma$, then for $n \geq \frac{1}{(\sigma-\sigma_1)|q|}$ $(n \in \mathbf{N})$, we set two functions $f_n(x)$ and $g_n(y)$ as in Lemma 1 and find

$$J_1 = \left[\int_{-\infty}^{\infty} |x|^{p(1-\sigma)-1} f_n^p(x)dx\right]^{\frac{1}{p}} \left[\int_{-\infty}^{\infty} |y|^{q(1-\sigma_1)-1} g_n^q(y)dy\right]^{\frac{1}{q}} = 2n.$$

We obtain

$$\begin{aligned}
I_2 &:= \int_{-\infty}^{\infty} f_n(x) \left(\int_{\{y;|y|\geq\frac{1}{|x|}\}} h(xy)g_n(y)dy\right) dx \\
&= \int_{-1}^{0} \left(\int_{\{y;|y|\geq\frac{-1}{x}\}} h(xy)|y|^{\sigma_1-\frac{1}{qn}-1}dy\right) (-x)^{\sigma+\frac{1}{pn}-1}dx \\
&\quad + \int_0^1 \left(\int_{\{y;|y|\geq\frac{1}{x}\}} h(xy)|y|^{\sigma_1-\frac{1}{qn}-1}dy\right) x^{\sigma+\frac{1}{pn}-1}dx \\
&= \int_0^1 \left[\int_{\{y;|y|\geq\frac{1}{x}\}} (h(-xy) + h(xy))|y|^{\sigma_1-\frac{1}{qn}-1}dy\right] x^{\sigma+\frac{1}{pn}-1}dx \\
&= \int_{\{u;|u|\geq1\}} (h(-u) + h(u))|u|^{(\sigma_1-\frac{1}{qn})-1}du \int_0^1 x^{\sigma-\sigma_1+\frac{1}{n}-1}dx,
\end{aligned}$$

and then by Fubini theorem (cf. Ref. [45]) and (2.7), it follows that

$$\begin{aligned}
&2K_2\left(\sigma_1-\frac{1}{qn}\right)\frac{1}{\sigma-\sigma_1+\frac{1}{n}}\\
&=\int_0^1 x^{\sigma-\sigma_1+\frac{1}{n}-1}dx\int_{\{u;|u|\geq 1\}}(h(-u)+h(u))|u|^{\sigma_1-\frac{1}{qn}-1}du\\
&=I_2=\int_0^\infty g_n(y)\left(\int_{\{x;|x|\geq\frac{1}{|y|}\}}h(xy)f_n(x)dx\right)dy\\
&\geq M_2J_1=2M_2n. \qquad (2.11)
\end{aligned}$$

Since $\sigma_1-\frac{1}{qn}\leq\sigma$, we have

$$\begin{aligned}
0&\leq K^{(2)}\left(\sigma_1-\frac{1}{qn}\right)=\int_1^\infty(h(-u)+h(u))u^{\sigma_1-\frac{1}{qn}-1}du\\
&\leq\int_1^\infty(h(-u)+h(u))u^{\sigma-1}du=K^{(2)}(\sigma)<\infty,
\end{aligned}$$

and then by (2.11), for $n\to\infty$, it follows that $\infty>2K_2(\sigma)\frac{1}{\sigma-\sigma_1}\geq\infty$, which is a contradiction.

Hence, we conclude the fact that $\sigma_1=\sigma$.

For $\sigma_1=\sigma$, we reduce (2.10) as follows:

$$K^{(2)}\left(\sigma-\frac{1}{pn}\right)=\int_1^\infty(h(-u)+h(u))u^{\sigma-\frac{1}{pn}-1}du\geq M_2. \qquad (2.12)$$

Since $\{(h(-u)+h(u))u^{\sigma-\frac{1}{pn}-1}\}_{n=1}^\infty$ $(u\in[1,\infty))$ is a non-negative and increasing function sequence, still by Levi theorem (cf. Ref. [45]), we have

$$\begin{aligned}
K^{(2)}(\sigma)&=\int_1^\infty\lim_{n\to\infty}(h(-u)+h(u))u^{\sigma-\frac{1}{pn}-1}du\\
&=\lim_{n\to\infty}\int_1^\infty(h(-u)+h(u))u^{\sigma-\frac{1}{pn}-1}du\geq M_2.
\end{aligned}$$

The lemma is proved. □

3. First Kind of the Reverse Hardy-Type Integral Inequalities

Theorem 1. *If $K^{(1)}(\sigma) < \infty$, then the following conditions are equivalent:*

(i) For any $f(x) \geq 0$, satisfying

$$0 < \int_{-\infty}^{\infty} |x|^{p(1-\sigma)-1} f^p(x)dx < \infty,$$

we have the following first kind of the reverse Hardy-type integral inequality with the non-homogeneous kernel:

$$J := \left[\int_{-\infty}^{\infty} |y|^{p\sigma_1 - 1} \left(\int_{\frac{-1}{|y|}}^{\frac{1}{|y|}} h(xy)f(x)dx\right)^p dy\right]^{\frac{1}{p}}$$
$$> M_1 \left[\int_{-\infty}^{\infty} |x|^{p(1-\sigma)-1} f^p(x)dx\right]^{\frac{1}{p}}. \tag{3.1}$$

(ii) For any $g(y) \geq 0$, satisfying

$$0 < \int_{-\infty}^{\infty} |y|^{q(1-\sigma_1)-1} g^q(y)dy < \infty,$$

we have the following first kind of the reverse Hardy-type integral inequality with the non-homogeneous kernel:

$$J_1 := \left[\int_{-\infty}^{\infty} |x|^{q\sigma - 1} \left(\int_{\frac{-1}{|x|}}^{\frac{1}{|x|}} h(xy)g(y)dy\right)^q dx\right]^{\frac{1}{q}}$$
$$> M_1 \left[\int_{-\infty}^{\infty} |y|^{q(1-\sigma_1)-1} g^q(y)dy\right]^{\frac{1}{q}}. \tag{3.2}$$

(iii) For any $f(x), g(y) \geq 0$, satisfying $0 < \int_{-\infty}^{\infty} |x|^{p(1-\sigma)-1} f^p(x)$ $dx < \infty$, and

$$0 < \int_{-\infty}^{\infty} |y|^{q(1-\sigma_1)-1} g^q(y)dy < \infty,$$

we have the following inequality:

$$I := \int_{-\infty}^{\infty} g(y)\left(\int_{\frac{-1}{|y|}}^{\frac{1}{|y|}} h(xy)f(x)dx\right)dy$$

$$> M_1\left[\int_{-\infty}^{\infty} |x|^{p(1-\sigma)-1}f^p(x)dx\right]^{\frac{1}{p}}\left[\int_{-\infty}^{\infty} |y|^{q(1-\sigma_1)-1}g^q(y)dy\right]^{\frac{1}{q}}. \tag{3.3}$$

(iv) $\sigma_1 = \sigma$, and $K^{(1)}(\sigma) \geq M_1$.

If Condition (iv) holds true, then the constant $M_1 = K^{(1)}(\sigma)$ ($\in \mathbf{R}_+$) in (3.1), (3.2) and (3.3) (for $\sigma_1 = \sigma$) is the best possible.

Proof. *(i)* => *(iii)*. By the reverse Hölder's inequality (cf. Ref. [44]), we have

$$I = \int_{-\infty}^{\infty}\left(|y|^{\sigma_1-\frac{1}{p}}\int_{\frac{-1}{|y|}}^{\frac{1}{|y|}} h(xy)f(x)dx\right)\left(|y|^{\frac{1}{p}-\sigma_1}g(y)\right)dy$$

$$\geq J\left[\int_{-\infty}^{\infty} |y|^{q(1-\sigma_1)-1}g^q(y)dy\right]^{\frac{1}{q}}. \tag{3.4}$$

Then by (3.1), we have (3.2).

(ii) => *(iii)*. By the reverse Hölder's inequality (cf. Ref. [44]), we have

$$I = \int_{-\infty}^{\infty}\left(|x|^{\frac{1}{q}-\sigma}f(x)\right)\left(|x|^{\sigma-\frac{1}{q}}\int_{\frac{-1}{|x|}}^{\frac{1}{|x|}} h(xy)g(y)dy\right)dx$$

$$\geq \left[\int_{-\infty}^{\infty} |x|^{p(1-\sigma)-1}f^p(x)dx\right]^{\frac{1}{p}} J_1. \tag{3.5}$$

Then by (3.1), we have (3.3).

(iii) => *(iv)*. By Lemma 1, we have $\sigma_1 = \sigma$, and $K^{(1)}(\sigma) \geq M_1 > 0$.

(iv) => *(i)*. We obtain the following weight function: For $y \neq 0$,

$$\omega_1(\sigma, y) := |y|^{\sigma}\int_{\frac{-1}{|y|}}^{\frac{1}{|y|}} h(xy)|x|^{\sigma-1}dx$$

$$= |y|^{\sigma}\int_{\frac{-1}{|y|}}^{0} h(xy)(-x)^{\sigma-1}dx + |y|^{\sigma}\int_{0}^{\frac{1}{|y|}} h(xy)x^{\sigma-1}dx$$

$$= |y|^{\sigma} \int_0^{\frac{1}{|y|}} h(-xy)x^{\sigma-1}dx + |y|^{\sigma} \int_0^{\frac{1}{|y|}} h(xy)x^{\sigma-1}dx$$

$$= |y|^{\sigma} \int_0^{\frac{1}{|y|}} (h(-x|y|) + h(x|y|)x^{\sigma-1}dx$$

$$= \int_0^1 (h(-u) + h(u))u^{\sigma-1}du = K^{(1)}(\sigma). \tag{3.6}$$

By the reverse Hölder's inequality with weight and (3.6), we have

$$\left(\int_{\frac{-1}{|y|}}^{\frac{1}{|y|}} h(xy)f(x)dx\right)^p$$

$$= \left\{\int_{\frac{-1}{|y|}}^{\frac{1}{|y|}} h(xy)\left[\frac{|y|^{(\sigma-1)/p}}{|x|^{(\sigma-1)/q}}f(x)\right]\left[\frac{|x|^{(\sigma-1)/q}}{|y|^{(\sigma-1)/p}}\right]dx\right\}^p$$

$$\geq \int_{\frac{-1}{|y|}}^{\frac{1}{|y|}} h(xy)\frac{|y|^{\sigma-1}f^p(x)}{|x|^{(\sigma-1)p/q}}dx\left[\int_{\frac{-1}{|y|}}^{\frac{1}{|y|}} h(xy)\frac{|x|^{\sigma-1}}{|y|^{(\sigma-1)q/p}}dx\right]^{p-1}$$

$$= \int_{\frac{-1}{|y|}}^{\frac{1}{|y|}} h(xy)\frac{|y|^{\sigma-1}}{|x|^{(\sigma-1)p/q}}f^p(x)dx \cdot \left[\omega_1(\sigma,y)|y|^{q(1-\sigma)-1}\right]^{p-1}$$

$$= (K^{(1)}(\sigma))^{p-1}|y|^{-p\sigma+1}\int_{\frac{-1}{|y|}}^{\frac{1}{|y|}} h(xy)\frac{|y|^{\sigma-1}}{|x|^{(\sigma-1)p/q}}f^p(x)dx. \tag{3.7}$$

If (3.7) takes the form of equality for a $y \in \mathbf{R}\backslash\{0\}$, then (cf. Ref. [44]), there exists constants A and B, such that they are not all zero, and

$$A\frac{|y|^{\sigma-1}}{|x|^{(\sigma-1)p/q}}f^p(x) = B\frac{|x|^{\sigma-1}}{|y|^{(\sigma-1)q/p}} \text{ a.e. in } \mathbf{R}.$$

We suppose that $A \neq 0$ (otherwise $B = A = 0$). It follows that

$$|x|^{p(1-\sigma)-1}f^p(x) = |y|^{q(1-\sigma)}\frac{B}{A|x|} \text{ a.e. in } \mathbf{R},$$

which contradicts the fact that

$$0 < \int_{-\infty}^{\infty} |x|^{p(1-\sigma)-1} f^p(x)dx < \infty.$$

Hence, (3.7) takes the form of strict inequality.

For $\sigma_1 = \sigma$, by the above results and Fubini theorem (cf. Ref. [45]), we have

$$\begin{aligned}
J &> (K^{(1)}(\sigma))^{\frac{1}{q}} \left\{ \int_{-\infty}^{\infty} \left[\int_{\frac{-1}{|y|}}^{\frac{1}{|y|}} h(xy) \frac{|y|^{\sigma-1}}{|x|^{(\sigma-1)p/q}} f^p(x)dx \right] dy \right\}^{\frac{1}{p}} \\
&= (K^{(1)}(\sigma))^{\frac{1}{q}} \left\{ \int_{-\infty}^{\infty} \left[\int_{\frac{-1}{|x|}}^{\frac{1}{|x|}} h(xy) \frac{|y|^{\sigma-1}}{|x|^{(\sigma-1)(p-1)}} dy \right] f^p(x)dx \right\}^{\frac{1}{p}} \\
&= (K^{(1)}(\sigma))^{\frac{1}{q}} \left[\int_{-\infty}^{\infty} \omega_1(\sigma, x)|x|^{p(1-\sigma)-1} f^p(x)dx \right]^{\frac{1}{p}} \\
&= K^{(1)}(\sigma) \left[\int_{-\infty}^{\infty} |x|^{p(1-\sigma)-1} f^p(x)dx \right]^{\frac{1}{p}}.
\end{aligned}$$

Since $M_1 \leq K^{(1)}(\sigma)$, we have

$$\begin{aligned}
J &> K^{(1)}(\sigma) \left[\int_{-\infty}^{\infty} |x|^{p(1-\sigma)-1} f^p(x)dx \right]^{\frac{1}{p}} \\
&\geq M_1 \left[\int_{-\infty}^{\infty} |x|^{p(1-\sigma)-1} f^p(x)dx \right]^{\frac{1}{p}},
\end{aligned}$$

namely, (3.1) follows.

(iv) => *(ii)*. In the same way, for $\sigma_1 = \sigma$, and $K^{(1)}(\sigma) \geq M_1$, by the reverse Hölder's inequality, we have

$$\left(\int_{\frac{-1}{|x|}}^{\frac{1}{|x|}} h(xy)g(y)dy \right)^q < \frac{(K^{(1)}(\sigma))^{q-1}}{|x|^{q\sigma-1}} \int_{\frac{-1}{|x|}}^{\frac{1}{|x|}} h(xy) \frac{|x|^{\sigma-1} g^q(y)}{|y|^{(\sigma-1)q/p}} dy,$$

then we can deduce (3.2).

Therefore, Conditions (i), (ii), (iii) and (iv) are equivalent.

When Condition (iv) follows, if there exists a constant $M_1 \geq K^{(1)}(\sigma)$, such that (3.3) is valid, then by Lemma 1, we have $K^{(1)}(\sigma) \geq M_1$. It follows that the constant factor $M_1 = K^{(1)}(\sigma)$ in (3.3) is the best possible.

The constant factor $M_1 = K^{(1)}(\sigma)(\in \mathbf{R}_+)$ in (3.1) is still the best possible. Otherwise, by (3.4) (for $\sigma_1 = \sigma$), we can conclude that the constant factor $M_1 = K^{(1)}(\sigma)$ in (3.3) is not the best possible. In the same way, by (3.5) (for $\sigma_1 = \sigma$), we can prove that the constant factor $M_1 = K^{(1)}(\sigma)$ in (3.2) is the best possible. □

Remark 1. If $K^{(1)}(\sigma) \in \mathbf{R}_+$, then we still have the results of Theorem 1.

In particular, for $\sigma = \sigma_1 = \frac{1}{p}$ in Theorem 1, we have the following:

Corollary 1. *If $K^{(1)}(\frac{1}{p}) < \infty$, then the following conditions are equivalent:*

(i) For any $f(x) \geq 0$, satisfying $0 < \int_{-\infty}^{\infty} |x|^{p-2} f^p(x)dx < \infty$, we have the following inequality:

$$\left[\int_{-\infty}^{\infty} \left(\int_{\frac{-1}{|y|}}^{\frac{1}{|y|}} h(xy)f(x)dx\right)^p dy\right]^{\frac{1}{p}} > M_1 \left(\int_{-\infty}^{\infty} |x|^{p-2} f^p(x)dx\right)^{\frac{1}{p}}. \tag{3.8}$$

(ii) For any $g(y) \geq 0$, satisfying $0 < \int_{-\infty}^{\infty} g^q(y)dy < \infty$, we have the following inequality:

$$\left[\int_{-\infty}^{\infty} |x|^{q-2} \left(\int_{\frac{-1}{|x|}}^{\frac{1}{|x|}} h(xy)g(y)dy\right)^q dx\right]^{\frac{1}{q}} > M_1 \left(\int_{-\infty}^{\infty} g^q(y)dy\right)^{\frac{1}{q}}. \tag{3.9}$$

(iii) For any $f(x), g(y) \geq 0$, satisfying $0 < \int_{-\infty}^{\infty} |x|^{p-2} f^p(x)dx < \infty$, and $0 < \int_{-\infty}^{\infty} g^q(y)dy < \infty$, we have the following inequality:

$$\int_{-\infty}^{\infty} g(y) \left(\int_{\frac{-1}{|y|}}^{\frac{1}{|y|}} h(xy)f(x)dx\right) dy > M_1 \left(\int_{-\infty}^{\infty} |x|^{p-2} f^p(x)dx\right)^{\frac{1}{p}} \left(\int_{-\infty}^{\infty} g^q(y)dy\right)^{\frac{1}{q}}. \tag{3.10}$$

(iv) $K^{(1)}(\frac{1}{p}) \geq M_1$.

If Condition (iv) holds true, then the constant $M_1 = K^{(1)}(\frac{1}{p})\,(\in \mathbf{R}_+)$ in (3.8), (3.9) and (3.10) is the best possible.

Setting $y = \frac{1}{Y}$, $G(Y) = g(\frac{1}{Y})\frac{1}{Y^2}$ in Theorem 1, and then replacing Y by y, we have the following:

Corollary 2. *If $K^{(1)}(\sigma) < \infty$, then the following conditions are equivalent:*

(i) For any $f(x) \geq 0$, satisfying

$$0 < \int_{-\infty}^{\infty} |x|^{p(1-\sigma)-1} f^p(x)dx < \infty,$$

we have the following inequality follows:

$$\left[\int_{-\infty}^{\infty} |y|^{-p\sigma_1-1} \left(\int_{-|y|}^{|y|} h\left(\frac{x}{y}\right) f(x)dx\right)^p dy\right]^{\frac{1}{p}}$$

$$> M_1 \left[\int_{-\infty}^{\infty} |x|^{p(1-\sigma)-1} f^p(x)dx\right]^{\frac{1}{p}}. \tag{3.11}$$

(ii) For any $g(y) \geq 0$, satisfying

$$0 < \int_{-\infty}^{\infty} |y|^{q(1+\sigma_1)-1} g^q(y)dy < \infty,$$

we have the following integral inequality:

$$\left[\int_{-\infty}^{\infty} |x|^{q\sigma-1} \left(\int_{\frac{-1}{|x|}}^{\frac{1}{|x|}} h\left(\frac{x}{y}\right) G(y)dy\right)^q dx\right]^{\frac{1}{q}}$$

$$> M_1 \left[\int_{-\infty}^{\infty} |y|^{q(1+\sigma_1)-1} G^q(y)dy\right]^{\frac{1}{q}}. \tag{3.12}$$

(iii) For any $f(x), G(y) \geq 0$, satisfying $0 < \int_{-\infty}^{\infty} |x|^{p(1-\sigma)-1} f^p(x)\, dx < \infty$, and

$$0 < \int_{-\infty}^{\infty} y^{q(1+\sigma_1)-1} G^q(y)dy < \infty,$$

we have the following inequality:

$$\int_{-\infty}^{\infty} G(y)\left(\int_{-|y|}^{|y|} h\left(\frac{x}{y}\right) f(x)dx\right) dy$$
$$< M_1\left[\int_{-\infty}^{\infty} |x|^{p(1-\sigma)-1} f^p(x)dx\right]^{\frac{1}{p}}\left[\int_{-\infty}^{\infty} |y|^{q(1+\sigma_1)-1} G^q(y)dy\right]^{\frac{1}{q}}. \tag{3.13}$$

(iv) $\sigma_1 = \sigma$, and $K^{(1)}(\sigma) \geq M_1$.

If Condition (iv) holds true, then the constant $M_1 = K^{(1)}(\sigma)\,(\in \mathbf{R}_+)$ in (3.11), (3.12) and (3.13) (for $\sigma_1 = \sigma$) is the best possible.

Note. $h(\frac{x}{y})$ is a homogeneous function of degree 0, namely, $h(\frac{x}{y}) = k_0(x, y)$.

Setting $h(u) = k_\lambda(u, 1)$, where $k_\lambda(x, y)$ is the homogeneous function of degree $-\lambda(\in \mathbf{R})$, for $g(y) = y^\lambda G(y)$ and $\mu = \lambda - \sigma_1$ in Corollary 2, we have the following:

Corollary 3. *If* $k_\lambda^{(1)}(\sigma) := \int_0^1 (k_\lambda(-u, 1) + k_\lambda(u, 1))u^{\sigma-1}du < \infty$, *then the following conditions are equivalent:*

(i) For any $f(x) \geq 0$, satisfying

$$0 < \int_{-\infty}^{\infty} |x|^{p(1-\sigma)-1} f^p(x)dx < \infty,$$

we have the following first kind of the reverse Hardy-type integral inequality with the homogeneous kernel:

$$\left[\int_{-\infty}^{\infty} y^{p\mu-1}\left(\int_{-|y|}^{|y|} k_\lambda(x, y)f(x)dx\right)^p dy\right]^{\frac{1}{p}}$$
$$> M_1\left[\int_{-\infty}^{\infty} |x|^{p(1-\sigma)-1} f^p(x)dx\right]^{\frac{1}{p}}. \tag{3.14}$$

(ii) For any $g(y) \geq 0$, satisfying

$$0 < \int_{-\infty}^{\infty} |y|^{q(1-\mu)-1} g^q(y)dy < \infty,$$

we have the following first kind of the reverse Hardy-type integral inequality with the homogeneous kernel:

$$\left[\int_{-\infty}^{\infty}|x|^{q\sigma-1}\left(\int_{-|x|}^{|x|}k_{\lambda}(x,y)g(y)dy\right)^{q}dx\right]^{\frac{1}{q}}$$

$$> M_1\left[\int_{-\infty}^{\infty}|y|^{q(1-\mu)-1}g^{q}(y)dy\right]^{\frac{1}{q}}. \tag{3.15}$$

(iii) For any $f(x), g(y) \geq 0$, satisfying $0 < \int_{-\infty}^{\infty}|x|^{p(1-\sigma)-1}f^p(x)dx < \infty$, and

$$0 < \int_{-\infty}^{\infty}|y|^{q(1-\mu)-1}g^{q}(y)dy < \infty,$$

we have the following inequality:

$$\int_{-\infty}^{\infty}g(y)\left(\int_{-|y|}^{|y|}k_{\lambda}(x,y)f(x)dx\right)dy$$

$$> M_1\left[\int_{-\infty}^{\infty}|x|^{p(1-\sigma)-1}f^{p}(x)dx\right]^{\frac{1}{p}}\left[\int_{-\infty}^{\infty}|y|^{q(1-\mu)-1}g^{q}(y)dy\right]^{\frac{1}{q}}. \tag{3.16}$$

(iv) $\mu + \sigma = \lambda$, and $k_{\lambda}^{(1)}(\sigma) \geq M_1$.

If Condition (iv) holds true, then the constant $M_1 = k_{\lambda}^{(1)}(\sigma)\,(\in \mathbf{R}_+)$ in (3.14), (3.15) and (3.16) is the best possible.

Remark 2. If $k_{\lambda}^{(1)}(\sigma) \in \mathbf{R}_+$, then we still have the results of Corollary 3.

In particular, for $\lambda = 1$, $\sigma = \frac{1}{q}$, $\mu = \frac{1}{p}$ in Corollary 3, we have the following:

Corollary 4. *If* $k_1^{(1)}(\frac{1}{q}) < \infty$, *then the following conditions are equivalent:*

(i) For any $f(x) \geq 0$, satisfying $0 < \int_{-\infty}^{\infty}f^p(x)dx < \infty$, we have the following inequality:

$$\left[\int_{-\infty}^{\infty}\left(\int_{-|y|}^{|y|}k_1(x,y)f(x)dx\right)^{p}dy\right]^{\frac{1}{p}} > M_1\left(\int_{-\infty}^{\infty}f^p(x)dx\right)^{\frac{1}{p}}. \tag{3.17}$$

(ii) For any $g(x) \geq 0$, satisfying $0 < \int_{-\infty}^{\infty} g^q(y)dy < \infty$, we have the following inequality:

$$\left[\int_{-\infty}^{\infty}\left(\int_{-|x|}^{|x|} k_1(x,y)g(y)dy\right)^q dy\right]^{\frac{1}{q}} > M_1\left(\int_{-\infty}^{\infty} g^q(y)dy\right)^{\frac{1}{q}}. \tag{3.18}$$

(iii) For any $f(x), g(y) \geq 0$, satisfying $0 < \int_{-\infty}^{\infty} f^p(x)dx < \infty$, and $0 < \int_{-\infty}^{\infty} g^q(y)dy < \infty$, we have the following inequality:

$$\int_{-\infty}^{\infty} g(y)\left(\int_{-|y|}^{|y|} k_1(x,y)f(x)dx\right)dy$$
$$> M_1\left(\int_{-\infty}^{\infty} f^p(x)dx\right)^{\frac{1}{p}}\left(\int_{-\infty}^{\infty} g^q(y)dy\right)^{\frac{1}{q}}. \tag{3.19}$$

(iv) $k_1^{(1)}(\frac{1}{q}) \geq M_1$.

If Condition (iv) holds true, then the constant $M_1 = k_1^{(1)}(\frac{1}{q})$ $(\in \mathbf{R}_+)$ in (3.17), (3.18) and (3.19) is the best possible.

4. Second Kind of the Reverse Hardy-Type Integral Inequalities and Examples

Similarly, in view of Lemma 2, we obtain the following weight function: For $y \neq 0$,

$$\omega_2(\sigma, y) := |y|^\sigma \int_{\{x;|x|\geq\frac{1}{|y|}\}} h(xy)|x|^{\sigma-1}dx$$
$$= \int_1^{\infty}(h(-u)+h(u))u^{\sigma-1}du = K^{(2)}(\sigma),$$

and in the same way, we have the following:

Theorem 2. *If $K^{(2)}(\sigma) < \infty$, then the following conditions are equivalent:*

(i) For any $f(x) \geq 0$, satisfying

$$0 < \int_{-\infty}^{\infty} |x|^{p(1-\sigma)-1} f^p(x)dx < \infty,$$

we have the following second kind of the reverse Hardy-type integral inequality with the non-homogeneous kernel:

$$\left[\int_{-\infty}^{\infty} |y|^{p\sigma_1 - 1} \left(\int_{\{x;|x|\geq \frac{1}{|y|}\}} h(xy)f(x)dx\right)^p dy\right]^{\frac{1}{p}}$$

$$> M_2 \left[\int_{-\infty}^{\infty} |x|^{p(1-\sigma)-1} f^p(x)dx\right]^{\frac{1}{p}}. \tag{4.1}$$

(ii) For any $g(y) \geq 0$, satisfying

$$0 < \int_{-\infty}^{\infty} |y|^{q(1-\sigma_1)-1} g^q(y)dy < \infty,$$

we have the following second kind of the reverse Hardy-type integral inequality with the non-homogeneous kernel:

$$\left[\int_{-\infty}^{\infty} |x|^{q\sigma - 1} \left(\int_{\{y;|y|\geq \frac{1}{|x|}\}} h(xy)g(y)dy\right)^q dx\right]^{\frac{1}{q}}$$

$$> M_2 \left[\int_{-\infty}^{\infty} |y|^{q(1-\sigma_1)-1} g^q(y)dy\right]^{\frac{1}{q}}. \tag{4.2}$$

(iii) For any $f(x), g(y) \geq 0$, satisfying $0 < \int_{-\infty}^{\infty} |x|^{p(1-\sigma)-1} f^p(x)dx < \infty$, and

$$0 < \int_{-\infty}^{\infty} |y|^{q(1-\sigma_1)-1} g^q(y)dy < \infty,$$

we have the following inequality:

$$\int_{-\infty}^{\infty} g(y) \left(\int_{\{x;|x|\geq \frac{1}{|y|}\}} h(xy)f(x)dx\right) dy$$

$$> M_2 \left[\int_{-\infty}^{\infty} |x|^{p(1-\sigma)-1} f^p(x)dx\right]^{\frac{1}{p}} \left[\int_{-\infty}^{\infty} |y|^{q(1-\sigma_1)-1} g^q(y)dy\right]^{\frac{1}{q}}. \tag{4.3}$$

(iv) $\sigma_1 = \sigma$, and $K^{(2)}(\sigma) \geq M_2$.

If Condition (iv) holds true, then the constant $M_2 = K^{(2)}(\sigma)$ ($\in \mathbf{R}_+$) in (4.1), (4.2) and (4.3) (for $\sigma_1 = \sigma$) is the best possible.

Remark 3. If $K^{(1)}(\sigma) \in \mathbf{R}_+$, then we still have the results of Theorem 2.

In particular, for $\sigma = \sigma_1 = \frac{1}{p}$ in Theorem 2, we have the following:

Corollary 5. *If $K^{(2)}(\frac{1}{p}) < \infty$, then the following conditions are equivalent:*

(i) For any $f(x) \geq 0$, satisfying $0 < \int_{-\infty}^{\infty} |x|^{p-2} f^p(x)dx < \infty$, we have the following inequality:

$$\left[\int_{-\infty}^{\infty}\left(\int_{\{x;|x|\geq\frac{1}{|y|}\}} h(xy)f(x)dx\right)^p dy\right]^{\frac{1}{p}} > M_2\left(\int_{-\infty}^{\infty}|x|^{p-2}f^p(x)dx\right)^{\frac{1}{p}}. \quad (4.4)$$

(ii) For any $g(y) \geq 0$, satisfying $0 < \int_{-\infty}^{\infty} g^q(y)dy < \infty$, we have the following inequality:

$$\left[\int_{-\infty}^{\infty}|x|^{q-2}\left(\int_{\{y;|y|\geq\frac{1}{|x|}\}} h(xy)g(y)dy\right)^q dx\right]^{\frac{1}{q}} > M_2\left(\int_{-\infty}^{\infty}g^q(y)dy\right)^{\frac{1}{q}}. \quad (4.5)$$

(iii) For any $f(x), g(y) \geq 0$, satisfying $0 < \int_{-\infty}^{\infty} |x|^{p-2} f^p(x)dx < \infty$, and $0 < \int_{-\infty}^{\infty} g^q(y)dy < \infty$, we have the following inequality:

$$\int_{-\infty}^{\infty} g(y)\left(\int_{\{x;|x|\geq\frac{1}{|y|}\}} h(xy)f(x)dx\right)dy$$
$$> M_2\left(\int_{-\infty}^{\infty}|x|^{p-2}f^p(x)dx\right)^{\frac{1}{p}}\left(\int_{-\infty}^{\infty}g^q(y)dy\right)^{\frac{1}{q}}. \quad (4.6)$$

(iv) $K^{(2)}(\frac{1}{p}) \geq M_2$.

If Condition (iv) holds true, then the constant $M_2 = K^{(2)}(\frac{1}{p})$ ($\in \mathbf{R}_+$) in (4.4), (4.5) and (4.6) is the best possible.

Setting $y = \frac{1}{Y}$, $G(Y) = g(\frac{1}{Y})\frac{1}{Y^2}$ in Theorem 2, and then replacing Y by y, we have the following:

Corollary 6. *If $K^{(2)}(\sigma) < \infty$, then the following conditions are equivalent:*

(i) For any $f(x) \geq 0$, satisfying

$$0 < \int_{-\infty}^{\infty} |x|^{p(1-\sigma)-1} f^p(x)dx < \infty,$$

we have the following inequality:

$$\left[\int_{-\infty}^{\infty} |y|^{-p\sigma_1-1} \left(\int_{\{x;|x|\geq|y|\}} h\left(\frac{x}{y}\right) f(x)dx\right)^p dy\right]^{\frac{1}{p}}$$
$$> M_2 \left[\int_{-\infty}^{\infty} |x|^{p(1-\sigma)-1} f^p(x)dx\right]^{\frac{1}{p}}. \tag{4.7}$$

(ii) For any $g(y) \geq 0$, satisfying

$$0 < \int_{-\infty}^{\infty} |y|^{q(1+\sigma_1)-1} g^q(y)dy < \infty,$$

we have the following integral inequality:

$$\left[\int_{-\infty}^{\infty} |x|^{q\sigma-1} \left(\int_{\{y;|y|\geq|x|\}} h\left(\frac{x}{y}\right) G(y)dy\right)^q dx\right]^{\frac{1}{q}}$$
$$> M_2 \left[\int_{-\infty}^{\infty} |y|^{q(1+\sigma_1)-1} G^q(y)dy\right]^{\frac{1}{q}}. \tag{4.8}$$

(iii) For any $f(x), G(y) \geq 0$, satisfying $0 < \int_{-\infty}^{\infty} |x|^{p(1-\sigma)-1} f^p$ $(x)dx < \infty$, and

$$0 < \int_{-\infty}^{\infty} |y|^{q(1+\sigma_1)-1} G^q(y)dy < \infty,$$

we have the following inequality:

$$\int_{-\infty}^{\infty} G(y) \left(\int_{\{x;|x|\geq|y|\}} h\left(\frac{x}{y}\right) f(x)dx\right) dy$$
$$> M_2 \left[\int_{-\infty}^{\infty} |x|^{p(1-\sigma)-1} f^p(x)dx\right]^{\frac{1}{p}} \left[\int_{-\infty}^{\infty} |y|^{q(1+\sigma_1)-1} G^q(y)dy\right]^{\frac{1}{q}}. \tag{4.9}$$

(iv) $\sigma_1 = \sigma$, and $K^{(2)}(\sigma) \geq M_2$.

If Condition (iv) holds true, then the constant $M_2 = K^{(2)}(\sigma)\,(\in \mathbf{R}_+)$ in (4.7), (4.8) and (4.9) (for $\sigma_1 = \sigma$) is the best possible.

Setting $h(u) = k_\lambda(u,1)$, where $k_\lambda(x,y)$ is the homogeneous function of degree $-\lambda(\in \mathbf{R})$, for $g(y) = y^\lambda G(y)$ and $\mu = \lambda - \sigma_1$ in Corollary 6, we have the following:

Corollary 7. *If* $k_\lambda^{(2)}(\sigma) := \int_1^\infty (k_\lambda(-u,1) + k_\lambda(u,1))u^{\sigma-1}du < \infty$, *then the following conditions are equivalent:*

(i) For any $f(x) \geq 0$, satisfying

$$0 < \int_{-\infty}^{\infty} |x|^{p(1-\sigma)-1} f^p(x)dx < \infty,$$

we have the following second kind of the reverse Hardy-type integral inequality with the homogeneous kernel:

$$\left[\int_{-\infty}^{\infty} |y|^{p\mu-1}\left(\int_{\{x;|x|\geq|y|\}} k_\lambda(x,y)f(x)dx\right)^p dy\right]^{\frac{1}{p}} > M_2\left[\int_{-\infty}^{\infty} |x|^{p(1-\sigma)-1} f^p(x)dx\right]^{\frac{1}{p}}. \tag{4.10}$$

(ii) For any $g(y) \geq 0$, satisfying

$$0 < \int_{-\infty}^{\infty} |y|^{q(1-\mu)-1} g^q(y)dy < \infty,$$

we have the following second kind of the reverse Hardy-type integral inequality with the homogeneous kernel:

$$\left[\int_{-\infty}^{\infty} |x|^{q\sigma-1}\left(\int_{\{y;|y|\geq|x|\}} k_\lambda(x,y)g(y)dy\right)^q dx\right]^{\frac{1}{q}} > M_2\left[\int_{-\infty}^{\infty} |y|^{q(1-\mu)-1} g^q(y)dy\right]^{\frac{1}{q}}. \tag{4.11}$$

(iii) For any $f(x), g(y) \geq 0$, satisfying $0 < \int_{-\infty}^{\infty} |x|^{p(1-\sigma)-1} f^p(x)dx < \infty$, and

$$0 < \int_{-\infty}^{\infty} |y|^{q(1-\mu)-1} g^q(y)dy < \infty,$$

we have the following inequality:

$$\int_{-\infty}^{\infty} g(y)\left(\int_{\{x;|x|\geq|y|\}} k_\lambda(x,y)f(x)dx\right)dy$$
$$> M_2\left[\int_{-\infty}^{\infty}|x|^{p(1-\sigma)-1}f^p(x)dx\right]^{\frac{1}{p}}\left[\int_{-\infty}^{\infty}|y|^{q(1-\mu)-1}g^q(y)dy\right]^{\frac{1}{q}}. \tag{4.12}$$

(iv) $\mu+\sigma=\lambda$, and $k_\lambda^{(2)}(\sigma)\geq M_2$.

If Condition (iv) holds true, then the constant $M_2=k_\lambda^{(2)}(\sigma)\,(\in \mathbf{R}_+)$ in (4.10), (4.11) and (4.12) is the best possible.

Remark 4. If $k_\lambda^{(1)}(\sigma)\in\mathbf{R}_+$, then we still have the results of Corollary 7.

In particular, for $\lambda=1,\sigma=\frac{1}{q},\mu=\frac{1}{p}$ in Corollary 7, we have the following:

Corollary 8. *If $k_1^{(2)}(\frac{1}{q})<\infty$, then the following conditions are equivalent:*

(i) For any $f(x)\geq 0$, satisfying $0<\int_{-\infty}^{\infty}f^p(x)dx<\infty$, we have the following inequality:

$$\left[\int_{-\infty}^{\infty}\left(\int_{\{x;|x|\geq|y|\}}k_1(x,y)f(x)dx\right)^p dy\right]^{\frac{1}{p}} > M_2\left(\int_{-\infty}^{\infty}f^p(x)dx\right)^{\frac{1}{p}}. \tag{4.13}$$

(ii) For any $g(y)\geq 0$, satisfying $0<\int_{-\infty}^{\infty}g^q(y)dy<\infty$, we have the following inequality:

$$\left[\int_{-\infty}^{\infty}\left(\int_{\{y;|y|\geq|x|\}}k_1(x,y)g(y)dy\right)^q dx\right]^{\frac{1}{q}} > M_2\left(\int_{-\infty}^{\infty}g^q(y)dy\right)^{\frac{1}{q}}. \tag{4.14}$$

(iii) For any $f(x), g(y) \geq 0$, satisfying $0 < \int_{-\infty}^{\infty} f^p(x)dx < \infty$, and $0 < \int_{-\infty}^{\infty} g^q(y)dy < \infty$, we have the following inequality:

$$\int_{-\infty}^{\infty} g(y) \left(\int_{\{x;|x|\geq|y|\}} k_1(x,y)f(x)dx \right) dy$$

$$> M_2 \left(\int_{-\infty}^{\infty} f^p(x)dx \right)^{\frac{1}{p}} \left(\int_{-\infty}^{\infty} g^q(y)dy \right)^{\frac{1}{q}}. \qquad (4.15)$$

(iv) $k_1^{(2)}(\frac{1}{q}) \geq M_2$.

If Condition (iv) holds true, then the constant $M_2 = k_1^{(2)}(\frac{1}{q}) (\in \mathbf{R}_+)$ in (4.12), (4.13) and (4.14) is the best possible.

Example 1. Setting $h(xy) = \frac{|\ln|xy||^\beta}{(\max\{|xy|,1\})^{\lambda-1}|xy-1|}$ and

$$k_\lambda(x,y) = \frac{|\ln|x/y||^\beta}{(\max\{|x|,|y|\})^{\lambda-1}|x-y|} \; (x,y \in \mathbf{R}),$$

for $\beta > 0, \sigma > 0, \sigma + \mu = \lambda$, it follows that $h(u) = \frac{|\ln|u||^\beta}{(\max\{|u|,1\})^{\lambda-1}|u-1|}$,

$$K^{(1)}(\sigma) = k_\lambda^{(1)}(\sigma) = \int_0^1 \frac{|\ln u|^\beta u^{\sigma-1}}{(\max\{u,1\})^{\lambda-1}} \left(\frac{1}{u+1} + \frac{1}{|u-1|} \right) du$$

$$= \int_0^1 (-\ln u)^\beta \left(\frac{1}{u+1} + \frac{1}{1-u} \right) u^{\sigma-1} du$$

$$= 2\int_0^1 (-\ln u)^\beta \frac{u^{\sigma-1}}{1-u^2} du = 2\int_0^1 (-\ln u)^\beta \sum_{k=0}^{\infty} u^{2k+\sigma-1} du.$$

By Lebesgue term by term integration theorem (cf. Ref. [45]), we have

$$K^{(1)}(\sigma) = k_\lambda^{(1)}(\sigma) = 2\sum_{k=0}^{\infty} \int_0^1 (-\ln u)^\beta u^{2k+\sigma-1} du$$

$$= 2\sum_{k=0}^{\infty} \frac{1}{(2k+\sigma)^{\beta+1}} \int_0^\infty v^\beta e^{-v} dv$$

$$= \frac{\Gamma(\beta+1)}{2^\beta} \zeta\left(\beta+1, \frac{\sigma}{2}\right) \in \mathbf{R}_+,$$

where $\zeta(s,a) = \sum_{k=0}^{\infty} \frac{1}{(k+a)^s}(\mathrm{Re}\, s > 1; a > 0)$ is the extended Riemann zeta function ($\zeta(s,1) = \sum_{k=1}^{\infty} \frac{1}{k^s}(\mathrm{Re}\, s > 1)$ is the Riemann zeta function) (cf. Ref. [21]).

In the same way, for $\beta > 0, \mu > 0, \sigma + \mu = \lambda$, we find

$$K^{(2)}(\sigma) = k_\lambda^{(2)}(\sigma) = \int_1^\infty \frac{|\ln u|^\beta u^{\sigma-1}}{(\max\{u,1\})^{\lambda-1}} \left(\frac{1}{u+1} + \frac{1}{|u-1|}\right) du$$
$$= \int_0^1 \frac{|\ln u|^\beta u^{\mu-1}}{(\max\{u,1\})^{\lambda-1}} \left(\frac{1}{u+1} + \frac{1}{|u-1|}\right) du$$
$$= \frac{\Gamma(\beta+1)}{2^\beta} \zeta\left(\beta+1, \frac{\mu}{2}\right) \in \mathbf{R}_+.$$

Example 2. Setting $h(xy) = \frac{1}{|xy-1|^\lambda}, k_\lambda(x,y) = \frac{1}{|x-y|^\lambda}$ $(x, y \in \mathbf{R})$, for $\sigma > 0, \sigma + \mu = \lambda < 1$, it follows that

$$K^{(1)}(\sigma) = k_\lambda^{(1)}(\sigma) = \int_0^1 \left(\frac{1}{(u+1)^\lambda} + \frac{1}{|u-1|^\lambda}\right) u^{\sigma-1} du$$
$$= \int_0^1 \frac{u^{\sigma-1}}{(u+1)^\lambda} du + \int_0^1 \frac{1}{(1-u)^\lambda} u^{\sigma-1} du$$
$$= \sum_{k=0}^\infty \binom{-\lambda}{k} \frac{1}{k+\sigma} + B(1-\lambda, \sigma) \in \mathbf{R}_+.$$

In the same way, for $\mu > 0, \sigma + \mu = \lambda < 1$, we find

$$K^{(2)}(\sigma) = k_\lambda^{(2)}(\sigma) = \int_1^\infty \left(\frac{1}{(u+1)^\lambda} + \frac{1}{|u-1|^\lambda}\right) u^{\sigma-1} du$$
$$= \int_0^1 \left(\frac{1}{(u+1)^\lambda} + \frac{1}{|u-1|^\lambda}\right) u^{\mu-1} du$$
$$= \sum_{k=0}^\infty \binom{-\lambda}{k} \frac{1}{k+\mu} + B(1-\lambda, \mu) \in \mathbf{R}_+.$$

Example 3. Setting $h(xy) = \frac{|\ln|xy||^{\beta}}{(\min\{|xy|,1\})^{\lambda-1}|xy-1|}$ and

$$k_{\lambda}(x,y) = \frac{|\ln|x/y||^{\beta}}{(\min\{|x|,|y|\})^{\lambda-1}|x-y|} \; (x, y \in \mathbf{R}),$$

for $\beta > 0, \lambda = \sigma + \mu, \mu < 1$, it follows that $h(u) = \frac{|\ln|u||^{\beta}}{(\min\{|u|,1\})^{\lambda-1}|u-1|}$, and

$$K^{(1)}(\sigma) = k_{\lambda}^{(1)}(\sigma) = \int_0^1 \frac{|\ln u|^{\beta} u^{\sigma-1}}{(\min\{u,1\})^{\lambda-1}} \left(\frac{1}{u+1} + \frac{1}{|u-1|} \right) du$$

$$= 2\int_0^1 (-\ln u)^{\beta} \frac{u^{\sigma-\lambda}}{1-u^2} du = 2\int_0^1 (-\ln u)^{\beta} \sum_{k=0}^{\infty} u^{2k+\sigma-\lambda} du.$$

By Lebesgue term by term integration theorem (cf. Ref. [45]), we have

$$K^{(1)}(\sigma) = k_{\lambda}^{(1)}(\sigma) = 2\sum_{k=0}^{\infty} \int_0^1 (-\ln u)^{\beta} u^{2k+\sigma-\lambda} du$$

$$= 2\sum_{k=0}^{\infty} \frac{1}{(2k+\sigma-\lambda+1)^{\beta+1}} \int_0^{\infty} v^{\beta} e^{-v} dv$$

$$= \frac{\Gamma(\beta+1)}{2^{\beta}} \zeta(\beta+1, \frac{\sigma-\lambda+1}{2}) \in \mathbf{R}_+.$$

In the same way, for $\beta > 0, \lambda = \sigma + \mu, \sigma < 1$, we find

$$K^{(2)}(\sigma) = k_{\lambda}^{(2)}(\sigma) = \int_1^{\infty} \frac{|\ln u|^{\beta} u^{\sigma-1}}{(\min\{u,1\})^{\lambda-1}} \left(\frac{1}{u+1} + \frac{1}{|u-1|} \right) du$$

$$= \int_0^1 \frac{|\ln u|^{\beta} u^{\mu-1}}{(\min\{u,1\})^{\lambda-1}} \left(\frac{1}{u+1} + \frac{1}{|u-1|} \right) du$$

$$= \frac{\Gamma(\beta+1)}{2^{\beta}} \zeta \left(\beta+1, \frac{\mu-\lambda+1}{2} \right) \in \mathbf{R}_+.$$

Remark 5. In view of Remarks 1–4, we can use Examples 1–3 to Theorems 1 and 2, Corollaries 3 and 7 in the particular kernels.

Acknowledgments

This work is supported by the National Natural Science Foundation (No. 61772140) and Science and Technology Planning Project Item of Guangzhou City (No. 201707010229). We are grateful for their help.

References

[1] G. H. Hardy, J. E. Littlewood and G. Pólya, *Inequalities*, (Cambridge University Press, Cambridge, USA, 1934).

[2] B. C. Yang, *The Norm of Operator and Hilbert-Type Inequalities*, (Science Press, Beijing, China, 2009).

[3] B. C. Yang, *Hilbert-Type Integral Inequalities*, (Bentham Science Publishers Ltd., The United Arab Emirates, 2009).

[4] B. C. Yang, On the norm of an integral operator and applications, *J. Math. Anal. Appl.*, 321, 2006, 182–192.

[5] J. S. Xu, Hardy-Hilbert's inequalities with two parameters, *Adv. Math.*, 36(2), 63–76.

[6] B. C. Yang, On the norm of a Hilbert's type linear operator and applications, *J. Math. Anal. Appl.*, 325(2007), 529–541.

[7] D. M. Xin, A Hilbert-type integral inequality with the homogeneous kernel of zero degree, *Math. Theory Appl.*, 30(2), 2010, 70–74.

[8] B. C. Yang, A Hilbert-type integral inequality with the homogenous kernel of degree 0, *J. Shandong Univ. (Nat. Sci.)*, 45(2), 2010, 103–106.

[9] L. Debnath and B. C. Yang, Recent developments of Hilbert-type discrete and integral inequalities with applications, *Int. J. Mathe. Math. Sci.*, 2012, 2012, Article ID 871845, p. 29.

[10] Y. Hong, On the structure character of Hilbert's type integral inequality with homogeneous kernal and applications, *J. Jilin Univ. (Sci. Edn.)*, 55(2), 2017, 189–194.

[11] M. Th. Rassias and B. C. Yang, On half-discrete Hilbert's inequality. *Appl. Math. Computat.*, 220, 2013, 75–93.

[12] B. C. Yang and M. Krnic, A half-discrete Hilbert-type inequality with a general homogeneous kernel of degree 0. *J. Meth. Inequal.*, 6(3), 2012, 401–417.

[13] Th. M. Rassias and B. C. Yang, A multidimensional half - discrete Hilbert - type inequality and the Riemann zeta function, *Appl. Math. and Computat.*, 225, 2013, 263–277.

[14] M. Th. Rassias and B. C. Yang, On a multidimensional half - discrete Hilbert - type inequality related to the hyperbolic cotangent function, *Appl. Math. Computat.*, 242, 2013, 800–813.

[15] M. Th. Rassias and B. C. Yang, A multidimensional Hilbert - type integral inequality related to the Riemann zeta function, *Applications of Mathematics and Informatics in Science and Engineering* (N. J. Daras, ed.), (Springer, New York, 2014), pp. 417–433.

[16] Q. Chen and B. C. Yang, A survey on the study of Hilbert-type inequalities. *J. Inequal. Appl.*, 2015, 2015, 302.

[17] B. C. Yang, M. Th. Rassias and A. Raigorodskii, On an extension of a Hardy-Hilbert-type inequality with multi-parameters. *Mathematics*, 9, 2021, 2432. https://doi.org/10.3390/math9192432.

[18] M. Th. Rassias, B. C. Yang and A. Raigorodskii, A new Hardy-CMulholland-type inequality with a mixed kernel. *Adv. Oper. Theory*, 6, 2021, 27. https://doi.org/10.1007/s43036-020-00123-0.

[19] M. Th. Rassias, B. C. Yang and A. Raigorodskii. Equivalent conditions of a multiple Hilbert -Type integral inequality with the non homogeneous kernel. *Rev. Real Acad. Cienc. Exact as Fis. Nat.Ser.A-Mat.*, 116, 2022, 107.

[20] B. C. Yang, A new Hilbert-type integral inequality, *Soochow J. Math.*, 33(4), 2007, 849–859.

[21] Z. Q. Wang and D. R. Guo, *Introduction to Special Functions*, (Science Press, Beijing, China, 1979).

[22] B. He and B. C. Yang, On a Hilbert-type integral inequality with the homogeneous kernel of 0-degree and the hypergeometrc function, *Math. Pract. Theory*, 40(18), 2010, 105–211.

[23] B. C. Yang, A new Hilbert-type integral inequality with some parameters, *J. Jilin Univ. (Sci. Edn.)*, 46(6), 2008, 1085–1090.

[24] B. C. Yang, A Hilbert-type integral inequality with a non-homogeneous kernel, *J. Xiamen Univ. (Nat. Sci.)*, 48(2), 2008, 165–169.

[25] Z. Zeng and Z. T. Xie, On a new Hilbert-type integral inequality with the homogeneous kernel of degree 0 and the integral in whole plane, *J. Inequal. Appl.*, 2010, 2010, Article ID 256796, p. 9.

[26] A. Z. Wang and B. C. Yang, A new Hilbert-type integral inequality in whole plane with the non-homogeneous kernel, *J. Inequal. Appl.*, 2011, 2011, 123.

[27] D. M. Xin and B. C. Yang, A Hilbert-type integral inequality in whole plane with the homogeneous kernel of degree −2, *J. Inequal. Appl.*, 2011, 2011, Article ID 401428, p. 11.

[28] B. He and B. C. Yang, On an inequality concerning a non-homogeneous kernel and the hypergeometric function, *Tamsul Oxford J. Inf. Math. Sci.*, 27(1), 2011, 75–88.

[29] B. Yang, A reverse Hilbert-type integral inequality with a non-homogeneous kernel, *J. Jilin Univ. (Sci. Edn.)*, 49(3), 2011, 437–441.

[30] Z. T. Xie, Z. Zeng and Y. F. Sun, A new Hilbert-type inequality with the homogeneous kernel of degree -2, *Adv. Appl. Math. Sci.*, 12(7), 2013, 391–401.
[31] Q. L. Huang, S. H. Wu and B. C. Yang, Parameterized Hilbert-type integral inequalities in the whole plane, *Sci. World J.*, 2014, 2014, Article ID 169061, p. 8.
[32] Z. Zhen, K. Raja Rama Gandhi and Z. T. Xie, A new Hilbert-type inequality with the homogeneous kernel of degree -2 and with the integral, *Bull. Math. Sci. Appl.*, 3(1), 2014, 11–20.
[33] M. Th. Rassias and B. C. Yang, A Hilbert - type integral inequality in the whole plane related to the hyper geometric function and the beta function, *J. Math. Anal. Appl.*, 428(2), 2015, 1286–1308.
[34] X. Y. Huang, J. F. Cao, B. He and B. C. Yang, Hilbert-type and Hardy-type integral inequalities with operator expressions and the best constants in the whole plane. *J. Inequal. Appl.*, 2015, 2015, 129.
[35] Z. H. Gu1 and B. C. Yang, A Hilbert-type integral inequality in the whole plane with a non-homogeneous kernel and a few parameters. *J. Inequal. Appl.*, 2015, 2015, 314.
[36] L, P. He, Y. Li and B. C. Yang, An extended Hilbert's integral inequality in the whole plane with parameters. *J. Inequal. Appl.*, 2018, 2018, 216.
[37] D. M. Xin, B. C. Yang and A. Z. Wang. Equivalent property of a Hilbert-type integral inequality related to the beta function in the whole plane. *J. Funct. Spaces*, 2018, 2018, Article ID 2691816, p. 8.
[38] Y. R. Zhong, M. F. Huang and B. C. Yang. A Hilbert-type integral inequality in the whole plane related to the kernel of exponent function. *J. Inequal. Appl.*, 2018, 2018, 234.
[39] M. Th. Rassias and B. C. Yang, On a Hilbert-type integral inequality in the whole plane related to the extended riemann zeta function. *Complex Anal. Oper. Theory*, 2018. https://doi.org/10.1007/s11785-018-0830-5.
[40] M. Th. Rassias and B. C. Yang. On a Hilbert-type integral inequality in the whole plane related to the extended Riemann zeta function. In Th. M. Rassias, P. M. Pardalos (eds.), *Mathematical Analysis and Applications*, Springer, 2019.
[41] M. Th. Rassias, B. C. Yang and A. Raigorodskii. On Hardy-type integral inequality in the whole plane related to the extended Hurwitz-zeta fanction. *J. Inequal. Appl.*, 2020, 2020, 94.
[42] M. Th. Rassias, B. C. Yang and A. Raigorodskii. On the reverse Hardy-type integral inequalities in the whole plane with the extended Riemann-Zeta function. *J. Math. Inequal.*, 14(2), 2020, 525–546.

[43] A. Z. Wang and B. C. Yang, Equivalent statements of a Hilbert-type integral inequality with the extended Hurwitz zeta function in the whole plane. *J. Math. Inequal.*, 14(4), 2020, 1039–1054.
[44] J. C. Kuang, *Applied Inequalities.* (Shangdong Science and Technology Press, Jinan, China, 2004).
[45] J. C. Kuang, *Introduction to Real Analysis.* (Hunan Education Press, Changsha, China, 1996).

https://doi.org/10.1142/9789819820498_bmatter

Index

C

D

O

P

Q